SEXUAL AND REPRODUCTIVE HEALTH:

A PUBLIC HEALTH PERSPECTIVE

SEXUAL AND REPRODUCTIVE HEALTH:

A Public Health Perspective

EDITOR-IN-CHIEF

Paul F A Van Look
Consultant in Sexual and Reproductive Health, Val-d'Illiez, Switzerland

EDITORS

H K Heggenhougen
Centre for International Health, University of Bergen, Norway; Department of International Health, Boston University School of Public Health, Boston, USA; and Department of Global Health and Social Medicine, Harvard Medical School, Boston, USA

Stella R Quah
Duke-NUS Graduate Medical School Singapore, Singapore

ELSEVIER

Amsterdam • Boston • Heidelberg • London • New York • Oxford
Paris • San Diego • San Francisco • Singapore • Sydney • Tokyo
Academic Press is an imprint of Elsevier

ACADEMIC
PRESS

Academic Press is an imprint of Elsevier
525 B Street, Suite 1900, San Diego, CA 92101-4495, USA
30 Corporate Drive, Suite 400, Burlington, MA 01803, USA
32 Jamestown Road, London NW1 7BY, UK
Radarweg 29, PO Box 211, 1000 AE Amsterdam, The Netherlands

The Chapter on *Worldwide Burden of Gynecologic Cancer: The Size of the Problem* by Sankaranarayanan and Ferlay was previously published under the title of *Worldwide burden of gynaecologic cancer: The size of the problem* in Best Practice & Research Clinical Obstetrics and Gynaecology 20[2]: 207–225, 2006. The chapter on *Cervical Cancer* by Zeferino and Derchain also previously appeared in the same journal (20[3]: 339–354, 2006) under the title *Cervical cancer in the developing world*, and the chapter on *Endometrial Cancer* by Sonoda and Barakat was also previously published by this journal (20[2]: 363–377, 2006) under the title *Screening and the prevention of gynecologic cancer: Endometrial cancer.* Minor updating and adaptation were made in all three articles for the purposes of the present book.

All other material originally appeared in *The International Encyclopedia of Public Health*, edited by Kris Heggenhougen and Stella Quah (Elsevier Inc., 2008)

Library of Congress Cataloging-in-Publication Data

Sexual and reproductive health: a public health perspective/editor-in-chief, Paul van Look; editors, H.K. Heggenhougen, Centre for International Health, University of Bergen, Norway, Department of International Health, Boston University School of Public Health, and Department of Global Health and Social Medicine, Harvard Medical School, Stella R. Quah, Duke-NUS Graduate Medical School Singapore, Singapore.
p. cm.
Includes bibliographical references and index.
ISBN 978-0-12-385009-6 (hardback : alkaline paper)
1. Reproductive health services. I. Look, Paul F. A. van, editor. II. Heggenhougen, Kris (Kristian), editor. III. Quah, Stella R., editor.
[DNLM: 1. Reproductive Health Services. 2. Family Planning Services. 3. Perinatal Care. 4. Pregnancy Complications–prevention & control. 5. Reproductive Rights. 6. Sexually Transmitted Diseases–prevention & control. WQ 200]
RA652.S49 2011
613.9–dc22

2010046204

British Library Cataloguing in Publication Data
A catalogue record for this book is available from the British Library

ISBN: 978-0-12-385009-6

For information on all Academic Press publications
visit our website at elsevierdirect.com

PRINTED AND BOUND IN USA
10 11 12 13 10 9 8 7 6 5 4 3 2 1

CONTENTS

SECTION 3 REPRODUCTIVE CANCERS

SECTION 4 SELECTED ETHICAL AND OTHER GENERAL ASPECTS

CONTRIBUTORS

V Alam
Merck Serono, Geneva, Switzerland

F Alvarez
PROFAMILIA, Santo Domingo, Dominican Republic

R A Ankeny
University of Adelaide, Adelaide, South Australia, Australia

A Auvinen
University of Tampere School of Public Health, Tampere, Finland

A Bajpai
Royal Children's Hospital, Parkville, VIC, Australia

L S Bakketeig
Norwegian Institute of Public Health, Oslo, Norway

R R Barakat
Memorial Sloan-Kettering Cancer Center, New York, NY, USA

H M Behre
Martin-Luther-University, Halle-Wittenberg, Halle, Germany

P Bergsjø
Norwegian Institute of Public Health, Oslo, Norway

P Boyle
International Agency for Research on Cancer, Lyon, France

C H Browner
University of California, Los Angeles, CA, USA

P S Chandra
National Institute of Mental Health and Neuro Sciences, Bangalore, India

R Choi
University of Washington, Seattle, WA, USA

J G Cleland
Centre for Population Studies, London School of Hygiene and Tropical Medicine, London, UK

R J Cook
University of Toronto, Toronto, Ontario, Canada

J C Cottingham
World Health Organization, Geneva, Switzerland

H B Croxatto
Chilean Institute for Reproductive Medicine, Santiago, Chile

L Dartnall
Medical Research Council, Pretoria, South Africa

S F Derchain
State University of Campinas, Campinas, Brazil

R J Edmondson
Queen Elizabeth Hospital, Tyne and Wear, UK

R Engelman
Worldwatch Institute, Washington, DC, USA

J N Erdman
University of Toronto, Toronto, Ontario, Canada

C Farquhar
University of Washington, Seattle, WA, USA

M F Fathalla
Assiut University, Assiut, Egypt

M M F Fathalla
Assiut University, Assiut, Egypt

A Faúndes
State University of Campinas, Campinas, Brazil

J Ferlay
International Agency for Research on Cancer, Lyon, France

C García-Moreno
World Health Organization, Geneva, Switzerland

M Garefalakis
University of Western Australia, Western Australia, Australia

A Gebbie
Family Planning and Well Woman Services, Edinburgh, UK

G G Giles
Cancer Epidemiology Centre, Carlton, Victoria, Australia

A Glasier
Family Planning and Well Woman Services, Edinburgh, UK

S R Grover
Royal Children's Hospital, Parkville, Victoria, Australia

S Haddock
Population Action International, Washington, DC, USA

M Hakama
University of Tampere School of Public Health, Tampere, Finland

M Hickey
University of Western Australia, Western Australia, Australia

E Huyghe
Hôpital Ranguel, Toulouse, France

R Jewkes
Medical Research Council, Pretoria, South Africa

N Johnson
University of Auckland, Auckland, New Zealand

V A S Krishna
Washington University School of Medicine, St. Louis, MO, USA

P A L Lancaster
Australian Health Policy Institute, University of Sydney, New South Wales, Australia

E Leahy
Population Action International, Washington, DC, USA

J S Mukherjee
Partners In Health, Boston, MA, USA

E Mulligan
Flinders University of South Australia School of Law, Adelaide, South Australia, Australia

R C Pattinson
Kalafong Hospital, Pretoria, Gauteng, South Africa

V N G P Raghunandan
National Institute of Mental Health and Neuro Sciences, Bangalore, India

T K S Ravindran
Achutha Menon Centre for Health Science Studies, Trivandrum, India

M Ripper
University of Adelaide, Adelaide, South Australia, Australia

W Rogers
Flinders University, Adelaide, South Australia, Australia

R Root
Baruch College, City University of New York, New York, NY, USA

R Sankaranarayanan
International Agency for Research on Cancer, Lyon, France

L Say
World Health Organization, Geneva, Switzerland

S Sayers
Menzies School of Health Research, Charles Darwin University, Darwin, Northern Territory, Australia

J-E Schwarze
Clinica las Condes, Santiago, Chile

Y Sonoda
Memorial Sloan-Kettering Cancer Center, New York, NY, USA

B P Stoner
Washington University in St. Louis, St. Louis, MO, USA

D W Sturdee
Solihull Hospital, West Midlands, UK

A R Todd
Northern Centre for Cancer Treatment, Newcastle General Hospital, Newcastle upon Tyne, UK

U Veronesi
European Institute of Oncology, Milan, Italy

K Wellings
London School of Hygiene and Tropical Medicine, London, UK

P H Wise
Stanford University, Stanford, CA, USA

L C Zeferino
State University of Campinas, Campinas, Brazil

F Zegers-Hochschild
Clinica las Condes, Santiago, Chile

PREFACE

This volume presents the highlights of current global thinking about sexual and reproductive health. Major changes have taken place in the last 15 years in the way decision-makers think about the subject and the manner in which programs deliver comprehensive sexual and reproductive health services. The turning point was the International Conference on Population and Development (ICPD) held in Cairo, Egypt, in 1994. ICPD was a watershed for several reasons. First, more than in any of the preceding United Nations population conferences, the issue of population was clearly placed as being central to sustainable development. Second, the narrow focus on population growth ('the population bomb'), which had been a neo-Malthusian concern and preoccupation ever since the Club of Rome published its 1972 report *Limits to Growth*, was replaced by the comprehensive concept of (sexual and) reproductive health. Third, and linked to the definition and introduction of the reproductive health concept, was the strong call for a paradigm shift away from a policy environment driven by demographic considerations (sometimes to the point of using coercion in family planning services in order to reach demographic targets) to an environment that recognized the right of individuals to make their own choices. Last but not least, ICPD as well as the Fourth World Conference on Women (FWCW) held the following year in Beijing, People's Republic of China, strongly emphasized that the rights of women and men to good sexual and reproductive health are firmly grounded in universal human rights.

At the same time, however, the debates that took place both at ICPD and at FWCW, and in many other fora since then, amply demonstrate the difficulties some people had, and continue to have, with internalizing the concept of sexual and reproductive health in all of its dimensions, given that it embodies some of the most intimate and personal aspects of our lives as human beings. Even in some of the most broad-minded and liberal societies, subjects such as sexual education for and the sexuality of young people, induced abortion, sexual violence against women, sexual orientation, and sexual dysfunction, to name but a few, continue to be avoided in public discourse and can be considered taboo subjects for discussion altogether. This may, to some extent, explain why we seem to have such difficulty in making headway with tackling some of the obstacles that still encumber the achievement of good sexual and reproductive health by all.

When looked at from a global perspective, as this book tries to do, there is another striking characteristic in the field of sexual and reproductive health, namely the immense gap that exists between the developed and developing countries, between the rich and the poor. Perhaps in no other branch of health is this inequity so great. Take, for instance, the risk of dying a woman has when she becomes pregnant. Globally, according to the latest figures from WHO, UNICEF, UNFPA, and The World Bank, some 358,000 women died in 2008 from causes related to pregnancy, delivery, and the immediate period after delivery; more than 99 per cent of these deaths occurred in developing countries. A sub-Saharan woman who becomes pregnant is nearly 100 times more likely to die from a pregnancy-related cause than a woman in a Western European country; for some developing countries such as Afghanistan, Chad, Guinea-Bissau, and Somalia, the risk is 140 times or more. Similarly, unmet needs for good-quality contraceptive methods and services remain huge in many areas of the world, particularly in sub-Saharan Africa, and sexual transmission of the human immunodeficiency virus (HIV), the cause of the acquired immune deficiency syndrome (AIDS), is also strongly concentrated in developing countries, again particularly in sub-Saharan Africa, where in several countries AIDS is having a major impact of shortening life expectancies.

Some 15 years have gone by since the global community met at ICPD and FWCW. Nevertheless, sexual and reproductive health and rights continue to catch the limelight, perhaps even more so in the last few years now that the overriding goal of ICPD ('achieve, by 2015, universal access to reproductive health') has finally been formally integrated into the Millennium Development Goals (MDGs) framework as one of the two targets under MDG 5 ('improving maternal health'). Thus, the present time seems most auspicious for this volume with its specific focus on sexual and reproductive health and rights looked at through a global public health lens.

This book, which we hope will become a valuable tool in your daily work, attempts to capture the comprehensive, and often complex and politicized, field of sexual and reproductive health and rights. It comprises 37 chapters; 34 of these are taken from the *International Encyclopedia of Public Health* published by Elsevier in 2008 and the remaining three (on the worldwide burden of gynecologic cancer, on cervical cancer, and on endometrial cancer) are slightly adapted and updated versions of papers published elsewhere. Given that all chapters address aspects within the theme of sexual and reproductive health, there is some inevitable overlap between some of the chapters and some authors go beyond the theme of sexual and reproductive health. For instance, the chapter on cancer screening also has some general information on the predictive value of screening tests and the organization of screening programs as well as information on screening for lung and colorectal cancers in addition to screening for cervical, breast, and prostate cancers. We believe, however, that this additional material, rather than being distractive, helps to put the information on sexual and reproductive health issues in a wider context.

The 37 chapters are divided into four sections. The first section has nine chapters dealing with human reproductive physiology (female and male reproductive function, puberty, and menopause) and the basic underpinnings of sexual and reproductive health. It also contains chapters on demography (population growth and trends in fertility) and perinatal epidemiology; the latter topic has been included since perinatal outcomes, in particular stillbirths and early neonatal deaths, are so much influenced by events happening during pregnancy and at the time of childbirth.

The second section focuses on the core elements that make up sexual and reproductive health as we understand it today. This section has, of course, chapters on the cornerstones of sexual and reproductive health, such as family planning, sexually transmitted infections, abortion, maternal mortality and morbidity, and infertility. But in this section you will also find chapters on sexual and reproductive health aspects that traditionally are not generally dealt with in a book of this nature. These aspects are, for instance, sexual violence and violence against women (often still considered taboo subjects), HIV/AIDS (which is, after all, very much a sexual issue in many parts of the world and linked closely to infant's health where there exists a risk of mother-to-child transmission of the virus), and the causes and outcomes of fetal growth retardation (which is often linked to the mother's condition during pregnancy).

The third section has eight chapters on the most common genital cancers in women (i.e., those affecting the uterine cervix, endometrium, breast, and ovary) and in men (cancers of prostate and testis). Of particular interest to readers may be the two chapters at the beginning of this section, namely those on cancer screening and on the global burden of gynecologic cancers, which is another telling example of the "North–South" divide although, in this instance, the picture is a mixed one with incidence rates for some cancers, notably cancer of the uterine body and ovarian cancer, greater in developed countries than in developing countries while the reverse is true for cancer of the uterine cervix.

The one single characteristic in sexual and reproductive health matters that has undoubtedly created the greatest public debate in recent times, ever since the birth of Louise Brown in 1978 following the successful *in vitro* fertilization of a human egg by a human spermatozoon, has been the ethical dimensions of applying various reproductive technologies in humans. Prime examples of these ethical debates in the recent past have been on the subject of the potential (mis)use of reproductive cloning (for instance to create a carbon-copy individual to serve as organ donor as evoked in the recent film *My sister's keeper* after the similarly named novel by Jodi Picoult) and on the harvesting of stem cells from human embryos for research with potential future clinical application in a range of degenerative and other ill-health conditions. It should not come as a surprise, therefore, that three of the eight chapters in the fourth and final section of this book are devoted to reproductive ethics. The other chapters in this section concentrate on some of the factors that can have a profound impact on sexual and reproductive health, namely, gender, the enjoyment (or lack thereof) of reproductive rights, and the cultural context in which women and men try to achieve their sexual and reproductive health goals. The final chapter in this section addresses the emerging recognition of the magnitude of women's mental ill-health, much of which is related to women's experience of sexual and reproductive events during the life course.

In compiling a book on an issue of such breadth, complexity, and sensitivity – at personal, community, national, and global levels – it is inevitable that a certain element of eclecticism enters the choice of articles. We do, nevertheless, hope that you will find this collection of papers most helpful in your work as an academician, policy maker, student, or other concerned professional with a passion for and dedication to improving the sexual and reproductive health and rights of our fellow citizens, wherever they may live. We are, of course, most interested to hear your comments and suggestions for possible future editions of this book.

Finally, in closing, we thank the plethora of distinguished authors who enthusiastically accepted the invitation to include their work in this volume. We also say a big 'thank you' to the Elsevier Editorial Team, in particular Nancy Maragioglio and Carrie Bolger, who made sure that we stayed on the "straight and narrow" of publishing deadlines and who were as committed as we were to make this book a valuable resource.

The Editors

SECTION 1

PHYSIOLOGY, GENERAL EPIDEMIOLOGY AND DEMOGRAPHY

Female Reproductive Function

H B Croxatto, Chilean Institute for Reproductive Medicine, Santiago, Chile

Introduction

The organs most involved in reproduction in the female are made up of the genital tract and ovaries located in the lower part of the abdomen, the hypothalamus, and pituitary gland located at the base of the brain and the breasts (**Figures 1** and **2**). The female reproductive system is formed during embryonic and fetal development, reaching incomplete development at the time of birth. Further growth and acquisition of full functional capacity is attained after puberty, which unfolds between 9 and 13 years of age. At puberty, the hypothalamic-pituitary-ovarian-uterine axis is activated and ovarian and endometrial cycles (**Figure 3**) become manifest by the occurrence of menstruation with a monthly periodicity. The first menstruation (or menarche) marks the onset of the fertile segment of a woman's life or reproductive years, in which women can conceive and give birth to one or more children. The number of oocytes (female gametes) present in the ovaries at birth is limited to a few hundred thousand or less and their number decreases thereafter continuously through a resorption process known as atresia. As a consequence, around the age of 50 years, the pool of oocytes becomes exhausted, ovarian and endometrial cycles cease to occur, and reproductive capacity wanes after the last menstruation (or menopause).

Menstruation and the Menstrual Cycle

The vast majority of vertebrates exhibit reproductive cyclicity as they can bear offspring repeatedly, but with a periodicity that is entrained with environmental phenomena determined by the Earth's rotation and yearly circling around the Sun. Other than parturition or egg-laying, an outstanding external signal of reproductive cyclicity is the female's behavior in response to male attempts to copulate. In most species, the female does not allow copulation except for a short period of hours or a few days preceding ovulation (the release of one or more oocytes from the ovary), enhancing in this way the probability of becoming pregnant. With the exception of the human and a few other species, mating is precisely timed to optimize the likelihood of fertilization (union of male and female gametes) and therefore reproduction.

This female receptive behavior toward mating is known as heat or estrus, hence the cycle in these animals is called the estrous cycle. This stereotyped receptive behavior toward mating is practically absent in the human female since women engage in sexual intercourse at any time of the menstrual cycle, during pregnancy, during lactational amenorrhea, and after the menopause. As a consequence, the vast majority of acts of sexual intercourse in the human are nonreproductive, affording only pleasant, uniting, and recreational experiences to the couple, without the invariable reproductive aim characteristic of most other animals.

The most prominent external sign of cyclicity during the reproductive years of the human female is menses or menstruation, which is blood mixed with tissue sloughed off from the superficial or functional layer of the endometrium (internal lining of the womb), being expelled through the vagina to the exterior. Thus, the cycle in women is called the menstrual cycle. Menses occur from menarche until menopause more or less regularly with a periodicity approaching the lunar cycle, and are temporarily suppressed during pregnancy and breastfeeding. While the occurrence of menstruation allows a woman to assume she is fertile, a delay in its occurrence, beyond its expected timing during the reproductive years in cycling, sexually active women, usually indicates that she has become pregnant.

At the time of impending menstruation or during menses, many women can experience a variety of unpleasant changes in mood and/or bodily sensations within a wide range of intensities, conditioned by the hormonal oscillations that cause menstruation. This, together with the sanitary requirements imposed by several days of vaginal bleeding, makes most women experience this monthly sign of cyclicity of their reproductive function with great awareness, although its importance and meaning varies considerably among them. Furthermore, different cultures assign a special meaning to menstruation and impose various rules on women that affect their social behavior or marital relationships while the bleeding episode is taking place. On the other hand, vaginal bleeding does not always reflect menstruation since there are diverse pathological conditions that are associated with blood loss from the endometrium or other segments of the female genital tract.

Menstruation is the culmination and external sign of the end of a nonconceptional cycle. It is immediately followed by the endometrial proliferative phase in which cells multiply to rebuild the functional layer that was sloughed off during menstruation (see **Figure 3**). The deeper or basal layer of the endometrium remaining after menstruation is no more than 3 mm in thickness

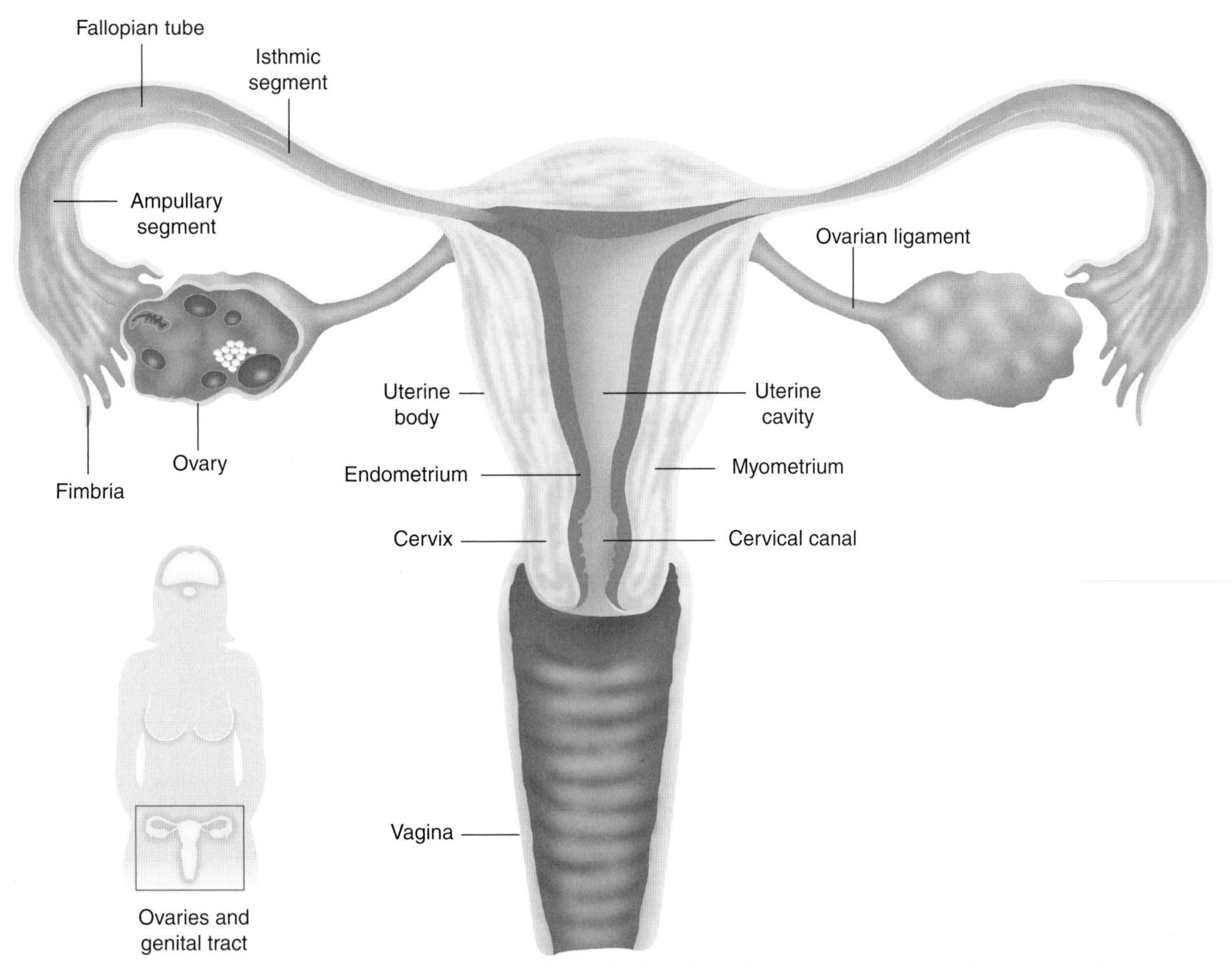

Figure 1 The female genital tract encompassing the ovaries, Fallopian tubes, uterus, and vagina. A partial frontal section allows appreciating the cavity of the tubular organs and the way they are connected.

and develops into a fully grown layer of up to 12–15 mm or more in about 2 weeks through stimulation by the increasing levels of estradiol (female hormone produced by the ovary) in blood. This proliferative phase is followed by a secretory phase in which tissue differentiation and remodeling and endometrial glandular secretion predominate under the influence of progesterone (another ovarian hormone) in preparation for nesting a fertilized egg (or zygote), if it should happen to be formed and arrive into the uterus in that cycle. The secretory phase also entails building the necessary mechanisms to produce menstruation if no zygote is formed during that cycle.

Recurrent sloughing off and rebuilding of the endometrium constitutes the endometrial cycle and takes place continuously for years until conception or the menopause occur. This process is not autonomous and is heavily dependent on estradiol and progesterone secreted by the ovary, therefore the endometrial cycle is precisely synchronized with the ovarian cycle (see **Figure 3**). Ovarian cycles are absent and ovarian hormone production is minimal or nil before puberty, during pregnancy and lactation, and after the menopause.

An outstanding feature of the menstrual cycle is its variability within and between women, not only in the intervals between menses but in their duration, in the timing of ovulation, the blood levels attained by the hormones involved, and several other parameters. The days of the cycle are usually counted taking the first day of menstruation as the first day of the cycle. Only for didactic purposes it is generally said that the menstrual cycle lasts 28 days and ovulation takes place on day 14 of the cycle. However, the normal cycle length varies from 21 to 35 days and follicular rupture can take place as early as day 10 or as late as day 22.

The Ovarian Cycle

The ovary plays a central role in female reproductive cyclicity and the menstrual cycle. Usually, the full ovarian cycle occurs in one of the two ovaries while the other exhibits incomplete waves of follicular growth for one or more cycles until the fully active side is reversed in a nonpredictable fashion.

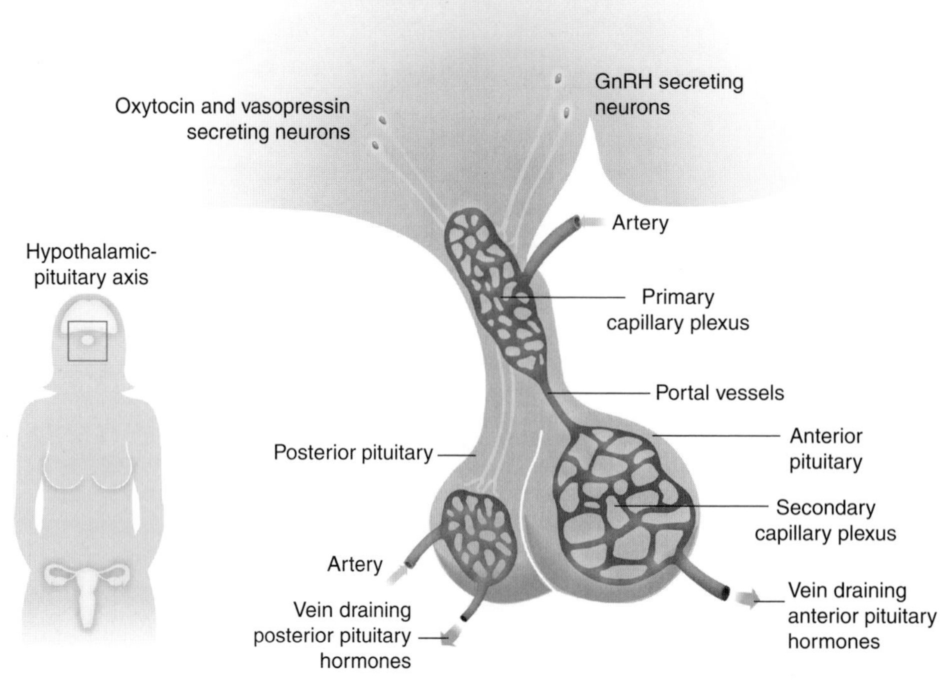

Figure 2 Components of the hypothalamic-pituitary axis most relevant to the reproductive process. Some neurons, which have their cell bodies located in hypothalamic nuclei, project their axons toward the primary capillary plexus of the hypothalamic-pituitary portal system where they deliver their hormonal secretions, notably gonadotropin-releasing hormone (GnRH). GnRH circulates down the portal vessels that resolve into the secondary capillary plexus in the anterior pituitary lobe. Here, GnRH diffuses out of the capillaries and acts on the gonadotrophs, stimulating the secretion of the gonadotropic hormones. Other hypothalamic neurons extend their axons all the way down into the posterior lobe of the pituitary where they release vasopressin and oxytocin, which are involved in parturition and milk ejection, respectively.

At the onset of each ovarian cycle, small antral follicles (tiny spheres containing each an oocyte and a fluid-filled cavity) grow rapidly stimulated by the gonadotropic hormones secreted by the anterior pituitary gland (**Figure 4**). Eventually, one of them becomes the leading or dominant follicle, which continues to grow to become a mature or Graafian follicle. A transient surge of the gonadotropic hormones in blood triggers the ovulatory process only in the mature follicle, leading to its rupture, release of the oocyte it contains and its transformation into a corpus luteum. The life span of the corpus luteum is self-limited to approximately 2 weeks, unless a developing zygote signals its presence. Functional demise of the corpus luteum allows both menstruation and the beginning of a new ovarian cycle.

The fundamental structures that sustain ovarian cyclicity come from a pool of thousands of primordial follicles, from which cohorts of about two dozen start growing every day in an autonomous manner. After a few months, nearly half a dozen reach the stage of antral follicle with a diameter of 2–5 mm while the rest undergo atresia at various stages of their development and disappear. The wall of antral follicles is formed by two layers of cells. The outermost layer is called the theca layer and is richly vascularized and innervated. The innermost layer, called granulosa, lacks vascular irrigation and innervation and is in direct contact with antral fluid. It is separated from the theca by a thin homogeneous lamina (basal membrane). Usually, each follicle contains a single oocyte that is immersed in the granulosa. The granulosa cells surrounding the oocyte form the cumulus oophorus during the ovulatory process.

Figure 5 shows a cross-section of a mature follicle which illustrates the layers forming its wall. From the outside to the inside: the external layer of cells forms the theca externa and interna. The theca externa is adjacent to the ovarian stroma (not shown). The theca interna is richly vascularized and innervated. A basal membrane separates the theca interna from the granulosa that lacks blood irrigation and innervation. The granulosa consists of several layers of cells housing a single oocyte and surrounding a cavity filled with follicular fluid. The cells immediately surrounding the oocyte (cumulus cells) behave differently from the rest of granulosa cells (mural granulosa). At ovulation, a mucinous material accumulates between them forming the cumulus

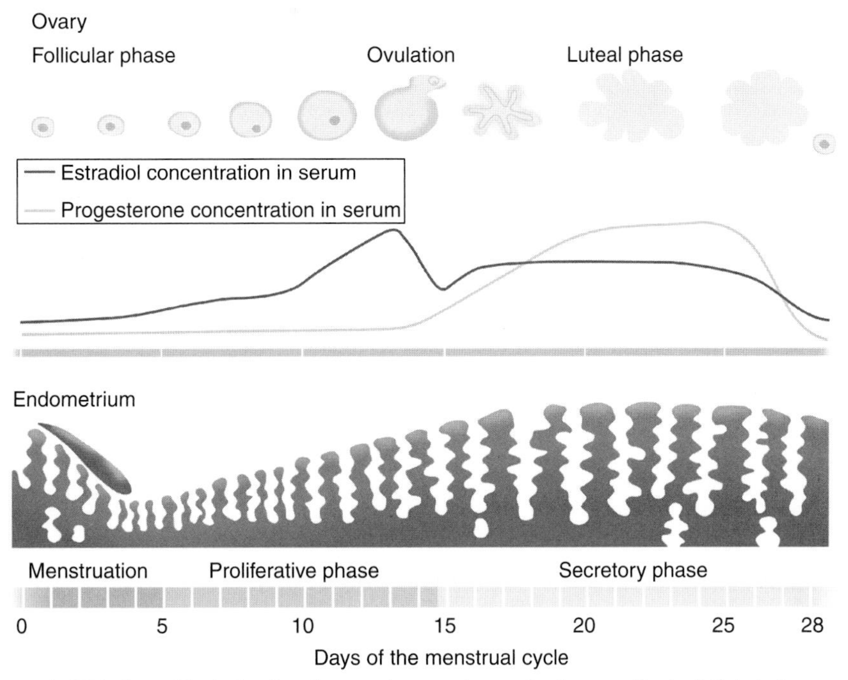

Figure 3 The upper part of this figure illustrates the phases of an ovarian cycle: the growth of a follicle in the ovary ending in ovulation and transformation into a corpus luteum that grows and then regresses and wanes. The middle part shows the oscillations in the blood serum concentration of estradiol and progesterone during the ovarian cycle. The lower part illustrates the phases of the endometrial cycle, beginning with menstruation. Increasing levels of estradiol secreted by the leading follicle as it grows stimulate cell proliferation and increasing thickness in the endometrium. The corpus luteum secretes both estradiol and progesterone in amounts proportional to its development. Progesterone stimulates glandular secretion and acquisition of receptivity to the blastocyst in the endometrium. Demise of the corpus luteum and associated decrease in estradiol and progesterone levels cause the onset of menstruation and of a new ovarian and endometrial cycle.

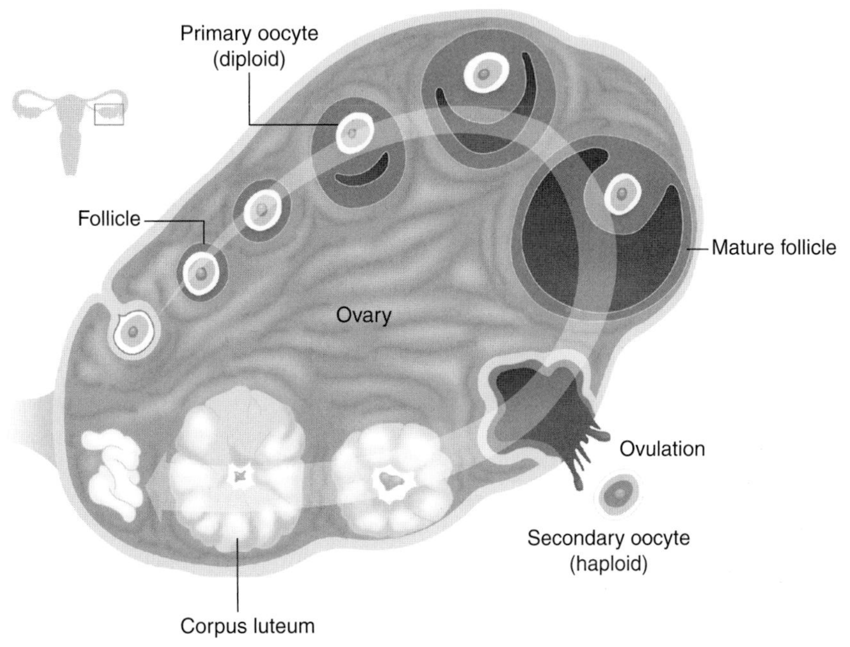

Figure 4 Life history of the follicle from a primary follicle destined to ovulate until ovulation, formation of a corpus luteum, and its demise in the ovary. See Figure 3 for the time frame of each stage.

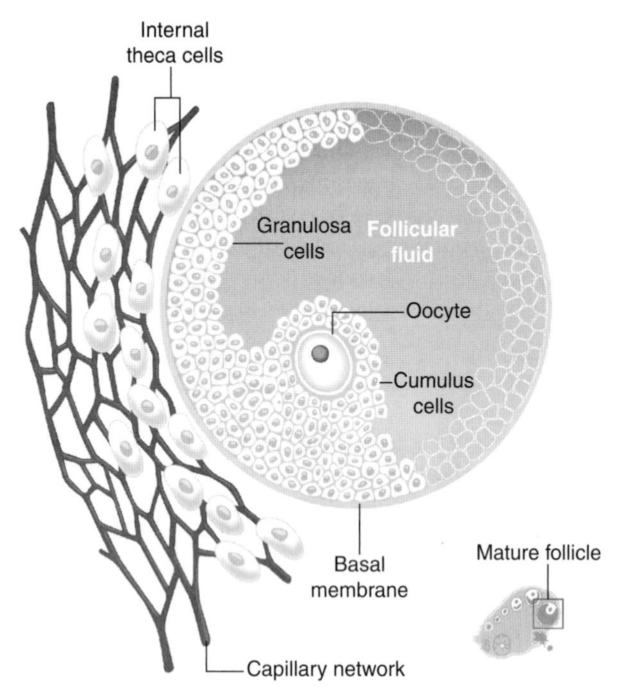

Figure 5 A cross-section of a mature follicle.

oophorus; they detach from the wall of the follicle and are released from the follicle surrounding the oocyte at ovulation. Theca cells respond primarily to luteinizing hormone (LH) by secreting androgens, whereas granulosa cells respond to follicle-stimulating hormone (FSH) and transform the androgens into estrogens.

At each transition from the end of one cycle to the beginning of the next, there is an elevation of FSH, a gonadotropic hormone in the circulation which acts to accelerate growth of the small antral follicles that began to develop months earlier. These follicles enter into a sort of competition until one of them becomes dominant and the others stop growing and undergo atresia. The dominant follicle continues growing by multiplication of its cells and accumulation of fluid in its cavity until reaching about 20 mm in diameter in 10–15 days. During this phase, the dominant follicle secretes increasing quantities of estradiol, resulting in an exponential rise in the blood level of this hormone until it surpasses a critical level that triggers a massive and transient release of the gonadotropic hormones, LH and FSH, from the pituitary gland. This surge of gonadotropic hormones acting on the mature follicle initiates the ovulatory process that will culminate 36 hours later in follicular rupture and release of the oocyte into a space that can be reached by spermatozoa to accomplish fertilization.

The ovulatory process is a complex series of parallel, synchronous, and, to some extent, independent changes affecting almost every component of the mature Graafian follicle. A crucial process is the resumption of oocyte meiosis that leads to extrusion of the first polar body (a small daughter cell resulting from oocyte division) containing one chromosome of each pair and, thus, the reduction of the number of maternal chromosomes within the oocyte to a single set. Another key process is the expansion of the intercellular matrix uniting the granulosa cells that surround the oocyte, to form the cumulus oophorus that detaches from the follicle wall so it can be extruded from the follicle at the time of follicular rupture and voiding. A third process is the activation of enzymes that erode the intercellular matrix of the follicle wall and allow the apex of the follicle to yield to the slightly positive pressure of the antral fluid, which is maintained throughout the entire process by the sustained contraction of the outer thecal cells at the base of the follicle. Both the granulosa and thecal cells begin a process called luteinization that entails major changes in steroid-synthesizing machinery leading to decreased synthesis of estradiol and increased synthesis of progesterone. The large amounts of progesterone to be produced in the following days impose increased energetic demands by these cells so granulosa cells secrete angiogenic factors that lead to disruption of the basal membrane and growth of newly formed blood capillary vessels into the luteinizing granulosa layer. Finally, the wall of the follicle ruptures at the apex, the follicle contracts, and the antral fluid containing the cumulus oophorus is voided toward the peritoneal cavity or directly to the fimbria of the Fallopian tube. Altogether these processes constitute ovulation.

Luteinized theca and granulosa cells remaining in the ruptured follicle reorganize to form the corpus luteum. They grow substantially in size, accumulate large amounts of lipids required for steroid hormone synthesis and, in the course of one week, they constitute the corpus luteum, a compact body in the superficial layer of the ovary that replaces the ruptured follicle (see **Figure 4**). During the second week after ovulation, the corpus luteum starts regressing and produces less and less progesterone and estradiol, which results in decreasing circulating levels of these hormones up to the point where the endometrium, lacking this hormonal support, starts to break down and menstruation ensues. The reduction in the circulating level of ovarian hormones weakens their negative feedback on gonadotropin secretion, allowing a transient small elevation of FSH that recruits a new cohort of antral follicles for initiating the next ovarian cycle.

The alternation of dominant follicle and corpus luteum as the predominant structure in the ovary gives rise to the denomination of follicular phase and luteal phase, respectively, to refer to these two phases of the ovarian cycle. These structural ovarian changes are accompanied by changes in the rate of secretion and blood levels of steroid and protein hormones produced by the ovary. The length of the follicular phase is more variable than that of the luteal phase and accounts for most of the variability in the length of the menstrual cycle observed within and between women. The length of the luteal phase, taking as the first day the day in which the echographic image of the leading follicle shows it has collapsed, and as the last day the day preceding the onset of menstruation, varies from 9 to 16 days being 13 to 15 days in nearly 70% of the cycles.

The Endometrial Cycle

As the dominant follicle grows and secretes increasing amounts of estradiol, the cells located in the basal layer of the endometrium proliferate, endometrial glands and surface epithelium are reconstructed, and blood vessels grow again side by side with the endometrial glands. At the time of ovulation, endometrial thickness is 12–15 mm and, as progesterone secretion by the corpus luteum increases, endometrial glands begin to secrete their products and the luminal epithelium (the single cell layer that covers the endometrial surface) exhibits transient swellings known as pinopodes. By the seventh day of exposure to elevated progesterone, the endometrium becomes receptive to embryo implantation and remains so for a few days, a period known as the window of implantation or receptive phase.

The success of embryo implantation depends on the quality of the blastocyst (a developmental stage of the zygote able to attach to the endometrium) and on endometrial receptivity. The latter depends on progesterone which, acting through its receptors on an estrogen-primed endometrium, changes the transcription rate of target genes. Hence, endometrial receptivity is associated with either enhanced or decreased expression of certain repertoires of genes in comparison with prereceptive stages of the endometrium. A morphological correlate of this receptivity is the predecidual transformation of the cells surrounding the endometrial blood vessels. The absence of a blastocyst in the mid-luteal phase allows luteolysis (breakdown of the corpus luteum) to proceed and menstruation to start a week later.

The Conceptional Cycle

The cycle in which a pregnancy begins is referred to as the conceptional cycle. In such a cycle, menstruation does not take place 2 weeks after ovulation as it does in an infertile cycle because the corpus luteum does not regress but continues to produce progesterone, thus menstruation does not occur. The conceptional cycle ends normally 9 months later with parturition, followed by lactational amenorrhea until menstrual cyclicity resumes. Ten percent or more of conceptional cycles will end prematurely due to death of the conceptus (developing zygote), the embryo or the fetus, or other causes, but before the product of conception is viable outside the womb. Loss of the embryo or fetus in these circumstances is referred to as abortion (or clinically often as 'miscarriage'). In addition, a variable number of conceptuses (estimated at 20–50%) die during or soon after they implant in the endometrium and menstruation may occur at the normal time or somewhat delayed without the woman being aware of this failed implantation.

The ovarian and endometrial cycles of the menstrual and the conceptional cycles do not differ until after ovulation, except for possible subtle changes in response to sexual intercourse and the presence of seminal plasma components and spermatozoa that interact with the epithelial cells lining the inner surface of the genital tract. The major differences between both cycles, before the delay in menses occurs, take place after ovulation and particularly at the time of implantation and thereafter.

A sequence of several fundamental processes is required for the natural occurrence of a conceptional cycle. The first two events are sexual intercourse and ovulation; the former must be close to, but not after, ovulation. They must be followed by the encounter of the gametes, oocyte, and spermatozoon, usually in the ampullary segment of the Fallopian tube (**Figure 6**), where fertilization takes place leading to the formation of a single cell, the zygote, representing the beginning of a new individual. This is followed by development of the zygote up to the morula stage, its transport to the uterine cavity where it continues to develop up to the blastocyst stage, which is the stage at which it is capable of implanting in the endometrium.

Following fertilization within the ampullary segment of the Fallopian tube, the zygote develops within this organ up to the early morula stage composed of 8–12 cells in the course of 3 days at the end of which it passes into the uterine cavity. In the following 3 days the morula develops into a blastocyst inside the uterus. At the end of this period, the blastocyst expands, loses the zona pellucida, and begins to implant in the endometrium.

Implantation of the blastocyst starts in the middle of the luteal phase when progesterone secretion by the corpus luteum is at its peak. As implantation begins, the trophoblastic cells of the blastocyst begin to secrete increasing amounts of the hormone human chorionic gonadotropin (hCG), which passes into the mother's bloodstream. Acting upon the corpus luteum, hCG prevents its involution and keeps it secreting high levels of progesterone and estradiol. In this manner, the hormonal support of the endometrium is maintained and menstruation is prevented so that the implanted blastocyst can continue its development in the endometrium.

Gamete Encounter and Fertilization

The vast majority of acts of sexual intercourse do not generate a pregnancy in fertile couples who are not

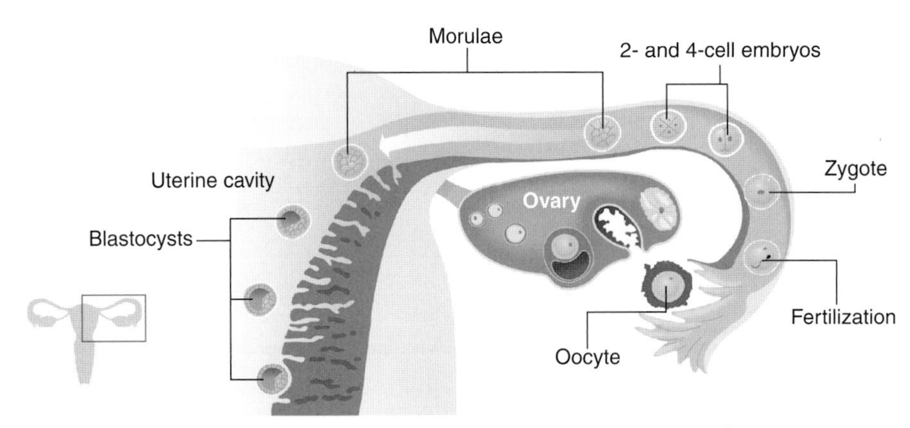

Figure 6 Preimplantation development of the zygote.

using any means to avoid it, indicating that the reproductive process in the human has rather low efficiency in comparison with other domestic and wild species in which almost every mating is followed by pregnancy. There are several reasons for this difference, one of them being that ovulation takes place in one of the 21–35 days of the menstrual cycle and intercourse can occur multiple times randomly distributed along that period and by chance not close enough to ovulation to allow the encounter of fresh gametes. By carefully monitoring the day of ovulation and days of intercourse in 625 cycles of 221 women seeking to become pregnant, Wilcox *et al.* (1995) were able to establish that a single act of sexual intercourse can generate a conceptional cycle if it takes place on the day of ovulation or in any of the 5 preceding days. These six days of the menstrual cycle are designated as the fertile period of the menstrual cycle. The probability of generating a conceptional cycle is highest when intercourse occurs on the last three days of the fertile period, when it is close to 35%. Wilcox *et al.* (1998) also established that 30% of conceptional cycles identified by detection of hCG in the urine ended before the pregnancy was clinically recognized. Since menstruation in such cycles occurs at about the expected time, women are not aware of these early conception losses and perceive the cycle as a normal, nonconceptional menstrual cycle.

Spermatozoa can survive in the genital tract of women and retain their fertilizing potential up to 5 days after intercourse. On the other hand, the oocyte needs to be fertilized within the first few hours after ovulation to generate a viable, healthy zygote with full developmental capacity. This explains why the fertile period is limited to six days asymmetrically distributed relative to the day of ovulation.

Ovulation is preceded by high estradiol and low progesterone levels in the circulation. Therefore, the endocrine milieu prevailing during the fertile period is optimal for the production of abundant, clear, more watery and less viscous cervical mucus that fills the cervical canal and is easily penetrable by spermatozoa.

When a man ejaculates during coitus, semen containing several hundred million spermatozoa is delivered to the vagina. If this occurs during the fertile period, sperm cells will easily pass into cervical mucus extruding from the external opening of the cervical canal. In fact, several hundred thousand spermatozoa will do so within the first hour and most of them will colonize the cervical crypts where they can remain viable for days, forming a sperm reservoir. In a short-lived first phase, some sperm cells are passively transported within minutes to the Fallopian tube while the others actively migrate from the cervical reservoir to the Fallopian tube continuously over the next hours and days. Animal experiments have demonstrated that passively transported sperm do not have the capacity to fertilize and that the fertilizing sperm come from the active phase.

Noncapacitated spermatozoa (not yet ready to fertilize an oocyte) reaching the Fallopian tubes bind for hours to the epithelial cells lining the lumen and subsequently are released in a capacitated state. Capacitated spermatozoa can readily fertilize an oocyte, but they soon lose viability if ovulation has not taken place. Fresh spermatozoa coming from the cervical reservoir can repeat this cycle until ovulation takes place or for a maximum of 5 days after intercourse. Once a capacitated sperm cell penetrates the cumulus oophorus, it binds to the proteinaceous layer covering the oocyte, called zona pellucida. Here it loses part of the outer cell membrane allowing the release of various enzymes (acrosome reaction) that help the sperm drilling a pathway through the zona until it reaches the oocyte cell-membrane. Again, the sperm binds to this membrane and fuses its own membrane with it. The sperm head carrying one set of paternal chromosomes then comes inside the oocyte cytoplasm. Meanwhile, the oocyte undergoes several processes in response to sperm signaling. It releases components (cortical granules) that turn the zona pellucida impenetrable to further spermatozoa, completes the second meiotic division keeping only one half of the maternal chromosomes (haploid condition) and becomes denuded from cumulus cells. At this stage, the chromosomes of the oocyte and those of the fertilizing spermatozoon are separately contained in membrane-bound structures called pronuclei. Afterward they come together so that the fertilized oocyte now contains the full endowment of chromosomes (diploid condition) becoming a single cell zygote, which represents the beginning of a new individual of the species.

Development and Transport of the Zygote to the Site of Implantation

The zygote develops steadily by successive divisions of its initial mass without accruing additional living matter. Thus, each division gives rise to equal-size daughter cells of smaller and smaller size. Initially, this happens inside the Fallopian tube up to approximately the 12-cell stage or morula in the course of 3 days (see **Figure 6**). During this time, the developing zygote resides within the ampullary segment of the Fallopian tube and, by the end of the third day, it is transported through the isthmic segment into the endometrial cavity where it continues to divide.

A fluid-filled cavity develops at the center of the developing zygote and enlarges, giving rise to a developmental stage known as blastocyst that continues to grow in size by fluid accumulation. At the same time, the blastocyst cells differentiate into two classes. The most numerous cells, called trophoblast cells, line the outer surface, and the less numerous, a cluster of 10–20 cells, is called the inner cell mass and lies inside, adjacent to one section of the trophoblast. At about the seventh day after fertilization, the blastocyst comprises some 200 cells, 10% of which correspond to the inner cell mass that will actually give rise to the embryo while the rest will form the placenta. At this stage, due to fluid accumulation, the volume of the blastocyst is at least twice as large as the oocyte from which it originated; the blastocyst then hatches from the zona pellucida and is ready to initiate implantation (**Figure 7**).

Following menstruation, cells of the basal layer proliferate, rebuilding the spongy and compact layers where glands and spiral arteries grow. Following ovulation and fertilization on day 14, the zygote begins its preimplantation development and the newly formed corpus luteum starts producing increasing amounts of progesterone that induce glandular secretion and receptivity to the embryo in the endometrium. By day 21, the blastocyst has attained full development and begins to implant. At the same time, it produces hCG that passes into the mother's bloodstream and acts upon the corpus luteum to sustain progesterone secretion and avoid menstruation. The endometrial stroma undergoes decidualization to contain the invading trophoblast. By day 30, the blastocyst has turned into an embryo that is entirely immersed in the endometrium. Trophoblast cells have differentiated in two cell lineages that interact with endometrial blood vessels to build the placenta, a nurturing system for the embryo whose postimplantation development is now ongoing.

Implantation

The particular distribution of the two cell types that constitute the blastocyst confers polarity to its structure. Trophoblast cells covering the inner cell mass attach to the epithelial cells lining the endometrial cavity and eventually their cytoplasmic projections penetrate between the epithelial cells, cross the basal membrane underlying the epithelial cells and begin to invade deeper endometrial layers. The conceptus immerses into the endometrium and becomes separated from the uterine cavity by the reconstructed epithelial lining (see **Figure 7**). This is followed by the establishment of a complex interphase between the embryo itself and the mother, allowing the exchange of nutrients and many other substances required for further development of the embryo in the months to follow. The placenta is the main component of this interphase and is derived from the trophoblast. The endometrium itself changes into a structure known as decidua

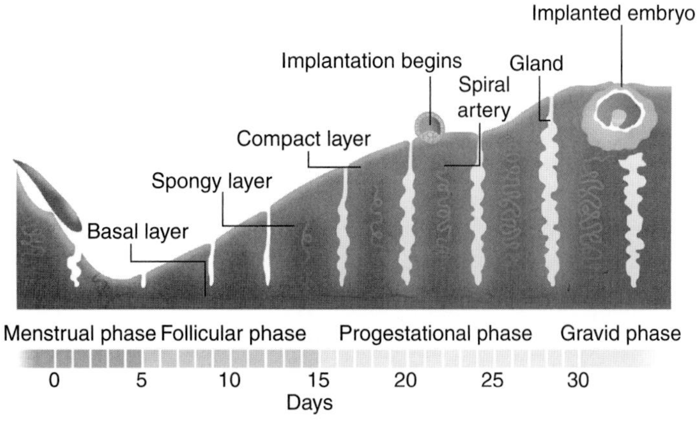

Figure 7 Schematic illustration of the first 30 days of an endometrial cycle in a conceptional cycle.

that can support the placenta while at the same time limiting the invasiveness of the cells derived from the trophoblast.

An important adaptation of the maternal side modulates the local immune system to prevent embryo rejection. The mother's immune system would otherwise recognize as nonself the embryonic proteins encoded by paternal-inherited genes and this would mobilize an allograft rejection. In fact, failure of this protective mechanism is thought to be responsible for a proportion of early embryo wastage.

As the conceptus establishes conduits to access maternal nutrient supplies, new living matter can be synthesized and real growth begins, resulting in a new phase of embryonic development where cells can keep their size after each division and new cell types can emerge that associate temporally and spatially in varying manners thereby forming the rudimentary organs that eventually shape the body. Once these organs are in place, fetal growth ensues until the fetus reaches enough maturity to trigger the onset of parturition and continue living outside the womb.

Regulation of the Hypothalamic-Pituitary-Ovarian Axis

The three organs forming the hypothalamic-pituitary-ovarian axis communicate with each other mainly by means of chemical signals that are the hormones they produce that travel from one organ to the other via the bloodstream. The hypothalamus located at the base of the brain contains neurons that produce a peptide composed of 10 amino acids known as gonadotropin-releasing hormone (GnRH). This molecule is secreted directly into the primary blood capillary plexus of the hypothalamic-pituitary portal system through which it reaches the pituitary gland in high concentrations (see **Figure 2**). This portal system conveys blood from the median eminence of the hypothalamus directly into the anterior lobe of the pituitary gland (adenohypophysis) where it resolves into a secondary blood capillary plexus, allowing hypothalamic hormones to reach the adenohypophysis without being diluted in the general circulation.

As in men, GnRH is released in a pulsatile fashion in women, with an interpulse interval varying within the range of 60–120 minutes. Specialized cells within the adenohypophysis known as gonadotrops synthesize and secrete the gonadotropins FSH and LH in response to GnRH pulses. Both gonadotropins consist of two subunits, designated α and β, linked together by disulfide bonds. The α-subunit is common to other glycoprotein hormones while the β-subunit is more specific. Nevertheless, the β-subunit of LH is highly homologous to the β-subunit of hCG. Both pituitary gonadotropins are released into the blood in response to GnRH, although there are other factors that modulate this response so that under certain conditions, the patterns of their secretion differ to some extent. Each GnRH pulse generates a gonadotropin pulse, being more pronounced in LH than in FSH.

FSH, acting on granulosa cells of antral follicles, stimulates the secretion of several growth-promoting factors and the expression of aromatase, an enzyme that converts androgens into estrogens. As a consequence, follicular growth is accompanied by increasing secretion of estradiol (see **Figure 3**). LH, in turn, through its action on thecal cells stimulates the synthesis of androgens (male-type hormones) that in part diffuse into the bloodstream, but also into the granulosa cells where they are converted into estrogens. As blood estradiol levels rise during the follicular phase, they exert negative feedback on gonadotropin secretion, keeping the response of the ovary in check so that usually only one follicle matures in each cycle. When the level of estradiol in blood surpasses a certain critical level, the negative feedback turns into positive feedback: pulsatile activity of GnRH neurons increases and the pituitary gonadotropic cells become highly sensitive to GnRH's stimulatory action resulting in a massive discharge of both gonadotropins, which causes a transitory increase in the blood concentrations of both hormones. The rise in gonadotropins is far more noticeable for LH than FSH. It is known as the preovulatory gonadotropin discharge, midcycle gonadotropin surge, or LH peak, and is responsible for selectively triggering the ovulatory process in the mature follicle. Both hormones return to their basal level approximately 48 h after the beginning of the discharge.

At the onset of luteinization induced by the LH peak, estradiol secretion drops at the same time as progesterone secretion begins to increase. Thereafter, the blood concentrations of both steroids increase progressively until the mid-luteal phase when they start decreasing again during the second half of this phase if no conceptus has signaled its presence. During the luteal phase, the combined effect of elevated estradiol and progesterone keeps the frequency of GnRH and gonadotropin pulses at a reduced rate and no follicular recruitment takes place. When the corpus luteum begins to involute and eventually ceases to produce the two steroid hormones, gonadotropins are released from their negative feedback and their basal levels rise, leading to the initiation of a new follicular phase.

Apart from sex steroids, peptide hormones – some of them produced by the ovaries – contribute to the regulation of gonadotropin secretion. Among them, inhibins, activins, and follistatin are functionally involved in the regulation of FSH secretion. Inhibins (both the A and B types) are produced by granulosa cells and inhibit selectively the secretion of FSH. Activins (of which there are A and B types too) are also produced by granulosa cells, as well as by cells of several other organs including the pituitary gland, but, in contrast to inhibins, activins stimulate the secretion of FSH. Follistatin is a protein produced in the pituitary gland that binds activins and prevents their stimulatory effect on the FSH-secreting gonadotropic cells. It is likely that activins and follistatin produced locally in the pituitary are at least as important for FSH regulation as their circulating levels. Another peptide hormone produced by granulosa cells is Müllerian inhibiting substance (MIS) (also known as anti-Müllerian hormone). In male fetuses it is produced by Sertoli cells of the testicles and it suppresses Müllerian duct development. In the female it is produced after birth by primordial, preantral, and very early antral follicles and its function is to restrain primordial follicles from entering into FSH-dependent growth. MIS serum levels are stable throughout the menstrual cycle and begin to decrease with age as the primordial follicle pool shrinks. Its synthesis ceases when these structures disappear in the menopause.

Conclusion

Knowledge and understanding of the normal, human female reproductive process has grown considerably in the last century aided by animal experimentation, clinical research, and technological developments that allow measuring minute amounts of hormones in body fluids, tracing their action in cells, and measuring cell responses under a variety of conditions. The reproductive process across mammals bears many features in common and the reproductive process in the human female has, if any, very few distinguishing features from the rest of mammals. The wealth of basic knowledge acquired has allowed bringing into practice diagnostic and therapeutic procedures for improving sexual and reproductive health. Moreover, methods to control fertility at times when childbearing is unwanted or to allow infertile couples to have babies have become possible and are now widespread and generally successful.

See also: Family Planning/Contraception; Gynecological Morbidity; Infertility; Male Reproductive Function; Menopause; Puberty.

Citations

Wilcox AJ, Weinberg CR, and Baird DD (1995) Timing of sexual intercourse in relation to ovulation effects on the probability of conception, survival of the pregnancy, and sex of the baby. *New England Journal of Medicine* 333: 1517–1521.

Wilcox AJ, Weinberg CR, and Baird DD (1998) Post ovulatory ageing of the oocyte and embryo failure. *Human Reproduction* 13: 394–397.

Further Reading

Bischof P and Campana A (1996) A model for implantation of the human blastocyst and early placentation. *Human Reproduction Update* 2: 262–270.

Burrows TD, King A, and Loke YW (1996) Trophoblast migration during human placental implantation. *Human Reproduction Update* 2: 307–321.

Croxatto HB (1995) Gamete transport. In: Adashi EY, Rock JA, and Rosenwaks Z (eds.) *Reproductive Endocrinology, Surgery, and Technology* vol. 1, pp. 385–402. Philadelphia, PA: Lippincott-Raven.

Dimitriadis E, White CA, Jones RL, and Salamonsen LA (2005) Cytokines, chemokines and growth factors in endometrium related to implantation. *Human Reproduction Update* 11: 613–630.

Evans JP (2002) The molecular basis of sperm-oocyte membrane interactions during mammalian fertilization. *Human Reproduction Update* 8: 297–311.

Giudice LC (2004) Microarray expression profiling reveals candidate genes for human uterine receptivity. *American Journal of Pharmacogenomics* 4: 299–312.

Goldfien A and Monroe SE (1997) Ovaries. In: Greenspan FS and Strewler GJ (eds.) *Basic and Clinical Endocrinology,* 5th edn., pp. 434–574. Stamford, CT: Appleton and Lange.

Hotchkiss J and Knobil E (1993) The menstrual cycle and its neuroendocrine control. In: Knobil E, Neill JD, Greenwald GS, Markert CL, and Pfaff DW (eds.) *The Physiology of Reproduction* vol. 1, pp. 711–750. New York: Raven Press.

Luisi S, Florio P, Reis FM, and Petraglia F (2005) Inhibins in female and male reproductive physiology: Role in gametogenesis, conception, implantation and early pregnancy. *Human Reproduction Update* 11: 123–135.

Messinis IE (2006) Ovarian feedback, mechanism of action and possible clinical implications. *Human Reproduction Update* 12: 557–571.

Norman R and Phillipson G (1998) The normal menstrual cycle; changes throughout life. In: Fraser IS, Jansen RPS, Lobo RA, Whitehead MI, and Mishell DR, Jr. (eds.) *Estrogens and Progestogens in Clinical Practice*, pp. 105–118. London: Churchill Livingstone.

Seifer DB and Maclaughlin DT (2007) Mullerian inhibiting substance is an ovarian growth factor of emerging clinical significance. *Fertility and Sterility* 88: 539–546.

Sirios J, Sayasith K, Brown KA, *et al.* (2004) Cyclooxygenase-2 and its role in ovulation: A 2004 account. *Human Reproduction Update* 10: 373–385.

Suarez SS and Pacey AA (2006) Sperm transport in the female reproductive tract. *Human Reproduction Update* 12: 23–37.

Veenstra van Nieuwenhoven AL, Heineman MJ, and Faas MM (2003) The immunology of successful pregnancy. *Human Reproduction Update* 9: 347–357.

Yanagimachi R (2003) Fertilization and development initiation in orthodox and unorthodox ways: From normal fertilization to cloning. *Advances in Biophysics* 37: 49–89.

Yanagimachi R (2005) Male gamete contributions to the embryo. *Annals of the New York Academy of Sciences* 1061: 203–207.

Relevant Websites

http://www.web-books.com/eLibrary/Medicine/Physiology/Reproductive/Female.htm – Female Reproductive System.

http://en.wikibooks.org/wiki/Human_Physiology/The_female_reproductive_system – Human Physiology/The Female Reproductive System.

http://biology.clc.uc.edu/courses/bio105/sexual.htm – Reproductive Physiology, Conception, Prenatal Development.

http://www.educypedia.be/education/reproductivesystem.htm – Reproductive System, Male and Female Reproductive Anatomy.

Male Reproductive Function

H M Behre, Martin-Luther-University, Halle-Wittenberg, Halle, Germany

Introduction

Male reproductive function depends on the integrity of the organs of the male reproductive tract, most relevantly, the testis, epididymis, deferent duct, seminal vesicle, prostate, and penis, as well as the various parts of the central nervous system, in particular the hypothalamus and the pituitary gland. In contrast to female reproductive function, hormone and gamete production in healthy men persists lifelong. However, various disorders of male reproductive function show a relatively high prevalence and are therefore of significant relevance for public health.

The Testis

The main male reproductive organ is the testis. The testis has an endocrine as well as an exocrine function. The predominant endocrine function is synthesis of testosterone, which is important for normal sperm production within the testis as well as – after entering the general circulation and binding to the androgen receptors of the various target organs – for initiating and maintaining androgen-dependent functions. The most important exocrine function of the testis is production of immature spermatozoa that are – after maturation in the epididymis – able to fertilize an egg and induce a pregnancy.

Endocrine testicular dysfunction is characterized by testosterone deficiency. A recent population-based, observational survey indicates that the overall prevalence of testosterone deficiency associated with relevant clinical symptoms in men aged 30–79 years is 5.6% (Araujo *et al.*, 2007). These data demonstrate the overt relevance of testosterone deficiency for public health. In primary endocrine testicular failure, low testosterone secretion is caused by a deficiency or absence of Leydig cell function. Clinically relevant diseases include anorchia, gonadal dysgenesis, testicular tumors, testicular maldescent, Leydig cell hypoplasia, enzymatic defects in testosterone synthesis, numerical or structural chromosome abnormalities, gene mutations, irradiation, and systemic diseases. Testosterone deficiency can also be caused by exogenous factors such as environmental toxins or medications (**Table 1**). In contrast to primary endocrine testicular failure, secondary endocrine testicular failure is caused by absent or insufficient bioactivity of gonadotropin-releasing hormone (GnRH) or luteinizing hormone (LH).

Men diagnosed with primary exocrine testicular dysfunction are infertile. Although exact figures are not available, it is assumed that about 7% of all men are confronted with impaired fertility in the course of their lives, resulting in a significant impact on public health (Nieschlag, 2000a). Infertility might be caused by impaired spermatogenesis due to deficiency of GnRH or gonadotropin secretion, anorchia, gonadal dysgenesis, varicocele, orchitis, structural abnormalities of spermatozoa, testicular tumors, testicular maldescent, Leydig cell hypoplasia, enzymatic defects in testosterone synthesis, irradiation, or systemic diseases, or by exogenous factors such as environmental toxins or medications, numerical chromosome abnormalities, structural chromosomal abnormalities, Y chromosome microdeletions, gene mutations, or other still unknown genetic defects (**Table 1**).

Hormonal Regulation of Testicular Function

Testicular steroidogenesis and spermatogenesis are controlled primarily by hypothalamic GnRH and the two pituitary gonadotropins, LH and follicle-stimulating hormone (FSH) (**Figure 1**). The production and secretion of these stimulating hormones are under negative feedback control of hormones produced by the testis, especially testosterone and its active metabolites and inhibin B - (Weinbauer *et al.*, 2000).

Gonadotropin-releasing hormone (GnRH)

GnRH is a decapeptide produced in specific neurons of the hypothalamus. It has a relatively short half-life of less than 10 min and is released into the portal blood in discrete pulses, which are at least partially controlled by stimulatory and inhibitory effects of various steroid hormones, neurotransmitters, and neuromodulators. GnRH stimulates LH and FSH synthesis and secretion from the pituitary gland. The nature of the pulsatile GnRH secretion has not been well-characterized in the human but seems to be important for GnRH action, as experimental long-term continuous GnRH release is not able to stimulate LH and FSH synthesis and secretion.

GnRH stimulates gonadotropin secretion following occupation and activation of specific GnRH receptors in the pituitary gland. These receptors belong to a family of G protein-coupled receptors having the typical seven-membrane domain structure. GnRH is capable of modulating the number and activity of its own receptor.

Table 1 Classification of disorders of male reproductive function based on localization of cause

Localization of disorder	*Disorder*	*Cause*
Hypothalamus/pituitary	Kallmann syndrome	Congenital disturbance of GnRH secretion, defect of the Kal-1 gene
	Idiopathic hypogonatropic hypogonadism	Congenital disturbance of GnRH secretion
	Prader-Labhart-Willi syndrome	Congenital disturbance of GnRH secretion
	Constitutionally delayed puberty	Delayed biological clock
	Secondary disturbance of GnRH secretion	Tumors, infiltrations, trauma, irradiation, disturbed circulation, malnutrition, systemic diseases
	Hypopituitarism	Tumors, infiltrations, trauma, irradiation, ischemia, surgery, GnRH receptor mutation
	Pasqualini syndrome	Isolated LH-deficiency
	Hyperprolactinemia	Adenomas, medications, drugs
Testes	Congenital anorchia	Fetal loss of testes
	Acquired anorchia	Trauma, torsion, tumor, infection, surgery
	Testicular maldescent	Testosterone or AMH deficiency, congenital anatomical hindrance
	Varicocele	Venous insufficiency
	Orchitis	Infection and destruction of germinal epithelium
	Sertoli-cell-only syndrome	Congenital/acquired (e.g., after radiotherapy)
	Spermatogenetic arrest	Congenital/acquired
	Globozoospermia	Absence of acrosome formation
	Immotile cilia syndrome	Absence of dynein arms
	Klinefelter syndrome	Meiotic nondysjunction
	46,XX-male	Translocalization of part of Y chromosome
	47,XYY-male	Meiotic nondysjunction
	Noonan syndrome	Gene mutation
	Structural chromosomal anomalies	Deletions, translocations
	Persistent oviduct	AMH receptor mutation
	Gonadal dysgenesis	Genetic disturbances of gonadal differentiation
	Leydig cell hypoplasia	LH receptor mutation
	Male pseudohermaphroditism	Enzymatic defects in testosterone synthesis
	True hermaphroditism	Genetic disturbance in gonadal differentiation
	Testicular tumors	Congenital/unknown
	Disorder caused by exogenous factors or systemic diseases	Medication, irradiation, heat, environmental and recreational toxins, liver cirrhosis, renal failure
	Idiopathic infertility	Cause(s) unknown
Excurrent seminal ducts and accessory sex glands	Obstructions	Congenital anomalies, infections, vasectomy, accidental damage during appendectomy, herniotomy, or kidney transplantation
	Cystic fibrosis	Mutation of the CFTR-gene
	CBAVD (congenital bilateral aplasia of the vas deferens)	Mutation of the CFTR-gene
	Young syndrome	Mercury poisoning?
	Disturbance of liquefaction	Cause(s) unknown
	Immunologic infertility	Autoimmunity
Disturbed semen deposition	Hypospadias, epispadias	Congenital
	Penis deformations	Congenital/acquired
	Erectile dysfunction	Multifactorial origin
	Disturbed ejaculation	Congenital/acquired
	Phimosis	Congenital/acquired
Androgen target organs	Testicular feminization	Complete androgen receptor defect
	Reifenstein syndrome	Incomplete androgen receptor defect
	Prepenile bifid scrotum + hypospadias	Incomplete androgen receptor defect
	Bulbospinal muscular atrophy	Androgen receptor defect
	Perineo-scrotal hypospadias with pseudovagina	5α-reductase deficiency
	Gynecomastia	Hormonal imbalance

AMH, anti-Müllerian hormone; CFTR, cystic fibrosis transmembrane conductance regulator; GnRH, gonadotropin-releasing hormone; LH, luteinizing hormone.

Modified from Nieschlag E (2000b) Classification of andrological disorders. In: Nieschlag E and Behre HM (eds.) *Andrology: Male Reproductive Health and Dysfunction*, 2nd ed, pp. 83–87. Berlin: Springer-Verlag. With kind permission of Springer Science and Business Media.

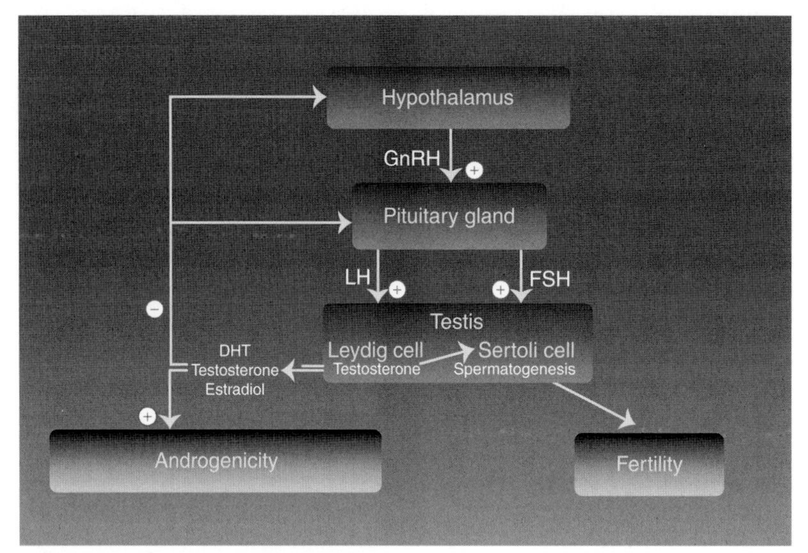

Figure 1 Endocrine regulation of male reproductive function.

Several disorders and diseases are known to cause testicular dysfunction by disturbances of GnRH synthesis, secretion, or function (**Table 1**). Diminished or absent hypothalamic secretion of GnRH can be found in patients with idiopathic hypogonadotropic hypogonadism (IHH) or Kallmann syndrome. Secondary GnRH deficiency might be caused by a tumor in the region of the diencephalon (craniopharyngioma or meningioma), metastases, granulomatous illnesses or hemochromatosis, fractures of the skull base, ischemic and hemorrhagic lesions, or radiotherapy. In patients with absent or diminished GnRH secretion, highly effective restoration of testicular function can be achieved by exogenous administration of GnRH intravenously or subcutaneously, in a pulsatile manner (Büchter *et al.*, 1999).

Several patients with inactivating mutations of the GnRH receptor causing testicular dysfunction have been described. In these men, exogenous GnRH is ineffective, but proper treatment of infertility can be achieved by exogenous gonadotropin administration (see 'Gonadotropins'). If fertility is not desired, patients should be treated with testosterone preparations.

Gonadotropins

LH and FSH, the gonadotropins regulating testicular function, are glycoprotein hormones secreted from the pituitary gland in response to GnRH (**Figure 1**). They consist of two subunits, alpha and beta. While the alpha subunit of LH and FSH is identical and is similar to the alpha subunit of other members of the glycoprotein hormone family, such as thyroid stimulating hormone or human chorionic gonadotropin (hCG), the specific actions of LH and FSH are caused by their specific beta subunits. Due to different terminal glycosylation, LH and FSH have different elimination half-lives of about 20 min and 120 min, respectively. Therefore, the pulsatility as measured in serum is more apparent for LH than for FSH. In addition to the pulsatile secretion of both gonadotropins, FSH is, to some degree, secreted constitutively from the pituitary gland.

Steroid hormones secreted from the testes, especially testosterone and its active metabolites, dihydrotestosterone and estradiol, exert a negative feedback control of both gonadotropins at the level of both the hypothalamus and pituitary. In addition, FSH secretion is under negative control of inhibin B secreted from the Sertoli cells of the seminiferous tubules in the testes.

Both gonadotropins act via specific receptors on different testicular cell types, with LH exerting its effects on Leydig cells and FSH binding to and activating receptors in Sertoli cells. The LH and FSH receptors belong to the family of G protein-coupled receptors, and both have a large extracellular hormone-binding domain. Normal function of both gonadotropins and of their receptors seems to be necessary for quantitatively and qualitatively normal initiation and development of spermatogenesis at puberty. Although some degree of spermatogenesis might be maintained thereafter by one normally functioning gonadotropin and its receptor, both LH and FSH activity seem to be mandatory for normal spermatogenesis.

A lack of normal physiological levels of gonadotropins in patients might be caused by primary tumors (gonadotropin, prolactin, growth hormone, thyroid hormone, ACTH-secreting, or endocrinologically inactive adenomas) or metastases of tumors in the pituitary or the hypophyseal stalk, or by brain surgery, radiotherapy, trauma, infections, hemochromatosis, or vascular disorders (**Table 1**). The most frequent cause of pituitary insufficiency is a prolactin-secreting pituitary adenoma (prolactinoma) that might destroy the gonadotropin-secreting pituitary cells or impair their function.

In most patients, prolactinomas can be treated effectively with dopamine agonistic drugs such as bromocriptine. Patients with irreversible destruction of gonadotropin-secreting pituitary cells can be successfully treated for their infertility with administration of exogenous gonadotropins. Instead of the natural LH, the gonadotropin hCG – which has LH activity – is commonly used because of its prolonged duration of action (Büchter *et al.*, 1999).

In some patients, inactivating mutations of the LH or FSH receptor gene have been detected as the cause for testicular dysfunction. In these patients, gonadotropin preparations with LH and FSH activity are generally not effective.

Endocrine Function: Testosterone Production

Testosterone is the most important sex hormone in men. More than 95% of the endogenous testosterone is produced in the testes, with 6–7 mg secreted every day. Testosterone is produced by the Leydig cells in the interstitial compartment in response to LH binding to its specific Leydig cell membrane receptor. As no testosterone can be stored in the Leydig cells, it has to be produced continuously *de novo*.

Testosterone is an important intratesticular factor for regulating spermatogenesis. In addition, it is released into the general circulation and transported in serum, with 54% bound to sex hormone binding globulin (SHBG), 44% bound to albumin, and 2% circulating as free or unbound steroid. The free and albumin-bound fractions are considered to be biologically active. However, the physiological role of SHBG-bound testosterone has not yet been fully elucidated. Several studies have shown that the testosterone concentration within the testis is about 100-fold higher than in the general circulation.

In serum, testosterone levels follow a circadian pattern, with morning concentrations 20–40% higher than evening values (Diver *et al.*, 2003). As normal ranges for serum testosterone levels are commonly based on morning levels, blood for diagnostic tests should be drawn in the morning. Short, intensive physical exercise can increase serum testosterone concentrations, whereas extended, exhausting physical exercise and the practice of high-performance sports can result in significant decreases.

Testosterone has two clinically relevant metabolites. It is 5 alpha-reduced to 5α-dihydrotestosterone (DHT), which is important for normal development of the external male genitalia and prostate gland. In a separate metabolic pathway, testosterone is aromatized to 17β-estradiol, which has essential functions in bone development.

Testosterone and its active metabolite DHT bind to an intracellular androgen receptor that belongs to the family of steroid hormone receptors. Once occupied, these receptors translocate to the cell's nucleus and bind to specific sequences of the genomic DNA to induce RNA and protein synthesis. In addition, a rapid nontranscriptional mode of action has been recently demonstrated for testosterone and DHT.

Various androgen receptor mutations have been described in infertile patients; these result in diminished or absent androgen action. In addition, polymorphisms of the androgen receptor, especially the so-called CAG repeat polymorphism in exon 1 of the androgen-receptor gene, have been shown to modify androgen action (Zitzmann and Nieschlag, 2007). In men with higher numbers of triplet residues, testosterone's effects in the various target organs seem to be attenuated.

Hypogonadism

Hypogonadism can be defined as a condition with absent or decreased biological action of testosterone. It might be caused by disorders at the testicular level and is then classified as primary hypogonadism. Hypogonadism due to disorders at the hypothalamic or pituitary level is classified as secondary hypogonadism. In addition, hypogonadism might be caused by androgen receptor dysfunction at the androgen target organs (**Table 1**).

The clinical symptoms of hypogonadism depend on the age of manifestation, in particular whether onset occurs before completed puberty or thereafter (**Table 2**). If androgen deficiency exists at the time of normal onset of puberty, an eunuchoid tall stature will result because of delayed epiphyseal closure. Arm span will exceed body length and the legs will become longer than the trunk. Onset of androgen deficiency after puberty will not result in a change of body proportions, although the musculature can be atrophic, depending on the duration and degree of androgen deficiency.

Long-standing testosterone deficiency may lead to osteoporosis, which can result in severe lumbago and pathological spine and hip fractures. Fat tissue distribution might have female characteristics, and lean body mass can be decreased and fat mass increased compared to normal.

Late-onset hypogonadism, defined as testosterone deficiency with clinical symptoms in the aging male, has become a topic of increasing interest throughout the world (Nieschlag *et al.*, 2005). Current demographic trends show a worldwide 'greying' of the population, with increasing percentage of the population aged 60–65 years and above. Average testosterone levels fall progressively with age, and a significant percentage of elderly men have serum testosterone levels lower than the normal ranges of young adults. A recent study estimated the overall prevalence of testosterone deficiency in combination with typical clinical symptoms of hypogonadism to be as high as 5.6% in men aged between 30 and 79 years. Prevalence was not related to race or ethnic group; however, it increased substantially with increasing age (Araujo *et al.*, 2007).

Table 2 Symptoms of hypogonadism relative to age of manifestation

Affected organ/ function	*Before completed puberty*	*After completed puberty*
Larynx	No voice mutation	No change
Hair	Horizontal pubic hairline, straight frontal hairline, diminished beard growth	Diminishing secondary body hair, decreased beard growth
Skin	Absent sebum production, lack of acne, pallor, skin wrinkling	Decreased sebum production, lack of acne, pallor, skin wrinkling
Bones	Eunuchoid tall stature, osteoporosis	Osteoporosis
Bone marrow	Low-degree anemia	Low-degree anemia
Muscles	Underdeveloped	Atrophy
Prostate	Underdeveloped	Atrophy
Penis	Infantile	No change of size
Testes	Small volume, possibly maldescended testes	Decrease of volume and consistency
Spermatogenesis	Not initiated	Decreased
Libido	Not developed	Loss
Erectile function	Decreased	Decreased

Adapted from Behre HM, Yeung CH, Holstein AF, *et al.* (2000) Diagnosis of male infertility and hypogonadism. In: Nieschlag E and Behre HM (eds.) *Andrology: Male Reproductive Health and Dysfunction*, 2nd edn., pp. 89–124. Berlin: Springer-Verlag. With kind permission of Springer Science and Business Media.

Late-onset hypogonadism is often caused by mixed primary and secondary endocrine testicular failure. Symptoms of late-onset hypogonadism are given in **Table 3**. However, the clinical diagnosis might escape detection for various reasons: Not all the signs and symptoms necessarily present together, they often progress slowly and are subtle in nature, and the unspecific signs and symptoms of late-onset hypogonadism might not be discernible from the unavoidable process of aging itself (Nieschlag *et al.*, 2004).

The most comprehensive review of the long-term clinical consequences and morbidity of late-onset hypogonadism has been compiled by the U.S. Institute of Medicine (Liverman and Blazer, 2004). Potential impacts of age-related testosterone decline on health have been identified for bone, body composition and strength, physical function, cognitive function, mood and depression, sexual function, health-related quality of life, cardiovascular and hematologic outcomes, prostate outcomes, sleep apnea, and other areas.

Apparent good health, defined as the absence of chronic illness, medication, obesity, or excessive drinking, attenuates the age-related decrease of testosterone (Feldman *et al.*, 2002). On the other hand, at all ages, serum testosterone concentrations may be transiently or permanently affected by comorbidities or their respective treatment (Kaufman and Vermeulen, 2005). Recently, it has been demonstrated in men aged ≥ 45 years and visiting primary care practices in the United States that odds ratios for having hypogonadism were significantly higher for men with hypertension, hyperlipidemia, diabetes, obesity, prostate disease, and asthma or chronic obstructive pulmonary disease than in men without these conditions (Mulligan *et al.*, 2006). Therefore, a proper diagnostic workup is mandatory in all men with known or suspected testosterone deficiency.

Table 3 Typical clinical symptoms of testosterone deficiency in late-onset hypogonadism

The easily recognized features of diminished sexual desire (libido) and erectile quality and frequency, particularly nocturnal erections
Changes in mood with concomitant decreases in intellectual activity, cognitive functions, spatial orientation ability, fatigue, depressed mood, and irritability
Sleep disturbances
Decrease in lean body mass with associated diminution in muscle volume and strength
Increase in visceral fat
Decrease in body hair and skin alterations
Decreased bone mineral density resulting in osteopenia, osteoporosis, and increased risk of bone fractures

Adapted from Nieschlag E, Swerdloff R, Behre HM, *et al.* (2005) Investigation, treatment and monitoring of late-onset hypogonadism in males: ISA, ISSAM, and EAU recommendations. *International Journal of Andrology* 28: 125–127.

In patients with disturbed endocrine testicular function and without present intention to induce a pregnancy, the therapy of choice is testosterone substitution. Numerous studies and more than 50 years of clinical experience have confirmed that testosterone substitution has beneficial effects on the symptoms of patients with primary or secondary hypogonadism, with acceptable adverse side effects. Today, several testosterone preparations are available for clinical use that can be applied via the intramuscular, transdermal, subcutaneous, oral, or buccal route. The various preparations have different pharmacokinetics, different adverse side effects, and different costs. As evidence from comparative studies is not available, the choice of the testosterone preparation is based on personal preferences and experience of the patient and the treating physician.

Men with late-onset hypogonadism may or may not benefit from testosterone treatment; the risks associated with such intervention are not well-described for this population (Nieschlag *et al.*, 2005). Results from recent studies show short-term beneficial effects of testosterone in older men that are similar to those in young men, but long-term data on the effects of such treatment in men with late-onset hypogonadism are limited. Future studies have to evaluate the specific risks on the prostate and cardiovascular system as well as the potential long-term benefits, including those that may retard frailty of elderly men.

Exocrine Function: Spermatogenesis

The second major function of the testis is production of spermatozoa. This process, termed spermatogenesis, takes place in the tubular compartment of the testis, while the testosterone production of the Leydig cells is located in the interstitial compartment (Weinbauer *et al.* 2000).

The tubular compartment of the testis represents 60–80% of the testicular volume. In total, each testis contains about 600 seminiferous tubules. The Sertoli cells within the seminiferous tubules are somatic cells and extend from the basal membrane to the lumen of the tubule. Sertoli cells support the structure of the seminiferous tubule and have a role in regulating the internal environment. The number of Sertoli cells in a testis determines the ultimate spermatogenic output. Sertoli cell function is regulated by occupation and activation of androgen and FSH receptors, and these hormonal signals are then transduced to the germinal cells.

Spermatogenesis takes place in three phases: (1) mitotic proliferation and differentiation of spermatogonia, (2) meiotic division, and (3) spermiogenesis, which represents a transformation of round haploid spermatids arising from the final division of meiosis into the complex structure of the spermatozoon (**Figure 2**). The overall duration of spermatogenesis is at least 64 days.

In most men with severely reduced sperm concentration in the ejaculate and no obstruction of the seminal ducts, various disturbances of spermatogenesis can be detected by testicular biopsy. In complete germ cell aplasia or Sertoli-cell-only syndrome (SCO), the tubules are reduced in diameter and contain Sertoli cells, but no other cells involved in spermatogenesis. Germ cell aplasia can also be focal with a variable percentage of tubules containing germ cells, but in these tubules spermatogenesis is often limited. Germ cell aplasia or SCO syndrome is one common cause of nonobstructive azoospermia (complete absence of spermatozoa in the ejaculate).

Spermatogenic arrest is a histopathological description of the interruption of normal germ cell maturation at the level of a specific cell type in the pathway leading from spermatogonia to spermatocytes to spermatids. Complete arrest of spermatogenesis results in azoospermia.

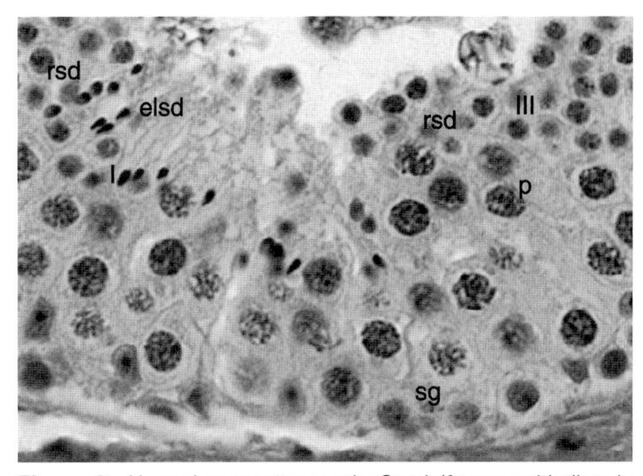

Figure 2 Normal spermatogenesis: Seminiferous epithelium in stage I (I) and stage III (III) of spermatogenesis showing spermatogonia (sg), pachytene spermatocytes (p), round step 1 and 3 spermatids (rsd), and elongating spermatids (elsd). Primary magnification, × 40. Reproduced from Behre HM and Bergmann M (2003) Primary testicular failure. In: McLachlan R (ed.) *Endocrinology of Male Reproduction*. South Dartmouth, MA: Endotext.com. http://www.endotext.org/male/index.htm (accessed February 2008), with permission from MOTEXT.com, INC.

Spermatogenetic arrest or germ cell aplasia might be caused by, for example, microdeletions of the Y chromosome and numerical or structural chromosome abnormalities.

Hypospermatogenesis is the third major classification of impaired spermatogenesis; it is characterized by a mildly, moderately, or severely reduced number of all germ cell types, including mature spermatids, in some or all tubules.

Seminal Ducts

Epididymis

After production within the seminiferous tubules of the testis, the immature testicular spermatozoa together with fluid secreted by the Sertoli cells pass the tubuli recti and the rete testis and enter into 12–18 efferent ducts that form the caput of the epididymis. Here, the efferent ducts unite and form the single tubule of the epididymis. The single tubule of the epididymis is then coiled into lobules and forms the corpus epididymis and finally the cauda epididymis. The total length of the epididymis is estimated to be 5–6 m. The time for spermatozoa to pass through the epididymis is between 2 to 11 days. The major function of the epididymis is the maturation of spermatozoa and storage prior to ejaculation.

Deferent Duct and Seminal Vesicle

During the emission phase of ejaculation, spermatozoa from the lower portion of the epididymis are transported via the deferent duct through its peristaltic action of muscle layers to the intraprostatic ejaculatory duct. Each deferent duct is about 30 cm long, and the ejaculatory duct about 2 cm. The seminal vesicle is a tubular gland of approximately 5 cm in length, located posteroinferior to the urinary bladder and surrounded by a thick coat of smooth muscle. The capacity of each seminal vesicle is approximately 2 ml, which contributes about 70% of the ejaculate volume. During the emission phase of ejaculation, sperm-containing fluid of the proximal deferent duct is diluted with fluids of the seminal vesicle, and the mixture transported via the intraprostatic ejaculatory duct into the prostatic urethra at the seminal colliculus.

Prostate

The prostate is a tubulo-alveolar exocrine gland surrounding the urethra just below the urinary bladder. Its main function is storage and secretion of clear alkaline fluid that constitutes up to 30% of the ejaculate volume. The prostatic urethra transports urine or, during ejaculation, the seminal fluid containing spermatozoa. During the ejection phase of ejaculation, the bladder neck becomes firmly closed, preventing reflux of the ejaculate into the bladder. By contraction of the prostatic musculature and pelvic floor muscles, the ejaculate is propelled into the anterior penile urethra and beyond. Ejaculatory disorders include retrograde, premature, retarded, or absent ejaculation, as well as various orgasmic disturbances. There are no good data on the prevalence of ejaculatory disorders; however, they seem to be as frequent as erectile dysfunction. For example, premature ejaculation has been found to affect approximately 13–18% of men aged between 40 and 80 years (Beutel *et al.*, 2006).

Penis

From the male reproductive function perspective, normal erectile function of the penis is important for sexual intercourse resulting in proper intravaginal sperm deposition for successful reproduction. Three hemodynamic factors in the penis are essential for normal erection: (1) relaxation of the muscle cells within the corpora cavernosa of the penis, (2) increase of arterial blood inflow by dilatation of the arterial vessels, and (3) restriction of venous blood outflow by compression of the intracavernosal and subtunical venous plexus.

According to the National Institutes of Health consensus, erectile dysfunction is defined as the inability to achieve and maintain an erection sufficient for satisfactory sexual performance. Erectile dysfunction can be caused by various disturbances. It is estimated that 50–80% of organically caused erectile dysfunction is due to arterial dysfunction. Other common reasons include neurogenic, endocrine, drug-induced, and – more often in younger men – psychogenic etiologies. In many patients, the reason for erectile dysfunction is multifactorial. Numerous epidemiological studies have been published on the prevalence of erectile dysfunction. In all these studies, the prevalence was strictly associated with aging. The overall prevalence has been estimated to be as low as 2.3% in men aged 30–39 years, but up to about 15–70% in men aged above 70 years (Beutel *et al.*, 2006).

Infertility

Infertility may be diagnosed when a couple fails to achieve a pregnancy within 1 year of regular unprotected intercourse. In contrast to other diseases or disorders, the diagnosis of infertility can be made only in a couple, and only in a couple with a desire for pregnancy. There is a significant interdependence of the reproductive function of the male and female partners. Disturbed male reproductive function, such as decreased sperm count, can be compensated for by optimal reproductive functions of the female partner, and vice versa.

Various diseases in the male might cause disorders of male reproductive function and infertility. These conditions can be classified by the topographic localization of the cause (**Table 1**). There are several causes for primary and secondary testicular failure. In many patients who have various degrees of abnormal spermatogenesis, the exact cause for testicular dysfunction is unknown. These patients are categorized as having idiopathic infertility and remain a majority of patients seen in modern andrology practices. For these patients, symptomatic treatment applying methods of assisted reproduction can be offered, such as intrauterine insemination, intracytoplasmic sperm injection with ejaculated sperm or sperm retrieved from testicular biopsies by testicular sperm extraction (TESE), depending on the site and degree of spermatogenic failure.

Acute or chronic infections of the epididymis, seminal vesicles, or prostate are curable conditions of male infertility. In patients with bilateral obstruction of the epididymis or deferent duct, microsurgical reanastomosis can be offered to restore fertility. When this is not possible, for instance in patients with congenital bilateral aplasia of the vas deferens/deferent duct (CBAVD), sperm can be retrieved from the epididymis by microsurgical sperm aspiration, or from testicular biopsies by TESE. For patients with CBAVD, a condition caused by mutations in the cystic fibrosis transmembrane conductance regulator gene, genetic counseling and testing should be offered.

In couples unable to conceive due to erectile dysfunction, various highly effective therapies can be offered, such as phosphodiesterase type V inhibitors, intracavernosal prostaglandin E1 injections, and hormonal, surgical, or psychological therapy. After proper diagnosis of the etiology of erectile dysfunction in the individual affected patient, the effectiveness of these rational therapies is high.

Infertility can be caused by retrograde ejaculation, for example, after surgery or trauma. In affected patients, the ejaculate – which would normally exit via the urethra – is redirected toward the urinary bladder. Patients with retrograde ejaculation can effectively be treated with various drugs, such as imipramine or chlorpheniramine in combination with phenylpropanalamine, which produce antegrade ejaculation and thereby restore fertility.

The multiplicity of diseases, disorders, or exposures resulting in infertility clearly indicates the need for a systematic andrological examination of the affected man, and not just an analysis of his ejaculate. Rational and preventive therapy of male infertility is based on proper clinical and laboratory diagnoses. These should include a comprehensive history, physical examination, application of ancillary methods such as ultrasonography of the testes (**Figure 3**), prostate, and seminal glands, and proper endocrine laboratory diagnosis and semen analysis, according to standard procedures recommended by the World Health Organization (World Health Organization Department of Reproductive Health and Research, 2010). In certain patients, examination might be completed by testicular biopsies and molecular and cytogenetic analyses.

Whenever possible, preventive therapy should be applied to maintain normal male reproductive function and integrity of the reproductive organs. For example, timely use of effective antibiotics in case of infections of the epididymis or prostate might prevent irreversible obstruction leading to azoospermia. Early hormonal treatment of maldescended testes or, if unsuccessful, surgical therapy might preserve fertility and decrease the risk of future testicular cancer (Pettersson *et al.*, 2007). Exogenous factors, such as radiation or environmental toxins, should be minimized or eliminated as soon as possible. Prior to compulsory chemotherapy or radiotherapy because of malignant diseases, patients should be offered cryopreservation of spermatozoa in order to maintain reproductive ability. From the public health perspective, it should be noted that smoking, obesity, abuse of alcohol or certain drugs such as anabolic steroids, and untreated systemic diseases are potentially avoidable factors leading to infertility (Handelsman, 2000; Brinkworth and Handelsman, 2000).

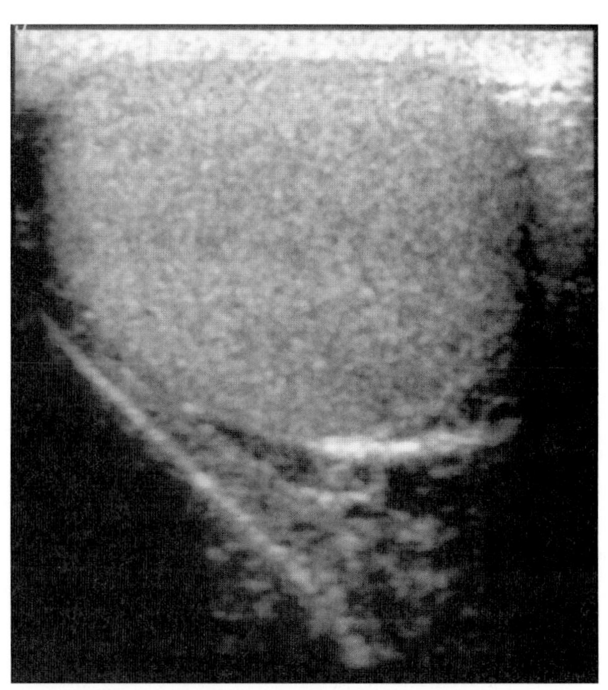

Figure 3 Scrotal sonography of a normal testis.

Conclusion

Male reproductive function depends on normal functioning of the various organs of the male reproductive tract and the respective regulatory mechanisms of the central nervous system. It should be noted that disturbances of reproductive function, such as hypogonadism, ejaculatory disorders, erectile dysfunction, and infertility, have a relatively high worldwide prevalence of more than 5% for each disorder. Therefore, proper diagnosis and treatment of male reproductive dysfunction as well as prevention whenever possible are of utmost relevance for public health.

See also: Family Planning/Contraception; Female Reproductive Function; Infertility; Prostate Cancer; Puberty; Sexual Health; Sexually Transmitted Infections: Overview; Testicular Cancer.

Citations

Araujo AB, Esche GR, Kupelian V, *et al.* (2007) Prevalence of symptomatic androgen deficiency in men. *Journal of Clinical Endocrinology and Metabolism* 92: 4241–4247.

Behre HM and Bergmann M (2003) Primary testicular failure. In: McLachlan R (ed.) *Endocrinology of Male Reproduction.* South Dartmouth, MA: Endotext.com. http://www.endotext.org/male/index.htm (accessed February 2008).

Behre HM, Yeung CH, Holstein AF, *et al.* (2000) Diagnosis of male infertility and hypogonadism. In: Nieschlag E and Behre HM (eds.) *Andrology: Male Reproductive Health and Dysfunction,* 2nd ed, pp. 89–124. Berlin: Springer-Verlag.

Beutel ME, Weidner W, and Brähler E (2006) Epidemiology of sexual dysfunction in the male population. *Andrologia* 38: 115–121.

Brinkworth MH and Handelsman DJ (2000) Environmental influences on male reproductive health. In: Nieschlag E and Behre HM (eds.) *Andrology: Male Reproductive Health and Dysfunction,* 2nd ed, pp. 253–270. Berlin: Springer-Verlag.

Büchter D, Behre HM, Kliesch S, and Nieschlag E (1999) Pulsatile GnRH or human chorionic gonadotropin/human menopausal gonadotropin as effective treatment for men with hypogonadotropic

hypogonadism: A review of 42 cases. *European Journal of Endocrinology* 139: 298–303.
Diver MJ, Imtiaz KE, Ahmad AM, Vora JP, and Fraser WD (2003) Diurnal rhythms of serum total, free and bioavailable testosterone and of SHBG in middle-aged men compared with those in young men. *Clinical Endocrinology* 58: 710–717.
Feldman HA, Longcope C, Derby CA, *et al.* (2002) Age trends in the level of serum testosterone and other hormones in middle-aged men: Longitudinal results from the Massachusetts male aging study. *Journal of Clinical Endocrinology and Metabolism* 87: 589–598.
Handelsman DJ (2000) Testicular dysfunction in systemic diseases. In: Nieschlag E and Behre HM (eds.) *Andrology: Male Reproductive Health and Dysfunction,* 2nd ed, pp. 241–251. Berlin: Springer-Verlag.
Kaufman JM and Vermeulen A (2005) The decline of androgen levels in elderly men and its clinical and therapeutic implications. *Endocrine Reviews* 26: 833–876.
Liverman CT and Blazer D (eds.) (2004) *Testosterone and Aging: Clinical Research Directions.* Washington, DC: The National Academies Press.
Mulligan F, Frick MF, Zuraw QC, *et al.* (2006) Prevalence of hypogonadism in males aged at least 45 years: the HIM study. *International Journal of Clinical Practice* 60: 762–768.
Nieschlag E (2000a) Scope and goals of andrology. In: Nieschlag E and Behre HM (eds.) *Andrology: Male Reproductive Health and Dysfunction,* 2nd ed, pp. 1–8. Berlin, Heidelberg, New York: Springer-Verlag.
Nieschlag E (2000b) Classification of andrological disorders. In: Nieschlag E and Behre HM (eds.) *Andrology – Male Reproductive Health and Dysfunction,* 2nd ed, pp. 83–87. Berlin: Springer-Verlag.
Nieschlag E, Behre HM, Bouchard P, *et al.* (2004) Testosterone replacement therapy: current trends and future directions. *Human Reproduction Update* 10: 409–419.
Nieschlag E, Swerdloff R, Behre HM, *et al.* (2005) Investigation, treatment and monitoring of late-onset hypogonadism in males: ISA, ISSAM, and EAU recommendations. *International Journal of Andrology* 28: 125–127.
Pettersson A, Richiardi L, Nordenskjold A, Kaijser M, and Akre O (2007) Age at surgery for undescended testis and risk of testicular cancer. *New England Journal of Medicine* 356: 1835–1841.
Weinbauer GF, Gromoll J, Simoni M, and Nieschlag E (2000) Physiology of testicular function. In: Nieschlag E and Behre HM (eds.) *Andrology: Male Reproductive Health and Dysfunction,* 2nd ed, pp. 23–61. Berlin, Heidelberg, New York: Springer-Verlag.
World Health Organization Department of Reproductive Health and Research (2010) *WHO Laboratory manual for the examination and processing of human semen*, 5th ed. Geneva: World Health Organization.
Zitzmann M and Nieschlag E (2007) Androgen receptor gene CAG repeat length and body mass index modulate the safety of long-term intramuscular testosterone undecanoate therapy in hypogonadal men. *Journal of Clinical Endocrinology and Metabolism* 92: 3844–3853.

Further Reading

Bhasin S (2007) Approach to the infertile man. *Journal of Clinical Endocrinology and Metabolism* 92: 1995–2004.
Gromoll J and Simoni M (2005) Genetic complexity of FSH receptor function. *Trends in Endocrinology and Metabolism* 16: 368–373.
Hafez ESE, Hafez B, and Hafez SD (2004) *An Atlas of Reproductive Physiology in Men*. New York: Parthenon Publishing.
Krausz C and Giachini C (2007) Genetic risk factors in male infertility. *Archives of Andrology* 53: 125–133.
Nieschlag E and Behre HM (eds.) (2000) *Andrology: Male Reproductive Health and Dysfunction,* 2nd edn. Berlin: Springer-Verlag.
Nieschlag E and Behre HM (eds.) (2004) *Testosterone: Action, Deficiency, Substitution,* 3rd edn. Cambridge, UK: Cambridge University Press.
Oehninger SC and Kruger TF (eds.) (2007) *Male Infertility: Diagnosis and Treatment.* London: Informa Healthcare.
Schill WB, Comhaire FH and Hargreave TB (eds.) (2006) *Andrology for the Clinician.* Berlin: Springer-Verlag.
Strauss JF and Barbieri RL (2004) *Yen & Jaffe's Reproductive Endocrinology: Physiology, Pathophysiology and Clinical Management,* 5th edn. Philadelphia, PA: Saunders.

Relevant Websites

http://www.uni-leipzig.de/~eaa/ – European Academy of Andrology.
http://www.andrology.org – International Society of Andrology.
http://www.who.int/reproductive-health/ – World Health Organization, Sexual and Reproductive Health.

Puberty

S R Grover and A Bajpai, Royal Children's Hospital, Parkville, Victoria, Australia

Definition

Puberty is the phase of life characterized by physical changes that mark the transition from childhood to adulthood. It includes a process involving accelerated growth and physical maturation of the respective reproductive systems to allow fertility in boys and girls. The end result is the constellation of physical features that distinguish males and females, including secondary sexual characteristics and the major differences of size, shape, body composition, and function in many body structures and systems.

When the processes of puberty fail to follow their usual time course, fail to occur, or are affected by an altered environment or physical anomalies, there can be long-term impacts on the achievement of future adult life and fertility.

The process of physical changes is generally known as puberty. In contrast, the slower process of psychosocial, emotional, and behavioral changes that occur between childhood and adulthood, combined with the physical changes, is generally known as adolescence. There has been increasing recognition that these broader adolescent changes are associated with risk behaviors (including smoking, drug usage, early sexual activity, and eating disorders), which may become the basis for significant long-term health problems in adulthood. However, there is an opportunity to promote and develop healthy habits and behaviors during adolescence as well. Public health professionals can play an important role in this respect.

Physiological Changes

Critical changes in the interaction of endocrine functions of childhood begin as puberty approaches. These changes include the activation of the hitherto quiescent hypothalamus–pituitary–gonadal axis resulting in gonadarche, i.e., the onset of gonadal functions. The increased production of adrenal androgens giving rise to the pubic and axillary hair is known as adrenarche.

Puberty in girls generally follows the sequence of accelerated growth, breast development, adrenarche, and menarche and takes an average 4.5 years (range, 1.5–6 years), although this sequence of events is not identical for all individuals (**Figure 1**). For 95% of girls, secondary sexual characteristics appear between the ages of 8.5 and 13 years. For boys, the sequence is different, with genital changes, body and facial hair, adrenarche, voice changes, and accelerated growth occurring in this order (**Figure 1**). Pubertal onset occurs in 95% of boys between the ages of 9.5 and 13.5 years (Tanner, 1989).

The initial endocrine changes in the hypothalamus–pituitary–gonadal axis involve an increase in the pulsatile secretion of luteinizing hormone (LH) from the pituitary in response to pulsatile nocturnal gonadotropin-releasing hormone (GnRH) secretion from the hypothalamus. This nocturnal LH release precedes breast development in girls and testicular volume changes in boys by several years (Mitamura *et al.*, 2000) and is sleep-dependent (Boyar *et al.*, 1974). There is a progressive increase in LH pulse size as well as a gradual shift to the adult pattern of an LH pulse occurring every 90 min throughout the day and night. In contrast, follicle-stimulating hormone (FSH) does not have a diurnal pattern with nocturnal release, but there is a progressive increase in pulse amplitude until mid-puberty.

The triggers for the changes in the interaction between the hypothalamus–pituitary–gonadal axis are not well understood. Importantly, the effects of induction factors are time-dependent and many factors have only a permissive and not a contributory role. There is some evidence

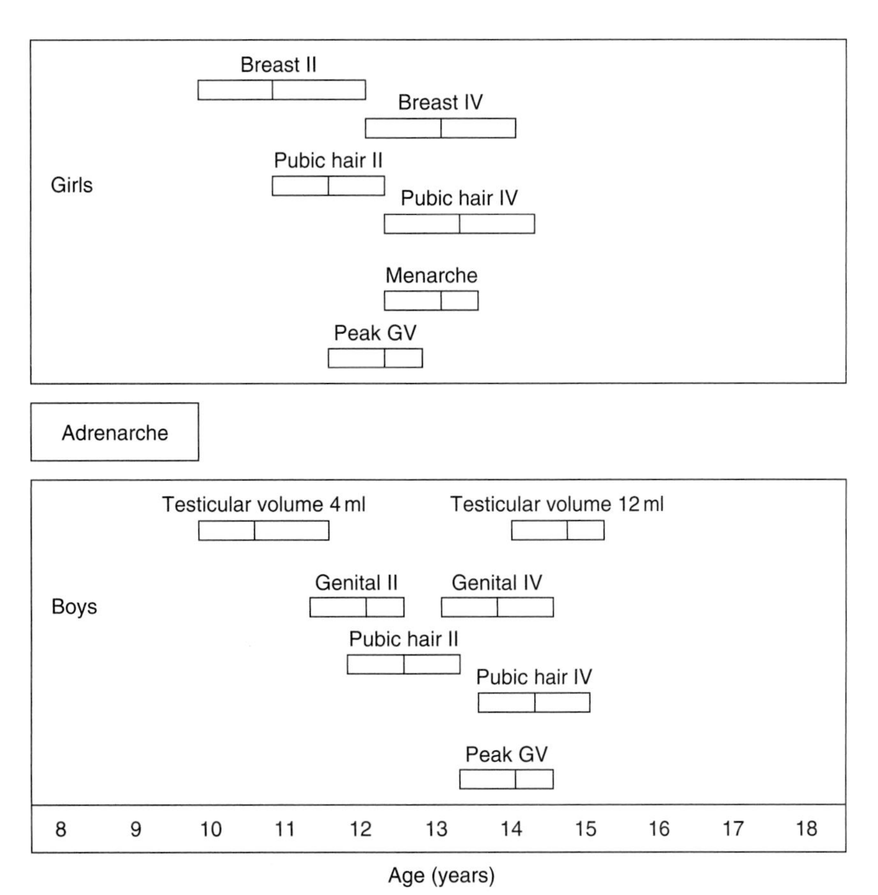

Figure 1 Line diagram demonstrating the pattern of pubertal progression observed in girls and boys. The boxes represent the 25th and 75th percentiles for the age at which individual pubertal stages are achieved. GV = growth velocity. Adapted from Patton GC and Viner R (2007) Pubertal transitions in health. *Lancet* 369: 1130–1139.

that leptin may play a role (Garcia-Mayor *et al.*, 1997), providing the link between reproductive function and energy balance. A possible role for leptin is supported by the presence of leptin receptors in the hypothalamus, especially on the neuropeptide Y (NPY) neurons. Neuropeptide Y has a role in regulating brain function, including food-intake behaviors (Terasawa and Fernandez, 2001), which ties closely with theories relating to critical weight, body mass index, or body fat content as necessary prerequisites for onset of puberty (see the section titled 'Secular Trends'). Before puberty, NPY appears to have an inhibitory role on GnRH secretion; this inhibition is thought to decline during and after puberty. Additionally, as puberty requires a functional growth hormone (GH) axis for full gonadal activation and the onset of puberty is associated with a dramatic increase in linear height growth, the role of the growth hormone–insulin-like growth factor-1 (IGF1) axis also appears to be critical. Recently, kisspeptin, a hypothalamic peptide, has been shown to be the key regulator of pubertal onset. Kisspeptin is also believed to provide the missing link between integration of nutritional signals (leptin and ghrelin) and pubertal onset (Seminara, 2006).

For girls, the altered LH pattern provokes a rise in estrogen production from the ovaries with a subsequent development of breast buds (Tanner stage 2), which is usually the first identified physical change of puberty. For boys, the first change is an increase in testicular volume (Mitamura *et al.*, 2000).

Maturation of the hypothalamic–pituitary–gonadal axis results in the achievement of fertility. For girls, ovulation – the mark of attaining potential fertility – occurs usually 1–2 years after menarche (i.e., the first episode of menstrual bleeding). It appears that the later the age of menarche the longer the delay before 50% of the cycles are ovulatory, with the interval being 12 months if menarche occurs before age 12 years, 3 years when menarche occurred at age 12–12.9 years, and 4.5 years when menarche was after age 13 years. For boys, fertility is indicated by the production of sperm. While this is often difficult to detect, examination of an early morning centrifuged urine sample for the presence of spermatozoa has been increasingly shown to be effective. Researchers from developed countries have found sperm in the urine of boys at ages 11–12 years, sometimes even before pubarche (pubic hair development) (Richardson and Short, 1978).

Pubic hair development usually occurs approximately 6 months after thelarche (breast development); however, in one-third of all girls pubic hair may be the first manifestation of puberty. This is the marker of adrenarche.

Although there appears to be a temporal relationship between these two hormonal systems (the hypothalamus–pituitary–gonadal system and the hypothalamic–pituitary–adrenal system), with both contributing to the development of secondary sexual characteristics, there is considerable evidence to suggest that their control mechanisms are independent (Sklar *et al.*, 1980; Counts *et al.*, 1987). Premature adrenarche (precocious appearance of pubic and axillary hair before 8 years of age) is not associated with premature gonadarche. In hypergonadotropic hypogonadism (gonadal dysgenesis), where there is a failure in gonadal hormone production, adrenarche still occurs. Similarly, gonadarche is unaffected in children with adrenal insufficiency.

The adolescent growth spurt happens in girls before boys (Gordon and Laufer, 2005). For girls, this may be the first marker of puberty, whereas in boys it invariably occurs after testicular enlargement. Estrogen has a gender nondimorphic effect on growth and is responsible for the growth spurt in both girls and boys. Local conversion of androgens to estrogens under the influence of the enzyme aromatase is crucial for these effects in boys. The growth-promoting effects of estrogen are largely related to growth hormone secretion with minor effects on IGF1 production. For girls, menarche coincides with a slowing of the growth velocity to approximately 4 cm/year. Prenatal growth significantly influences pubertal development as demonstrated by normal pubertal maturation but smaller pubertal height and weight gain in babies born small for gestational age.

The development of peak bone mass occurs during the teenage years, modulated by sex steroids and growth hormone. The achievement of optimum bone density also requires adequate dietary calcium, vitamin D, and exercise, particularly weight-bearing exercise. Several studies suggest that 60% of adult bone mass is acquired during the pubertal growth spurt (Gordon and Laufer, 2005). Peak bone mass is achieved by the age of 16 years in girls and 18 years in boys.

Secular Trends

The age of onset of puberty varies considerably between individuals and populations and seems to be influenced by both genetic and environmental factors. Johann Sebastian Bach recorded the age of choir boys transitioning from singing treble to alto in three churches in Leipzig between the years 1723 and 1750. It averaged around 17 years (Daw, 1970). Today, the boys of the choir in King's College Chapel, Cambridge, United Kingdom, usually break their voices at around 13 years (Potts and Short, 1999). For girls, the onset of menarche is well documented, with records from 1850 in Scandinavia that show that the average age of menarche was about 17.5 years. Throughout the Western world, there has been a well-documented fall in age of first menstruation over the last few centuries and by as much as 6 months in a decade over much of the last century. The age appears to have reached a plateau at between 12.5 and 13 years in Europe and the United States.

This secular trend has been attributed to improved nutrition and health. Whether environmental factors such as endocrine disrupters having an estrogenic effect have an influence remains to be shown. There are indications that breast development is also occurring earlier in American girls. This had led to the recommendation of lowering the age cut-off for the diagnosis of precocious puberty by the Lawson Wilkins Pediatric Endocrine Society to 6 years in black and 7 years in white girls (Kaplowitz and Oberfield, 1999). The earlier breast development has, however, not been associated with a similar trend in the age of menarche, indicating that early thelarche is associated with slower pubertal progression. The critical role of improved nutrition in the development of a secular trend is exemplified by the lack of such a trend in developing countries. While there are limited data on the pattern of pubertal development in resource-poor countries, anecdotal reports indicate later pubertal onset compared to industrialized countries. In a recent population-based prospective study, Senegalese girls were found to achieve sexual maturity on average 3 years later than their Western counterparts (Garnier *et al.*, 2005).

Genetic factors are clearly a major influence as children with a family history of early onset of puberty will also start puberty at an earlier than average age. Also, black American girls achieve all stages of pubertal development 6 months earlier than white American girls. Other factors that appear to have an influence on the onset and rate of progression of puberty include geographic location, exposure to light, general health and nutrition, and psychological factors. Obesity and residence closer to the equator, at lower altitudes, and in an urban area are associated with earlier onset of puberty.

It has been argued that a critical body mass (47.8 kg or 105 pounds) is required to achieve menarche for girls. Rather than total weight *per se*, the more crucial measure is probably body composition and the percentage of body fat (thought to be approximately 16–23%). Thus, the moderately overweight (20–30% over normal weight) girls have earlier menarche, and anorexic girls and intense exercisers have delayed menarche. Nevertheless, there are inconsistencies in the data, as it has been reported in a longitudinal study that there is no change in body fat composition until after menarche. Support for the theory that body weight and puberty are linked was provided by the identification of leptin and its role (Garcia-Mayor *et al.*, 1997). Studies have demonstrated that leptin levels increase in puberty and are low in athletes and those with anorexia, and that they delayed puberty. Additionally, rodent and monkey experiments demonstrate a role for leptin in the onset of sexual maturation.

Given the important relationship of nutrition and pubertal development, it is not surprising that a wide variety of nutritional disorders are associated with disordered pubertal development. Thus, puberty is usually delayed in individuals with malnutrition (including those with eating disorders), chronic infections, malabsorption, and anemia.

Variation Within the Normal Range

As there is considerable individual variation in onset and duration of puberty, there can be concern that the pubertal development of a young person may be abnormal, either delayed or precocious. Often this is not the case, and the boy or girl simply represents either end of the normal spectrum.

Delayed puberty can result in a young person being physically immature and short compared to peers. Since adolescence is a time when the young person is becoming more independent, and the establishment of a peer group is an important step in this process, delayed puberty can have a negative impact on overall well-being. This response varies considerably among individuals but boys in particular may be more concerned by their short stature than by any delay in the development of their secondary sexual characteristics. Early pubertal development may also have a negative psychosocial impact. These children appear more physically mature than their level of intellectual development and are prone to sexual abuse.

Abnormal Pubertal Development

Pubertal development can be precocious or delayed. Precocious puberty is defined as pubertal onset (stage II breast development in girls and testicular volume more than 4 ml in boys) before the age of 8 years in girls and 9.5 years in boys. No underlying cause is identified in most girls with precocious puberty, in contradistinction to boys who usually have an identifiable etiology. Precocious puberty may be caused by increased gonadotropin production (gonadotropin-dependent) or autonomous gonadal hormone production (gonadotropin-independent). Gonadotropin-dependent precocious puberty may be related to a wide variety of intracranial pathologies including brain tumors, congenital malformation, and neurological insults emphasizing the need for thorough neurological evaluation (including neuroimaging). Gonadotropin-independent precocious puberty is rare in girls and most commonly related to estrogen-producing ovarian cysts. This form of puberty is much more common in boys and caused by increased adrenal or testicular androgen production. Besides the significant psychosocial impact, the most important consequence of precocious puberty is short stature. While these children are tall compared to their peers at presentation, their bone age is disproportionately advanced, resulting in compromised final height. The most effective treatment for gonadotropin-dependent precocious puberty is gonadotropin-releasing hormone (GnRH) analogs. Prolonged stimulation by these analogs results in the desensitization of the pituitary and reversal of pubertal changes. These agents are expensive, however, and should be used only in individuals with progressive puberty and compromised growth potential. Medroxyprogesterone acetate and cyproterone acetate are economical and effective in preventing menarche but have no effect on final height. They may be used in girls with intellectual disability and early puberty if height is not a major issue of concern. Gonadotropin-independent precocious puberty is treated by agents that decrease sex hormone production (aromatase inhibitors) or action (antiandrogens in boys and selective estrogen receptor modulators in girls).

Delayed puberty is defined by the absence of pubertal development by the age of 13 years in girls and 14 years in boys. In addition, inappropriate progress of puberty with no menarche within 5 years of thelarche indicates delayed puberty. Delayed puberty may be caused by disorders of the hypothalamic–pituitary axis (hypogonadotropic hypogonadism) or of the gonadal functions (primary gonadal failure or hypergonadotropic hypogonadism). Hypogonadotropic hypogonadism is most commonly reversible and related to systemic diseases, nutritional disorders, and increased physical activity. On rare occasions, it is caused by intracranial tumors, congenital malformations, and neurological insult such as radiotherapy and surgery. Hypergonadotropic hypogonadism is most commonly associated with chromosomal disorders (with Turner's syndrome in girls being the most common). With increasing numbers of young people surviving childhood malignancies, the potentially negative effects of chemotherapy and radiotherapy on the gonads, including possible gonadal failure, need to be remembered. Importantly, delayed puberty may represent a normal variation of pubertal development (constitutional delay of puberty and growth). The condition is particularly common in boys and associated with delayed skeletal maturation and paternal history of delayed puberty. Treatment of delayed puberty should be started at an appropriate age (13 years in girls and 14 years in boys). Low-dose estrogens (10% of adult doses) should be started in girls and gradually increased over 2 years to adult doses. Progesterone should be added after the onset of withdrawal bleeding. In boys, parenteral testosterone treatment should be started at low doses (100 mg monthly) and increased to adult doses (300 mg every 3 weeks) over 2 years. Boys with constitutional delay of puberty and growth respond well to three to six doses of testosterone.

Abnormal adrenal function can result in early adrenarche (independent of gonadarche) with resultant pubic and axillary hair, acne, and an increase in growth rate with advance in skeletal height. The condition is usually nonprogressive and does not require extensive evaluation. Features of virilization (change in voice, clitoromegaly, and increased muscularity) should prompt evaluation for hyperandrogenic conditions such as congenital adrenal hyperplasia and virilizing adrenal or ovarian tumors.

The psychosocial repercussions of achieving physical maturation well ahead of the cognitive and psychosocial development, and thus being out of synchrony with peers, can have major consequences in increasing health risk behaviors in these young people.

Disorders of Sexual Development

Although some of the conditions associated with disorders of sexual development are diagnosed at birth or in early childhood (for example, girls with ambiguous genitalia at birth due to congenital adrenal hyperplasia, girls with descended testes due to the complete androgen insensitivity syndrome, children with ambiguous genitalia at birth due to mixed gonadal dysgenesis, partial androgen insensitivity, or true hermaphroditism), some of these conditions are not diagnosed until pubertal development is noted to be abnormal. It is estimated that genital anomalies occur in 1 in 4500 births. Likewise, structural anomalies of the urogenital tract may have required initial corrective surgery in the neonatal period, but further interventions relating to the altered anatomy nevertheless may be required when the young person passes through puberty. This is particularly the case for young women with obstructive genital tract anomalies that may not have been appreciated until menstruation begins. Delays in making these diagnoses can have a negative impact on future fertility.

Virilization at puberty, delayed puberty, or primary amenorrhea may all be presentations of disorders of sexual development. Correct identification of the underlying cause is important to ensure an optimal long-term outcome.

The impact of having a disorder of sexual development also needs careful psychological attention – because issues of fertility and infertility, being normal or abnormal, i.e., being different from one's peers – are important issues for adolescents in terms of their psychosocial development (Liao, 2004). Thus, even for those young people in whom the diagnosis and management of a disorder of sexual development was commenced in infancy, issues regarding genital appearance, sexual function, and general well-being are ideally addressed by a multidisciplinary team (Warne *et al.*, 2005). A shift in health-care provision is required in adolescence to ensure that the young person – rather than the parents on behalf of the child – is the focus of care.

The issues for teenage boys and girls with disorders of sexual development are similar. Body image and self-identity, satisfaction with genital appearance and function, and prospects of being fertile or infertile are important. Although there are cultural differences in response to these issues, increasing evidence across cultures reveals that disclosure is important for all individuals.

Conclusions

The timing of puberty has changed over time, with a range of factors influencing its onset and course. Although the knowledge regarding these processes has improved, there are still many triggers and influences that are poorly understood. The end result of a physically mature, fertile individual is the outcome of a developmental process that is coupled closely to the cognitive and psychosocial maturation process. When the pubertal processes fail to progress normally, there are often clear consequences in terms of failure to achieve full sexual and reproductive health as well as mental health. Conversely, failure to achieve psychosocial maturity also has implications for achieving the physical endpoint of a healthy and successful sexual and reproductive life.

See also: Female Reproductive Function; Gynecological Morbidity; Male Reproductive Function; Sexual Health.

Citations

Boyar RM, Rosenfeld RS, Kapen S, *et al.* (1974) Human puberty: Simultaneous augmented secretion of luteinising hormone and testosterone during sleep. *Journal of Clinical Investigation* 54: 609–618.

Counts DR, Pescovitz OH, Barnes KM, *et al.* (1987) Dissociation of adrenarche and gonadarche in precocious puberty and in isolated hypogonadotrophic hypogonadism. *Journal of Clinical Endocrinology and Metabolism* 64: 1174–1178.

Daw SF (1970) Age of puberty in Leipzig,1727–49, as indicated by voice breaking in JSBach's choir members. *Human Biology* 42: 87–89.

Garcia-Mayor RV, Andrade MA, Rios M, *et al.* (1997) Serum leptin levels in normal children: Relationship to age, gender, body mass index, pituitary-gonadal hormones, and pubertal stage. *Journal of Clinical Endocrinology and Metabolism* 82: 2849–2855.

Garnier D, Simondon KB, and Bénéfice E (2005) Longitudinal estimates of puberty timing in Senegalese adolescent girls. *American Journal of Human Biology* 17: 718–730.

Gordon CM and Laufer MR (2005) The physiology of puberty. In: Emans SJ, Laufer MR, and Goldstein DP (eds.) *Pediatric and Adolescent Gynecology*, pp. 120–155. Philadelphia, PA: Lippincott, Williams, and Wilkins.

Kaplowitz PB and Oberfield SE (1999) Reexamination of the age limit for defining when puberty is precocious in girls in the United States: Implications for evaluation and treatment. Drug and Therapeutics and Executive Committees of the Lawson Wilkins Pediatric Endocrine Society. *Pediatrics* 104: 936–941.

Liao LM (2004) Development of sexuality: Psychological perspectives. In: Balen AH, Creighton SM, Davies MC, MacDougall S, and Stanhope R (eds.) *Paediatric and Adolescent Gynaecology – A Multidisciplinary Approach*, pp. 77–93. Cambridge, UK: Cambridge University Press.

Mitamura R, Yano K, Suzuki N, *et al.* (2000) Diurnal rhythms of luteinizing hormone, follicle-stimulating hormone, testosterone, and estradiol secretion before the onset of female puberty in short children. *Journal of Clinical Endocrinology and Metabolism* 85: 1074–1080.

Patton GC and Viner R (2007) Pubertal transitions in health. *Lancet* 369: 1130–1139.

Potts M and Short RV (1999) Growing up. In: Potts M and Short RV (eds.) *Ever Since Adam and Eve*, pp. 161–163. Cambridge, UK: Cambridge University Press.

Richardson DW and Short RV (1978) The time of onset of sperm production in boys. *Journal of Biosocial Science* 5: 15–25.

Seminara SB (2006) Mechanisms of disease: The first kiss – A crucial role for kisspeptin-1 and its receptor G-protein-coupled receptor 54, in puberty and reproduction. *Nature Clinical Practice. Endocrinology and Metabolism* 2: 328–334.

Sklar CA, Kaplan SL, and Grumbach MM (1980) Evidence for dissociation between adrenarche and gonadarche: Studies in patients with idiopathic precocious puberty, gonadal dysgenesis, isolated gonadotroph deficiency, and constitutionally delayed growth and adolescence. *Journal of Clinical Endocrinology and Metabolism* 51: 548–556.

Tanner JM (1989) *Fetus into Man: Physical Growth from Conception to Maturity*. Cambridge MA: Harvard University Press.

Terasawa E and Fernandez DL (2001) Neurobiological mechanisms of the onset of puberty in primates. *Endocrine Reviews* 22: 111–151.

Warne G, Grover S, Hutson J, *et al.* (2005) A long-term outcome study of intersex conditions. *Journal of Pediatric Endocrinology and Metabolism* 18: 555–567.

Further Reading

Balen AH, Creighton SM, Davies MC, MacDougall S, and Stanhope R (2004) *Paediatric and Adolescent Gynaecology – A Multidisciplinary Approach*. Cambridge, UK: Cambridge University Press.

National Center for Health Statistics (2007) *2000 CDC Growth Charts*. United States Department of Health and Human Services. http://www.cdc.gov/growthcharts (accessed February 2008).

Potts M and Short RV (1999) *Ever Since Adam and Eve*. Cambridge, UK: Cambridge University Press.

Rogol AD, Clark PA, and Roemmich JN (2000) Growth and pubertal development in children and adolescents: Effects of diet and physical activity. *American Journal of Clinical Nutrition* 72: S521–S528.

Sperling MA (2002) *Pediatric Endocrinology*. 2nd edn. Philadelphia, PA: Saunders.

Speroff LGR and Kase N (1999) *Clinical Gynecologic Endocrinology and Infertility*, 6th edn. Philadelphia, PA: Lippincott Williams and Wilkins.

Menopause

D W Sturdee, Solihull Hospital, West Midlands, UK

Introduction

One of the earliest references to menopause in the world literature comes from the Bible: "Abraham and Sarah were old and well stricken in age; and it ceased to be with Sarah after the manner of women" (Gen. 18:11), although three chapters later Sarah gives birth to Isaac, so in reality she could not have had her menopause at that time. In the whole of nature only the human female experiences a physical menopause. While other animals remain potentially fertile until death, the woman of today, in the developed parts of the world at least, will spend on average one-third of her life without the benefits of the female sex hormone – estrogen. By contrast, a man slides without abrupt change from one stage of his life to the next maintaining the ability to father children well into old age. The ovary is the only organ in the body that naturally ceases to function in midlife. For many reasons, however, women will readily appreciate the merit of not being fertile into old age, as the human child is more dependent on its mother than any other animal.

It is only relatively recently that menopause has attracted much attention. Indeed, at the beginning of the twentieth century a woman in the Western world could only expect to live to about 48 years, but now life expectancy is about 80 years, with the number of women in the United Kingdom aged over 85 years expected to rise by 50% in the next 20 years. The United Nations projects that by the year 2050 women aged 60 years and over will constitute 40% of the female population of Europe and 23% of the entire world. In comparison, in 1997, the life expectancy in sub-Saharan Africa was only 52 years and, in many African countries, life expectancy has decreased by 3 years as a result of acquired immune deficiency syndrome (AIDS), so that menopause is less likely to be of much relevance for many women.

Menopause, or last menstrual period, is a natural biological event resulting from failure of the ovaries to continue producing eggs (ova) and the associated hormones estrogen and progesterone. The term is derived from the Greek *menos*, meaning month, and *pausos*, meaning an ending. The median age at which menopause occurs is 52 years and has changed little over the years. It tends to occur earlier in communities of low socioeconomic status, those living at high altitudes, and 1–2 years earlier in women who smoke. The timing of the natural menopause tends to be inherited, with daughters experiencing menopause around the same time as their mothers.

Types of Menopause

Premature

A premature menopause occurs in about 1% of women, being defined as menopause before the age of 40 years. This may be due to genetic factors or be iatrogenic as a result of a medical or gynecological condition. The associated infertility can be very distressing when premature menopause occurs unexpectedly before intended pregnancies have been achieved.

Surgical

A surgical menopause occurs when functioning ovaries are removed, usually together with a hysterectomy (removal of the uterus). The sudden cessation of ovarian function and rapid decline in circulating estrogen is more likely to result in menopausal symptoms than a more gradual decrease in ovarian activity associated with a natural menopause.

Natural

The mechanism of a natural menopause is not fully understood but is related to the ovaries running out of suitable follicles from which ova are produced. The ovaries have a finite number of follicles whose numbers are maximal during intrauterine life, after which there is a logarithmic reduction until about 50 years later when the stock is exhausted (**Table 1**), or those remaining are unable to respond to stimulation by endogenous or exogenous gonadotropins. Prior to menopause and during a variable number of months or years there is a gradual reduction in the control mechanism for the ovarian cycle with accompanying endocrine changes (Burger *et al.*, 2004). In particular, there is a decrease in the secretion of estradiol, the main estrogen, from the ovaries accompanied by a rise in circulating follicle-stimulating hormone (FSH) as the pituitary gland tries to stimulate the increasingly resistant remaining ovarian follicles to develop (**Figure 1**). This loss of control of the cycle causes disturbances of menstruation with heavier, painful, irregular, and sometimes erratic periods that are frequent causes of medical consultation. In the majority of cases, there is no underlying disease process and the abnormal periods are labeled 'dysfunctional uterine bleeding.' However, benign conditions, such as uterine fibroids and endometriosis, are also more common at this time and may

Table 1 Follicular attrition with age[a]

Number of follicles	*Age*
7 million	5–7 months in utero
2 million	At birth
155 000	18–24 years
59 000	25–31 years
8300	40–44 years

[a]Only 0.02% of primordial follicles ever ovulate. Adapted from Bloch E (1952) Quantitative morphological investigations of the follicular system in women. Variations at different ages. *Acta Anatomica* 14: 108–123; Faddy MJ, *et al.* (1992) Accelerated disappearance of ovarian follicles in mid-life: Implications for forecasting menopause. *Human Reproduction* 10: 1342–1346.

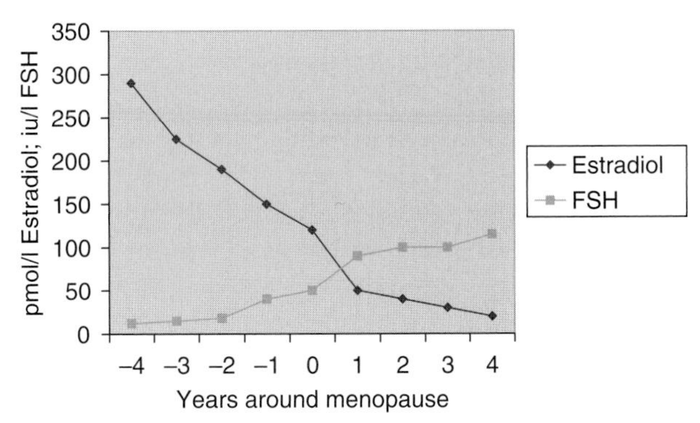

Figure 1 Geometric means of FSH and estradiol in relation to the final menstrual period; the horizontal axis represents time with respect to the final menstrual period at point 0. Adapted from Burger HG, Dudley EC, Hopper JL, *et al.* (1999) Prospectively measured levels of serum follicle-stimulating hormone, estradiol and the dimeric inhibins during the menopausal transition in a population-based cohort of women. *Journal of Clinical Endocrinology and Metabolism* 84: 4025–4030.

exacerbate such disturbances that can severely affect the quality of life.

The period of time encompassing the transition from normal regular menstrual cycles through menopause and into the postmenopause and postreproductive phase is called the climacteric – from the Greek meaning 'a ladder.' A woman is said to be postmenopausal when there has been a gap of 12 months without menstruation. This is mainly of medical significance, as any bleeding after this time is abnormal and requires investigation to exclude cancer of the endometrium (lining of the womb), which will be found in about 10% of women.

Effects of Menopause

Menopause has different implications for women depending on their culture and tradition. In societies in which menstruating women are considered to be unclean and may thus be excluded from praying in public, the release from such inferior status allows them to assume an important and rewarding role in the family and society (in most Indian languages there is no word for menopause). In contrast, the modern, Western societal values and attitudes are more likely to emphasize the negative aspects, such as loss of femininity and sexual attractiveness. This time of life may also coincide with domestic changes such as children leaving home (empty nest syndrome), parents becoming aged and infirm, husband and wife at a stressful time in their careers, all of which may influence how a woman adjusts to the climacteric and menopause. It is therefore not without good reason that the phrase 'change of life' has often been applied to this time.

Symptoms

Aging in itself is accompanied by various inevitable physical changes and it can be difficult to distinguish which clinical symptoms are a result of the loss of ovarian function and estrogen, and which are due to the aging process itself. About 80% of Caucasian women will experience some symptoms considered to be due to estrogen deficiency during the climacteric. Hot flushes and sweats are the most common symptoms and of those experiencing a natural menopause 60–75% will be affected. In Japan, however, flushes are much less common and there is not even a word in the Japanese language for them. Cultural, genetic, and dietary factors with, in particular, the high intake of soy protein in the Japanese diet, which is rich in phytoestrogens, may contribute to the low prevalence of hot flushes in Japan. The Mayan Indian women in Mexico are said not to experience hot flushes at all.

The Hot Flush

The hot flush is a unique and disagreeable sensation of heat that can occur at any time of the day or night. It may be triggered by various situations such as sudden temperature change; drinking tea, coffee, or alcohol; eating spicy foods; or embarrassment, although in the latter it is definitely different from a blush. Commonly, the sensation starts in the face, neck, head, or chest and the initial focal point may be very specific, such as an earlobe, forehead, or between the breasts. Subsequent spread of the sensation of heat may be in any direction and some experience it all over the body. The frequency of flushing varies between individuals ranging from a few per month to several per hour, and similarly the duration may vary from a few seconds to one hour, although the mean is around 3 minutes (Sturdee *et al.*, 1978). The more intense and severe flushing episodes may be followed by sweating, though some women will also sweat without an apparent initial flush. When these disturbances occur at night, there is a resultant domino effect causing disturbed sleep, lethargy, poor concentration and irritability, all of which improve with successful treatment of the flushes. Hot flushes are commonly experienced for around 1 to 2 years but in 25% of women they may continue for more than 5 years and rarely for 25–40 years. Hot flushes can therefore have a very significant effect on quality of life and cause great embarrassment, especially in the workplace or socially. Although no one will ever die from a hot flush, the impact on general well-being and self-esteem should not be underestimated.

While men do not experience a comparable climacteric, following the uncommon situation of testicular failure or bilateral orchidectomy (removal of the testes), severe hot flushes and sweats may occur, which have similar features and associated physiological changes to those in women.

The mechanism of flushing is incompletely understood and difficult to investigate, but it involves a disturbance of the temperature-regulating center (thermostat) situated in the middle of the brain (Freedman, 2001). The decline in estrogen during the climacteric, and probably the fluctuating levels of estrogen in particular, alter the sensitivity of this center such that it initiates inappropriate mechanisms to reduce body temperature by suddenly increasing the blood flow to the skin, which produces the sensation of heat and causes sweating (**Figure 2**). These vasomotor disturbances, referring to regulation of peripheral blood vessel tone, are common in the years before and after menopause when estrogen levels are erratic but cease in later years when estrogen levels, although very low, have stabilized. Priming with estrogen seems to be an essential prerequisite for flushing, as young women with Turner's syndrome (a condition in which the ovaries have not developed or are inactive) have very low circulating levels of estrogen, but do not have flushes. However, if these women are given estrogen replacement therapy, which is later stopped, hot flushes can occur and are often a very distressing, new problem. The link with estrogen is also evident in women being treated for breast cancer with tamoxifen, who often have distressing hot flushes due to the antiestrogenic actions of the drug also affecting the brain. There are complex neuroendocrine actions also involved in the temperature-regulating mechanism in which the hormone serotonin has a key role. Serotonin has many functions in the body including controlling mood and recently it has been found that drugs that are used for relieving depression can also suppress hot flushes, although not as effectively as estrogen replacement therapy.

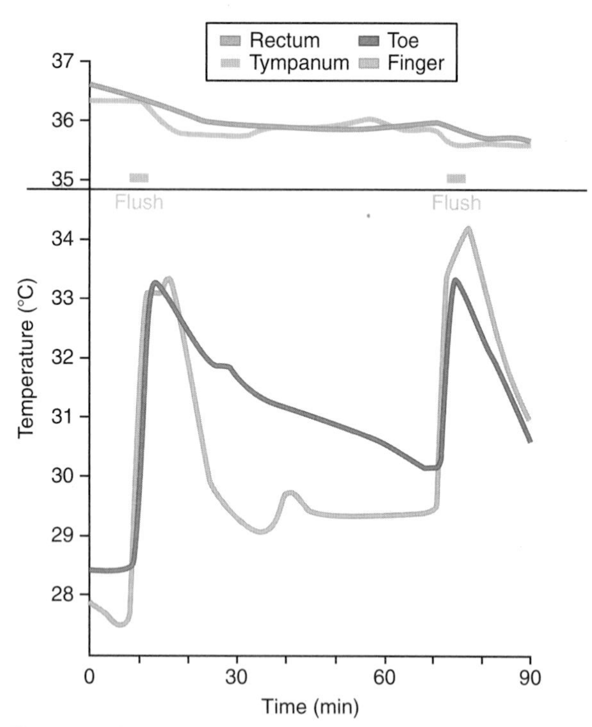

Figure 2 Changes in body temperature during a hot flush. Core body temperature recorded in the rectum and tympanum (eardrum) decrease as a result of the increased blood flow and temperature rise with each flush as recorded in the toe and finger. Adapted from Molnar GW (1975) Body temperatures during menopausal hot flashes. *Journal of Applied Physiology* 38: 499–503.

Other Symptoms

While the hot flush is the most characteristic symptom and experienced by most women, the prevalence of other symptoms that can be attributed to the decline in circulating estrogen is very variable and less easily determined. Women who have experienced severe premenstrual symptoms during the premenopausal years are more likely to suffer from symptoms during the climacteric. Many problems at this time are attributed to 'the change' (**Table 2**) but in reality the only genuine estrogen deficiency symptoms are:

- Hot flushes and sweats
- Vaginal dryness
- Disturbed sleep
- Poor memory/concentration
- Depression.

It is frequently thought that depression is more common in the climacteric, but the term has now been used to cover a wide variety of symptoms and degrees of illness. A depressed mood is a normal human emotion that everyone experiences at some time, and feelings of sadness and disappointment are part of normal life experience. Individual circumstances will influence whether the arrival of menopause is accompanied by psychological distress or, for some women, a feeling of relief and improved general well-being. For those who are depressed during menopause transition, the main predictors seem to be a history of depression, low socioeconomic status, stressful life events, and chronic ill health. Various rating scales and

Table 2 Frequency of 23 symptoms in 135 women attending a menopause clinic who had experienced a natural menopause

Symptom	%
Hot flushes	75
Sweats	71
Lethargy	68
Loss of libido	65
Tension, irritability	65
Insomnia	64
Anxiety	62
Headaches	62
Muscular/joint pains	61
Depression	60
Hair/skin changes	58
Loss of memory/concentration	53
Dry vagina	48
Fears of aging/health	47
Loss of confidence	47
Weight gain	45
Indigestion/nausea	45
Discomfort with intercourse	45
Palpitations/dizziness	44
Backache	39
Formication[a]	33
Loss of femininity	32
Urinary symptoms	30

[a]Formication; sensation as if ants crawling on the skin, possibly related to stimulation of sweat glands.

instruments have been used to evaluate the symptoms associated with menopause and estrogen deficiency. A study of women before and after menopause has shown significant differences in anxiety, somatic physical complaints, flushes and sweats (vasomotor symptoms), and sexual interest, but not in depression (**Figure 3**).

Women who have had a surgical menopause and those who hold negative beliefs about menopause, are more likely to become depressed. Moreover, psychosocial factors have been found to account for much more of the variation in psychological symptoms than stage of menopause (Hunter, 1996). Nevertheless, it remains possible that hormonal changes might influence mood for some women. Improvements in some psychological symptoms with estrogen therapy may just be due to a mental tonic effect of estrogen. It is well recognized that women will often state that they never felt so well as when they were pregnant, when the very high levels of circulating estrogen may be an important factor.

Libido

Loss of sexual desire may be reported by about 45% of postmenopausal women. Various physical factors related to aging in both men and women will affect both desire and performance, but vaginal dryness and atrophic changes due to estrogen deficiency can be a major factor that is easily corrected by estrogen replacement therapy, given either locally to the vagina or systemically. In men, sex drive is strongly correlated with testosterone, both being at their peak in men in their 20s. Women also produce testosterone and other androgens from their ovaries and adrenal glands. The reduction in ovarian function at menopause results in decreased testosterone levels. Women who have had their functioning ovaries surgically removed, resulting in a precipitous drop in both estrogen and testosterone, report a reduction in well-being, energy, and libido. These can be restored more effectively by therapy with a combination of estrogen and testosterone than by estrogen alone. However, the link between libido and testosterone in women is not as clear as might be expected. Measurement of circulating testosterone levels is generally unhelpful and in many women testosterone therapy does not help the libido, thus indicating that other factors are responsible.

Physical Effects

Urogenital Tract

Many tissues are sensitive to estrogen and will change as a result of the decrease in circulating estrogen levels. The urogenital tract is especially well endowed with estrogen receptors and the vaginal skin shows the most marked changes, which usually are evident a year or more after menopause. Estrogen deficiency causes gradual atrophy of the vagina and urinary tract. The earliest symptom is usually vaginal dryness, which may cause pain or discomfort during intercourse. Estrogen promotes a good blood supply to the vagina and stimulates glands in the cervix and at the entrance to the vagina to produce lubricating secretions. These secretions are fermented by lactobacillus bacteria in the vagina, and produce an acid environment, which protects against infection. Thus, in the absence of estrogen, the vagina becomes less acid and loses some resistance to infection. Itching of the vagina and vulva is also common.

Sexual activity has an impact on the health of the vagina. Comparisons of sexually active and abstinent postmenopausal women have shown less vaginal atrophy in those who are active, despite similar blood levels of estrogen in both groups (Bachmann *et al.*, 1984). Estrogen also helps to maintain a healthy lining of the bladder and urethra and, as is the case with the vagina, lack of estrogen leads to an increase in urinary tract infections. Estrogen deficiency may also make the bladder muscle more irritable, which could explain why postmenopausal women can suffer from urge incontinence (overactive bladder). This is characterized by a very strong desire to empty the bladder, which may not be resisted causing incontinence. However, although a direct role of estrogen in this process has not been confirmed, urge incontinence is

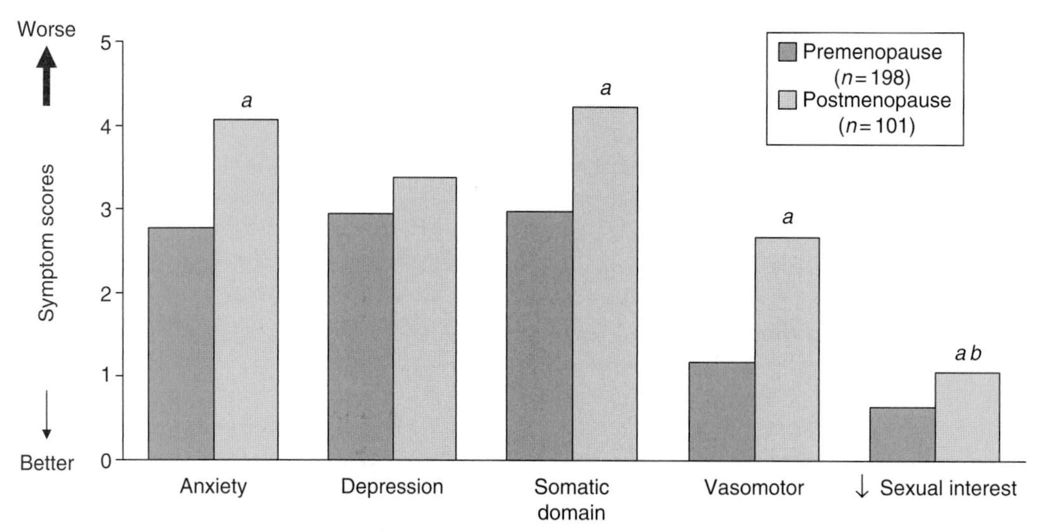

Figure 3 Effect of menopause on quality of life using the Greene climacteric scale: A study of 493 women, age range 45–65 years. [a]These results are from a population-based survey that assessed climacteric symptoms among a group of 493 Dutch women. Participants were divided according to menopausal status: pre-, peri-, and postmenopausal, and posthysterectomy. The Greene Climacteric Scale measures 21 symptoms and is divided into subscales. Anxiety and depression together constitute the psychological subscale: anxiety, 1–6; depression, 7–11; somatic, 12–18; vasomotor, 19–20; sexual dysfunction, 21. To complete the evaluation, the participant assigns a score to each symptom by using a 4-point scale: 0 = not at all; 1 = a little; 2 = quite a bit; and 3 = extremely. The total score on the Greene Scale is the sum of all 21 scores. The graph represents mean scores for pre- and postmenopausal women. Except for the depression subscale, all scores were significantly worse for peri- and postmenopausal women than for premenopausal women, indicating more severe symptoms among peri- and postmenopausal women (data for perimenopausal women not shown). However, scores were not significantly different between peri- and postmenopausal women for most parameters. For the sexual interest evaluation, postmenopausal women scored significantly worse than either pre- or perimenopausal women. [b]$P < 0.05$ vs. perimenopause also. Adapted from Barentsen R, *et al.* (2001) Climacteric symptoms in a representative Dutch population sample as measured with the Greene Climacteric Scale. Maturitas 38: 123–128.

exacerbated by urinary infections, which are reduced by estrogen. The other common type of incontinence is stress incontinence, in which leakage results from a sudden rise in abdominal pressure, such as in coughing, sneezing, laughing, or exercising. Up to one-third of women over the age of 60 years suffer from this condition in which many factors are likely to be involved including the effects of previous childbirth, being overweight, smoking, age, and muscular weakness, but estrogen deficiency does not seem to be a significant factor.

Bone

Bone density and strength change considerably over a lifetime, but especially in women, when after menopause there is an accelerated loss of bone density due to estrogen deficiency. In many women this will lead to osteoporosis, which is defined as a skeletal disorder characterized by compromised bone strength predisposing to an increased risk of fracture. In both sexes, bone density increases during adolescence reaching a peak around 30 years of age, which is sustained for some years and begins to decline during the mid 40s (**Figure 4**). After menopause, an accelerated period of bone loss occurs lasting for up to 10 years followed by a much slower rate of bone loss.

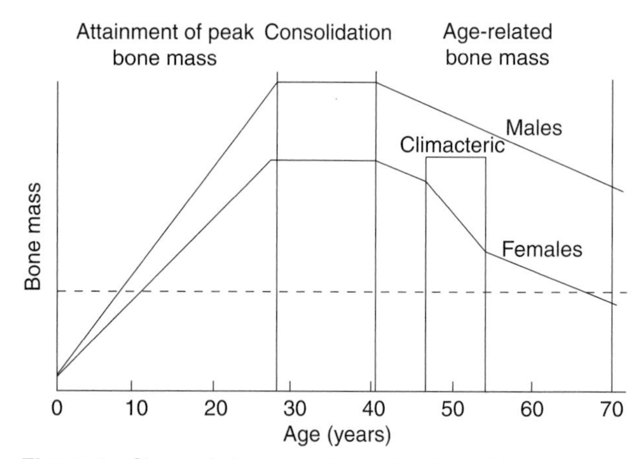

Figure 4 Change in bone density (skeletal mass) in men and women during lifetime, showing that at every age men have stronger bones and in women the loss of bone mass with age is accelerated for a few years after menopause. Reproduced from Compston JE (1990) Osteoporosis. *Clinical Endocrinology* 33: 653–682.

The risk of developing osteoporosis largely depends on the level of the peak bone mass as well as the rate of postmenopausal bone loss and longevity. Men generally develop osteoporosis only late in life because they start losing bone from a much higher peak bone mass and do not have the accelerated loss due to the sudden decrease in sex hormone production as occurs at menopause. By the age of 70 years, women have lost 50% of their skeleton while men lose only 25% by the age of 90 years. Without estrogen, female bone density typically decreases by about 2% per year in the spine and 1% per year in the hip, although in some women the loss is as much as 10 and 5%, respectively. Osteoporosis does not cause any symptoms until a fracture occurs, but every 10% loss of bone doubles the risk of fracture.

The different incidences of fractures in men and women with age are well recognized (**Figure 5**). Gradual loss of height with advancing years is part of the normal aging process but, in addition, the loss of bone in the vertebral bodies in women accelerates the process. In women with severe osteoporosis, repeated vertebral compression fractures may result in thoracic kyphosis, commonly known as a dowager's hump. There is also some evidence that oral bone and attending tooth loss are associated with menopausal estrogen deficiency and osteoporosis (Lindsay *et al.*, 2002).

The incidence of osteoporosis is high and is predicted to increase, constituting one of the most serious health problems currently faced by industrialized countries. Some statistics concerning osteoporosis fractures are the following:

- In the United States at least 1.3 million fractures per year are attributable to osteoporosis.
- In the United Kingdom the annual cost to the National Health Service is around UK£500 million.
- In some countries, 20% of hospital beds are occupied by people with hip fractures.
- One in four hip fracture patients needs long-term nursing care and one in three will never regain full independence.

Women with a premature menopause should be given hormone replacement therapy (HRT) at least until the normal age of menopause to avoid premature loss of bone. Also, prevention of osteoporosis should start in early life by encouraging young women to develop optimum bone strength through lifestyle measures, such as healthy diet, regular exercise, and not smoking.

Cardiovascular Disease

Cardiovascular disease (CVD), of which coronary heart disease (CHD) is the most frequent, is the most common cause of death in women over the age of 60 years, and is relatively uncommon prior to menopause. About 50% of women develop CHD in their lifetime and 30% will die from the disease, with a further 20% developing a stroke. Functioning ovaries seem to provide some degree of protection, which is lost at menopause, particularly if this is premature. The postmenopausal decline in

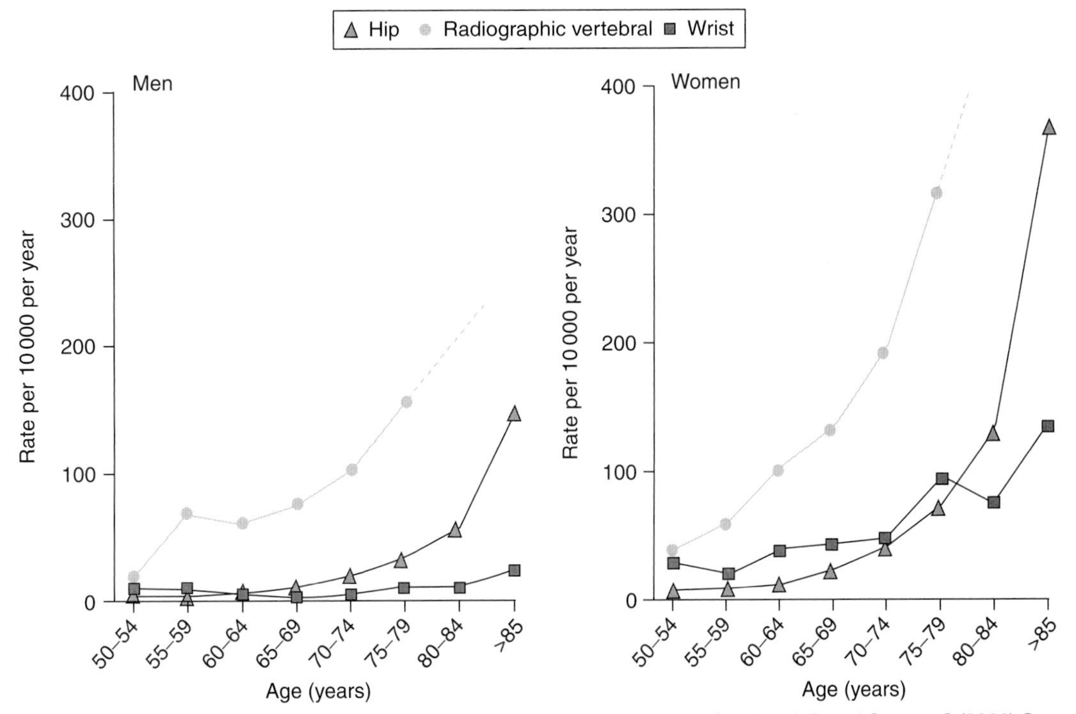

Figure 5 Incidence of osteoporotic fractures in men and women. Reproduced from Sambrook P and Cooper C (2006) Osteoporosis. *The Lancet* 367: 2010–2018.

estrogen results in several metabolic changes, which can promote cardiovascular disease. Notable among these changes are an increase in blood total cholesterol levels of about 20% and triglycerides of 10–15%, an increase in insulin resistance, and a redistribution of fat from the hourglass figure of the younger woman with fat around the breasts and hips to more abdominal fat, all of which are risk factors for CHD (Stevenson, 2004). Nevertheless, the impact of menopause on CHD remains controversial among cardiologists and physicians, and particularly the potential role of HRT remains a matter of much debate. Recent, much-publicized studies from the United States, such as those from the Women's Health Initiative (WHI) (Writing Group for the WHI Investigators, 2002) have demonstrated no benefit from HRT in women with established CHD and those who are several years post-menopause, but there is increasing evidence of a therapeutic window of opportunity whereby HRT, if started around the time of menopause, may delay the onset of CHD. This is supported by animal studies, but more research in women is required to clarify the situation and before there could be a recommendation of using HRT to prevent CHD.

Dementia

Dementia affects 10% of the population over the age of 65 years and 50% over the age of 85 years. Alzheimer's disease, which is the most common cause of dementia, is more prevalent in women and there is some limited evidence of a preventive effect of HRT but, as with CVD, there is no benefit or even a worsening of the condition in those with established disease. Whether there is a potential benefit if started around the time of menopause also has yet to be established.

Skin and Hair

The skin is the largest organ of the body and like all tissues it changes with age, but in addition it is much affected by solar radiation and smoking. After menopause, there is a gradual reduction in thickness due to loss of collagen, the main constituent of the dermis, and loss of elasticity, and the skin becomes dry. Some 30% of the skin collagen is lost in the first 5 years after menopause, some of which can be replaced by estrogen replacement therapy, but there is no evidence that wrinkles are prevented.

Changes in hair growth with age are very variable but among postmenopausal women the most common complaints concern the decrease in body and pubic hair and the increase in facial hair. The latter may be partly due to a change in the ratio of androgens to estrogens.

Treatments

Hormone Replacement Therapy

HRT is intended to replace the deficiency of estrogen associated with the climacteric and is suitable for women experiencing menopausal symptoms that are affecting the quality of life and for some women at particular risk of developing osteoporosis. Estradiol, which is the main estrogen produced by the ovary before menopause, or conjugated equine estrogens (CEE), for example, Premarin®, are the most commonly prescribed worldwide. Estradiol may be given by tablet, transdermal patch or gel, vaginal ring or subcutaneous pellet implant, and CEE by tablet or vaginal cream. Women who have had a hysterectomy may take estrogen alone, otherwise a progestin, a synthetic form of progesterone (the other main hormone produced by the ovary during the normal menstrual cycle) is required in addition to protect the endometrium from overstimulation by estrogen, which could cause cancer. For women who are experiencing symptoms related to the urogenital tract only (see the section 'Urogenital tract'), local, low-dose estrogen given by tablet, vaginal pessary, cream, or ring may be sufficient. These low-dose preparations act locally and do not have any beneficial effects on other menopausal symptoms or cause stimulation of other tissues such as the breasts or skeleton.

The WHI study (Writing Group for the WHI Investigators, 2002) reported that the use of HRT, with estrogen and progestin combined, for more than 5 years may be associated with an increased risk of breast cancer, venous thromboembolism (venous thrombosis and/or pulmonary embolism), and stroke of the order of about eight extra cases per 10 000 women per year. However, in women taking estrogen alone for up to 7 years, there was no evidence of an increased risk of breast cancer, though the other risks remained. Conversely, this study also confirmed a reduction in all osteoporotic fractures and a reduced risk of colorectal cancers.

The timing of the initiation of HRT seems to be critical, so that if started within 10 years of menopause there may be some protection of the coronary arteries from the development of atherosclerosis (Rossouw *et al.*, 2007); as well, the risks of stroke and thromboembolism are also less than in older women. This finding has suggested that there is a window of opportunity for giving HRT in the early years after menopause, but starting therapy after the age of about 60 years, as in the WHI study in which the mean age of women on entering the trial was 63 years, is associated with a significantly increased risk of cardiovascular complications in particular.

In response to the WHI and other studies, drug regulatory authorities in many countries have issued guidelines for the use of HRT, which in summary advise the use of HRT only for the relief of significant menopausal symptoms, and to use the lowest dose for the shortest time, although neither of these criteria have been defined.

Alternative Treatments

Nonhormonal alternative treatments for menopausal symptoms and dietary supplements have received widespread use especially since the publicity in the media about the results of the WHI and other studies. There are limited, randomized controlled data suggesting some benefit in the relief of hot flushes with the use of plant-derived preparations containing phytoestrogens. However, none can compare with the effectiveness of HRT and some herbal remedies such as black cohosh (*Cimicifuga racemosa*) may cause significant side effects including liver damage.

Conclusion

Menopause is a natural and inevitable part of aging in women occurring at about two-thirds of the way through a normal life span. The loss of ovarian function provides some benefits and disadvantages to varying degrees. Greater awareness and knowledge of these changes will allow women to accept and cope with this transition. For those whose quality of life is significantly affected, the merits of HRT can be considered (Hickey *et al.*, 2005).

See also: Family Planning/Contraception; Female Reproductive Function; Gynecological Morbidity; Reproductive Ethics: Ethical Issues and Menopause.

Citations

Bachmann GA, Leiblum SR, Kemmann E, *et al.* (1984) Sexual expression and its determinance in the postmenopausal woman. *Maturitas* 16: 19–29.

Barentsen R, van de Weijer PH, van Gend S, and Foekema H (2001) Climacteric symptoms in a representative Dutch population sample as measured with the Greene Climacteric Scale. *Maturitas* 38: 123–128.

Bloch E (1952) Quantitative morphological investigations of the follicular system in women. Variations at different ages. *Acta Anatomica* 14: 108–123.

Burger HG, Dudley EC, Hopper JL, *et al.* (1999) Prospectively measured levels of serum follicle-stimulating hormone, estradiol and the dimeric inhibins during the menopausal transition in a population-based cohort of women. *Journal of Clinical Endocrinology and Metabolism* 84: 4025–4030.

Burger H, Robertson D, Landgren B-M, and Dennerstein L (2004) Endocrinology of the menopause and menopause transition. In: Critchley H, Gebbie A, and Beral V (eds.) *Menopause and Hormone Replacement*, pp. 3–14. London: RCOG Press.

Compston JE (1990) Osteoporosis. *Clinical Endocrinology* 33: 653–682.

Faddy MJ, Gosden RG, Gougeon A, *et al.* (1992) Accelerated disappearance of ovarian follicles in mid-life: Implications for forecasting menopause. *Human Reproduction* 10: 1342–1346.

Freedman RR (2001) Physiology of hot flashes. *American Journal of Human Biology* 13: 453–464.

Hickey M, Davis SR, and Sturdee DW (2005) Treatment of menopausal symptoms: What shall we do now? *The Lancet* 366: 409–421.

Hunter MS (1996) Mental changes: Are they due to estrogen deficiency? In: Birkhäuser MH and Rozenbaum H (eds.) *Menopause European Consensus Development Conference*, pp. 73–77. Paris, France: Editions Eska.

Lindsay R, Ettinger B, McGowan JA, Redford M, and Barrett-Connor E (2002) Osteoporosis and oral bone loss in aging women: Risks and therapy. In: Wenger N, *et al.* (eds.) *Women's Health and Menopause: A Comprehensive Approach*, pp. 181–202. Bethesda, MD: NIH Publication No. 02-3284.

Molnar GW (1975) Body temperatures during menopausal hot flashes. *Journal of Applied Physiology* 38: 499–503.

Rossouw J, Prentice RL, Manson JE, *et al.* (2007) Postmenopausal hormone therapy and risk of cardiovascular disease by age and years since menopause. *Journal of the American Medical Association* 297: 1465–1477.

Sambrook P and Cooper C (2006) Osteoporosis. *The Lancet* 367: 2010–2018.

Stevenson JC (2004) Metabolic effects of the menopause and hormone replacement therapy. In: Critchley H, Gebbie A, and Beral V (eds.) *Menopause and Hormone Replacement*, pp. 15–22. London: RCOG Press

Sturdee DW, Wilson KA, Pipili E, and Crocker AD (1978) Physiological aspects of menopausal hot flush. *British Medical Journal* 2: 79–80.

Writing Group for the Women's Health, Initiative (WHI) Investigators (2002) Risks and benefits of estrogen plus progestin in healthy postmenopausal women. *Journal of the American Medical Association* 288: 321–333.

Further Reading

Birkhäuser MH, Barlow DH, Notelovitz M, and Rees M for the International Menopause Society (2005) *Health Plan for the Adult Woman; Management Handbook*. Oxford, UK: Taylor and Francis.

Critchley H, Gebbie A, and Beral V (2004) *Menopause and Hormone Replacement*. London: RCOG Press.

Genazzani AR (2006) *Postmenopausal Osteoporosis; Hormones and Other Therapies*. Oxford, UK: Taylor and Francis.

Gordon T, Kannel WB, Hjortland MC, *et al.* (1978) Menopause and coronary heart disease: The Framingham Study. *Annals of Internal Medicine* 89: 157–161.

Lauritzen C and Studd JWW (2005) *Current Management of the Menopause*. Oxford, UK: Taylor and Francis.

Nedrow A, Miller J, Walker M, Nygren P, Huffman LH, and Nelson HD (2006) Complementary and alternative therapies for the management of menopause related symptoms: A systematic evidence review. *Archives of Internal Medicine* 166: 1453–1465.

Pines A, Sturdee DW, Birkhäuser MH, Schneider HPG, Gambacciani M, and Panay N on behalf of the Board of the International Menopause Society (2007) IMS Updated Recommendations on postmenopausal hormone therapy. *Climacteric* 10: 181–194.

Rees M and Purdie DW (2006) *Management of the Menopause*. London: Royal Society of Medicine Press and British Menopause Society Publications.

Sturdee DW (2004) *The Facts of Hormone Therapy for Menopausal Women*. London: Parthenon.

Tomlinson J (2005) *Sexual Health and the Menopause*. London: Royal Society of Medicine Press and British Menopause Society Publications.

Wenger N, Paoletti R, Lenfant CJM, and Pinn VW (2002) International position paper on women's health, menopause: A comprehensive approach. Bethesda, NJ: NIH Publication No. 02-3284, July (2002) National Heart, Lung and Blood Institute, Office of Research on Women's Health, National Institutes of Health and Giovanni Lorenzini Medical Science Foundation.

Relevant Websites

http://www.the-bms.org – British Menopause Society.

http://www.rcplondon.ac.uk/pubs/wp/wp_osteo_update.htm – Guidelines for prevention and treatment from the Royal College of Physicians, London.
http://www.imsociety.org – International Menopause Society.
http://www.pofsupport.org – International Premature Ovarian Failure Association.
http://www.menopausematters.co.uk – Menopause Matters.
http://www.nlm.nih.gov/medlineplus/menopause.html – MedlinePlus: Menopause.
http://www.nof.org/osteoporosis – National Osteoporosis Foundation (United States).
http://www.nos.org.uk – National Osteoporosis Society (NOS) (UK).
http://www.menopause.org – North American Menopause Society.

Sexual and Reproductive Health: Overview

M F Fathalla and M M F Fathalla, Assiut University, Assiut, Egypt

Introduction

The second half of the twentieth century witnessed a vast expansion of health technologies and health services to provide people with certain elements of health care related to sexuality and reproduction. The services were, however, fragmented and not oriented to respond to the totality of health needs. Certain elements received disproportionate attention; other elements were relatively neglected. The concept of reproductive health has evolved to offer a comprehensive and integrated approach to health needs related to sexuality and reproduction.

Another shortcoming in health services was in the mindset with which services were provided. Women were often considered as means in the process of reproduction, and as targets in the process of fertility control. Services were not provided to women as ends in themselves. Women benefited from the process, but were not at its center. They were objects, and not subjects.

The needs of women in reproduction have been traditionally addressed within the concept of maternal and child health. The needs of women were submerged in their needs as mothers. Maternal and child health programs and services have played, and continue to play, an important role in promotive, preventive, and curative health care of mothers and children. Maternal and child health services, however, tended to focus on a healthy child as the desired outcome. While mothers care very much for this desired outcome because of the investment they make in the process of reproduction, this focus resulted in less emphasis being put on caring about the health risks to which mothers are liable during pregnancy and childbirth, and on putting in place the essential obstetric services and facilities to deal with them. As a result, the tragedy of maternal mortality in developing countries has reached dimensions that can no longer be ignored.

Despite all their benefits to the quality of life of women, family planning programs have left women with some genuine concerns as well as unmet needs. Women have more at stake in fertility control than do men. Contraceptives are meant to be used by women to empower themselves by maximizing their choices. Family planning, however, has sometimes been used to control rather than to empower women. The family planning movement has been largely demographically driven. Some governments were so short-sighted as not to see that, when women are given a real choice, and the information and means to implement their choice, they will make the most rational decisions for themselves, for their families and communities, and ultimately for the world at large.

Sexual and Reproductive Health: Concepts and Dimensions

Reproductive Health

The concept of reproductive health was endorsed at the United Nations International Conference on Population and Development (ICPD), held in Cairo in 1994 (UN, 1994). At ICPD, representatives of 179 countries reached a consensus that responding to the needs of individuals is the way to address the aggregate problem of rapid population growth; population policies should address social development beyond family planning, with particular emphasis on advancement of the status and empowerment of women; and family planning should be provided in the context of comprehensive reproductive health care. Moreover, ICPD set the goal that all countries should strive to make accessible, through the primary health-care system, reproductive health to all individuals of appropriate ages as soon as possible and no later than the year 2015. Realizing that laws differ among countries, the ICPD emphasized that methods to be used are those that are not against the law in the country.

Definition of reproductive health

The following definition was adopted in the Programme of Action of the ICPD, and at the Fourth World

Conference on Women, also sponsored by the United Nations, in Beijing in 1995. The full definition reads:

> Reproductive health is a state of complete physical, mental and social well-being and not merely the absence of disease or infirmity, in all matters relating to the reproductive system and to its functions and processes. Reproductive health therefore implies that people are able to have a satisfying and safe sex life and that they have the capability to reproduce and the freedom to decide if, when and how often to do so. Implicit in this last condition are the right of men and women to be informed and to have access to safe, effective, affordable and acceptable methods of family planning of their choice, as well as other methods of their choice for regulation of fertility which are not against the law, and the right of access to appropriate health-care services that will enable women to go safely through pregnancy and childbirth and provide couples with the best chance of having a healthy infant. In line with the above definition of reproductive health, reproductive health care is defined as the constellation of methods, techniques and services that contribute to reproductive health and well-being by preventing and solving reproductive health problems. It also includes sexual health, the purpose of which is the enhancement of life and personal relations, and not merely counselling and care related to reproduction and sexually transmitted diseases. (UN, 1994: 30, paragraph 7.2)

Sexual Health

In the evolution of our species, the temporal relationship between sex and reproduction has been severed. This dissociation of the act of sex from reproduction appears to be a purposeful act of nature aimed at strengthening the pair bond. Sex has been elevated from an act of reproductive instinct to an expression of love and a confirmation of human bonding. With the worldwide trend for the adoption of a small family norm, reproduction has been receding further into the background.

As with reproductive health, sexual health should be defined in light of the positive definition of health as a state of complete physical, mental, and social well-being and not merely the absence of disease or infirmity.

> Sexual health is a state of physical, emotional, mental and social well-being in relation to sexuality: it is not merely the absence of disease, dysfunction or infirmity. Sexual health requires a positive and respectful approach to sexuality and sexual relationships, as well as the possibility of pleasurable and safe sexual experiences, free of coercion, discrimination and violence. For sexual health to be attained and maintained, the sexual rights of all persons must be respected, protected and fulfilled. (Glasier *et al.*, 2006: 1596)

The Human Rights Dimension

The Declaration and Platform for Action adopted by 187 UN Member States in the 1995 Fourth World Conference on Women makes explicit that

> [t]he human rights of women include their right to have control over and decide freely and responsibly on matters related to their sexuality, including sexual and reproductive health, free of coercion, discrimination and violence. Equal relationships between women and men in matters of sexual relations and reproduction, including full respect for the integrity of the person, require mutual respect, consent and shared responsibility for sexual behavior and its consequences. (UN, 1995: 39, paragraph 96)

Application of human rights law, until recently, has been gender blind. For too long, it focused on the public arena of men and neglected the private sphere of women. Sexual and reproductive rights embrace certain human rights that are already recognized in national laws, international human rights documents, and other consensus documents. Human rights in relation to sexuality and reproduction impose on the Member States the obligation to respect, protect, and implement sexual and reproductive rights.

The Development Dimension

Sexual and reproductive health is not only a public health and a human rights issue. It is a development concern. It should be noted that the concept of reproductive health was adopted in a conference on population and development, the ICPD. This was also highlighted in the 2005 Millennium Project report to the UN Secretary-General titled *Investing in Development: A Practical Plan to Achieve the Millennium Development Goals.* The report calls for sexual and reproductive health issues to be included in national, regional, and international poverty-reduction strategies, and states that sexual and reproductive health is essential for reaching development goals.

The Burden of Sexual and Reproductive Ill-Health

Global Situation of Sexual and Reproductive Health

This article, rather than dealing with a discussion regarding the different components of sexual and reproductive health, provides a summary overview based on data presented in the World Health Organization (WHO) Reproductive Health Strategy to accelerate progress toward the attainment of international development goals and targets (WHO Dept. of Reproductive Health and Research, 2004), together with more recently available data. It highlights not only the magnitude of the problems, but also the glaring inequities in sexual and reproductive health in the world.

Pregnancy, Childbirth, and Health of Newborns

Each year, some 8 million of the estimated 210 million women who become pregnant, suffer life-threatening complications related to pregnancy, many experiencing long-term morbidities and disabilities. In 2005, an estimated 535 900 women died during pregnancy and childbirth from largely preventable causes (Hill *et al.*, 2007). Globally, the maternal mortality ratio has not changed substantially over the past decade. Regional inequities are extreme, with 99% of these maternal deaths occurring in developing countries. At the national level, estimated maternal mortality ratios range from below 10 deaths per 100 000 live births in most developed countries to as high as 2000 deaths per 100 000 live births in some developing countries.

In the developing world, a third of all pregnant women receive no health care during pregnancy, 60% of deliveries take place outside health facilities, and only about 60% of all deliveries are attended by trained staff (Glasier *et al.*, 2006).

Over 130 million babies are born every year, and more than 10 million infants die before their fifth birthday, and almost 8 million before their first (WHO Dept. of Making Pregnancy Safer, 2006). More than half of the infant deaths occur in the first 4 weeks after birth. Of these neonatal deaths, 98% occur in developing regions. It is estimated that in developing regions, the risk of death in the neonatal period is more than six times that of developing countries; in the least-developed countries, it is more than eight times higher. Most deaths in the neonatal period occur in the first few days after birth. Additionally, an estimated 3.3 million infants are stillborn. Many of these deaths are related to the poor health of the woman and inadequate care during pregnancy, childbirth, and the postpartum period. Furthermore, a mother's death can seriously compromise the survival of her children.

Family Planning

Contraceptive use has substantially increased in many developing countries and in some is approaching that practiced in developed countries. Yet, surveys indicate that in developing countries and countries in transition, more than 120 million couples have an unmet need for safe and effective contraception despite their expressed desire to avoid or to space future pregnancies. Between 9 and 39% of married women (including women in union) have this unmet need for family planning. Data suggest that unmarried, sexually active adolescents and adults also face this unmet need. About 80 million women every year have unintended or unwanted pregnancies, some of which occur through contraceptive failure, as no contraceptive method is 100% effective.

Unsafe Abortion

Some 42 million unintended pregnancies are terminated each year, an estimated 20 million of which are unsafe; two-thirds of all unsafe abortions occur among women aged between 15 and 30 years (WHO, 2007). Unsafe abortions kill an estimated 65 000 to 70 000 women every year, representing 13% of all pregnancy-related deaths. In addition, they are associated with considerable morbidity; for instance, more than 3 million suffer from the effects of reproductive tract infection, and almost 1.7 million will develop secondary infertility.

Sexually Transmitted Infections, including HIV and Reproductive Tract Infections

An estimated 340 million new cases of sexually transmitted bacterial infections, most of which are treatable, occur annually. Many are untreated because they are difficult to diagnose and because competent, affordable services are lacking. Among other consequences, they can be a leading cause for infertility in women. If cases of mostly incurable viral infections are added, the yearly number of sexually transmitted infections acquired may well exceed 1 billion (Glasier *et al.*, 2006).

Sexually transmitted human papillomavirus infection is closely linked with cancer of the cervix, which is the second most common cancer in women worldwide, with about 500 000 new cases and 250 000 deaths each year (WHO Dept. of Making Pregnancy Safer, 2006). Almost 80% of cases occur in low-income countries, where programs for screening and treatment are seriously deficient or lacking.

The 2007 AIDS epidemic update estimates that of the 30.8 million adults living with AIDS, 50% are women (UNAIDS, 2007). In sub-Saharan Africa, about 61% are women. In 2007, it is estimated that 2.5 million people were newly infected, including 420 000 children under age 15.

Additionally, reproductive tract infections, such as bacterial vaginosis and genital candidiasis, which are not sexually transmitted, are known to be widespread, although the prevalence and consequences of these infections are not well documented.

Together, these aspects of sexual and reproductive ill health (maternal and perinatal mortality and morbidity, cancers, sexually transmitted infections, and HIV/AIDS) account for nearly 20% of the global burden of ill health for women and some 14% for men (WHO Dept. of Reproductive Health and Research, 2004). These statistics do not capture the full burden of ill health, however. Gender-based violence, and gynecological conditions such as severe menstrual problems, urinary and fecal incontinence due to obstetric fistulae, uterine prolapse, pregnancy loss, and sexual dysfunction – all of which have major social, emotional, and physical consequences – are currently not estimated, or underestimated, in global burden of disease estimates.

The Unfair Burden on Women

Men have their own sexual and reproductive health needs relating to sexuality, protection against sexually transmitted infections, infertility prevention and management, and fertility regulation. Protection against prostate hypertrophy and prostate cancer is another concern. These needs must be met. But the burden of sexual and reproductive ill health, for biological and social reasons, falls largely on women.

Maternity is a unique privilege and a unique health burden for women. In some other aspects of sexual and reproductive health, where the responsibility is shared between men and women, the burden, for both biological and social reasons, falls heavily on women. This applies to the burden of sexually transmitted infections, fertility regulation, infertility, and perinatal mortality and morbidity.

For a mix of biological and social reasons, the burden of sexually transmitted infections falls disproportionately on women. Women are more likely to be infected, are less likely to seek care, are more difficult to diagnose, are at greater risk for severe disease sequelae, and are more subject to social discrimination. The most widely used effective method available for protection against sexually transmitted infections, the male condom, is controlled by men. Although a female condom is available, a simple and effective vaginal microbicide that a woman can use without the need or necessity of her partner's cooperation does not yet exist. The past few years have witnessed a major infusion of funding for this field, but it is difficult to predict when an effective, safe, acceptable, and affordable vaginal microbicide will be available to women.

The modern revolution in contraceptive technology provided women with reliable methods of birth control, which they can use independently of cooperation of their partners. This independence was at a price. Women had to assume the inconveniences and risks involved. The role and responsibility of the male partner receded as contraception has come to be considered a woman's business. Not only do women have an undue burden of responsibility in fertility regulation, but the effective methods that women have available for use are those associated with potential health hazards.

Responsibility for infertility is commonly shared by the couple. The burden of infertility, however, for biological and social reasons, is unequally shared. The infertility investigation of the female partner is much more elaborate and is associated with more inconvenience and risk. The burden of treatment also falls mostly on the female partner. Even for male infertility, the promise of successful management is now shifting to assisted conception technologies, where the female assumes the major burden. The psychological and social burden of infertility in most societies is much heavier on the woman. A woman's status is often identified with her fertility, and failure to have children can be seen as a social disgrace or a cause for divorce. The suffering of the infertile woman can be very real.

Perinatal (fetal and early neonatal) morbidity and mortality should be recognized as part of the disease burden on women. Health statistics classify perinatal mortality as a category on its own, a condition that affects both males and females. Perinatal mortality and morbidity are outcomes of pregnancy and delivery by women. Women make major investments of themselves in pregnancy and childbirth. The unfavorable perinatal outcome of a pregnancy can be a frustration or an additional burden on the woman for the care of an unhealthy infant and child.

Sexual and Reproductive Health Indicators

For the comprehensive sexual and reproductive health approach, there is a need for a new set of indicators. A WHO interagency meeting proposed the following short list of 17 indicators, and WHO provided guidelines for their generation, interpretation, and analysis for global monitoring (WHO Dept. of Reproductive Health and Research, 2006a):

- Total fertility rate (TFR): total number of children a woman would have by the end of her reproductive period if she experienced the currently prevailing age-specific rates throughout her childbearing life.
- Contraceptive prevalence (CP): percentage of women of reproductive age (15–49 years) who are using (or whose partner is using) a contraceptive method at a particular point in time.
- Maternal mortality ratio (MMR): the number of maternal deaths per 100 000 live births.
- Percentage of women attended, at least once during pregnancy, by skilled health personnel (excluding trained or untrained traditional birth attendants) for reasons related to pregnancy.
- Percentage of births attended by skilled health personnel (excluding trained or untrained traditional birth attendants).
- Number of facilities with functioning, basic, essential obstetric care per 500 000 population.
- Number of facilities with functioning, comprehensive, essential obstetric care per 500 000 population.
- Perinatal mortality rate: Number of perinatal deaths per 1000 total births.
- Percentage of live births of low birth weight (less than 2500 g).

- Positive syphilis serology prevalence in pregnant women attending for antenatal care: percentage of pregnant women, 15 to 24 years old, attending antenatal clinics, whose blood has been screened for syphilis, with positive serology for syphilis.
- Prevalence of anemia in women: percentage of women of reproductive age (15–49 years old) screened for hemoglobin levels who are anemic (with levels below 110 g/dl for pregnant women, and below 120 g/dl for nonpregnant women).
- Percentage of obstetric and gynecological admissions owing to abortion: percentage of all cases admitted to service delivery points, providing inpatient obstetric and gynecological services, which are due to abortion (spontaneous and induced, but excluding planned termination of pregnancy).
- Reported prevalence of women with female genital mutilation (FGM): percentage of women interviewed in a community survey, reporting themselves to have undergone FGM.
- Prevalence of infertility in women: percentage of women of reproductive age (15–49 years old) at risk of pregnancy (not pregnant, sexually active, noncontracepting, and nonlactating) who report trying for a pregnancy for two years or more (this indicator measures infertility whether it is caused by a male or female factor).
- Reported incidence of urethritis in men: percentage of men aged 15 to 49 years interviewed in a community survey reporting episodes of urethritis in the last 12 months.
- HIV prevalence in pregnant women: percentage of pregnant women aged 15 to 24 years, attending antenatal clinics, whose blood has been tested for HIV, who are sero-positive for HIV.
- Knowledge of HIV-related prevention practices: percentage of all respondents who correctly identify all three major ways of preventing the sexual transmission of HIV and who reject three major misconceptions about HIV transmission or prevention.

In selecting these indicators, the WHO took into consideration that a good indicator must be ethical, useful at the national and international levels, scientifically robust, representative, understandable, and accessible.

Inequity in Sexual and Reproductive Health

Measurement of overall population coverage rates can mask inequities. Unless a special effort is made, scaling up of services is likely to favor better-off groups. Managers pressed to increase overall population coverage rates quickly may focus on reaching people who can most easily be covered, rather than the poor and disadvantaged.

Inequalities in health are a fact of life. Inequalities in health become inequitable when they are unfair. They are unfair when they stem from factors that society can do something about, such as differential availability and access to health-care services and differential exposure to risk factors. Health equity means the provision of equal health services for all those with equivalent needs, and the provision of more or enhanced services to those with greater needs.

Determinants of Sexual and Reproductive Health

Sexual and reproductive health, and health in general, is promoted or undermined by the individual's lifestyle behavior. It is partly predetermined by the social and economic conditions in the community in which people are born and in which they live. It can be improved by the availability and accessibility of health-care services. The relative importance of these determinants varies from country to country, and their relevance varies in the different components of sexual and reproductive health. The important point about these determinants is that health system interventions alone are not enough to improve sexual and reproductive health. Interventions beyond the health system are needed.

Behavioral Determinants

Health is a human right. But health cannot be granted. Health should be pursued. In fact, health is more in the hands of the individual than in the hands of the physician. This cannot be truer than in sexual and reproductive health, where unsafe sex is one of the most important global risk factors to health, and where sexual and reproductive behavior has such far-reaching consequences for health. But for individuals to pursue their health, they need to be equipped with the information, and they need to be empowered to act on that information. Also, it can be the behavior of others that impacts adversely on their health. Culture plays an important role as a behavioral determinant.

Adolescents are often denied sex education when their evolving capacities are ready for it and they need it. When this education is not available from school or from home, adolescents get the information, or more commonly misinformation, from their peers.

Powerlessness of women is a serious health hazard. Empowerment of women has a two-way relationship to sexual and reproductive health. A majority of women and girls in the world today still live under conditions that limit educational attainment, restrict economic participation, and fail to guarantee them equal rights and freedoms as compared to men. Yet, the empowerment of women through education and employment opportunities is known to be among the most powerful determinants of their sexual and reproductive health. And one of the most effective ways of empowering women is to recognize and enforce their human right of control over their own sexual and reproductive life, enabling them to protect themselves against infection and disease, as well as against unwanted pregnancies.

In the case of sexual and reproductive health, it may not be the woman's behavior but the behavior of others that matters more. There is evidence that in many developing countries women get sexually transmitted infections, including HIV, from their husbands who have multiple sexual partners, and not through their own behavior. Women can be victimized by violence, including intimate partner violence and sexual violence and abuse. Men can play a positive or negative role in women's sexual and reproductive health. Shared responsibility and mutual respect between women and men in matters related to sexual and reproductive behavior are essential to improving women's sexual and reproductive health. Another case in point is the damage inflicted by female genital mutilation on a young girl, who does not have a choice.

Economic and Social Determinants

In poor societies, women are the poorest of the poor. Feminization of poverty is real. Pregnant women and children are the first to suffer under poor economic conditions.

Economic growth is important for improving health, but is not enough on its own. It has to be properly managed, fully exploited, and equitably distributed. Some developing countries have been more successful than others in translating economic growth into health gains. Sexual and reproductive health is a major concern for women. Where women are not in the position to make or to influence policy decisions, sexual and reproductive health may fall between the cracks.

Social norms also matter in sexual and reproductive health. Women's sexual and reproductive health is not simply determined by their sex, but also and more importantly by their gender. Biological characteristics identify the sex of an individual as male or female. Gender, on the other hand, is a social construct. Gender is what it means in a particular society to be male or female. Gender attributes include codes of behavior considered appropriate to each gender and a division of labor between the two genders. Imposed codes of behavior are often embedded in the law and historically have been claimed to provide a justification for discrimination against women. Gender division of labor does not only segregate tasks. It also often assigns different values to these tasks. This differential value of tasks often works against the equality between men and women, underestimating the value of women's contribution and hence the value of women.

Barriers to improving women's sexual and reproductive health are thus often rooted in economic, social, cultural, legal, and related conditions that transcend health-care considerations.

Health System Determinants

Healthy behavior and good socioeconomic conditions are not enough to ensure sexual and reproductive health. There is a need for health services. Nowhere is this more apparent than in the area of maternal health. Where no health intervention is done to avert maternal death, natural mortality is around 1000–1500 per 100 000 births, an estimate based on historical studies and data from contemporary religious groups who do not intervene in childbirth (WHO, 2005). Life-threatening complications of pregnancy and childbirth are not always predictable or preventable.

Sexual and reproductive health services are still deficient for a majority of the world's population. Where they are available, there can be an imbalance. imbalance between urban and rural services; between curative and preventive care; between building infrastructure and health manpower development and training; and between roles of different health professionals. In addition, services may be of poor quality, or may not be accessible. Services, even in resource-poor settings, may be underused if they are not culturally acceptable.

Affordability is another important determinant. Consumer spending as measured by out-of-pocket expenditure represents a large, if not the largest, share of resources spent on sexual and reproductive health in developing countries, raising concern about the accessibility of services by those who cannot afford to pay. This concern applies to health services in general, with the increasing adoption of schemes for health sector reform. Sexual and reproductive health is particularly vulnerable where women are the poorest of the poor.

Enough information and tools are available to improve sexual and reproductive health. But there is always room and need for more research. The frontiers of science are continuously expanding and should be exploited to develop new and improved cost-effective interventions that can be implemented in resource-poor settings. Topics considered taboos before are now opening up for legitimate scientific inquiry. These include sexual behavior among adolescents and adults, commercial sex, unsafe abortion, female genital mutilation, and intimate partner violence against women. There is also a need for research to improve the effectiveness and expand the capacity of health systems. But research should not be used as an excuse for inaction. To ensure that the outcome of the research can be applied and scaled up, the research community is called upon to walk the extra mile, to ensure that research is responsive to the real needs of people and that it is communicated to where the action is, and to where it can have an impact.

The Sexual and Reproductive Health Package

The Five Core Aspects of Sexual and Reproductive Health

The global WHO Reproductive Health Strategy (WHO Dept. of Reproductive Health and Research, 2004) defines the following five core aspects: improving antenatal, perinatal, postpartum, and newborn care; providing high-quality services for family planning, including infertility services; eliminating unsafe abortion; combating sexually transmitted infections, including HIV, reproductive tract infections, cervical cancer, and other gynecological morbidities; and promoting sexual health.

Sexual and Reproductive Health: An Integrated Package

The different needs in sexual and reproductive health are not isolated from each other. They are simultaneous or consecutive related needs. People cannot be healthy if they have one element of the sexual and reproductive health package but are missing the others. Moreover, improvements in one element can result in potential improvements in other elements. Similarly, lack of improvement in one element can hinder progress in other elements.

Pelvic infection is a major cause for infertility worldwide. The resultant infertility is also the most difficult to treat. The magnitude of the problem of infertility will not be ameliorated except by a reduction of sexually transmitted infections, by safer births that avoid postpartum infection, and by decreasing the need for and the resort to unsafe abortion practices.

Infant and child survival, growth, and development cannot be improved without good maternity care. Proper planning of births, including adequate child spacing, is a basic ingredient of any child survival package. Unless adequately controlled, sexually transmitted infections, and in particular HIV infection, can impede further progress in child survival.

Family Planning

Family planning is a basic component of the sexual and reproductive health package. Fertility by choice, not by chance, is a basic requirement for women's health. A woman who does not have the means to regulate and control her fertility cannot be considered in a 'state of complete physical, mental and social well-being,' the definition of health (shown previously) in the WHO constitution. She cannot have the joy of a pregnancy that is wanted, avoid the distress of a pregnancy that is unwanted, plan her life, pursue her education, undertake a productive career, and plan her births to take place at optimal times for childbearing, ensuring more safety for herself and better chances for her child's survival and healthy growth and development. A woman with an unwanted pregnancy cannot be considered in good health, even if the pregnancy is not going to impair her physical health, and even if she delivers the unwanted child alive and with no physical disability.

Fertility regulation is a major element in any safe motherhood strategy. It reduces the number of unwanted pregnancies, with a resultant decrease in the total exposure to the risk as well as a decrease in the number of unsafe abortions. Proper planning of births can also decrease the number of high-risk pregnancies.

Family planning improves the quality of life not only for women, but also for the family as a whole and particularly for children. The quality of child care – including play and stimulation as well as health and education – inevitably rises as parents are able to invest more of their time, energy, and money in bringing up a smaller number of children.

In the sexual and reproductive health approach, family planning services are not 'demographic posts.' Women are not 'targets' for contraception, for which policymakers and administrators set 'quota' for services to accomplish. As the ICPD Programme of Action states, "Family planning programmes work best when they are part of or linked to broader reproductive health programmes that address closely related health needs" (UN, 1994: 33).

Integration of Sexual and Reproductive Health Services

Related sexual and reproductive health interventions often come to be delivered over time as separate and discrete activities, rather than as a comprehensive response to people's needs at different stages of their lives. For example, family planning clinics often function independently of those catering to other aspects of maternal and child health. A woman attending a clinic session for child immunization may not, at the same time, be able to obtain care for her own health needs; she will be asked to come back on a different day, or even to go to a different health facility.

The approach to integration should be pragmatic. Different situations in countries should be judged in their own context. There is much room for broadening the primary health-care package to include sexual and reproductive health interventions beyond the traditional maternal and child health approach. There is also a potential for expanding the range of services offered in community-based and clinic-based family planning services to include other selected specific interventions in sexual and reproductive health care. Specific elements in the strategy for prevention and control of sexually transmitted infections can be incorporated in maternal and child health (including prenatal care) and family planning services. The guiding principle is that no opportunity should be missed, where a capacity exists, for meeting all sexual and reproductive health needs.

The need for comprehensive sexual and reproductive health care should not translate into an all-or-none situation. Providing people with some elements of the service is better than providing no services. Services could be built up as resources become available and according to level of need and demand. Nor should integration result in dilution of available resources. Rather, it should result in more effective use of resources.

Implementation of the Sexual and Reproductive Health Package

Components in the sexual and reproductive health package pose different challenges in implementation (Fathalla, 2002).

One part of the package includes traditional services with which we already have experience. These include maternal and child health services and family planning. The challenge will be how to expand the coverage and improve the quality of the services.

In other parts of the package, we are challenged to meet sensitive and emerging needs. These include the pandemics of sexually transmitted infections and HIV, the tragedy of unsafe abortion, and gender-based violence and sexual abuse in all its offensive forms. The challenge here is how to approach these sensitive and socially divisive issues, how to overcome deep-rooted traditions and social barriers, and how to influence human behavior.

In another part of the package, we are challenged to serve new customers. Among the new customers we need to reach and serve are adolescents, men, and the growing number of displaced persons and refugees. Services, including information and education, need to be tailored to serve the needs of these new customers. The biggest-ever generation of young people between 15 and 24 years is rapidly expanding in many countries and their special needs have to be met. The long-neglected sexual and reproductive health needs of refugees are now being increasingly recognized.

Another part of the package includes services needed that are less affordable. These include treatment of HIV-positive pregnant women to prevent transmission of the infection to the fetus and newborn; infertility management; and detection and management of reproductive cancers. The challenge is to develop and test new, cost-effective interventions that can be implemented in resource-poor settings.

The appropriate balance in the sexual and reproductive health package can only be designed and developed at the national level. Each country needs to identify problems, set priorities, and formulate its own strategies in light of its needs and capacities.

Special Considerations of Sexual and Reproductive Health Care

Sexual and reproductive health care is a part of general health care. There are, however, different considerations and challenges. Sexual and reproductive health-care providers deal with mostly healthy people. They often have to consider the interests of more than one 'client' at the same time. They deal mostly with women. And they have to deal and interact with society. These special considerations have ethical, legal, and human rights implications.

Dealing with Healthy People

Sexual and reproductive health-care providers respond to needs of healthy subjects, related to their physiological, sexual, and reproductive functions (including the prevention of unwanted pregnancy and the prevention of sexually transmitted infections). Sexual and reproductive health care is more health-oriented than disease-oriented. Promotive and preventive health care are major components of sexual and reproductive health care.

Dealing with healthy people implies a change in the provider–patient relationship, from a giver–recipient relationship to a more participatory type of health care. Counseling is the description for a good part of sexual and reproductive health care. There is no other field of medicine in which participation of the patients in health-care decisions is that necessary. The general ethical principles of respect and autonomy dictate that patients are required to give their informed consent to the treatment proposed by the health-care provider, freely and without undue pressure or inducement. In the case of sexual and reproductive health care, clients have to make informed choices and decisions.

The risk/benefit ratio is different when a drug or device is used for a healthy subject, to prevent a condition that may or may not happen, from what it is when a drug or device is used by a person suffering from a disease that in itself carries a risk. The risk/benefit ratio is also different from another aspect. Often, very large numbers of people are involved, so that rare, adverse effects assume more importance. Safety and efficacy of drugs and devices are regulated by government agencies, and are subject to laws of product liability. Sexual and reproductive health products, and particularly contraceptives, are recognized as different from products developed and marketed to treat patients who have disease conditions.

In dealing with healthy people, the temptation to overmedicate for normal life events would be resented, but in sexual and reproductive health care of women, and particularly in maternity care, it has often been accepted, sometimes without valid scientific evidence, by the health profession. A WHO report cites four such interventions as particularly subject to overuse: cesarean section, routine episiotomy, routine early amniotomy, and abuse of oxytocin (WHO, 2005).

Dealing with More than One Client

Different from other health professionals, sexual and reproductive health-care providers often deal with, or have to consider, more than one client at the same time. This other party to a management decision could be a woman's male partner, or her fetus. The interests of the different parties may coincide and may diverge, or even conflict. The ethical principles of beneficence and nonmaleficence, to do good and to do no harm, pose a challenge when, for example, the interest of the mother conflicts with the interest of the fetus.

Dealing Mostly with Women

Providers of sexual and reproductive health care deal mostly with women, who in many societies are still subordinated and undervalued. Apart from respecting women and treating them as equals, providers must also be sensitive to their concerns and perceptions. Respect for women should be shown in providing them with confidentiality, privacy, and access to all information they need to make well-informed decisions about their health. Health-care professionals in developing countries know that even though the majority of women in many parts of the world are illiterate this does not mean that they are not fully capable of making sound decisions. Poor people have a very narrow safety margin for error in making decisions about their lives and, consciously or subconsciously, they know that. Women, literate or illiterate, rich or poor, given the information and the right to choose and decide, will make the right decisions for themselves and their families.

Lack of respect for confidentiality for women seeking sexual and reproductive health care can deter them from seeking advice and treatment and thereby adversely affect their health and well-being. Women will be less willing, for the reason of unreliable confidentiality, to seek health care for diseases of the genital tract, for contraception, or for incomplete abortion, and in cases where they have suffered sexual or physical violence.

Women have been excluded from historical sources of moral authority, and are still underrepresented in learned professions of medicine and law, and in legislative assemblies. The voices of women, and their perspectives, often have not been taken into consideration in laws, policies, and regulations governing sexual and reproductive health care for women.

Dealing with Society

No society, no culture, no religion, and no legal code has been neutral about sexual and reproductive life. No other health profession has to deal with such emotionally charged health issues as sexuality and abortion. As new health technologies develop, for example, in the area of infertility management, new issues arise for which society may not be well prepared. Enforcing perceived interests of the society may violate women's sexual and reproductive health rights.

International Commitment to Sexual and Reproductive Health

Sexual and Reproductive Health and the Millennium Development Goals (MDGs)

The Millennium Development Goals (MDGs), which grew out of the United Nations Millennium Declaration adopted by 189 Member States in 2000, provide the new international framework for measuring progress toward sustaining development and eliminating poverty (UN Millennium Project, 2005). Although there was no specific goal about sexual and reproductive health, out of the eight MDGs, three – improve maternal health, reduce child mortality, and combat HIV/AIDS, malaria, and other diseases – are directly related to sexual and reproductive health. Specific targets under these goals are to reduce by three-quarters, between 1990 and 2015, the maternal mortality ratio; to reduce by two-thirds, between 1990 and 2015, the under-five mortality rate; and to have halted by 2015, and begun to reverse, the spread of HIV/AIDS. Moreover, at the 2005 World Summit review of the MDGs at the United Nations, world leaders committed themselves to "achieving universal access to reproductive health by 2015, as set out at the ICPD, integrating this goal in strategies to attain the internationally agreed development goals, including those contained in the Millennium Declaration" (UN, 2005: 16).

WHO Reproductive Health Strategy to Accelerate Progress Toward the Attainment of MDGs

A reproductive health strategy to accelerate progress toward the attainment of MDGs was adopted by the 57th World Health Assembly in May 2004. The guiding principles for the strategy are the internationally agreed instruments and global consensus declarations on human rights (WHO Dept. of Reproductive Health and Research, 2004).

The strategy recognizes that each country needs to identify problems, set priorities, and formulate its own strategies for accelerated action through consultative processes involving all stakeholders. Five overarching activities are identified: strengthening health systems capacity; improving information for priority setting; mobilizing political will; creating supportive legislative and regulatory frameworks; and strengthening monitoring, evaluation, and accountability. A framework for implementing the WHO Reproductive Health Strategy was also developed (WHO Dept. of Reproductive Health and Research, 2006b).

Conclusion

A global overview of sexual and reproductive health provides reasons for public health, development, and human rights concern. Improving the situation, alleviating the burden, and addressing the glaring inequity have been on the international and national agendas for many years now. Although progress has been made, it has been uneven, and major parts of the world still fall short of desired goals. The know-how is available. Cost-effective interventions are affordable. A sustained collaborative effort supported by political commitment, together with mobilization and rational allocation of resources, can bring about a brighter future for sexual and reproductive health.

See also: Female Reproductive Function; Gynecological Morbidity; Maternal Mortality and Morbidity; Reproductive Rights; Trends in Human Fertility.

Citations

Fathalla MF (2002) Implementing the reproductive health approach. In: Sadik N (ed.) *An Agenda for People – The UNFPA through Three Decades*, pp. 24–46. New York and London: New York University Press.

Glasier A, Gulmezoglu M, Schmid GP, Moreno CG, and Van Look PFA (2006) Sexual and reproductive health: A matter of life and death. *The Lancet* 368: 1595–1607.

Hill K, Thomas K, Abou Zahr C, *et al.* on behalf of the Maternal Mortality Working Group(2007) Estimates of maternal mortality worldwide between 1990 and 2005: An assessment of available data. *The Lancet* 370: 1311–1319.

Joint United Nations Programme on HIV/AIDS (UNAIDS) (2007) *AIDS Epidemic Update*. Geneva, Switzerland: UNAIDS and WHO.

United Nations (1994) *Report of the International Conference on Population and Development*, Cairo, 5–13 September. (with Anastasion D) New York: UN.

United Nations (1995) *Fourth World Conference on Women Platform for Action and the Beijing Declaration*, Beijing, China, 4–15 September. New York: UN.

United Nations (2005) World summit outcome. Document A/RES/60/1. New York: UN.

United Nations Millennium Project (2005) *Investing in Development: A Practical Plan to Achieve the Millennium Development Goals*. London: United Nations Development Programme.

World Health Organization (WHO) (2005) *The World Health Report 2005. Make Every Mother and Child Count*. Geneva, Switzerland: WHO.

World Health Organization (2007) *Unsafe Abortion – Global and Regional Estimates of the Incidence of Unsafe Abortion and Associated mortality in 2003*. Geneva, Switzerland: WHO.

World Health Organization Department of Making Pregnancy Safer (2006) *Neonatal and Perinatal Mortality: Country, Regional, and Global Estimates*. Geneva, Switzerland: WHO.

World Health Organization Department of Reproductive Health and Research (2004) *Reproductive Health Strategy to Accelerate Progress towards the Attainment of International. Development Goals and Targets*. Geneva, Switzerland: WHO.

World Health Organization Department of Reproductive Health and Research (2006a) *Reproductive Health Indicators – Guidelines for their Generation, Interpretation and Analysis for Global Monitoring*. Geneva, Switzerland: WHO.

World Health Organization Department of Reproductive Health and Research (2006b) *Accelerating Progress Towards the Attainment of International Reproductive Health Goals – a Framework for Implementing the WHO Reproductive Health Strategy*. Document WHO/RHR/06.3. Geneva, Switzerland: WHO.

World Health Organization and UN Population Fund (UNFPA) (2006) *Preparing for the Introduction of HPV Vaccines – Policy and Programme Guidance for Countries*. Geneva, Switzerland: WHO.

Further Reading

Cook RJ, Dickens BM, and Fathalla MF (2003) *Reproductive Health and Human Rights: Integrating Medicine, Ethics, and Law*. Oxford, UK: Oxford University Press.

Fathalla MF (2000) Women, the profession and the reproductive revolution. In: Special Millennium edn. 223–231O'Brien PMS (ed.) *The Yearbook of Obstetrics and Gynecology,* Special Millennium edn. Vol. 8, pp. 223–231London: RCOG Press.

Fathalla MF, Sinding SW, Rosenfield A, and Fathalla MMF (2006) Sexual and reproductive health for all: A call for action. *The Lancet* 368: 2095–2100.

Murray CJL and Lopez AD (eds.) (1998) *Health Dimensions of Sex and Reproduction,* Global Burden of Disease Series, (vol. 3). Boston, MA: Harvard School of Public Health on behalf of the World Health Organization and the World Bank.

Potts P and Short R (1999) *Ever Since Adam and Eve – The Evolution of Human Sexuality*. Cambridge, UK: Cambridge University Press.

Sadik N (ed.) (2002) *An Agenda for People – The UNFPA through Three Decades.* New York and London: New York University Press.

The Lancet Sexual and Reproductive Health Series (October 2006) www.thelancet.com.

United Nations (2006) Economic and Social Council, Commission on Population and Development, 39th session. Follow-up actions to the recommendations of the International Conference on Population and Development. Flow of financial resources for assisting in the implementation of the Programme of Action of the International Conference on Population and Development. *Report of the Secretary-General.* Document E/CN.9/2006/5. New York: UN.

United Nations Population Division Department of Economic and Social Affairs (2004) *World Population Monitoring 2002 – Reproductive Rights and Reproductive Health*. New York: UN.

UN Millennium Project (2006) *Public Choices, Private Decisions: Sexual and Reproductive Health and the Millennium Development Goals*. New York: UN Development Programme.

Relevant Websites

http://www.measuredhs.com – Demographic and Health Surveys (DHS).

http://www.who.int/reproductive-health – Department of Reproductive Health and Research (WHO).
http://www.engenderhealth.org – EngenderHealth.
http://www.familycareintl.org – Family Care International (FCI).
http://www.fhi.org – Family Health International (FHI).
http://www.guttmacher.org – Guttmacher Institute.
http://www.figo.org – International Federation of Gynecology and Obstetrics (FIGO).
http://www.ippf.org – International Planned Parenthood Federation (IPPF).
http://www.iwhc.org – International Women's Health Coalition (IWHC).
http://www.ipas.org – Ipas.
http://www.unaids.org – Joint United Nations Programme on HIV/AIDS (UNAIDS).
http://www.popcouncil.org – Population Council.
http://www.prb.org – Population Reference Bureau (PRB).
http://www.unicef.org – United Nations Children's Fund (UNICEF).
http://www.unpopulation.org – United Nations Population Division.
http://www.unfpa.org – United Nations Population Fund (UNFPA).
http://www.worldbank.org – World Bank.

Sexual Health

K Wellings, London School of Hygiene and Tropical Medicine, London, UK

Introduction

Defining the Area of Sexual Health

Sexual health is often seen in terms of the prevention of adverse outcomes of sexual behavior, such as unplanned pregnancy and sexually transmitted infections (STIs). The definition of sexual health formulated by the World Health Organization as a guide for those working in this public health field, however, incorporates more positive and pleasurable aspects of sexual health, so broadening the public health remit to include the enhancement of life and personal relations, and not merely counseling and care related to reproduction and STIs:

> Sexual health is a state of physical, emotional, and social well-being in relation to sexuality; it is not merely the absence of disease, dysfunction or infirmity. Sexual health requires a positive and respectful approach to sexuality and sexual relationships, as well as the possibility of having pleasurable and safe sexual experiences, free of coercion, discrimination and violence. For sexual health to be attained and maintained, the sexual rights of all persons must be respected, protected and fulfilled. (WHO, 2006)

The past decade has seen growing attention in the international policy arena to sexual rights, and new standards have been formulated for the creation and maintenance of a sexually healthy society, invoking values of respect and choice (Miller and Vance, 2004).

Barriers to the Achievement of Sexual Health

Sexual health presents particular challenges to public health. The human behaviors involved are not only for the most part personal and private, but are often stigmatized and discriminated against. As a result, those who practice them are often hidden or difficult to reach. This has consequences for sexual health status at a number of levels. Men and women may feel unable to talk about safer sex; they may feel disinclined to seek help; politicians may be unwilling to support provision of services for some populations; and service providers may feel unable to reach people in need of help, or may have negative attitudes toward them.

The Public Health Burden

Efforts to improve sexual health are vital to the achievement of the Millennium Development Goals on gender equality, maternal health, and HIV/AIDS. Sexual ill-health contributes significantly to the global burden of disease (Ezzati *et al.*, 2003). The impact of the HIV epidemic is well documented. By the end of 2007, more than 30 million people worldwide were living with HIV, 15 million of whom were women, and 2.5 million, children (UNAIDS and WHO, 2007). HIV/AIDS is responsible for 6% of the global burden of disease. In developing countries, STIs and their complications are among the top five diseases for which adults seek care and are a major cause of infertility.

Sexual and reproductive ill health accounts for one-third of the global burden of disease among women of reproductive age and one-fifth of the burden of disease among the population overall (Ezzati *et al.*, 2003). Unintended pregnancy often leads to unsafe abortion, which in turn may lead to other health consequences, such as reproductive tract infections, pelvic pain, and infertility.

The burden of sexual ill health due to STIs varies markedly between populations, since transmission patterns relate to both patterns of sexual behavior and access

to treatment. Congenital syphilis, for example, is rare in Western countries but is a major cause of infant mortality and morbidity in resource-poor countries lacking effective screening and treatment programs. Similarly, the advent of highly active anti-retroviral therapy (HAART) has considerably increased survival rates for HIV in developed countries, while prognoses in poor settings where HAART is unavailable are still poor.

Sexual health problems also include sexual dysfunction, sexual violence, and discrimination on the basis of sexual orientation. Global estimates of lifetime prevalence of sexual violence by an intimate partner range from 10–50%. Population-based estimates of the prevalence of sexual dysfunction vary greatly. In developed countries, the range is from 1 in 12 to 1 in 3 of the adult population (Laumann *et al.*, 1999) and a pan-European study of women (Graziottin, 2007) also showed large differences between countries. The variation stems largely from differences in defining and measuring the problem. Persistent lack of interest in or desire for sex is classified as a disorder of sexual function both by ICD-10 (WHO, 1992) and the DSM-IV-TR (American Psychiatric Association, 2000). Yet sexual interest varies across age, gender, cultural context, and sexual orientation, and attempts to define 'abnormal' must be made with caution.

The concept of 'psychosexual dysfunction,' which first appeared in the third edition of the DSM (DSM-III) (American Psychiatric Association, 1980), was based on a biomedical model of sex involving physical processes such as lubrication and orgasm, and stages of desire, arousal, and orgasm. Subsequent revisions of the DSM have reflected advances in thinking about the nature of sexual difficulties, in particular, the concern to avoid labeling temporary alterations in sexual performance as 'sexual dysfunction' and to broaden the diagnostic criteria to include duration of, and distress caused by, the problem, and the extent to which it reflects the relationship context (Balon *et al.*, 2007). Increasing use of sildenafil citrate (Viagra) has helped many people, but there is growing recognition that pharmacological solutions are likely to prove effective only if psychosocial issues are also dealt with, and if medical treatment and sex therapy are combined.

Risk Behaviors

Trends and Patterns

An understanding of trends and patterns in sexual behavior is essential to the design of public health interventions to improve sexual health status. Marked shifts in sexual behavior have occurred in the past half century. These include demographic changes in the age structures of populations, in the timing of marriage, and in the scale of mobility and migration between and within countries. The increase in intra- and international travel, particularly, has played its part in increasing possibilities for the transmission of STIs, including HIV. The HIV epidemic has significantly influenced the context in which sexual behavior occurs, providing the impetus for public discussion of sexual matters, research on the subject, and innovation in interventions designed to improve sexual health.

Policy and legislation governing health-care systems and public health strategies have also brought about changes. The advent of reliable contraception has increasingly freed sexual expression from its reproductive consequences. Access to family planning services has increased, and few areas have been unaffected by efforts to prevent HIV transmission. Attitudes toward sexual behavior have altered in many countries (Wellings *et al.*, 2006). Global communications, including the Internet, have had a bearing on social norms, transporting Western sexual images to more conservative societies, particularly those in which advances in information technology have been rapid.

Given these trends, surveys of sexual behavior show perhaps less change over time than might have been supposed. Considerable public health interest surrounds onset of sexual activity, since early sexual intercourse is more likely to be nonconsensual, to be regretted, to be unprotected against unplanned pregnancy and infection, and to be associated with larger lifetime numbers of sexual partners. Trends toward early and premarital sex are less marked than is sometimes assumed. In countries in which first intercourse still occurs predominantly within marriage, the trend toward later marriage has been accompanied by a trend toward later sex among young women. In industrialized countries, sexual activity before age 15 has become more common in recent decades but the prevalence is low. The trend toward later marriage has led to an increase in the prevalence of premarital sex in some, but not all, countries.

Multiple sexual partnerships increase the risk of STIs and HIV. The majority of people report only one sexual partner in the past year. The prevalence of multiple partnerships varies regionally but is notably higher in industrialized countries. However, rates of condom use are lower in countries with lower sexual health status, and this is likely to be attributable to factors relating to access and service provision.

Risk Factors and Risk Groups

Sexual ill health affects all age groups and sections of society, but some groups are more vulnerable than others. Young people, minority ethnic groups, and those affected by poverty and social exclusion are disproportionately affected. Globally, young, sexually active people are at highest risk of unplanned pregnancy and STIs. Preventive efforts also focus on men who have sex with men, a particularly high-risk group for HIV and other STIs.

Mean partner numbers are higher among male and female sex workers. (In addition to the formal trading of sex for money, sex may be exchanged for gifts or favors within personal relationships, and hence the term 'transactional sex' is sometimes preferred.) However, the primary risk for STIs among commercial sex workers is unprotected sex with high-risk regular partners. In countries with wide gender differences between men and women in the prevalence of premarital sex, young men are more likely to report sex with sex workers. Clients of sex workers are important bridging groups in the transmission of STIs and HIV to wider sexual networks.

A focus on behaviors in addition to groups is needed in a public health context, since risk behaviors are not necessarily limited to the traditional high-risk groups. Many men who have sex with men, for example, have also had sex with women and, except in the West, are unlikely to identify as 'gay.' An individual's risk depends also on his or her partner's risk. Monogamous women in many parts of the world may be more susceptible to sexually transmitted infection on account of their partner's risk behavior, yet may be unable to negotiate condom use (Clark, 2004). Similarly, simply identifying the 'multi-partnered' will not help target men and women at greatest risk. Increasingly, high rates of concurrent partnerships are thought to contribute substantially to the global HIV epidemic (Kretzschmar and Morris, 1996). Concurrent sexual partnerships (those that overlap in time) allow more rapid spread of STIs, including HIV, than the same rate of new sequential partnerships.

Social and Environmental Factors

Social and environmental factors are powerful influences shaping sexual behavior and its consequences for sexual health. Poverty, deprivation, and unemployment, for example, contribute markedly to sexual ill health. Economic adversity limits the power of men and women to take control over their health; unemployment may drive men and women to sell sex or travel greater distances to work. Being away from home is associated in both developed and developing countries with concurrent partnerships, disruption of existing partnerships, and an increase in risk behaviors (**Figure 1**).

There are striking differences between men and women in sexual behavior, and these are most pronounced in the less industrialized countries (Wellings *et al.*, 2006). Men report more premarital sex and multiple partnerships than women in all but the more industrialized countries. The sexual 'double standard,' whereby restraint is expected of women while excesses are tolerated for men, compounds the problems for both men and women. Reporting bias doubtless accounts for some of these differences, but in some areas, for example, in Africa, the difference can be largely explained by the age structure and patterns of age mixing, that is, older men having sex with younger women. Women may be disadvantaged in protecting their sexual health when their partner is senior to them in age and/or superior in status, and when they are beholden to a man for favors, goods, or money in return for sex.

Possibly the most powerful influences on human sexuality are the social norms governing its expression. Morals, taboos, laws, and religious beliefs employed by societies the world over circumscribe and radically determine the sexual behavior of their citizens. Such strictures may have been instituted to protect well-being and rights, yet they may also hinder attempts by both men and women to protect their sexual health, and they strongly influence the selection of acceptable public health messages. In some countries, such as Brazil, condoms are available to young people in schools; in others, such as parts of Indonesia, possession is a criminal offence.

Nowhere are social norms more strongly felt than in the area of homosexual activity. In some parts of the world, sex between men can be celebrated in public parades of pride; in others, it carries the death penalty. Whether sexual orientation is innate is an issue that is hotly contested in scientific circles, but more important from a public health perspective is the issue of sexual identity, a term used to refer to the way in which an individual sees him- or herself and is seen by others. Where cultures stigmatize homosexual behavior and relationships, men and women may be wary of assuming an openly gay identity. Behaviors that are discriminated against may be driven underground, thwarting public health initiatives to protect sexual health. Homophobic attitudes may have a negative impact on mental health, including a higher incidence of depression and suicide. In its extreme form, homophobia can lead to violence; studies have shown that 80% of gay men and lesbians had experienced verbal or physical harassment on the basis of their orientation.

Effective Public Health Strategies

Interventions Aimed at Individual Behavior Change

Men and women have sex for different reasons and in different ways in different settings. This diversity needs to be respected in a range of approaches tailored to specific populations and to particular groups and individuals within them. The repertoire of public health approaches is broad and includes health promotion, social marketing, media advocacy, health service provision, legislative activities, and community empowerment.

Approaches based on voluntary individual behavior change continue to be a central plank of public health policy. Interventions, however, focusing on provision of

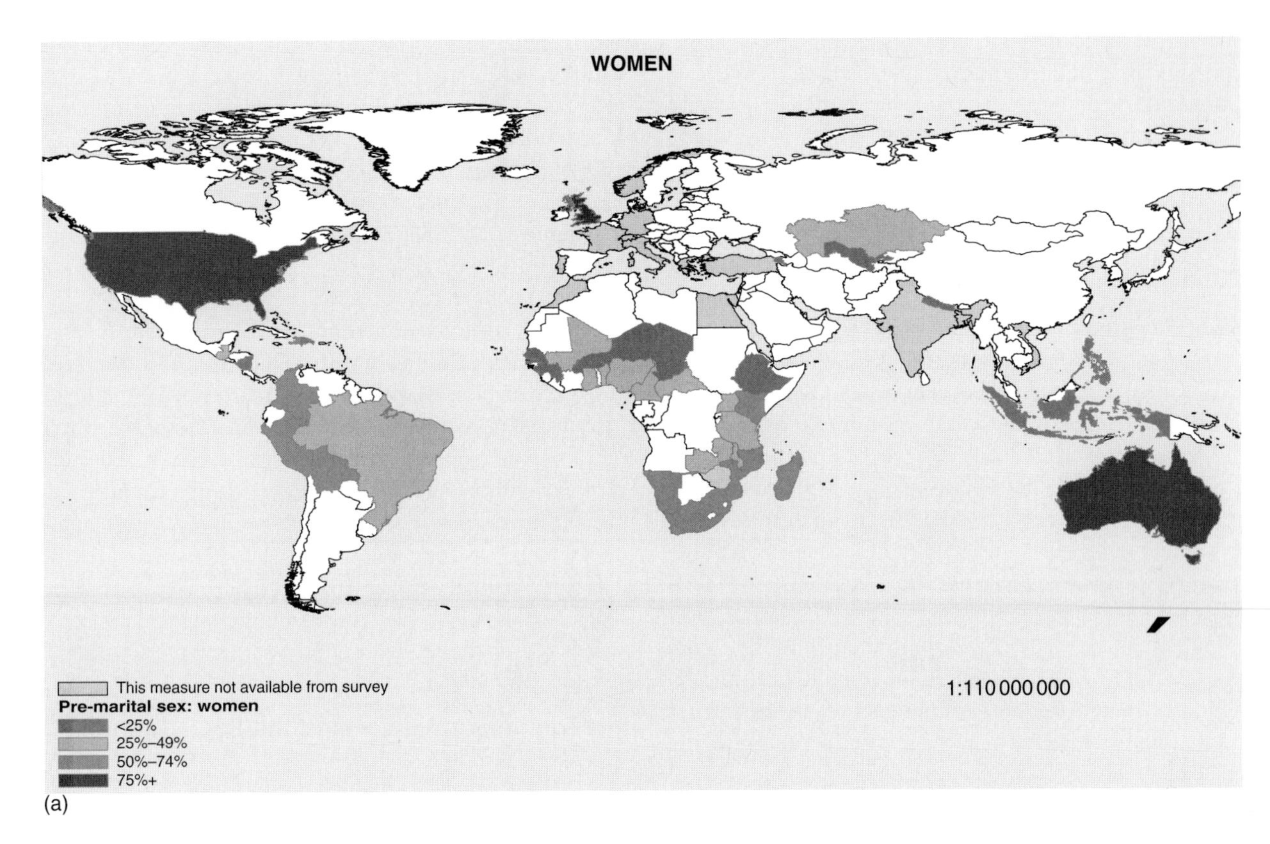

(a)

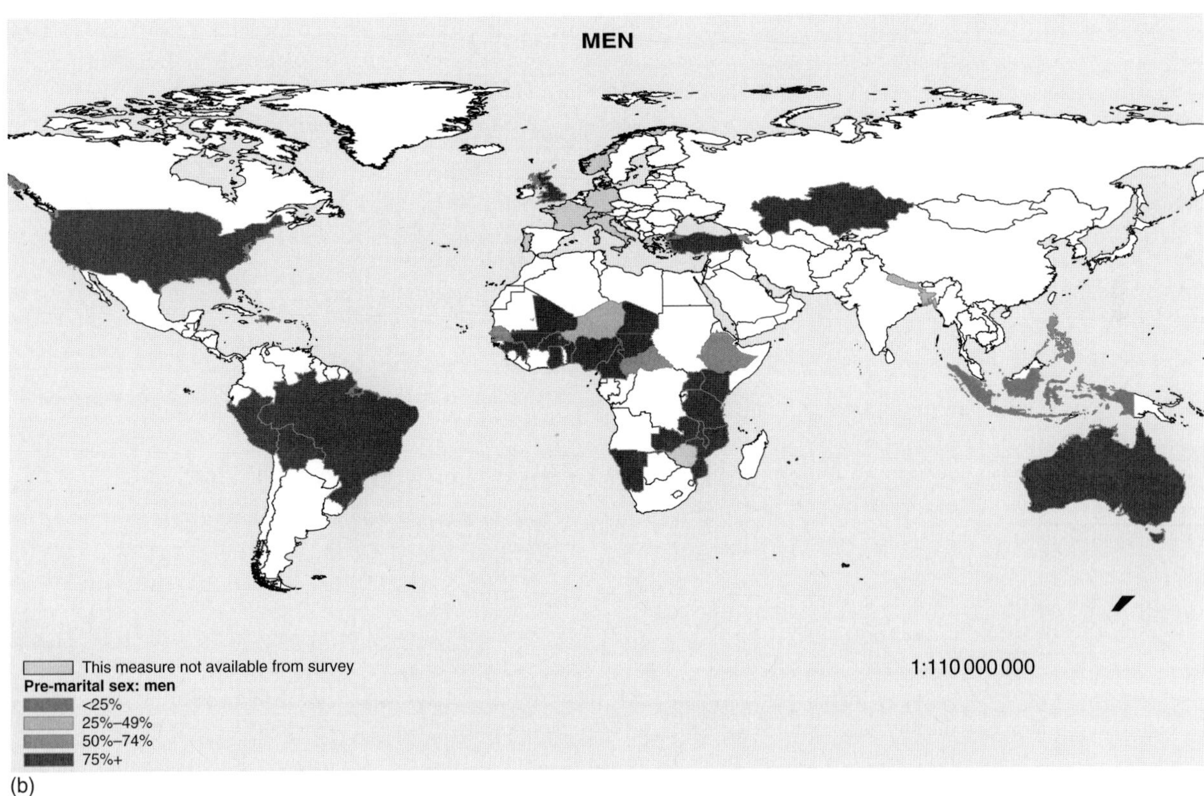

(b)

Figure 1 Global prevalences of premarital sex. Reproduced with permission from Wellings K, Collumbien M, Slaymaker E, *et al.* (2006) Sexual behaviour in context: a global perspective. *Lancet* 368: 1706–1728.

information alone have been shown to be less effective than those including the development of skills relating to condom use and negotiation of safer sex that use multiple delivery methods, and take account of the social context and the need for sustainability (Rotheram-Borus *et al.*, 2000).

Mass media campaigns can be effective in bringing sexual health issues to the attention of the broad mass of the public, in increasing knowledge, and in influencing social norms, but they are less effective in increasing self-efficacy and personal skills (Bertrand *et al.*, 2006).

To be successful, individual-based interventions also need to be targeted, in the workplace, for example, and in school settings. The evidence is that curriculum-based sex education does not increase risky sexual behavior as is sometimes feared, but instead is likely to increase the practice of safer sex and delay, rather than hasten, the onset of sexual activity (Santelli *et al.*, 2007).

Risk-reduction messages need to take account of the diversity of sexual behavior and the need for people to be able to act on them. Options in public health may include advice to abstain from sex, to have sex with only one partner, or to practice safer sex. Considerable debate takes place over the relative merits of each of these. Research has shown that a significant proportion of young women and men first have sex before they are ready, and so efforts to help young people optimize the timing of first intercourse, and to resist unwanted sex, are justified. Yet abstinence may not be an option where first sexual relations may be forced, where the sexual abuse of adolescents is common, and where financial circumstances force young people to sell sex. A WHO study has shown that over 30% of women who first had sex before the age of 15 years described having been forced to do so (Garcia-Moreno *et al.*, 2005). Similarly, advice to an individual to be monogamous may not reliably safeguard sexual health where men and women are put at risk by their partners' risk behavior. In Uganda, married women constitute the population group among whom HIV transmission is increasing most rapidly. Married women find it more difficult than single women to negotiate safer sex, and fewer use condoms for family planning.

Interventions also need to address the social norms influencing sexual health. Efforts on the part of professional educators and the medical profession need to be supported by those working at the grassroots level. Community-level interventions have been effective in mobilizing local groups in support of preventive strategies. Information gained through social networks is more salient, and more likely to lead to behavior change, than that conveyed by more impersonal agencies (McIntyre, 2005). The prompt response from gay communities to the prevention of HIV/AIDS in Western countries in the early 1980s owed much to the preexistence of infrastructures in the nongovernmental sector and to the visibility and mobilization of gay men. Preventive programs using naturally occurring social networks have been used to reduce levels of risky behavior among gay men in Russia and to increase contraceptive use among married women in Bangladesh and condom use among sex workers in India; and have proved more effective in changing norms than more orthodox approaches using conventional health care and field workers.

Sexual behavior is subject to extensive legislation, and this also has implications for sexual health. In general, laws protect the young and those vulnerable to coercion and exploitation, but they may also impede safer sex practices. Where sexual practices are illegal, they are more likely to be engaged in in a furtive or clandestine manner, and people may have fewer opportunities to protect themselves. For example, condom use is uncommon among sex workers in India, where commercial sex is heavily socially proscribed, but is near-universal among those in Kampala and Mexico, where public health agencies have been able to adopt a more open approach.

Intervention at the Level of the Social Context

Patterns of risk behavior vary between cultures and regions, as do patterns of sexual health status, yet often there is no obvious association between the two. The link between sexual behavior and rates of STIs/HIV and unplanned pregnancy is not always seen. It has been difficult to explain differences in patterns of HIV transmission in Africa and in the West, for example, by reference to differences in sexual behavior, and evaluation studies have not always shown a link between behavior change and health outcomes. The prevalence of multiple partnerships is comparatively high in the richer countries of the West, yet it is the poorer countries of the world that have higher rates of STIs and HIV (Wellings *et al.*, 2006). The impact of poor sexual and reproductive health falls hardest on the most disadvantaged groups, and disproportionately affects women in low-income countries (**Figure 2**).

This has led to pleas for a public health focus on the broader determinants of sexual health such as poverty, mobility, and gender. Interventions need to be pitched not only at the individual level but also at the level of the social context within which sex occurs. Individual-level intervention may have less chance of success in settings in which people have less control over their lives. In particular, individual-level interventions may have less chance of success where sex is tied to livelihoods, duty, and survival, and individual agency is more limited. In wealthier countries, personal choice is greater, yet power inequalities persist.

Addressing the broader social influences on sexual behavior, however, presents considerable challenges for public health. Structural factors like poverty,

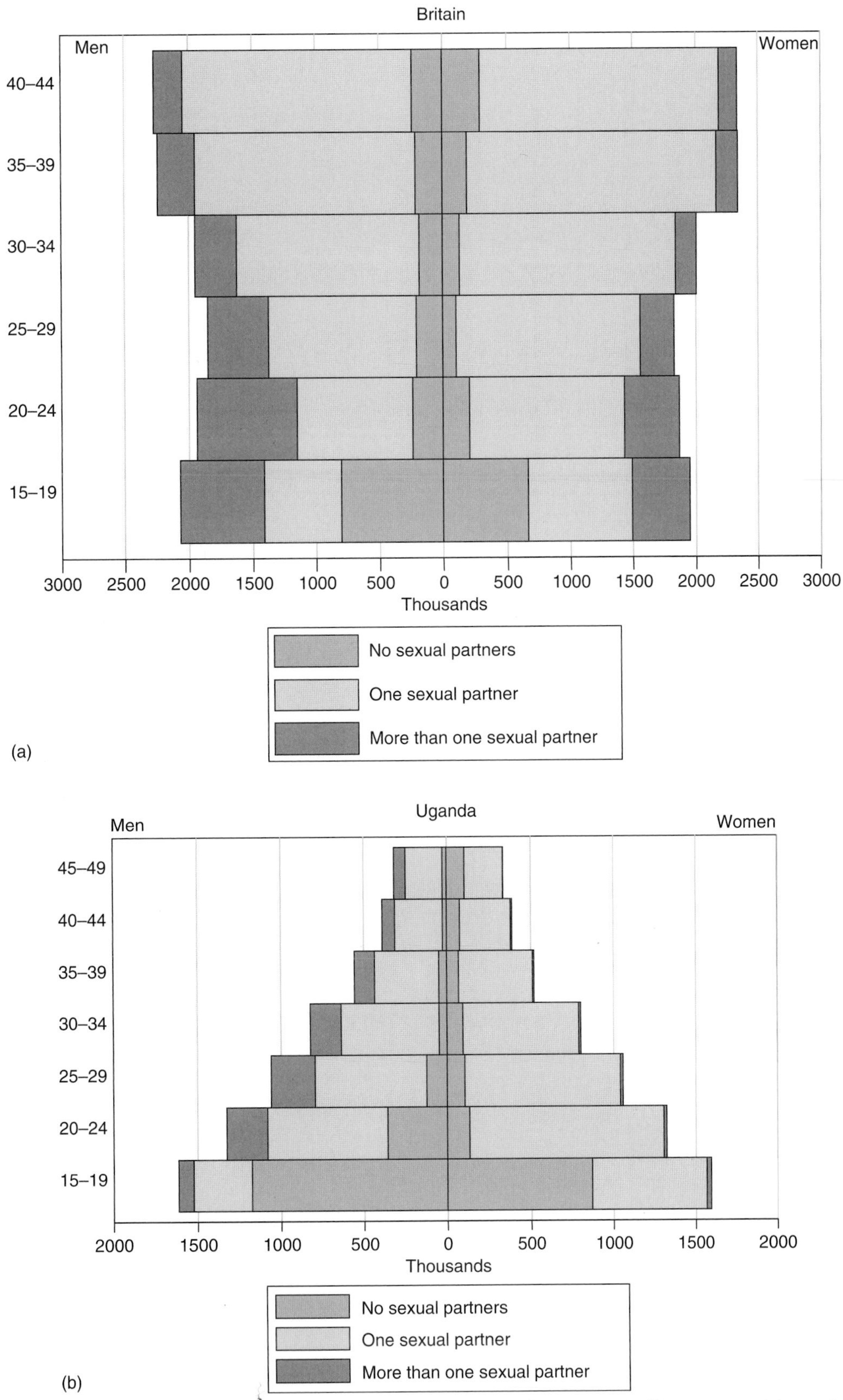

Figure 2 Comparison of sexual partnership patterns in a resource-poor country compared with a resource-rich country. (a) Britain. (b) Uganda. Reproduced with permission from Wellings K, Collumbien M, Slaymaker E, *et al.* (2006) Sexual behaviour in context: a global perspective. *Lancet* 368: 1706–1728.

unemployment, and gender are hard to modify, and social and political norms are slow to shift. Yet efforts are being made to address macro-level forces such as gender and poverty in innovative approaches linking health with development, for example, by empowering sex workers to avoid sex work through business and information technology training. In South Africa, a current sexual rights advocacy campaign is focused on achieving a more comprehensive perspective of sexual health and rights into prevention work, involving working with nongovernmental and community-based organizations to address HIV/AIDS, violence against women, and adolescent sexual health in an integrated manner.

Tackling determinants of sexual health, such as poverty, requires coordination and collaboration across health and social sectors and agencies. Partnerships need to be forged between public health personnel and, for example, economists, politicians, industry, the judiciary, and nongovernmental organizations. Improvements in sexual health are generally achieved as a result of a combination of preventive policies and strategies mounted by different agencies, with strong partnerships between the media, government, nongovernmental organizations, sex workers, the AIDS community, and international and local public health agencies, endorsed at the highest political level.

The need for preventive approaches that are multilevel and multipartnered, that have strong support from government, and that promote messages carefully tailored to the local context is demonstrated in many of the success stories in public health: the successes of Thailand and Uganda in reducing HIV rates, for example, and of the United States in reducing teenage pregnancy rates. The 100% condom-use program in Thailand, which achieved increasing rates of condom use and reduced STI and HIV rates and sex work activity, incorporated a variety of components, such as free condom provision; mass media campaigns and workplace programs; regular check-ups for STIs and encouragement of testing and treatment among sex workers; extensive contact tracing; and legal action against brothel owners not implementing the program (Nelson *et al.*, 1996). Coordination between national and local governments, public health officials, and brothel owners was a strong characteristic of the program and the approach was pragmatic, involving tacit acceptance of sex work.

By the same token, Uganda's past success in reducing HIV prevalence and improving reproductive health status compared with neighboring countries has been attributed to the careful tailoring of messages to cultural context; the successful assimilation of scientific knowledge about modes of transmission; restoration of stability following the end of the Ugandan civil war in 1986; and two decades of strong policy support (Parkhurst, 2002). Uganda was the first African nation to establish a national AIDS program, and the response to the epidemic went beyond the individual-level factors, incorporating mobilization of local groups; collaborative partnerships with religious groups and community activists; and commitment to tackling the problem at the highest political level.

Rates of teenage pregnancy in the United States fell by 30% during the 1990s. Some studies have suggested that the decline resulted from increased uptake of abstinence and increased use of reliable methods of contraception. But other studies also provide evidence that the explanation may lie in changes in economic factors, notably the welfare reforms introduced by the Clinton administration. A number of factors influence adolescents' motivation to prevent pregnancy, and no one approach satisfactorily accounts for the decline that has occurred so far.

Conclusion

Sexuality is an essential part of human nature, and its expression needs to be affirmed rather than denied, if public health messages are to be heeded. No general public health approach will work everywhere and, by the same token, no single-component intervention is likely to work anywhere. The evidence is that the interventions that work best are complex, multilevel, multipartner behavioral interventions that take account of the social context in mounting individual-level programs; attempt to modify social norms to facilitate uptake and support maintenance of behavior change; and tackle the structural factors contributing to risky sexual behavior. Public health strategies limited to voluntary individual behavior change are unlikely to achieve such radical change without sociostructural changes. The mix of components in national programs needs to be tailored to the local context. The emphasis must be not merely on which approaches work, but on why and how they do so in particular social contexts.

See also: Abortion; Family Planning/Contraception; Violence Against Women.

Citations

American Psychiatric Association (1980) *Diagnostic and Statistical Manual of Mental Disorders*. 3rd edn. Washington, DC: American Psychiatric Press.

American Psychiatric Association (2000) *Diagnostic and Statistical Manual of Mental Disorders*. 4th edn. Washington, DC: American Psychiatric Press.

Balon R, Segraves RT, and Clayton A (2007) Issues for DSM-V: Sexual dysfunction, disorder, or variation along normal distribution: Toward rethinking DSM criteria of sexual dysfunctions. *American Journal of Psychiatry* 164: 198–200.

Bertrand JT, O'Reilly K, Denison J, Anhang R, and Sweat M (2006) Systematic review of the effectiveness of mass communication programs to change HIV/AIDS-related behaviors in developing countries. *Health Education Research* 21(4): 567–597.

Clark S (2004) Early marriage and HIV risks in sub-Saharan Africa. *Studies in Family Planning* 35: 149–160.
Ezzati M, Vander Hoorn S, and Rodgers A (2003) Estimates of global and regional potential health gains from reducing multiple major risk factors. *Lancet* 362: 271–280.
Garcia-Moreno C, Jansen HA, Ellsberg M, Herse L, and Watts C (2005) *WHO Multi-Country Study on Women's Health and Domestic Violence Against Women*. Geneva, Switzerland: World Health Organization.
Graziottin A (2007) Prevalence and evaluation of sexual health problems: HSDD in Europe. *Journal of Sexual Medicine* 4(supplement 3): 211–219.
Kretzschmar M and Morris M (1996) Measures of concurrency in networks and the spread of infectious disease. *Mathematical Biosciences* 133: 165–195.
Laumann EO, Paik A, and Posen RC (1999) Sexual dysfunction in the United States: Prevalence and predictors. *Journal of the American Medical Association* 281: 237–244.
McIntyre JA (2005) Sex, pregnancy, hormones, and HIV. *Lancet* 366: 1141–1142.
Miller AM and Vance CS (2004) Sexuality, human rights, and health. *Health and Human Rights* 7: 5–16.
Nelson KE, Celentano DD, Eiumtrakol S, *et al.* (1996) Changes in sexual behavior and a decline in HIV infection among young men in Thailand. *New England Journal of Medicine* 335: 297–303.
Parkhurst JO (2002) The Ugandan success story? Evidence and claims of HIV-1 prevention. *Lancet* 360(9326): 78–80.
Rotheram-Borus MJ, Cantwell S, and Newman PA (2000) HIV prevention programs with heterosexuals. *AIDS* 14(Supplement 2): S59–S67.
Santelli JS, Lindberg LD, Singh S, and Finer LB (2007) Recent declines in adolescent pregnancy in the United States: more abstinence or better contraceptive use? *American Journal of Public Health* 97(1): 150–156.
UNAIDS and World Health Organization (2007) *AIDS Epidemic Update: December 2007*. Geneva, Switzerland: UNAIDS. http://data.unaids.org/pub/EPISlides/2007/2007_epiupdate_en.pdf (accessed February 2008).
Wellings K, Collumbien M, Slaymaker E, *et al.* (2006) Sexual behaviour in context: a global perspective. *Lancet* 368: 1706–1728.
World Health Organization (1992) ICD–10: *International Statistical Classification of Diseases and Related Health Problems*, 10th edn. Geneva, Switzerland: World Health Organization.
World Health Organization (2006) *Defining Sexual Health: Report of a Technical Consultation on Sexual Health, 28–31 January 2002, Geneva*. Geneva, Switzerland: WHO. http://www.who.int/reproductive-health/publications/sexualhealth/index.html (accessed February 2008).

Further Reading

Bancroft J, Graham CA, and McCord C (2001) Conceptualizing sexual problems in women. *Journal of Sex and Marital Therapy* 27: 95–103.
Coates TJ, Aggleton P, Gutzwiller F, *et al.* (1996) HIV prevention in developed countries. *Lancet* 348: 1143–1148.
Fathalla MF, Sinding SW, Rosenfield A, and Fathalla MM (2006) Sexual and reproductive health for all: A call for action. *Lancet* 368: 2095–2100.
Glasier A, Gülmezoglu AM, Schmid GP, Garcia Moreno CG, and VanLook PFA (2006) Sexual and reproductive health: A matter of life and death. *Lancet* 368: 1595–1607.
Graham CA and Bancroft J (2007) The sexual dysfunctions. In: Gelder M, Lopez-Ibor J, Andreasen N and Geddes J (eds.) *New Oxford Textbook of Psychiatry*, 2nd edn. Oxford, UK: Oxford University Press.
Marston C and King E (2006) Factors that shape young people's sexual behaviour: A systematic review. *Lancet* 368: 1581–1586.
Parker RG, Easton D, and Klein CH (2000) Structural barriers and facilitators in HIV prevention: A review of international research. *AIDS* 14: S22–S32.
Tiefer L (1996) The medicalization of sexuality: Conceptual, normative, and professional issues. *Annual Review of Sex Research* 7: 252–282.
UN Millennium Project (2006) *Public Choices, Private Decisions: Sexual and Reproductive Health and the Millennium Development Goals*. Geneva, Switzerland: United Nations Development Programmehttp://www.unmillenniumproject.org/reports/srh_main.htm (accessed February 2008).

Relevant Websites

http://www.jr2.ox.ac.uk/bandolier/booth/booths/std.html – Bandolier, Sexual Health. Provides critical reviews and evidence-based reports on sexual health issues.
http://www.guttmacher.org – The Guttmacher Institute, a nonprofit organization focused on sexual and reproductive health research, policy analysis, and public education.
http://www.ippf.org/ – International Planned Parenthood Foundation, a global service provider and a leading advocate of sexual and reproductive health and rights for all.
http://www.unaids.org/ – The United Nations Joint Programme on HIV/AIDS.
http://www.who.int/topics/sexual_health/en/ – World Health Organization, Sexual Health. Provides links to descriptions of activities, reports, news, and events; contacts and cooperating partners in the various WHO programs and offices working on this topic; and links to related websites and topics.

Population Growth

S Haddock and E Leahy, Population Action International, Washington, DC, USA
R Engelman, Worldwatch Institute, Washington, DC, USA

Dynamics of Population Growth

In the last half-century, population growth has proceeded at a rate unprecedented in the history of the planet. The human population took hundreds of thousands of years to grow to 2.5 billion people in 1950. Then, within the time span of just 50 years, the world's population more than doubled to exceed six billion people. Most of this growth has occurred in developing countries where advances in public health have contributed to lower mortality at all ages. Until recently, death rates fell faster than birthrates, resulting in rapid population growth.

While growth rates have fallen from their all-time high, human numbers are still increasing. The world population is growing by about 1.2% a year, down from a peak of roughly 2% in the late 1960s. However, even at this lower growth rate, the world's population still experiences a net increase of about 76 million people each year; many more than the 50 million or so added annually in the 1950s when the term population explosion first gained currency.

Population Projections

World population could reach anywhere between 7.7 billion and 10.6 billion by the mid-21st century, based on the 2004 United Nations (UN) projections, depending mostly on future birthrates (UN, 2007). A population projection is a conditional forecast based on assumptions about current and future fertility, mortality, and migration. **Figure 1** shows the UN's high, medium, and low projections for 2300. The variances between these three projections stem from differences in projected fertility; that is, the average number of children a woman has in the study population. In the high projection, global fertility averages around 2.6 children per woman. In the medium, it averages about 2.1 children per woman. The lowest fertility projection assumes an average of about 1.6 children per woman. Most users of these projections tend to cite the medium projection as the most likely. It is important, however, to be aware of the low and high projections as a kind of outer bounds of demographic possibility.

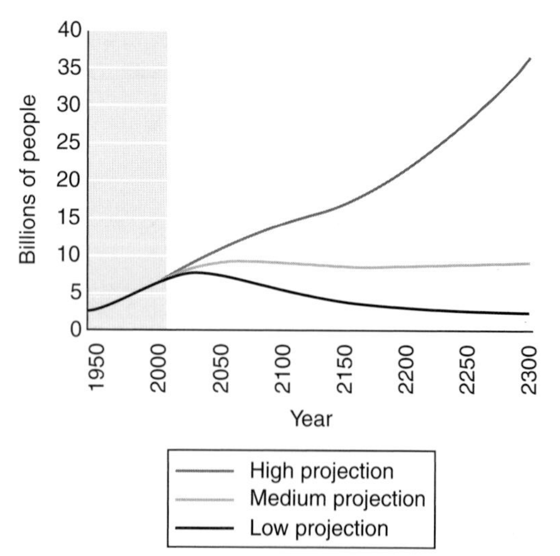

Figure 1 Population projections (high, medium and low to 2300). The United Nations high, medium, and low population projections are based on assumptions about current and future fertility, mortality, and migration. Data from United Nations Population Division (2004) *World Population to 2300*. New York, United Nations Population Division.

Recent UN projections are lower than most previous projections for the same period, reflecting earlier-than-expected declines in family size in some countries. Yet even under the UN's slow-growth scenario, the population would continue to grow for at least four more decades, to about 7.8 billion people, before declining gradually. Under the UN's high-growth scenario, the world population would exceed 35 billion late in the 23rd century and continue growing rapidly. The long-term medium projection suggests that the population would peak just above 9.1 billion around 2075, then drift downward to around 8.5 billion before growing again late in the 23rd century.

However, there is no certainty in projecting future population trends. All population projections are based on fairly rigid assumptions and therefore cannot provide completely reliable guides to the world's demographic future.

The Demographic Transition

The transformation of a population that is characterized by high birth and death rates to one in which people tend to live longer lives and raise smaller families is called the demographic transition. This transition typically shifts population age structure from one dominated by children and young adults up to age 29 to one dominated by middle-aged and older adults age 40 and above. About one-third of the world's countries, including most of Europe, Russia, Japan, and the United States, have effectively completed this transition. These 65 or so countries contain half of the world's population.

In **Figure 2(a)** and **2(b)**, two population pyramids geometrically display age structures: The proportion of people per age group, relative to the population as a whole. Afghanistan is an example of a country in the early phase of the demographic transition. Such countries typically have age structures composed predominantly of young adults and children (commonly called a youth bulge). As countries advance into the late phase of transition, as shown by the population age structure of South Korea, the proportion of children begins to decrease, but a youth bulge persists for a decade or two, moving upward into the older age cohorts.

Fertility Rates

Improvements in the education of girls and the overall status of women, together with increased access to contraceptive methods, have played important roles in advancing the demographic transition and reducing the global average fertility rate from 4.9 children in the first half of the 1960s. Nevertheless, the world's average fertility rate is currently 2.7 children, considerably higher than the level of fertility that would ultimately stabilize the growth of most populations (UN, 2005). This level, called

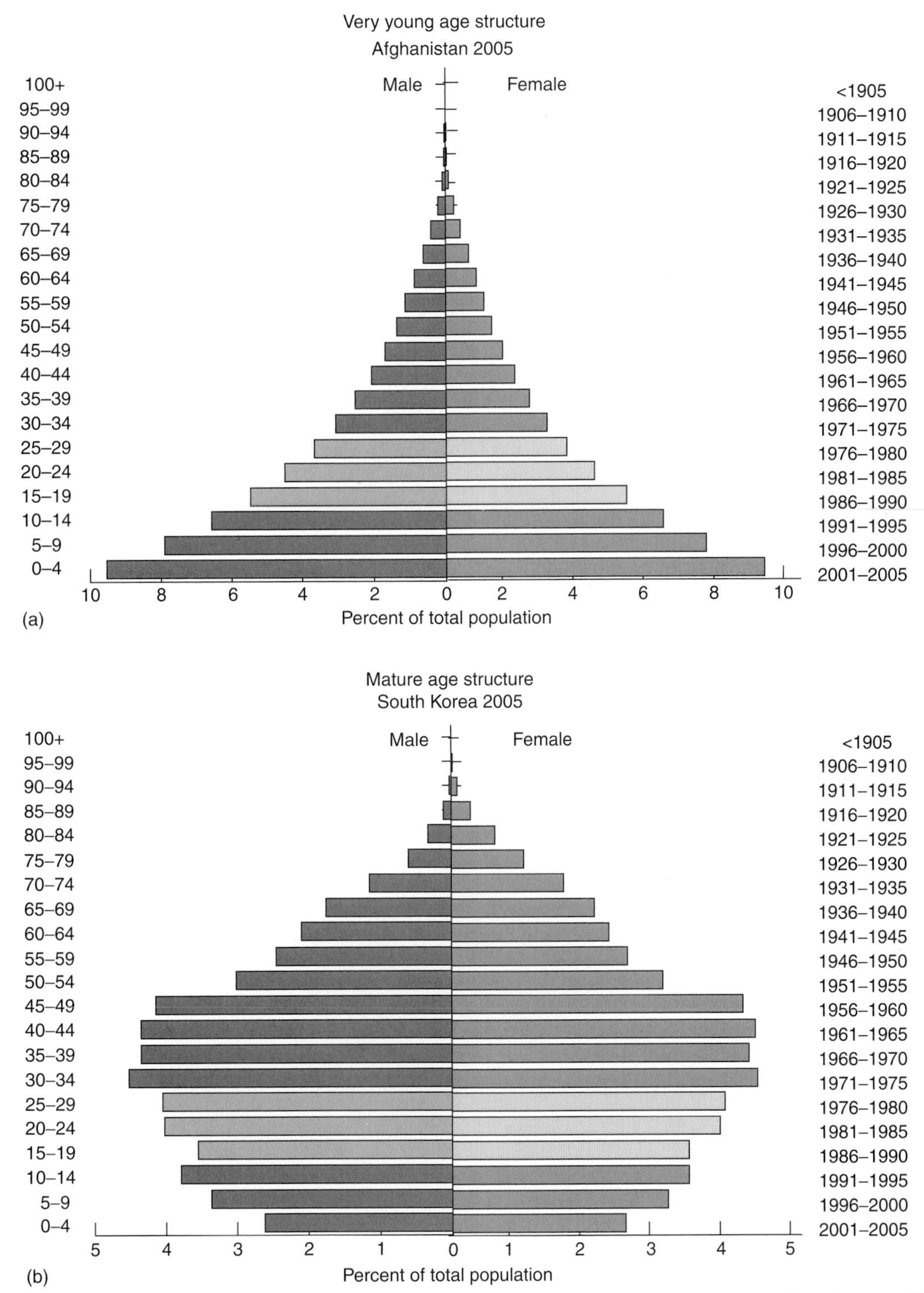

Figure 2 Population age structures: Afghanistan (a) and South Korea (b). Afghanistan is in the early phase of the demographic transition, with a large youth bulge. South Korea, which is in the late phase of transition, has a decreasing proportion of youth. Data from United Nations Population Division (2005) *World Population Prospects: The 2004 Revision*. New York: United Nations Population Division.

replacement fertility, represents the number of children a couple needs to have in order to replace themselves in the population. The values for replacement fertility are surprisingly diverse and vary by country, reflecting mostly differential rates of survival of children to their own reproductive ages. Replacement fertility rates range today from a low of 2.04 for Réunion Island to a high of 3.35 in Swaziland (Engelman and Reahy, 2006).

In nearly all developing countries, the average number of childbirths per woman has decreased in recent decades. While a number of factors have contributed to this revolution in childbearing, improved access to modern contraceptive methods has played a key role, in most of sub-Saharan Africa, in many countries of the Middle East, and in parts of South Asia. Growing at rates of 2.5–3.5% annually, these countries could double in size in 20 to 30 years. Almost all industrialized countries now have an average family size of fewer than two children and are growing relatively slowly, or even declining in population. In some eastern and southern European countries, women have an average of 1.2 or 1.3 children. These nations comprise less than 5% of the world's population, however, and thus have little influence on overall global trends (**Figure 3**).

Population Aging

Due to low fertility rates in some developed countries, concerns are emerging about population aging and the possibility of population decline. Population aging is defined as a rising average age within a population and is an inevitable result of longer life expectancies and lower birthrates. In many societies, population aging is a positive social and environmental development, but its potential persistence to extreme average ages is worrisome to governments concerned with the funding of social security programs and general economic growth. However, like population growth itself, population aging cannot continue indefinitely and will eventually end, even in stabilized populations.

Population decline is occurring in some industrialized countries. European population growth, which fueled immigration to the Americas for three centuries, has ended and begun to reverse course, despite significant streams of migration from developing countries. If low fertility rates persist, the populations of Germany, Italy, Russia, and Spain could shrink by 5–10% by 2025-an average reduction of 0.5% per year. In East Asia, the populations of Japan, China, and South Korea are likely to peak and begin a gradual decline in size before the middle of the 21st century. However, major declines outside Europe and East Asia are unlikely for decades.

Factors That Influence Population Growth

Contraceptive Prevalence

Demographers attribute much of the recent decline in global fertility to improved access to and use of family

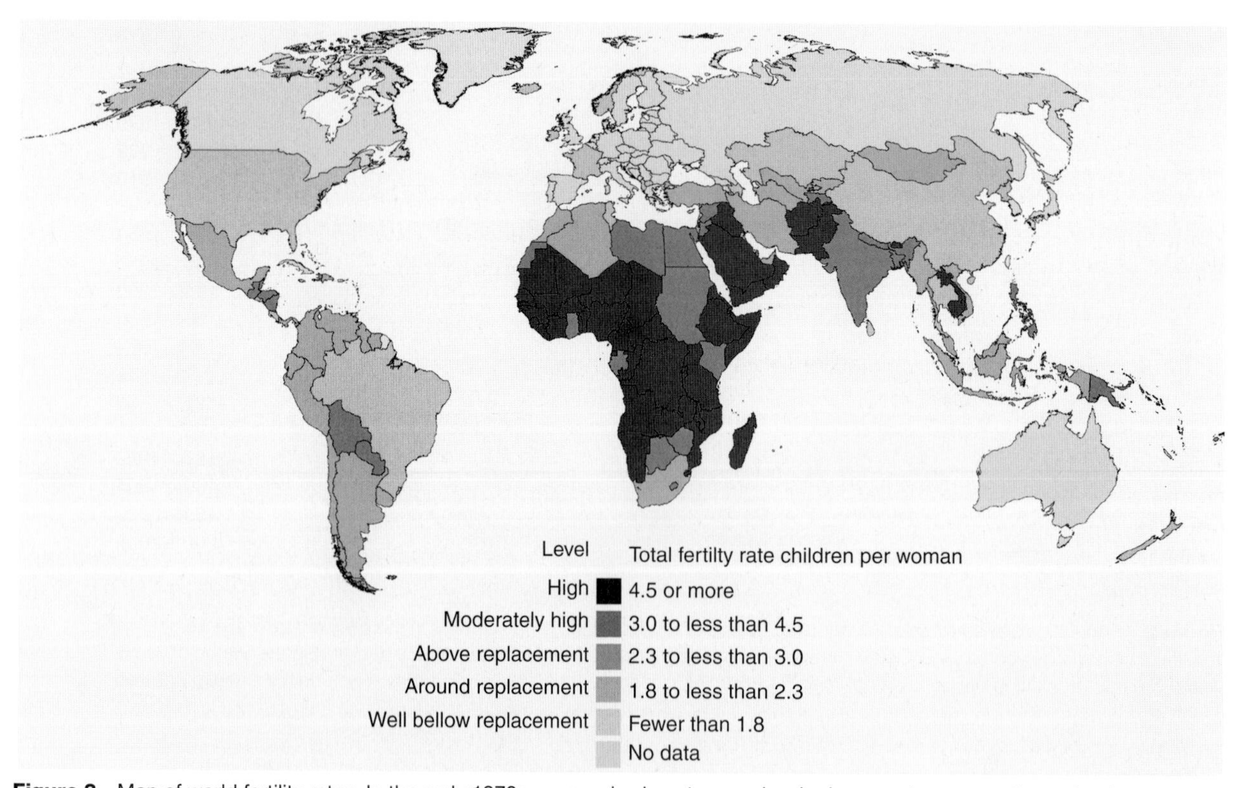

Figure 3 Map of world fertility rates. In the early 1970s, women in almost every developing country were estimated to have, on average, more than four children during their lives. In the three decades that followed, a revolution in childbearing spread across the world. Yet progress along the demographic transition has been uneven and lags in several regions. Data from United Nations Population Division (2005) *World Population Prospects: The 2004 Revision*. New York: United Nations Population Division.

planning information and methods. More than half of all married women in developing countries now use family planning, compared to 10% in the 1960s. Across the developing world, use of modern methods of contraception ranges from less than 8% of married women in Western and Central Africa to 65% in South America. The highest contraceptive prevalence rate in the world is in Northern Europe, where 75% of married women are currently using a modern method of family planning (**Figure 4**) (UN, 2004).

Despite these gains in availability, contraceptive services are still difficult to obtain, unaffordable, or of poor quality in many countries. Approximately 201 million married women worldwide who would prefer to delay or avoid having another child are not using modern methods of contraception (Alan Guttmaches Institute and UNFPA, 2003). Unmet need is highest in sub-Saharan Africa, where 46% of women at risk of unintended pregnancy are not using any contraceptive, presumably reflecting lack of access or other barriers. Meanwhile, in developing countries, the number of women in their childbearing years is increasing by about 22 million women a year. For fertility rates to continue their global decline, family planning services will have to expand rapidly to keep up with both population growth and rising demand.

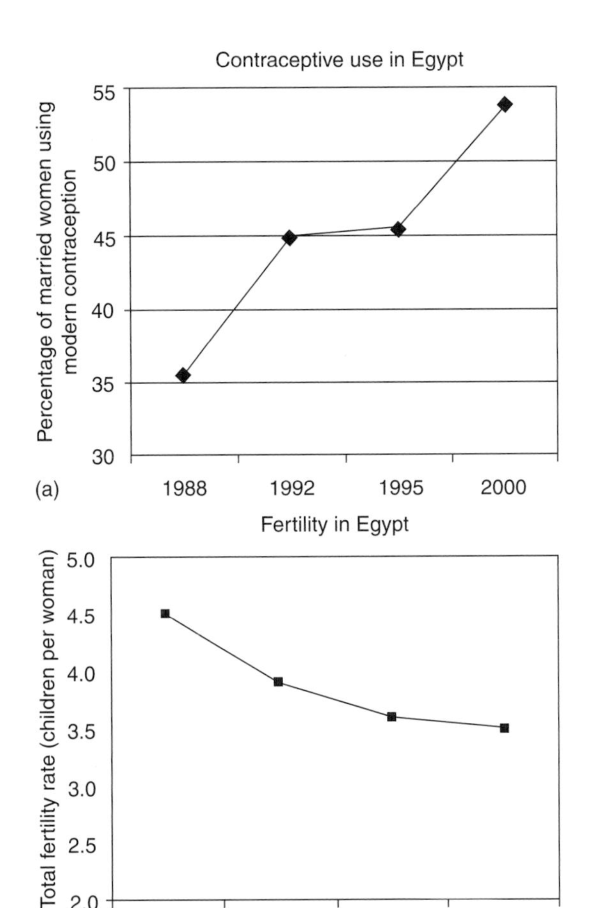

Figure 4 Contraceptive prevalence and fertility decline in Egypt. These figures illustrate the negative correlation between contraceptive use and total fertility rates that occurred over a 12-year period in Egypt. As an additional 20% of married women began using modern contraception between 1988 and 2000 (a), the country's total fertility rate (b) decreased from 4.5 to 3.5 children per woman. El-Zanaty F and Way A (2001) *Egypt Demographic and Health Survey 2000.* Calverton, MD: Ministry of Health and Population (Egypt), National Population Council and ORC Macro.

Mortality from HIV/AIDS and Other Infectious Diseases

Like no other disease, AIDS debilitates and kills people in their most productive years. Ninety percent of HIV-associated fatalities occur among people of working age, who leave behind large numbers of orphans – currently 11 million in sub-Saharan Africa alone – with few means of supporting themselves. A projected 10–18% of the working-age population will be lost in the next 5 years in nine central and southern African countries, primarily due to AIDS-related illnesses (Cincotta *et al.*, 2003).

High rates of AIDS mortality lead to a bottle-shaped population age structure, with very high proportions of young people and many fewer older adults. As birth rates remain high while people of reproductive age die from AIDS-related causes, the share of young dependants to each working-age adult rises dramatically. Without significant advances in HIV prevention or in access to life-saving drugs in poor countries, AIDS-related mortality rates could increase significantly. UN population projections suggest that some countries will experience lower population growth rates from AIDS, but high fertility rates mean that population size in these countries is likely to still increase significantly (**Figure 5**).

Population growth also threatens global health through increased vulnerability to other infectious diseases. A growing population size living in and moving to and from densely populated areas creates expanded opportunities for disease to spread and intensify. At the country and community levels, governments often lack the resources to improve sanitation and public health services at the same rate that populations are increasing. At the household level, evidence from demographic surveys suggests that children born after several siblings tend to receive fewer immunizations and less medical attention than children born earlier or in smaller families (Rutstein, 2005). The cumulative effect of all these influences is a greater risk of disease with higher birthrates and rapid population growth.

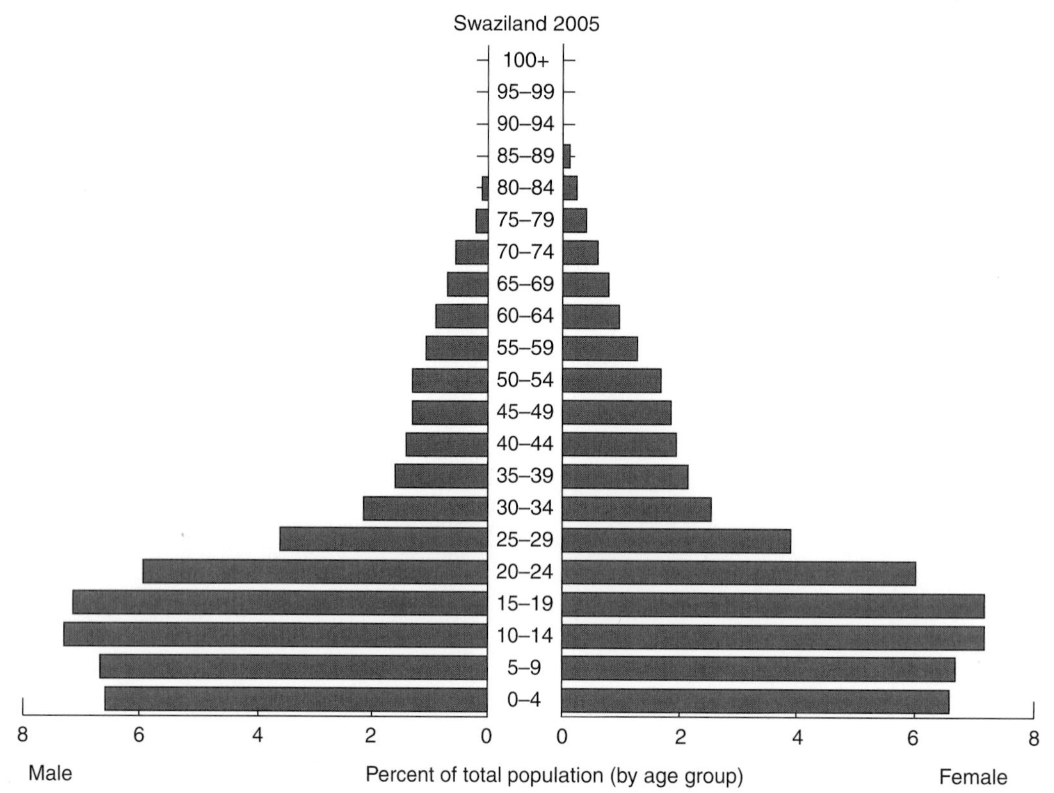

Figure 5 The Impact of HIV/AIDS on population age structure in Swaziland. The age structure of Swaziland's population in 2005 shows the decimating impact of the HIV/AIDS pandemic. As the disease leads to high mortality rates among working-age adults, high fertility rates create a youthful population. Data from United Nations Population Division (2005) *World Population Prospects: The 2004 Revision*. New York: United Nations Population Division.

Gender Equity

When women are able to determine the size of their families, fertility rates are generally lower than in settings where women's status is poor. In particular, there is a strong correlation between female school enrollment rates and fertility trends. In most countries, girls who are educated, especially those who attend secondary school, are more likely to delay marriage and childbearing. Women with some secondary education commonly have between two and four fewer children than those who have never been to school (Conly, 1998).

Early childbearing often limits educational and employment opportunities for women. When women have an education, their children tend to be healthier; in India, a baby born to a woman who has attended primary school is twice as likely to survive as one born to a mother with no education. By delaying marriage and childbearing, educating girls also helps lengthen the span between generations and slow the momentum driving future population growth.

Migration

The movement of people from one country to another – emigration in the case of those who leave their native country and immigration to describe the increase in a country's foreign-born population – continues to increase in both scale and frequency. International migration has doubled in the past 25 years, with about 200 million people – 3% of the world's population – today living in a country different from the one in which they were born. Approximately 60% of international migrants have chosen to live in developed countries, but migration within developing countries remains significant. Asia has three times as many international migrants as any other region of the developing world.

Although there are many reasons for increases in migration, the dramatic population growth of the past few decades has been a primary impetus. This has led, with a 15- to 20-year time lag, to the rapid growth of the world's labor force, especially among the young adults who make up the age group most likely to migrate. Tens of millions of people are added to the labor force each year, and the search for decent jobs is the leading reason people migrate. In addition deteriorating environmental conditions related to population growth – water and food shortages, for example, or human-induced climate change – can also spur large movements of population across international borders. Lower rates of population growth can help ease the pressures to migrate and improve the underlying conditions that force many people to seek a better life elsewhere.

Government Policies

Throughout history, government regulations on fertility – either by promoting or restricting childbirth – have directly affected population dynamics. Authoritarian regimes have pursued pro-natalist policies to increase the size of a population for militaristic, nationalist, and/or economic reasons. A few countries, including the world's most populous – China and India – have placed direct and indirect regulations on their citizens' fertility as a means of reducing population growth.

All state regulation of fertility, whether pro-natalist or intended to slow population growth, has been strongly criticized by human rights advocates. The importance of individual freedom of choice concerning fertility and reproductive health was outlined and agreed upon by 19 countries in 1994 at the International Conference on Population and Development.

Worldwide, more than one-fifth of all pregnancies are terminated annually. The prevalence of abortion is one indication of the high level of unintended pregnancy worldwide. Political disagreements over the highly contentious issue of abortion have in recent years hampered the funding of population and reproductive health programs in both the United States and in many developing countries.

As in the case of abortion, government policy related to international migration cannot really be called population policy, since in no cases is it expressly designed to remedy perceived deficiencies in national population trends. Nonetheless, government policies on migration and the process itself do influence the pace and distribution of population change, and it remains possible that some countries with especially low fertility may design future migration policies specifically to slow either the aging or the decline of their populations, or both.

Impacts of Population Growth

Health

High fertility rates can have an adverse effect on the health of women and children, especially in developing countries that lack a health infrastructure. Every year, an estimated 529 000 women die in pregnancy or childbirth (WHO, 2004), and for every woman who dies as many as 30 others suffer chronic illness or disability (Ashford 2002). Moreover, every year more than 46 million women resort to an induced abortion; 18 million do so in unsafe circumstances. Each year, approximately 68 000 women die from unsafe abortions and tens of thousands more suffer serious complications leading to chronic infection, pain and infertility (WHO, 2005).

Pregnancy-related deaths overall account for one-quarter to one-half of deaths to women of childbearing age. For both physiological and social reasons, pregnancy is the leading cause of death for young women aged 15–19 worldwide, with complications of childbirth and unsafe abortion being the major factors (WHO, 2004). Furthermore, each year 3.3 million babies are stillborn, more than 4 million die within 28 days of birth, and another 6.6 million children die before age 5 (UN, 2005).

The timing of births has an impact on child health. When a pregnant woman has not had time to fully recover from a previous birth, the new baby is often born underweight or premature, develops too slowly, and has an increased risk of dying in infancy or contracting infectious diseases during childhood (Rustein, 2005). Research shows that children born less than 2 years after the previous birth are about 2.5 times as likely to die before age 5 than children who are born 3–5 years apart (Setty-Venugopal and Upadhyay, 2002).

Chronic hunger, which affects an estimated 852 million undernourished people worldwide, is another critical health issue that is directly linked to population growth. The number of chronically hungry people in developing countries increased by nearly 4 million per year in the latter part of the 1990s. Globally, the number of food emergencies each year has doubled over the past two decades, totaling more than 30 per year. Africa has the highest number and proportion of countries facing food emergencies. The average number of food emergencies in Africa has almost tripled since the 1980s, and one-third of the population of sub-Saharan Africa is undernourished. A 1996 study by the Food and Agriculture Organization estimated that Africa's food supply would need to quadruple by 2050 just to meet people's basic caloric needs under even the lowest, and most optimistic, population growth projections (Food and Agriculture Organization, 2004).

Poverty

Improving reproductive health care, including family planning, can have economic benefits. The expanded provision of family planning services was responsible for as much as a 43% reduction in fertility between 1965 and 1990 in developing countries; falling fertility rates, in turn, have been associated with an average 8% decrease in the incidence of poverty in low-income countries. Countries with poverty rates above 10% have fertility rates that are double those in richer countries (Leete and Schoch, 2003).

Lowering fertility rates through access to family planning information and contraceptive supplies has both macro- and microeconomic benefits. Just as national governments are able to spend more per capita on social welfare when population growth slows, parents can invest more in health and education costs for each child if they have fewer children. Couples who have smaller families are often able to have women work outside the home and to save more of their income, thus increasing the national labor supply, investment, and growth.

Economists credit declining fertility over the past four decades as one contributor to sustained economic growth among the East Asian Tiger countries, which include South Korea, Taiwan, Thailand, Singapore, Indonesia, Malaysia, and Hong Kong. Research indicates that shifts to smaller family size and slower rates of population growth in the region helped create an educated workforce, expand the pool of household and government savings, raise wages, and dramatically increase investments in manufacturing technology (Williamson and Higgins, 1997; Williamson, 2001). A shift to smaller families produced three important demographic changes: Slower growth in the number of school-age children, a lower ratio of dependants to working-age adults (dependency ratio), and a reduced rate of labor force growth. As fertility declined, household and government investments per child increased sharply, as did household savings, while governments reduced public expenditures. Domestic savings rose to displace foreign funds as the leading source of private sector investment and countries that had been major foreign aid recipients emerged to become a significant part of industrialized countries' export market. Although this model of fertility decline and economic growth has not yet been replicated in other regions of the world, the East Asian experience demonstrates how important a smaller average family size can be to economic development.

Natural Resources

Greater numbers of people and higher per capita resource consumption worldwide are multiplying humanity's impacts on the environment and natural resources. While population is hardly the only force applying pressure to the environment, the challenges of sustainability and conservation become harder to address as the number of people continues to increase. Currently, more than 1.3 billion people live in areas that conservationists consider the richest in nonhuman species and the most threatened by human activities. While these areas comprise about 12% of the planet's land surface, they hold nearly 20% of its human population. The population in these biodiversity hotspots is growing at a collective rate of 1.8% annually, compared to the world population's annual growth rate of 1.3% (Engelman, 2004). Human pressure on renewable freshwater, cropland, forests, fisheries, and the atmosphere is unprecedented and continually increasing.

Changes in ecosystems (natural systems of living things interacting with the nonliving environment) can greatly impact human prosperity, security, and public health. Food supply, fresh water, clean air, and a stable climate all depend upon healthy ecosystems. Human-induced or human-accelerated disturbances to ecosystems can increase the risk of acquiring infectious diseases, either directly or through the impact made to biodiversity. For example, air and water pollution are linked to increases in asthma and other respiratory problems, as well as cancer and heart disease. The emergence of Lyme disease in the United States is thought to be associated with habitat loss and diminished biodiversity. Many experts believe that climate change is also contributing to a rise of infectious disease worldwide, as warmer temperatures bring pathogens to new geographical areas (Epstein *et al.*, 2003).

Currently, more than 745 million people face water stress or scarcity (Engelman, 2004). Much of the fresh water now used in water-scarce regions originates from aquifers that are not being refreshed by the natural water cycle. In most of the countries where water shortage is severe and worsening, high rates of population growth exacerbate the declining availability of renewable fresh water. Cultivated land is at similar levels of critical scarcity in many countries. Despite the Green Revolution and other technological advances, agriculture experts continue to debate how long crop yields will keep up with population growth. The food that feeds future generations will be grown mostly on today's cropland, which must remain fertile to keep food production secure. Easing world hunger would be even more difficult if population growth resembles demographers' higher projections.

Likewise, future declines in the per capita availability of forests, especially in developing countries, are likely to pose major challenges for both conservation and for the provision of food, fuel, and shelter necessary for human well-being. Based on the United Nations' medium population projection and current deforestation trends, by 2025 the number of people living in forest-scarce countries could swell to 3.2 billion in 54 countries (Engelman, 2004). Most of the world's original forests have been lost to the expansion of human activities.

In 2002, per capita emissions of CO_2 continued a 3-year upward climb that appeared unaffected by the global economic jolt from the September 2001 terrorist attacks on the United States. In the 1999–2002 period, population growth and increasing per capita fossil-fuel consumption joined forces on an equal basis to boost global CO_2 emissions rapidly. With about 4.7% of the world's population, the United States accounted for almost 23% of all emissions from fossil-fuel combustion and cement manufacture, by far the largest CO_2 contributor among nations. (China, with more than four times the U.S. population, is number two.) Emissions remained grossly inequitable, with one-fifth of the world's population accounting for 62% of all emissions in 2002, while another — and much poorer — fifth accounted for 2%. Population growth and the corresponding increase in global CO_2 emissions are of crucial importance to scientists and policy makers concerned with the potential effects of global climate change.

Conflict and Security

During the last three decades of the twentieth century, the demographic transition was correlated with continuous declines in the vulnerability of countries to civil conflicts, including ethnic wars, insurgencies, and terrorism. Progress through the demographic transition gives countries a more mature and less volatile age structure, slower workforce growth, and a more slowly growing school-age population. It reduces urban growth and gives countries additional time to expand infrastructure, meet the demand for public services, and conserve dwindling natural resources. Countries in the early and middle phases of demographic transition have cumulatively been significantly more vulnerable to civil conflict than late-transition countries.

The likelihood of civil conflict steadily decreased for high-risk countries as they experienced overall declines in birth and death rates. From the 1970s to the 1990s, a decline in a country's annual birth rate of ten births per 1000 people corresponded to a decrease of approximately 10% in the likelihood of an outbreak of civil conflict.

The demographic factors most closely associated with the likelihood of a new outbreak of civil conflict during the 1990s were a high proportion of young adults ages 15–29 years and a rapid rate of urban population growth. When coupled with a large youth bulge, countries with a very low availability of cropland and/or renewable fresh water, plus a rapid rate of urban population growth, had a roughly 40% probability of experiencing an outbreak of civil conflict in the 1990s. Migration and differential rates of population growth among ethnic groups competing for political and economic power have also played important roles in political destabilization.

Over the past 40 years, the demographic transition has been progressing impressively in nearly all of the world's regions. Most of the world's countries are moving toward a distinctive range of population structures that make civil conflict less likely. Yet this progress through the demographic transition is not universal. Continuing declines in birth rates and increases in life expectancy in the poorest and worst-governed countries will be required.

Conclusion

In summary, the world is demographically complex, diverse, and divided in ways that have no precedent in human history. No one can say with confidence how problematic this diversity is or how humanity's demographic future will unfold. However, it is certain that given the linkages between world population growth and disease and mortality, environmental resources, gender equity, and civil conflict, demography will play a major role in the future of global public health.

See also: Abortion; AIDS: Epidemiology and Surveillance; Family Planning/Contraception; Gender Aspects of Sexual and Reproductive Health; Infant Mortality/Neonatal Disease; Maternal Mortality and Morbidity; Mother-to-Child Transmission of HIV; Reproductive Ethics: Perspectives on Contraception and Abortion; Reproductive Rights; Sexual and Reproductive Health: Overview; Trends in Human Fertility.

Citations

The Alan Guttmacher Institute (1999) *Induced Abortion Worldwide.* New York: The Alan Guttmacher Institute.

The Alan Guttmacher Institute and UNFPA (2003) *Adding it up: The Benefits of Investing in Sexual and Reproductive Health Care.* New York: AGI and UNFPA.

Ashford L (2002) *Hidden Suffering: Disabilities from Pregnancy and Childbirth in Less Developed Countries.* Washington, DC: Population Reference Bureau.

Conly S (1998) *Educating Girls: Gender Gaps and Gains.* Washington, DC: Population Action International.

El-Zanaty F and Way A (2001) *Egypt Demographic and Health Survey 2000.* Calverton, MD: Ministry of Health and Population (Egypt), National Population Council and ORC Macro.

Engelman R (with Anastasion D) (2004) *Methodology. People in the Balance: Update 2004.* http://www.populationaction.org/resources/publications/peopleinthebalance/downloads/AcknowlAndMethodo.pdf.

Engelman R and Leahy E (2006) *How Many Children Does It Take to Replace Their Parents? Variation in Replacement Fertility as an Indicator of Child Survival and Gender Status.* Paper presented at the Population Association of America annual conference Los Angeles, 30 March–1 April.

Epstein PR, Chivian E, and Frith K (2003) Emerging diseases threaten conservation. *Environmental Health Perspectives.* 111(10): A506–A507.

Food Agriculture Organization (2004) *The State of Food Insecurity in the World 2004.* Rome: FAO.

Global Commission on International Migration (2005) *Migration in an Interconnected World: New Directions for Action.* Geneva, Switzerland: GCIM.

Joint United Nations Programme on HIV/AIDS (2005) *AIDS Epidemic Update: December 2005.* Geneva, Switzerland: UNAIDS.

Leete R and Schoch M (2003) Population and poverty: Satisfying unmet need as the route to sustainable development. *Population and Poverty: Achieving Equity, Equality and Sustainability* 8: 9–38.

Rutstein S (2005) Effects of preceding birth intervals on neonatal, infant and under-five years mortality and nutritional status in developing countries: Evidence from the Demographic and Health Surveys. *International Journal of Gynecology and Obstetrics* 89 (supplement 1): s7–s24.

Setty-Venugopal V and Upadhyay UD (2002) *Three to Five Saves Lives. Population Reports. L13: 2.* Baltimore, MD: Johns Hopkins University.

UNFPA. State of World Population (2004) *Reproductive Health and Family Planning.* http://www.infpa.org/swp/2004/english/ch6/page3.html (accessed February 2006).

United Nations Children's Fund (2005) *State of the World's Children 2005.* New York: UNICEF.

United Nations Population Division (2004) *World Population to 2300.* New York, United Nations Population Division.

United Nations Population Division, Department of Economic and Social Affairs (2004) *World Contraceptive Use 2003*. New York: United Nations.
United Nations, Department of Economic and Social Affairs, Statistics Division (2005) *Word Population Prospects: The 2004 Revision*. New York: United Nations.
United Nations Population Division (2005) *World Population Prospects: The 2004 Revision.* New York: United Nations Population Division.
United Nations Population Division (2007) *World Population Prospects: The 2006 Revision*. New York: United Nations Population Division.
Williamson JG and Higgins M (1997) The accumulation and demography connection in East Asia. *Proceedings of the Conference on Population and the East Asian Miracle*. Honolulu, Hawaii: East-West Center.
Williamson JG (2001) Demographic change, growth, and inequality. In: Birdsall N, Kelley AC, and Sinding SW (eds.) *Population Matters: Demographic Change, Economic Growth, and Poverty in the Developing World*, pp. 107–136. Oxford, UK: Oxford University Press
World Health Organization (2004) *Maternal Mortality in 2000: Estimates Developed by WHO, UNICEF and UNFPA.* Geneva, Switzerland: WHO.
World Health Organization (2005) *The World Health Report 2005: Make Every Mother and Child Count.* Geneva, Switzerland: WHO.

Further Reading

Ashford L (1995) *New Perspectives on Population: Lessons from Cairo.* Washington, DC: Population Reference Bureau.
Birdsall N, Kelley A, and Sinding S (eds.) (2001) *Population Matters: Demographic Change, Economic Growth, and Poverty in the Developing World.* New York: Oxford University Press.
Bongaarts J (1994) Population policy options in the developing world. *Science* 263: 771–776.
Chivian E (ed.) (2002) *Biodiversity: Its Importance to Human Health: Interim Executive Summary.* Cambridge, MA: Center for Health and the Global Environment Harvard Medical School.
Cincotta R, Wisnewski J, and Engelman R (2000) Human population in the biodiversity hotspots. *Nature* 404: 990–992.
Coale A and Hoover E (1958) *Population Growth and Economic Development in Low-Income Countries.* Princeton, NJ: Princeton University Press.
Cohen JE (1995) *How Many People Can the Earth Support?* New York: Norton.
Engelman R and LeRoy P (1993) *Sustaining Water: Population and the Future of Renewable Water Supplies.* Washington, DC: Population Action International.
Food and Agriculture Organization (1996) *Food Requirements and Population Growth.* World Summit Background Document No. 4. Rome: FAO.
Livi-Bacci M (1992) *Concise History of World Population.* Cambridge, MA: Blackwell Publishers.
Lutz W, O'Neill B, and Scherbov S (2003) Europe's population at a turning point. *Science* 299: 1991–1992.
Mason A (ed.) (2002) *Population Change and Economic Development in East Asia: Challenges Met, Opportunities Seized.* Stanford, CA: Stanford University Press.
National Research Council (1986) *Population Growth and Economic Development: Policy Questions.* Washington, DC: National Academy of Science Press.
Robey B, Rutstein S, and Morris L (1993) The fertility decline in developing countries. *Scientific American* 269: 60–67.
Smil V (1991) Population growth and nitrogen: An exploration of a critical existential link. *Population and Development Review* 17: 569–601.
United Nations (1994) *Programme of Action Adopted at the International Conference on Population and Development.* New York: United Nations.

Relevant Websites

http://www.measuredhs.com – Demographic and Health Surveys.
http://www.popact.org – Population Action International.
http://www.un.org/esa/population/unpop.htm – United Nations Population Division.
http://www.who.int – World Health Organization.

Trends in Human Fertility

J G Cleland, Centre for Population Studies, London School of Hygiene and Tropical Medicine, London, UK

Introduction

The number of children born per woman and the timing of births are directly relevant to public health and health services in many ways. The level of childbearing, for instance, determines the demand for obstetric and child health services and has a direct effect on the maternal mortality rate. The age pattern of childbearing influences the incidence of obstetric complications, because pregnancies in the early teenage years and at ages over 35 years pose an increased risk to the mother. Fecundity also declines after age 35, and thus postponement of births will increase the need for assisted reproduction. The spacing of births has important health implications; conceptions occurring within 24 months of a previous live birth are at elevated risk of fetal death, prematurity, low birth weight for gestational age, and infant mortality. Unintended pregnancies may result in induced abortion, which in many countries is restricted and unsafe.

Fertility also affects public health indirectly through socioeconomic pathways. Birth rates are the crucial determinant of population growth (or decline) and the age structure of populations, and both factors have profound socioeconomic implications. Countries growing at 2% per year or more (implying a doubling in population size every 36 years or less), because mortality has declined but fertility remains high, face greater difficulties in escaping from poverty and illiteracy than other countries, mainly because nearly half the population is aged under 15 years, thus placing a heavy dependency burden on the adult population. When fertility falls, an era of several decades follows when the labor force is proportionately large, the dependency burden is atypically low, and prospects for making rapid socioeconomic progress are especially bright. This era is inevitably followed by a return to a high-dependency burden because of an increase in the elderly population, which poses a strain on governments' health and welfare budgets.

The sequence is well illustrated by the case of the Republic of Korea (**Figure 1**). By 1960, mortality had already declined but fertility remained high at about six births per woman, and the population was growing at 2.8% per year. At that time, 42% of the total population was aged under 15 years but only 3% were aged 65 or more. In the next 40 years, fertility fell sharply and by 2000 the number of births per woman was about 1.4 and the growth rate had abated to 0.6% per year. Between 1960 and 2000 the number of working-age adults (15–64 years) per 100 dependants rose from 120 to 250. Between 2000 and 2040 it is projected that the proportion of Korea's population aged 65 or more will rise from 7.4% to 30.5% and the ratio of workers to dependants will have fallen from 250 to 137 per 100 dependants. All industrialized countries now face similar problems of population aging.

In the absence of any constraints, it is estimated that the average number of births per woman would be about 15. In all societies, fertility is held well below this biological maximum by a blend of four main factors: restrictions on sexual intercourse, typically operating through marriage systems; lactation that inhibits ovulation; contraception; and induced abortion. The highest fertility recorded for a national population was 8.7 births per woman in Yemen between 1970 and 1985. In premodern societies, fertility was typically in the range of four to six births.

Between 1950 and 2005, the global fertility rate halved from about five to 2.5 births. Under conditions of moderate to low mortality, a little over two births per woman is required to bring about long-term stabilization of population size. Thus, the world as a whole may be approaching the end of an era of sustained growth, from 1 billion in 1830 to 6.5 billion in 2005 and a projected total of 9.2 billion in 2050 (United Nations, 2006). However, these global figures mask huge differences between regions

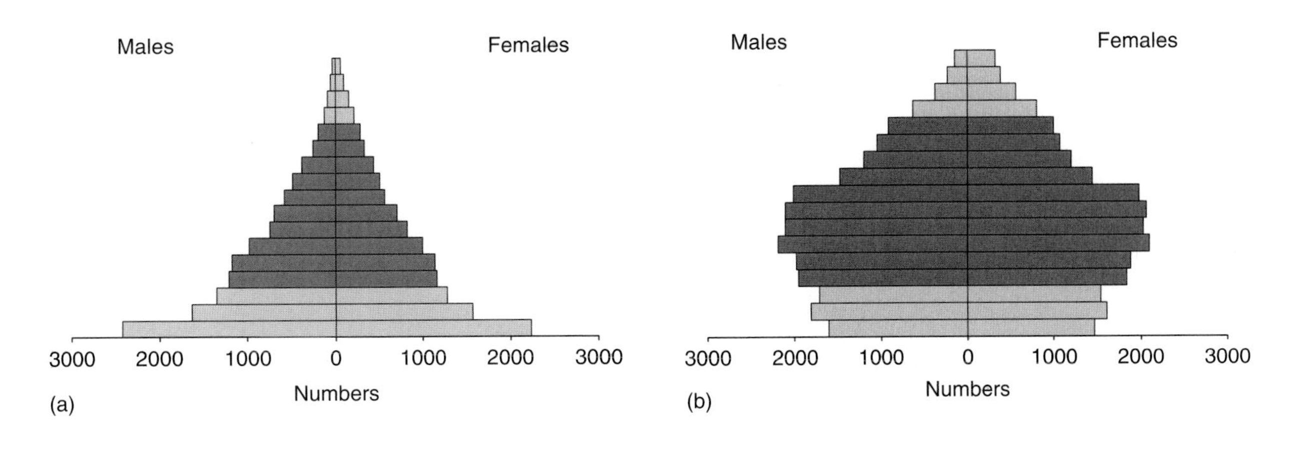

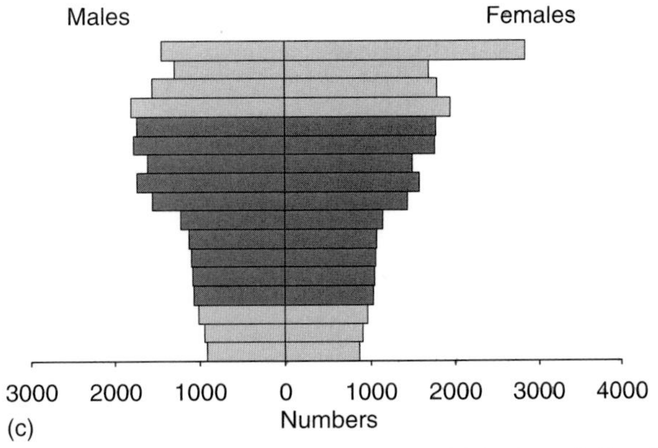

Figure 1 Age structure in Republic of Korea: 1960 (a), 2000 (b), and 2040 (c) (projected). Population aged 15–64 years is darkly shaded. Source: UN Department of Economic and Social Affairs, Population Division (2007) *World Population Prospects: The 2006 Revision*. New York: United Nations.

and countries. The level of childbearing in most industrialized countries has fallen well below the two-child mark and therefore these countries face the possibility of population decline, combined with population aging. Conversely, many of the poorest countries in the world still retain buoyant fertility levels and can expect substantial increases in population size. In this article, fertility trends since 1950 will be described and the underlying causes and implications discussed.

Trends in Industrialized Countries

In most of Europe and in North America, fertility decline started in the late 19th century (well before the development of modern contraceptives) and birth rates fell to low levels in the Great Depression of the 1930s, giving rise to concerns about population decline. These concerns were short-lived because, following the end of World War II, fertility rose in most industrialized countries and in some it continued to rise throughout the 1950s (**Figure 2**). This postwar baby boom was most pronounced in the United States, where fertility climbed from 2.9 births in 1946 to peak at 3.7 births in 1957. Japan is the clearest exception: This country experienced a dramatic decline from 4.5 births in 1947 to 2.0 births a decade later, partly in response to a shift from pro- to antinatalist policies and liberalization of abortion laws.

The mid-1960s marked the start of a second and unforeseen phase of fertility decline. By 1980, fertility in most (developed) countries had fallen below the replacement level of two births. In 2005, childbearing was below 1.5 births in Italy, Spain, Germany, Austria, the Russian Federation, and much of Eastern Europe, and also in the economically advanced East Asian states and territories (Japan, Hong Kong, Singapore, Taiwan, Republic of Korea). Some of this decline can be attributed to increased efficiency in the prevention of unintended births. The advent of oral contraception in the 1960s represented a decisive break of the sex–reproduction nexus. Access to legal abortion was also made easier in many countries. In 2000, about 20% of known pregnancies were legally terminated in France, Norway, Denmark, Italy, the United Kingdom, and Sweden. This percentage exceeded 40% in many countries of the former Soviet bloc and the Russian Federation itself (United Nations, 2005). In the United States, the fraction of births reported by women as unwanted fell from 20% in the early 1960s to 7% by the late 1970s, and the same trend no doubt occurred in many industrialized countries, though is less well documented.

However, most commentators have sought explanations in more fundamental changes than improved birth

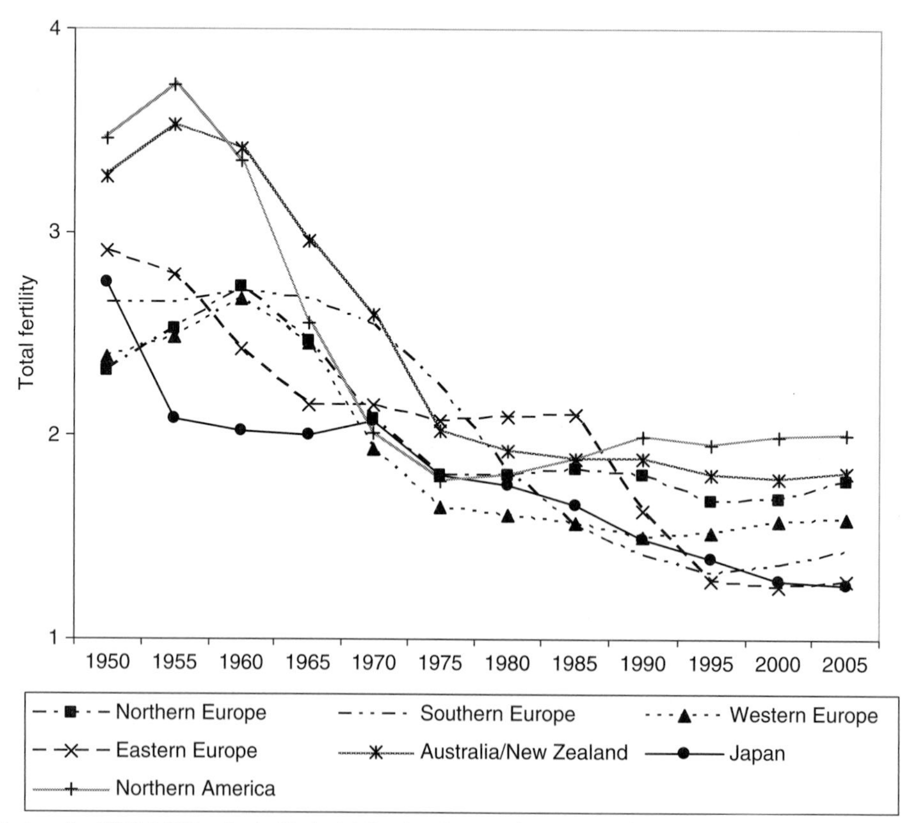

Figure 2 Fertility trends, 1950–2005: selected industrialized regions and countries. Source: UN Department of Economic and Social Affairs, Population Division (2007) *World Population Prospects: The 2006 Revision*. New York: United Nations.

control. The fertility decline in many industrialized countries has been accompanied and partly propelled by postponement of marriage and parenthood, rises in cohabitation, nonmarital births, and divorce, increased acceptance of diverse sexual lifestyles, and a growing independence of women. These interwoven features, dubbed the 'second demographic transition,' represent an appreciable departure from marriage and parenthood as the central pillars of adult life. The key underlying cause is identified by some as the changing roles of women in society, together with the sluggish adaptation of men to this emancipation, for instance, by reluctance to shoulder a more equal share of the burden of housekeeping and child-rearing. The shift away from marriage and motherhood has been called 'the revenge of women on men.' Paradoxically, however, the level of childbearing is lowest in countries where women's labor force participation is also very low: Japan, Greece, Italy, and Spain. It is also of note that these same countries have low proportions of nonmarital births. Other experts, such as Ronald Lesthaeghe and Dirk van de Kaa, have sought an explanation in the development of broader 'postmodern' values of individualism, secularism, and the desire for self-fulfillment. Compelling region-specific explanations abound. For instance, the turbulence and insecurity caused by the break-up of the Soviet bloc coincides with the period of sharpest decline in these countries. Given the economic, social, and cultural diversity of the very-low-fertility countries, it seems unlikely that there is a single underlying cause.

A sustained fertility rate of 1.5 births implies a halving of population size approximately every 65 years. While this prospect of population shrinkage is welcomed by some environmental groups, it is regarded with alarm by many European and Asian governments. International migration on a sufficient scale to offset low birth rates and prevent population decline does not appear to be politically feasible. Hence, the main policy responses have been aimed at stimulating reproduction and have included generous maternity/paternity allowances (Sweden), child allowances that increase with parity (France), and cash payments at birth (Australia, Italy). Many other countries have shunned explicit pronatalist policies but have attempted to make family building and work more compatible, for instance, by better provision of infant care centers. None of these policies can claim long-term effectiveness at raising birth rates, and the demographic futures of industrialized countries is uncertain (Gauthier, 1996). While most experts foresee the continuation of very low fertility, the United Nations envisages a slight but steady increase over the next 40 years. One factor favoring an increase concerns postponement of births, which depresses period rates but may not affect the number of children that women have over their life course. Sooner or later, the trend toward delayed childbearing must end and, when this happens, period fertility rates will increase, typically by an average of 0.2–0.4 births per woman (Lutz *et al.*, 2003). It is also true that a two-child family remains a widespread aspiration despite downward shifts in some countries (Goldstein *et al.*, 2003).

Trends in Developing Regions

In Asia, Latin America, and to a lesser extent Africa, life expectancy increased sharply while birth rates remained high in the two decades following World War II. The resulting acceleration of rates of population growth rang alarm bells because of evidence that socioeconomic progress would be jeopardized. Key U.S. leaders, such as President Lyndon Johnson and Robert MacNamara, president of the World Bank, became convinced of the need to reduce birth rates through the promotion of family planning. In 1969, the United Nations Fund for Population Activities (now the United Nations Population Fund – UNFPA) was created, with the shrewd choice of a Filipino Roman Catholic as its first executive director. Knowledge, attitude, and practice (KAP) surveys indicated the existence of favorable attitudes toward smaller family sizes and contraception in many poor countries. Pilot programs in Taiwan and the Republic of Korea showed that a ready demand existed for modern contraception, specifically the intrauterine device. Thus, the stage was set for a novel form of social engineering, the reduction of fertility through government-sponsored family planning programs. The number of developing countries with official policies to support family planning rose from two in 1960 to 115 by 1996.

Between 1950 and 2005, fertility in both Asia and Latin America fell from a little under six births per woman to about 2.5 births (**Figure 3**). The main exceptions are Afghanistan, the Lao People's Democratic Republic and Pakistan in Asia, and Guatemala, Bolivia, and Paraguay in Latin America. In the Arab states of North Africa, the corresponding decline was from 6.8 births to 3.2 births. Only in sub-Saharan Africa does fertility remain high at 5.5 births.

The relative influence of family planning promotion and socioeconomic development (in particular, increased life expectancy and education, which raise the number of surviving children and the costs of rearing them) on fertility trends is hotly contested. In most Latin American countries, the role of state intervention has been minor. Governments in this region were hesitant to promote birth control partly because of the influence of the Roman Catholic church, and early efforts to popularize contraception were spearheaded by nongovernmental organizations with the prime objective of reducing illegal and unsafe abortions. The imprint of government actions can be more clearly discerned in Asia, notably in China,

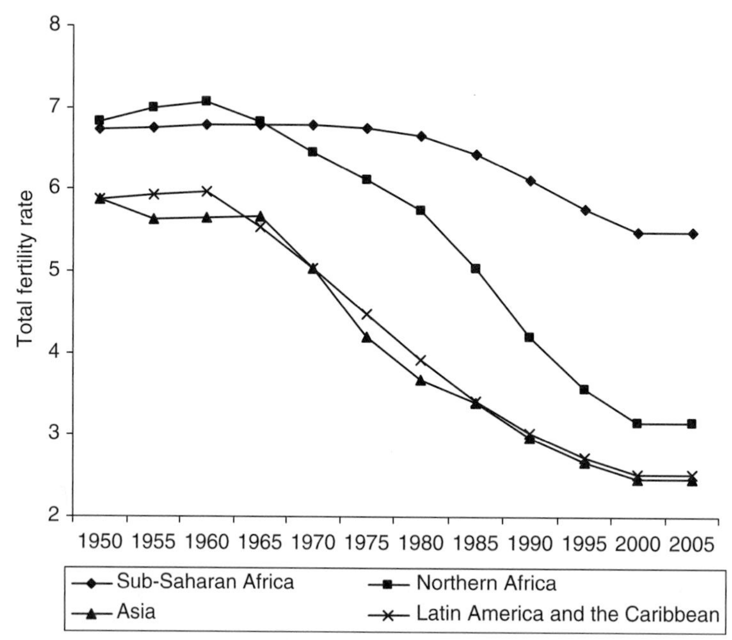

Figure 3 Fertility trends, 1950–2005: Selected developing regions. Source: UN Department of Economic and Social Affairs, Population Division (2007) *World Population Prospects: The 2006 Revision*. New York: United Nations.

Bangladesh, and Indonesia. In China, programs to reduce population growth started in 1972 and, in the next seven years, fertility fell sharply – but not sufficiently, in the opinion of government planners. In 1979, the one-child policy was enacted and enforced rigorously in urban areas, where fertility fell quickly to one child. In rural areas, however, there was entrenched resistance and in the 1980s policy implementation in many provinces was relaxed to permit two children, particularly if the first born was a daughter (Gu *et al.*, 2007). China's fertility is currently estimated to be 1.7 births per woman. Because of the vast size of China's population, government policies in this country have made a major contribution to global stabilization but the price has also been high: denial of reproductive freedom, sex-selective abortion, abandonment of daughters, and instances of female infanticide.

In Bangladesh, one of the poorest and least literate countries in Asia, governments, faced with a highly visible population problem, had little choice but to address it as a top priority. Starting in 1975, a comprehensive family planning service was created, accompanied by incessant publicity through the mass media and other channels. Between 1975 and the 2000, fertility halved from six to three births per woman – a vivid demonstration that poverty and illiteracy are not incompatible with small families.

The demographic histories of the Philippines and Indonesia also demonstrate that high educational levels and reasonably high status of women, factors thought to be particularly conducive to low fertility, do not always have this effect. In 1960, the Philippines was one of the wealthiest countries of Southeast Asia. The level of adult literacy was 72%, compared with only 39% in Indonesia. Income per head was almost double the Indonesian level at that time, life expectancy was 10 years longer, and infant mortality correspondingly lower. It came as no surprise, therefore, that fertility decline started earlier in the Philippines than in Indonesia. However, since the mid-1970s, the pace of decline has been much greater in Indonesia; the level of fertility in 2000–05 was estimated to be 2.5 births, one birth lower than in the Philippines.

The explanation for this unexpected outcome lies well beyond the realm of statistical evidence, but almost certainly involves the intertwined factors of religion and government policy. The Indonesian government skillfully evaded the potential danger of opposition from Islamic leaders by eschewing abortion and contraceptive sterilization. It mounted a very forceful family planning program, with considerable community pressure on couples to adopt birth control (Warwick, 1986). In the Philippines, no compact between church and state was reached, and Roman Catholic leaders remain strongly and openly opposed to modern birth control. This opposition has both inhibited the evolution of comprehensive family planning services and no doubt influenced the climate in which reproductive decisions are taken.

The persistence of rather high fertility in the Philippines is made even more surprising in view of the relatively high status of women in that country. Educational levels for women are exceptionally high, as is labor force participation. However, the example of the Philippines suggests that the status of women – if defined in terms of participation in public life, including paid employment – is not such an important precondition for sustained fertility decline as so often claimed. Indeed, it is probably more a consequence of decline than a cause.

Though fertility in sub-Saharan Africa remains high on average, the subregion is demographically diverse. In the Republic of South Africa and Zimbabwe (then Rhodesia), vigorous family planning promotion started before the advent of majority rule and fertility in both countries, and in Botswana is now at a relatively low level. HIV epidemics in southern Africa are especially severe and thus falling fertility has been accompanied by rapidly rising mortality. These countries face an exceptionally abrupt end to an era of population growth.

In eastern Africa, fertility decline has started in most countries, but most markedly in Kenya. In that country, a vigorous family planning effort was initiated in the early 1980s and, in the next 15 years, fertility fell from nearly eight to 4.8 births. Unexpectedly, the rate then plateaued, one likely reason being that funds and energy were diverted from family planning to HIV/AIDS. Between 1988 and 2003, the proportion of contraceptive users relying on government services dropped from 68% to 53% and the percentage of births reported by mothers as unwanted rose from 11% to 21% (Westoff and Cross, 2006). Both trends imply a deterioration of government services. In 2002, the United Nations projected Kenya's population in 2050 at 44 million. In 2004, this projection was raised to 83 million, mainly in response to the fertility stall but also to a reduction in expected AIDS-related mortality.

This sequence of events in Kenya may be an extreme example of a more pervasive trend throughout much of the region. Rather than gathering pace in the past 15 years, the trend toward lower fertility has faltered. In west and central Africa, fertility typically remains close to six births per woman, contraceptive use prevalence among married women remains below 10%, and desired family sizes are still high. The United Nations projects that fertility in sub-Saharan Africa will decline steadily to reach 2.5 births by midcentury. Even if this projection is accurate, Africa's population is set to rise from 0.77 billion in 2005 to 1.76 billion in 2050. Most countries will double in population size in the next 40 or so years, and some will triple in size.

Is the persistence of high fertility throughout much of sub-Saharan Africa simply a reflection of low socioeconomic development or of distinctive features of culture and social organization that set the region apart from Asia and Latin America? Certainly, standards of living for many Africans worsened in the 1990s, but equally high levels of poverty and illiteracy did not stifle fertility transition in Asia, as trends in Bangladesh and Nepal since the 1980s show. It is also true that countless surveys reveal that Africans attach a higher value to large families than citizens elsewhere. According to Caldwell, the explanation for this pronatalism lies in the subordination of the nuclear family to the lineage. For lineage leaders, the patriarchs, high fertility is advantageous because it enhances their prestige, power, and patronage. Thus, they see children as sources of wealth rather than as drains on emotional and financial resources. A related explanation stems from the multiplicity of ethnic and linguistic groups in Africa, with the inevitable tension and conflict over resources that this diversity implies. A buoyant birth rate and numerical strength may well have conferred advantage in these circumstances, thus engendering strongly pronatalist values.

Since the most recent International Conference on Population and Development in Cairo, in 1994, policies to reduce population growth through family planning promotion have fallen from fashion and been replaced by a broader agenda of women's reproductive health and rights. The topic of population has been marginalized in key reports on development and was omitted from the Millennium Development Goals. International funding for family planning has fallen. This neglect may prove to be mistaken. **Figure 4** shows a display of the 76 poorest countries with a current population of 5 million or more, by fertility rate (2000–05) and by the level of unmet need for family planning – defined as the percent of married women who want no child for at least 2 years but are using no method of contraception. Over one-third of these countries, mostly in Africa, still have high fertility of five births or more per woman, and the vast majority of these countries also have a high unmet need for family planning. In a further 25 countries, fertility is in the range of three to five births, well above replacement level.

Conclusions

For millennia, the human population grew at a miniscule rate because moderate fertility was matched by high, albeit fluctuating, mortality. The scientific and technological revolution of the past 200 years broke this demographic balance and gave rise to an unprecedented surge in human numbers. The past 50 years has seen a necessary and welcome return toward balance; fertility has fallen in most countries, and world population may stabilize in the latter half of this century. Thus, the prospect of the Malthusian nightmare of famine and warfare, so prominently proclaimed in the 1960s by Paul Ehrlich and others, has receded.

No consensus on the ideal level of fertility exists, but a range of 1.7 to 2.3 births per woman has much to recommend it, as it implies modest growth or decline. As shown in this article, the world is still far away from such a benign outcome. Fertility rates in many industrialized countries have plunged well below 1.7 while many poor countries have rates well above 2.3. Indeed, the

Total fertility (births per woman) 2000–2005	Low (<10%)	Medium (10%–19%)	High (20%+)
High (5.0 or more)		Mozambique [5.52] Nigeria [5.85] Niger [7.45] Nicaragua [8.00]	Senegal [5.22] Togo [5.37] Zambia [5.65] Ethiopia [5.78] Rwanda [6.01] Yemen [6.02] Malawi [6.03] Burkina Faso [6.36] Sierra Leone [6.50] Mali [6.70] D R Congo [6.70] Uganda [6.75] Kenya [5.00] Cote d'Ivoire [5.06] Madagascar [5.28] Tanzania [5.66] Guinea [5.84] Benin [5.87] Chad [6.54] Somalia [6.43] Angola [6.75] Burundi [6.80] Afghanistan [7.50]
Medium (3.0–4.99)	Egypt [3.17]	Sri Lanka [3.02] India [3.11] Bangladesh [3.22] Paraguay [3.48] Jordan [3.53] Philippines [3.54] Zimbabwe [3.56] Hounduras [3.72]	Syria [3.48] Laos [3.59] Cambodia [3.64] Nepal [3.68] Tajikistan [3.81] Bolivia [3.96] Pakistan [3.99] Haiti [4.00] Papua New Guinea [4.32] Ghana [4.59] Guatemala [4.60] Sudan [4.82] Iraq [4.86] Cameroon [4.92]
Low (1.0–2.9)	Romania [1.29] Cuba [1.63] China [1.70] Thailand [1.83] Korea, Dem Rep [1.92] Kazakhstan [2.01] Vietnam [2.32] Indonesia [2.38] Brazil [2.35] Colombia [2.47]	Ukraine [1.15] Belarus [1.24] Azerbaijan [1.67] Tunisia [2.04] Kyrgyz Republic [2.50] Morocco [2.52] Algeria [2.53] Peru [2.70] Uzbekistan [2.74] Ecuador [2.82] El Salvador [2.88] Dominican Republic [2.95]	Bulgaria [1.26] Serbia [1.75] Myanmar [2.25]

Unmet need for family planning

Figure 4 Classification of low-income countries by fertility rate (2000–05) and unmet need for family planning. Figures in parentheses show the fertility rate.

fertility of nations has rarely been so diverse. Our demographic future is still uncertain. Will birth rates in Africa fall as fast and pervasively as in Asia and Latin America, and will fertility edge steadily up in countries such as Japan and Italy? What happens in Africa is partly a matter of political priorities because a large body of successful experience at reducing fertility has accumulated. Policies to raise fertility do not have a successful track record, and so trends in low-fertility countries are particularly difficult to predict.

See also: Abortion; Family Planning/Contraception; Infertility; Reproductive Rights.

Citations

Gauthier AN (1996) *The State and the Family: A Comparative Analysis of Family Policies in Industrialised Countries*. Oxford, UK: Clarendon Press.
Goldstein J, Lutz W, and Testa MR (2003) The emergence of sub-replacement family size ideals in Europe. *Population Research and Policy Review* 22(5–6): 479–496.
Gu B, Wang F, Guo Z, and Zhang E (2007) China's local and national fertility policies at the end of the twentieth century. *Population and Development Review* 33: 129–147.
Lutz W, O'Neill BC, and Scherbov S (2003) Europe's population at a turning point. *Science* 299: 1991–1992.
United Nations (2005) *The New Demographic Regime: Population Challenges and Policy Responses*. Geneva, Switzerland: Economic Commission for Europe and United Nations Population Fund.
United Nations (2006) *World Population Prospects: The 2006 Revision*. New York: Department of Social and Economic Affairs, Population Division.
Warwick DP (1986) The Indonesian family planning program: Government influence and client choice. *Population and Development Review* 12: 453–490.
Westoff C and Cross A (2006) *The Stall in the Fertility Transition in Kenya*. Demographic and Health Surveys Analytical Study No. 9. Calverton, MD: ORC Macro.

Further Reading

Caldwell JC and Caldwell P (1987) The cultural context of high fertility in sub-Saharan Africa. *Population and Development Review* 13: 409–437.
Caldwell JC and Schidlmayr T (2003) Explanation of the fertility crisis in modern societies: A search for commonalities. *Population Studies* 57: 241–264.
Cleland J, Bernstein S, Ezeh A, Faundes A, and Innis J (2006) Family planning: The unfinished agenda. *Lancet* 368: 1810–1827.
Davis K, Bernstam MS, and Ricardo-Campbell R (eds.) (1986) Below-replacement fertility in industrial societies: Causes, consequences, policies. *Population and Development Review*, 12 (supplement).
Guzmán JM, Singh S, Rodríguez G, and Pantelides EA (eds.) (1996) *The Fertility Transition in Latin America.* Oxford, UK: Clarendon Press.
Kirk D (2000) The demographic transition. *Population Studies* 50: 361–388.
Leete R and Alam I (eds.) (1993) *The Revolution in Asian Fertility: Dimensions, Causes and Implications.* Oxford, UK: Clarendon Press.
Lesthaeghe R and Meekers D (1986) Value changes and the dimension of familism in the European community. *European Journal of Population* 2: 225–268.
van de Kaa DJ (2001) Post modern fertility preferences: From changing value orientation to new behavior. In: Bulatao RA and Casterline JB (eds.) Global Fertility Transition. *Population and Development Review,* pp. 290–331.

Relevant Websites

http://www.populationaction.org – Population Action International.
http://www.prb.org – Population Reference Bureau.
http://www.un.org/esa/population/unpop.htm – United Nations Department of Economic and Social Affairs, Population Division.
http://unstats.un.org/unsd/demographic/default.htm – United Nations Statistics Division.

Perinatal Epidemiology

L S Bakketeig and P Bergsjø, Norwegian Institute of Public Health, Oslo, Norway

Introduction

It is well established that events during pregnancy and childbirth influence the health and development of the newborn baby. In recent years, we have come to realize that events occurring in pregnancy and early childhood may also affect health into adult life. Even events occurring before pregnancy and intergenerationally can affect reproduction. Population-based studies of these relationships are called perinatal epidemiology, which has evolved into a major subspecialty of epidemiology. Events or exposures may affect both mother and child; those affecting the mother are labeled maternal, while the term perinatal refers to the child.

Perinatal surveillance has been established through a system of medical registration of births in a number of countries. In Norway, medical registration of births was established in 1967. The main reason was the thalidomide catastrophe in Europe some years earlier. The other Nordic countries followed suit: Denmark in 1968, Iceland in 1972, and Sweden in 1973. Finland established its registry somewhat later, in 1987. The registries have served several other purposes than monitoring the frequency and type of congenital malformations. The data, which contain civic and medical information on the child, parents, and family, have been extensively used for perinatal research, of which several examples will be given. In Norway, mother, father, and child are identified by unique individual registration numbers. This facilitates linkage of successive births to the same mother with the same father or different fathers and linkage across generations. The medical birth registries can also be linked to other health-related registries, for example the cancer registries.

Traditionally, perinatal epidemiologists have focused on exposures and outcomes that occur in the perinatal period, which for research purposes covers the period from conception through birth and the first part of life. For perinatal surveillance and statistical comparisons of mortality and morbidity between regions and countries, a more specific definition is used.

Definitions and Definitional Pitfalls

According to *Webster's New World Medical Dictionary*, the perinatal period "starts at the 20th to 28th week of gestation and ends 1 to 4 weeks after birth." By international convention, strict limits are defined in the tenth revision of the *International Statistical Classification of Diseases and Related Health Problems* (ICD-10), stating that the perinatal period commences at 22 completed weeks (154 days) of gestation (the time when birth weight is normally 500 g), and ends 7 completed days after birth. The lower limit is arbitrarily chosen to represent the time when a fetus is potentially viable; before this time a terminated pregnancy would be labeled miscarriage or abortion, after this time it is a birth. However, the ICD-10 definition of birth does not take gestational age into account; it only distinguishes between live birth and fetal death (deadborn fetus).

The ICD-10 definition is not as straightforward as it looks at first glance. Gestational age is variously defined. According to ICD-10, it should be based on the date of the first day of the last normal menstrual period, or on the best clinical estimate if this date is not available (or uncertain). With the advent of second-trimester ultrasound dating, sonographic biometry has increasingly replaced menstrual dates in countries where ultrasound is an integral part of prenatal care. Seven completed days has three logical interpretations: 168 h from the hour of birth, the date of birth plus 6 completed days and the date of birth plus 7 completed days, which are all in use. Since the incidence of early neonatal death tapers off toward a minimum during the first week after birth, these different definitions do not seriously affect international comparisons.

In many countries, civil notification of stillbirth starts at 28 completed gestational weeks, which precludes the enumeration of stillbirths between 22 and 28 weeks. International perinatal comparisons are therefore most often restricted to births beyond 28 weeks. To avoid enigmatic definitions of gestational length, age is in some instances replaced by birthweight of 1000 g or more. The different limits for compulsory notification of stillbirths entails another source of error in comparisons, since births shortly past the time limit tend to be underreported. In a concerted European comparison of perinatal mortality, it was found that adjustment for different cut-off points for birthweight and gestational age would reduce the differences in perinatal mortality by between 14% and 40% in the different countries and regions (Graafmans *et al.*, 2001).

Perinatal Mortality

Perinatal mortality is defined as the number of fetal deaths past 22 (or 28) completed weeks of pregnancy plus the number of deaths among live-born children up to 7 completed days of life, per 1000 total births (live births and stillbirths). A joint interagency expert meeting on global indicators of sexual and reproductive health organized by WHO, UNICEF, and UNFPA in 2000 recommended perinatal mortality rate (PNMR) to be one of 17 chosen indicators. To avoid the problem of determining gestational age, the expert group advised that for comparative purposes 500 g and 1000 g should replace the time limits (World Health Organization, 2001). It goes without saying that PNMR will be higher when the 22-week limit or 500-g limit is used, because of the added number of stillbirths between 22 and 27 weeks. **Figure 1** shows perinatal death rates in Norway in four time periods by gestational age. There has been substantial improvement in survival of extremely preterm births since 1967; in 1997–2004 there were 48% perinatal deaths among those born between 22 and 27 weeks, compared to 91% in 1967–76 (Skjærven, 2006). Norway was chosen for the example because there is compulsory notification of all births (stillbirths and live births) from 16 weeks onwards, which ensures nearly complete coverage. The remarkably enhanced survival over a span of 40 years is largely ascribed to better neonatal intensive care, which is of course related to those born alive. During the same period, the stillbirth rate in Norway past 28 weeks fell from 11.2 in 1967 to 2.9 per 1000 births in 2004. It is difficult to sort out the causes, but better overall maternal health status is thought to play an important part.

Using PNMR for global monitoring pinpoints the wide gap in reproductive health between rich and poor countries. Perinatal death is more common than maternal death, by a factor of about 100. Local registration will serve as an impetus to perinatal audit, which in turn may induce measures for improvement. **Figure 2** shows perinatal mortality in the WHO regions of the world over a span of 18 years and **Figure 3** gross country estimates in the year 2000. With a PNMR of 62 per 1000 births, Africa at present ranks higher than the Nordic countries did more than 100 years ago. However, a local population-based study in rural Northern Tanzania came up with an optimistically lower figure, 27 per 1000 births, for which no scientific explanation was found (Hinderaker *et al.*, 2003).

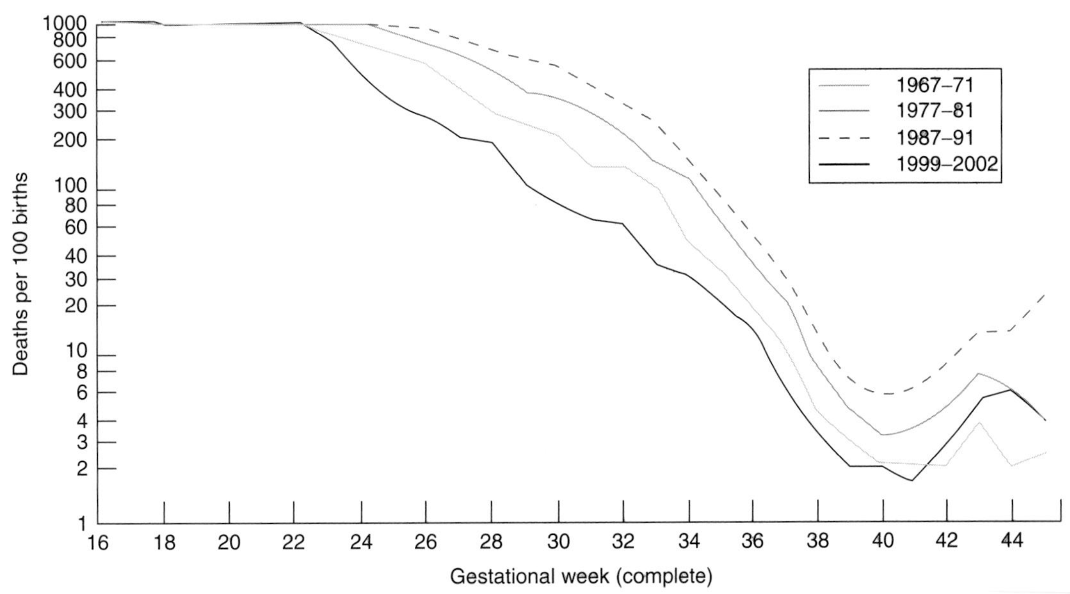

Figure 1 Perinatal mortality rates in Norway in four time periods during 1967–2002, by gestational age. Data from the Medical Birth Registry of Norway.

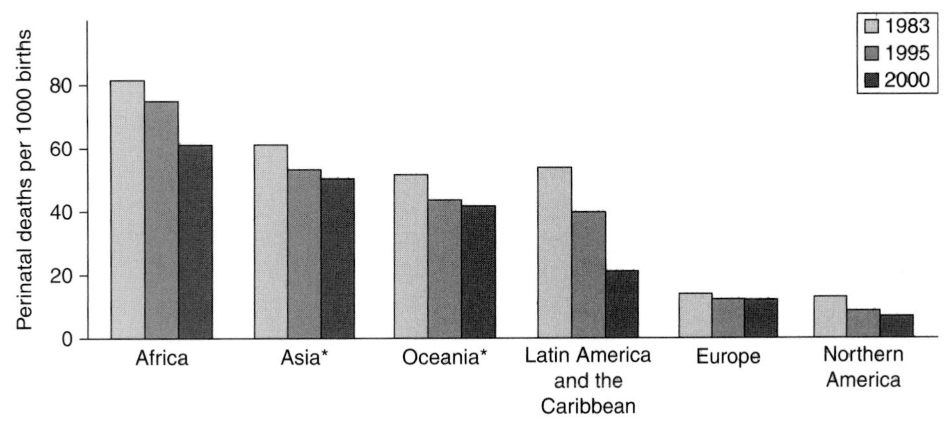

Figure 2 Perinatal mortality rates in three time periods by WHO Regions. *Australia, New Zealand, and Japan have been excluded from regional estimates but are included in the total for developed countries. From World Health Organization (2006) *Neonatal and Perinatal Mortality. Country, Regional and Global Estimates.* Geneva: World Health Organization.

In developing countries, the ratio of stillbirths to first-week deaths is approximately 50/50. In Europe, stillbirths have the higher share, while in Northern America it is lower, the PNMR in both regions being around 10 per 1000 births (**Table 1**). In all likelihood, these latter differences are caused by registration shortfalls.

Causes of Perinatal Death

To monitor perinatal mortality over time and make comparisons between countries and regions, one needs to know the causes of death. This implies a classification system that is clearly defined and fit for the purpose of introducing corrective measures. Various systems have been presented, with different emphasis on maternal and fetal or newborn conditions and complications. To determine the cause of death in a stillborn fetus as a rule requires an autopsy; this in practice is an exception rather than a standard procedure.

Table 2 is an example of causes of perinatal death in districts of rural Tanzania in 1995–96, grouped according to a classification system developed by the International Collaborative Effort (ICE) on Birth Weight, Plurality, Perinatal and Infant Mortality (Cole *et al.*, 1989). The data derive from a prospective population-based study. Of the deaths recorded, 46% were stillbirths. Some of the neonates died more than 1 week after birth. Infection-related deaths

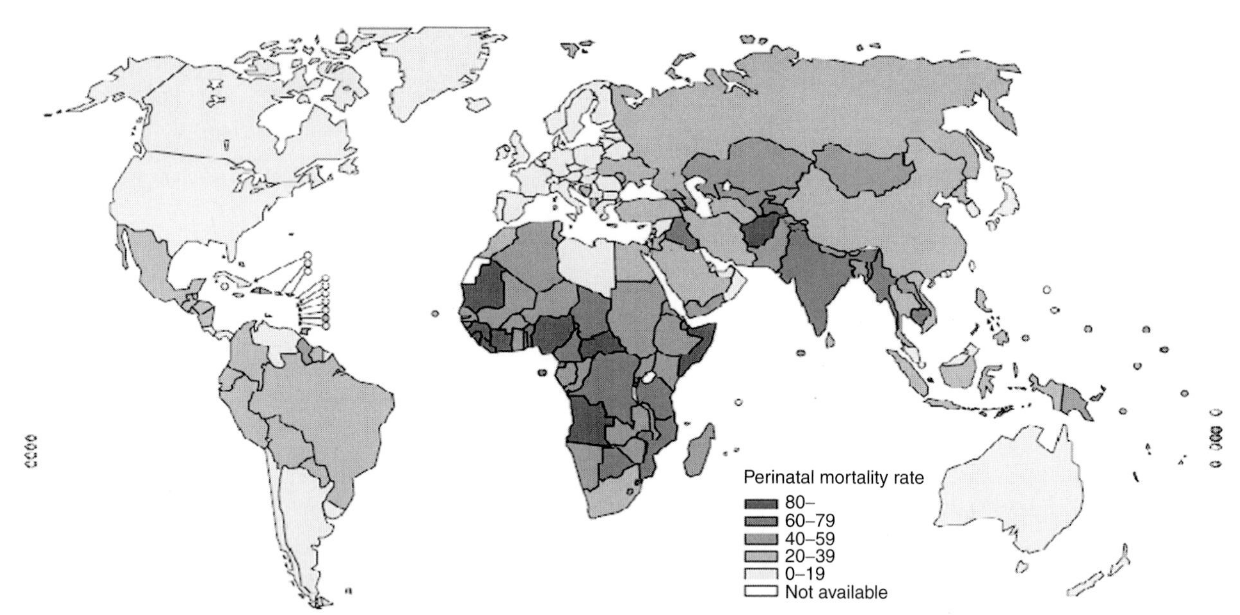

Figure 3 Perinatal mortality rates by country, 2000. From World Health Organization (2006) *Neonatal and Perinatal Mortality. Country, Regional and Global Estimates.* Geneva: World Health Organization.

Table 1 Global estimates of stillbirths, early neonatal deaths and perinatal mortality by WHO region in 2000

Region	*PNM rate*	*Still birth rate*	*ENMR*
Africa	62	32	31
Asia	50	27	24
Europe	13	8	4
Latin America	21	10	12
North America	7	3	4
Oceania	42	23	19

From World Health Organization (2006) *Neonatal and Perinatal Mortality. Country, Regional and Global Estimates.* Geneva: World Health Organization.

Table 2 Causes of death in 136 cases of perinatal death in rural Tanzania, 1995–96

	All cases	*Avoidable*
Total	136	34
Infections	53	19
Asphyxia-related conditions	32	9
Immaturity-related conditions	20	1
Other causes/unclassifiable	16	2
Congenital conditions	9	0
External causes of death	2	0
Sudden death	1	0
Specific conditions	1	1
No information	2	2

Adapted from Hinderaker SG, Olsen BE, Bergsjø PB, *et al.* (2003) Avoidable stillbirths and neonatal deaths in rural Tanzania. *International Journal of Obstetrics and Gynecology* 110: 616–623.

dominated. Malaria was diagnosed in 20 out of 53 cases (Hinderaker *et al.*, 2003).

Besides the direct cause of death, a number of other factors have been shown to influence the incidence of perinatal mortality. Gestational age and birth weight are obvious. Maternal age, parity, and a number of maternal conditions and diseases such as diabetes, the antiphospholipid syndrome, preeclampsia, and smoking are risk factors to varying degrees. In multiple pregnancies, the second in a set of twins carries the higher risk. More boys than girls die in the perinatal period. Isoimmunization due to blood group incompatibility has largely been eliminated following the introduction of anti-rhesus-gammaglobulin, but is still a risk factor. Intrapartum events (breech and transverse fetal presentation, mode of delivery, maternal hemorrhage) may threaten the baby's life. In rich countries with decreasing PNMRs, congenital conditions take an increasing share of perinatal deaths.

Perinatal Morbidity

Perinatal epidemiologists also study diseases, injuries, and genetic aberrations diagnosed *in utero*, at birth, or in the early neonatal period, and their short- or long-term consequences. They can be classified in much the same way as the causes of death in **Table 2**. The Medical Birth Registry of Norway routinely reports incidence rates of a number of specified neonatal conditions, shown in **Table 3**. The most common of these conditions was treated neonatal hyperbilirubinemia (or jaundice), which occurred in almost 10% of the newborns. Treatment for the majority of these was phototherapy. Systemic antibiotic treatment was given to 2.4%, while hip joint dysplasia and respiratory distress syndrome were diagnosed each in about 1% of the newborns (Skjærven, 2006).

Perinatal Audit

Perinatal audit is defined as "the systematic, critical analysis of the quality of perinatal care, including the procedures used for diagnosis and treatment, the use of resources and the resultant outcome and quality of life for women and their babies" (Dunn and McIlwaine, 1996). Peer review is another label. The definition covers a variety of quality assessment activities, with the implicit understanding that audit feedback should lead to changes to improve services. In this context, audit should be a formal procedure, with a multidisciplinary panel consisting of professionals of high standing, chosen among people of the same ranks and positions as those who cared for the cases to be audited, but preferably with no links to their departments. We have previously given a more detailed account of our experiences of perinatal audit through 20 years (Bergsjø *et al.*, 2003) and will only give a few examples in this brief outline.

Audits of perinatal deaths have been used to advantage in both developing and developed countries. An example of the former is given in **Table 2**, which provides the added information that one of four cases was considered to be possibly avoidable. In a similar exercise covering ten European countries or regions, a multinational and multidisciplinary panel decided that between 35% and 54% of selected cases of perinatal death were potentially avoidable (**Table 4**). The cases had been selected among groups in which care and treatment were most likely to have a significant impact on the outcome, and the figures therefore only indicate relative differences in standards between countries and regions (Richardus *et al.*, 2003). If all of the perinatal deaths within each birth population had been reviewed, the proportion of potentially avoidable cases would have been considerably lower.

Table 3 Selected neonatal conditions and interventions in Norway 2004. Numbers and proportions per 1000 births. The total number of births was 58 041

Condition	*Number*	*Proportion per 1000*
Respiratory distress syndrome	563	9.7
Intracranial hemorrhage	134	2.3
Fractured limbs	234	4.0
Facial paresis	35	0.6
Brachial plexus injury	136	2.3
Hip joint dysplasia, treated	572	9.9
Conjunctivitis, treated	442	7.6
Systemic antibiotics treatment	1363	23.5
Hyperbilirubinemia, treated	5431	93.6

Perinatal Risk Factors

Birth registries serve as important surveillance systems allowing description of relevant factors associated with pregnancies and childbirth. For example, the frequencies of interventions such as cesarean section and adverse outcome of birth such as perinatal death. The associations between maternal age and parity and preterm birth are shown in **Table 5**. With the exception of teenaged mothers, the incidence of preterm birth is consistently lower among those who have given birth previously.

Another example is the association between the length of maternal education and preterm birth in different time periods, as shown in **Table 6**. Mothers with a low

Table 4 Summary results of audit of selected types of perinatal deaths in some Western European countries and regions

	Unavoidable		*Potentially avoidable*		*Total graded*
Country	*Number*	*%*	*Number*	*%*	*Number*
Belgium	92	48.9	96	51.1	188
Denmark	127	48.8	133	51.2	260
England	100	46.5	115	53.5	215
Finland	111	68.1	52	31.9	163
Greece	51	48.6	54	51.4	105
The Netherlands	81	51.6	76	48.4	157
Norway	84	60.4	55	39.6	139
Scotland	42	49.4	43	50,6	85
Spain	57	55.9	45	44.1	102
Sweden	83	64.3	46	35.7	129
Total	828	53.7	715	46.3	1543

Most countries are represented by regions (provinces).
Adapted from the EuroNatal study, Richardus JH, Graafmans WC, Verloove-Vanhorick SP, and Mackenbach JP (2003) Differences in perinatal mortality and suboptimal care between 10 European regions: Results of an international audit. *BJOG: An International Journal of Obstetrics and Gynaecology* 110: 97–105.

Table 5 Age, parity, and preterm birth

Maternal age and preterm birth		*Proportion of preterm births (%)*
Para 0	Age < 20 years	7.4
	Age 20–24	5.3
	Age 25–29	5.4
	Age 30–34	6.5
	Age 35+	8.9
	All ages	5.8
Para 1+	Age < 20 years	9.0
	Age 20–24	4.8
	Age 25–29	4.0
	Age 30–34	4.5
	Age 35 +	6.1
	All ages	4.6

Based on data from Skjærven R (2006) Forekomst av og overlevelse blant ekstremt premature barn. In: *Fødsler i Norge 2003–2004. Årsrapport/Annual Report.* Bergen, Norway: Medisinsk fødselsregister/Medical Birth Registry of Norway, pp. 199–205 ($n = 1\,680\,994$).

Table 6 Frequency of preterm birth (%) by level of maternal education in three periods in Norway

	1967–76		*1977–86*		*1987–98*	
Maternal education	*%*	*RR*	*%*	*RR*	*%*	*RR*
Low	5.9	1.40	5.6	1.47	6.6	1.43
Medium	4.8	1.20	4.6	1.21	5.4	1.17
High	4.0	1.00	3.8	1.00	4.6	1.00
Unknown	6.8	1.70	6.7	1.76	7.0	1.52
Total	5.1		4.7		5.4	

Based on data from Skjærven R (2006) Forekomst av og overlevelse blant ekstremt premature barn. *Fødsler i Norge 2003–2004. Årsrapport/Annual Report*, pp. 199–205. Bergen, Norway: Medisinsk fødselsregister/Medical Birth Registry of Norway ($n = 1\,680\,994$).

education level had increased risk of preterm birth in all three time periods, 1967–76, 1977–86, and 1987–98.

Table 7 shows the association between the risk of preterm birth by sex of the fetus, previous preterm birth, and previous fetal death. It can be seen that male fetuses have 20% increased risk of being born preterm (6.5% vs. 5.4%). A previous preterm birth increases the incidence from 5.2% to 11.4% (RR = 2.19) and a previous fetal death increases the risk from 5.6 to 8.1 (RR = 1.45).

Time trends of preterm births of less than 37 completed weeks' gestational age in Norway 1967–98 are shown in **Table 8**. It appears that little change in the frequency has occurred over this time period, in spite of considerable changes in habits of living (diet and smoking), in vitro fertilization, ultrasound dating of pregnancies, and a more interventional obstetric practice. Perinatal survival has improved considerably for these tiny babies, as shown in **Figure 1**. The causation of preterm birth is mainly unknown, although a significant part of the causation is linked to genes from both parents. The genetic and environmental components of the causation call for innovative research in order for decision makers to develop effective preventive strategies.

Table 7 Percentage preterm births by sex of the fetus and by previous pregnancy outcomes

Sex and previous outcomes		*Preterm birth (%)*
Sex of fetus	Male	6.5
	Female	5.4
Previous preterm birth	Yes	11.4
	No	5.2
Previous fetal death	Yes	8.1
	No	5.6

Based on data from Skjærven R (2006) Forekomst av og overlevelse blant ekstremt premature barn. In: *Fødsler i Norge 2003–2004. Årsrapport/Annual Report.* Bergen, Norway: Medisinsk fødselsregister/Medical Birth Registry of Norway, pp. 199–205, 1967–98 ($n = 1\,680\,994$).

Table 8 Time trends of preterm births of less than 37 completed weeks' gestational age (%) in Norway, 1967–98

	Time periods		
Gestational age (weeks)	*1967–76*	*1977–86*	*1987–98*
16–27	0.5	0.5	0.6
28–32	0.8	0.7	0.8
33–36	3.8	3.6	4.0
Total <37	5.1	4.8	5.4
Total number of births <37 completed weeks	390 021	467 752	623 221

Based on data from Skjærven R (2006) Forekomst av og overlevelse blant ekstremt premature barn. In: *Fødsler i Norge 2003–2004. Årsrapport/Annual Report.* Bergen, Norway: Medisinsk fødselsregister/Medical Birth Registry of Norway, pp. 199–205, 1967–98 ($n = 1\,680\,994$).

Analytical Studies

In searching for risk factors, two major approaches are employed within perinatal epidemiology. The individuals may be defined according to a specific outcome variable, for example preterm birth. Risk factors can be related to this outcome by the case–control approach. Is the risk factor more prevalent among the cases than among the controls? Control cases can be matched to the cases for one or more variables, such as age and parity. Alternatively, the total cohort can be used for the comparisons, in which case it is called a case–cohort study. The other approach involves identification of individuals according to the presence of the risk factor under study and following the individuals over time and recording the outcome(s) in question. This is called a prospective, or cohort, study. Recently, huge cohorts of pregnancies have been established allowing cohort analyses, or so-called nested case–control studies, which mean case–control studies within the cohort material.

During the 1960s, diethylstilbestrol (DES) was widely used to prevent abortion or preterm birth (on scanty evidence, later shown to be invalid). An investigation that started in Boston in 1971 (Herbst *et al.*, 1977) is a classical example of a case–control study in perinatal epidemiology. Three physicians at the Vincent Memorial Hospital reported a striking association between maternal use of DES during the first trimester of pregnancy and the development 15–20 years later of vaginal cancer in daughters born of these mothers. Seven young women were diagnosed with this cancer that was rarely seen in young women. Because of this clustering of cases (seven within 4 years), it was decided to conduct a case–control study comparing these seven cases plus an additional one from another hospital and their families with an appropriate comparison group (controls) in order to identify factors that might explain the appearance of the cancers. Four matched controls were selected among young women who did not have vaginal cancer (matched for age and place of birth). Thus the study consisted of eight cases and 32 controls. As shown in **Table 9**, seven of the eight women with vaginal cancer had mothers who had been exposed to DES in their pregnancies compared to none of the 32 controls. No other differences between cases and controls were found. This was such a strong observation that causality was suspected. Thus a previously unknown cause of a disease was disclosed.

Most studies that use data from the medical birth registries use a cohort design where both exposure and outcome variables are collected from the registry. A recent example concerns fetal and infant survival following preeclampsia. The study showed that fetal and infant survival in preeclamptic pregnancies have improved over the past 35 years in Norway. However, the selective risk of neonatal death following a preeclamptic pregnancy has not changed over time (Basso *et al.*, 2006).

Table 9 Exposure to diethylstilbestrol (DES) among mothers of eight young women with cancer of the vagina, compared to 32 matched controls

		Cancer vaginalis	
		+	–
DES	+	7	0
	–	1	32
Total		8	32

Based on data from Herbst AL, Ulfelder H, and Poskanzer D (1971) Adenocarcinomas of the vagina association of maternal stilbestrol therapy with tumor appearance in young women. *New England Journal of Medicine* 284: 878–881.

Some researchers combine data from a medical birth registry with data from other registries. Fosså *et al.* linked data from the patient registry of the Norwegian Radium Hospital and from the Norwegian Cancer Registry with data from the Birth Registry of Norway. They examined the parenthood in successive births after adult cancer. Female cancer survivors delivered more preterm births and low-birth-weight infants than women in a matching noncancer population. Male cancer survivors, on the other hand, did not differ from noncancer fathers in the outcome of successive births (Fosså *et al.*, 2005).

Recurrence Risk

Sibling studies became possible because the Norwegian Medical Birth Registry employs the unique individual identification system, whereby it is possible to link successive births to the same mothers. Bakketeig (1977) presented the first data showing the strong tendency to repeat low birth weight, preterm birth, and small for gestational age (SGA) birth. **Table 10** shows the relative prediction of adverse outcomes among second births based on the outcome of the mothers' first birth. For example, the prediction of low birth weight (LBW) is 5.6 times as strong if the elder sibling was also LBW as opposed to having appropriate birth weight. The risk of preterm birth was 4.0 times higher if the mother's first birth was also a preterm delivery, compared to a nonpreterm birth. A similar risk was shown for a SGA second birth where the first birth was also SGA compared to not being SGA. This tendency to repeat was later confirmed in much larger data sets (Bakketeig *et al.*, 1979). Mothers somehow appeared to be programmed to produce births of a certain gestational age and size. If they departed from

Table 10 The cumulative tendency to repeat preterm birth (< 37 weeks)

Previous births		*Subsequent births preterm birth*		
First	*Second*	*No*	*%*	*RR*
N		461 052	3.9	1.0
P		29 278	16.3	4.18
N	N	155 417	4.1	1.0
N	P	7081	14.8	3.61
P	N	9912	9.2	2.24
P	P	1960	32.3	7.88

Based on the first 7-year period of the Medical Birth Registry of Norway, 1967–1973 (a total of 464 067 births). Skjærven R (2006) Forekomst av og overlevelse blant ekstremt premature barn. In: *Fødsler i Norge 2003–2004. Årsrapport/Annual Report.* Bergen, Norway: Medisinsk fødselsregister/Medical Birth Registry of Norway, pp. 199–205. P, gestational age <37 weeks; N, gestational age ≥37 weeks.

Table 11 Preterm birth (<37 weeks) by parental gestational age at birth

Mother	*Father*	*Total number of births*	*Outcome preterm births (%)*
N	N	83 191	6.5
P	N	3310	9.1
N	P	4441	7.1
P	P	204	11.3
Total		91 146	

P, preterm; N, not preterm.

this pattern the offspring was at a greater risk of perinatal death (Bakketeig and Hoffman, 1983).

The tendency to report birth outcomes was found to be cumulative. As shown in **Table 10**, two preterm births increased the risk of a subsequent preterm birth nearly eightfold compared to the two first births not being preterm.

The risk of preterm birth is dependent on whether the mother or the father, or both parents, were born preterm, as shown in **Table 11**.

Skjærven *et al.* (1988) examined whether the birth weight and perinatal mortality of second birth were conditional on the weight of the first birth. They showed that a woman's successive offspring tend to have similar birth weights. Mean birth weights among second births differ by as much as 1000 g depending on the weight of the first birth. Also, the survival of the second baby at a given weight is strongly affected by its weight relative to the first baby's weight. A baby may be average size compared to the population norm, but small compared to its elder siblings. Such a baby has increased mortality that goes with being relatively small. For example, an infant weighing 3250 g is relatively large if the mother's previous baby weighed 2250 g, but relatively small if the previous birth weighed 4250 g. In the first case, the perinatal mortality risk of the 3250-g baby is 2.2 per 1000 while in the second case the baby with the same weight has perinatal mortality of 9.0 per 1000 (or four times higher).

Focusing on mothers with two births Vatten and Skjærven (2003) showed that women who changed partners between their pregnancies had an increased risk of preterm birth, birth with low birth weight, and increased risk of infant mortality, compared with women with the same partner in their pregnancies.

Mjelve *et al.* (1999) studied mothers' two first singleton births. They showed that siblings' gestational ages were significantly correlated ($r = 0.26$) and that the risk of having a preterm birth was nearly ten times higher among mothers whose firstborn had been delivered before 32 weeks' gestation compared to mothers whose first birth had been at 40 weeks' gestation. They also found that the perinatal mortality in preterm second births was significantly higher among mothers whose first infant was born at term compared with mothers whose first born child was delivered at 32–37 weeks. They concluded that since perinatal mortality among preterm births is dependent on gestational age in the mother's previous birth, a common threshold of 37 weeks' gestation for defining preterm birth as a risk factor for perinatal death may not be appropriate for all births.

Generational Studies

Skjærven *et al.* (1997) focused on whether a baby's survival is related to its mother's birth weight. They linked births during 1981–94 to data on all mothers born from 1967 onwards, thereby forming 105 014 mother–offspring units. The mother's birth weight was strongly associated with the weight of her baby. Mortality among small babies was much higher where the mothers were born large. For example, babies weighing between 2500 and 3000 g had a threefold higher perinatal mortality if their mothers' birth weight was 4000 g or higher, compared to those whose mothers had been small at birth (between 2500 and 3000 g).

Magnus *et al.* (1993) examined the correlation of gestational age and birth weight across generations. Based on linkage of 1967–69 births and 1986–89 births, they obtained 11 072 pairs of mother–firstborn offspring. A low correlation coefficient (0.086) was found for gestational age across generations, whereas the correlation between maternal and offspring birth weight was 0.242. Mothers with birth weight below 2500 g had a significantly increased risk (OR = 3.03, 95% CI 1.79–5.11) of having a low weight baby compared with mothers with birth weight above 4000 g. On the other hand, if the mother was born before 37 weeks' gestation, the risk of having a preterm birth was not significantly increased (OR = 1.48, 95% CI 0.96–2.21) compared with mothers born at term. The authors concluded that in contrast to birth weight, variation in human gestational age does not appear to be influenced by genetic factors to any large degree.

Record Linkage to Registries Outside Medical Birth Registries

In a study focusing on preeclampsia and subsequent breast cancer Vatten *et al.* (2002) showed that women with preeclampsia and/or hypertension in their first pregnancy had a 19% (95 CI, 9%–29%) lower risk of breast cancer compared to other parous women. This was based on data from the Medical Birth Registry of Norway linked to the Cancer Registry of Norway. The results indicate that pathophysiology surrounding preeclampsia and gestational hypertension plays a role in breast cancer etiology.

Maternal and Child Cohorts

Large cohorts of pregnancies have been established for research purposes in Denmark and Norway. Pregnant women have been included from early on in their pregnancies. Information has been collected as well as biological material (blood, urine) from the women and partly from their partners (Norway). These cohorts are important additional sources for further studies on maternal and child health. To exemplify, one can establish case–control (or case–cohort) studies within the cohort on its scientific strength. We may want to test a specific hypothesis on causation of a certain type of cancer. Each of these cohorts contains a variety of information on, for example, lifestyle of the mother when the patient was a fetus or baby, in addition to biological material collected during pregnancy, which would allow for testing of a specific hypothesis on causation.

See also: Infant Mortality/Neonatal Disease; Maternal Mortality and Morbidity.

Citations

Bakketeig LS (1977) The risk of repeated preterm or low weight delivery. In: Reed DM and Stanley FJ (eds.) *The Epidemiology of Prematurity.* Baltimore, MD: Urban and Schwarzenberg.

Bakketeig LS, Hoffman HJ, and Harly EE (1979) The tendency to repeat gestational age and birth weight in successive births. *American Journal of Obstetrics and Gynecology* 135: 1086–1103.

Bakketeig LS and Hoffman HJ (1983) The tendency to repeat gestational age and birth weight in successive birth related to perinatal survival. *Acta Obstetricia et Gynecologica Scandinavica* 62: 85–92.

Basso O, Rasmussen S, Weinberg CR, Wilcox AJ, Irgens LM, and Skjærven R (2006) Trends in fetal and infant survival following preeclampsia. *Journal of the American Medical Association* 296: 1357–1362.

Bergsjø P, Bakketeig LS, and Langhoff-Roos J (2003) The development of perinatal audit: Twenty years' experience. *Acta Obstetricia et Gynecologica Scandinavica* 82: 780–788.

Cole S, Hartford RB, Bergsjø P, and McCarthy B (1989) International Collaborative Effort (ICE) on Birth Weight, Plurality, Perinatal, and Infant Mortality: III. A method of grouping underlying causes of infant death to aid international comparisons. *Acta Obstetricia et Gynecologica Scandinavica* 68: 113–117.

Dunn PM and McIlwaine G (eds.) (1996) Perinatal audit. *Prenatal and Neonatal Medicine* 1: 160–194.

Fosså SD, Magelssen H, Melve K, Jacobsen AB, Langmark F, and Skjærven R (2005) Parenthood in survivors after adulthood cancer and perinatal health in their offspring. A preliminary report. *Journal of the National Cancer Institute Monographs* 34: 77–82.

Graafmans WC, Richardus J-H, Macforlane A, *et al.* (2001) Comparability of published perinatal mortality rates in Western Europe: the quantitative impact of differences in gestational age are birthweight criteria. *BJOG: An International Journal of Obstetrics and Gynecology* 108: 1237–1245.

Herbst AL, Ulfelder H, and Poskanzer D (1971) Adenocarcinomas of the vagina association of maternal stilbestrol therapy with tumor appearance in young women. *New England Journal of Medicine* 284: 878–881.

Hinderaker SG, Olsen BE, Bergsjø PB, *et al.* (2003) Avoidable stillbirths and neonatal deaths in rural Tanzania. *International Journal of Obstetrics and Gynecology* 110: 616–623.

Magnus P, Bakketeig LS, and Skjærven R (1993) Correlations of birth weight and gestational age across generations. *Annals of Human Biology* 20: 231–238.

Mjelve KK, Skjærven R, Gjessing HK, and Øyen N (1999) Recurrence of gestational age in sibships: Implication of perinatal mortality. *American Journal of Epidemiology* 150: 756–762.

Richardus JH, Graafmans WC, Verloove-Vanhorick SP, and Mackenbach JP (2003) Differences in perinatal mortality and suboptimal care between 10 European regions: Results of an international audit. *BJOG: An International Journal of Obstetrics and Gynaecology* 110: 97–105.

Skjærven R (2006) Forekomst av og overlevelse blant ekstremt premature barn. *Fødsler i Norge 2003–2004. Årsrapport/Annual Report*, pp. 199–205. Bergen, Norway: Medisinsk fødselsregister/ Medical Birth Registry of Norway.

Skjærven R, Wilcox AJ, and Russel D (1988) Birthweight and perinatal mortality of second births conditional on weight of the first. *International Journal of Epidemiology* 17: 830–838.

Skjærven R, Wilcox AJ, Øyen N, and Magnus P (1997) Mothers' birth weight and survival of their offspring: Population-based study. *British Medical Journal* 314: 1376–1380.

Vatten LJ and Skjærven (2003) Effects on pregnancy outcome of changing partner between first two births: Prospective population study. *British Medical Journal* 327: 1138.

Vatten LJ, Romundstad PR, Triohopulos D, and Skjærven R (2002) Pre-eclampsia in pregnancy and subsequent risk of breast cancer. *British Journal of Cancer* 87(9): 971–973.

World Health Organization (2001) *Reproductive Health Indicators for Global Monitoring. Report of the second Interagency meeting*. WHO, Geneva 17–19 July 2000. WHO/RHR/01.19. Geneva, Switzerland: World Health Organization.

World Health Organization (2006) *Neonatal and Perinatal Mortality. Country, Regional and Global Estimates*. Geneva, Switzerland: World Health Organization.

SECTION 2
CORE ELEMENTS

Family Planning/Contraception

A Glasier and A Gebbie, Family Planning and Well Woman Services, Edinburgh, UK

Introduction

Until the twentieth century, the prevention of pregnancy was largely achieved by abstinence, avoiding coitus during the fertile period or having it only infrequently, or by coitus interruptus. These methods of contraception served well the populations going through the Industrial Revolution. In developed countries, birth rates had fallen long before the introduction of modern methods of contraception. Before 1950, condoms (first described in 1350 BC), female barrier methods (available from the 1800s), and the intrauterine device (IUD, first developed in 1909 with modern copper devices produced in 1969) were the only artificial methods available. The advent of the oral contraceptive pill in 1960, developed by Pincus and Rock and colleagues, is widely regarded as heralding a revolution in contraception; the pill is currently used by over 60 million women worldwide. The uptake of modern methods of contraception (hormonal and intrauterine contraceptives) has led to a further fall in total fertility rates (TFRs) in the developed world and a marked fall in some less-developed countries. In 2005, 60% of married women were using contraception (all methods) and 53% were using modern methods, resulting in a TFR for the world of 2.7. These impressive statistics hide enormous variation between countries. In China, 87% of married women use contraception (86%, modern methods) while in Chad only 11% are using contraception (2%, modern methods). Overall contraceptive use in developed and developing countries is compared in **Figure 1**.

Modern contraceptives can be divided into irreversible methods (male and female sterilization) and reversible methods, made up of hormonal methods (delivering either a combination of estrogen and progestogen or progestogen alone), IUDs, and barrier methods, the latter of which are the only methods that also protect against sexually transmitted infections (STIs). Periodic abstinence, withdrawal, and lactational amenorrhea also have a major role to play in some countries.

A wide range of factors affect contraceptive choice and uptake, including availability (not all methods are licensed everywhere and in some parts of the world supplies of even those that are licensed are precarious), cost, effectiveness, ease of use, risks, and side effects. Providers and cultural acceptability also play a large part in determining method use and there are wide variations between countries in patterns of use. For example, while the IUD is used by 17% of women using contraception in Sweden, it accounts for less than 1% of contraceptive use in the USA. Injectable, progestogen-only contraception is the most popular method in South Africa and Thailand but hardly used in France, Italy, or Spain. In Turkey, 24% of couples rely on withdrawal. In Latin America, female sterilization is common whereas vasectomy is hardly ever performed.

Most modern methods of contraception are highly effective when used perfectly, that is, correctly and with every act of intercourse (**Table 1**), but those methods that require the user to do something every day (e.g., take a pill) or something with every act of intercourse (e.g., put on a condom) have relatively high failure rates in typical use (i.e., when a method is used sometimes incorrectly and/or inconsistently). In the developed world, unintended pregnancy most commonly arises because of inconsistent or incorrect use of a method including discontinuation of the method. Contraceptives that require a visit to a health professional for discontinuation (IUDs and implants) tend to have higher continuation rates than those that the user can simply stop when he or she feels like it.

All methods are extremely safe in healthy women. Methods that contain estrogen are associated with an increased risk of cardiovascular disease and breast cancer but the absolute risk of serious side effects is very small. For all methods, particularly in less-developed countries, the risk of pregnancy far outweighs the risks of using contraception. The World Health Organization (WHO) has developed a simple system to help assess who can use which method safely – the Medical Eligibility Criteria for Contraceptive Use (WHOMEC) (**Table 2**). Medical problems or special conditions affecting eligibility to use a method are classified into one of four categories, which allow for unrestricted use (category 1), use when the advantages outweigh the theoretical or proven risks (category 2), use when the risks outweigh the advantages (category 3), and use when the risk of using the method is unacceptable (category 4). The system allows for providers with limited clinical judgment to provide contraception to women with conditions that fall into categories 1 and 2. WHO has also developed international guidance for effective use and clinical management of contraception – the Selected Practice Recommendations for Contraceptive Use (SPR) (World Health Organization 2005).

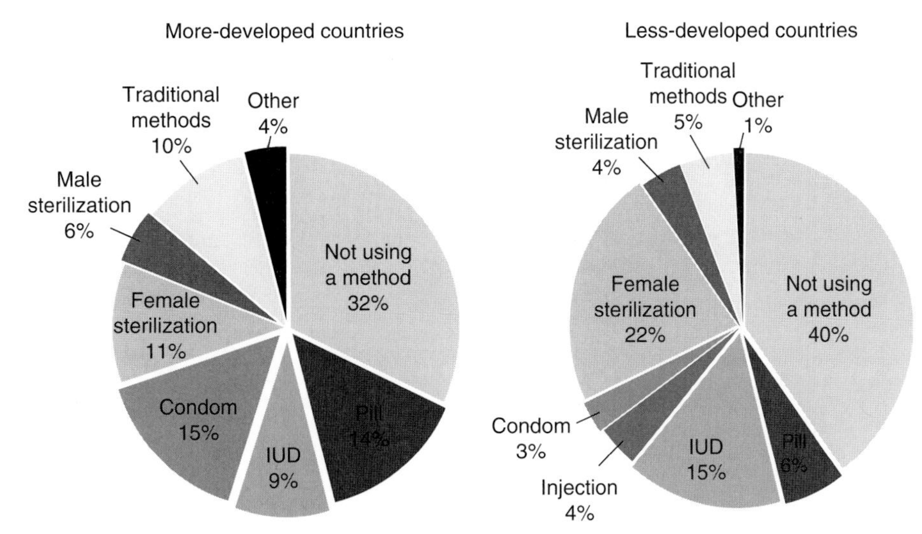

Figure 1 Contraceptive use in more-developed and less-developed countries. Source: Population Reference Bureau (2004) Transitions in world population. *Population Bulletin* 59: 1–40.

Table 1 Effectiveness of contraceptive methods

		Pregnancy rate (%)[a]	
	Method	*Typical use*	*Perfect use*
More effective	Male sterilization	0.15	0.10
	Female sterilization	0.5	0.5
	Implant	0.05	0.05
	LNG-IUS[c]	0.1	0.1
	Copper IUD	0.8	0.6
	Progestogen injectable	3	0.3
	Combined injectable	3	0.05
Effective	Combined pill	8	0.3
	Combined patch	8	0.3
	Combined vaginal ring	8	0.3
	Progestogen-only pill	8	0.5
Less effective	Male condom	15	2
	Female condom	21	5
	Diaphragm	16	6
	Cervical cap (parous)	32	26
	(nulliparous)	16	9
	Periodic abstinence	20	1–9[b]
	Spermicides	26	6
	Withdrawal	19	4
No contraception		85%	

[a]Pregnancy rates (in %) among USA women during first year of use of a method.
[b]Depending on the precise method used.
[c]Levonorgestrel intrauterine system.
Adapted from Trussell J (2004) Contraceptive efficacy. In: Hatcher RA, Trussell J, Stewart F, *et al.* (eds.) *Contraceptive Technology*, 18th rev. edn., pp. 773–845. New York: Ardent Media.

Combined Hormonal Contraception

The combined oral contraceptive pill was first developed in the late 1950s and since then has become widely available throughout the world. It contains a combination of synthetic estrogen and progestogen hormones and since it was first introduced the dose of both hormones has been significantly reduced.

The combined pill offers highly effective and convenient contraception that is independent of intercourse. Alternative delivery systems of combined hormonal contraception have been developed and may offer advantages in terms of convenience and compliance. Most evidence relating to oral, combined hormonal contraception can be extrapolated to the other routes of combined hormonal administration for which there are as yet few data.

Mode of Action

The main mode of action of combined hormonal contraception is to inhibit ovulation by suppressing gonadotropin release from the anterior pituitary, thereby inhibiting ovarian follicular development and ovulation. Additional contraceptive mechanisms include effects on cervical mucus and atrophy of the endometrium.

Combined Oral Contraceptives

The estrogen contained in almost all combined pills is ethinylestradiol and the dose ranges between 15 and 50 μg, with 30 μg being the most common dose. There is a wider selection of progestogens within combined pills

Table 2 WHO Medical Eligibility Criteria (WHOMEC)

Category	*With clinical judgment*	*With limited clinical judgment*	*Example condition in relation to use of combined pill*[a]
1	Use method in any circumstances	Yes (Use the method)	Depressive disorders
2	Generally use the method	Yes (Use the method)	Obesity (BMI > 30)
3	Use of method not usually recommended unless other more appropriate methods are not available or not acceptable	No (Do not use the method)	Adequately controlled hypertension
4	Method not to be used	No (Do not use the method)	Breast cancer

[a]This column gives examples of how the medical eligibility criteria are used in practice for women wishing to use combined hormonal contraception. If the woman suffers from a depressive disorder there is no restriction on her using combined contraceptive pills. Women who are obese generally can also use combined pills but women with breast cancer should not use such pills and women with hypertension, even if adequately controlled, should also not use such pills unless no other method is available or acceptable. Reproduced from World Health Organization (2004) *Medical Eligibility Criteria for Contraceptive Use*, 3rd edn. Geneva: World Health Organization.

Table 3 Classification of progestogens within combined pills

Classification	*Progestogen*
2nd generation	Norethisterone/norethisterone acetate Levonorgestrel Ethynodiol
3rd generation	Gestodene Desogestrel
2nd/3rd generation	Norgestimate
With antimineralocorticoid activity	Drospirenone
With anti-androgenic activity	Cyproterone acetate

and these vary in potency and dosage (**Table 3**). Most pill preparations contain identical combinations of hormones throughout the packet, that is, they are known as monophasic preparations, although some contain a mixture of two (biphasic) or three (triphasic) different pill combinations in the same pack.

Most combined pill preparations contain 21 active hormone tablets and the user takes a tablet daily followed by a pill-free interval of 7 days. Some combined pills have seven placebo pills in the blister pack so that the user can continue with regular, daily pill-taking. Some women choose to run packets of combined pill preparations together, without the pill-free interval, often three at a time. This 'tricycling' of pills or continuous pill use avoids the regular, monthly withdrawal bleed and a marketed preparation is now available in the USA, which has 84 pills packaged for continuous use followed by a 7-day, pill-free interval.

Contraindications

The WHOMEC Category 4 conditions (which represent absolute contraindications to prescribing of the combined pill) fall under the broad headings of past or present circulatory disease, diseases of the liver, and history of a serious condition known to be affected by sex steroids (e.g., breast cancer). Many common medical conditions are not adversely affected by use of the combined pill but lack of conclusive evidence makes most of them WHOMEC Category 2 (the advantages of the method generally outweigh the theoretical or proven risks).

Side effects and risks

Most side effects of the combined pill are mild and often settle with time. Common complaints include irregular bleeding, including breakthrough bleeding, breast tenderness, bloatedness, mood changes, weight gain, and nausea. These troublesome symptoms can result in poor compliance and subsequent discontinuation of the combined pill in some users.

Serious complications in combined pill users are very rare and the main risks relate to thrombotic events and cancer of the breast and cervix. Long-term data from the Royal College of General Practitioners' Study in the United Kingdom showed no excess mortality in users of the combined pill compared with users of other methods (Beral *et al.*, 1999).

Venous thromboembolism (VTE): Estrogen within combined pills has a prothrombotic effect and the risk of VTE for users is increased compared to nonusers (**Table 4**). The risk is greater in women taking combined pills containing the newer progestogens – desogestrel and gestodene – rather than the older progestogens, but overall VTE risk in users (15–30 cases per 100 000 users) is still only about half of the risk associated with pregnancy (60 cases per 100 000 pregnant women). Risk of VTE in individual users of combined oral contraceptives is also influenced by factors such as obesity, immobility, and presence of any inherited thrombophilia.

Table 4 Risk of venous thromboembolism with combined pills

Risk of venous thromboembolism	*Per 100 000 women*
Nonusers of combined pills	5
2nd-generation combined pill users	15
3rd-generation combined pill users	30
Pregnancy	60

Arterial disease: The WHO Scientific Group on Cardiovascular Disease and Steroid Hormone Contraception (1998) reviewed risk of myocardial infarction and stroke in combined pill users and their conclusions are reassuring. Both conditions are very uncommon in young women of reproductive age so the absolute risks are extremely small. In brief summary, there is virtually no effect of combined pill use on risk of myocardial infarction in women with normal blood pressure who do not smoke. Risk of hemorrhagic stroke similarly is not increased in women under the age of 35 years who do not smoke and have normal blood pressure. The risk does, however, increase with age and this effect is magnified in users of the combined pill. The relative risk of thrombotic stroke is increased in combined pill users to around 1.5 and further increased in pill users who smoke or have high blood pressure.

Cancer: The relationship of the combined pill with both breast and cervical cancers is complex. Data from the Collaborative Group on Hormonal Factors in Breast Cancer (1996) found that current or recent users of the combined pill have an increased relative risk of breast cancer of around 1.24. This increased risk was not related to duration of pill use or hormone dose and disappeared 10 years after discontinuing the pill. In a recent reanalysis of worldwide data from 24 epidemiological studies, the relative risk of cervical cancer among women currently using oral contraceptives for 5 years or more was twice that of never-users and the risk increased with increasing duration of use (International Collaboration of Epidemiological Studies of Cervical Cancer, 2007).

Noncontraceptive benefits

In addition to highly effective protection against pregnancy, the combined pill offers significant noncontraceptive benefits for the user. Withdrawal bleeds are usually lighter and less painful than normal menstrual periods. Women are less likely to become anemic and iron stores rise. Premenstrual mood changes often improve and women using the combined pill are less likely to develop functional ovarian cysts and pelvic inflammatory disease.

The combined pill offers important long-term protection against both ovarian and endometrial cancers with a three- to fourfold reduction in the risk of both malignancies, which lasts for at least 15 years after the combined pill is stopped.

Combined Hormonal Patches, Injectables, and Rings

All these delivery systems have been developed relatively recently and it may be of benefit for some individuals to avoid the oral route of hormonal administration in the presence of gastrointestinal disorders or if nausea and vomiting are a problem. Women may find these methods more convenient to use and thus compliance may also be improved.

Injectable: One monthly injection contains a natural estrogen ester, estradiol cypionate (5 mg), combined with medroxyprogesterone acetate (25 mg). Women experience a regular monthly bleed with this preparation and the return to fertility on discontinuation is rapid. Various other combinations have been used, particularly in the People's Republic of China and Latin America, with varying dosage regimens and generally high efficacy.

Transdermal patch: A combined hormonal patch is now available in many countries. It releases ethinylestradiol (20 μg) and norelgestromin (150 μg) daily. The monthly schedule consists of three patches, each worn for 7 days followed by a patch-free week during which a withdrawal bleed occurs. Efficacy is similar to the combined pill.

Vaginal ring: The combined contraceptive ring is a soft silastic structure that is inserted high into the vagina. It measures 5.4 cm in diameter and releases ethinylestradiol (15 μg) and etonorgestrel (120 μg) daily. It is placed in the vagina for 21 days and then removed for a 7-day interval at which time a withdrawal bleed occurs.

Progestogen-Only Contraception

Progestogen-only methods avoid the estrogen-related side effects and risks of combined hormonal contraception. They provide highly effective contraception in a range of dosages and delivery systems (**Table 5**). The individual characteristics of each progestogen-only method vary quite considerably and they offer a variety of options, including pills, implants, and injectables. A progesterone-releasing vaginal ring is marketed in South America for lactating women.

In contrast to the estrogen-containing, combined methods, low-dose, progestogen-only methods have no significant prothrombotic effect and therefore can be used by women with risk factors for cardiovascular disease when combined hormonal contraception is contraindicated. High-dose, progestogen-only methods may exert some effect on coagulation so should be avoided in women with a personal history of thrombosis. All progestogen-only methods are associated with significant menstrual change ranging from total amenorrhea to daily bleeding, although the latter is rarely heavy.

Table 5 Progestogen-only contraception

Mode of delivery	*Method*	*Type of progestogen*	*Dose*	*Effective for*
IM injection	DMPA Depo Provera	Medroxyprogesterone acetate	150 mg	3 months
IM injection	NET-EN Noristerat	Norethisterone enanthate	200 mg	2 months
Subdermal implant	Norplant	Levonorgestrel	25–80 μg/day	5 years
Subdermal implant	Jadelle	Levonorgestrel	25–80 μg/day	5 years
Subdermal implant	Implanon	Etonorgestrel	25–70 μg/day	3 years
Intrauterine	Progestasert	Progesterone	65 μg/day	1 year
Intrauterine	Mirena	Levonorgestrel	20 μg/day	5 years
Oral		Norethisterone	350 μg/day	Daily
Oral	Various brands	Levonorgestrel	30 μg/day	Daily
Oral		Ethynodiol	500 μg/day	Daily
Oral	Cerazette	Desogestrel	75 μg/day	Daily

Mode of Action

Low-dose progestogen methods act mainly by interfering with cervical mucus and endometrial development, thereby affecting sperm transport and implantation. Higher-dose methods reliably suppress ovulation in addition.

Progestogen-only pills

The progestogen-only pill (POP) is a low-dose method that requires a meticulous pill-taking regimen to ensure effectiveness. Extra contraceptive measures are required if a POP is taken more than 3 hours late. A small range of pills is available. One pill is taken every day without a break. Menstruation may be regular but is more likely to be erratic and unpredictable. Depending on the brand of POP up to 20% of women (in whom ovarian activity is suppressed) develop amenorrhea.

Efficacy shows a clear relationship to age, with the POP being around 10 times more effective in women over 40 years compared to women aged 25–29 years. For this reason, the POP is rarely recommended for adolescents, who often find it difficult to achieve good compliance with the required, strict pill-taking. Many women take the POP while lactating and efficacy is very high in this situation. There is a recommendation in some countries that, if a woman weighs more than 70 kg, she should then take a double dose of the POP to improve efficacy. If a woman taking the POP does become pregnant, there is a higher chance than normal that she might have an ectopic pregnancy.

A new, more potent POP containing the progestogen desogestrel (75 μg) is available in some countries and reliably inhibits ovulation in virtually all cycles. As a result, it is theoretically more effective than the older preparations and can prove helpful in the management of conditions such as menstrual migraine and premenstrual syndrome.

Injectables

Two injectable progestogen-only methods are available. Depot medroxyprogesterone acetate (DMPA) (150 mg) is the more commonly used and is given by deep, intramuscular injection every 3 months. Norethisterone enanthate (NET-EN) (200 mg) is given every 8 weeks. Amenorrhea is common with both methods; many women welcome the lack of menstruation provided they are well counseled about this in advance.

Long-term use of DMPA appears to cause a slight reduction in bone mineral density compared to nonusers. Most studies suggest that this is reversible on discontinuation of DMPA but it is of concern that adolescents using DMPA may fail to reach their peak bone mass and that women approaching menopause may not recover their bone mineral density and so may be at increased risk of osteoporosis in later life. DMPA is often associated with slight initial weight gain and, in long-term users, there is a delay in return of fertility of up to 6 months in comparison with other hormonal methods.

Implants

Contraceptive implants consist of nonbiodegradable, flexible rods that are inserted subdermally into the upper arm. The active hormone is contained within the core of each rod and is released by steady diffusion into the circulation. Insertion and removal have to be undertaken by trained health professionals. Although the up-front cost of implants is higher than for other methods, they are a highly cost-effective method of contraception if the user continues for the full duration of the implant's lifespan. The most commonly used implants are listed here:

- Norplant has six rods that release levonorgestrel. It is effective for at least 5 years.
- Jadelle is a two-rod system releasing levonorgestrel with a similar profile to Norplant.

- Implanon consists of a single rod that releases etonorgestrel, a precursor of desogestrel. It is effective for 3 years.

Implants are commonly associated with menstrual disruption and women should be carefully counseled prior to insertion about the menstrual change and risk of other minor hormonal side effects. Reinsertion of a new implant can be performed at the time of removal of the old implant. Fertility is immediately restored on removal of the implant.

Intrauterine Devices and Systems

The modern copper-bearing IUD consists of a plastic frame with copper wire around the stem and, in some models, has copper sleeves on the arms (**Figure 2**). The copper-bearing IUDs are smaller and more effective than the older, inert plastic devices. Both efficacy and duration of use are related to the amount of copper the device bears and many devices will provide up to 10 years of contraception. Insertion of an IUD should be performed by trained health-care staff.

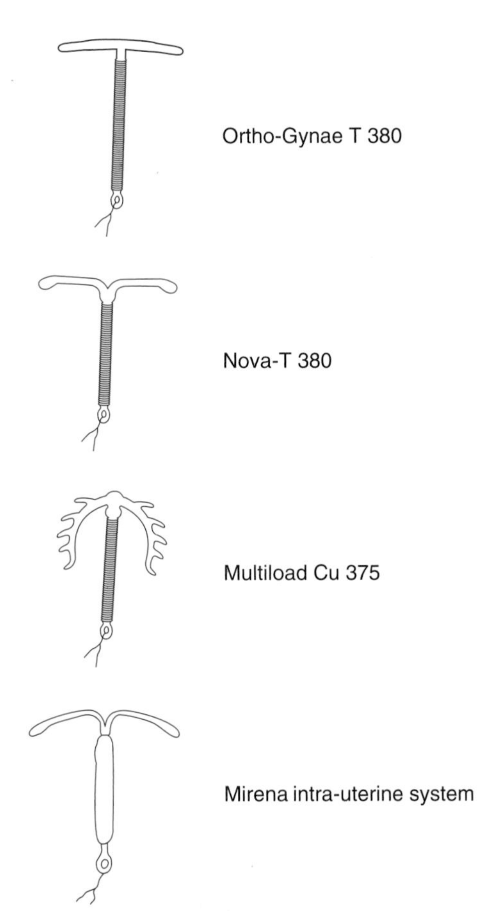

Figure 2 Commonly used intrauterine methods of contraception. Source: Glasier A, Gebbie A (eds.) (2008) *Handbook of Family Planning and Reproductive Healthcare*, 5th edn. Edinburgh, UK: Churchill Livingstone.

Mode of Action

The main mode of action of a copper-bearing IUD is to prevent fertilization by the toxic effects of copper ions on gametes. In addition, an IUD causes an inflammatory reaction of the endometrium that inhibits implantation.

Side Effects and Risks

The risk of uterine perforation at the time of insertion is around 1 in 1000. The risk of spontaneous expulsion of an IUD is around 1 in 20 in the first 3 to 6 months after insertion. An IUD is associated with a menstrual change, with around a 25% increase in menstrual blood loss and increased dysmenorrhea.

Pelvic infection: There is a slight increase in risk of pelvic infection in the first few weeks following insertion of an IUD. Thereafter, the overall risk of pelvic infection in IUD users is not increased significantly compared to women using no contraception. If a woman with an IUD contracts a sexually transmitted infection (STI) then the IUD may enhance the spread of ascending infection in the pelvic organs. The decision whether women are to be screened for STIs prior to insertion of an IUD should be based on prevalence of infection within the community and the individual's own risk factors.

Ectopic pregnancy: Women with an IUD are protected against all pregnancies, including ectopic pregnancies (pregnancies implanted abnormally, outside the uterine cavity). However, should the method fail, the risk of ectopic pregnancy is around three times higher than observed in pregnancies that occur without an IUD.

Intrauterine System

The intrauterine system (IUS), also known as the levonorgestrel-releasing IUD, releases 20 μg of levonorgestrel per day from a central hormone reservoir in the stem of the device (see **Figure 2**). It exerts its potent contraceptive effect by causing profound endometrial atrophy and making cervical mucus hostile to sperm transport. Ovulation is not inhibited. The IUS is associated with a profound reduction in menstrual blood loss and most long-term users will be oligo-amenorrheic (infrequent and scant menstrual bleeds). In addition to contraception, the IUS is also licensed in many countries for treatment of menorrhagia and as the progestogen component of hormone replacement regimens.

Barrier Methods

Male Condom

The male condom has been used in various forms for centuries. Most condoms are made of latex rubber but condoms made of both plastic and animal material are available. Allergy to latex condoms is a not uncommon problem in either partner. Various sizes and shapes of condoms are manufactured. Some individuals dislike the altered sensation of wearing a condom during sexual intercourse and the necessity to interrupt the sexual act to put the condom on. It is important that condoms are manufactured to stringent standards and many countries use a recognizable mark indicating production to an accepted quality. Out-of-date condoms are more likely to tear or burst during use.

Failure rates of condoms vary and are related to how consistently and diligently couples use them. The main benefit of using condoms as a method of contraception is the additional protection they offer against STI and, provided the condom is used prior to any genital contact, then the level of protection against infection is very high. Condoms have been heavily promoted in Safe Sex campaigns across the globe, particularly with respect to preventing HIV infection.

Female Condom

The female condom (**Figure 3**) is a lubricated polyurethane sheath that is designed to lie loosely within the vagina. There is a flexible polyurethane ring at the open, outer end and a second polyurethane ring inside the closed, inner end of the condom, which helps to insert the condom and anchor it within the vagina. It offers significant protection against STIs. Designed for single use, it can be inserted either immediately or several hours prior to intercourse. Occasionally, the condom can be displaced or pushed right up into the vagina with penetration taking place outside the condom (i.e., between the vaginal wall and wall of the female condom). It is relatively expensive and many individuals dislike the concept and aesthetics of the female condom. If no fresh condom is available, a used female condom may be reused for a limited number of times provided it is carefully washed and disinfected.

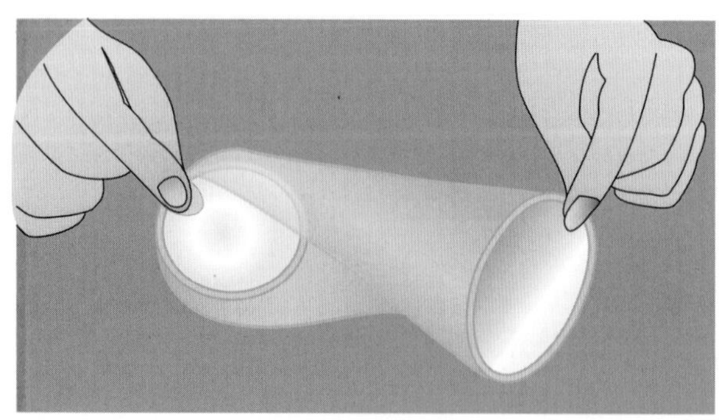

Figure 3 Female condom. Source: Glasier A, Gebbie A (eds.) (2008) *Handbook of Family Planning and Reproductive Healthcare*, 5th edn. Edinburgh, UK: Churchill Livingstone.

Diaphragms

A contraceptive diaphragm consists of a thin latex rubber hemisphere with a reinforced rim of a flat or coiled metal spring (**Figure 4**). It is designed to cover the cervix by lying across the upper vagina. Diaphragms are made in various sizes ranging from 50–95 mm in diameter and a woman should be fitted with a diaphragm by a trained health-care professional. A diaphragm should always be used in conjunction with spermicide to improve efficacy as it does not have a tight seal around it in the vagina. Diaphragms offer protection against pelvic inflammatory disease and cervical cancer but do not protect against HIV or other lower genital tract infections.

Cervical Caps

The traditional cervical cap is a thimble-shaped rubber device that is smaller than the diaphragm and is secured directly onto the cervix with a suction effect (see **Figure 4**). It comes in various sizes and should also be used with a spermicide to increase efficacy. It may be dislodged during intercourse without the woman noticing. Other cervical caps (e.g., Lea's Shield and FemCap) are available in different countries and may be manufactured in plastic, which allows them to remain in position for longer periods.

Spermicides

A range of spermicidal products is available comprising creams, gels, pessaries, foams, sponges, and dissolving squares of film. The active constituent is the chemical nonoxynol-9, which has toxic effects on sperm cell membranes. Spermicides also offer some degree of protection against most STIs but not against HIV. In general, they should be used in conjunction with other barrier methods as they have an unacceptably high failure rate if used on their own. There is concern that spermicidal preparations

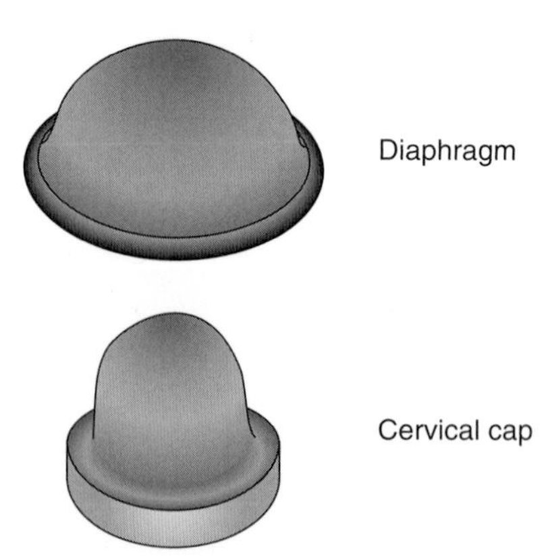

Figure 4 Diaphragm and cervical cap. Source: Glasier A, Gebbie A (eds.) (2008) *Handbook of Family Planning and Reproductive Healthcare*, 5th edn. Edinburgh, UK: Churchill Livingstone.

can have an irritant effect on vaginal and penile tissues, which may enhance the risk of HIV transmission. As a result, it is recommended that individuals at high risk of HIV infection do not use spermicides. In addition, for this reason, condoms that are prelubricated with spermicide are no longer recommended.

Fertility Awareness Methods

There are only a limited number of days in each menstrual cycle when fertilization can occur and, using this knowledge, it is possible to predict when intercourse should be avoided if a couple wishes to avoid pregnancy. The success of this approach is totally dependent on very careful adherence to the period of abstinence that many individuals find restrictive. Alternatively, individuals can choose to use a barrier method of contraception during the fertile period.

These methods may be practiced by couples who have religious or moral objections to other contraceptive methods, have no consistent access to modern methods, or who prefer 'natural' methods. They require careful teaching of the method and good comprehension by the couple of the significance of the symptoms and signs of ovulation.

There are various ways of detecting the fertile period:

Calendar or rhythm method: This method calculates fertile days by subtracting 18 from the length of the shortest cycle and 11 from the length of the longest cycle. These numbers then represent the start and the end of the fertile period, respectively. In some countries a set of colored beads is used to help women keep track of the cycle and fertile times. Each color-coded bead represents a single day in the cycle, and the color of a particular bead on a particular day indicates whether the woman is likely to be fertile or not (**Figure 5**).

Temperature method: This method monitors the small rise in basal body temperature that occurs after ovulation. Couples should abstain from intercourse during the first part of the cycle until 3 days after this rise; this method does not identify the beginning of the fertile period.

Cervical mucus method or Billings method: This method involves monitoring changes in cervical mucus that occur in the normal menstrual cycle. Fertile cervical mucus is clear and 'stretchy.' Abstinence should be maintained from the day the fertile mucus is first seen or felt in the vagina until 3 days after the final day it is detected.

Multiple index methods: These methods combine various indicators of fertility and, while this undoubtedly improves efficacy, the method does require particular diligence on the part of the user.

Ovulation predictor kits: These kits are available and monitor urinary hormone levels. The computer detects 6–10 days each month when intercourse should be avoided.

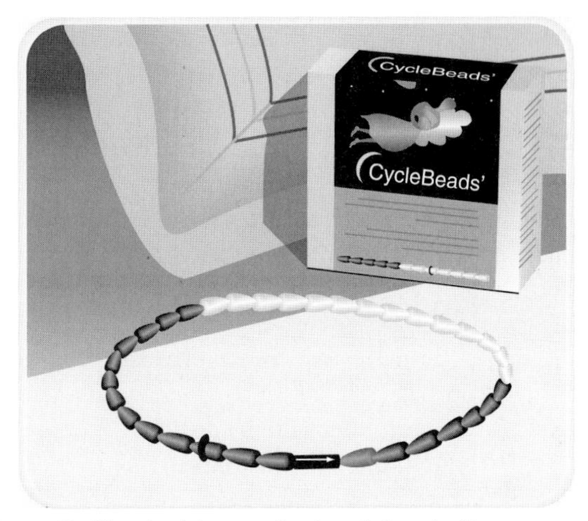

Figure 5 Standard days method cycle beads. Source: Glasier A, Gebbie A (eds.) (2008) *Handbook of Family Planning and Reproductive Healthcare*, 5th edn. Edinburgh, UK: Churchill Livingstone.

Lactational Amenorrhea Method

Women who are exclusively breastfeeding, are amenorrheic, and have a baby under the age of 6 months have an extremely low risk of pregnancy. Women in this situation can rely on lactational amenorrhea as a contraceptive method but should start using another method once supplemental foods are introduced for the baby, once 6 months have passed since delivery, or if menstruation returns. The uptake and success of this method depend on the pattern and duration of breastfeeding and, in many developed countries, women introduce formula milk and solid food at an early stage.

Withdrawal

This is the oldest method of contraception, first described in the Bible. Withdrawal (or coitus interruptus) involves withdrawing the erect penis from the vagina before ejaculation occurs, thus avoiding the passage of sperm into the female genital tract. The method requires discipline on the part of the male and can be reasonably effective if the man can recognize the imminence of ejaculation. Pre-ejaculatory secretions containing sperm may escape from the urethra before withdrawal.

Sterilization Procedures

Female Sterilization

Female sterilization involves a surgical procedure to occlude the Fallopian tubes thereby preventing sperm reaching and fertilizing the ovum. It must be performed by trained health-care professionals and can be carried out under local or general anesthesia. The choice of technique and type of anesthesia are often dictated by local availability of equipment and training of personnel more than by patient preference.

Timing of the procedure

The procedure can be performed at any time in the menstrual cycle, although care should be taken to ensure that the woman is not at risk of having been fertilized just before the procedure. Immediate postpartum sterilization can be performed and may be the only option available in developing countries or remote communities. Post-abortion sterilization can also be undertaken, but this is associated with a higher rate of regret and is therefore generally best deferred until the woman has an opportunity to plan her reproductive future at a less stressful time.

Techniques

Various methods can be used to block the Fallopian tubes (**Figure 6**). Clips made of metal and plastic, silastic rings, diathermy, and laser ablation are all in common use. The Fallopian tubes can also be surgically divided with excision of a small portion of tube.

A female sterilization procedure can be carried out by both laparoscopy and mini-laparotomy. Laparoscopy is usually done under general anesthesia but can be done under spinal or local anesthesia. Mini-laparotomy involves a 2–3 cm suprapubic incision and avoids the need for expensive laparoscopy equipment. It is the procedure of choice in the immediate postpartum period when the uterus is large.

Essure is a nonsurgical sterilization technique that is available in North America and some European countries.

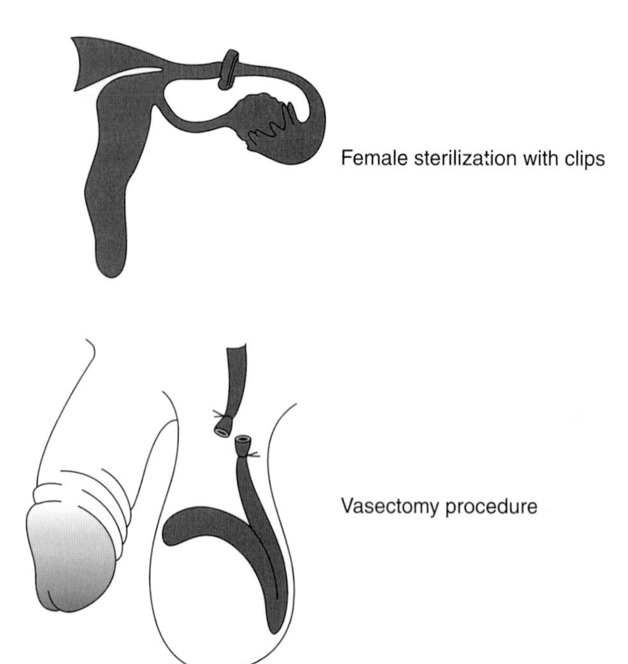

Figure 6 Female sterilization and vasectomy. Source: Glasier A, Gebbie A (eds.) (2008) *Handbook of Family Planning and Reproductive Healthcare*, 5th edn. Edinburgh, UK: Churchill Livingstone.

Under hysteroscopic guidance, a metal coil is inserted into each Fallopian tube via a slim catheter. Scar tissue develops around the coils and results in occlusion of the Fallopian tubes. It takes 3 months to ensure occlusion of the tubes and initial failure rates appear very low.

Clinical management

Women undergoing sterilization must be carefully counseled about the irreversibility of the procedure and the very small failure rate. A consent form explaining these important points and requiring the woman's signature is recommended. Reversal of female sterilization is technically possible but success rates vary according to the type of procedure performed initially. Successful reversal rates are highest when the tubes have been occluded with clips or rings. Reversal procedures are technically complex and expensive operations.

Failure of female sterilization is associated with an increased risk of ectopic pregnancy; therefore, women with symptoms of pregnancy following sterilization must seek medical advice to exclude an ectopic pregnancy.

Male Sterilization

Male sterilization (vasectomy) involves division or occlusion of both vasa deferentia and is a more effective procedure than female sterilization. It is also much safer and simpler to perform and is undertaken under local anesthesia in most men.

Technique

The standard technique involves making a single midline or two incisions, one on each side, to expose the vas. The vas on each side is then divided surgically or occluded by clips or diathermy (see **Figure 6**). No-scalpel vasectomy uses a special instrument to puncture the skin and expose the vas. It is quicker and less traumatic to the surrounding tissues so is associated with less bruising and discomfort.

Clinical management

Signed consent is usually obtained for the procedure, explaining the permanent nature of vasectomy and its very small failure rate. In contrast to female sterilization, vasectomy is not effective immediately and it takes around 3 months to clear sperm from the vas beyond the site of the division or occlusion. Contraception must be continued during this time and until two consecutive semen samples have been found to be devoid of sperm.

Uncommonly, men may experience chronic pain following vasectomy which may be the result of sperm granulomas developing as a local inflammatory response at the cut ends of the vas. Late canalization of the vas deferens can occasionally occur many years following vasectomy but is rare.

Some studies have suggested an association between vasectomy and subsequent development of both testicular and prostatic cancer and of cardiovascular disease.

This has not been substantiated and these small studies have not influenced clinical practice.

Emergency Contraception

Emergency contraception is a back-up method used following unprotected intercourse or when another method has failed. It is commonly used when condoms have burst, but can also be used when contraceptive pills have been omitted.

Hormonal Emergency Contraception

The most modern regimen involves a single dose of 1.5 mg of the hormone levonorgestrel up to 72 hours following the unprotected episode of intercourse (some practitioners recommend it up to 120 hours but efficacy is highest when taken in the first 24 hours following intercourse and declines over time). Side effects are uncommon. It is available in more than 100 countries on medical prescription and also over-the-counter in some countries. The mode of action of hormonal emergency contraception is not clear but disruption of ovulation is the most likely mechanism. Hormonal emergency contraception probably only prevents around three-quarters of pregnancies that would otherwise have occurred so should not be used as a regular method of contraception.

The older Yuzpe regimen using combined contraceptive pills involved giving two doses, each of ethinylestradiol (100 μg) and levonorgestrel (0.5 mg), 12 hours apart and was associated with a significant degree of nausea and vomiting. In China, the antiprogestogen mifepristone is marketed as an additional emergency contraceptive.

Postcoital Intrauterine Device

A copper-bearing IUD can be inserted following unprotected intercourse and appears to be highly effective in preventing pregnancy. The effectiveness of the levonorgestrel-releasing system (LNG-IUS) when inserted postcoitally has not yet been studied and hence this device should not be used for this indication. A device can be inserted up to 5 days after the anticipated day of ovulation, despite multiple episodes of unprotected intercourse, without fear of causing abortion. The device can be removed following the next menstrual period or can remain *in situ* as the ongoing method of contraception. Women should be carefully assessed for risk of STI prior to insertion of a postcoital IUD.

Conclusion

A range of methods exists that allow couples to regulate their fertility. While permanent sterilization is available for both men and women, there are only a few reversible methods for use by men, namely condoms and withdrawal, and both are among the less effective methods of contraception. For women, a range of reversible methods allows a choice between barriers, intrauterine contraceptives, and hormonal methods, with or without estrogen, and deliverable by a variety of routes (oral, injectable, etc.).

All methods are very safe for healthy women and a robust classification of medical eligibility helps providers choose suitable methods for women with preexisting medical conditions.

Method choice is influenced by a long list of factors but is vital to successful use of contraception. While methods vary in their effectiveness, and while those that rely little or not at all on compliance are the most effective, correct, consistent, and continued use is most likely when people use the method they find most acceptable.

Use of contraception allows couples (and individuals) to choose whether, and when, to have children and contributes enormously to sexual and reproductive health. In a recent article, Cleland *et al.* (2006) described family planning as being:

> unique among medical interventions in the breadth of its potential benefits: reduction of poverty, and maternal and child mortality; empowerment of women by lightening the burden of excessive childbearing; and enhancement of environmental sustainability by stabilising the population of the planet (Cleland *et al.*, 2006: 1810).

See also: Female Reproductive Function; Gender Aspects of Sexual and Reproductive Health; Infertility; Male Reproductive Function; Reproductive Ethics: Perspectives on Contraception and Abortion; Reproductive Rights; Sexual and Reproductive Health: Overview; Trends in Human Fertility.

Citations

Beral V, Hermon C, Kay C, *et al.* (1999) Mortality associated with oral contraceptive use: 25-year follow up of cohort of 46,000 women from Royal College of General Practitioners' oral contraceptive study. *British Medical Journal* 318: 96–100.

Cleland J, Bernstein S, Ezeh A, *et al.* (2006) Family planning: The unfinished agenda. *Lancet* 368: 1810–1827.

Collaborative Group on Hormonal Factors in Breast Cancer (1996) Breast cancer and hormonal contraceptives: Collaborative reanalysis of individual data of 53,297 women with breast cancer and 100, 239 women without breast cancer from 54 epidemiological studies. *Lancet* 347: 1713–1727.

Glasier A and Gebbie A (eds.) (2008) *Handbook of Family Planning and Reproductive Healthcare,* 5th edn. Edinburgh, UK: Churchill Livingstone.

International Collaboration of Epidemiological Studies of Cervical Cancer (2007) Cervical cancer and hormonal contraceptives: Collaborative reanalysis of individual data for 16,573 women with cervical cancer and 35,509 women without cervical cancer from 24 epidemiological studies. *Lancet* 370: 1609–1621.

Population Reference Bureau (2004) Transitions in world population. *Population Bulletin* 59: 1–40.

Trussell J (2004) Contraceptive efficacy. In: Hatcher RA, Trussell J, Stewart F, *et al.* (eds.) *Contraceptive Technology,* 18th rev. edn., pp. 773–845. New York: Ardent Media.

World Health Organization (1998) *Cardiovascular Disease and Steroid Hormone Contraception. Technical Report Series 877*. Geneva, Switzerland: World Health Organization.
World Health Organization (2004) *Medical Eligibility Criteria for Contraceptive Use*. 3rd edn. Geneva, Switzerland: World Health Organization.
World Health Organization (2005) *Selected Practice Recommendations for Contraceptive Use*. 2nd edn. Geneva, Switzerland: World Health Organization.
World Health Organization Department of Reproductive Health and Research (WHO/RHR) and Johns Hopkins Bloomberg School of Public Health/Center for Communications Programs (CCP), INFP Project (2007) *Family Planning: A Global Handbook for Providers*. Baltimore, MD and Geneva, Switzerland: CCP and WHO.

Further Reading

Glasier A (ed.) (2006) Contraception and hormone replacement therapy. *Medicine International* 34(1): 1–34.
Glasier A and Winikoff B (2005) *Fast Facts Contraception*. 2nd edn. Oxford, UK: Health Press.
Hatcher RA, Trussell J, Stewart F, *et al.* (2004) *Contraceptive Technology*. 18th rev. edn. New York: Ardent Media.
Mishell D (ed.) (2007) A special edition on intrauterine devices. *Contraception* 75(supplement 1): S1–S117.

Relevant Websites

http://www.ffprhc.org.uk – Faculty of Family Planning and Reproductive Healthcare in the UK.
http://www.infoforhealth.org/cire/cire_pub.pl – The INFP Project and Continuous Identification of Research Evidence (CIRE) system.
http://www.popcouncil.org – Population Council in the USA.
http://www.who.int/reproductive-health/publications/MEC/index – World Health Organization Medical Eligibility Criteria for Contraceptive Use (WHOMEC).
http://www.who.int/reproductive-health/publications/spr/index – World Health Organization Selected Practice Recommendations for Contraceptive Use.

Gynecological Morbidity

M Garefalakis and M Hickey, University of Western Australia, Western Australia, Australia
N Johnson, University of Auckland, Auckland, New Zealand

Introduction

Vaginal bleeding is considered abnormal if the bleeding is prolonged (greater than 7 days' duration), the flow is greater than 80 ml per cycle (or subjective impression of heavier than normal flow), and the cycle length less than 24 days or greater than 35 days. Intermenstrual, postcoital, and postmenopausal bleeding are also considered abnormal.

Abnormal vaginal bleeding is usually attributed to a uterine source, but may arise from problems at multiple anatomic sites including the lower genital tract (vulva, vagina, cervix) and upper genital tract (Fallopian tubes and ovaries, as well as the uterus). In addition, the origin of bleeding may include nongynecologic sites, such as the urethra, bladder, perianal region, and bowel.

It may be due to normal physiologic processes (such as the anovulatory cycles, which are common in the first postmenarchal years and during the perimenopause) or may be a manifestation of pregnancy and its complications, endocrine dysfunction, trauma, infection (e.g., endometritis), medicines (e.g., anticoagulants), contraceptives (e.g., hormonal contraception, intrauterine devices), hemostatic defects, systemic illnesses (e.g., uremia, advanced liver disease, acute leukemia), or neoplasia. Dysfunctional uterine bleeding describes abnormal bleeding that cannot be attributed to anatomic, organic, or systemic lesions or disease. The likelihood of the various causes is dependent upon the age of the woman and the pattern of bleeding (ovulatory or anovulatory).

Coagulation disorders may present with abnormal uterine bleeding, particularly menorrhagia. In one series of women aged 18–45 years, bleeding disorders were diagnosed in 10.7% of women with menorrhagia (von Willebrand disease, coagulation factor deficiencies, and platelet dysfunction) but in only 3.2% of controls without abnormal vaginal bleeding (Dilley *et al.*, 2001).

von Willebrand disease is a common inherited bleeding disorder and should be suspected in women who present with menorrhagia from menarche, if associated with anemia, requirement for hospitalization, or if there is a family history of coagulopathy.

Amenorrhea

Amenorrhea is the condition of having no menstrual periods. Primary amenorrhea describes the absence of menses (before menarche), and secondary amenorrhea describes the cessation of menses (after menarche). Oligomenorrhea is the condition of having infrequent menstrual periods.

The current recommendations for the timing of the evaluation of primary amenorrhea recognizes the trend to earlier age at menarche and is therefore indicated when there has been a failure to menstruate by age 15 years in the presence of normal secondary sexual development or within 5 years after breast development if that occurs before age 10 years.

The absence of development of secondary sexual characteristics by age 13 years also requires investigation.

It is advised that secondary amenorrhea lasting three months and oligomenorrhea involving less than nine cycles a year should be investigated (Practice Committee of the American Society for Reproductive Medicine, 2006).

There is a wide range of causative factors for amenorrhea (see **Figure 1**), but often no clear cause can be determined. The possibility of pregnancy should always be considered, and a comprehensive history and thorough examination should be conducted. Age-appropriate amenorrhea over the age of 45 years is consistent with menopause (average age 51 years).

Laboratory tests (including gonadotropins, prolactin, and thyroid-stimulating hormone) and imaging (pelvic or sella turcica) may be indicated. If there is evidence of hyperprolactinemia or thyroid dysfunction, appropriate investigation and management should restore menses.

Low gonadotropin levels (i.e., hypogonadotropic hypogonadism) suggest an abnormality of the pituitary or hypothalamus but a common cause in teenagers is constitutional delay, which is often familial. Other rare causes include central nervous system tumors, cranial irradiation, hypothalamic or pituitary destruction, Sheehan syndrome, and Kallmann syndrome (which is associated with anosmia). Hypothalamic amenorrhea is often caused by excessive weight loss, strenuous exercise, stress, or chronic illness. Young athletes may develop a combination of conditions called the 'female athlete triad' that includes eating disorder, amenorrhea, and osteoporosis. Menses will usually return after a healthy body weight is achieved with a modest increase in caloric intake and/or a decrease in training. It is important for adolescents and young women to avoid bone loss during this period of peak bone mass development.

Elevated gonadotropin levels (follicle-stimulating hormone (FSH) and luteinizing hormone (LH)) suggest an ovarian abnormality (hypergonadotropic hypogonadism). Repeatedly elevated gonadotropin levels in women aged under 40 years are defined as premature ovarian failure (POF), which affects 0.1% of women by 30 years and 1% of women by 40 years of age. It may be caused by chemotherapy, radiotherapy, galactosemia, and mumps oophoritis or be associated with autoimmune disorders (hypothyroidism, diabetes mellitus, Addison disease/adrenal insufficiency) and genetic disorders (including gonadal dysgenesis such as Turner syndrome (45,XO karyotype) and fragile X syndrome). Women under 30 years should have karyotype analysis. Some 5–10% of women with POF may achieve a natural conception, but women need to be made aware of options for assisted conception including ovum donation. Women with POF may be at increased risk of osteoporosis and heart disease.

Amenorrheic women with normal gonadotropin levels may have androgen insensitivity syndrome (XY karyotype; importantly, the undescended testes in this condition should be removed because of the high risk of malignant transformation after puberty), obstruction of menstrual blood outflow (congenital or acquired), or hyperandrogenic chronic anovulation (most common cause is polycystic ovarian syndrome (PCOS); see below). Müllerian (genital tract) abnormalities causing amenorrhea include congenital absence or abnormal development of the vagina and uterus. Imperforate hymen or transverse vaginal septum may be suspected if there is cyclic abdominal pain from accumulation of blood in the uterus and vagina. Asherman syndrome (i.e., intrauterine synechiae and scarring, usually

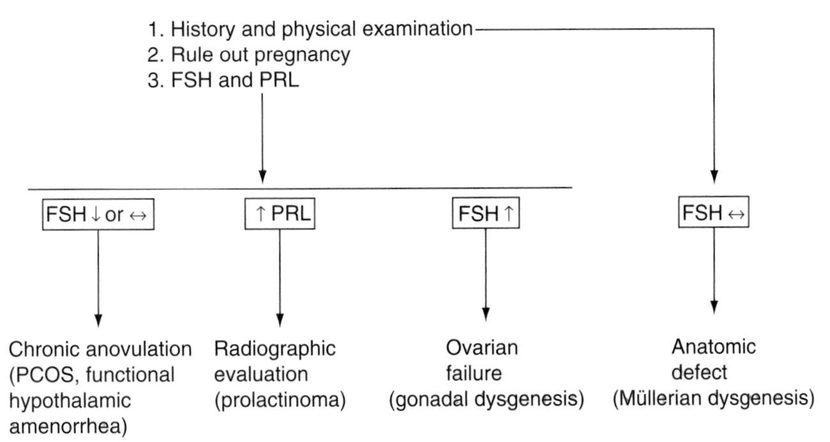

Figure 1 Suggested flow diagram aiding in the evaluation of women with amenorrhea. Adapted from Practice Committee of the American Society of Reproductive Medicine (2006) Current evaluation of amenorrhea. *Fertility and Sterility* 86 (supplement 5): S148–S155.

from endometrial curettage or infection) and cervical stenosis are the most common causes of acquired outflow obstruction. Diagnosis of Asherman syndrome is made by hysteroscopy, hysterosalpingography or sonohysterography, and management involves hysteroscopic resection of the adhesions under ultrasonographic, fluoroscopic, or laparoscopic guidance. This is followed by treatment aimed at reducing the reformation of adhesions, including insertion of an intrauterine device or catheter balloon and/or estrogen therapy, although there is limited evidence for efficacy.

Causes of hyperandrogenic anovulation include polycystic ovarian syndrome (PCOS), androgen-secreting ovarian or adrenal tumors, Cushing disease, acromegaly, non-classic congenital adrenal hyperplasia, and administration of exogenous androgens.

Polycystic Ovarian Syndrome (PCOS)

Polycystic ovarian syndrome (PCOS) is the most common reproductive disorder affecting women, thought to affect about 6% of women in the reproductive age group, and varies according to ethnicity (Azziz *et al.*, 2004). The rate of incidental finding of polycystic ovaries on ultrasound has been reported as 22% (Polson *et al.*, 1988; Lowe *et al.*, 2005).

Demographic data about the prevalence of PCOS and research into this common disorder have been hampered by the existence of conflicting definitions in past years. Controversy persists regarding the criteria for PCOS diagnosis, but a unifying definition was proposed in 2003 at a joint consensus meeting between the American Society of Reproductive Medicine and the European Society of Human Reproduction and Embryology. This definition encompasses a description of the ultrasound appearance of polycystic ovaries, infrequent or absent ovulation, and clinical or biochemical evidence of excess androgen levels (such as hirsutism and acne) while emphasizing the importance of ruling out other disorders with a similar presentation (e.g., congenital adrenal hyperplasia, Cushing syndrome, hyperprolactinemia, and androgen-secreting tumors) (Rotterdam ESHRE/ASRM-Sponsored PCOS Consensus Workshop Group, 2004).

It is being increasingly recognized that infrequent menstrual periods in adolescence may be one of the first manifestations of PCOS. Although PCOS is generally not diagnosed until more than two to three years after menarche, it is believed that its origins lie in fetal life or childhood, since excess androgen exposure of primates *in utero* produces PCOS-like features. Due to the familial clustering of PCOS cases, it is believed that there is a genetic predisposition to PCOS whereby genomic variants act under the influence of environmental factors, particularly obesity, and produce insulin resistance.

PCOS has significant implications for women, including the increased risk of anovulatory infertility, increased risk of miscarriage and pregnancy-related complications, and, in later life, increased risk of Type II diabetes mellitus, cardiovascular disease, and endometrial hyperplasia and cancer.

Evidence suggests that the adverse features of PCOS can be ameliorated with lifestyle intervention, such as diet and exercise.

Available treatments for women desiring contraception include the combined oral contraceptive pill (those with less androgenic progestogens may be desirable) and anti-androgenic medications such as cyproterone acetate and spironolactone. Treatments for subfertility include ovulation induction with anti-estrogens (e.g., clomiphene citrate) gonadotropins and ovarian drilling. Ovarian drilling may be useful when there has been an exaggerated response to fertility drugs.

While benefits related to ovulation may be derived from treatment with insulin-altering agents such as metformin, there is current uncertainty about long-term use of metformin in women with PCOS to protect against adverse cardiovascular outcomes and to reduce pregnancy-related complications.

Heavy Menstrual Bleeding (Menorrhagia)/ Dysfunctional Uterine Bleeding

Heavy menstrual bleeding (HMB) or menorrhagia is clinically defined as greater than or equal to 80 ml blood loss per menstrual cycle. In clinical practice it is impractical to measure blood loss, and the woman's perception of her menstrual loss is the key determinant in management decisions. HMB has profound social, financial, and medical implications. It is a common cause of anemia and lost working days and it may impair quality of life.

Underlying factors may include uterine abnormalities and other pelvic pathology, endocrine abnormalities, and, occasionally, coagulation disorders. These should be excluded before decisions are made about treatment as they may require different management. In most cases no pathological cause can be identified and the abnormal bleeding is labeled 'dysfunctional uterine bleeding.'

Although patient satisfaction with hysterectomy is high, this major surgical procedure is not without significant morbidity and occasional mortality. Effective medical therapy is an attractive alternative. A wide variety of medications are available, but few can guarantee a prolonged reduction in bleeding for the majority of women. The National Institute for Health and Clinical Excellence guideline on heavy menstrual bleeding (2007) makes recommendations on a range of treatment options.

Medical Treatments

Anti-fibrinolytic drugs (plasminogen activator inhibitors), such as tranexamic acid, work by inhibiting the breakdown of blood clots (fibrinolysis). These drugs reduce menstrual loss by 40–50% but do not generally alleviate cramping, which commonly accompanies HMB. They are taken only during menstruation and are usually well tolerated but can cause mild nausea and diarrhea (Lethaby *et al.*, 2000).

Non-steroidal anti-inflammatory drugs (NSAIDs) such as mefenamic acid and naproxen act by inhibiting the production of prostaglandins. Limited data suggest that they are more effective than placebo in reducing heavy menstrual bleeding but less effective than tranexamic acid, danazol, or the levonorgestrel-releasing intrauterine system (LNG-IUS). They can cause gastrointestinal disturbances, though serious side effects are unlikely as they are taken only for a short period during menstruation (Lethaby *et al.*, 2007).

There is some evidence that the combined oral contraceptive pill (COCP) significantly reduces menstrual blood loss and relieves cramping; in addition to inhibiting ovulation, combined contraceptive pills induce endometrial regression. Common side effects include nausea, headache, breast tenderness, and alterations in libido and mood.

Progestogens such as norethisterone and medroxyprogesterone acetate are ineffective if taken only during the luteal phase of the menstrual cycle (between ovulation and menstruation), but may reduce HMB if taken continuously.

The LNG-IUS releases a low dose of progestogen into the endometrium for at least five years. This profoundly suppresses the endometrium but does not inhibit ovulation (**Figure 2**). The device also provides contraception, with fertility returning when the system is removed. It is as effective as endometrial ablation (see below) and will reduce menstrual blood loss by 94% after three months (Lethaby *et al.*, 2005a).

The device is well accepted by most women but frequently causes irregular vaginal bleeding or spotting, especially during the first few months of use. Some women experience hormonal side effects such as breast tenderness and bloating and, occasionally, the device is expelled spontaneously. Functional (follicular) ovarian cysts may occur transiently.

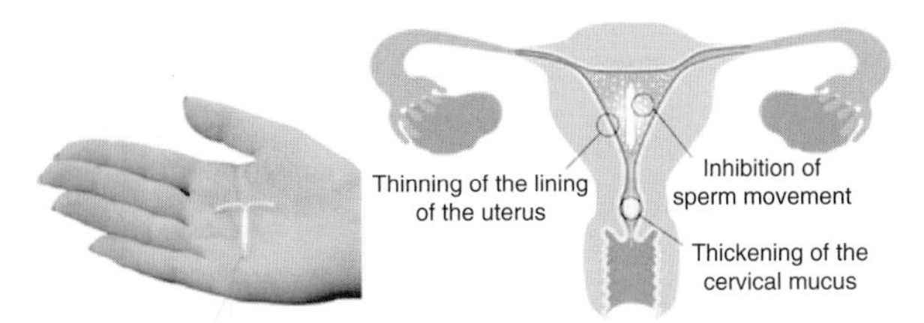

Figure 2 LNG-IUS shown in hand; LNG-IUS *in utero* and mechanisms of contraceptive action. Reproduced with permission from Bayer Schering Pharma AG. All rights reserved by Bayer Schering Pharma AG 2008.

Other medications such as danazol (a synthetic testosterone derivative) and Zoladex (a gonadotropin-releasing hormone (GnRH) analogue) may effectively treat HMB but are generally used only in short-term treatment due to the high incidence of side effects such as weight gain and acne (with danazol) and headache, nausea, tiredness, and menopause-like symptoms (with Zoladex). Prolonged use of GnRH analogues such as Zoladex may also lead to bone demineralization and has cost implications.

Surgical Treatments

Surgery may be indicated for women who have completed childbearing for whom medical treatment is ineffective or intolerable, or it may be chosen as first-line therapy. Surgery may aim to conserve or remove the uterus.

Hysterectomy (removal of the uterus) has traditionally been regarded as the definitive surgical treatment for HMB and is one of the most commonly performed surgical procedures. Menstrual disorders are the leading indication for hysterectomy. Hysterectomy can be performed abdominally, vaginally, or laparoscopically; when possible, the vaginal route is preferred since it leads to shorter recovery time and fewer complications than the abdominal or laparoscopic routes. However, hysterectomy by any route has a relatively high incidence of short-term complications such as hemorrhage, infection, thromboembolism, and wound healing problems, and often requires a lengthy postoperative recovery period. Nevertheless, hysterectomy is 100% successful in treating HMB and for most women any problems are relatively short-term. Satisfaction rates after hysterectomy are very high, at over 95% up to three years after surgery (Lethaby *et al.*, 1999).

There are widely varying hysterectomy rates among different regions, countries, and gynecologists. It has been reported that the rate of hysterectomy for benign uterine conditions is decreasing in some countries, which may be associated with the widespread use of LNG-IUS (Reid and Mukri, 2005; Jacobson *et al.*, 2006).

Less-invasive surgical options, which conserve the uterus, include endometrial resection or ablation, and involve the destruction by various means of the endometrium (the inner lining of the uterus) and the underlying basal glands. The procedure may be preceded by a course of hormonal medication to thin the endometrium in order to facilitate its removal. Some techniques for endometrial destruction utilize a surgical telescope (hysteroscope) to aid viewing of the uterus, along with a variety of electrosurgical or laser tools. These techniques usually require a general or regional anesthetic, specialized surgical skill, and sometimes a short hospital admission. They are significantly safer than hysterectomy but still involve a small risk of uterine perforation, hemorrhage, fluid overload, and infection, the short-term complication rate being around 4% (Lethaby *et al.*, 2005b). Non-hysteroscopic 'second generation' techniques have been developed that utilize the controlled application of heat, cold, microwave, or other forms of energy to destroy the endometrium. They require sophisticated equipment but can usually be performed as day surgery under local anesthetic. However, women often continue to have some bleeding and, in the longer term (around four years), over one-third go on to receive further surgical treatment of some kind, usually hysterectomy, for continued bleeding and/or pain.

Endometrial destruction is quicker than hysterectomy, recovery is faster, and there are fewer postoperative complications. These techniques are not contraceptive, and if pregnancy occurs it is often complicated. Endometrial ablation is similar to the LNG-IUS in terms of efficacy and satisfaction, but the LNG-IUS is the most cost-effective in the short term.

Benign Gynecological Tumors

Endometrial Polyps

Endometrial polyps are nonmalignant, pedunculated or sessile growths in the lining of the uterus. They occur in up to 10% of women and are most commonly diagnosed in the 5th decade. There is uncertainty about their natural history and pathological significance. Established risk factors for endometrial polyps are late menopause and the use of unopposed estrogen or tamoxifen.

Endometrial polyps are generally small (a few mm) but may enlarge to fill the uterine cavity. They may be asymptomatic or be associated with abnormal bleeding, particularly intermenstrual bleeding, menorrhagia, and abnormal perimenopausal bleeding. Up to 25% of women with abnormal bleeding patterns have an endometrial polyp. They may be detected on direct examination (if they protrude through the cervix), ultrasound, saline infusion sonohysterography (a technique utilizing endovaginal ultrasound and instillation of sterile saline simultaneously through the cervical canal), hysteroscopy, or curettage. They are managed through removal by curettage or operative hysteroscopy. All polyps should be examined histologically, as up to 0.5% will be hyperplastic or malignant.

Cervical Polyps

Cervical polyps are benign, pedunculated growths of varying size that extend from the ectocervix or endocervical canal.

The etiology of cervical polyps is unknown. They are believed to result from chronic inflammation and 99% are benign. Cervical polyps are most common among multiparous women in their 30s and 40s. They are usually asymptomatic, but women may present with vaginal discharge and/or postcoital, intermenstrual, or postmenopausal bleeding. Cervical polyps should be removed and sent for histological evaluation.

Fibroids (Uterine Leiomyomata)

Fibroids are the most common benign tumors in women of childbearing age. Fibroids are lumps made of muscle cells and other tissue that grow within, or are attached to, the wall of the uterus (**Figure 3**). They may be asymptomatic but may also lead to heavy or painful periods, prolonged bleeding, pelvic pain, or feelings of pressure and 'fullness' in the lower abdomen as well as reproductive problems, such as infertility, recurrent miscarriage, or preterm labor. Around 30% of women of childbearing age have clinically symptomatic uterine fibroids.

The cause of uterine fibroids is not known. However, genetic, hormonal, immunological, and environmental factors may play a role in starting the growth of fibroids and/or in promoting that growth. Some racial groups are at increased risk, such as those of African or Asian origin. The incidence increases with age, but fibroids regress after menopause.

Fibroids may be detected on gynecological examination by the palpation of an enlarged, irregular uterus or with the aid of imaging techniques, such as ultrasonography, sonohysterography, or magnetic resonance imaging (MRI).

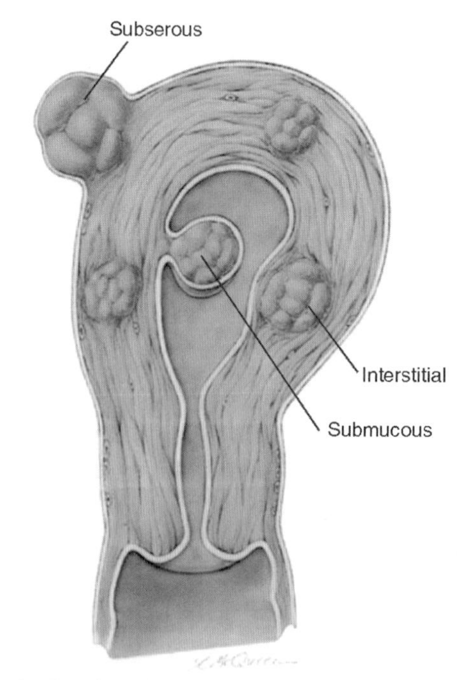

Figure 3 Drawing of cut surface of uterus showing characteristic whorl-like appearance and varying locations of leiomyomas. Reproduced with permission from Novak ER and Woodruff JD (eds.) (1967) *Novak's Gynecologic and Obstetric Pathology*, 6th edn., p. 215. Philadelphia, PA: WB Saunders.

Treatment depends on the severity of symptoms, desire for future pregnancies, general health, and characteristics of the fibroid(s). Asymptomatic fibroids generally do not require treatment. Mild symptoms may be helped by non-steroidal anti-inflammatory drugs (NSAIDs), such as ibuprofen and naproxen. Major symptoms may be treated medically or surgically. Medical treatments are only effective for the duration of treatment and may include GnRH agonists (GnRHa), LNG-IUS, or antiprogestogens, such as mifepristone. Surgical therapy may be performed transvaginally, laparoscopically, or as an open procedure.

Uterine artery embolization (UAE) aims to shrink fibroids by removing or reducing their blood supply. UAE reduced heavy menstrual bleeding associated with fibroids by around 85% (range 60–90%) and reduced fibroid volume by around 30–46% (Gupta *et al.*, 2006). New conservative treatments being studied include cryomyolysis (using liquid nitrogen to 'freeze' the fibroids to make them shrink) and myolysis (using electric current to destroy the fibroids and reduce their blood supply).

Adenomyosis

Uterine adenomyosis is defined as the presence of endometriumlike tissue in the muscular layer of the uterus (myometrium). The incidence of adenomyosis is unclear, but the condition is more common in women with endometriosis (see below). Diagnosis has traditionally been histopathological (following hysterectomy) but may now be possible using newer imaging techniques, such as MRI.

It is unclear to what extent adenomyosis causes symptoms. The commonly quoted presenting symptoms are painful and heavy menstrual periods, although many women are asymptomatic. Pelvic examination may reveal a symmetrically enlarged, tender uterus, but endometrial assessment should be considered in all women with abnormal bleeding.

Because the disease has been infrequently diagnosed prior to hysterectomy, there are few well-designed studies of medical or surgical management. Hysterectomy or insertion of the LNG-IUS are the most common treatments.

Endometriosis

Endometriosis is the presence of endometriumlike tissue at sites outside the uterus. The ectopic endometrial tissue responds cyclically to estrogen by proliferating and producing local inflammation.

The etiology of endometriosis is unknown, although several theories have been proposed. Implantation of eutopic endometrium that migrated from the uterine cavity via the Fallopian tubes is a popular theory.

The prevalence of endometriosis is unclear, but it may affect around 7–10% of women in the general population and around 38% of infertile women, and up to 80% of women with chronic pelvic pain. Endometriosis shows familial clustering, but a clear genetic cause has not yet been determined.

Endometriosis may be asymptomatic, but commonly presents with dysmenorrhea, pelvic pain, dyspareunia, intermenstrual bleeding, and infertility. Other symptoms include perimenstrual back pain, dyschezia (pain on passage of feces), abdominal pain, urinary symptoms, or cyclic rectal bleeding. Pain is the most troublesome symptom, but the severity of pain relates poorly to the visible extent of the disease. The reason why endometriosis causes pain is incompletely understood, but may be related to the location and depth of the endometriotic implant, with deep implants in highly innervated areas most consistently associated with pain.

Pelvic examination may appear completely normal if lesions are small and few. In more advanced disease, cervical displacement of 1 cm or more to the left or right of the midline and nodularity of the uterosacral ligaments and posterior cul-de-sac may be detected. Clinical examination during menstruation may better enable the practitioner to identify tender nodules of endometriosis. Fixed retroversion of the uterus and adnexal masses may also be palpated.

Definitive diagnosis requires inspection +/− histological confirmation from biopsy of endometriotic lesions. Disease severity may be classified using a standardized system, such as the classification system of the American Society for Reproductive Medicine, which was most recently revised in 1996. Unfortunately, this system does not correlate well with symptom severity or prognosis (American Society for Reproductive Medicine, 1997).

There are a variety of medical and surgical therapeutic options. Medical management should be considered initially in women with mild to moderate symptoms who wish to maintain their fertility potential and/or avoid surgery. NSAIDs during menses may relieve dysmenorrhea, and there are various hormonal medical therapies aimed at interrupting the cycle of stimulation and bleeding. These include the combined oral contraceptive pill, progestogens including the LNG-IUS, gestrinone or danazol (a synthetic testosterone derivative now rarely used owing to its high incidence of androgenizing side effects), and GnRH agonists (GnRHa) with add-back therapy (including hormone replacement, progestogens, tibolone, and bisphosphonates). The various medications are comparable in their ability to provide relief from dysmenorrhea, dyspareunia, and pelvic pain associated with endometriosis, but are not of proven value in improving fertility.

The preoperative and postoperative use of medical therapies such as GnRHa and danazol to decrease the extent of endometriosis and the size of endometriomas (ovarian endometriosis) is of unproven benefit, and is associated with side effects (e.g., hot flushes or vaginal dryness). Growing evidence suggests that the postoperative use of the LNG-IUS reduces the risk of recurrent symptoms and of recurrence of endometriotic lesions following surgical removal.

Surgical therapy to remove endometriotic deposits improves symptoms and fertility rates. The aim is to remove all visible disease and to restore normal anatomy by procedures such as the division of adhesions and removal of scar tissue. The laparoscopic approach is preferable and is associated with fewer complications than open surgery.

Total abdominal hysterectomy and bilateral salpingo-oophorectomy has previously been considered the definitive treatment of endometriosis, but the enlightened view of surgical therapy is that removal of endometriotic lesions is what is required. Presacral neurectomy has been shown to be an effective adjunct to surgical removal of endometriosis (but carries a high complication rate) for relief of pain; laparoscopic uterine nerve ablation (LUNA) has not been proven to significantly add to the benefit of laparoscopic removal of endometriosis.

Unless the ovaries are removed and the patient rendered permanently hypoestrogenic, endometriosis may recur, and appropriate follow-up intervals depend on the choice of therapy and the woman's needs.

Adnexal Pathologies

The adnexa is the region adjoining the uterus that contains the ovary and Fallopian tube, as well as associated vessels, ligaments, and connective tissue. Pathology in this area may also be related to the uterus, bowel, or retroperitoneum, or metastatic disease from another site.

Adnexal Masses

Masses in the adnexa occur at all ages and may be symptomatic or discovered incidentally. Some will regress spontaneously; others require a surgical procedure for histologic diagnosis and treatment. The differential diagnosis of an adnexal mass discovered in women of reproductive age is broad and includes physiologic or functional cysts, ectopic pregnancy, inflammatory etiologies such as a tubo-ovarian abscess, endometrioma, benign and malignant ovarian neoplasms, or neoplasms metastatic to the ovary.

The reported prevalence varies widely depending upon the population studied and the criteria employed. A large series examining 1101 subjects with adnexal masses found that 80% of neoplasms in women younger than 55 years of age were hormone-dependent (functional cysts or endometriomas), whereas 8% were benign neoplasms (teratomas, cystadenomas, or leiomyomas) and only 4 (0.4%) were carcinomas (Nezhat *et al.*, 1992).

In one study of 1769 asymptomatic postmenopausal women, 116 (6.6%) had simple ovarian cysts on ultrasound. Women with persistent cysts underwent surgery and no malignancies were identified (Conway *et al.*, 1998).

The ovaries are organs that are anatomically difficult to access during routine physical examination. Bimanual pelvic examination is not an effective method for identifying cases of adnexal mass at an early stage. Ultrasound is currently the most efficacious and practical study for evaluating an adnexal mass.

The most serious concern when an adnexal mass is discovered is the possibility that it is malignant. There is currently no good screening test to distinguish a benign from a malignant adnexal mass. Characteristics that increase the likelihood of malignancy include: age less than 15 years or postmenopausal status, a complex or solid-appearing mass on ultrasound, presence of ascites, elevated tumor markers in peripheral blood (e.g., CA 125) or a history of nongenital cancer (e.g., breast or gastric cancer). Benign adnexal masses typically have the following ultrasonographic features: low echogenicity, a thin cyst wall, unilocular or, if septated, a thin septation, and absence of internal papillary excrescences. Surgery and histopathological examination is often required for definitive diagnosis and therapy.

Adnexal Torsion

Adnexal torsion is the twisting of part or all of the adnexa on its mesentery. It usually involves the ovary but may also involve the Fallopian tube.

Torsion accounts for 2–3% of gynecological emergencies and usually occurs in the reproductive age group (including during pregnancy), but is not uncommon before menarche or in postmenopausal women. It occurs because of the mobility of the supporting pedicles, infundibulopelvic and ovarian ligaments. During torsion, venous blood flow followed by arterial flow is impaired. This leads to congestion, adnexal edema, ischemia, and necrosis.

Torsion of a normal-sized ovary or of an ovary with cysts smaller than 5 cm is rare. Risk factors include the presence of large and heavy ovarian cysts, such as benign cystic teratoma (dermoid cyst), and tubal pathology including hematosalpinx, hydrosalpinx, and para-ovarian cysts.

Presentation is usually due to a sudden onset of continuous lower abdominal pain. Nausea and vomiting are quite common, occurring in 70% of those affected, and fever may also be present. Delay in diagnosis is common, with the time span from the onset of symptoms to operation reported to vary from hours to several days.

There are no specific laboratory findings but the white cell count may be elevated. Ultrasound findings and Doppler flow studies may be helpful but, even though reduced or absent adnexal vascular flow is suggestive of torsion, the diagnosis cannot be based solely on the absence or presence of blood flow on color Doppler sonography (Mazouni *et al.*, 2005).

Awareness and suspicion is needed in patients who present with the classical findings of lower abdominal pain associated with nausea, vomiting, and presence of an adnexal mass. Urgent surgical intervention (usually by the laparoscopic approach) is indicated to prevent irreversible ovarian and/or tubal damage.

In most cases an ovarian tumor is present. Complete arterial obstruction does not usually occur, and some blood supply is obtained from either the ovarian or uterine arteries. The misleading ischemic-hemorrhagic, black-bluish appearance of the adnexa is a result of venous and lymphatic stasis rather than gangrene. The duration of ischemia causing irreversible damage is unknown.

Despite the 'necrotic' appearance of the twisted ischemic ovary, detorsion is the only procedure that should be performed at surgery, and removal of the injured organ should only be performed if there are obvious signs of adnexal disruption, such as ligament detachment or ovarian tissue decomposition. Several studies have shown that estimation of the degree of necrosis during surgery is inaccurate. Studies assessing subsequent ovarian function by observing follicular development on ultrasound, the macroscopic appearance of the adnexa at subsequent surgery, and fertilization of oocytes retrieved during *in vitro* fertilization have demonstrated that ovarian function is preserved in 88–100% of cases (Oelsner *et al.*, 2003).

Ovarian cystectomy during detorsion of the black-bluish ischemic adnexa should also be avoided because handling of the edematous, friable, and ischemic adnexa may cause additional damage to the ovary, and a high proportion of the ovarian masses are functional cysts, which should not be removed. Patients with true ovarian tumors should be operated upon electively four to six weeks later. Recurrent torsion may be managed with ovarian fixation. The incidence of pulmonary embolism after torsion is 0.2% and is not increased after detorsion.

Bartholin Gland Cyst/Abscess

The Bartholin glands are two pea-sized, mucus-secreting glands located on either side of the vaginal vestibule (vaginal opening). A Bartholin gland cyst results from obstruction of the duct leading to the outside, which results in dilatation of the gland. Obstruction of the duct may occur as a result of infection, inspissated mucus, or trauma; abscess formation may occur.

Bartholin gland cysts may be asymptomatic or present with unilateral swelling of the labia, tenderness, purulent discharge, dyspareunia, pain when walking or sitting, and enlarged inguinal lymph nodes. Secondary infection is common and microbiological swabs of any discharge should be taken and tests for bacterial infection including *Neisseria gonorrheae* should be performed. Abscess formation may be recurrent.

Differential diagnosis includes inclusion cyst, sebaceous cyst, primary cancer of Bartholin gland or duct (especially among women older than 40 years), cancer metastasis, lipoma, fibroma, hernia, and hydrocele.

Treatment of inflammation or abscess formation should include analgesia and intervention to effect permanent duct drainage and prevention of further abscess formation. Appropriate antibiotic treatment should be provided if gonorrhea is identified. Surgical interventions include incision, drainage, and insertion of a Word catheter (a bulb-tipped inflatable catheter that remains in the cyst for several weeks) or marsupialization (the creation of a pouch).

Chronic Pelvic Pain (CPP)

Chronic pelvic pain (CPP) in women is a common and difficult problem. The true incidence and prevalence is difficult to ascertain and the number of population-based studies from less-developed countries is low (Latthe *et al.*, 2006). Many gynecological pathological conditions are more frequent in women with CPP, but the development of a chronic pelvic pain syndrome is often multifactorial.

Chronic pelvic pain syndromes are usually characterized by a duration of 6 months or longer; incomplete relief with most treatments; significantly impaired function at home and/or work; signs of depression, such as early awakening, weight loss, or anorexia; and altered family roles.

There are many recognized antecedent causes of CPP including both gynecological (e.g., endometriosis, chronic pelvic infections, ovarian cysts, ovarian remnant syndrome, residual ovary syndrome, fibroids, adenomyosis, vulvodynia) and nongynecological (e.g. interstitial cystitis, chronic urinary tract infections, irritable bowel syndrome, constipation, myofascial pain, disc disease, nerve entrapment syndromes, psychological stress, substance abuse, history of physical or sexual abuse).

Clinical evaluation and management should include medical, surgical, and psychological assessment. Pathology affecting the reproductive tract, surrounding viscera, and musculoskeletal system may coexist. Because of the strong association of sexual abuse with reported pelvic pain, there should be screening for the possibility of past or present physical or emotional abuse. Depression, sleep disturbance, and sexual dysfunction may develop in response to prolonged pain and need to be addressed. CPP may best be managed within a multidisciplinary model.

More detailed information on the pathophysiology, etiologies, investigation, and management of CPP is available in the Society of Obstetricians and Gynecologists of Canada clinical practice guidelines (2005).

Premenstrual Syndrome

Premenstrual syndrome (PMS) and the more severe variant, premenstrual dysphoric disorder (PMDD), are characterized by the presence of physical, behavioral, and/or emotional symptoms that occur repetitively in the second half of the menstrual cycle (luteal phase), resolve at or soon after the beginning of the menstrual period, and are severe enough to interfere with some aspects of the woman's life. Premenstrual symptoms are common, affecting up to 75% of women; however, clinically significant PMS occurs in 20–30% of women and PMDD in 3–8% of women with regular cycles.

The most common physical manifestations are abdominal bloating, which occurs in 90% of women with this disorder, and breast tenderness and headaches, which occur in more than 50% of cases. PMDD can be differentiated from PMS by the presence of at least one affective symptom, such as anger, irritability, or internal tension.

The origin of PMS is not fully understood but appears to arise from an interaction of cyclic changes in ovarian steroids with central neurotransmitters.

First-line management may include regular exercise, relaxation techniques, and vitamin and mineral supplementation, although there is little evidence to support the effectiveness of these interventions.

Controlled clinical trials of progesterone, spironolactone, evening primrose oil, essential fatty acids, and *Ginkgo biloba* have failed to show any consistent benefit for the treatment of physical or emotional symptoms of PMS. Two recent placebo-controlled trials reported that a combined oral contraceptive (containing drospirenone and ethinylestradiol) administered in a 24-day regimen rather than a 21-day regimen was more effective than placebo in reducing some premenstrual symptoms. Further studies are needed (Pearlstein *et al.*, 2005).

Selective serotonin reuptake inhibitors (SSRIs) have been shown to reduce or eliminate disturbing emotional symptoms for many women, and are generally well-tolerated, often at doses significantly lower than those required to treat depression. Drug therapy is generally recommended as first-line management of women with PMDD because of the severity of symptoms associated with this condition.

Diethylstilbestrol (DES) Exposure

Diethylstilbestrol (DES) was used during the mid-1900s to treat pregnancy complications, including bleeding, premature labor, diabetes, and preeclampsia. DES use was discontinued in 1971 when *in utero* exposure was linked to the occurrence of vaginal clear cell adenocarcinoma. It has been estimated that 5–10 million women and infants were prescribed DES during pregnancy or were exposed to the drug *in utero.*

Women who took DES appear to have an increased incidence of breast cancer and are advised to follow current guidelines for breast cancer screening.

Women exposed to DES *in utero* may develop sequelae including structural genital tract changes, such as transverse vaginal and cervical ridges, an abnormally shaped uterine cavity, uterine hypoplasia, vaginal adenosis, and an increased incidence of miscarriage, ectopic pregnancy, infertility, and preterm birth due to cervical incompetence and premature rupture of membranes. There is also an increased incidence of clear cell adenocarcinoma of the cervix or vagina (incidence rises at age 15 years and median age at diagnosis is 19 years) and an increased incidence of breast cancer. It is advised that these women be educated about these possible sequelae and encouraged to have yearly pelvic examinations from menarche onward, including careful inspection of the cervix and vagina with cervical and vaginal cytology. Any abnormalities warrant referral to an expert colposcopist. Multigenerational studies are being conducted, but there is no good evidence that the grandchildren of women who received DES during pregnancy are at increased risk for abnormalities (Palmer *et al.*, 2005; Rubin, 2007).

Pelvic Organ Prolapse

Pelvic organ prolapse (POP) is defined as the descent of one or more of the following: anterior vaginal wall, posterior vaginal wall, or the apex of the vagina (cervix or uterus) or vault (after hysterectomy).

The pelvic organs are supported by a complex interplay between the uterosacral/cardinal ligament complex, the levator ani muscles, and the endopelvic fascia. These structures attach the pelvic organs to the bony pelvis and form a continuous and interdependent organ complex. Damage to any of the muscles and fibrous structures results in descent of the vaginal walls and pelvic organs. Different types of prolapse may exist, and the site at which the muscle or fascia is damaged determines the type of prolapse present.

Rectocele, enterocele, cystocele, and urethrocele, respectively, describe protrusion of the rectum, small bowel/peritoneum, bladder, and urethra into the vaginal lumen. Uterine prolapse is the downward protrusion of the cervix and uterus, with procidentia describing total prolapse beyond the hymen. Vault prolapse is eversion of the vagina and can occur after hysterectomy.

Some 10–15% of all women and as many as 50% of women older than 50 years have some degree of POP. The Women's Health Initiative reported 34% of women had anterior vaginal wall prolapse, 19% had posterior vaginal wall prolapse, and 14% had uterine prolapse (Hendrix *et al.*, 2002).

POP results from the loss of normal support mechanisms, involving weakening of the pelvic support connective tissue and muscles as well as nerve damage. Risk factors include multiparity, obesity, chronic cough, heavy lifting, and intrinsic tissue weakness.

POP may be asymptomatic or cause significant symptoms, such as a dragging sensation, a lump ('something coming down'), or protrusion of pelvic organs into or through the vagina. The dragging sensation is usually relieved by lying down, is less noticeable in the morning, and worsens as the day progresses. Women may also describe lower backache and dyspareunia and the need to digitally replace the prolapse to void or defecate.

Vaginal examination allows the description of observed and palpable abnormalities and assessment of pelvic floor muscle function. Severity may be graded using the Pelvic Organ Prolapse Quantification system (POP-Q), a standardized system that can facilitate communication between health practitioners and aid research.

Treatment is individualized according to the severity of symptoms, degree of prolapse, the woman's previous treatment, activity level, and health status, and whether or not sexual function is to be maintained. Nonsurgical treatment options include pelvic floor muscle training, behavioral modification, and the use of intravaginal devices such as pessaries. Follow-up after fitting a pessary involves re-evaluation of the appropriate size and type of pessary and ensuring that ulceration, irritation, or exacerbation of incontinence has not occurred.

Various surgical methods have been described and are performed via an open abdominal, laparoscopic, or vaginal approach. The aims are to relieve symptoms and, in most cases, to restore vaginal anatomy. Prognosis depends upon the severity of symptoms, extent of the prolapse, surgeon's experience, and patient expectations. Surgery has traditionally been associated with a recurrence/re-operation rate of up to 30%. Simultaneous surgery for concomitant urinary or fecal incontinence may also be considered.

Urinary and Fecal Incontinence

Continence depends upon having intact lower urinary tract, anorectal, pelvic, and neurologic components and the functional ability to access and use toilet facilities. This includes sufficient mobility and manual dexterity, and the cognitive ability to recognize and react appropriately to sensations of bladder and rectum filling. The presence of incontinence often reflects multiple factors.

Urinary Incontinence

Urinary incontinence is defined as 'the complaint of any involuntary leakage of urine' and is further characterized into the following categories:

- Stress urinary incontinence is the complaint of involuntary leakage on effort or exertion, or on sneezing or coughing (i.e., during periods of increased intra-abdominal pressure).
- Urge urinary incontinence is the complaint of involuntary leakage accompanied by or immediately preceded by urgency. Urgency is the complaint of a sudden compelling desire to pass urine, which is difficult to defer. Women may also describe increased daytime frequency and nocturia.
- Mixed urinary incontinence is the complaint of involuntary leakage associated with urgency and also with exertion, effort, sneezing, or coughing.
- Continuous urinary incontinence is the complaint of continuous leakage.
- Nocturnal enuresis is the complaint of loss of urine occurring during sleep.
- Other types may be situational.
- Overflow incontinence symptoms include difficulty in initiating urination, a weak or intermittent stream, and post-void dribbling.

Half of young women report occasional minor leakage of urine, while 5–10% of adult women have daily leakage that interferes with their daily activities. The severity tends to increase with age, and urge incontinence becomes more common. Symptoms should be sought because many of those affected do not report the problem to their health-care providers.

Risk factors for urinary incontinence in women include increasing parity, increasing age, obesity, high-impact physical activities, chronic cough, constipation, medications (e.g., sedatives, diuretics, anticholinergics, alpha-adrenergic blockers), diabetes mellitus, and functional impairment. Acute onset of symptoms suggests a urinary tract infection.

Complications include vulval cellulitis, perineal candidial infection, urinary tract infections, falls, disturbed sleep, social isolation, and psychological morbidity.

Investigations to identify the lower urinary tract pathology and any reversible exacerbating factors are performed according to symptoms and signs and may include urinalysis, urine cytology, frequency and volume charts, pad tests, and specialized tests called urodynamic studies.

Urodynamic investigations are a group of tests used to assess the function of the lower urinary tract and include cystometry, urethral pressure profile measurements, leak point pressure measurements, uroflowmetry, pressure-flow studies, electromyography, video urodynamics, and ambulatory monitoring. They may be invasive and time-consuming and are not universally available. They are not performed routinely and must be interpreted in the context of the entire clinical picture (Glazener and Lapitan, 2002).

It is preferable to commence treatment of incontinence with less invasive therapies. General management strategies that have been found helpful in many patients include adequate fluid intake (up to 2L per day), avoidance of caffeinated beverages and alcohol, minimizing evening fluid intake (if nocturnal incontinence), avoidance of constipation, weight loss, smoking cessation, and treatment of respiratory disease if cough is exacerbating stress or mixed-urge incontinence.

Various therapies, such as bladder retraining and pelvic muscle exercise, may be effective for several types of urinary incontinence, including urge, stress, and mixed incontinence. Other therapies include medications (e.g., antimuscarinic drugs for urge and mixed incontinence; side effects associated with the use of these drugs can be reduced by slow titration of dose and extended release preparations), continence pessaries (particularly if there is pelvic organ prolapse), various surgical techniques, and indwelling catheterization or diversion procedures as a last resort. Treatment with oral hormone replacement therapy was associated with increased incontinence in the Women's Health Initiative trial (Hendrix *et al.*, 2005) and the Heart and Estrogen/Progestin Replacement Study (Grady *et al.*, 2001), but vaginal estrogen may be helpful.

Surgery offers the highest cure rates for stress urinary incontinence, even in elderly women. Methods available include bladder neck suspension procedures, periurethral bulking injections, and sling procedures, such as tension-free vaginal tape.

Fecal Incontinence

Fecal incontinence is defined as the recurrent, involuntary loss of fecal material (solid feces, liquid stool, or flatus) in an individual older than 4 years of age. It can be an embarrassing and debilitating condition, which may lead to social isolation and loss of self-esteem. Prevalence has been reported between 1–21%, and this variation may be due to differences in symptom definition and the age of the population studied. Many women do not report their symptoms to their health-care providers (Boreham *et al.*, 2005).

Loss of fecal continence may result from dysfunction of the internal and/or external anal sphincters, abnormal rectal compliance, reduced rectal sensation, local benign and neoplastic pathologies, inflammatory bowel conditions, diarrhea, and fecal impaction. Risk factors include increasing age, female gender, vaginal delivery, anal sphincter injuries, anorectal surgery, and neurologic disorders such as diabetes mellitus, demyelinating conditions, and spinal cord injury.

Evaluation is guided by history and physical examination and may include stool cultures, mucosal evaluation, and specialized tests. Anorectal functional tests are not universally available and may include anorectal manometry, endo-anal ultrasonography, defecography, pudendal nerve terminal latency, and anorectal electromyography.

Medical treatment is aimed at reducing stool frequency and improving stool consistency. This may involve dietary changes, use of bulking agents, or antidiarrheal and other medications. Other therapies such as biofeedback and bowel training are recommended but have limited data to support their efficacy.

Surgical treatment may be considered in those who do not respond to conservative treatment. Approaches that may be considered include various methods of sphincter repair and sphincter replacement, surgery for rectal or hemorrhoidal prolapse, sacral nerve root stimulation, and fecal diversion procedures (Rudolph and Galandiuk, 2002).

Female Genital Tract Fistulae

A fistula is an abnormal connection or passageway between organs or vessels that are not normally connected.

Vesicovaginal fistulae (located between the bladder and the vagina) are the most common female genital tract fistulae, but they may occur between the vagina, uterus, bowel, urethra, ureters, or bladder.

Worldwide, the most common cause of pelvic fistulae is obstructed labor. Robust data are lacking but the World Health Organization (WHO) estimates that approximately 2 million women worldwide currently live with obstetric fistulae and that another 50 000 to 100 000 new cases occur each year (Department of Making Pregnancy Safer, 2006).

If cephalopelvic disproportion occurs during delivery, constant pressure from the fetal head on the soft tissues of the pelvis leads to tissue necrosis. This results in formation of fistulae, which are often large and complex. Access to family planning, skilled birth attendants, and modern obstetric care can prevent this problem, and the most common causes of female genital tract fistulae in developed countries are surgical injury, malignancy, and radiation therapy rather than obstructed delivery. Female genital circumcision and traumatic injury (e.g., as a result of botched abortions) may also cause fistulae. In recent years, thousands of women in eastern Congo have presented with traumatic fistula caused by systematic, sexual violence that occurred during the country's war.

Symptoms include uncontrolled loss of urine, flatus or feces, ulceration, infection, renal damage, and psychological morbidity. Many of these women are ostracized by family members and communities.

Pelvic examination with a speculum should always be performed in an attempt to locate the fistula(e) and assess the size and number of fistulae and associated inflammation or other pelvic pathology.

The presence of a vesicovaginal fistula can be confirmed by instilling a blue dye or sterile milk into the bladder per urethra and observing for discolored vaginal drainage. Various imaging techniques may be used to clarify the number and position of fistulae.

Genital tract fistulae are managed expectantly until the inflammatory process resolves. Spontaneous closure may occur but persistent fistulae require management by surgical repair or diversional procedures (e.g., urinary conduit or colostomy). Rates of success (defined as closure of the fistula) after reconstructive surgery have been reported to be as high as 90% for uncomplicated cases but less than 60% for complicated cases. Counseling and support are also important to address psychological morbidity and to facilitate social reintegration. Cesarean section is recommended for future pregnancies to prevent recurrence.

Female Genital Mutilation (FGM)

Female genital mutilation (FGM), also referred to as 'female circumcision' or 'female genital cutting,' comprises all procedures involving partial or total removal of the external female genitalia or other injury to the female genital organs, whether for cultural, religious, or other nontherapeutic reasons.

Most of the girls and women who have undergone FGM live in Africa, although some live in the Middle East and Asia. They are also increasingly found in Europe, Australia, and North America, primarily among immigrant populations.

To date, the number of girls and women who have undergone FGM is estimated at between 100–140 million. It is estimated that, each year, a further 2 million girls are at risk of undergoing FGM (Department of Reproductive Health and Research, 2006).

The WHO classification was first developed through a technical consultation in 1995. Such classification is valuable in research on the consequences of different forms of FGM, estimates of prevalence and trends of change, gynecological examination and management of health consequences, and for classification in legal cases. Limitations in the applicability of the 1995 classification were reported and the typology of classification was modified to better capture the variety of procedures (**Table 1**).

Immediate complications include severe pain, shock, hemorrhage, urinary retention, and death. Possible long-term consequences include cyst, abscess, fistula and keloid scar formation, voiding difficulties, urinary incontinence, urinary tract infections, dysmenorrhea, dyspareunia, psychological morbidity, sexual dysfunction, infertility, and obstructed vaginal delivery, leading to higher rates of cesarean delivery and greater need for resuscitation of the newborn, as well as increased perinatal mortality (WHO Study Group on Female Genital Mutilation and Obstetric Outcome, 2006).

Defibulation may be performed prior to coitus to prevent dyspareunia or prior to pregnancy or delivery to prevent

Table 1 Extract from the revised Joint Statement on the Elimination of Female Genital Mutilation. Geneva: World Health Organization (in press)

WHO FGM typology modified, 2007	*WHO FGM typology, 1995*
Type I: Partial or total removal of the clitoris and/or the prepuce (clitoridectomy)[a]	Type I: Excision of the prepuce, with or without excision of part or the entire clitoris
Type II: Partial or total removal of the clitoris and the labia minora, with or without excision of the labia majora (excision)[b]	Type II: Excision of the clitoris with partial or total excision of the labia minora
Type III: Narrowing of the vaginal orifice by creating a covering seal through the cutting and apposition of the labia minora and/or labia majora, with or without excision of the clitoris (infibulation)[c]	Type III: Excision of part or all of the external genitalia and stitching/narrowing of the vaginal opening (infibulation)
Type IV: Unclassified: All other harmful procedures to the female genitalia for nonmedical purposes, for example: pricking, piercing, incision, cauterization, and scraping	Type IV: Unclassified: pricking, piercing or incising of the clitoris and/or labia; stretching of the clitoris and/or labia; cauterization by burning of the clitoris and surrounding tissue; scraping of tissue surrounding the vaginal orifice (angurya cuts) or cutting of the vagina (gishiri cuts); introduction of corrosive substances or herbs into the vagina to cause bleeding or for the purpose of tightening or narrowing it; and any other procedure that falls under the definition given above

[a]Where it is important to distinguish between the major variations of Type I, typically for research purposes, we propose the following subdivisions: Type Ia, Removal of the clitoral hood/prepuce only; Type Ib, Removal of the clitoris with or without the prepuce.
[b]Where it is important to distinguish between the major variations that have been documented, typically for certain research purposes, we propose the following subdivisions: Type IIa, Removal of the labia minora only; Type IIb, Partial or total removal of the clitoris and of the labia minora; Type IIc, Partial or total removal of clitoris, labia minora, and labia majora. Note also that in French, the term 'excision' is often used as a general term covering all types of female genital mutilation.
[c]Where it is important to distinguish between variations in infibulation for research purposes, we propose the following subdivisions: Type IIIa, Removal and apposition of labia minora; Type IIIb, Removal and apposition of labia majora.

problems with vaginal birth. A major objective of WHO's work on FGM is to generate knowledge and develop interventions to promote the elimination of FGM.

See also: Female Reproductive Function; Gender Aspects of Sexual and Reproductive Health; Menopause; Reproductive Rights; Sexual and Reproductive Health: Overview.

Citations

American Society for Reproductive Medicine (1997) Revised American Society for Reproductive Medicine classification of endometriosis: 1996. *Fertility and Sterility* 67: 817–821.

Azziz R, Woods KS, Reyna R, *et al.* (2004) The prevalence and features of the polycystic ovary syndrome in an unselected population. *Journal of Clinical Endocrinology and Metabolism* 89: 2745–2749.

Boreham MK, Richter HE, Kenton KS, *et al.* (2005) Anal incontinence in women presenting for gynecologic care: Prevalence, risk factors, and impact upon quality of life. *American Journal of Obstetrics and Gynecology* 192: 1637–1642.

Conway C, Zalud I, Dilena M, *et al.* (1998) Simple cyst in the postmenopausal patient: Detection and management. *Journal of Ultrasound in Medicine* 17: 369–372.

Department of Making Pregnancy Safer (2006) *Obstetric Fistula: Guiding Principles for Clinical Management and Programme Development*. Geneva, Switzerland: World Health Organization. http://www.who.int/making_pregnancy_safer/publications/obstetric_fistula.pdf (accessed November 2007).

Department of Reproductive Health and Research (2006) Female genital mutilation: New knowledge spurs optimism. *Progress in Sexual and Reproductive Health Research* 72: 1–7. http://www.who.int/reproductive-health/hrp/progress/72.pdf (accessed November 2007).

Dilley A, Drews C, Miller C, *et al.* (2001) von Willebrand disease and other bleeding disorders in women with diagnosed menorrhagia. *Obstetrics and Gynecology* 97: 630–636.

Glazener CMA and Lapitan MC (2002) Urodynamic investigations for management of urinary incontinence in children and adults. *Cochrane Database of Systematic Reviews* 3: CD003195 [DOI: 10.1002/14651858.CD003195].

Grady D, Brown J, Vittinghoff E, *et al.* (2001) Postmenopausal hormones and incontinence: The Heart and Estrogen/Progestin Replacement Study. *Obstetrics and Gynecology* 97: 116–120.

Gupta JK, Sinha AS, Lumsden MA, and Hickey M (2006) Uterine artery embolization for symptomatic uterine fibroids. *Cochrane Database of Systematic Reviews* Issue 1: CD005073. DOI: 10.1002/14651858.CD005073.pub2.

Hendrix SL, Clark A, Nygaard *et al.* (2002) Pelvic organ prolapse in the Women's Health Initiative: Gravity and gravidity. *American Journal of Obstetrics and Gynecology* 186: 1160–1166.

Hendrix SL, Cochrane BB, Nygaard IE, *et al.* (2005) Effects of estrogen with and without progestin on urinary incontinence. *Journal of the American Medical Association* 293: 935–948.

Jacobson GF, Shaber RE, Armstrong MA, and Hung Y (2006) Hysterectomy rates for benign indications. *Obstetrics and Gynecology* 107: 1278–1283.

Latthe P, Latthe M, Say L, *et al.* (2006) WHO systematic review of prevalence of chronic pelvic pain: A neglected reproductive health morbidity. *BMC Public Health* 6: 177–183. http://www.pubmedcentral.nih.gov/articlerender.fcgi.

Lethaby A, Augood C, Duckitt K, and Farquhar C (2007) Nonsteroidal anti-inflammatory drugs for heavy menstrual bleeding. *Cochrane Database of Systematic Reviews* 4: CD000400 [DOI: 10.1002/14651858.CD000400.pub2].

Lethaby A, Cooke I, and Rees M (2005a) Progesterone or progestogens-releasing intrauterine systems for heavy menstrual bleeding. *Cochrane Database of Systematic Reviews* 4: CD002126 [DOI: 10.1002/14651858.CD002126.pub2].

Lethaby A, Farquhar C, and Cooke I (2000) Antifibrinolytics for heavy menstrual bleeding. *Cochrane Database of Systematic Reviews* 4: CD000249 [DOI: 10.1002/14651858.CD000249].

Lethaby A, Hickey M, and Garry R (2005b) Endometrial destruction techniques for heavy menstrual bleeding. *Cochrane Database of Systematic Reviews* 2: CD001501 [DOI: 10.1002/14651858.CD001501.pub2].

Lethaby A, Shepperd S, Cooke I, and Farquhar C (1999) Endometrial resection and ablation versus hysterectomy for heavy menstrual bleeding. *Cochrane Database of Systematic Reviews* 2: CD000329 [DOI: 10.1002/14651858.CD000329].

Lowe P, Kovacs G, and Howlett D (2005) Incidence of polycystic ovaries and polycystic ovary syndrome amongst women in Melbourne, Australia. *Australian and New Zealand Journal of Obstetrics and Gynaecology* 45: 17–19.

Mazouni C, Bretelle F, Menard JP, *et al.* (2005) Diagnosis of adnexal torsion and predictive factors of adnexal necrosis. *Gynécologie, Obstétrique et Fertilité* 33: 102–106.

National Institute for Health and Clinical Excellence (2007) Clinical guideline. *Heavy Menstrual Bleeding: Investigation and Treatment*. http://www.nice.org.uk/guidance/CG44/?c=91520 (accessed November 2007).

Nezhat F, Nezhat C, Welander CE, *et al.* (1992) Four ovarian cancers diagnosed during laparoscopic management of 1011 women with adnexal masses. *American Journal of Obstetrics and Gynecology* 167: 790–796.

Novak ER and Woodruff JD (eds.) (1967) *Novak's Gynecologic and Obstetric Pathology*, 6th edn., p. 215. Philadelphia, PA: WB Saunders.

Oelsner G, Cohen SB, Soriano D, *et al.* (2003) Minimal surgery for the twisted ischaemic adnexa can preserve ovarian function. *Human Reproduction* 18: 2599–2602.

Palmer JR, Wise LA, Robboy SJ, *et al.* (2005) Hypospadias in sons of women exposed to diethylstilbestrol in utero. *Epidemiology* 16: 583–586.

Pearlstein TB, Bachmann GA, Zacur HA, and Yonkers KA (2005) Treatment of premenstrual dysphoric disorder with a new drospirenone-containing oral contraceptive formulation. *Contraception* 72: 414–421.

Polson DW, Adams J, Wadsworth J, and Franks S (1988) Polycystic ovaries: A common finding in normal women. *Lancet* 1: 870–872.

Practice Committee of the American Society for Reproductive Medicine (2006) Current evaluation of amenorrhea. *Fertility and Sterility* 86 (supplement 5): S148–S155.

Reid PC and Mukri F (2005) Trends in number of hysterectomies performed in England for menorrhagia: Examination of health episode statistics, 1989 to 2002–3. *British Medical Journal* 330: 938–939.

Rotterdam ESHRE/ASRM-Sponsored PCOS Consensus Workshop Group (2004) Revised 2003 consensus on diagnostic criteria and long-term health risks related to polycystic ovary syndrome. *Fertility and Sterility* 81: 19–25.

Rubin MM (2007) Antenatal exposure to DES: Lessons learned . . . future concerns. *Obstetrical and Gynecological Survey* 62: 548–555.

Rudolph W and Galandiuk S (2002) A practical guide to the diagnosis and management of fecal incontinence. *Mayo Clinic Proceedings* 77: 271–275.

Society of Obstetricians and Gynaecologists of Canada (2005) SOGC Clinical Practice Guidelines: Consensus guidelines for the management of chronic pelvic pain, parts 1 and 2. *Journal of Obstetrics and Gynaecology Canada* 27: 781–801 and 869–887.

WHO Study Group on Female Genital Mutilation and Obstetric Outcome (2006) Female genital mutilation and obstetric outcome: WHO collaborative prospective study in six African countries. *Lancet* 367: 1835–1841.

Further Reading

Abrams P, Cardozo L, Fall M, *et al.* (2002) The standardisation of terminology of lower urinary tract function: Report from the standardisation sub-committee of the International Continence Society. *Neurology and Urodynamics* 21: 167–178.

Bhatia J and Cleland J (2000) Methodological issues in community-based studies of gynecological morbidity. *Studies in Family Planning* 31: 267–273.
Bump RC, Mattiasson A, Bo K, *et al.* (1996) The standardization of terminology of female pelvic organ prolapse and pelvic floor dysfunction. *American Journal of Obstetrics and Gynecology* 175: 10–17.
Cochrane Collaboration (2007) Menstrual disorders and subfertility. In: *Cochrane Reviews.* Oxford, UK: Wiley Interscience. http://www.cochrane.org/reviews/en/topics/76_reviews.html (accessed November 2007).
Department of Women's Health, Family and Community Health, WHO (2000) *A Systematic Review of the Health Complications of Female Genital Mutilation Including Sequelae in Childbirth*. Geneva, Switzerland: World Health Organization. http://www.who.int/gender/other_health/en/systreviewFGM.pdf (accessed November 2007).
Donnay F and Ramsey K (2006) Eliminating obstetric fistula: Progress in partnerships. *International Journal of Gynecology and Obstetrics* 94: 254–261.
Jarrell JF, Vilos GA, Allaire C, *et al.* (2005) Consensus guidelines for the management of chronic pelvic pain: Part one of two. *Journal of Obstetrics and Gynaecology Canada* 27: 781–826. http://www.sogc.org/guidelines/public/164E-CPG1-August2005.pdf (accessed November 2007).
Jarrell JF, Vilos GA, Allaire C, *et al.* (2005) Consensus guidelines for the management of chronic pelvic pain: Part two of two. *Journal of Obstetrics and Gynaecology Canada* 27: 869–887. http://www.sogc.org/guidelines/public/164E-CPG2-September2005.pdf (accessed November 2007).
Kurman RJ (ed.) (2002) *Blaustein's Pathology of the Female Genital Tract,* 5th edn. New York: Springer-Verlag.
Master-Hunter T and Heiman DL (2006) Amenorrhea: Evaluation and treatment. *American Family Physician* 73: 1374–1382.
Reslova T, Tosner J, Resl M, *et al.* (1999) Endometrial polyps: A clinical study of 245 cases. *Archives of Gynecology and Obstetrics* 262: 133–139.
Roenneburg ML, Genadry RG, and Wheeless CR (2006) Repair of obstetric vesicovaginal fistulas in Africa. *American Journal of Obstetrics and Gynecology* 195: 1748–1752.
Speroff L and Spitz MA (2005) *Clinical Gynecologic Endocrinology and Infertility*. 7th edn. Philadelphia, PA: Lippincott Williams & Wilkins.

Relevant Websites

http://www.nice.org.uk/guidance/CG44/?c=91520 – National Institute for Health and Clinical Excellence, Heavy Menstrual Bleeding, Clinical Guideline.
http://www.sogc.org/guidelines/public/164E-CPG1-August2005.pdf – Society of Obstetricians and Gynaecologists of Canada, SOGC Clinical Practice Guidelines, Consensus Guidelines for the Management of Chronic Pelvic Pain, Part 1 of 2.
http://www.sogc.org/guidelines/public/164E-CPG2-September2005.pdf – Society of Obstetricians and Gynaecologists of Canada, SOGC Clinical Practice Guidelines, Consensus Guidelines for the Management of Chronic Pelvic Pain, Part 2 of 2.

Sexually Transmitted Infections: Overview

B P Stoner, Washington University in St. Louis, St. Louis, MO, USA

Introduction

Sexually transmitted infections (STIs) are those conditions that are primarily spread through person-to-person sexual contact. They are of profound importance in terms of the immediate suffering caused to affected individuals, but also in terms of the public health impact and overall cost to humanity. Their magnitude is immense and underappreciated in policy-making circles. Even in areas where STIs are considered 'under control' they tend to persist in risk groups stratified by poverty, inequality, and general access to health and preventive services.

In general, experts distinguish between curable STIs caused by bacteria or protozoa (such as syphilis, gonorrhea, chlamydia, and trichomoniasis), and chronic STIs caused by viruses (such as genital herpes and human papillomavirus infection), which are treatable but not curable in the classic sense. Epidemiologically, the various STIs inhabit different sorts of infection networks that overlap to greater or lesser degree, but are slightly different from one another. Research investigators, clinicians, and public health officials now increasingly prefer the term STIs over the previously accepted and widely used term sexually transmitted diseases (STDs), since many STIs are asymptomatic and do not cause overt symptoms in the infected individual. The term STI captures the notion that persons may be infected, and therefore may represent a transmission risk to others, without suffering from a 'disease' in the classic sense.

By and large, Western, developed nations have developed and implemented STI surveillance and control programs to deal with community-level spread of infection, and these have been successful to greater or lesser degree depending on the particular STI and aspects of the social, cultural, and political environment in which they occur. Above all, STIs are malleable and adaptable to changing social norms of sexual contact and associated sexual behaviors.

Human immunodeficiency virus (HIV) is transmitted predominantly as an STI, but also is characterized by other routes of transmission such as bloodborne exposure or through breastfeeding. Two types of HIV are distinguished, HIV-1, which has a worldwide distribution and is responsible for the great majority of HIV infections, and HIV-2, which is geographically linked to West African origins. Recent research continues to emphasize the ways in which genital ulcer and mucosal inflammatory diseases facilitate and reinforce HIV transmission and acquisition across populations at risk (Fleming and Wasserheit, 1999). For this reason, many experts see STI prevention and control as an important part of a comprehensive HIV prevention strategy.

History and Epidemiology of STIs

Formerly called venereal diseases (after Venus, the goddess of love), STIs have afflicted human populations for centuries. Ancient Greek medical texts accurately described the clinical manifestations of gonorrhea (literally, 'flowing seed') and recognized a sexual route of transmission. Sixteenth-century Europe was ravaged by the Great Pox, later determined to be syphilis, and now widely believed to have been acquired from New World populations and transmitted across the Atlantic shortly after European contact. Military campaigns are intimately associated with STIs, often linked to young male soldiers acquiring infections through engagement with prostitution. Modern STI control strategies developed following World War I, as public health took a greater interest in community health preservation and the prevention of person-to-person spread. In the USA, this effort was most effectively spearheaded by Surgeon General Thomas Parran, whose monograph, *Shadow on the Land* (1937), helped establish surveillance systems for STIs and formal partner notification programs in local public health jurisdictions. Other countries also implemented surveillance and control programs for STIs, focusing on those conditions for which primary curative therapy could be implemented and distributed more widely. Bacterial STIs, such as syphilis, gonorrhea, and chlamydia, were added to the roster of reportable diseases, both to ensure that infected individuals receive appropriate therapy, and also to ensure that their sex partners are appropriately identified and treated as well.

Surveillance of STIs across time and space is a major public health responsibility. Tracking of STIs occurs through a combination of active and passive surveillance, and is most effectively implemented in developed countries with strong governmentally based public health structures in place. Nevertheless, STI incidence and prevalence remain incompletely tracked under current surveillance systems. In the USA, for example, chlamydia is the most commonly reported infectious condition, with nearly 1 million cases reported annually (Centers for Disease Control and Prevention, 2006a). However, this figure is a gross underestimate of the true burden of infection due to the asymptomatic nature of the disease, incomplete screening, and underreporting. A recent study estimated the incidence of STIs in the USA at approximately 18.9 million cases annually (Weinstock *et al.*, 2004). Globally, the situation is much bleaker. Focusing solely on curable STIs (syphilis, gonorrhea, chlamydia, and trichomoniasis), the World Health Organization (WHO) estimated an incidence of 340 million infections occurring annually among men and women 15–49 years of age (World Health Organization, 2001).

In general, viral STIs other than HIV are not reportable conditions in most jurisdictions, and surveillance for these infections is undertaken through population-based serologic testing. Viral STIs tend to be chronic, long-standing infections as they are not curable in the classic sense with antibiotic treatment. In the USA alone, prevalence estimates for genital tract infection with herpes simplex virus type 2 (HSV-2) and human papillomavirus (HPV) range from 11 to 33% of the sexually active population. Based on these figures, experts believe the true burden of viral STIs worldwide may exceed several hundred million persons. Evidence is also increasingly mounting that HSV-2 is a powerful cofactor for HIV transmission (Freeman *et al.*, 2006), and studies are underway to determine the extent to which treatment of STIs (including genital herpes) serves to reduce sexual transmission of HIV.

Unfortunately, surveillance for STIs other than HIV in many developing countries is rudimentary, so the true prevalence of STIs in most areas of the world is unknown.

Transmission Dynamics of STIs

A great conceptual leap forward in thinking about STIs was the development of the transmission dynamics model, first promoted in a series of articles in the late 1980s by Roy Anderson and Robert May (May and Anderson, 1987; Anderson and May, 1988). Building on prior work by Hethcote, Yorke, and other early modeling pioneers, the transmission dynamics model states that the reproductive rate of infection within a population (R_o) is a function of three parameters: the efficiency of transmission of the infectious organism during sexual contact (β), the rate of sex partner change among members of the population (c), and the duration of infectiousness (D). At equilibrium, $R_o = 1$, meaning that each case of infection generates one new case of infection and infection rates overall do not change. When the rate of infection is increasing within a population, then $R_o > 1$ (more people are entering the infected pool of individuals than are leaving it). Conversely, when infection rates are receding within a population, then $R_o < 1$ and more people are leaving the pool of infected individuals than are entering the pool (presumably through early detection, effective treatment, reducing risk behaviors, etc.).

The transmission dynamics model clarified two basic tenets which are essential to the public health approach to

STIs: (1) reducing STI rates is achieved by implementing program interventions to reduce β, c, and D of the model components; and (2) when interventions are implemented, the infection will contract toward hard-to-reach populations, which are relatively resistant to the effects of the intervention and which serve to perpetuate and sustain infection within the community at large. These hard-to-reach populations are termed core groups, and the emergence of core group theory is specifically linked to the conceptual model that Anderson and May first formulated.

Public health structures focus primarily on minimizing β and D from the transmission dynamics model. In general, β is a measure of how efficiently STI pathogens are transmitted from an infected person to a susceptible partner within sexual partnerships. Using condoms correctly and consistently for all sexual contacts is a way of reducing the likelihood of transmission of STI pathogens during sexual partnerships, and thus serves to reduce β at the population level when this behavior is adopted widely throughout the population. Similarly, use of vaccination to prevent HPV acquisition effectively reduces β by decreasing the partner's susceptibility to infection. D is reduced through early screening to detect indolent or asymptomatic infection, and provision of effective medication to cure or treat these infections. The screening of reproductive-age women to detect and treat chlamydial infection serves to reduce D when implemented broadly. Also, public awareness campaigns encouraging individuals to get tested for STI generally work to detect, and presumably treat, ongoing infections and reduce D by limiting these individuals' ability to transmit infection to others. Tackling parameter c has been quite a different issue.

The transmission dynamics model suggests that reducing the rate of partner change (later refined by Anderson and May to the effective mean rate of partner change) should lead to reduced spread of STI pathogens by limiting the number of susceptible individuals who are potentially exposed through sexual contact. However, public health officials have generally shied away from promoting what amounts to social engineering in the realm of sexual behavior and decision making, and have tended to avoid overt recommendations to reduce the rate at which persons acquire new partners.

The concept of core groups, facilitated by STI networks, helps explain how infections can persist within subsets of the population even as effective disease control and prevention strategies are implemented widely within a community. Aggressive screening, detection, and treatment of infection during an epidemic serves to drive the fewer remaining cases into harder-to-access populations, often marginal to mainstream society and with limited access to social service organizations. Core groups are subsegments of the general population for whom $R_o > 1$, that is, who generate more than one case of infection for each infected individual. Such persons may sustain sexual partnerships with high rates of transmission of infectious agents (elevated β), high rates of partner change (elevated c), and long durations of infectiousness (elevated D), thereby serving at the population level to keep R_o at equilibrium, even as the infection recedes throughout lower-risk segments of the community. Research in this area has also begun to clarify the nature of overlapping social and sexual networks, and the role of social networks in the spread of STI pathogens. Network studies of STI patients argue for the inclusion of social (nonsexual) network members in disease control investigations and outbreak interventions, presumably because these individuals are more likely to be engaged in similar sexual risk behaviors as the STI index case, even though they are not direct sexual contacts.

Classification of STI Pathogens

More than 30 sexually transmitted pathogens have been described, although the majority of STIs are caused by just a few different agents. New pathogens (e.g., *Mycoplasma genitalium*) continue to be identified or confirmed as sexually transmitted organisms, based on advances in laboratory testing methodologies. A useful classificatory scheme is to distinguish between curable STIs, generally caused by bacteria or protozoa and which can be treated with antimicrobial therapy, and chronic STIs caused by viruses, which are often treatable but not curable in a classic sense. Some of the most important STIs are included in **Table 1**. Asymptomatic carriage is common, and infected individuals may or may not experience the full range of symptoms listed. Moreover, the data in **Table 1** are incomplete inasmuch as a variety of other pathogens can be sexually transmitted.

Complete discussion of all STI pathogens is beyond the scope of this chapter. Presented here is basic background information, common clinical manifestations, and long-term complications associated with the most common STIs: four bacterial infections (syphilis, gonorrhea, chlamydia, chancroid), two viral infections (genital herpes and HPV infection), and one parasitic infection (trichomoniasis). The reader is referred to a more comprehensive textbook source for additional information on these agents, and other STI pathogens not discussed here.

Syphilis

Syphilis is a bacterial infection caused by *Treponema pallidum*. It is one of the most studied STIs, particularly because of its rich historical background. Most experts agree that syphilis was unknown in Europe until the late fifteenth century, when it suddenly appeared as a major scourge affecting the European population. Syphilis is classified as a genital ulcer disease, and in its primary phases causes the development of painless ulcerations in

Table 1 The major curable and chronic sexually transmitted infections

Curable STIs		
Condition	*Pathogen*	*Examples of signs/symptoms*
Syphilis	*Treponema pallidum*	Genital ulcers, skin rash, neurological damage, cardiovascular damage
Gonorrhea	*Neisseria gonorrhoeae*	Urethritis, cervicitis, pharyngitis, proctitis, epididymitis, pelvic inflammatory disease, disseminated infection
Chlamydia	*Chlamydia trachomatis*	Urethritis, cervicitis, proctitis, epididymitis, pelvic inflammatory disease, lymphogranuloma venereum
Trichomoniasis	*Trichomonas vaginalis*	Vaginitis, urethritis
Chancroid	*Haemophilus ducreyi*	Genital ulcers, lymphadenitis
Chronic STIs		
Condition	*Pathogen*	*Examples of signs/symptoms*
Genital herpes	Herpes simplex virus (HSV) type 1 and type 2	Genital ulcers, neonatal herpes infection, aseptic meningitis, encephalitis
HPV infection	Human papillomavirus (HPV)	Anogenital warts, intraepithelial neoplasia and carcinoma
HIV infection	Human immunodeficiency virus (HIV) type 1 and type 2	Acquired immune deficiency syndrome (AIDS)

the genital tract. The infection then progresses to a secondary phase, which results in cutaneous eruptions typically described as a rash but which can be very aggressive pustules or poxlike lesions. Untreated, the disease develops a latent state and can remain inactive for many years. However, a substantial proportion (25–35%) of those with latent infection may progress to develop long-term complications affecting the central nervous, cardiovascular, or other body organ systems. In the preantibiotic era, neurological syphilis was a major cause of long-term disability and early death. Upon invading the central nervous system, the organism affects the dorsal columns of the spinal cord leading to difficulties in walking and balance (tabes dorsalis), and also causes premature difficulties with mentation and cognition through to the development of general paresis (also known as dementia paralytica). Syphilis can also be transmitted transplacentally from infected mother to fetus (congenital syphilis). Up to 50% of pregnancies in women with early syphilis may end in stillbirth; those infants born with syphilis suffer developmental abnormalities affecting the bones, teeth, brain, and other organ systems.

The Columbian theory of syphilis transmission holds that *T. pallidum* infections were prevalent in New World populations but generally absent from Europe prior to Columbus's voyage in the late fifteenth century. These New World treponemal infections may have been nonvenereal dermatoses that afflicted native populations of North, Central, and South America and the Caribbean, and substantial paleopathological and historical evidence backs up the assertion that syphilis, or some variant thereof, was present in the New World for centuries prior to European contact. Upon arriving in the New World, Europeans quickly acquired venereal forms of syphilis and brought the infection back to Europe, where population-level herd immunity was absent. The infection spread quickly and virulently through European communities, often following lines of battle and the advancing armies of warring nations. A common clinical presentation was the development of aggressive, large, poxlike lesions affecting the face and body. These lesions were larger and different in character from other pox-causing diseases of the day, leading to the ascription of the term Great Pox to describe early pustular forms of secondary syphilis in the late fifteenth and early sixteenth centuries. Artistic renditions of persons with syphilitic lesions appear in woodcuttings and paintings starting at about this time, and are virtually unknown prior to this period. The current name by which we know this disease was ultimately taken from a poem by the early sixteenth-century Italian poet Fracastoro, in which the protagonist named Syphilis offended the god Apollo and was afflicted with the disease as punishment.

Over the years, several treatments have been proposed for syphilis, including the use of heavy metals such as arsenic (salvarsan), or intentional infection of patients with malaria-causing organisms (championed by the Austrian physician Julius Wagner-Jauregg, who won the 1927 Nobel Prize in Physiology or Medicine for this work). All of these treatments were generally ineffective and were associated with long-term toxicities of their own. Syphilis treatment was revolutionized in the 1940s with the discovery and clinical use of penicillin. To this day, penicillin remains the drug of choice for treating syphilis.

In the USA, the history of syphilis is inextricably tied to the notorious Tuskegee Syphilis experiment, a natural history study of syphilis conducted by the U.S. Public Health Service from the mid-1930s to the mid-1960s. This was a nontreatment study of poor, rural, African-American men in Tuskegee, Alabama, ostensibly to monitor the long-term effects of untreated syphilis in a time when the treatments carried substantial side effects and toxicities of their own. Men who participated in the study were told they had 'bad blood' and were monitored serologically over an extended time period. Treatment was withheld from the study participants long after the curative effects of penicillin were well recognized. Today, reference to the Tuskegee study resonates strongly as a touchstone for governmental deception, abuse of power, and disenfranchisement of minority populations at the hands of a powerful medical/public health apparatus.

Today, syphilis continues to exert a major toll on human populations. Many developing populations continue to suffer from syphilis due to lack of surveillance systems, screening laboratory tests and infrastructure, and required treatment medication. The persistence of congenital syphilis is particularly unfortunate, inasmuch as serotesting and treatment of infected women during pregnancy could significantly reduce or eliminate this condition. In recent years, increasing rates of syphilis have been seen among men who have sex with men (MSM), and syphilis is increasingly implicated as a cofactor in sexual transmission of HIV among high-risk communities.

Gonorrhea

Gonorrhea is caused by infection with the bacterium *Neisseria gonorrhoeae*, a spherical-shaped organism (coccus) with Gram-negative staining characteristics. The bacterium is often referred to as the gonococcus, and the organisms appear in nature as paired sets of bacteria attached side-by-side, or diplococci. It is an ancient disease, with references to its clinical manifestations appearing in early texts from Greek, Roman, and biblical times. Among males, uncomplicated gonococcal infection often causes copious amounts of thick green or yellow urethral discharge, coupled with intense burning on urination (dysuria). This finding may have given rise to its name, which is derived from Greek roots ('flow of seed'). In fact, the flowing material is a mixture of bacteria and pus, with the body's immune system producing large amounts of infection-fighting white corpuscles (leukocytes) to combat the offending bacterial agent. Among females, gonorrhea typically infects the uterine cervix, rather than the urethra, and symptoms of uncomplicated infection are generally milder. Most women do not experience urinary symptoms; some may notice a mild vaginal discharge, and many are asymptomatic altogether.

Unfortunately, untreated gonococcal infection can cause serious and significant complications. In males, the bacteria can ascend the urethral passage and establish infection in the epididymis and testicles (epididymitis and epididymo-orchitis). In females, gonorrhea can cause infection of the Bartholin's and Skene's glands near the vaginal introitus, leading to painful local abscess formation. Gonococcal infection (along with chlamydia) is also one of the most important causes of pelvic inflammatory disease (PID), characterized by upper genital tract infection of the uterus (endometritis), Fallopian tubes (salpingitis), and the abdominal-pelvic lining (peritonitis). Classically, gonococcal PID leads to irritation around the area of the liver in the right upper quadrant of the abdomen (peri-hepatitis or Fitz-Hugh-Curtis syndrome) and the development of adhesions in this area. The brisk inflammatory response triggered by gonococcal infection also causes scarring of other infected structures, particularly the Fallopian tubes, leading to tubal-factor infertility, ectopic pregnancy, and tubo-ovarian abscess as a result. Chronic pelvic pain is also a consequence of peritoneal scarring and adhesions. In both males and females, disseminated gonococcal infection (DGI) occurs when the bacterium gains access to the bloodstream and causes severe systemic disease, typically causing small-joint arthritis and the development of pustules on the skin surface (arthritis-dermatitis syndrome). Infection of the heart valves (endocarditis) has also been described. Gonococcal infection is also an important cause of eye infection in the newborn. Neonatal conjunctivitis (also known as ophthalmia neonatorum) due to untreated maternal gonorrhea manifests as a purulent discharge from the conjunctival tissues developing within a few days after delivery, and can lead to blindness in the neonate if left untreated.

Antibiotic resistance has become, and continues to be, a major problem in treating gonorrhea. Over time, the organism has acquired resistance (both plasmid-mediated and chromosomally mediated) to many antibiotic classes, including penicillins, tetracyclines, and more recently fluoroquinolone antibiotics. Currently, quinolone-resistant *Neisseria gonorrhoeae* (QRNG) is a significant public health concern in developed as well as developing countries; rates of QRNG are high in Asian and Pacific Island settings, and are increasing across North America and Western Europe, further limiting the effective antibiotic armamentarium for treating these infections (Centers for Disease Control and Prevention, 2006b). In general, cephalosporin antibiotics are considered the drugs of choice for treating gonorrhea. Coinfection with chlamydia is common, so antichlamydial therapy is also often prescribed for patients with gonorrhea.

Chlamydia

Chlamydia is caused by infection with *Chlamydia trachomatis*, an obligate intracellular pathogen that affects the mucous membrane tissues of the genital tract in males and females. In developing countries, the organism is also responsible for causing trachoma, a devastating ocular infection leading to corneal scarring and blindness; however, the chlamydial serovars (bacterial subspecies) that cause trachoma are different from those that cause genital tract infection. In many ways, sexually transmitted chlamydial infection behaves much like gonorrhea in terms of the body sites affected (male urethra and female cervix) as well as the long-term complications it causes (PID leading to tubal scarring and infertility). Chlamydia is a relatively newly recognized pathogen, coming to public health consciousness in the late twentieth century with the advent of effective laboratory testing procedures to detect its presence. In general, genital tract chlamydial infections tend to be less aggressive and less pyogenic than gonococcal infections, inasmuch as the inflammatory response is not as brisk and symptoms are generally milder. However, chlamydia is much more widely distributed than gonorrhea in developed countries and the numbers of infected individuals are much higher, such that chlamydia has become the single most important cause of tubal-factor infertility in developed countries. Implementation of widespread screening programs for chlamydia among reproductive-age women in developed nations is an important component of infertility prevention strategies. The true prevalence of genital tract chlamydial infection in developing countries is unknown, due to a lack of effective surveillance systems and laboratory testing strategies.

Many males and females with genital tract chlamydial infections are asymptomatic. Some males develop symptoms of urethritis, including burning on urination (dysuria) and a mucoid or sticky urethral discharge. Clinically, this condition looks much like gonococcal urethritis, although in general the discharge is less purulent and less aggressive, and microbiologic staining of the discharge material fails to reveal the classic Gram-negative diplococci of gonorrhea (hence the term NGU or nongonococcal urethritis to describe cases of chlamydial infection among symptomatic males). Females with cervical chlamydial infection are generally detected through screening, as symptoms are subtle (mild vaginal discharge or mild cervical bleeding) and may not be perceived by the infected individual. Infection of the newborn may occur when a child is born to a mother with untreated genital tract chlamydia. Neonatal chlamydia manifests as conjunctival infection (ophthalmia neonatorum) occurring within 1–2 weeks, or pneumonia that may not become clinically evident for several weeks after delivery.

Antibiotic treatment for chlamydia is achieved through the use of macrolide, tetracycline, or fluoroquinolone antibiotics. Single-dose treatment with azithromycin, a macrolide agent, is preferred by many clinicians for ease of treatment and likelihood of medication compliance. At the present time, antibiotic resistance is less of a concern for chlamydia than for gonorrhea, and widespread systematic resistance has not yet occurred. Infection rates remain high in North America and other developed nations despite aggressive screening of reproductive-age women.

Chancroid

Chancroid is an infectious disease caused by the bacterium *Haemophilius ducreyi*. Individuals with acute infection develop lesions that resemble the 'chancre' of primary syphilis, hence the name chancroid. In fact, chancroid is markedly different from syphilis, causing the development of deep genital ulcerations that frequently are complicated by infection of the inguinal lymph nodes with subsequent nodal suppuration.

Transmission of chancroid is exclusively by way of sexual contact, with the development of clinical symptoms occurring approximately 3–10 days after exposure. Infected individuals generally develop painful ulcerations in the genital tract, associated with swelling or pain in the inguinal region due to infection of regional lymph nodes. Chancroid ulcers typically have ragged, irregular edges with undermined borders (soft chancre), and the ulcer base is generally necrotic and purulent. The ulcers may mimic those of syphilis, but they typically lack the firm, indurated border that is characteristic of syphilitic lesions (hard chancre). Infected inguinal lymph nodes (buboes) may ultimately rupture as a consequence of ongoing infection, and pus can often be expressed from ruptured nodes on direct palpation.

Chancroid is more common in developing-country settings, or among individuals with a recent travel history to endemic areas. Studies have documented higher rates of infection in marginalized populations, such as commercial sex workers and their clients. Uncircumcised men have also been shown to have a higher risk of acquisition of chancroid than circumcised men, presumably because the foreskin serves as a bacterial reservoir that can harbor infectious bacteria after exposure. Localized urban outbreaks in developed countries are often linked to illicit drug use or commercial sex work. Chancroid and other genital ulcers have been shown to be important cofactors that facilitate HIV acquisition and transmission in developing countries.

Diagnosing chancroid can be difficult. The causative organism (*H. ducreyi*) is a small, facultative, anaerobic coccobacillus that requires hemin (X-factor) for growth, and clinical samples often grow poorly in culture even using selective media. DNA amplification methods to detect chancroid, while promising, are not yet clinically available. Therefore, a high index of clinical suspicion for chancroid is required to ensure that appropriate antibiotic therapy is rendered in a timely fashion to persons who may be infected. A number of antibiotic agents are effective against chancroid, including members of the macrolide, cephalosporin, and fluoroquinolone classes. Buboes may require drainage by aspiration if large or painful, and extended treatment courses may be required to ensure complete bubo resolution. Higher rates of treatment failure have been reported among uncircumcised men, and close clinical follow-up is required to ensure treatment adequacy, particularly among HIV-infected individuals.

Genital Herpes

Genital herpes is caused by herpes simplex virus (HSV). The virus is classically divided into two viral types (type 1 and type 2), which can be distinguished molecularly, and which historically have been held to infect different parts of the body. HSV-1 commonly causes infection of the mouth, lips, and oral mucous membranes (orolabial herpes), while HSV-2 commonly causes infection of the penis, scrotum, labia, vulva, vagina, and other genital tract structures (genital herpes). Serologic and virologic testing demonstrates a significant degree of overlap with regard to herpes types and body sites infected: some orolabial infections are caused by HSV-2, and increasing numbers of genital tract infections are caused by HSV-1. Type-specific serologic tests for HSV are now available in many developed-country settings and are being increasingly used to identify individuals with HSV infection. By and large, HSV-2 infections are generally held to be sexually transmitted, and serologic studies of HSV-2 infection confirm high rates of infection in general population samples. Recent studies estimate overall age-adjusted seroprevalence of HSV-2 in the USA at 17.0% of the general population (Xu *et al.*, 2006).

Clinical manifestations of genital herpes can range from aggressive, painful blisters and ulcerations (clinical infection), to subtle but perceivable irritations, rashes, fissures, and other manifestations (subclinical infection), to a complete absence of identifiable clinical symptoms (asymptomatic infection). The virus is believed to enter the body through microabrasions that occur during sexual contact. Clinical symptoms, which occur in up to 20% of persons with genital tract HSV-2 infection, may develop within a few days of exposure and include painful genital lesions (blisters, ulcerations, rashes), local edema, and systemic symptoms such as fever, malaise, and fatigue. The virus is neurotropic and ascends local nerve roots, establishing latency in the dorsal root ganglion of the spinal column; infection is lifelong and is not curable in the classic sense. Viral reactivation is common, and causes local recurrence of genital lesions, usually in a unilateral (dermatomal) distribution.

Subclinical infection and asymptomatic infection are actually more common than classic clinical infection. Many individuals experience recurrent rashes, irritations, fissures, or other genital tract lesions which they do not recognize or perceive as herpes, but which can later be so identified through laboratory testing. Infants born to infected mothers are at risk for developing neonatal herpes, a devastating condition affecting the newborn that is associated with serious, long-term, neurological consequences. Pregnant women with active genital-tract herpes lesions at the time of labor are delivered abdominally by cesarean section to prevent or limit neonatal exposure to herpes virus in the birth canal. Research studies have identified the highest risk of neonatal herpes among infants whose mothers acquire primary genital tract herpes infection during pregnancy (Brown *et al.*, 2003).

Untreated clinical infection ultimately resolves over several days or weeks. Early treatment with oral acyclovir (or related medications such as famciclovir or valacyclovir) significantly reduces the duration of clinical lesions and reduces viral shedding in primary infection as well as in outbreaks of recurrent infection. Management of recurrent genital herpes infections includes patient education to identify and recognize clinical symptoms, coupled with antiviral treatment to reduce or prevent morbidity. Episodic treatment of recurrences involves use of antiviral medication at the time lesions begin to develop, or as soon thereafter as possible, so as to limit the duration of the clinical recurrence. Suppressive treatment involves daily use of antiviral medication to prevent the onset of recurrent clinical lesions, and is generally recommended for persons with frequent or traumatic recurrences. Some studies suggest that antiviral medications may be beneficial for preventing sexual transmission of HSV within serodiscordant partnerships. Recent data suggest a strong link between genital-tract herpes infection and the spread of HIV in areas of high HIV prevalence such as sub-Saharan Africa (Freeman *et al.*, 2006). Research studies are underway to determine whether herpes treatment reduces the likelihood of HIV acquisition in high-prevalence settings.

Human Papillomavirus (HPV) Infection

More than 80 different viral types of human papillomavirus (HPV) have been identified, and more than 20 of these have been found to infect the human genital tract. HPV has long been recognized as the cause of external warts in the anogenital area, and aggregations of warts are commonly referred to as condylomata acuminata. These manifestations are typically caused by HPV types 6 and 11, although other HPV types can also cause anogenital warts. Warts are generally treated with local therapies designed to eliminate the wart itself, rather than the viral infection. After treatment, wart recurrence rates are high because the virus remains viable within dermal and subdermal tissues.

Other HPV types are oncogenic and have been closely linked to development of cancer of the cervix, vulva, vagina, and anus. These 'high-risk' HPV types include 16 and 18, and to a lesser extent several others (e.g., 31, 33, 35, 45, 56), and are implicated as the cause of most cervical and anal cancers. For this reason, aggressive screening for high-risk HPV cervical infection has been advocated among reproductive-age women in developed countries for early detection and treatment of HPV-related dysplasias and neoplasias, in an effort to prevent cervical cancer morbidity and death. (In practice, this generally takes the form of routine cervical screening for dysplasia by Papanicolaou (Pap) smear, followed by reflex HPV testing of Pap smear specimens only if dysplastic or suspicious cells are identified.) By and large, the wart-causing HPV types 6 and 11 are not oncogenic, and individuals with external genital warts caused by HPV are not necessarily at risk for the development of cervical or other genital tract neoplasia. However, coinfection with other high-risk HPV types can occur, so individuals with anogenital warts should be counseled to follow routine screening guidelines for cervical cancer prevention. Two preventive vaccines against HPV have been recently developed and are being introduced for clinical use. One is a quadrivalent vaccine targeting cancer-causing HPV types 16 and 18 as well as wart-causing HPV types 6 and 11, while the other is a bivalent vaccine targeting HPV types 16 and 18 alone. Widespread use of the vaccine among adolescent and young adult females may help prevent further spread of HPV and the development of HPV-associated diseases (Bosch *et al.*, 2006).

Trichomoniasis

Trichomoniasis, also known as trichomonal infection, is a clinical condition caused by infection with the parasite *Trichomonas vaginalis*. It is commonly diagnosed in women, in whom it typically causes a profuse, malodorous vaginal discharge with vaginal inflammation. Male infection also occurs but is much more likely to be asymptomatic. Trichomonal infection is the most common curable STI worldwide, accounting for more than half of the estimated 340 million bacterial or parasitic infections occurring annually (World Health Organization, 2001). Trichomoniasis has been linked to premature rupture of membranes in pregnancy and preterm delivery, and has been identified in epidemiological studies as a cofactor in HIV transmission.

The causative agent, *T. vaginalis*, is a flagellated anaerobic protozoan that is transmitted during sexual contact. Up to 50% of infected women are initially asymptomatic, although many of these will proceed to develop clinical symptoms within several months of infection. Because the organism can survive for months or years in epithelial crypts and periglandular areas, subclinical infection may occur and diagnosis may be delayed. Most commonly, women develop a frothy gray or yellow-green vaginal discharge, associated with vaginal itching (pruritus) indicative of a brisk vaginal inflammatory response. Some women classically develop small petechial hemorrhages on the cervix, which may be visible on inspection during a pelvic examination ('strawberry cervix'), although this occurs in a minority of cases. Infection in males is most commonly asymptomatic, although symptomatic urethritis may occur in a minority of cases.

Most cases of trichomonal vaginitis are diagnosed by direct visualization of the parasitic organisms on a microscopic examination of vaginal secretions. Unfortunately, this method of diagnosis is relatively insensitive (40–70%) and is dependent on the technique of specimen collection as well as the experience of the microscopist. Diagnostic accuracy can be increased by incubating vaginal secretions in commercially available culture media, although these tests are not widely available in many clinical sites. Diagnosis in resource-poor settings is generally based on syndromic presentation and clinical history.

Metronidazole or tinidazole are the treatments of choice for trichomonal infection. Asymptomatic male partners must also be treated to prevent reinfection of treated females. Drug resistance, although unusual, should be suspected in cases of trichomoniasis that do not respond to standard antimicrobial therapy. Individuals should be counseled not to drink alcohol while taking these medications, because they can cause a disulfiram-type reaction (flushing, shortness of breath, and severe nausea and vomiting).

Consequences of STIs

Sexually transmitted infections have immediate as well as long-term consequences. Many STIs cause pain, discomfort, and immediate suffering to infected individuals. Genital herpes causes painful, recurrent ulcerations in the genital tract. Early treatment with antiviral medication can reduce the intensity and duration of the lesions, but is not curative of the infection. Trichomonal infection often causes an uncomfortable, malodorous vaginal discharge in women, and is known to cause urethral irritation and inflammation in some men as well. Gonorrhea in men classically causes intense burning on urination and copious amounts of urethral discharge, and both syphilis and chancroid cause ulcerations in the genital tract along with other associated symptoms. The immediate pain and suffering of STIs is an important public health consequence, for these symptoms often trigger a health-seeking response among affected individuals in need of care for their medical condition.

In addition to causing substantial physical discomfort, STIs are often stigmatizing conditions in society, and commonly lead to feelings of shame, guilt, or regret among infected individuals. Substantial research is underway to clarify the extent to which stigma and shame help structure the STI experience at the level of the individual. For example, persons at risk for STI may be reluctant to seek care if they believe their privacy or confidentiality could be compromised through the care-seeking process. As a result, lack of early detection and of early treatment can further facilitate the transmission of infection by prolonging the duration of infectiousness. Chronic infections, such as genital herpes, may be more stigmatizing than curable infections, although limited data are available to address the magnitude of these effects.

From a public health standpoint, the long-term medical and economic consequences of STIs are profound, particularly with regard to reproductive tract infections in women. STIs are the most important preventable cause of human infertility: in industrialized countries, untreated chlamydial and gonococcal infections are a major cause of PID, resulting in infertility from Fallopian tube scarring, as well as other complications such as ectopic (tubal) pregnancy, tubo-ovarian abscess, and chronic pelvic pain. Untreated STIs among pregnant women are a major cause of spontaneous abortion, stillbirth, premature labor, and low birthweight, and can lead to serious infections in the newborn including neonatal herpes, congenital syphilis, and ophthalmia neonatorum. STIs are important facilitators of HIV transmission as well. Studies in developed and developing countries demonstrate that STIs, particularly those causing genital ulcerations, enhance the likelihood of HIV acquisition among HIV-negative individuals (Fleming and Wasserheit, 1999) and also promote genital-tract shedding of HIV among persons who are already HIV-infected. Moreover, some STIs have oncogenic potential: oncogenic strains of HPV are widely recognized as the most important cause of anogenital cancers, particularly cancer of the cervix in women. And hepatitis B, which like HIV may be transmitted through sexual contact as well as through bloodborne routes, is a major cause of liver cancer worldwide. Finally, from an economic perspective, research suggests that STIs other than HIV account for up to 5% of the total discounted healthy life-years lost in sub-Saharan Africa. STIs consistently rank among the top disease categories for which adults seek health care in developing countries, and remain among the top causes of disease, death, and healthy life lost among women of childbearing age.

STI Control and Prevention

The public health approach to STIs is based on the twin tenets of control and prevention. Control is the term used to describe the implementation of basic programmatic strategies at the national level to detect infections and limit their spread throughout the wider population. Prevention speaks to the systematic application of public health principles to inhibit the primary acquisition of STIs, and to limit further spread of infection in the event that primary acquisition does occur. Often, control and prevention go hand in hand. For example, treatment of an incident bacterial STI such as gonorrhea serves a control function by eliminating the infection within the affected individual, and also serves a prevention function by eliminating the opportunity for the individual to spread the infection to others in the future.

In general, three different levels of prevention for STIs can be distinguished (**Table 2**). Primary prevention is the implementation of programmatic activities to prevent direct exposure to an STI pathogen. Educational programs promoting safer sex and correct and consistent condom use, or programs promoting delay of first sexual contact among teenagers, are examples of primary STI prevention programs. (Vaccination against STI pathogens such as HPV constitutes another example of primary prevention.) Secondary prevention is the implementation of strategies to prevent the onset of disease in those who have been exposed and may be infected with an STI pathogen. Provision of empiric antibiotic therapy to sex partners of persons with a bacterial STI such as syphilis is an example of secondary STI prevention. Tertiary prevention is the effort to limit morbidity and restore health to those who are infected. This is the most basic form of STI control, and is really a form of harm-reduction inasmuch as treatment is designed to prevent further tissue damage or long-term complications from an ongoing clinical condition. Clinical care for symptomatic curable STIs is an example of tertiary prevention. (Some argue that STI treatment also serves a primary prevention function by reducing the number of infected individuals within the community, thereby reducing the risk for others as well.)

Table 2 Levels of prevention for STI

Level of prevention	*Example*
Primary prevention	Educational programs promoting consistent and correct condom use
Secondary prevention	Empiric antibiotic therapy provided to persons who are exposed to bacterial STI, but who do not yet show signs of infection
Tertiary prevention	Treatment of incident STI to decrease suffering and prevent complications

Partner Notification and Referral

Identifying, tracking down, and providing treatment to the 'source case' of the infection, as well as to secondary cases who may also have been infected by the index case, is another aspect of STI control. In developed countries, this process is typically implemented through systems of partner notification and referral for medical treatment. Two types of partner notification have been described. The first form, patient referral, occurs when an STI patient is instructed to notify his or her own partners that they have been exposed to an STI, and that they should seek medical care for testing and treatment. The second form, provider referral, occurs when a health-care provider or health agency member interviews an STI patient to elicit the names and contact information of recent sex partners, and then notifies these partners of their STI exposure and need for testing and treatment. Unfortunately, these methods of partner notification are inadequate and incomplete. Individuals with an STI are often unwilling or unable to identify their sex partners or release contact information to a health department investigator when asked. Some individuals do not wish to notify sex partners who they believe infected them in the first place. And STI network studies suggest that infections occur among social as well as sexual network members, such that only following direct chains of sexual contact inevitably misses some cases of STI occurring in social (but not sexual) networks of interaction with infected individuals.

Recently, advocates of expedited partner therapy (EPT) have demonstrated enhanced outcomes in partner notification and referral through the provision of antimicrobial therapy to partners outside of the usual channels of notification and referral (Golden *et al.*, 2005). These sorts of novel strategies for partner treatment often obviate the need for a physical examination of the partner, since treatment is offered outside of clinical care settings. In some cases, EPT can be achieved through patient-delivered mechanisms, whereby persons with an STI are given extra doses of medication to give to their partners. In other settings, pharmacy-delivered mechanisms have been used, whereby partners are referred to local pharmacies who then give antimicrobial treatment. Advocates point to increased numbers of partners treated by way of EPT strategies compared with traditional partner notification and referral patterns, and also to decreased rates of reinfection or persistent infection of the initial STI patient, presumably due to effective treatment of sex partners through EPT who otherwise would remain untreated and infectious.

Passive and active surveillance systems are also important aspects of STI control and prevention programs in developed and some developing countries. Passive surveillance occurs when health authorities track and take note of case reports of disease that are submitted by treating clinicians and laboratory-testing facilities. These case reports form the basis of STI intervention planning, for they direct health authorities to areas of ongoing concern or changing epidemiologic patterns. Active surveillance implies the intentional use of screening or directed testing among large sections or subsections of the population, with the intent of finding asymptomatic or subclinical infection that might otherwise remain undetected without aggressive outreach. The implementation of chlamydia screening among women of reproductive age is an example of an active surveillance system. Both methods are important components of comprehensive STI control and prevention strategies.

The Future of STIs

New developments in STI research and programmatic activities hold promise for reducing the public health burden of these infections. The use of field-based test strips for diagnosing syphilis is one such development. Historically, syphilis has been diagnosed using standard serologic tests, which require a certain amount of laboratory processing that is often unavailable in resource-poor settings. Blood samples must be collected and centrifuged to separate blood cells from serum, and then testing is performed on these serum samples. Recently, several manufacturers have developed field-based, immunochromatographic, syphilis test strips that can be used with whole blood samples, thereby obviating the need for additional specimen processing in the laboratory. A number of these test strips have been evaluated by WHO and other health authorities to determine sensitivity, specificity, and performance characteristics. Production and distribution of the test strips at low cost will advance syphilis detection and may serve to facilitate early treatment, particularly among pregnant women for prevention of congenital syphilis infection.

Additional research is needed in the realm of STI treatment, particularly with regard to effective, single-dose, oral therapies. Such treatments, if available at low cost, could be distributed widely to populations at risk in developing- and developed-country settings alike to reduce STI-associated suffering. Unfortunately, few such agents are currently available which meet these criteria. Single-dose azithromycin has activity against chlamydia (and at higher doses, gonorrhea) but recent studies have demonstrated high levels of resistance to azithromycin among syphilis strains in North America and Europe, and have cast a shadow on the utility of this agent for oral treatment of syphilis. Fluoroquinolone antibiotics, once a mainstay of treatment for gonorrhea, are now limited in their utility by antimicrobial resistance to quinolone antibiotics. STI control and prevention will depend to a great degree on the availability of field-based treatments that can be administered easily and safely by public health personnel in a variety of clinical and nonclinical settings.

Finally, the use of social and sexual network paradigms will continue to advance STI prevention goals by shedding light on the ways in which sex partners are selected within and outside of networks of social interaction. Degrees of network embeddedness, overlap, and separation affect STI risk of network members, and additional research in these areas is warranted. Network analyses provide insight with regard to the dynamic nature of human sexual behavior, and the relationship between sexual behavior and STI risk. As public health incorporates network methods into standard STI control activities, enhanced case finding and disease prevention goals are advanced.

Conclusion

STIs are widely distributed across human societies and are a significant cause of suffering worldwide. A variety of microbial agents can be sexually transmitted, including bacteria, viruses, and parasites. Bacterial and parasitic

STIs are curable with antibiotic therapy, and viral STIs can now be effectively treated with appropriate antiviral agents. STIs represent an important public health problem because of their propensity to spread asymptomatically across sectors of the population and because of their substantial impact on reproductive and neonatal health. Surveillance and testing of populations at risk are essential elements of STI control and prevention in developed and developing countries alike. Further research is needed in STI biology, epidemiology, and prevention to identify more effective tools to prevent the spread of STI pathogens and reduce the burden of these pathogens on human populations.

See also: AIDS: Epidemiology and Surveillance; Cervical Cancer; Female Reproductive Function; Gender Aspects of Sexual and Reproductive Health; Infertility; Male Reproductive Function; Mother-to-Child Transmission of HIV; Sexual and Reproductive Health: Overview.

Citations

Anderson RM and May RM (1988) Epidemiological parameters of HIV transmission. *Nature* 333: 514–519.
Bosch FX, Cuzick J, Schiller JT, *et al.* (eds.) (2006) HPV vaccines and screening in the prevention of cervical cancer. *Vaccine* 24(Suppl 3): S251–S261.
Brown ZA, Wald A, Morrow RA, *et al.* (2003) Effect of serologic status and cesarean delivery on transmission rates of herpes simplex virus from mother to infant. *Journal of the American Medical Association* 289: 203–209.
Centers for Disease Control and Prevention (CDC) (2006a) Sexually transmitted disease surveillance, 2005. Atlanta, GA: U.S. Department of Health and Human Services.
Centers for Disease Control and Prevention (CDC) (2006b) Sexually transmitted diseases treatment guidelines, 2006. *MMWR Morbidity and Mortality Weekly Report* 55: No. RR-11.
Fleming DT and Wasserheit JN (1999) From epidemiological synergy to public health policy and practice: The contribution of other sexually transmitted diseases to sexual transmission of HIV infection. *Sexually Transmitted Infections* 75: 3–17.
Freeman EE, Weiss HA, Glynn JR, *et al.* (2006) Herpes simplex virus 2 infection increases HIV acquisition in men and women: Systematic review and meta-analysis of longitudinal studies. *AIDS* 20: 73–83.
Golden MR, Whittington WLH, Handsfield HH, *et al.* (2005) Effect of expedited treatment of sex partners on recurrent or persistent gonorrhea or chlamydial infection. *New England Journal of Medicine* 352: 676–685.
May RM and Anderson RM (1987) Transmission dynamics of HIV infection. *Nature* 326: 137–142.
Parran T (1937) *Shadow on the Land*. New York: Reynal and Hitchcock.
Weinstock H, Berman S, and Cates W, Jr. (2004) Sexually transmitted diseases among American youth: Incidence and prevalence estimates, 2000. *Perspectives on Sexual and Reproductive Health* 36: 6–10.
World Health Organization (WHO) (2001) *Global Prevalence and Incidence of Selected Curable Sexually Transmitted Infections: Overview and Estimates*. Geneva, Switzerland: World Health Organization.
Xu F, Sternberg MR, Kottiri BJ, *et al.* (2006) Trends in herpes simplex virus type 1 and type 2 seroprevalence in the United States. *Journal of the American Medical Association* 296: 964–973.

Further Reading

Anderson RM and May RM (1991) *Infectious Diseases of Humans: Dynamics and Control*. Oxford, UK: Oxford University Press.
Brandt AM (1987) *No Magic Bullet: A Social History of Venereal Disease in the United States since 1880*. New York: Oxford University Press.
Holmes KK, Sparling PF, Mardh PA, *et al.* (eds.) (1999) *Sexually Transmitted Diseases,* 3rd edn. New York: McGraw-Hill.
Jones JH (1981) *Bad Blood: The Tuskegee Syphilis Experiment*. New York: The Free Press.
Stoner BP (2007) Current controversies in the management of adult syphilis. *Clinical Infectious Diseases* 44: S130–S146.
World Health Organization (2005) *Sexually Transmitted and Other Reproductive Tract Infections – A Guide to Essential Practice*. Geneva, Switzerland: World Health Organization.
World Health Organization (2007) *Global Strategy for the Prevention and Control of Sexually Transmitted Infections: 2006–2015*. Geneva, Switzerland: World Health Organization.

Relevant Websites

http://www.astda.org – American Sexually Transmitted Diseases Association (ASTDA).
http://www.ashastd.org – American Social Health Association (ASHA).
http://www.bashh.org – British Association for Sexual Health and HIV (BASHH).
http://www.cdc.gov/std – Sexually Transmitted Diseases (CDC).
http://www.who.int/topics/sexually_transmitted_infections/en – Sexually Transmitted Infections (WHO).

AIDS: Epidemiology and Surveillance

R Choi and C Farquhar, University of Washington, Seattle, WA, USA

Introduction

During the last 25 years, human immunodeficiency virus (HIV) has claimed the lives of millions of men, women, and children across the globe and has developed into an international public health crisis. The HIV/AIDS (acquired immune deficiency syndrome) epidemic has spared few regions in the world and has been particularly devastating in sub-Saharan Africa, where more than 60% of all HIV-infected adults and 90% of HIV-infected infants reside (Joint United Nations Programme on HIV/AIDS [UNAIDS]/World Health Organization [WHO], 2006).

At the time of the first reports of AIDS cases among gay men in the USA during the early 1980s, it would have been difficult to envision such a calamitous outcome. Even after identification of the human immunodeficiency virus in 1983 and diagnosis of HIV infection among heterosexual men and women, infants and children, intravenous drug users, hemophiliacs, and recipients of other blood products, the enormity of the situation was not fully recognized. As a result, the global response to HIV was initially slow and in many parts of the world remains inadequate, despite in-depth knowledge of the salient risk factors for HIV transmission and improved surveillance data defining the nature of the epidemic.

Overall Prevalence of HIV

Despite significant advances in HIV prevention and treatment, HIV remains a disease without a cure and continues to threaten the social and economic stability of many developing nations. According to the *AIDS Epidemic Update 2006* by UNAIDS and WHO, an estimated 39.5 million (34.1–47.1) people worldwide were living with HIV in 2006 (**Figure 1**). Approximately 17.7 million (15.1–20.9) of HIV-infected were women, an increase of over 1 million between 2004 and 2006. In the same year, an estimated 4.3 million (3.6–6.6) became newly infected with HIV, and approximately 2.9 million (2.5–3.5) lost their lives to AIDS (**Figures 2** and **3**). HIV has infected over 65 million people and claimed over 25 million lives worldwide since 1981 (UNAIDS/WHO, 2006).

Distinctions between HIV Types 1 and 2

Two human immunodeficiency viruses have been identified and characterized in humans: HIV type 1 (HIV-1) and HIV type 2 (HIV-2). While HIV-1 and HIV-2 share the same modes of transmission, HIV-2 has a less efficient rate of transmission than HIV-1 (Kanki *et al.*, 1994). Studies conducted in Senegal estimated the HIV-2 transmission rate per sexual act with an infected partner to be 3.4- to 3.9-fold lower than that of HIV-1 (Gilbert *et al.*, 2003). Other studies conducted in The Gambia, Ivory Coast, and Senegal have demonstrated rates of mother-to-child HIV-2 transmission to be approximately 6- to 20-fold lower than those among women with HIV-1 infection (Andreasson *et al.*, 1993; Abbott *et al.*, 1994; Adjorlolo-Johnson *et al.*, 1994; O'Donovan *et al.*, 2000). Studies have also demonstrated that progression to AIDS is slower

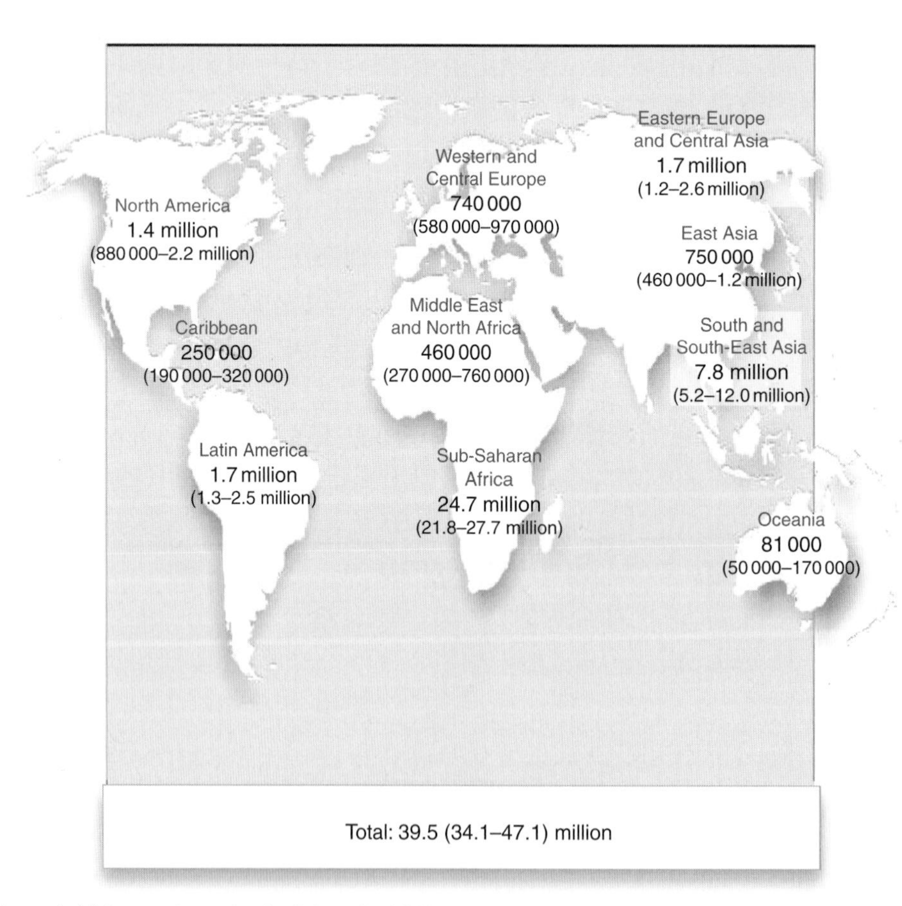

Figure 1 Adults and children estimated to be living with HIV in 2006. Reproduced with kind permission from UNAIDS/WHO (2006) *AIDS Epidemic Update 2006*. Geneva: UNAIDS/WHO.

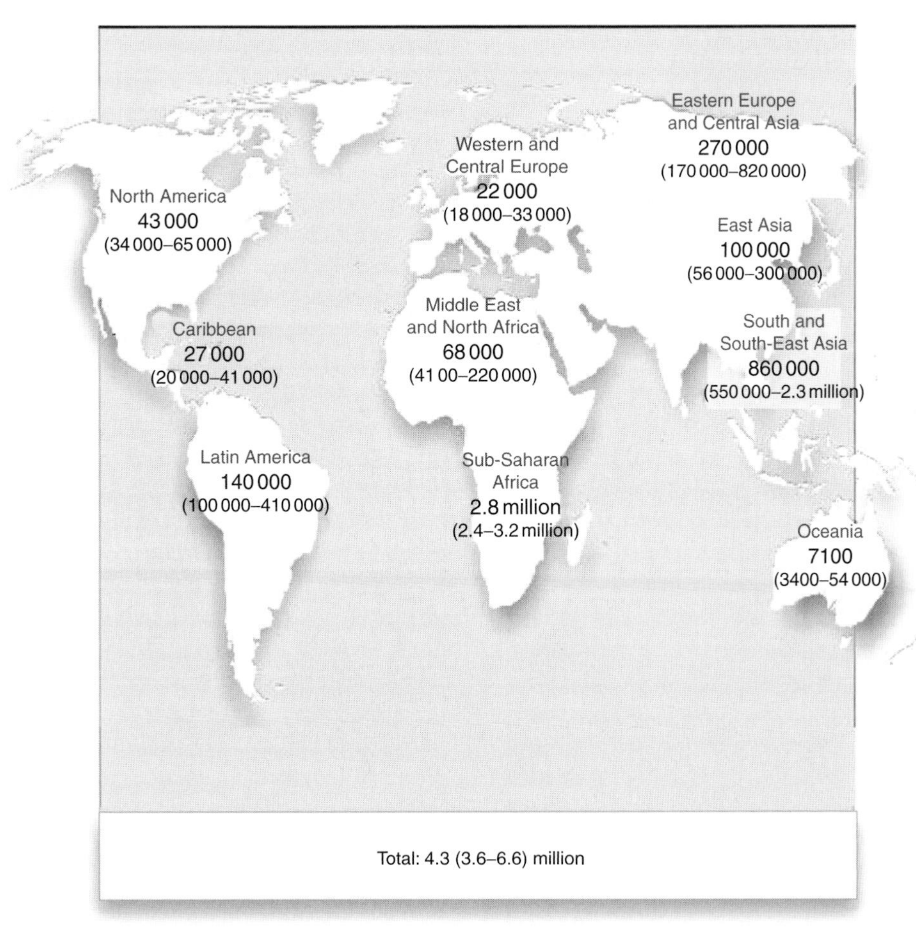

Figure 2 Estimated number of adults and children newly infected with HIV during 2006. Reproduced with kind permission from UNAIDS/WHO (2006) *AIDS Epidemic Update 2006*. Geneva: UNAIDS/WHO.

among individuals with HIV-2 compared to HIV-1, although the consequences of immunosuppression and the propensity for opportunistic infections remain similar at lower CD4 counts (Marlink *et al.*, 1994).

Another difference between HIV-1 and HIV-2 is in their antiretroviral susceptibility, which may be important in future delivery of antiretroviral therapy to Africa and management of the epidemic. The current antiretroviral therapy for HIV-1 includes various combinations of nucleotide or nucleoside reverse transcriptase inhibitors (NRTIs), nonnucleoside reverse transcriptase inhibitors (NNRTIs), and protease inhibitors (PIs). While NRTIs have been found to be as effective against HIV-2 as they are against HIV-1 *in vitro* (Mullins *et al.*, 2004), NNRTIs have been shown to be largely ineffective against HIV-2 *in vitro* (Witvrouw *et al.*, 1999). PIs have shown mixed effectiveness against HIV-2 (Pichova *et al.*, 1997).

Given lower rates of transmission and progression to disease with HIV-2, it is not surprising that HIV-1 is responsible for the majority of HIV infections and AIDS cases worldwide. HIV-2 is restricted to relatively small geographic areas and is the predominant HIV type in the West African countries of Guinea-Bissau, The Gambia, Cape Verde, and Senegal, where it was first described in 1985 (**Figure 4**) (Barin *et al.*, 1985). HIV-2 is also found in Portugal and former Portuguese colonies, such as Mozambique, Angola, southwestern India, and Brazil, although it is less prevalent in these regions than HIV-1 infection (Kanki, 1997).

While these differences between HIV-1 and HIV-2 promote a better understanding of HIV, it is clear that HIV-1 drives the HIV/AIDS epidemic globally, and HIV-1 will therefore be the focus of the remainder of this article.

HIV-1 Groups and Subtypes

The great genetic diversity of HIV is reflected in the many genotypes or clades that have been identified. Currently, HIV-1 genotypes are divided into Groups M, N, and O, with Group M being further subdivided into subtypes A to K because it is the most common. Groups N and O are limited to West Africa. Group O represents the outlier strains, while Group N represents 'new group' or non-O and non-M strains. Generally, intragenotype strain variation is less than 15%, while intergenotypic variation is 20–30% (Burke *et al.*, 1997).

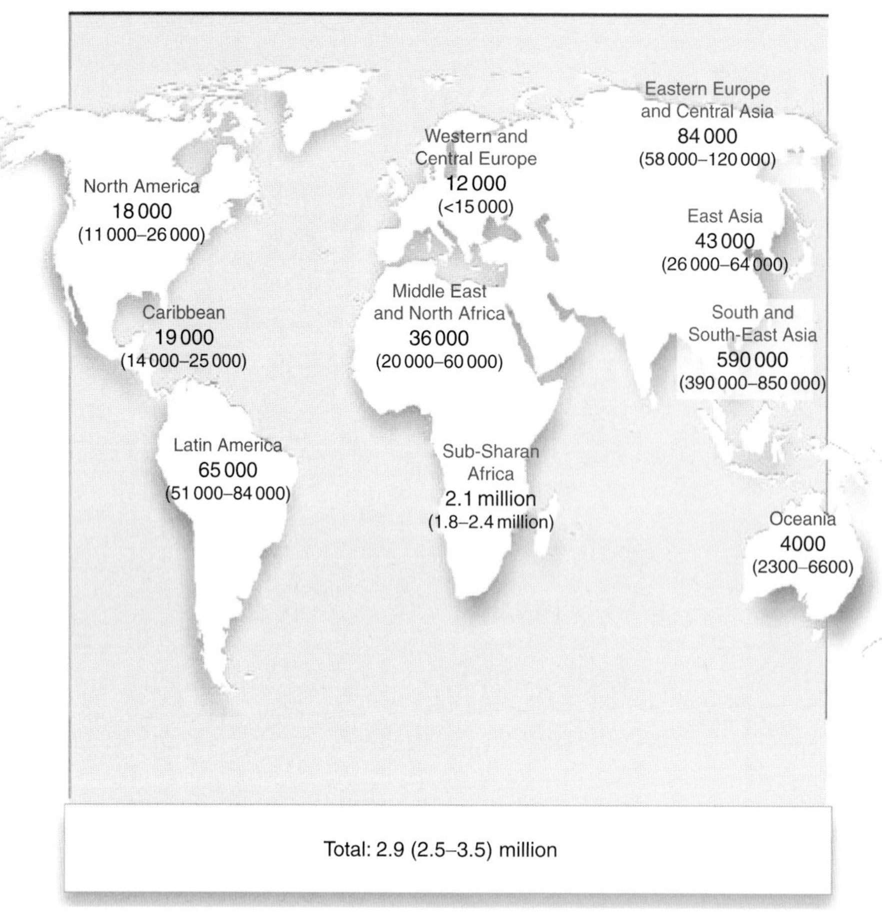

Figure 3 Estimated adult and child deaths from AIDS during 2006. Reproduced with kind permission from UNAIDS/WHO (2006) *AIDS Epidemic Update 2006.* Geneva: UNAIDS/WHO.

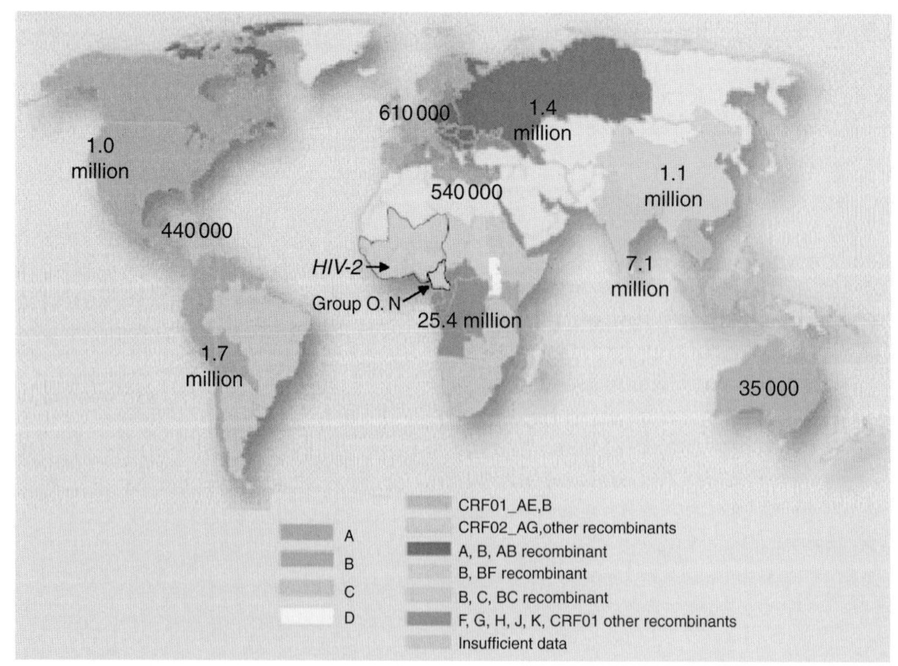

Figure 4 Global distribution of HIV-1 subtypes and HIV-2. Mc Cutchan (2006) Global epidemiology of HIV. *Journal of Medical Virology* 78: s7–s12. Copyright 2006 Wiley & Sons, Inc. Reprinted with permission of Wiley-Liss, Inc., a subsidiary of John Wiley & Sons, Inc.

Subtypes A to K are defined by significant clustering across the genome and are often more prevalent in particular regions of the world (see **Figure 4**) (McCutchan, 2006). Of these subtypes, subtype C is currently the most common subtype worldwide, and is found in the eastern and southern parts of Africa and in India (McCutchan, 2000). Subtype B predominates in the USA and Europe, subtypes A and D are found in eastern parts of Africa, and subtype E in South-East Asia.

Several studies suggest that subtypes A to K differ in their primary mode of transmission, virulence, and infectivity. In Thailand, where subtypes B and E were both prevalent at the start of the epidemic, subtype E was initially the major subtype among female sex workers, while subtype B was associated with intravenous drug use (IDU) (Nelson *et al.*, 1993; Nopkesorn *et al.*, 1993). However, subtype E subsequently became more common than subtype B among IDUs, indicating that mode of transmission may be less important than other biological or behavioral factors (Nopkesorn *et al.*, 1993). In Kenya, subtypes A, C, and D all circulate and individuals infected with HIV-1 subtype C have been found to have higher plasma levels of viral RNA and lower CD4 counts compared to individuals with subtypes A or D (Neilson *et al.*, 1999). In Uganda and Kenya, HIV infection with subtype D has been associated with more rapid progression when compared to infection with subtype A (Kaleebu *et al.*, 2001; Baeten *et al.*, 2007). However, in Sweden, studies showed no differences among subtypes A, B, C, or D when examining clinical HIV disease progression (Alaeus *et al.*, 1999).

Individuals may also be coinfected with 2 strains of HIV-1 and these may be the same or different viral subtypes. Coinfection occurs when a person acquires two different viral strains simultaneously or when a chronically HIV-infected individual is reinfected with HIV-1, a phenomenon also known as super-infection. This has led to recombinant forms of HIV-1, which may in turn be transmitted and have the potential to become major strains within populations. For example, a recombinant form comprised of subtypes A and E is a major circulating strain throughout South-East Asia and a recombinant strain of subtypes A and G is a major strain in West Africa. Recombinant strains will undoubtedly continue to emerge and challenge those involved in HIV vaccine development, surveillance efforts, and clinical care.

Modes of HIV Transmission: Rates and Risk Factors

Understanding the different modes of transmission and appreciating associated risk factors play an essential role in grasping the epidemiology of the HIV/AIDS epidemic. HIV may be transmitted via sexual intercourse (either male to female, female to male, or male to male), via IDU, vertically from mother to child, through transfusions of blood or other blood products, or via occupational exposure to HIV-infected bodily fluids.

Sexual Transmission

Sexual transmission accounts for 75–85% of HIV infections worldwide and heterosexual intercourse accounts for the vast majority of these transmission events (Royce *et al.*, 1997). The likely explanation for this is that heterosexual transmission predominates in those regions of sub-Saharan Africa with the highest HIV prevalence, rather than that heterosexual intercourse is a more efficient means of transmission. The average rate of infection for vaginal intercourse is estimated to be 0.0011 per contact and several studies suggest vaginal transmission from an infected male to an uninfected female is more efficient than when the female partner is HIV-infected (**Figure 5**) (Gray *et al.*, 2001). This is in comparison to significantly higher rates for unprotected anal sex, which is the most efficient mode of sexual transmission and has an average rate of infection of 0.0082 per contact (Vittinghoff *et al.*, 1999).

One of the most significant and well-established risk factors in sexual HIV-1 transmission is the quantity of HIV-1 in the blood of the infected partner. In a rural

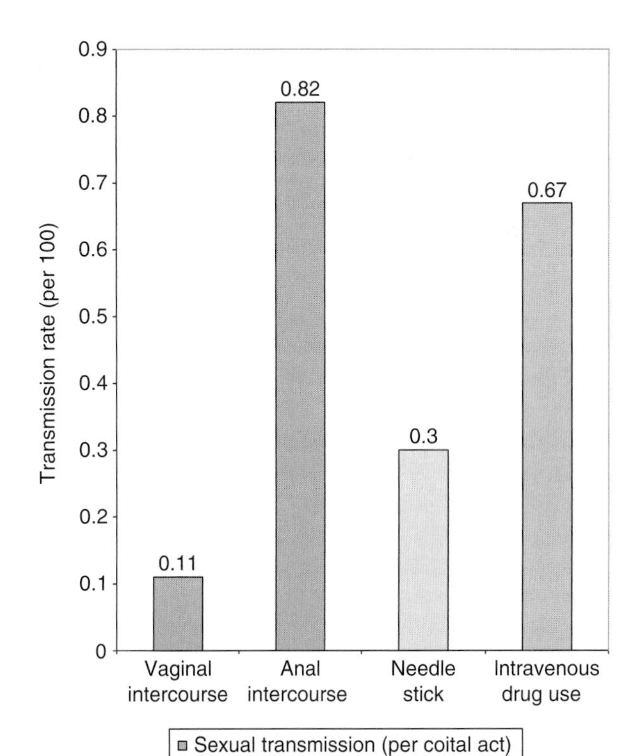

Figure 5 Rates for different modes of transmission.

district of Uganda, one study demonstrated a significant dose–response relationship between increased HIV-1 RNA viral load in plasma and increased risk of transmission. For every $\log_{10}$ increment in the viral load, there was a 2.5-fold increase in HIV-1 transmission and no transmission occurred in this cohort when the infected partner had a plasma HIV-1 RNA level less than 1500 copies per milliliter (Quinn *et al.*, 2000).

The highest HIV-1 viral load occurs in the setting of end-stage AIDS and during primary HIV infection, making these periods of high transmission risk. The latter is a particular concern from a public health standpoint because when a person is first infected with HIV, he or she may not know his or her HIV status, and may unknowingly spread HIV at high rates. A study in Uganda demonstrated that the highest HIV transmission rates were within 2.5 months after seroconversion of the infected partner, which represents primary infection (0.0082 per coital act), and during the period 6–26 months before the death of the infected partner, which represents advanced AIDS (0.0028 per coital act). The average rates of HIV transmission were 0.0007 per coital act among chronically infected partners (Wawer *et al.*, 2005).

Genital shedding of HIV-1 may also increase HIV transmission risk and some have suggested that it may be a better predictor of HIV transmission than systemic HIV-1 viral load (Baeten and Overbaugh, 2003). A number of studies have reported strong correlation between plasma and genital tract HIV-1 levels; however, only one published study has found a significant association between genital HIV shedding and increased transmission (Pedraza *et al.*, 1999).

Another factor that appears to increase risk of transmitting HIV, as well as risk of acquiring HIV in the exposed, uninfected partner, is the presence of sexually transmitted infections (STIs). STIs may facilitate HIV shedding in the genital tract, cause local recruitment of susceptible inflammatory cells, or disrupt genital mucosal surfaces. Increased risk of both HIV transmission and acquisition has been shown for both ulcerative and nonulcerative STIs. Early studies identified syphilis as increasing risk of acquisition 3.1- to 12.8-fold (Otten *et al.*, 1994). A separate study found a fivefold increase in acquisition of HIV in the presence of genital ulcers, where 89% of the ulcers were chancroid (Cameron *et al.*, 1989). The ulcerative STI that has received the most attention recently has been herpes simplex virus type 2 (HSV-2). Two meta-analyses have demonstrated that having HSV-2 increases the risk of transmission approximately threefold (Freeman *et al.*, 2006) and increases risk of HSV-2 acquisition twofold (Wald, 2004). The association between HSV-2 infection and HIV systemically and in genital secretions has been firmly established in a recent trial in Burkina Faso which showed that HSV suppression therapy with valacyclovir significantly reduced genital and plasma HIV-1 RNA level in women who are coinfected with HSV-2 and HIV-1 (Nagot *et al.*, 2007). Randomized clinical trials (RCTs) testing the hypotheses that HSV-2 suppression will reduce risk of HIV transmission and acquisition are underway.

In terms of nonulcerative STIs, several studies have shown significantly increased risk in the acquisition of HIV in the presence of gonorrhea, chlamydia, trichomoniasis, and bacterial vaginosis (Plummer *et al.*, 1991; Laga *et al.*, 1993; Craib *et al.*, 1995; Sewankambo *et al.*, 1997; Taha *et al.*, 1998). These data and others led to four community-level RCTs conducted in Mwanza, Tanzania; Rakai, Uganda; Masaka, Uganda; and Manicaland, Zimbabwe, to see the effect of STI treatment on HIV transmission. The study in Mwanza showed a 38% reduction in HIV incidence among those receiving the intervention to prevent STIs when compared with the nonintervention group (Grosskurth *et al.*, 1995); however, none of the other three studies was able to demonstrate a positive impact of STI treatment on HIV transmission (Wawer *et al.*, 1999; Kamali *et al.*, 2003). The main explanation for these discordant findings is that Mwanza represented a community that was in an early phase of the HIV epidemic, while the other three communities were in a more mature phase of the epidemic (Korenromp *et al.*, 2005). Therefore, new HIV epidemics characterized by high-risk behavior may benefit significantly more from interventions than HIV epidemics in a more mature phase.

Vaginal microbicides targeting STIs and HIV have also been evaluated for a protective effect against HIV transmission. Nonoxynol-9 has been studied in several RCTs, and to date, none has shown any benefit against HIV transmission (Kreiss *et al.*, 1992; Wilkinson *et al.*, 2002). In fact, some of these studies have demonstrated increased risk of HIV transmission among those women using the microbicide, probably secondary to increased genital inflammation and ulceration. Such risks are being closely monitored in ongoing RCTs examining other microbicides to determine their impact on HIV transmission.

Hormonal contraception is another potential risk factor for HIV acquisition that has been extensively studied. Hormonal contraception could increase a woman's susceptibility via hormonally induced changes in the vaginal mucosa or through associated cervical ectopy (Chao *et al.*, 1994; Daly *et al.*, 1994; Sinei *et al.*, 1996). However, recent studies, including a trial following 6109 women in Uganda, Zimbabwe, and Thailand, have not found an association between hormonal contraceptive use and HIV acquisition (Mati *et al.*, 1995; Morrison *et al.*, 2007; Myer *et al.*, 2007).

Increased transmission rates have also been associated with specific sexual practices and several partnership factors. The different modes of sexual intercourse differ in their efficiency of HIV transmission, with anal intercourse having the highest rates of transmission, followed by vaginal, and then oral sex (Royce *et al.*, 1997). In addition, traumatic sex (fisting, rape), sex while under the influence of alcohol or cocaine, sex during menses, sex during pregnancy, the use of vaginal desiccants, and the use of vaginal tightening agents have been associated with increased HIV transmission risk in several studies (Lazzarin *et al.*, 1991; Henin *et al.*, 1993; Seidlin *et al.*, 1993). Partnership factors that increase HIV transmission risk include high number of partners, high frequency of sexual contact, and concurrency. Concurrency is a partnership timing factor defined as being engaged in more than one sexual relationship at a single time point. In contrast, serial monogamy occurs when each partnership ends before the start of the next sexual relationship (Morris and Kretzschmar, 1997).

One important intervention to prevent HIV transmission is condom use. In several longitudinal studies, consistent condom use was 80–87% effective in preventing HIV transmission (Davis and Weller, 1999; Weller and Davis, 2002). Condom use is also associated with decreased transmission of gonorrhea, chlamydia, and HSV-2, all important HIV risk factors (Shafii *et al.*, 2004). Despite these compelling data, increasing condom use within at-risk populations has been a challenge. Several behavioral studies designed to reduce risky behavior and increase condom use have had no impact on HIV incidence (Hearst and Chen, 2004).

Strong data also support circumcision of adult men as a means to protect against HIV transmission. A number of observational studies have demonstrated a significant association between male circumcision and HIV transmission (Quinn *et al.*, 2000). These were recently validated in three large clinical trials, which randomized men to circumcision or no circumcision. One biological explanation is that keratinization of the glans penis when circumcised prevents tears and abrasions and decreases the presence of Langerhans cells and other HIV target cells (Quinn, 2006). The first RCT was completed in South Africa and showed 60% protection against HIV infection in the intention-to-treat analysis and 75% protection in the per-protocol analysis (Auvert *et al.*, 2005). Two other concurrent RCTs in Kenya and Uganda were subsequently published and confirmed that male circumcision reduced HIV incidence in men with similar risk reduction as the RCT in South Africa (Bailey *et al.*, 2007; Gray *et al.*, 2007).

Transmission via Injection Drug Use (IDU)

HIV transmission from IDU accounts for 15–25% of HIV infection globally and approximately one-third of HIV infection outside sub-Saharan Africa. The rate of HIV transmission for IDU is 0.0067 per exposure (see **Figure 5**) (Kaplan and Heimer, 1992). The majority of HIV-infected intravenous drug users live in Southern China, East Asia, the Russian Federation, and the Middle East. In some regions, such as Eastern Europe and Central Asia, approximately 80% of incident HIV cases result from IDU (UNAIDS/WHO, 2006). As there are approximately 13 million intravenous drug users globally, and 8.8 million in Eastern Europe and Asia, HIV prevention efforts targeting IDU are essential in curbing the rapid spread of HIV in these regions.

The greatest risk factor for HIV transmission via IDU is the use of contaminated syringes, needles, and other paraphernalia. This is most often associated with the practice of sharing drug paraphernalia, which can occur for cultural, social, legal, or economic reasons. Sharing with multiple injection partners is an added risk factor. This can take place at sites where drugs are sold and injection equipment is made available for rental, such as in 'shooting galleries' (Marmor *et al.*, 1987). Practices that increase risk when sharing include backloading, frontloading, and booting. The first two occur when a drug solution is squirted from a donor syringe into another by removing the plunger (backloading) or needle (frontloading) from the receiving syringe. Booting is the practice of repeatedly drawing blood into the syringe and mixing it with drug.

In addition to drug practices, the type of drug may be important. The use of cocaine has shown higher risk of HIV transmission when compared to heroin (Anthony *et al.*, 1991), perhaps due to the need for greater injection frequency and also the more frequent practice of booting in cocaine drug injection. Also, cocaine is associated with high-risk sexual practices.

Another risk factor that is intimately linked with IDU is commercial sex work (Astemborski *et al.*, 1994). Drug users often have to fund their habits by working as sex workers and drug-using commercial workers are less likely to use condoms than nonintravenous drug users. HIV epidemics in China, Indonesia, Kazakhstan, Ukraine, Uzbekistan, and Viet Nam are fuelled by the overlap of commercial sex work and IDU (UNAIDS/WHO, 2006).

One way to decrease the risk of HIV transmission among intravenous drug users is through cessation of drug injection. Methadone or buprenorphine replacement has been associated with a decrease in risk of HIV infection among users as a result of decreases in drug use (Sullivan *et al.*, 2005). However, pharmacotherapy has overall been unsuccessful in complete cessation of IDU (Amato *et al.*, 2004). Other methods to decrease the risk of HIV include syringe and needle exchange programs. These programs offer clean syringes and needles in exchange for used injection equipment, with the goal of decreasing needle and syringe sharing. Several studies in the USA, the United Kingdom, and the Netherlands have shown them to be effective (Buning *et al.*, 1988; Donoghoe *et al.*, 1989; Watters *et al.*, 1994). Despite studies that show decreased needle sharing, there has been much controversy surrounding these programs due to concerns that the programs are condoning drug use and may increase the prevalence of drug use by making syringes and needles more available. These concerns have not been validated in other studies in the USA, Britain, France, Sweden, and the Netherlands, which showed no increase in the prevalence of drug use in the setting of needle exchange (Brickner *et al.*, 1989; Oliver *et al.*, 1994; Paone *et al.*, 1994). An additional benefit of needle exchange programs is that they provide an opportunity for counselors to educate drug users about preventive interventions.

Mother-to-Child Transmission

Vertical transmission rates in many developing countries remain as high as 10–20% even though the likelihood an HIV-1-infected mother will transmit HIV-1 to her infant has been reduced to less than 1% in the USA and Europe. As a result, more than 500 000 HIV infections occur in infants each year in sub-Saharan Africa and other resource-limited settings (UNAIDS/WHO, 2006). Differences in transmission rates may be attributed in part to poor access and uptake of HIV diagnostic testing and prevention interventions in those regions with the highest HIV prevalence among women of reproductive age. A number of other factors, including infant feeding practices, rates of coinfections, and breast pathology, also contribute.

HIV may be transmitted from mother to infant during pregnancy (*in utero*), during delivery (intrapartum), or through breastfeeding. In the absence of antiretroviral therapy, the overall transmission rate among non-breastfeeders is approximately 15–30% (**Figure 6**). Among non-breastfeeding women, approximately two-thirds of mother-to-child HIV-1 transmission events occur during delivery when the baby passes through the birth canal and is exposed to infected maternal blood and genital secretions. The remaining infant infections are attributable to *in utero* transmission.

In resource-limited settings, the majority of HIV-1-infected women breastfeed their infants due to lack of clean water or safe alternatives to breast milk, or due to stigma associated with not breastfeeding in regions where

Breastfeeders

In utero 5–10%	Intrapartum 10–20%	Breast milk 10–20%	Overall 30–45%

Non-breastfeeders

In utero 5–10%	Intrapartum 10–20%	Overall 15–30%

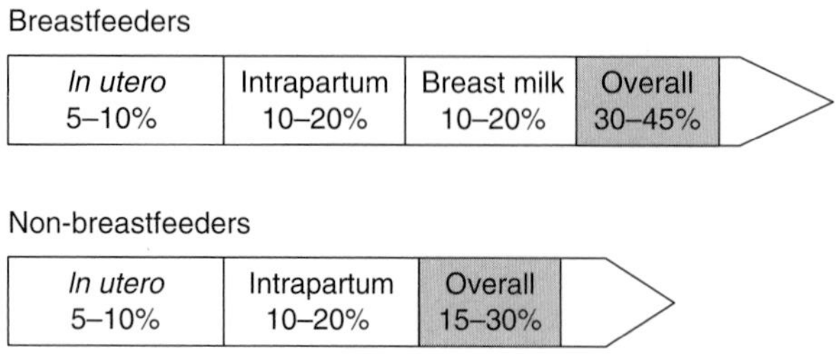

Figure 6 Timing and absolute rates of vertical HIV-1 transmission.

breastfeeding is the norm. Consequently, overall rates of transmission in the absence of antiretroviral prophylaxis increase to 30–45% (John and Kreiss, 1996) and several studies suggest that breast milk exposure accounts for one-third to one-half of these transmission events. Transmission during the *in utero* and intrapartum periods contributes the remainder of infections in proportions similar to those among non-breastfeeders (see **Figure 6**).

Valuable data on breast milk transmission rates were collected in a randomized clinical trial of breastfeeding versus formula feeding in Kenya where 16% of infants acquired HIV-1 via breast milk during the 2-year study period (Nduati *et al.*, 2000). The proportion of transmission events attributable to breastfeeding was 44%. A subsequent meta-analysis of nine cohorts of HIV-1-infected breastfeeding mothers found similar results, reporting that the proportion of infant infections attributable to breastfeeding after the first month postpartum was 24–42% (Coutsoudis *et al.*, 2004). Risk of breast milk transmission after the first month appears to remain relatively constant over time, with transmission events continuing to accumulate with longer duration of breastfeeding (Coutsoudis *et al.*, 2004). During this period, the risk of infant HIV-1 acquisition per liter of breast milk ingested is similar to the risk of an unprotected sex act among heterosexual adults (Richardson *et al.*, 2003).

High maternal plasma HIV-1 RNA load has been consistently associated with increased transmission and is considered the most important predictor of vertical transmission risk (Dickover *et al.*, 1996). Several studies have demonstrated that reduction of HIV-1 viral load with antiretroviral drugs decreases overall rates of transmission, and this has become the mainstay of global prevention efforts. HIV-1 viremia is highest in the setting of advanced disease and is usually associated with severe immunosuppression. High HIV-1 levels also occur during acute HIV infection, a time when a woman may not realize she is at risk for transmitting HIV-1 to her child. This has been examined in one study of women acquiring HIV-1 postpartum which found that there was a twofold increased risk of transmitting HIV-1 via breast milk during acute HIV (Dunn and Newell, 1992).

In addition to HIV-1 viral load, several biological and behavioral factors contribute to increased transmission risk. Ascending bacterial infections that cause inflammation of the placenta, chorion, and amnion have been associated with increased *in utero* transmission in several observational studies (Taha and Gray, 2000). Sexually transmitted infections (STIs), such as syphilis and gonorrhea, are considered to increase risk of transmission via this mechanism (Mwapasa *et al.*, 2006), as well as chorioamnionitis resulting from infection with local vaginal or enteric flora. However, the role of chorioamnionitis was not confirmed in a randomized clinical trial of more than 2000 women comparing antibiotic treatment of ascending infections to placebo which found no difference in transmission events between the two arms (Taha *et al.*, 2006).

Other risk factors include infant gender and malarial infection of placental tissue, which has been inconsistently reported to increase mother-to-child HIV-1 transmission (Brentlinger *et al.*, 2006). Strong associations have been found between infant gender and *in utero* transmission in several cohorts (Galli *et al.*, 2005; Taha *et al.*, 2005; Biggar *et al.*, 2006). Female infants have a twofold increased risk of infection at birth when compared to male infants, perhaps because *in utero* mortality is higher for male HIV-1-infected infants than for females (Galli *et al.*, 2005; Taha *et al.*, 2005; Biggar *et al.*, 2006).

Risk factors influencing the intrapartum period include genital tract HIV-1 levels, genital ulcer disease, delivery complications, and breaks in the placental barrier that may cause maternal–fetal microtransfusions. Genital ulcer disease has been shown to increase mother-to-child HIV-1 transmission at least twofold, both in the presence and absence of antiretrovirals (John *et al.*, 2001; Chen *et al.*, 2005; Drake *et al.*, 2007). The majority of genital ulcers are caused by HSV-2, and in regions with high HIV-1 seroprevalence, HSV-2 seroprevalence is greater than 70% among women of reproductive age. Other risk factors for intrapartum transmission include prolonged duration of ruptured membranes and cervical or vaginal lacerations that occur during delivery. To restrict exposure to maternal HIV-1 in blood and mucosal secretions, cesarean section has been used effectively in many settings.

Two of the most important determinants of breast milk transmission risk are duration of breastfeeding and HIV-1 viral levels in breast milk. Introduction of food other than maternal milk may also exert an effect on HIV-1 transmission risk, potentially by compromising infant mucosal surfaces in the oropharynx and gut. Breast milk viral load increases with increased plasma viremia and also as a result of local factors. These include breast inflammation resulting from mastitis, breast abscess, or other breast pathology, and inflammation within the breast milk compartment in the absence of clinical symptoms, which is known as subclinical mastitis (John *et al.*, 2001). Rates of subclinical mastitis are reported to be as high as 30% among HIV-1-infected breastfeeding women, and several studies have found it to be associated with increased HIV-1 levels in breast milk, as well as greater risk of infant HIV-1 acquisition (Willumsen *et al.*, 2003).

Transmission via Exposure to Blood Products

HIV transmission through blood transfusion and blood products has become rare after much progress in instituting careful screening and limiting the use of transfusions. Stringent screening of blood and blood products was a high priority at the start of the epidemic when it became known that transfusion with HIV-infected blood was an extremely efficient mode of transmission. In retrospective studies, approximately 90% of transfusion recipients were infected per single contaminated unit of blood (Donegan *et al.*, 1990), and 75–90% of recipients of Factor VIII concentrate acquired HIV (Caussy and Goedert, 1990).

In developed countries, blood product screening for HIV includes serologic and nucleic acid donor testing. In Canada, the estimated risk of HIV-infected donation being accepted was 1 per 7.8 million donations in the period from 2001 to 2005 (O'Brien *et al.*, 2007), and in the USA, the estimated risk of an HIV-infected donation being accepted was 1 per approximately 2.1 million donations in 2001 (Dodd *et al.*, 2002). While great strides have been made to improve the safety of the blood supply in many countries, antibody screening of blood donors is not universally done in many developing countries due to the lack of resources. In addition, some developing nations still rely on paid donors for their blood banks, and these paid donors have been shown to be at high risk for bloodborne infections (Volkow and Del Rio, 2005).

Occupational Exposure and HIV Transmission Risk

Health-care workers and laboratory personnel have a higher risk of HIV infection than individuals in other occupations. The average risk of HIV transmission is approximately 0.3% following percutaneous exposure to HIV-infected blood (see **Figure 5**) (Bell, 1997). Accidental percutaneous exposure carries the highest risk of subsequent infection and is the most important cause of occupational HIV transmission. Needlesticks are the most common cause of a percutaneous accident, and factors that increase transmission from a needlestick accident include deep puncture with a large-bore needle, larger quantity of blood, and exposure to blood from an HIV-infected individual with a high HIV viral load (Cox and Hodgson, 1988). Contaminated surgical instruments have also been associated with percutaneous injuries. Nonpenetrating accidents that involve intact skin are far less risky. Retrospective data suggest that immediate initiation of effective antiretroviral drugs can decrease risk of acquiring HIV after an occupational exposure (Cardo *et al.*, 1997). Currently, recommended postexposure prophylaxis comprises a two- or three-drug antiretroviral regimen, depending on the nature and severity of the exposure (Centers for Disease Control and Prevention, 2005).

Regional HIV Surveillance Data

The HIV epidemic has developed distinctly in different regions. Generalized epidemics occur when the general population HIV prevalence reaches 1% or more, and concentrated epidemics are described as an HIV prevalence of 5% or more in high-risk groups.

Sub-Saharan Africa: Overview

In 2006, HIV and AIDS continue to have a devastating effect on sub-Saharan Africa which is home to 24.7 million people living with HIV and 63% of all HIV-infected individuals in the world. Approximately 2.8 million people became newly infected in sub-Saharan Africa in 2006, and approximately 2.1 million people died from HIV in this same year. Women are more likely to be infected than men in this region, and 59% of all HIV-infected persons are estimated to be female (**Figure 7**) (UNAIDS/WHO, 2006).

Southern Africa

Within sub-Saharan Africa, Southern Africa has the highest burden of disease, accounting for 32% of HIV-infected individuals and 34% of AIDS deaths in the world. All countries of Southern Africa continue to increase in prevalence, except Zimbabwe, where HIV seroprevalence among adults has reduced to 20.1% from 22.1% in 2003. A decrease in HIV prevalence was reported in antenatal clinics in Zimbabwe from 30–32% in 2000 to 24% in 2004. Swaziland has the highest adult prevalence of HIV globally at 33.4%, and South Africa has the second greatest number of HIV-infected persons among all nations in the world after India. Across Southern Africa, the death rate from AIDS significantly increased in both men and women between 1997 and 2004, despite increased antiretroviral availability and use.

East Africa

Overall, HIV prevalence has been stabilizing or decreasing in this region of Africa. Kenya, Tanzania, and Rwanda have all shown a decline in prevalence over the last several years. In Kenya, adult HIV prevalence decreased from 10% in the late 1990s to approximately 6% in 2005 (Kenya Ministry of Health, 2005), and in Tanzania, adult HIV prevalence decreased from 8.1% to 6.5% between 1995 and 2004 (Somi *et al.*, 2006). There is concern that a similar trend may not continue in Uganda where there is a decrease in consistent condom use and increase in the number of men with more than one partner (Uganda Ministry of Health, 2006). In addition, there is concern that the growing number of intravenous drug users, especially in Kenya, Tanzania, and Zanzibar may significantly modify the course of the epidemic in these regions (Ndetei, 2004; Odek-Ogunde, 2004).

West and Central Africa

The epidemic has been less severe overall in West Africa compared to East and Southern Africa. There is declining HIV prevalence in Ghana, Côte d'Ivoire, and Burkina Faso; however, prevalence appears to be increasing in Mali (World Health Organization, 2005). In Mali, the HIV seroprevalence using antenatal clinic data has increased from 3.3% in 2002 to 4.1% in 2005. In Nigeria, which is known to have the third largest number of HIV-infected persons in the world, as a result of its large population, HIV prevalence appears to have stabilized at approximately 4.4% in 2005.

Asia: Overview

In 2006, there were 8.6 million people infected with HIV in Asia. Of these, 960 000 people were newly HIV-infected in 2006, and approximately 630 000 people died from AIDS. Compared to countries in sub-Saharan Africa, seroprevalence in most Asian countries is lower; however, generalized epidemics are found in Thailand, Cambodia, and Myanmar.

South-East Asia

South-East Asian countries have the highest HIV prevalence in Asia due to the commercial sex industry, widespread IDU, and transmission among men who have sex

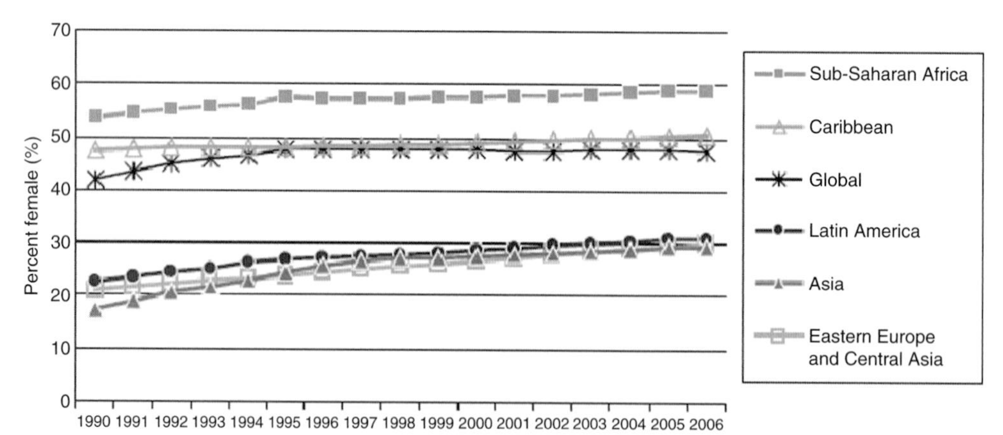

Figure 7 Percentage of adults greater than 15 years of age living with HIV who are female, 1990–2006. Reproduced with kind permission from UNAIDS/WHO (2006) *AIDS Epidemic Update 2006.* Geneva: UNAIDS/WHO.

with men (MSM) (**Figure 8**). While the epidemic appears to be stabilizing in Thailand, Cambodia, and Myanmar, in Viet Nam the number of infected individuals has almost doubled in number between 2000 and 2005. Despite the overall stabilizing trend in Thailand, there is a growing concern that the epidemic is increasing within the general population. One-third of newly infected individuals in 2005 were married women. In Cambodia, the stabilizing trend may have been partly due to behavioral changes in the commercial sex work industry. Consistent condom use increased from 53% to 96% among brothel-based sex workers in the major cities of Cambodia (Gorbach *et al.*, 2006) and this was followed by a decline in the HIV prevalence among sex workers from 43% to 21% between 1995 and 2003 (National Center for HIV/AIDS, Dermatology and STDs, 2004).

People's Republic of China

Currently, there are 650 000 people living with HIV in the People's Republic of China (PRC) China, and approximately 45% of these individuals are intravenous drug users (China Ministry of Health, 2006). The epidemic initially occurred in rural provinces along the major drug-trafficking routes, and major outbreaks from HIV-1 commercial plasma donors worsened the epidemic (Wu *et al.*, 2001). Sexual transmission has further fuelled the spread of HIV to urban areas in the People's Republic of China, which from 1998 to 2005 experienced an exponential increase in new HIV cases (Wu *et al.*, 2007). New cases were estimated to climb from 3300 to 36 000 during this period, with a greater proportion due to sexual transmission in populations considered at lower risk. As a result, there is concern that the epidemic is now spreading from high-risk populations of intravenous drug users and commercial sex workers to the general population.

India

The second most populous nation, India, had approximately 5.7 million HIV-infected people in 2005, more than any other country in the world. Most of these reported HIV cases are concentrated in 6 of the 28 states of India where transmission is occurring through unprotected heterosexual intercourse. Among those infected, 38% are women and HIV prevalence among women from 206 antenatal clinics has declined from 1.7% in 2000 to 1.1% in 2004 (Kumar *et al.*, 2006). The epidemics in the northeast part of India are mainly due to IDU.

Eastern Europe and Central Asia

Currently, there are 1.7 million people living with HIV in Eastern Europe and Central Asia, where new cases of HIV have grown significantly in the past decade and continue to increase. Ukraine and the Russian Federation comprise 90% of HIV infections of this region (EuroHIV, 2006), and approximately 80% of those infected in the Russian Federation are 15 to 30 years of age. The primary mode of transmission in this area is IDU, which makes up approximately 67% of cases in Eastern Europe and Central Asia (see **Figure 8**). However, the number of newly infected individuals through unprotected sex is also growing.

Latin America and the Caribbean

Approximately 250 000 people are living with HIV in the Caribbean. The epidemic was initially due to the commercial sex industry and eventually spread into the general population. In addition, 10% of those who are HIV-infected are MSM. The HIV infection rates have stabilized in the Caribbean, where Haiti and the Dominican Republic account for approximately 75% of HIV-infected cases.

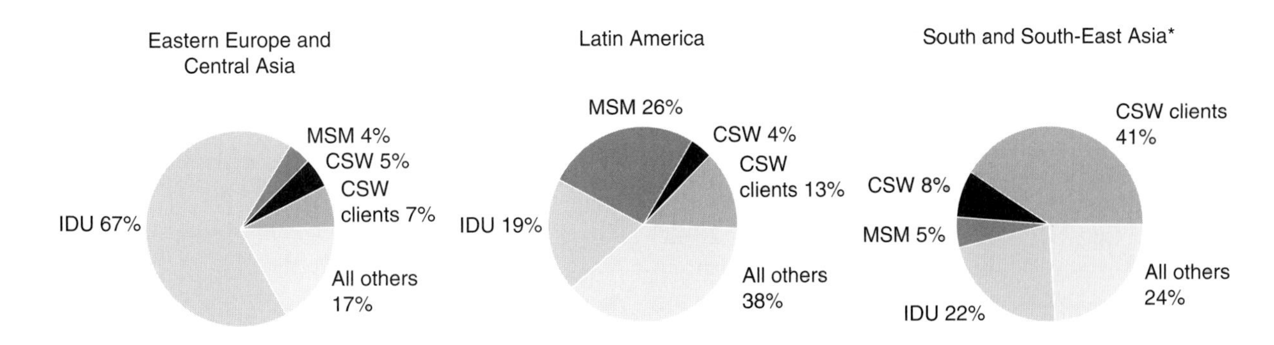

* India was omitted from this analysis because the scale of its HIV epidemic (which is largely heterosexual) masks the extent to which other at-risk populations feature in the region's epidemics.

Figure 8 Proportions of HIV infections in different population groups by region, 2005. Reproduced with kind permission from UNAIDS/WHO (2006) *AIDS Epidemic Update 2006.* Geneva: UNAIDS/WHO.

In Latin America, the epidemic has stabilized overall with 1.7 million people living with HIV and approximately 140 000 new infections annually. The majority of the cases are due to unsafe IDU and MSM (see **Figure 8**), and more than 60% of HIV-infected persons reside in Argentina, Brazil, Colombia, and Mexico. There has been much international praise for the efforts of the Brazilian government for HIV prevention and treatment efforts during the last decade.

North America, Western, and Central Europe

In North America and Western and Central Europe, the number of living HIV-infected persons is 2.1 million, a number that continues to increase. The USA accounts for 1.2 million people living with HIV, which places it among the 10 countries in the world with the highest number living with HIV or AIDS. This is because individuals with HIV are living longer as a result of increased use of antiretroviral drugs in these parts of the world. In North America and Western and Central Europe, there were 65 000 new infections and approximately 30 000 deaths from AIDS in 2006.

The main populations at risk in the USA and Europe are MSM and intravenous drug users. However, women, African-Americans, and immigrants are increasingly affected by the HIV epidemic. The proportion of women newly infected with HIV increased from 15% before 1995 to 27% in 2004 in the USA, and between 2001 and 2004, half of all AIDS diagnoses were in African-Americans. In 2004, African-Americans had a significantly higher risk of HIV infection than among Whites in the USA, 7 times greater in men and 21 times greater in women. In Western and Central Europe, three-quarters of all heterosexually acquired HIV are among migrants and immigrants (EuroHIV, 2006).

Middle and North Africa

There are an estimated 450 000 people living with HIV, with 68 000 newly infected, in Middle and North Africa. The major modes of transmission are IDU, commercial sex, and sex among MSM. In this region, Sudan is the only nation with a generalized epidemic with a prevalence of 1.6%. Other nations in this region have poor surveillance, and therefore it may be difficult to understand fully the nature of the epidemic. Greater efforts are needed to improve surveillance and access to high-risk populations to control the HIV epidemic.

Oceania

In Oceania, an estimated 81 000 persons were infected with HIV and 7100 were newly infected in 2006. Three-quarters of the infected in this region reside in Papua New Guinea, where prevalence has increased to 1.8%. Risk factors for transmission in Papua New Guinea include inconsistent condom use, sex with commercial sex workers, high rates of concurrent sexual relationships, early sexual initiation, high rates of sexual violence against women, and high rates of STIs (Brouwer *et al.*, 1998; Asian Development Bank, 2006). In Australia and New Zealand, transmission among MSM continues to drive the epidemic, accounting for 66% and 90% of all new HIV cases, respectively.

See also: Mother-to-Child Transmission of HIV; Sexually Transmitted Infections: Overview.

Citations

Abbott RC, NDour-Sarr A, Diouf A, *et al.* (1994) Risk factors for HIV-1 and HIV-2 infection in pregnant women in Dakar, Senegal. *Journal of Acquired Immune Deficiency Syndromes* 7: 711–717.

Adjorlolo-Johnson G, De Cock KM, Ekpini E, *et al.* (1994) Prospective comparison of mother-to-child transmission of HIV-1 and HIV-2 in Abidjan, Ivory Coast. *Journal of the American Medical Association* 272: 462–466.

Alaeus A, Lidman K, Bjorkman A, Giesecke J, and Albert J (1999) Similar rate of disease progression among individuals infected with HIV-1 genetic subtypes A-D. *AIDS* 13: 901–907.

Amato L, Minozzi S, Davoli M, Vecchi S, Ferri M, and Mayet S (2004) Psychosocial combined with agonist maintenance treatments versus agonist maintenance treatments alone for treatment of opioid dependence. *Cochrane Database of Systematic Reviews* 4, CD004147.

Andreasson PA, Dias F, Naucler A, Andersson S, and Biberfeld G (1993) A prospective study of vertical transmission of HIV-2 in Bissau, Guinea-Bissau. *AIDS* 7: 989–993.

Anthony JC, Vlahov D, Nelson KE, Cohn S, Astemborski J, and Solomon L (1991) New evidence on intravenous cocaine use and the risk of infection with human immunodeficiency virus type 1. *American Journal of Epidemiology* 134: 1175–1189.

Asian Development Bank (2006) www.adb.org/Documents/RRPs/PNG/39033-PNG-RRP.pdf (accessed April 2008).

Astemborski J, Vlahov D, Warren D, Solomon L, and Nelson KE (1994) The trading of sex for drugs or money and HIV seropositivity among female intravenous drug users. *American Journal of Public Health* 84: 382–387.

Auvert B, Taljaard D, Lagarde E, Sobngwi-Tambekou J, Sitta R, and Puren A (2005) Randomized, controlled intervention trial of male circumcision for reduction of HIV infection risk: The ANRS 1265 Trial. *Public Library of Science Medicine* 2: e298.

Baeten JM, Chohan B, Lavreys L, *et al.* (2007) HIV-1 subtype D infection is associated with faster disease progression than subtype A in spite of similar plasma HIV-1 loads. *Journal of Infectious Diseases* 195: 1177–1180.

Baeten JM and Overbaugh J (2003) Measuring the infectiousness of persons with HIV-1: Opportunities for preventing sexual HIV-1 transmission. *Current HIV Research* 1: 69–86.

Bailey RC, Moses S, Parker CB, *et al.* (2007) Male circumcision for HIVprevention in young men in Kisumu, Kenya: A randomised controlled trial. *Lancet* 369: 643–656.

Barin F, M'Boup S, Denis F, *et al.* (1985) Serological evidence for virus related to simian T-lymphotropic retrovirus III in residents of west Africa. *Lancet* 2: 1387–1389.

Bell DM (1997) Occupational risk of human immunodeficiency virus infection in healthcare workers: An overview. *American Journal of Medicine* 102: 9–15.

Biggar RJ, Taha TE, Hoover DR, Yellin F, Kumwenda N, and Broadhead R (2006) Higher in utero and perinatal HIV infection risk in girls than boys. *Journal of Acquired Immune Deficiency Syndromes* 41: 509–513.

Brentlinger PE, Behrens CB, and Micek MA (2006) Challenges in the concurrent management of malaria and HIV in pregnancy in sub-Saharan Africa. *Lancet Infectious Diseases* 6: 100–111.

Brickner PW, Torres RA, Barnes M, *et al.* (1989) Recommendations for control and prevention of human immunodeficiency virus (HIV) infection in intravenous drug users. *Annals of Internal Medicine* 110: 833–837.

Brouwer EC, Harris BM, and Tanaka S (1998) *Gender Analysis in Papua New Guinea*. Washington, DC: The World Bank.

Buning EC, van Brussel GH, and van Santen G (1988) Amsterdam's drug policy and its implications for controlling needle sharing. *NIDA Research Monographs* 80: 59–74.

Burke DS, McCutchan FE, and Francine E (1997) *Global Distribution of Human Immunodeficiency Virus-1 Clades*. Philadelphia, PA: Lippincott-Raven.

Cameron DW, Simonsen JN, D'Costa LJ, *et al.* (1989) Female to male transmission of human immunodeficiency virus type 1: Risk factors for seroconversion in men. *Lancet* 2: 403–407.

Cardo DM, Culver DH, Ciesielski CA, *et al.* (1997) A case-control study of HIV seroconversion in health care workers after percutaneous exposure. Centers for Disease Control and Prevention Needlestick Surveillance Group. *New England Journal of Medicine* 337: 1485–1490.

Caussy D and Goedert JJ (1990) The epidemiology of human immunodeficiency virus and acquired immunodeficiency syndrome. *Seminars in Oncology* 17: 244–250.

Centers for Disease Control and Prevention (2005) Updated U.S. Public Health Service Guidelines for the Management of Occupational Exposures to HIV and Recommendations for Postexposure Prophylaxis. *MMWR* 1–17.

Chao A, Bulterys M, Musanganire F, *et al.* (1994) Risk factors associated with prevalent HIV-1 infection among pregnant women in Rwanda. National University of Rwanda-Johns Hopkins University AIDS Research Team. *International Journal of Epidemiology* 23: 371–380.

Chen KT, Segu M, Lumey LH, *et al.* (2005) Genital herpes simplex virus infection and perinatal transmission of human immunodeficiency virus. *Obstetrics and Gynecology* 106: 1341–1348.

China Ministry of Health (MoH) (2006) 2005 update on the HIV/AIDS epidemic and response in China, ed. China MoH. UNAIDS Beijing, China: WHO.

Coutsoudis A, Dabis F, Fawzi W, *et al.* (2004) Late postnatal transmission of HIV-1 in breast-fed children: An individual patient data meta-analysis. *Journal of Infectious Diseases* 189: 2154–2166.

Cox J and Hodgson L (1988) Hepatitis B, AIDS and the neurological pin. *Medical Education* 22: 83.

Craib KJ, Meddings DR, Strathdee SA, *et al.* (1995) Rectal gonorrhoea as an independent risk factor for HIV infection in a cohort of homosexual men. *Genitourinary Medicine* 71: 150–154.

Daly CC, Helling-Giese GE, Mati JK, and Hunter DJ (1994) Contraceptive methods and the transmission of HIV: Implications for family planning. *Genitourinary Medicine* 70: 110–117.

Davis KR and Weller SC (1999) The effectiveness of condoms in reducing heterosexual transmission of HIV. *Family Planning Perspectives* 31: 272–279.

Dickover RE, Garratty EM, Herman SA, *et al.* (1996) Identification of levels of maternal HIV-1 RNA associated with risk of perinatal transmission. Effect of maternal zidovudine treatment on viral load. *Journal of the American Medical Association* 275: 599–605.

Dodd RY, Notari EP, and Stramer SL (2002) Current prevalence and incidence of infectious disease markers and estimated window-period risk in the American Red Cross blood donor population. *Transfusion* 42: 975–979.

Donegan E, Stuart M, Niland JC, *et al.* (1990) Infection with human immunodeficiency virus type 1 (HIV-1) among recipients of antibody-positive blood donations. *Annals of Internal Medicine* 113: 733–739.

Donoghoe MC, Stimson GV, Dolan K, and Alldritt L (1989) Changes in HIV risk behaviour in clients of syringe-exchange schemes in England and Scotland. *AIDS* 3: 267–272.

Drake AL, John-Stewart GC, Wald A, *et al.* (2007) Herpes simplex virus type 2 and risk of intrapartum human immunodeficiency virus transmission. *Obstetrics and Gynecology* 109: 403–409.

Dunn D and Newell ML (1992) Vertical transmission of HIV. *Lancet* 339: 364–365.

EuroHIV (2006) *HIV/AIDS Surveillance in Europe: End-year Report 2005* vol. 73. Saint-Maurice: Institut de Veille Sanitaire.

Freeman EE, Weiss HA, Glynn JR, Cross PL, Whitworth JA, and Hayes RJ (2006) Herpes simplex virus 2 infection increases HIV acquisition in men and women: Systematic review and meta-analysis of longitudinal studies. *AIDS* 20: 73–83.

Galli L, Puliti D, Chiappini E, *et al.* (2005) Lower mother-to-child HIV-1 transmission in boys is independent of type of delivery and antiretroviral prophylaxis: The Italian Register for HIV Infection in Children. *Journal of Acquired Immune Deficiency Syndromes* 40: 479–485.

Gilbert PB, McKeague IW, Eisen G, *et al.* (2003) Comparison of HIV-1 and HIV-2 infectivity from a prospective cohort study in Senegal. *Statistics in Medicine* 22: 573–593.

Gorbach PM, Sopheab H, Chhorvann C, Weiss RE, and Vun MC (2006) Changing behaviors and patterns among Cambodian sex workers: 1997–2003. *Journal of Acquired Immune Deficiency Syndromes* 42: 242–247.

Gray RH, Wawer MJ, Brookmeyer R, *et al.* (2001) Probability of HIV-1 transmission per coital act in monogamous, heterosexual, HIV-1 discordant couples in Rakai, Uganda. *Lancet* 357: 1149–1153.

Gray RH, Kigozi G, Serwadda D, *et al.* (2007) Male circumcision for HIV prevention in men in Rakai, Uganda: A randomised trial. *Lancet* 369: 657–666.

Grosskurth H, Mosha F, Todd J, *et al.* (1995) Impact of improved treatment of sexually transmitted diseases on HIV infection in rural Tanzania: Randomised controlled trial. *Lancet* 346: 530–536.

Hearst N and Chen S (2004) Condom promotion for AIDS prevention in the developing world: Is it working? *Studies in Family Planning* 35: 39–47.

Henin Y, Mandelbrot L, Henrion R, Pradinaud R, Coulaud JP, and Montagnier L (1993) Virus excretion in the cervicovaginal secretions of pregnant and nonpregnant HIV-infected women. *Journal of Acquired Immune Deficiency Syndromes* 6: 72–75.

John GC and Kreiss J (1996) Mother-to-child transmission of human immunodeficiency virus type 1. *Epidemiologic Reviews* 18: 149–157.

John GC, Nduati RW, Mbori-Ngacha DA, *et al.* (2001) Correlates of mother-to-child human immunodeficiency virus type 1 (HIV-1) transmission: Association with maternal plasma HIV-1 RNA load, genital HIV-1 DNA shedding, and breast infections. *Journal of Infectious Diseases* 183: 206–212.

Kaleebu P, Ross A, Morgan D, *et al.* (2001) Relationship between HIV-1 Env subtypes A and D and disease progression in a rural Ugandan cohort. *AIDS* 15: 293–299.

Kamali A, Quigley M, Nakiyingi J, *et al.* (2003) Syndromic management of sexually-transmitted infections and behaviour change interventions on transmission of HIV-1 in rural Uganda: A community randomised trial. *Lancet* 361: 645–652.

Kanki P (1997) *Epidemiology and Natural History of Human Immunodeficiency Virus Type 2*. Philadelphia, PA: Lippincott-Raven.

Kanki PJ, Travers KU, Marlink RG, *et al.* (1994) Slower heterosexual spread of HIV-2 than HIV-1. *Lancet* 343: 943–946.

Kaplan EH and Heimer R (1992) A model-based estimate of HIV infectivity via needle sharing. *Journal of Acquired Immune Deficiency Syndromes* 5: 1116–1118.

Kenya Ministry of Health (MoH) (2005) AIDS in Kenya. 7th edn, ed. (NASCOP) NAaSCP Nairobi, Kenya: Ministry of Health, Kenya.

Korenromp EL, White RG, Orroth KK, *et al.* (2005) Determinants of the impact of sexually transmitted infection treatment on prevention of HIV infection: A synthesis of evidence from the Mwanza, Rakai, and Masaka intervention trials. *Journal of Infectious Diseases* 191 (supplement 1): S168–S178.

Kreiss J, Ngugi E, Holmes K, *et al.* (1992) Efficacy of nonoxynol 9 contraceptive sponge use in preventing heterosexual acquisition of

HIV in Nairobi prostitutes. *Journal of the American Medical Association* 268: 477–482.
Kumar R, Jha P, Arora P, *et al.* (2006) Trends in HIV-1 in young adults in south India from 2000 to 2004: A prevalence study. *Lancet* 367: 1164–1172.
Laga M, Manoka A, Kivuvu M, *et al.* (1993) Non-ulcerative sexually transmitted diseases as risk factors for HIV-1 transmission in women: Results from a cohort study. *AIDS* 7: 95–102.
Lazzarin A, Saracco A, Musicco M, and Nicolosi A (1991) Man-to-woman sexual transmission of the human immunodeficiency virus. Risk factors related to sexual behavior, man's infectiousness, and woman's susceptibility. Italian Study Group on HIV Heterosexual Transmission. *Archives of Internal Medicine* 151: 2411–2416.
Marlink R, Kanki P, Thior I, *et al.* (1994) Reduced rate of disease development after HIV-2 infection as compared to HIV-1. *Science* 265: 1587–1590.
Marmor M, DesJarlais DC, Cohen H, *et al.* (1987) Risk factors for infection with human immunodeficiency virus among intravenous drug abusers in New York City. *AIDS* 1: 39–44.
Mati JK, Hunter DJ, Maggwa BN, and Tukei PM (1995) Contraceptive use and the risk of HIV infection in Nairobi, Kenya. *International Journal of Gynaecology and Obstetrics* 48: 61–67.
McCutchan FE (2000) Understanding the genetic diversity of HIV-1. *AIDS* 14(supplement 3): S31–S44.
McCutchan FE (2006) Global epidemiology of HIV. *Journal of Medical Virology* 78(supplement 1): S7–S12.
Morris M and Kretzschmar M (1997) Concurrent partnerships and the spread of HIV. *AIDS* 11: 641–648.
Morrison CS, Richardson BA, Mmiro F, *et al.* (2007) Hormonal contraception and the risk of HIV acquisition. *AIDS* 21: 85–95.
Mullins C, Eisen G, Popper S, *et al.* (2004) Highly active antiretroviral therapy and viral response in HIV type 2 infection. *Clinical Infectious Diseases* 38: 1771–1779.
Mwapasa V, Rogerson SJ, Kwiek JJ, *et al.* (2006) Maternal syphilis infection is associated with increased risk of mother-to-child transmission of HIV in Malawi. *AIDS* 20: 1869–1877.
Myer L, Denny L, Wright TC, and Kuhn L (2007) Prospective study of hormonal contraception and women's risk of HIV infection in South Africa. *International Journal of Epidemiology* 36: 166–174.
Nagot N, Ouedraogo A, Foulongne V, *et al.* (2007) Reduction of HIV-1 RNA levels with therapy to suppress herpes simplex virus. *New England Journal of Medicine* 356: 790–799.
National Center for HIV/AIDS, Dermatology and STDs (2004) *HIV Sentinel Surveillance (HSS) in Cambodia, 2003*. Phnom Penh, Cambodia: National Center for HIV/AIDS, Dermatology and STDs.
Ndetei D (2004) *Study on the assessment of the linkages between drug abuse, injecting drug abuse and HIV/AIDS in Kenya: A rapid situation assessment 2004*. Nairobi, Kenya: United Nations Office on Drugs and Crime.
Nduati R, John G, Mbori-Ngacha D, *et al.* (2000) Effect of breastfeeding and formula feeding on transmission of HIV-1: A randomized clinical trial. *Journal of the American Medical Association* 283: 1167–1174.
Neilson JR, John GC, Carr JK, *et al.* (1999) Subtypes of human immunodeficiency virus type 1 and disease stage among women in Nairobi, Kenya. *Journal of Virology* 73: 4393–4403.
Nelson KE, Celentano DD, Suprasert S, *et al.* (1993) Risk factors for HIV infection among young adult men in northern Thailand. *Journal of the American Medical Association* 270: 955–960.
Nopkesorn T, Mastro TD, Sangkharomya S, *et al.* (1993) HIV-1 infection in young men in northern Thailand. *AIDS* 7: 1233–1239.
O'Brien SF, Yi QL, Fan W, Scalia V, Kleinman SH, and Vamvakas EC (2007) Current incidence and estimated residual risk of transfusion-transmitted infections in donations made to Canadian Blood Services. *Transfusion* 47: 316–325.
O'Donovan D, Ariyoshi K, Milligan P, *et al.* (2000) Maternal plasma viral RNA levels determine marked differences in mother-to-child transmission rates of HIV-1 and HIV-2 in The Gambia. MRC/The Gambia Government/University College London Medical School working group on mother-child transmission of HIV. *AIDS* 14: 441–448.
Odek-Ogunde M (2004) *World Health Organization phase II drug injecting study: Behavioural and seroprevalence (HIV, HBV, HCV) survey among injecting drug users in Nairobi*. Nairobi, Kenya: WHO.
Oliver KMS, Friedman S, Maynard H, *et al.* (1994) *Behavioral and Community Impact of the Portland Syringe Exchange Program*. Washington, DC: National Academy Press.
Otten MW Jr, Zaidi AA, Peterman TA, Rolfs RT, and Witte JJ (1994) High rate of HIV seroconversion among patients attending urban sexually transmitted disease clinics. *AIDS* 8: 549–553.
Paone D, Des Jarlais DC, *et al.* (1994) *New York City Syringe Exchange: An Overview*. Washington, DC: National Academy Press.
Pedraza MA, del Romero J, and Roldan F (1999) Heterosexual transmission of HIV-1 is associated with high plasma viral load levels and a positive viral isolation in the infected partner. *Journal of Acquired Immune Deficiency Syndromes* 21: 120–125.
Pichova I, Weber J, Litera J, *et al.* (1997) Peptide inhibitors of HIV-1 and HIV-2 proteases: A comparative study. *Leukemia* 11(supplement 3): 120–122.
Plummer FA, Simonsen JN, Cameron DW, *et al.* (1991) Cofactors in male-female sexual transmission of human immunodeficiency virus type 1. *Journal of Infectious Diseases* 163: 233–239.
Quinn TC (2006) Male circumcision as a preventive measure limiting HIV transmission. *Retrovirology* 3(supplement 1): S109.
Quinn TC, Wawer MJ, Sewankambo N, *et al.* (2000) Viral load and heterosexual transmission of human immunodeficiency virus type 1. Rakai Project Study Group. *New England Journal of Medicine* 342: 921–929.
Richardson BA, John-Stewart GC, Hughes JP, *et al.* (2003) Breast-milk infectivity in human immunodeficiency virus type 1-infected mothers. *Journal of Infectious Diseases* 187: 736–740.
Royce RA, Sena A, Cates W Jr., and Cohen MS (1997) Sexual transmission of HIV. *New England Journal of Medicine* 336: 1072–1078.
Seidlin M, Vogler M, Lee E, Lee YS, and Dubin N (1993) Heterosexual transmission of HIV in a cohort of couples in New York City. *AIDS* 7: 1247–1254.
Sewankambo N, Gray RH, Wawer MJ, *et al.* (1997) HIV-1 infection associated with abnormal vaginal flora morphology and bacterial vaginosis. *Lancet* 350: 546–550.
Shafii T, Stovel K, Davis R, and Holmes K (2004) Is condom use habit forming? Condom use at sexual debut and subsequent condom use. *Sexually Transmitted Diseases* 31: 366–372.
Sinei SK, Fortney JA, Kigondu CS, *et al.* (1996) Contraceptive use and HIV infection in Kenyan family planning clinic attenders. *International Journal of STD and AIDS* 7: 65–70.
Somi GR, Matee MI, Swai RO, *et al.* (2006) Estimating and projecting HIV prevalence and AIDS deaths in Tanzania using antenatal surveillance data. *BMC Public Health* 6: 120.
Sullivan LE, Metzger DS, Fudala PJ, and Fiellin DA (2005) Decreasing international HIV transmission: The role of expanding access to opioid agonist therapies for injection drug users. *Addiction* 100: 150–158.
Taha TE, Brown ER, Hoffman IF, *et al.* (2006) A phase III clinical trial of antibiotics to reduce chorioamnionitis-related perinatal HIV-1 transmission. *AIDS* 20: 1313–1321.
Taha TE and Gray RH (2000) Genital tract infections and perinatal transmission of HIV. *Annals of the New York Academy of Sciences* 918: 84–98.
Taha TE, Hoover DR, Dallabetta GA, *et al.* (1998) Bacterial vaginosis and disturbances of vaginal flora: Association with increased acquisition of HIV. *AIDS* 12: 1699–1706.
Taha TE, Nour S, Kumwenda NI, *et al.* (2005) Gender differences in perinatal HIV acquisition among African infants. *Pediatrics* 115: e167–e172.
Uganda Ministry of Health (MoH) (2006) Uganda HIV/AIDS Sero-behavioural Survey 2004/2005, Ministry of Health and ORC Macro, ed. Calverton, MD: Ministry of Health and ORC Macro.

UNAIDS/WHO (2006) *AIDS Epidemic Update 2006*. Geneva, Switzerland: UNAIDS/WHO.
Vittinghoff E, Douglas J, Judson F, McKirnan D, MacQueen K, and Buchbinder SP (1999) Per-contact risk of human immunodeficiency virus transmission between male sexual partners. *American Journal of Epidemiology* 150: 306–311.
Volkow P and Del Rio C (2005) Paid donation and plasma trade: Unrecognized forces that drive the AIDS epidemic in developing countries. *International Journal of STD and AIDS* 16: 5–8.
Wald A (2004) Synergistic interactions between herpes simplex virus type-2 and human immunodeficiency virus epidemics. *Herpes* 11: 70–76.
Watters JK, Estilo MJ, Clark GL, and Lorvick J (1994) Syringe and needle exchange as HIV/AIDS prevention for injection drug users. *Journal of the American Medical Association* 271: 115–120.
Wawer MJ, Gray RH, Sewankambo NK, *et al.* (2005) Rates of HIV-1 transmission per coital act, by stage of HIV-1 infection, in Rakai, Uganda. *Journal of Infectious Diseases* 191: 1403–1409.
Wawer MJ, Sewankambo NK, Serwadda D, *et al.* (1999) Control of sexually transmitted diseases for AIDS prevention in Uganda: A randomised community trial. Rakai Project Study Group. *Lancet* 353: 525–535.
Weller S and Davis K (2002) Condom effectiveness in reducing heterosexual HIV transmission. *Cochrane Database Systematic Reviews* 3, CD003255.
World Health Organization (2005) HIV/AIDS epidemiological surveillance report for the WHO African region-2005 update. Harare, Zimbabwe: WHO Regional Office for Africa.
Wilkinson D, Tholandi M, Ramjee G, and Rutherford GW (2002) Nonoxynol-9 spermicide for prevention of vaginally acquired HIV and other sexually transmitted infections: Systematic review and meta-analysis of randomised controlled trials including more than 5000 women. *Lancet Infectious Diseases* 2: 613–617.
Willumsen JF, Filteau SM, Coutsoudis A, *et al.* (2003) Breastmilk RNA viral load in HIV-infected South African women: Effects of subclinical mastitis and infant feeding. *AIDS* 17: 407–414.
Witvrouw M, Pannecouque C, Van Laethem K, Desmyter J, De Clercq E, and Vandamme AM (1999) Activity of non-nucleoside reverse transcriptase inhibitors against HIV-2 and SIV. *AIDS* 13: 1477–1483.
Wu Z, Rou K, and Detels R (2001) Prevalence of HIV infection among former commercial plasma donors in rural eastern China. *Health Policy Plan* 16: 41–46.
Wu Z, Sullivan SG, Wang Y, Rotheram-Borus MJ, and Detels R (2007) Evolution of China's response to HIV/AIDS. *Lancet* 369: 679–690.

Relevant Websites

http://www.aidsinfo.nih.gov – AIDSinfo, HIV/AIDS Treatment Information, Department of Human and Health Services Guidelines.
http://www.cdc.gov – Centers for Disease Control and Prevention (CDC).
http://www.hivma.org/Content.aspx?id=1922 – HIV Medicine Association (HIVMA) Practice Guidelines.
http://www.unaids.org – Joint United Nations Programme on HIV/AIDS (UNAIDS).
http://www.iasusa.org/pub/arv_2006.pdf – *Treatment for Adult HIV Infection: 2006 Recommendations of the International AIDS Society – USA Panel.*

Mother-to-Child Transmission of HIV

J S Mukherjee, Partners In Health, Boston, MA, USA

Introduction

Since the first case reports of immune deficiency 25 years ago, the human immunodeficienty virus, or HIV, is ever more concentrated in vulnerable, impoverished, and marginalized populations. Differential access to effective therapies and the ideas that matter, and the power to use them deepens health inequities. Today, most of the 2.3 million children under 15 years of age infected with HIV (UNICEF, 2005) live in sub-Saharan Africa. As the majority of children under 15 years with HIV were infected by their mothers *in utero*, peripartum, or during breastfeeding, prevention of mother-to-child transmission (pMTCT) using antiretroviral therapy (ART) in the developed countries of Europe and North America has made pediatric HIV infection extremely rare there. In stark contrast, in resource-poor settings, less than 5% of pregnant women have access to HIV testing and even fewer have access to the drugs that prevent mother-to-child transmission. In the absence of testing, prevention, and treatment, approximately 1800 new pediatric HIV infections occur each day. But children under 15 years of age are not the only group placed at high risk for HIV. The situation is even more dire among adolescents and young adults aged 15–24 years. Among them, 6000 new infections happen each day. Girls are at extreme and perilous risk as the sexual debut for many occurs in the context of poverty, gender inequity, and violence that make prevention methods difficult if not utterly impossible.

Once infected with HIV, children in resource-poor settings fare poorly as they are exquisitely susceptible to aggressive opportunistic infections such as *Mycoplasma tuberculosis* and *Salmonella typhi,* in addition to syndromes commonly associated with infant mortality. Since 2002, there has been a massive effort to scale up ART in heavily HIV-burdened countries. However, HIV care for children has been slower to advance than among adults in resource-poor settings. Lack of health personnel trained in the management of children with HIV and the lack of pediatric formulations of AIDS medicines stack the odds against children.

This chapter addresses the public health importance of preventing HIV transmission to infants born to HIV-positive mothers, the risk of HIV among adolescents, and the important issue of pediatric HIV treatment in resource-poor settings.

Prevention of Transmission of HIV from HIV-Positive Mothers to Their Infants

Prevention of Mother-to-Child Transmission (pMTCT)

Women of childbearing age represented nearly half of the 39.4 million adults living with HIV/AIDS worldwide in 2004. In regions where ART prophylaxis is not readily available, rates of mother-to-child transmission (MTCT) of HIV range from 25 to 40% (Joint United Nations Programme on HIV/AIDS, 2004). Programs to prevent MTCT will have the greatest impact if the vulnerability of the whole family is taken into account. Ideally, HIV prevention and care, including pMTCT, should be integrated with primary health care and women's health services. If a pregnant woman is found to be HIV-positive, it is critical to preserve and improve the mother's health as well as to decrease the risk of transmission of the virus to the infant. HIV-positive pregnant women should receive medical care and extended social support, including an assessment of the family's living situation, nutritional assistance, and testing of any children and sexual partner (s). If the mother's immunologic state is deemed poor based on CD4 monitoring or clinical status, three antiretroviral medications should be started as soon as possible as per normal guidelines for the management of HIV infection in adults. Alternatively, if the mother is not ill, ART can be deferred until the 28th week of gestation, and then administered to prevent MTCT of HIV.

Nutritional supplementation should be provided to pregnant women beginning at the 12th week of gestation and continued through the pregnancy. This supplementation should be in the form of a daily multivitamin preparation containing B1 (20 mg), B2 (20 mg), B6 (25 mg), niacin (100 mg), B12 (50 μg), C (500 mg), E (30 mg), and folic acid (0.8 mg). This regimen has been found to reduce the risk of fetal death, severe preterm birth, small size for gestational age, and low birthweight among children of HIV-positive women. Because vitamin A has been shown to increase the risk of MTCT of HIV, particularly in women who take it during breastfeeding, nutritional supplements that include vitamin A should not be given.

The presence of sexually transmitted infections (STIs), anemia, increased viral load, and low CD4 count have all been associated with increased rates of HIV transmission from mother to child and should be addressed accordingly (Bobat *et al.*, 1996). Obstetric risk factors for transmission include prolonged rupture of amniotic membranes, vaginal rather than elective cesarean delivery (in particular, among women not receiving ART), chorioamnionitis, and obstetric complications. Thus, the provision of obstetrical services should be improved concurrently with the scale-up of ART if maternal mortality and vertical transmission of HIV are to be decreased. Prenatal testing, prenatal care, and appropriate screening and treatment for STIs are clearly linked to better obstetric and pediatric outcomes. However, of all known factors, high maternal viral load is the strongest predictor of vertical HIV transmission. The risk of MTCT is increased 2.4-fold for every log increase in viral load at the time of delivery (Ioannidis *et al.*, 2001). ART is thus the most important intervention for pMTCT.

ART in Pregnant Women

The use of maternal ART has led to perinatal transmission rates of less than 2%, compared to rates as high as 36% in the absence of ART (Perinatal HIV Guidelines Working Group, 2007). The literature regarding choice of ART during pregnancy, labor, and delivery continues to evolve rapidly. The landmark AIDS Clinical Trials Group (ACTG) 076 study used a three-part zidovudine (AZT) monotherapy regimen – antepartum and intrapartum for the mother, postpartum for the newborn – and reduced MTCT rates from 26% to 8% (Connor *et al.*, 1994). Since that trial, a number of other studies have attempted to determine whether shorter courses of monotherapy for the mother and/or infant, or combinations of ART, have equal or greater efficacy. Efficacy has been shown for regimens involving AZT alone, AZT and lamivudine (3TC), nevirapine (NVP) alone (single-dose to mother and infant), AZT with single-dose NVP, and AZT and 3TC with single-dose NVP (Guay *et al.*, 1999; Petra Study Team, 2002; Lallemant *et al.*, 2003).

In wealthy countries, the current standard of care for pMTCT is triple-drug maternal ART (European Collaborative Study, 2001). In a multivariate analysis, adjusting for maternal viral load and duration of therapy, the odds-ratio of MTCT for women receiving potent triple therapy compared with AZT monotherapy was 0.27, supporting the benefit of using three drugs (Cooper *et al.*, 2002). Data strongly support that transmission rates are significantly lower when multiagent therapy is being administered to an HIV-positive pregnant woman and when lower maternal plasma HIV-RNA levels are observed. Viral load is more significant a predictor of HIV transmission than mode of delivery. In addition to lowering the risk of MTCT, combination therapy also diminishes the risk of developed drug resistance in the mother. After the use of single-dose NVP for pMTCT, strains of HIV resistant to non-nucleoside reverse transcriptase inhibitors (NNRTIs) have been found (at least temporarily) in just under 50% of babies and a little more than 50% of women (Magder *et al.*, 2005). The clinical significance of this finding with regard to future pregnancies and future management of maternal and pediatric disease is currently unknown.

Recommended ART regimens for pMTCT

Decisions regarding maternal ART should be made based on the timing of presentation for care and maternal indications for therapy. Women who are already receiving ART at the time of conception should continue on the same regimen, unless it includes efavirenz (EFV), in which case NVP or a protease inhibitor (PI) should be substituted. When a pregnant woman is newly identified as HIV-positive and has clinical and/or immunologic reasons for ART (i.e., either clinical symptoms or CD4 below 350 cells/mm^3), she should immediately start combination therapy with three antiretroviral drugs (**Figure 1**). The preferred – and most widely available – regimen is AZT, 3TC, and NVP. Women presenting in pregnancy who are not already receiving ART and who do not have maternal indications for therapy (i.e., CD4 above 350 cells/mm^3) may be started on therapy at the 28th week of gestation for pMTCT.

Triple therapy is superior to monotherapy for pMTCT, yet NVP, the most widely used third agent in resource-poor settings, may be associated with maternal liver toxicity in women with CD4 counts greater than 250 cells/mm^3. Thus, the preferred triple therapy with AZT, 3TC, and a PI is preferred. HIV-positive women who present after the 28th week of gestation should be started on triple therapy as soon as possible, even before the results of CD4 testing are available. If NVP is used, liver function tests should be closely monitored. Women who present in labor and have not yet received ART should receive single-dose NVP (Guay *et al.*, 1999). In addition, based on reports of developed NVP resistance, any woman who received any NVP for pMTCT is now also given 1 week of AZT and 3TC. In some settings, an aggressive approach to maximizing adherence has been piloted, continuing AZT and 3TC for 2 weeks after the postpartum cessation of NVP.

All HIV-exposed infants should be given postexposure prophylaxis with ART. Studies have found that the critical factor is the time of initiation of maternal prophylaxis (28 weeks' gestation is more efficacious at preventing MTCT than 34 weeks) and that there is little difference between 1 and 6 weeks of infant prophylaxis with AZT. However, the addition of single-dose NVP to the postpartum infant within 72 hours of birth improves the prevention of transmission. Thus, it is reasonable to add single-dose NVP to infant prophylaxis of AZT and use the longer duration of AZT for a total of 6 weeks of therapy if the mother did not receive ART prior to the 34th week of gestation (**Figure 2**). If the mother is suspected of having resistant virus, alternative drug combinations may be considered for the infant. Infant protocol is determined by the available information about the mother's prior exposure to antiretrovirals, viral load, and presence or absence of resistance mutations.

ART toxicity during pregnancy

Cohorts of HIV-positive pregnant women given ART (specifically, AZT and 3TC) have demonstrated little in the way of maternal or fetal toxicity. However, several antiretroviral agents and combinations should be avoided or used with caution during pregnancy. The use of EFV in pregnant monkeys has been associated with abnormalities in their offspring; a single case of myelomeningocele has also been reported in a human infant exposed to EFV *in utero* (Fundaro *et al.*, 2002). Generally, pregnant women and women of childbearing age who are not using contraception should not be given EFV. If a woman becomes pregnant while receiving an EFV-containing regimen, NVP or a PI should be substituted. While this particular use has not been studied, EFV may possibly be given in the third trimester, as the development of the infant's neural tube occurs during the first trimester.

NVP has been associated with an increased risk of hepatotoxicity in women with a CD4 count above 250 cells/mm^3; thus, a regimen containing AZT, 3TC, and NVP may not always be recommended for pMTCT in women with high CD4 counts. However, NVP is the most widely available nonteratogenic third agent in resource-poor settings and in Zanmi Lasante clinics (in Haiti) is used in the first-line regimen for ART and pMTCT regardless of maternal CD4. A recent study from Brazil demonstrated minimal toxicity from this approach (Custódio João *et al.*, 2005). When NVP-based triple therapy is given to women with a CD4 count above 250 cells/mm^3, close monitoring of liver function tests – 1 week after the initiation of ART and then every 2 weeks subsequently, or if symptoms develop – is strongly recommended. Another reasonable alternative is to initiate AZT at the 28th week of gestation, with the addition of maternal NVP at labor and after birth for the infant (Lallemant *et al.*, 2004).

The use of stavudine (d4T) and didanosine (ddI) in combination is associated with increased mitochondrial toxicity during pregnancy and should be avoided. In addition, tenofovir (TDF) should not be used during pregnancy, due to concerns about osteopenia in infants and a general lack of safety data.

Mode of Delivery

Elective cesarean sections prior to rupture of membranes can reduce MTCT in HIV-positive women who are not receiving ART. For women receiving ART, however,

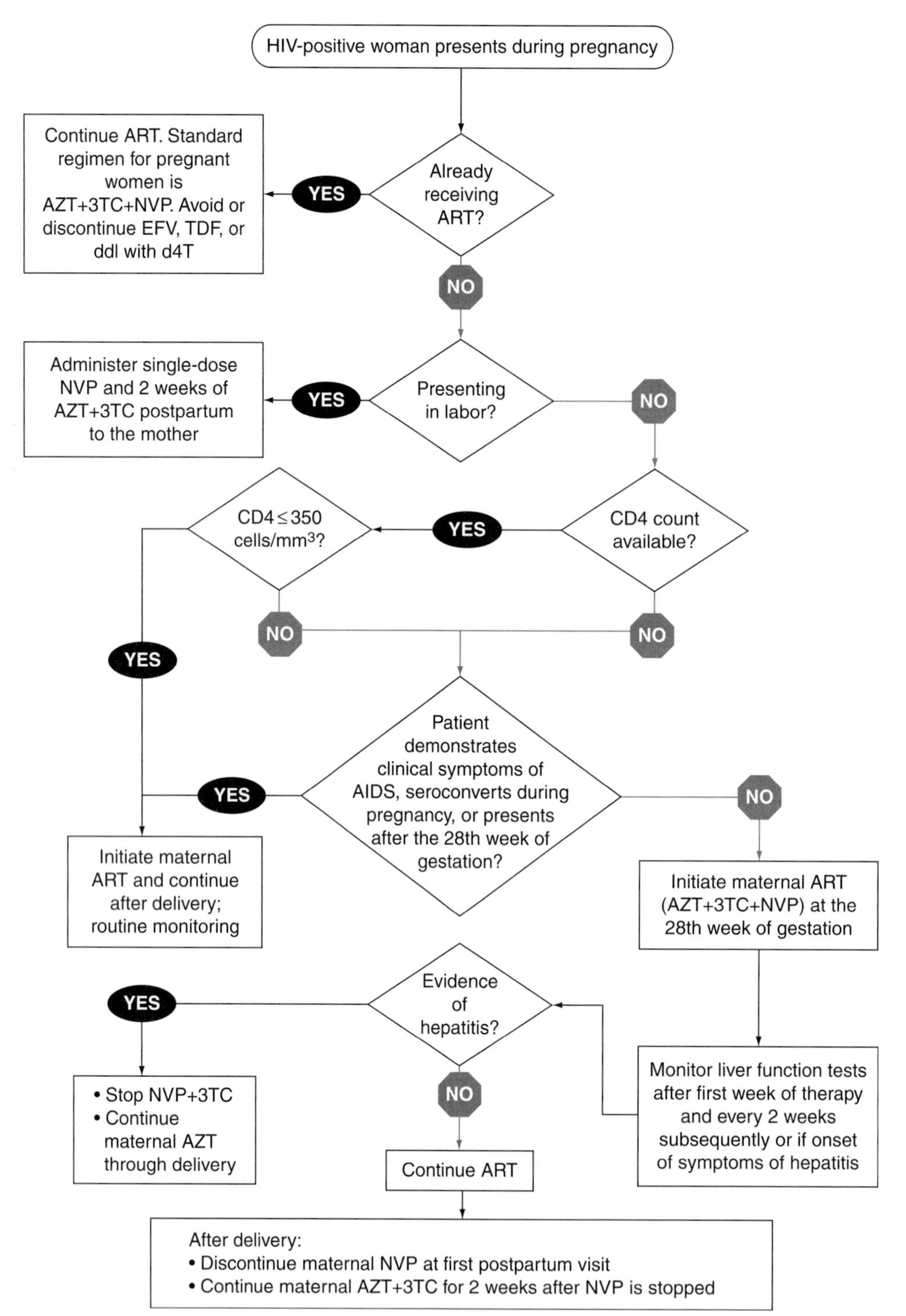

Figure 1 Antiretroviral therapy for pregnant women based on WHO guidelines for resource-poor settings. Reproduced with permission from Partners in Health (2006) *PIH Guide to the Community-Based Treatment of HIV in Resource-Poor Settings,* 2nd edn. Boston: Partners In Health.

cesarean sections do not decrease the risk of MTCT. Recent data from PACTG 367 indicate that transmission rates are significantly lower when multiagent therapy is being administered and when lower maternal plasma HIV RNA levels are observed; rates do not differ significantly according to mode of delivery (Shapiro *et al.*, 2004). Cesarean sections should therefore be reserved for women who are likely to have a detectable viral load at the time of labor – that is, those women who are not receiving effective ART or who demonstrate nonadherence

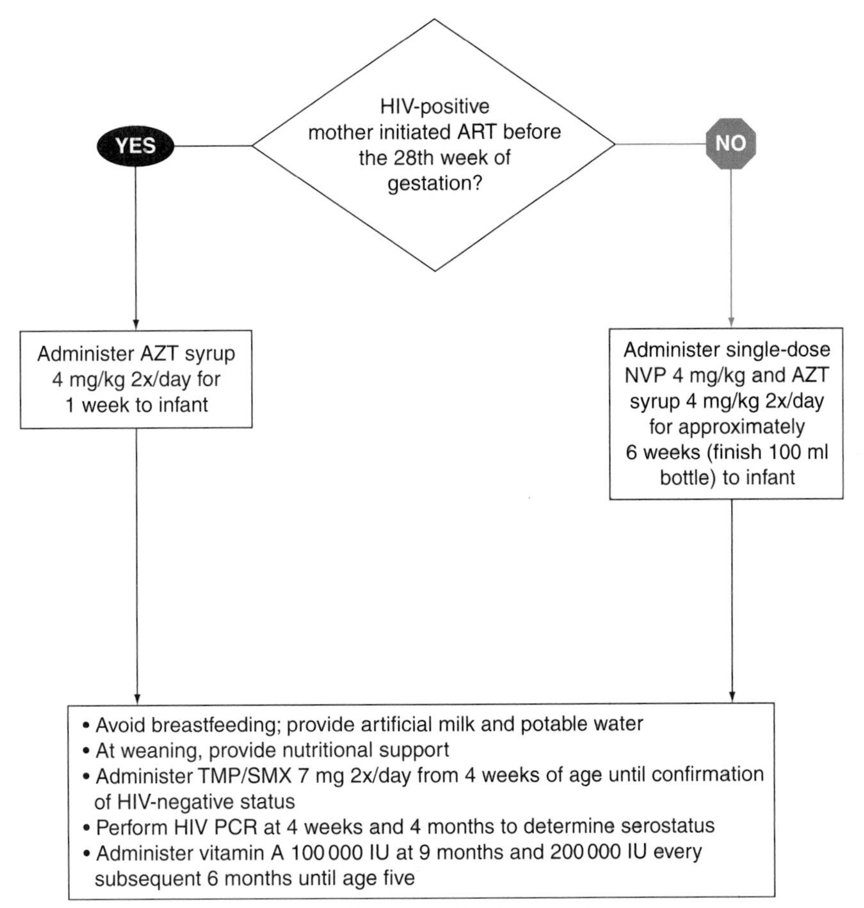

Figure 2 ART for infants born to HIV-positive mothers. Reproduced with permission from Partners in Health (2006) *PIH Guide to the Community-Based Treatment of HIV in Resource-Poor Settings,* 2nd edn. Boston: Partners In Health.

during pregnancy – and women who have an obstetric indication for a cesarean section. This stipulation is particularly important given that women with HIV infection may be at increased risk of febrile complications after a cesarean section.

Breastfeeding and MTCT Risk

The rate of MTCT of HIV through breastfeeding can be as high as 0.7% per month. In breastfeeding populations, 30 to 50% of MTCT is attributable to breastfeeding. Although the U.S. Centers for Disease Control and Prevention (CDC) has recommended since 1986 that women with HIV infection avoid breastfeeding (Centers for Disease Control and Prevention, 1985), breastfeeding is heavily implicated in fuelling ongoing vertical transmission in resource-poor settings. The Petra study (in which AZT and 3TC were given during pregnancy only) revealed that most of the impact of preventive ART was negated when infants breastfed for 18 months, with comparable rates of transmission among interventions and placebo (Petra Study Team, 2002). Risk of transmission via breast milk has been found to be dependent on factors such as maternal viral load, maternal immune status, and infant feeding patterns, as well as on the presence of infant oral candidiasis or maternal breast pathologies such as mastitis or fissure (Coutsoudis *et al.*, 1999). Debate over breastfeeding is especially fierce with regard to resource-poor settings, where the availability of infant formula and potable water is limited. However, obtaining formula and improving water sources is less complicated than administering lifelong care to HIV-infected infants. The provision of clean water also has a positive impact on the health of the mother, the family, and the community at large.

In settings such as the Partners in Health programs in Haiti and Rwanda, where assistance with the provision of clean water and aggressive prevention and treatment of diarrhea and dehydration are an integral part of primary health care and HIV programs, formula feeding for infants born to HIV-infected mothers makes sense both practically and ethically. In certain circumstances, the absolute lack of clean water or cooking fuel to boil water or the fear of HIV status disclosure with the use of bottle feeding, may mean that HIV-positive women have no alternative but to breastfeed. UNICEF and other organizations recommend exclusive breastfeeding for 6 months, as some studies have demonstrated increased MTCT through mixed (i.e., breast and formula) feeding. However, studies have found that 75% of HIV infections from breastfeeding occurred during the first 6 months. In addition, if early weaning is indeed promoted and nutritional supplementation is not provided, the infant is also at high risk for diarrhea and kwashiorkor. For women who choose to breastfeed, weaning at 6 months is encouraged and nutritional support is given during the weaning period. Because malnutrition, mastitis, and breast lesions are all associated with increased risk of HIV transmission through breast milk, a multivitamin supplement (containing vitamins B, C, and E), as well as basic instructions on the prevention of mastitis and breast lesions, should be provided to all breastfeeding women. Single-dose NVP for the mother during labor and for the infant within 72 hours of birth has been associated with a sustained reduction in the transmission of HIV in breastfed babies through 18 months of age. Studies are currently ongoing to determine if triple-drug maternal ART can reduce the risk of viral transmission to infants who continue to breastfeed.

The Care of HIV-Exposed Infants

Without preventive measures cumulative risk of transmission of HIV from infected mother to child is 25–45%. Yet, with antenatal and postnatal care, ART, and breast milk substitution, the transmission of HIV from mother to infant may be reduced to less than 2% (Connor *et al.*, 1994; Guay *et al.*, 1999; De Cock *et al.*, 2000; European Collaborative Study, 2001; Cooper *et al.*, 2002; Petra Study Team, 2002; Lallemant *et al.*, 2003). Because babies born to HIV-positive mothers receive passive transmission of maternal HIV antibody, a simple HIV serology will not determine whether the infant is infected with HIV. After 18 months of age, a positive HIV antibody can be assumed to be the child's response to HIV infection and is considered definitive evidence of pediatric infection. However, prior to 18 months of age, a virologic test must be performed to definitively diagnose infant infection. Thus, HIV-positive infants born to HIV-positive mothers, but with unknown HIV infection status are considered 'HIV-exposed' during this period.

Early virologic diagnosis of HIV infection in infants is performed routinely in developed countries as it is crucial to know whether a child is infected to optimize medical management and decisions related to infant feeding. Additionally, the psychological stress on the family is reduced significantly if parents can be informed of the infant's status as early as possible. Early knowledge of the infant's status also has been found in resource-poor settings to be cost effective (Sherman *et al.*, 2005a). Yet, virologic testing is not available in many resource-poor settings, which places infants in the nebulous category of HIV-exposed for up to 18 months. The following description of care of the HIV-exposed infant is geared toward such settings.

HIV-exposed infants should receive routine immunizations (including countries endemic for tuberculosis [TB], and Bacillus Calmette-Guerin [BCG] at birth) and should be given trimethoprim/sulfamethoxazole (TMP/SMX) (7 mg per kg twice daily) to prevent *Pneumocystis carinii* pneumonia and bacterial infections from 4–6 weeks of age until they are confirmed to be HIV-negative (World Health Organization/Joint United Nations Programme on HIV/AIDS/United Nations Children's Fund, 2004). (Prophylaxis of HIV-infected children is discussed in the section 'Care of the Child with HIV.') In addition to routine care, all HIV-exposed infants should receive 100 000 IU of vitamin A at 9 months of age and 200 000 IU every subsequent 6 months until the age of 5 years (Villamor *et al.*, 2000, 2002). When feasible to offer safe breast milk substitution, HIV-exposed infants should not be breastfed as breastfeeding is associated with 0.7% HIV transmission per month. Primary health care of the HIV-exposed child should include routine growth and development monitoring. HIV-exposed infants with weight loss or poor weight gain should be assessed carefully for opportunistic infections, especially TB. If HIV infection is suspected in infants younger than 18 months of age in whom virologic diagnosis is not available, the WHO staging criteria (**Table 1**) and CD4 count or percentage (**Table 2**), if available, may be sufficient criteria to initiate therapy. Note that clinical diagnosis in these infants remains uncertain because all signs and symptoms indicated – even those for WHO Pediatric Clinical Stage III – are also frequently seen in HIV-negative infants; this is particularly true in the developing world. For all infants less than 18 months of age in whom ART is initiated based solely on clinical criteria, HIV serum antibody should be measured after 18 months, and ART should be continued only in those infants who have a positive antibody test.

Diagnosing HIV Infection in Infants

Maternal antibody to HIV may be present in infants until 18 months of age; infants may therefore record a false-positive HIV antibody test up until clearance of these maternal antibodies. This period of uncertainty as to whether the infant is HIV-infected can be fatal. Intrauterine transmission accounts for 25%–40% of infection, labor and delivery 60%–75% of transmission and 20%–25% of perinatal infections are believed to be due to breastfeeding (CDC 2001). Studies have shown that the risk of disease progression is inversely correlated with the age of the child, with the youngest children at greatest risk of rapid disease progression (Newell *et al.* 2006). Thus, improving access to virological testing is currently being promoted. Virologic testing, real-time polymerase chain

Table 1 WHO clinical staging of HIV/AIDS for children with confirmed HIV

Clinical stage 1
Asymptomatic
Persistent generalized lymphadenopathy
Clinical stage 2
Unexplained[a] persistent hepatosplenomegaly
Papular pruritic eruptions
Extensive wart virus infection
Extensive molluscum contagiosum
Fungal nail infections
Recurrent oral ulcerations
Unexplained persistent parotid enlargement
Lineal gingival erythema
Herpes zoster
Recurrent or chronic upper respiratory tract infections (otitis media, otorrhea, sinusitis, or tonsillitis)
Clinical stage 3
Unexplained moderate malnutrition not adequately responding to standard therapy
Unexplained persistent diarrhea (14 days or more)
Unexplained persistent fever (above 37.5 °C intermittent or constant, for longer than 1 month)
Persistent oral candidiasis (after first 6–8 weeks of life)
Oral hairy leukoplakia
Acute necrotizing ulcerative gingivitis or periodontitis
Lymph node tuberculosis
Pulmonary tuberculosis
Severe recurrent bacterial pneumonia
Symptomatic lymphoid interstitial pneumonitis
Chronic HIV-associated lung disease including brochiectasis
Unexplained anemia (<8 g/dl), neutropenia ($<0.5 \times 10^9$ per liter) and/or chronic thrombocytopenia ($<50 \times 10^9$ per liter)
Clinical stage 4[b]
Unexplained severe wasting, stunting, or severe malnutrition not responding to standard therapy
Pneumocystis pneumonia
Recurrent severe bacterial infections (such as empyema, pyomyositis, bone or joint infection, or meningitis but excluding pneumonia)
Chronic herpes simplex infection (orolabial or cutaneous of more than 1 month's duration or visceral at any site)
Extrapulmonary tuberculosis
Kaposi sarcoma
Esophageal candidiasis (or candidiasis of trachea, bronchi, or lungs)
Central nervous system toxoplasmosis (after 1 month of life)
HIV encephalopathy
Cytomegalovirus infection: retinitis or cytomegalovirus infection affecting another organ, with onset at age older than 1 month
Extrapulmonary cryptococcosis (including meningitis)
Disseminated endemic mycosis (extrapulmonary histoplasmosis, coccidiomycosis)
Chronic cryptosporidiosis
Chronic isosporiasis
Disseminated nontuberculous mycobacterial infection
Cerebral or B-cell non-Hodgkin lymphoma
Progressive multifocal leukoencephalopathy
Symptomatic HIV-associated nephropathy or HIV-associated cardiomyopathy

[a]Unexplained refers to when the condition is not explained by other causes.
[b]Some additional specific conditions can also be included in regional classifications (such as reactivation of American trypanosomiasis [meningoencephalitis and/or myocarditis] in the WHO region of the Americas, penicilliosis in Asia, and HIV-associated rectovaginal fistula in Africa).
Adapted from World Health Organization (2007) WHO case definitions of HIV for surveillance and revised clinical staging and immunological classification of HIV-related disease in adults and children. Geneva: World Health Organization.

reaction (RT-PCR) of HIV-RNA and HIV-DNA and ultrasensitive p24 (Up24Ag) assays are being used in some resource-poor settings and can definitively diagnose HIV in infants after 4 weeks of age with sensitivity approaching 98%. Because blood may be difficult to collect from young infants the use of dried blood spots (DBSs) for both HIV-DNA or HIV-RNA testing and Up24Ag assay has been used in a variety of settings

Table 2 CD4 criteria for severe HIV immune deficiency in children

Immunological marker[a]	*Age-specific recommendation to initiate ART*[b]			
	≤11 months	*12 months to 35 months*	*36 months to 59 months*	*≥5 years*
%CD4+[c]	<25%	<20%	<15%	<15%
CD4 count[c]	<1500 cells/mm^3	<750 cells/mm^3	<350 cells/mm^3	<200 cells/mm^3

[a]Immunological markers supplement clinical assessment and should therefore be used in combination with clinical staging. CD4 is preferably measured after stabilization of acute presenting conditions.
[b]ART should be initiated by these cut-off levels, regardless of clinical stage; a drop of CD4 below these levels significantly increases the risk of disease progression and mortality.
[c]%CD4+ is preferred for children aged < 5 years.
Reproduced from World Health Organization (WHO) (2007) Antiretroviral therapy of HIV infection in infants and children in resource-poor settings: Toward universal access. Recommendations for a public health approach. Geneva: World Health Organization.

(Sherman *et al.*, 2005b). Dried blood spots do not require venipuncture but can be obtained by using blood from a finger-stick or heel-stick. While HIV antibody testing cannot be used to diagnose HIV infection definitively in infants under 18 months of age, a negative test is helpful in excluding HIV infection between 9 and 12 months of age in children 6 weeks after they were last breastfed (WHO, 2006) (**Figure 3**).

HIV Prevention and the Adolescent

HIV prevention should be targeted toward adolescents before the debut of sexual intercourse. School-based and out-of-school-based education programs have been shown to increase knowledge of how HIV is transmitted and how it can be prevented. However, many young people lack the agency needed to carry out real choice in terms of protection against unwanted sex, unwanted pregnancy, and sexually transmitted infections including HIV. Poverty, gender inequalities, and lack of child rights often result in coercion, rape, or economic necessity triggering the onset of sexual activity. The HIV/AIDS pandemic has fuelled the stresses on children resulting in massive losses of teachers (Grassly *et al.*, 2003), worsening homelessness, lack of education, and poverty among adolescents by leaving them orphaned or without parents who are well enough to care for them. Without an education, home, or means to support oneself, adolescents, particularly girls, are put in the position of using sex as a commodity toward achieving a better life. Several studies have documented the HIV risk associated with adolescent girls having older sexual partners. The much discussed 'sugar daddy' phenomenon is driven by economics and gender, and, therefore, power dynamics (Luke, 2005). Not only is AIDS education critical, but education itself is critical to protecting adolescents against HIV infection. Studies have linked condom use with higher levels of education as well as economic indicators (Gavin *et al.*, 2006).

Pediatric HIV Infection

Care of the Child with HIV

Once a child has been diagnosed with HIV infection, careful follow-up is critical. Growth and developmental monitoring should be performed monthly with careful attention to rule out opportunistic infections, particularly TB, which in children are often due to primary infections rather than reactivation of disease. Children are infected with largely the same pathogens as adults including (in developing countries) ubiquitous and aggressive pathogens such as *Mycoplasma tuberculosis, Pneumococcus pneumoniae, Salmonella typhi,* and more classically 'opportunistic pathogens' such as *Pneumocystis carinii, Toxoplasma gondii, Cryptococcus neoformans, Cryptosporidium parvum,* cytomegalovirus, herpes simplex, varicella zoster, and hepatitis B and C.

Children who are diagnosed with HIV infection should be given prophylaxis for opportunistic infections based on CD4 count or clinical staging of III or IV. Prophylaxis with trimethoprim/sulfamethoxazole (TMP/SMX) for *P. carinii* and bacterial infections is the mainstay of prophylaxis and also protects against *T. gondii.* Isoniazid for children with latent TB infection purified protein derivative (PPD) greater than 5 mm or a household contact with TB is important in endemic areas. Newborns who are known to be HIV-infected benefit from prophylaxis against *P. carinii* pneumonia and bacterial infections from birth regardless of CD4 count or percentage as the immune system is immature. A recent study from South Africa suggests that newborns greater than 8 weeks of age may have a survival benefit from isoniazid prophylaxis (Zar *et al.*, 2007).

Starting ART in Infants and Children

When to start ART in an HIV-infected child is an important decision. Perinatally infected children under 12 months of age have a 40% mortality if untreated. The classification of immune suppression in HIV-exposed

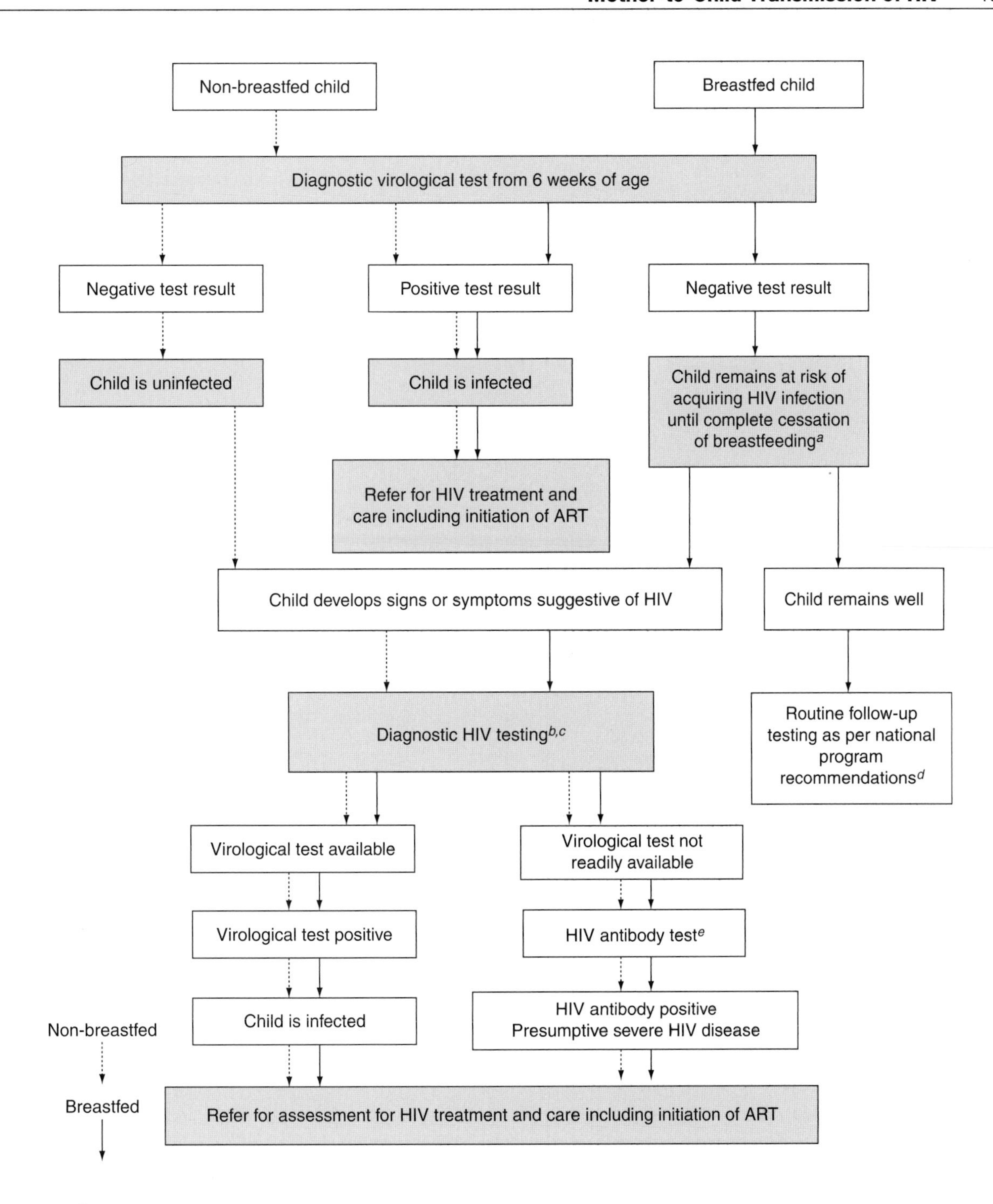

[a]The risk of HIV transmission remains if breastfeeding continues beyond 18 months of age.

[b]Infants over 9 months of age can be tested initially with HIV antibody test, as those who are HIV Ab negative are not HIV-infected, although still at risk of acquring infection if still breastfeeding.

[c]In children older than 18 months antibody testing is definitive.

[d]Usually HIV antibody testing from 9–18 months of age.

[e]Where virological testing is not readily available, HIV antibody testing should be performed; it may be necessary to make a presumptive clinical diagnosis of severe HIV disease in HIV-seropositive children. Confirmation of diagnosis should be sought as soon as possible.

Figure 3 Establishing the presence of HIV infection in infants born to HIV-positive mothers. Reproduced from World Health Organization (2006) Antiretroviral therapy of HIV infection in infants and children in resource-poor settings: Toward universal access. Recommendations for a public health approach. Geneva: World Health Organization.

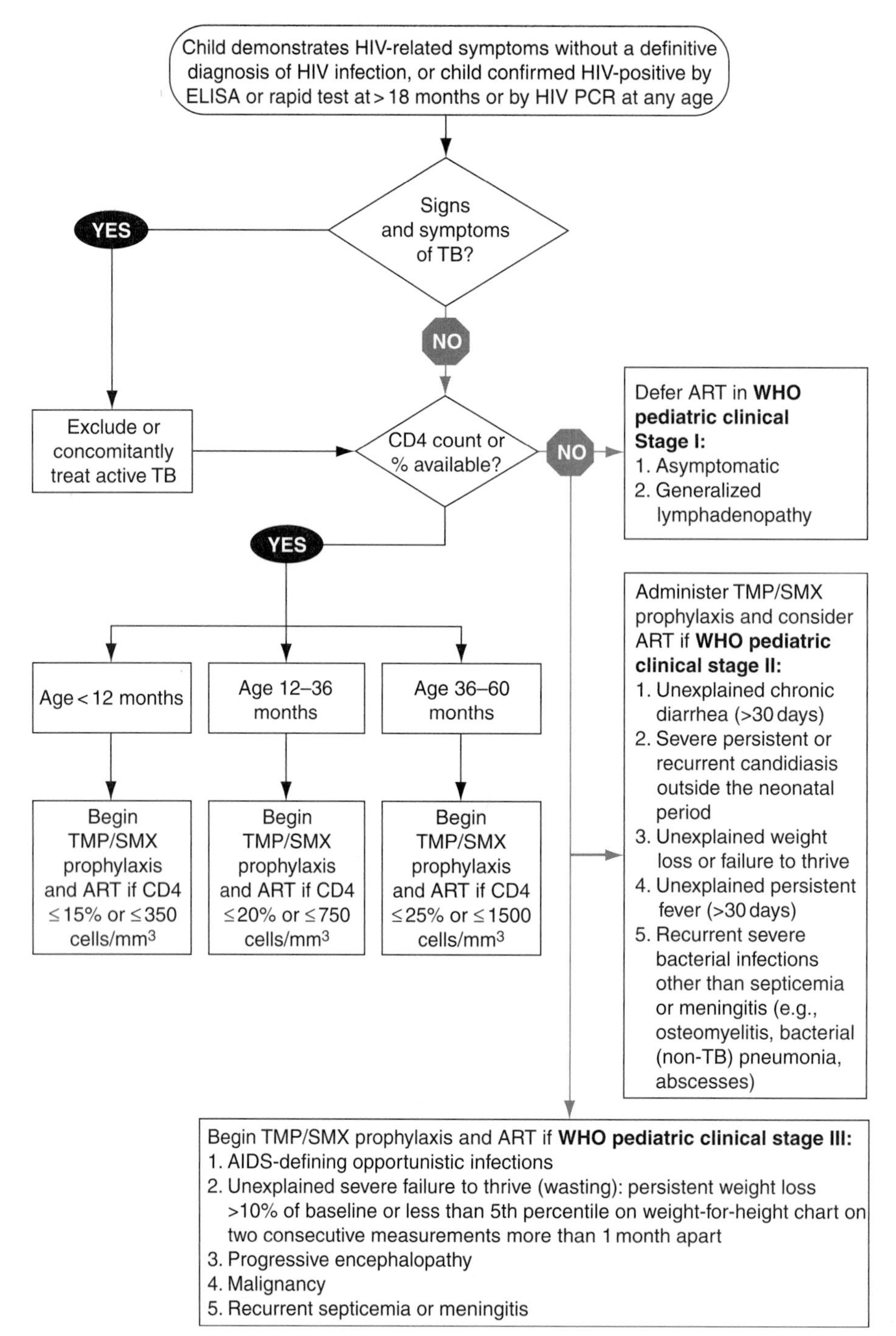

Figure 4 Initiation of ART in children less than 5 years of age in resource-poor settings. Reproduced with permission from Partners in Health (2006) *PIH Guide to the Community-Based Treatment of HIV in Resource-Poor Settings,* 2nd edn. Boston: Partners In Health.

children under 5 years of age is optimally assessed by percentage of CD4 cells of all T-lymphocytes (all CD3-positive cells) rather than by absolute CD4 count. **Figure 4** summarizes the approach to deciding when to initiate ART in children. If children are not yet eligible for ART, they should be monitored monthly for growth, development, and with CD4 every 6 months. If a child meets immunologic or clinical criteria for ART, the choice of first-line ART for children follows the same general principles as for adults – two NRTIs and either a NNRTI or a PI is used as the third agent. In children, the additional considerations regarding available formulations and certain pharmacokinetic factors must also be taken into account. First-line NRTIs generally used for children are 3TC and either AZT or d4T. The third agent is typically an NNRTI, usually NVP. Once a child weighs more than 10 kilograms, it is possible to administer tablets and capsules, which are cheaper and simpler to procure and store, rather than syrup formulations.

For children coinfected with TB, ART is generally deferred until after at least 2 months of TB therapy, or (if possible) until TB treatment is completed. Delaying ART avoids adverse drug interactions with rifampin and helps limit the number of medicines a child must take at any one time. In children above 3 years of age who are receiving rifampin as part of concomitant TB treatment, the third ART agent should be EFV instead of NVP. The bioequivalence for EFV in children under 3 years of age has not been determined, however. Thus, for HIV/TB coinfected children under 3 years of age who are receiving rifampin, abacavir (ABC) should be used as the third agent in lieu of EFV. Note that known perinatal exposure to NVP also warrants consideration in determining an appropriate pediatric ART regimen.

Conclusions

The rise or fall of health inequities, like the risks and outcomes of HIV among children, is neither random nor inevitable. Rather, it is the consequence of policy choices made by the privileged and powerful. In resource-poor settings, the number of children with HIV continues to rise because of their mothers' disproportionate burden of risk. Poor women's lack of access to HIV testing and ART to prevent vertical transmission, and continued breast-feeding in the absence of safe water and affordable infant formula drive the rise of the pandemic among the world's children. Moreover, adolescents and young adults are the fastest growing group of newly HIV-positive persons. And while HIV prevention education regarding condom use, delayed onset of sexual intercourse, and limiting partner number is important among this vulnerable group, the bulk of their risk is also related to social and economic inequities and to other human rights abuses that sharply constrain their ability to make choices and take action.

Active case finding of HIV among children living in HIV endemic areas is important, both for children known to be born to HIV-positive mothers and for children presenting with malnutrition, TB, chronic diarrhea, and other illness that may be associated with HIV or AIDS. When children are infected with HIV, they do respond well to treatment with ART if started promptly and managed appropriately. Here again, differential access to effective therapies exacerbates health inequities and inflicts harm, suffering, and death. Increasing access to virologic testing for infants born to HIV-positive mothers and to ART among children with HIV infection are critical if the scale-up of HIV treatment is to reach those in places where it is most acutely required.

See also: AIDS: Epidemiology and Surveillance; Infant Mortality/Neonatal Disease.

Citations

Bobat R, Coovadia H, Coutsoudis A, *et al.* (1996) Determinants of mother-to-child transmission of human immunodeficiency virus type 1 infection in a cohort from Durban, South Africa. *Pediatric Infectious Disease Journal* 15: 604–610.

Centers for Disease Control and Prevention (1985) Recommendations for assisting in the prevention of perinatal transmission of human T lymphotropic virus type III/lymphadenopathy-associated virus and acquired immunodeficiency syndrome. *Morbidity and Mortality Weekly Report* 34: 731–732.

Centers for Disease Control and Prevention (CDC) (2001) Revised guidelines for HIV counseling, testing, and referral and Revised recommendations for HIV screening of pregnant women. *Morbidity and Mortality Weekly Report* 50: RR-19, pp. 1–58 and pp. 59–86.

Connor EM, Sperling RS, Gelber R, *et al.* (1994) Reduction of maternal-infant transmission of human immunodeficiency virus type 1 with zidovudine treatment. Pediatric AIDS Clinical Trials Group Protocol 076 Study Group. *New England Journal of Medicine* 331: 1173–1180.

Cooper ER, Charurat M, Mofenson L, *et al.* (2002) Combination antiretroviral strategies for the treatment of pregnant HIV-1-infected women and prevention of perinatal HIV-1 transmission. *Journal of Acquired Immune Deficiency Syndromes* 29: 484–494.

Coutsoudis A, Pillay K, Spooner E, *et al.* (1999) Influence of infant-feeding patterns on early mother-to-child transmission of HIV-1 in Durban, South Africa: A prospective cohort study. South African Vitamin A Study Group. *Lancet* 354: 471–476.

Custódio João E, Calvet GA, Menezes JA, *et al.* (2005) Nevirapine toxicity in a cohort of HIV-1-infected pregnant women. *American Journal of Obstetrics and Gynecology* 194: 199–202.

De Cock KM, Fowler MG, Mercier E, *et al.* (2000) Prevention of mother-to-child HIV transmission in resource-poor countries: Translating research into policy and practice. *Journal of the American Medical Association* 283: 1175–1182.

European Collaborative Study (2001) HIV infected pregnant women: Vertical transmission in Europe since 1986. *AIDS* 15: 761–770.

Fawzi WW, Msamanga GI, Hunter D, *et al.* (2002) Randomized trial of vitamin supplements in relation to transmission of HIV-1 through breastfeeding and early child mortality. *AIDS* 16: 1935–1944.

Fundaro C, Genovese O, Rendeli C, *et al.* (2002) Myelomeningocele in a child with intrauterine exposure to efavirenz. *AIDS* 16: 299–300.

Gavin L, Galavotti C, Dube H, *et al.* (2006) Factors associated with HIV infection in adolescent females in Zimbabwe. *Journal of Adolescent Health* 39: 596.e11–596.e18.

Grassly NC, Desai K, Pegurri E, *et al.* (2003) The economic impact of HIV/AIDS on the education sector in Zambia. *AIDS* 17: 1039–1044.

Guay LA, Musoke P, Fleming T, *et al.* (1999) Intrapartum and neonatal single dose nevirapine compared with zidovudine for prevention of mother-to-child transmission of HIV-1 in Kampala, Uganda: HIVNET 012 randomised trial. *Lancet* 354: 795–802.

Ioannidis JP, Abrams EJ, and Ammann A (2001) Perinatal transmission of human immunodeficiency virus type 1 by pregnant women with RNA virus loads <1000 copies/ml. *Journal of Infectious Diseases* 183: 539–545.

Joint United Nations Programme on HIV/AIDS (2004) *Report on the Global HIV/AIDS Epidemic*. Geneva, Switzerland: UNAIDS.

Lallemant M, Jourdain G, LeCoeur S, *et al.* (2003) Multicenter, randomised controlled trial, assessing the safety and efficacy of nevirapine in addition to zidovudine for the prevention of perinatal HIV in Thailand, PHPT-2 update. *Antiviral Therapy* 8(Suppl. 1): S199(Abstract 62).

Lallemant M, Jourdain G, Le Coeur S, *et al.* (2004) Single-dose perinatal nevirapine plus standard zidovudine to prevent mother-to-child transmission of HIV-1 in Thailand. *New England Journal of Medicine* 351: 217–228.

Luke N (2005) Confronting the 'sugar daddy' stereotype: Age and economic asymmetries and risky sexual behavior in urban Kenya. *International Family Planning Perspectives* 31: 6–14.

Magder LS, Mofenson L, Paul ME, *et al.* (2005) Risk factors for in utero and intrapartum transmission of HIV. *Journal of Acquired Immune Deficiency Syndromes* 38: 87–95.
Newell ML, Patel D, Goetghebuer T, Thorne C, European Collaborative Study (2006) CD4 cell response to antiretroviral therapy in children with vertically acquired HIV infection: Is it associated with age at initiation? *Journal of Infectious Diseases* 193: 954–962.
Partners in Health (2006) *PIH Guide to the Community-Based Treatment of HIV in Resource-Poor Settings*, 2nd edn. Boston: Partners in Health.
Perinatal HIV Guidelines Working Group (2007) Public health service task force recommendations for use of antiretroviral drugs in pregnant HIV-infected women for maternal health and interventions to reduce perinatal HIV transmission in the United States. November 2, 2007, 1–96. http://aidsinfo.nih.gov/ContentFiles/PerinatalGL.pdf (accessed February 2008).
Petra Study Team (2002) Efficacy of three short-course regimens of zidovudine and lamivudine in preventing early and late transmission of HIV-1 from mother to child in Tanzania, South Africa, and Uganda (Petra study): A randomised, double-blind, placebo-controlled trial. *Lancet* 359: 1178–1186.
Shapiro D, Tuomala R, Pollack H, *et al.* Mother-to-child HIV transmission risk according to antiretroviral therapy, mode of delivery and viral load in 2895 U.S. women (PACTG 367). http://retroconference.org/2004/cd/Abstract/99.htm (accessed February 2008).
Sherman GG, Matsebula TC, and Jones SA (2005a) Is early HIV testing of infants in poorly resourced prevention of mother to child transmission programmes unaffordable? *Tropical Medicine and International Health* 10: 1108–1113.
Sherman GG, Stevens G, Jones SA, Horsfield P, and Stevens WS (2005b) Dried blood spots improve access to HIV diagnosis and care for infants in low-resource settings. *Journal of Acquired Immune Deficiency Syndromes* 38: 615–617.
UNICEF (2005) *Children: The Missing Face of AIDS*. New York: UNICEF.
Villamor E, Mbise R, Spiegelman D, Hertzmark E, Fataki M, Peterson KE, and Fawzi WW (2002) Vitamin A supplements ameliorate the adverse effect of HIV-1, malaria, and diarrheal infections on child growth. *Pediatrics* 109: E6.
Villamor E and Fawzi WW (2000) Vitamin A supplementation: Implications for morbidity and mortality in children. *Journal of Infectious Diseases* 182: S122–S133.
World Health Organization (2006) Antiretroviral therapy of HIV infection in infants and children in resource-poor settings: Toward universal access. Recommendations for a public health approach. Geneva, Switzerland: World Health Organization.
World Health Organization (2007) *WHO Case Definitions of HIV for Surveillance and Revised Clinical Staging and Immunological Classification of HIV-Related Disease in Adults and Children*. Geneva, Switzerland: World Health Organization.
World Health Organization, Joint United Nations Programme on HIV/AIDS, United Nations Children's Fund (2004) Joint WHO/UNAIDS/UNICEF statement on use of cotrimoxazole as prophylaxis in HIV exposed and HIV infected children. Geneva, Switzerland: World Health Organization. http://www.who.int/3by5/mediacentre/en/Cotrimstatement.pdf (accessed February 2008).
Zar HJ, Cotton MF, Strauss S, *et al.* (2007) Effect of isoniazid prophylaxis on mortality and incidence of tuberculosis in children with HIV: Randomised controlled trial. *British Medical Journal* 334: 136–139.

Further Reading

Brahmbhatt H, Kigozi G, Wabwire-Mangen F, *et al.* (2006) Mortality in HIV-infected and uninfected children of HIV-infected and uninfected mothers in rural Uganda. *Journal of Acquired Immune Deficiency Syndromes* 41: 504–508.
Centers for Disease Control and Prevention (CDC) (2001) Revised guidelines for HIV counseling, testing, and referral and Revised recommendations for HIV screening of pregnant women. *Morbidity and Mortality Weekly Report* 50: RR-19, pp. 1–58 and pp. 59–86.
Chaisilwattana P, Chokephaibulkit K, Chalermchockcharoenkit A, *et al.* (2002) Short-course therapy with zidovudine plus lamivudine for prevention of mother-to-child transmission of human immunodeficiency virus type 1 in Thailand. *Clinical Infectious Diseases* 35: 1405–1413.
Coll O, Hernandez M, Boucher C, *et al.* (1997) Vertical HIV-1 transmission correlates with a high maternal viral load at delivery. *Journal of Acquired Immune Deficiency Syndromes* 14: 26–30.
Coutsoudis A, Pillay K, Kuhn L, *et al.* (2001) Method of feeding and transmission of HIV-1 from mothers to children by 15 months of age: Prospective cohort study from Durban, South Africa. *AIDS* 15: 379–387.
Crowe S, Turnbull S, Oelrichs R, and Dunne A (2003) Monitoring of human immunodeficiency virus infection in resource-constrained countries. *Clinical Infectious Diseases* 37(Suppl 1): S25–S35.
Dabis F, Ekouevi D, Rouet F, *et al.* (2003) Effectiveness of a short course of zidovudine and lamivudine and peripartum nevirapine to prevent HIV-1 mother-to-child transmission. The ANRS DITRAME Plus trial, Abidjan, Côte d'Ivoire. *Antiviral Therapy* 8: S236–S237.
Dabis F, Msellati P, Meda N, *et al.* (1999) 6-month efficacy, tolerance, and acceptability of a short regimen of oral zidovudine to reduce vertical transmission of HIV in breastfed children in Côte d'Ivoire and Burkina Faso: A double-blind placebo-controlled multicentre trial. DITRAME Study Group. *Lancet* 353: 786–792.
Damond F, Descamps D, Farfara I, *et al.* (2001) Quantification of proviral load of human immunodeficiency virus type 2 subtypes A and B using real-time PCR. *Journal of Clinical Microbiology* 39: 4264–4268.
De Santis M, Carducci B, De Santis L, *et al.* (2002) Periconceptional exposure to efavirenz and neural tube defects [letter]. *Archives of Internal Medicine* 162: 355.
Desire N, Dehee A, Schneider V, *et al.* (2001) Quantification of human immunodeficiency virus type 1 proviral load by a TaqMan realtime PCR assay. *Journal of Clinical Microbiology* 4: 1303–1310.
Eshleman SH, Cunningham SP, Jones D, *et al.* (2003) Analysis of nevirapine resistance seven days after single-dose nevirapine prophylaxis: HIVNET 012. http://www.retroconference.org/2003/cd/Abstract/856.htm (accessed February 2008).
Eshleman S, Mracna M, Guay L, *et al.* (2001) Selection and fading of resistance mutations in women and infants receiving nevirapine to prevent HIV-1 vertical transmission (HIVNET 012). *AIDS* 15: 1951–1957.
European Mode of Delivery Collaboration (1999) Elective caesarean section versus vaginal delivery in prevention of vertical HIV-1 transmission: A randomized clinical trial. *Lancet* 353: 1035–1039.
Fawzi WW, Msamanga GI, Spiegelman D, *et al.* (1998) Randomised trial of effects of vitamin supplements on pregnancy outcomes and T cell counts in HIV-1-infected women in Tanzania. *Lancet* 351: 1477–1482.
Fawzi W, Msamanga G, Spiegelman D, *et al.* (2002) Transmission of HIV-1 through breastfeeding among women in Dar es Salaam, Tanzania. *Journal of Acquired Immune Deficiency Syndromes* 31: 331–338.
Fowler MG and Newell ML (2002) Breast-feeding and HIV-1 transmission in resource-limited settings. *Journal of Acquired Immune Deficiency Syndromes* 30: 230–239.
Garcia PM, Kalish LA, Pitt J, *et al.* (1999) Maternal levels of plasma human immunodeficiency virus type 1 RNA and the risk of perinatal transmission. Women and Infants Transmission Study Group. *New England Journal of Medicine* 341: 394–402.
Gerard Y, Maulin L, Yazdanpanah Y, *et al.* (2000) Symptomatic hyperlactatemia: An emerging complication of antiretroviral therapy. *AIDS* 14: 2723–2730.
Gibellini D, Vitone F, Gori E, La Placa M, and Re MC (2004) Quantitative detection of human immunodeficiency virus type 1 (HIV-1) viral load by SYBR green real-time RT-PCR technique in HIV-1 seropositive patients. *Journal of Virology Methods* 115: 183–189.
Grubert TA, Reindell D, Kastner R, *et al.* (1999) Complications after caesarean section in HIV-1 infected women not taking antiretroviral treatment. *Lancet* 354: 1612–1613.
Gueudin M, Plantier JC, Damond F, Roques P, Mauclere P, and Simon F (2003) Plasma viral RNA assay in HIV-1 group O infection by real-time PCR. *Journal of Virology Methods* 113: 43–49.

Hitti J, Frenkel LM, Stek AM, *et al.* (2004) Maternal toxicity with continuous nevirapine in pregnancy: Results from PACTG 1002. *Journal of Acquired Immune Deficiency Syndromes* 36: 772–776.

International Perinatal HIV Group (1999) The mode of delivery and the risk of vertical transmission of human immunodeficiency virus type 1 – A meta-analysis of 15 prospective cohort studies. *New England Journal of Medicine* 340: 977–987.

International Perinatal HIV Group (2001) Duration of ruptured membranes and vertical transmission of HIV-1: A meta-analysis from fifteen prospective cohort studies. *AIDS* 15: 357–368.

Jackson JB, Musoke P, Fleming T, *et al.* (2003) Intrapartum and neonatal single-dose nevirapine compared with zidovudine for prevention of mother-to-child transmission of HIV-1 in Kampala, Uganda: 18-month follow-up of the HIVNET 012 randomised trial. *Lancet* 362: 859–868.

Joy S, Poi M, Hughes L, *et al.* (2005) Third-trimester maternal toxicity with nevirapine use in pregnancy. *Obstetrics and Gynecology* 106: 1032–1038.

Lallemant M, Jourdain G, Le Coeur S, *et al.* (2000) A trial of shortened zidovudine regimens to prevent mother-to-child transmission of human immunodeficiency virus type 1. Perinatal HIV Prevention Trail (Thailand) Investigators. *New England Journal of Medicine* 343: 982–991.

Leroy V, Newell ML, Dabis F, *et al.* (1998) International multicentre pooled analysis of late postnatal mother-to-child transmission of HIV-1 infection. Ghent International Working Group on Mother-to-Child Transmission of HIV. *Lancet* 352: 597–600.

Leroy V, Sakarovitch C, Cortina-Borja M, *et al.* (2005) Is there a difference in the efficacy of peripartum antiretroviral regimens in reducing mother to child transmission of HIV in Africa? *AIDS* 19: 1865–1875.

Luzzati R, Del Bravo P, Di Perri G, *et al.* (1999) Riboflavine and severe lactic acidosis. *Lancet* 353: 901–902.

Miotti PG, Taha TE, Kumwenda NI, *et al.* (1999) HIV transmission through breastfeeding: A study in Malawi. *Journal of the American Medical Association* 282: 744–749.

Mofenson L (1995) A critical review of studies evaluating the relationship of mode of delivery to perinatal transmission of human immuno deficiency virus. *Journal of Infectious Diseases* 14: 169–176.

Mofenson LM, Harris DR, Moye J, *et al.* (2003) NICHD IVIG Clinical Trial Study Group. Alternatives to HIV-1 RNA concentration and CD4 count to predict mortality in HIV-1 infected children in resource-poor settings. *Lancet* 362: 1625–1627.

Mofenson LM, Oleske J, Serchuck L, Van Dyke R, and Wilfert C (2004) Treating opportunistic infections among HIV-exposed and infected children: Recommendations from CDC, the National Institutes of Health, and the Infectious Diseases Society of America. *Morbidity and Mortality Weekly Report* 53(RR-14): 1–92.

Moodley D, Moodley J, Coovadia H, *et al.* (2003) A multicenter randomized controlled trial of nevirapine versus a combination of zidovudine and lamivudine to reduce intrapartum and early postpartum mother to child transmission of human immunodeficiency virus type 1. *Journal of Infectious Diseases* 187: 725–735.

Nduati R, John G, Mbori-Ngacha D, *et al.* (2000) Effect of breastfeeding and formula feeding on transmission of HIV-1: A randomized clinical trial. *Journal of the American Medical Association* 283: 1167–1174.

Newell ML, Coovadia H, Cortina-Borja M, Rollins N, Gaillard P, and Dabis F (2004) Mortality of infected and uninfected infants born to HIV infected mothers in Africa: A pooled analysis. *Lancet* 364: 1236–1243.

Patton JC, Sherman GG, Coovadia AH, Stevens WS, and Meyers TM (2006) The Ultrasensitive HIV-1 p24 Antigen assay modified for use in dried whole blood spots as a reliable, affordable test for infant diagnosis. *Clinical and Vaccine Immunology* Jan: 152–155.

Pitt J, Brambilla D, Reichelderfer P, *et al.* (1997) Maternal immunologic and virologic risk factors for infant human immunodeficiency virus type 1infection: Findings from the Women and Infants Transmission Study. *Journal of Infectious Diseases* 175: 567–575.

Prata N, Vahidnia F, and Fraser A (2005) Gender and relationship differences in condom use among 15–24-year-olds in Angola. *International Family Planning Perspectives* 31: 192–199.

Rouet F, Ekouevi DK, Chaix ML, *et al.* (2005) Transfer and evaluation of an automated, low-cost real-time reverse transcription-PCR test for diagnosis and monitoring of human immunodeficiency virus type 1 infection in a West African resource-limited setting. *Journal of Clinical Microbiology* 43: 2709–2717.

Rouet F, Elenga N, Msellati P, *et al.* (2002) Primary HIV-1 infection in African children infected through breastfeeding. *AIDS* 16: 2303–2309.

Rouzioux C, Costagliola D, Burgard M, *et al.* (1995) Estimated timing of mother-to-child human immunodeficiency virus type 1 (HIV-1) transmission by use of a Markov model. The HIV Infection in Newborns French Collaborative Study Group. *American Journal of Epidemiology* 142: 1330–1337.

Sanne I, Mommeja-Marin H, Hinkle J, *et al.* (2005) Severe hepatoxicity associated with nevirapine use in HIV-infected subjects. *Journal of Infectious Diseases* 191: 825–829.

Schutten M, van den Hoogen B, van der Ende ME, Gruters RA, Osterhaus AD, and Niesters HG (2000) Development of a real-time quantitative RT-PCR for the detection of HIV-2 RNA in plasma. *Journal of Virology Methods* 88: 81–87.

Semba RD, Kumwenda N, Hoover DR, *et al.* (1999) Human immunodeficiency virus load in breast milk, mastitis, and mother-to-child transmission of human immunodeficiency virus type 1. *Journal of Infectious Diseases* 180: 93–98.

Semprini AE, Castagna C, Ravizza M, *et al.* (1995) The incidence of complications after cesarean section in 156 HIV-positive women. *AIDS* 9: 913–917.

Shaffer N, Chuachoowong R, Mock PA, *et al.* (1999) Short-course zidovudine for perinatal HIV-1 transmission in Bangkok, Thailand: A randomized controlled trial. Bangkok Collaborative Perinatal HIV Transmission Study Group. *Lancet* 353: 773–780.

Shapiro D, Tuomala R, Samelson R, *et al.* (2002) Mother-to-child HIV transmission rates according to antiretroviral therapy, mode of delivery, and viral load (PACTG 367). http://www.retroconference.org/2002/Abstract/12953.htm (accessed February 2008).

Stern JO, Robinson PA, Love J, *et al.* (2003) A comprehensive hepatic safety analysis of nevirapine in different populations of HIV infected patients. *Journal of Acquired Immune Deficiency Syndromes* 34: S21–S33.

Taha TE, Kumwenda NI, Hoover DR, *et al.* (2004) Nevirapine and zidovudine at birth to reduce perinatal transmission of HIV in an African setting: A randomized controlled trail. *Journal of the American Medical Association* 292: 202–209.

Tarantal AF, Castillo A, Ekert JE, *et al.* (2002) Fetal and maternal outcome after administration of tenofovir to gravid rhesus monkeys (Macaca mulatta). *Journal of Acquired Immune Deficiency Syndromes* 29: 207–220.

Taylor GP and Low-Beer N (2001) Antiretroviral therapy in pregnancy: A focus on safety. *Drug Safety* 24: 683–702.

Toni TD, Masquelier B, Lazaro E, *et al.* (2005) Characterization of nevirapine (NVP) resistance mutations and HIV type 1 subtype in women from Abidjan (Côte d'Ivoire) after NVP single-dose prophylaxis of HIV type 1 mother-to-child transmission. *AIDS Research and Human Retroviruses* 21: 1031–1034.

Tovo P, Gabiano C, and Tulisso S (1997) Maternal clinical factors influencing HIV-1 transmission. *Acta Pediatrica* Suppl 421: 52–55.

Wabwire-Mangen F, Gray R, Mmiro F, *et al.* (1999) Placental membrane inflammation and risks of maternal-to-child transmission of HIV-1 in Uganda. *Journal of Acquired Immune Deficiency Syndromes* 22: 379–385.

Wiktor SZ, Ekpini E, Karon JM, *et al.* (1999) Short-course oral zidovudine for prevention of mother-to-child transmission of HIV-1 in Abidjan, Côte d'Ivoire: A randomised trial. *Lancet* 353: 781–785.

Infertility

F Zegers-Hochschild and J-E Schwarze, Clinica las Condes, Santiago, Chile
V Alam, Merck Serono, Geneva, Switzerland

Glossary

Anovulation Lack of ovulation.
Azoospermia Lack of sperm in the semen.
Blastocyst An embryo with a fluid-filled blastocele cavity (usually developing by 5 or 6 days after fertilization).
Clinical abortion An abortion of a clinical pregnancy that takes place between the diagnosis of pregnancy and 20 completed weeks' gestational age.
Clinical pregnancy Evidence of pregnancy by clinical or ultrasound parameters (ultrasound visualization of a gestational sac). It includes ectopic pregnancy. Multiple gestational sacs in one patient are counted as one clinical pregnancy.
Embryo Product of conception from the time of fertilization to the end of the embryonic stage 8 weeks after fertilization (the term 'preembryo,' or dividing conceptus, has been replaced by 'embryo').
Endometriosis An abnormal gynecological condition characterized by the growth and function of endometrial tissue outside of the uterus.
Endometrium The mucous membrane lining the uterine cavity, which undergoes cellular changes during the monthly menstrual cycle.
Fecundity The ability to produce offspring frequently and in large numbers.
Fertilization The penetration of the ovum by the spermatozoon and fusion of genetic materials, resulting in the development of a zygote.
Fetus The product of conception starting from completion of embryonic development (at 8 completed weeks after fertilization) until birth or abortion.
Gamete Specialized reproductive cells (ova in females; spermatozoa in males), the union of which forms a new individual.
Gamete intrafallopian transfer (GIFT) ART procedure in which both gametes (oocytes and sperm) are transferred to the Fallopian tubes.
Gestational age Age of an embryo or fetus calculated by adding 14 days (2 weeks) to the number of completed weeks since fertilization.
Hysterosalpingography (HSG) Radiologic procedure involving the insertion of radio-opaque solution through the cervix to investigate the shape of the uterine cavity and the shape and patency of the uterine tubes.
Implantation The attachment and subsequent penetration by the zona-free blastocyst (usually in the endometrium), which starts 5–7 days following fertilization.
Intracytoplasmic sperm injection (ICSI) IVF procedure in which a single spermatozoon is injected through the zona pellucida into the oocyte.
Laparoscopy Examination of the abdominal cavity by means of an instrument called a laparoscope, which is introduced through a small abdominal incision, allowing the physician to see the reproductive organs; used especially when Fallopian tube damage or endometriosis is suspected.
Live birth A birth in which a fetus is delivered with signs of life after complete expulsion or extraction from its mother, beyond 20 completed weeks of gestational age. Live births are counted as birth events (e.g., a twin or triplet live birth is counted as one birth event).
Ovulation The release of a mature egg from the ovarian follicle. Both gonadotropic hormones, follicle-stimulating hormone (FSH) and luteinizing hormone (LH), are required for this process.
Preimplantation genetic diagnosis (PGD) Screening of cells from preimplantation embryos for the detection of genetic and/or chromosomal disorders before embryo transfer.
Spontaneous abortion Spontaneous loss of a clinical pregnancy before 20 completed weeks of gestation or, if gestational age is unknown, of a fetus with weight of 500 g or less.
Stillbirth A birth in which the fetus does not exhibit any signs of life when completely removed or expelled from the birth canal at or after 20 completed weeks of gestation. Stillbirths are counted as birth events (e.g., a twin or triplet stillbirth is counted as one birth event).
Zygote The diploid cell resulting from the fertilization of an oocyte by a spermatozoon, which subsequently develops into an embryo.

But you must come, sweet love, my baby, because water gives salt, the earth fruit, and our wombs guard tender infants, just as a cloud is sweet with rain.

(Federico Garcia Lorca, *Yerma*, 1934)

Introduction

For many couples, the inability to bear children develops into a life crisis. The confluxes of personal, interpersonal, social, and religious expectations bring a sense of failure, loss, and exclusion to those unable to conceive. Furthermore, relationships can become very strained when children are not forthcoming: one partner may seek to blame the other as being defective or unwilling.

Most societies, especially in developing countries, are structured to rely on children for the future care and maintenance of older family members. Even in developed countries with social support systems, it is the family, including the children, who are expected to provide most of the care for the elderly.

Their treatment varying enormously in different ethnic groups and cultures, childless couples can be excluded from taking part in important family functions and events such as birthdays, christenings, confirmations, bar mitzvahs, and other social gatherings. Furthermore, many religions assign important ceremonial tasks to the couple's children. Peer or parental pressure to 'have a baby' (Fonttis *et al.*, 2004) may lead couples to avoid participating in family or other social gatherings, affecting their family and social life.

Only recently has the international community acknowledged a comprehensive concept of reproductive health that encompasses more than just childbearing, contraception, and the prevention of sexually transmitted infections. In 1994, the United Nations Programme of Action formally expanded the meaning of reproductive health to include the right of men and women to choose the number, timing, and spacing of their children. The expended reproductive health concept also includes the recommendation to support, in reproductive health programs, the prevention and appropriate treatment of infertility (World Health Organization, 1994). Furthermore, since 2006, infertility has been considered a reproductive health indicator, alongside maternal mortality ratio, perinatal mortality rate, prevalence of anemia, and so forth (World Health Organization, 2006).

In general, 1 in 10 couples experiences primary or secondary infertility, thus affecting more than 80 million people in the world (United Nations, 1995). Infertility rates vary among countries, from less than 5% to more than 30% of couples in the reproductive age group (Vayena *et al.*, 2002).

The objective of this chapter is to provide a comprehensive review of the magnitude of the problem of infertility in different regions – its main causes as well as the impact of this condition on individuals, their families, and society as a whole. It is also the purpose of this chapter to describe the various treatments available today, and the accessibility worldwide to these treatment options.

Definitions

The terms 'infertility,' 'sterility,' 'infecundity,' and 'childlessness' are often used with little consideration of their precise meaning. The understanding of these terms may differ substantially when they are used by demographers or by clinicians dealing with infertile couples. The demographers' main focus is to describe the effect of reproduction on the structure of a certain population; therefore, their primary metric is the actual number of births taking place in a certain population. In this context, fertility rate is expressed as the ratio between the numbers of births and the number of women exposed to pregnancy; infertility is defined as the absence of births irrespective of whether pregnancy had occurred. On the other hand, for clinicians dealing with fertility treatments, the definition of infertility is the failure to conceive (become pregnant) after at least 1 year of unprotected coitus. Although, from the couple's perspective, the delivery of a live birth is the only significant event, for a given treatment, pregnancy is the first direct measure of the quality and efficacy of that intervention. For this reason, when evaluating efficacy or comparing different treatment modalities, both the clinical pregnancy rate and the delivery rate are used. Infertility can also be subclassified as primary or secondary, depending on whether the woman has been pregnant before, irrespective of the outcome of that pregnancy.

From a different perspective, the concept of childlessness is often applied to women who, by the end of their reproductive age, have remained, either voluntarily or involuntarily, without children.

Table 1 Demographic and reproductive medicine definitions

	Demography	*Reproductive medicine*
Infertility	Percentage of women who have been married for the past 5 years, who have ever had sexual intercourse, who have not used contraception during the past 5 years, and who have not had any births	Failure to conceive after at least 1 year of unprotected coitus
Childlessness	Percentage of women who are currently married, have been so for at least 5 years, and who have no living children	

Table 2 Primary and secondary infertility, and childlessness

Country	*Primary infertility*[a]	*Secondary infertility*[b]	*Childlessness*[c]
Sub-Saharan Africa			
Benin	3.3	22.0	3.2
Burkina Faso	1.8	23.2	1.4
Cameroon	3.5	26.4	7.3
Central African Republic	4.8	35.8	10.5
Chad	3.1	32.8	4.4
Comoros	6.9	24.7	5.5
Cote d'Ivoire	3.7	37.3	3.7
Eritrea	4.0	26.9	1.2
Ghana	4.1	24.3	1.1
Guinea	3.4	33.1	3.0
Kenya	2.5	24.0	1.5
Madagascar	5.2	26.3	4.7
Malawi	1.9	24.1	2.8
Mali	2.3	27.9	3.3
Mozambique	4.6	35.2	3.1
Niger	2.9	26.6	4.4
Nigeria	5.6	30.7	2.1
Senegal	3.5	20.3	3.8
Tanzania	3.2	24.1	2.8
Togo	4.4	18.1	1.4
Uganda	3.6	37.7	3.2
Zambia	3.1	22.0	2.3
Zimbabwe	2.8	19.5	2.7
North Africa/ West Asia			
Egypt	3.6	17.0	2.8
Jordan	3.5	13.5	2.4
Morocco	17.8	15.2	2.0
Turkey	8.9	18.9	1.8
Yemen	2.5	24.5	2.5
Central Asia/ South and Southeast Asia			
Bangladesh	1.9	22.7	1.8
Cambodia	2.6	38.4	2.1
India	2.9	27.5	2.6
Indonesia	3.4	26.1	3.7
Kazakhstan	4.1	15.7	2.0
Krygyz Rep.	2.0	17.0	0.5
Nepal	2.2	26.9	3.2
Philippines	2.9	19.8	1.6
Turkemenistan	2.1	18.7	1.3
Uzkebistan	1.9	16.6	0.9
Viet Nam	1.7	9.3	0.9
Latin America/ Caribbean			
Bolivia	4.1	17.5	1.1
Brazil	7.7	12.6	3.3
Colombia	7.7	10.3	2.5
Dominican Republic	4.0	18.1	3.7
Guatemala	1.2	20.2	1.7
Haiti	5.6	24.5	3.9
Nicaragua	2.5	17.7	1.4
Peru	5.8	15.7	1.4

[a]Primary infertility: Percentage of women aged 25–49 years who have had sexual intercourse but have never had a birth.

[b]Secondary infertility: Percentage of women aged 25–49 years who have had a delivery, but report themselves as infecund or have not had a live birth in the past five years while not using contraception during this period.

[c]Childlessness: Percentage of currently married women aged 40–44 years who have been married for at least five years and have no living children.

Source: Rutstein SO and Shah IH (2004) *Infecundity, Infertility, and Childlessness in Developing Countries*. DHS Comparative Reports No. 9. Calverton, MD: ORC Macro and the World Health Organization.

In this chapter, the 'demographic' definitions will be used when analyzing population data, and the 'clinical' definitions when analyzing the outcome of infertility treatments (see **Table 1**).

Prevalence of Infertility

The prevalence of primary infertility (percentage of women aged 25–49 years who have had unprotected sexual intercourse but have never had a birth) and secondary infertility (women who have had one or more live births and are unable to have another child) ranges among different countries and continents from 5–30% (Vayena *et al.*, 2002).

Consensus estimation is that approximately 10% of married and cohabiting couples will not be able to conceive after 1 year of unprotected coitus. The reasons for this broad range have to do with a variety of biological and sociocultural components, such as:

- the mean age of the female population,
- the age and mode of sexual initiation,
- the number of sexual partners (before women express their desire to become mothers),
- the incidence of sexually transmitted infections (STIs),
- the use and type of contraception, and
- the incorporation of women into academic education and the workforce.

Once sexual relations are established on a regular basis, pregnancy soon follows for the majority of women unless contraception is used. **Table 2** shows the proportions of women with primary and secondary infertility. Secondary infertility increases sharply with age, from about 5% at ages 20–24 years to about 62% at ages 45–49 years (Rutstein and Shah, 2004).

Many developing countries, especially in Africa, show low incidence of primary infertility, partly due to the fact that marriage and pregnancy take place at a younger age than in developed countries. This is confirmed in several reports that indicate that sexual activity begins at an increasingly younger age in different countries in Africa (Bambra, 1999). Furthermore, younger birth cohorts with first coitus occurring at younger age have been reported in several countries. However, due to lack of proper assistance in pregnancy and delivery, and especially the high incidence of STIs (Bambra, 1999), the rate of secondary infertility can be as high as 30% in Central Africa and 27% in Eastern Africa (Rutstein and Shah, 2004).

On the other hand, in many highly industrialized countries the exposure to pregnancy is postponed, allowing more sexual partners before seeking pregnancy. Therefore, the frequency of primary infertility is higher in developed countries when compared to developing countries, with a prevalence ranging from 3.6–14.3% (Schmidt and Münster, 1995).

Prevalence of infertility from an international perspective confronts two paradoxes (Nachtigall, 2006). The first paradox is that, due to its population structure, the developing world generally aims at controlling and containing population growth, while most of the industrialized nations of the developed world have fertility rates below those required to maintain their current populations. The second demographic paradox is that countries with the highest overall fertility are also those with the highest prevalence of infertility. Countries in Northern Africa, Southern Asia, and Latin America all report high incidence of secondary infertility, ranging from 15% to greater than 25% (Rutstein and Shah, 2004). In the so-called 'infertility belt' of sub-Saharan Africa, the percentage of couples with secondary infertility exceeds 30% in some countries, and, in Zimbabwe, it has been reported that almost 2 out of 3 women over the age of 25 years are infertile (Larsen, 2000).

Childlessness at the end of the reproductive years is most effectively studied by using women in the oldest age cohort, that is women aged 45–49 years. However, there are well-known reporting problems with this age group that involve both the determination of age and fertility. Data from the Demographic and Health Surveys give a proportion of childless women at ages 40–49 years ranging from less than 2% (Malawi, Tanzania, Zambia) to 9% (Philippines), with a mean close to 4% (Rutstein and Shah, 2004).

Table 2 presents data on women in a younger age cohort (40–44 years) than the one described above which measures lifetime childlessness. Among currently married women aged 40–44 years who have been married for at least 5 years, about 96% have one or more surviving children in the majority of countries. In most developing countries, even when fertility has already declined significantly, the proportion of childless women is low. Many developed countries experienced a baby boom in the 1950s and 1960s. Consequently, the percentage of childless women in some cohorts reached very low levels. Fertility and family surveys carried out in a number of European countries in the 1990s determined the proportion of childless women. These women, aged 40 to 49 years at the time of the survey, were born on average around 1950, and thus were post-baby-boom cohorts. The proportion of childless women is low in some countries (5–8% in Norway, Poland, Romania, Czech Republic, Ukraine), but exceeds 10% in Finland and Sweden, and peaks at 18% in Switzerland. The proportions for men are usually higher, sometimes substantially: 17% in Finland, and more than 21% in Switzerland (Rutstein and Shah, 2004).

Consequences of Infertility for Individuals and the Family

Infertility is an unexpected event in the life cycle of any person, and coming to terms with this reality is a stressful process for both the individuals and the couple. Although men and women use different strategies to manage stress associated with infertility, this condition has an effect on both partners, regardless of which one appears to be biologically responsible for the couple's infertility.

The magnitude of stress generated by infertility is only comparable to somatic diseases such as cancer. Furthermore, the stress generated by infertility treatment is ranked second to the stress involving the death of a family member or the stress involving divorce.

When confronted with the diagnosis of infertility, couples move through different psychological stages, often beginning with denial and disbelief, followed by an acute phase of anxiety. Later, many couples experience feelings of loss, grief, and depression (Covington, 1995). Many infertile couples also resent the lack of support from their families and social environment. Family and social environment actually can become an important source of stress due to people inquiring about a couple's childlessness and spreading myths about infertility and its solution. When couples decide to undergo an assisted reproductive technology (ART), which is often seen as the last recourse, anxiety increases even more. If treatment fails, the associated stress leads many couples to abandon treatment, even if additional treatment is available and affordable. On the other hand, the lack of access to treatment due to economic reasons can be an aggravating factor.

Infertility challenges the person's feelings about gender roles, self-worth, self-esteem, and self-perception of the body. While men and women may experience a variety of sexual demands in connection with this 'procreative' rather than 'relational' or 'recreational' sex, sexual dysfunction is commonly observed during and/or after infertility treatment. For couples whose previous sexual adjustment was difficult, infertility can negatively impact on their sexual response and the feeling of pleasure, decreasing libido in women and generating erectile dysfunction or other forms of sexual dysfunction in men.

There is substantial variability in how different couples are affected by infertility. Some couples' relationships deteriorate, whereas other couples emerge closer and more satisfied. In general, differences in gender perspective often generate conflicts among infertile couples. Within infertile couples, women see having children as more important than men do, and are more involved in the process of trying to conceive. Women want to talk and share their feelings about trying to have children more than their male partners do, and are more likely to experience emotional distress and loss of self-esteem as a result of experiencing infertility. In developing countries, the pressures on infertile women are, if anything, greater than the ones seen in developed countries. This is especially so in Africa and some countries in Latin America, where women unable to bear children are often rejected by their husbands and ostracized, being perceived as inferior and useless to society.

Etiology

The frequency of different etiologic factors varies between countries and is greatly dependent on cultural, social, and economic determinants. Furthermore, differences in the availability of modern and more sophisticated diagnostic tools can affect the capacity to reach certain diagnoses. The absence of an updated international standardization of diagnostic categories makes it difficult to compare the relative frequency of various etiologic factors in different regions of the world.

A report by Smith *et al.* (2003) in women from North America demonstrates that ovulatory disorders account for 17.6% of the female factors, tubal diseases 23.1%, endometriosis 6.6%, and male factors 25.9%. Unexplained infertility presented with a very high incidence, 25.6%. One of the few studies conducted in a geographically isolated but demographically and socioeconomically representative area of an industrialized Western society is the Walcheron study in The Netherlands. The authors followed a cohort of 726 couples during 2 to 10 years without interventions. Almost 10% (9.9%) of them presented with fertility complaints at least once during the cohort follow-up. Thirty percent were due to male factors, 25.9% had ovulation defects, 13% were tubal factors, and again there was a high incidence (30%) of unexplained infertility.

While in highly industrialized countries the delay in exposure to pregnancy is a cause of infertility, in poorly industrialized countries primary infertility is low, but secondary infertility is very high due to pelvic infections associated with STIs. Pelvic infections due to *Neisseria gonorrhea* and *Chlamydia trachomatis*, acquired through sex or after abortions and deliveries, remain a major public health challenge. Their prevalence has constantly increased in the last 20 years, despite governmental and public efforts to control them. In many African countries, some of the factors responsible for this increase in STIs are: (1) high birth rates, resulting in increasingly larger sexually active populations, (2) wars and civil unrest altering the male-to-female population rate, leading to increased numbers of sex workers, (3) urbanization and male migration, promoting male promiscuity and commercial sex, (4) poor medical services, leading to lack of treatment of common STIs and hence their spread, and (5) general poverty in female communities and low income (Bambra, 1999).

With STIs in developing communities and a delay in exposure to pregnancy in developed societies as the major conditions causing infertility, treatment modalities need to deal with either damaged Fallopian tubes or poor-quality oocytes, respectively. Both conditions require access to modern reproductive technology, which is available to only a small proportion of infertile couples in the world.

Although less prevalent, dietary or environmental factors may be responsible for some cases of infertility (Hamberger and Janson, 1997). Iodine and selenium deficiencies have been linked to infertility, as have exposures to the alatoxins that commonly contaminate foods in tropical countries. Men and women in developing countries may also face exposure to environmental and occupational toxicants such as arsenic, lead, solvents, pesticides, and industrial chemicals. Highly prevalent habits, such as smoking and alcohol and caffeine consumption, have also been linked to decreased fertility in both men and women.

Overview of Infertility Evaluation and Treatment Alternatives

Since 85% of couples will achieve pregnancy within 1 year of unprotected intercourse, medical intervention is indicated in couples attempting pregnancy who fail to conceive within that year. Earlier interventions can be initiated in women over 35 years of age, as natural fecundity and therefore the success rate of any medical procedure decrease dramatically as years pass by. Evaluation of both partners should begin simultaneously. The goal of the initial infertility evaluation is to start formulating a diagnosis based on a comprehensive anamnesis and physical examination, together with specific tests to evaluate the reproductive systems of the male and female partner, both as individuals and as reproductive couples (**Tables 3** and **4**).

The first requirement for a spontaneous conception to take place is an anatomical and functional integrity of the reproductive organs, allowing the male partner to deposit semen inside the vagina during coitus. This can be affected by various reasons, such as hymen obliteration

Table 3 Basic infertility workup

Man	Physical exam
	Semen analysis
Woman	Physical exam
	Pelvic ultrasound
	Mid-luteal progesterone
	Hysterosalpingography
	Postcoital test

Table 4 Diagnostic categories

Female infertility	Mechanical Endocrine Immunological Genetic Endometriosis Low ovarian reserve
Male infertility	Impaired sperm production Impaired sperm motility Impaired sperm morphology
Unknown, i.e., idiopathic	

and scars in the vulva as a consequence of trauma and infection, and sometimes as a sequel of clitoridectomy or other forms of female genital mutilation practiced in certain Muslim and African communities (Obermeyer, 2005). Developmental abnormalities in the penis such as Peyronie disease can also interfere with normal coitus. Apart from these structural abnormalities, appropriate release of semen in the vagina can be impaired by ejaculatory dysfunction and lack of orgasm in the male partner.

If normal coitus takes place during the fertile period, a sequence of sequential events needs to follow for pregnancy to occur:

1. The release by the ovary of a mature oocyte.
2. Sufficient number and quality of motile sperm reaching the Fallopian tube.
3. Encounter of female and male gametes in the oviduct, followed by fertilization.
4. The resulting embryo must then be transported from the oviduct toward the uterine cavity where it will continue development.
5. At the end of the 5th day after fertilization, the developing blastocyst will attach and penetrate the endometrium. It is only then that conception starts.

Infertility Evaluation of the Female Partner

A basic infertility investigation in the female partner should include: (1) evaluation of ovarian reserve, (2) assessment and characterization of the ovulatory cycle, (3) assessment of uterine cavity, and (4) evaluation of tubal indemnity and pelvic integrity.

Evaluation of ovarian reserve

Faddy *et al.* (1992) proposed a mathematical model describing the age-related decline in the population of small follicles in the human ovary. The model suggests that the rate of follicle disappearance almost doubles when the number of follicles falls to a critical figure of 25 000 at the age of 37.5 years. *In vitro* studies have further shown that the remaining follicles have fewer granulosa cells, decreased mitosis, and increased apoptosis. It can be speculated that in women who experience menopause by the age of 45 years or earlier, the critical point of 25 000 remaining follicles had been reached 13 years earlier, with a rapid decline of fertility starting at 32 years and possibly total loss of fertility by the age of 36 years. Nevertheless, age is not the only predictor of ovarian reserve, or of the pool of follicles in the ovary. Previous pelvic surgeries, smoking habits, and genetic factors can also affect ovarian reserve.

When matched by age, women with decreased ovarian reserve have a lower fertility rate when attempting pregnancy, lower pregnancy rate per cycle when undergoing low-complexity ART, and lower oocyte yield when undergoing high-complexity ART.

Several markers have been used to determine ovarian reserve: basal follicle-stimulating hormone (FSH) measured in plasma between days 3 and 5 of the menstrual cycle; inhibin B; and the number of antral follicles in the early follicular phase assessed by ultrasound. On the other hand, ovarian challenge tests offer the theoretical benefit of measuring ovarian reserve in a dynamic and clinically meaningful way. The exogenous FSH ovarian reserve test was designed to screen for good or poor response during *in vitro* fertilization (IVF). Although expensive, it has less intercycle variability than basal FSH alone. The clomiphene citrate challenge test has also been used with success. It is especially useful to identify a subgroup of women with decreased ovarian reserve in spite of having normal basal FSH.

Assessment and characterization of the ovulatory cycle

A clinical assessment of the ovulatory cycle starts with a medical history recording both the frequency and characteristics of the menstrual cycle. Regular menstrual periods (25 to 35 days) preceded by the loss of transparent cervical mucus secretion are usually associated with ovulation, especially if the "fertile" mucus is produced 12 to 15 days prior to each menstrual episode. This can be further confirmed by the mid-luteal measurement in plasma of progesterone or of its urinary metabolite. Ultrasound assessment of follicular growth and rupture, together with the visualization of changes in endometrium growth and texture, are also important objective markers of ovulation.

Ovulatory disorders include anovulation, which is frequently but not necessarily associated with amenorrhea, and ovulatory dysfunction, which includes a variety of menstrual disorders such as the loss of menstrual cycles, oligomenorrhea, and luteal phase defects, frequently associated with inadequate corpus luteum function. When characterizing ovulatory disorders, the first objective is to determine whether the cause is located in the pituitary gland, in the ovaries, or in other systems. The measurement of FSH, estradiol, and prolactin early in the follicular phase should allow for a topographic localization of the problem. If FSH is high (>12 IU/l), the cause is most probably the result of decreased ovarian reserve and there are very few treatment alternatives. On the other hand, if FSH is normal or low, the cause is most probably a pituitary failure or dysfunction, in which case there are many treatment alternatives aimed at restoring ovulation. In order to prevent multiple gestation, treatments should be directed at recruiting and maturating one or at the most two dominant follicles. Cumulative pregnancy rates of 40–70% are usually achieved after three ovulatory attempts with different treatment options.

The treatments most frequently used are: (1) clomiphene citrate, (2) human menopausal gonadotropins (HMG), extracted from urine of postmenopausal women, and (3) recombinant gonadotropins (rFSH and rLH) manufactured by recombinant DNA technology. The main complication when treating anovulation is the high incidence of multiple pregnancies, especially in women with pituitary dysfunction or failure. Once ovulation is restored in these women, cumulative conception rates are similar to normal women; therefore, exposure to multiple oocytes increases the risk of multiple gestations. This is especially so when HMG or rFSH is used without appropriate ultrasound monitoring of the number of follicles recruited and maturing. Multiple gestations derived from aggressive induction of ovulation should be avoided by following guidelines of best clinical practices, including appropriate decision to abandon the stimulation cycle.

Assessment of uterine cavity and evaluation of tubal indemnity and pelvic integrity

Abnormalities in the uterus can be the result of developmental abnormalities, such as bicornuate, septate, or arcuate uterus and other similar conditions, which can affect embryo implantation as well as the continuation of pregnancy and delivery, or acquired abnormalities, such as infection and trauma to the epithelial and/or muscular layer, usually associated with chlamydia infection or tuberculosis, among others. Trauma due to repeated dilation and curettages (D&Cs) and uterine perforation by contraceptive intrauterine devices (IUDs) can also damage the uterine cavity. Other acquired conditions are fibroids located near the endometrium, which can interfere with embryo implantation.

The Fallopian tubes can also be damaged, interfering with either oocyte pickup and/or embryo transport into the uterine cavity. What causes more damage to this organ are infections associated with STIs and adhesions generated after septic procedures and pelvic surgery. The uterine cavity can be assessed by ultrasound, hysterosalpingography (HSG), and hysteroscopy. Evaluation of tubal patency can only be obtained by either HSG or laparoscopy.

Obstructions, distortions, adhesions, infections, and other pathologies account for 20–40% of the causes of infertility, depending on the geographical region. The goal of assessing the female genital tract is to demonstrate tubal patency and absence of adhesions and/or endometriosis. HSG has a sensitivity of 65% and specificity of 83% for diagnosis of tubal patency. Positive HSG findings are confirmed by laparoscopy in 75.5% of the cases; however, accuracy in detecting peritubal disease is poor. Moreover, endometriosis, regardless of its severity, rarely causes abnormalities detectable by HSG. Therefore, endometriosis can only be diagnosed by laparoscopy. Laparoscopy is considered the gold standard in evaluating tubal patency, as demonstrated in a large prospective study. The superiority of laparoscopy over HSG in assessing adhesions and endometriosis has been shown in several studies with reported abnormal findings ranging between 21–68% of cases after a normal HSG. Given that laparoscopy is an invasive surgical procedure, it should be left as a final step in the infertility workup in women with normal HSG and no history of pelvic surgery or previous pelvic inflammatory disease. Moreover, laparoscopy is not required when ART is indicated, unless there are ultrasound and/or radiographic signs suggesting hydrosalpinx.

Microsurgery and laparoscopic surgery have been the treatments of choice for tubal diseases and adhesions, with success rates in the range of 20–40% according to the extent of tubal damage and the surgeon's experience. Nowadays, the usual treatment for tubal occlusions in developed countries is IVF. It is estimated that up to 50–60% of women with tubal factor as the only cause will give birth after four cycles of IVF. These results compare favorably with the best outcomes of reconstructive tubal surgery; but very few people in the world can afford four IVF cycles.

Endometriosis

A special condition affecting the anatomical integrity of the female pelvis is endometriosis. Endometriosis is defined as the presence of endometrium tissue (gland and stroma) outside the uterine cavity. The disease is mainly characterized by pelvic pain and infertility. The incidence of endometriosis is 30–50% among women consulting for infertility. Diagnosis can be suspected in women complaining of severe dysmenorrhea and the generation of pain during gynecologic examination. The presence of ovarian cysts diagnosed by ultrasound can also contribute to the diagnosis. However, it is only after direct observation through laparoscopy and histology that the final diagnosis can be confirmed. Available treatments include surgical and/or medical procedures. A recent review combined the results of two randomized controlled trials into a meta-analysis and showed that surgical treatment is better than medical management (odds ratio for pregnancy 1.7; 95% confidence interval 1.1–2.5).

Although the association of infertility with minimal and mild endometriosis remains unclear, treatment with controlled ovarian stimulation and intrauterine insemination enhances monthly pregnancy rates compared with expectant management. In cases of severe tubal damage and adhesions, the recommended treatment is ART.

Infertility Evaluation of the Male Partner

Semen analysis is the main method to assess the male partner. Although azoospermia is an undisputed and easily diagnosed cause of male infertility, the relative weight of other causes of male infertility is more difficult to establish. A WHO manual for the examination of semen has been used internationally. There are well-defined normal values of sperm density, motility, and morphology to discriminate between normal and abnormal semen (see **Table 5**). Nevertheless, the more fertile the female partner, the fewer normal sperm are needed to conceive. Inversely, no sperm will swim across the cervix if there is no mucus, and no fertilization will take place if the tubes are not patent.

An important abnormality in the male partner could be identified in approximately 30% of infertile couples. In another 20–30%, abnormalities could be detected in both partners. Thus, in over half of all infertile couples, there is a male factor contributing to infertility.

Unexplained Infertility

Despite the increased variety and precision of diagnostic procedures and the knowledge conveyed after more than 20 years of ART, 25–30% of fully investigated couples remain without a definitive cause for their infertility. A diagnosis of unexplained infertility is usually made after it has been demonstrated that the female partner ovulates regularly, has patent Fallopian tubes, shows no evidence of peritubal adhesions, fibroids, or endometriosis, and has a partner with normal sperm production.

Table 5 Normal semen analysis parameters

Volume	2–5 ml
Sperm concentration	20 million/ml or more
Amount	>40 million/ml
Motility	50% or more with forward progressive or 25% or more with rapid linear progression within 60 min after collection
Morphology	>50% or more with normal morphology
pH	7.2 to 7.8
Total sperm count	40 million or more
White blood cells	Fewer than 1×10^6 per ml
Zinc (total)	2.4 mol or more per ejaculate
Citric acid (total)	52 mol (10 mg) or more per ejaculate
Fructose (total)	13 mol or more per ejaculate
Viability	75% or more alive

Furthermore, it is not always possible to establish a direct relation between what can be considered an abnormal test and the cause of infertility. In fact, a study done in normal fertile couples reports that at least one infertility factor was present in two-thirds of cases.

In the absence of an evident cause of infertility, treatment is recommended to start 2 or more years after diagnosis. This is supported by the observation that approximately 60% of couples with unexplained infertility of less than 3 years' duration will become pregnant within 3 years of expectant management. This observation needs to consider the age of the female partner, since the spontaneous monthly fecundity rate declines with age and with increasing duration of unexplained infertility.

Overview of Assisted Reproductive Technology

A definition of ART would read as follows:

> all treatments or procedures that include the *in vitro* handling of human oocytes and sperm or embryos for the purpose of establishing a pregnancy. This includes, but is not limited to, *in vitro* fertilization and transcervical embryo transfer, gamete intrafallopian transfer, zygote intrafallopian transfer, tubal embryo transfer, gamete and embryo cryopreservation, oocyte and embryo donation and gestational surrogacy. ART does not include assisted insemination (artificial insemination) using sperm from either a woman's partner or a sperm donor.
>
> (Zegers-Hochschild *et al.*, 2006)

The last world survey on ART relates to procedures performed during 2000. Around 400 000 cycles of ART are reported every year. Today, it is estimated that 800 000 to 1 million cycles are initiated every year, with enormous regional variation. While Europe contributes approximately 300 000 cycles, Latin America contributes only 18 000–20 000 cycles per year. The overall delivery rate per oocyte aspiration is 13–20%, irrespective of the technology used to fertilize oocytes.

Although *in vitro* fertilization was initially indicated for women with damaged and/or irreparable Fallopian tubes, the current indications have been extended to cases of severe male factors, endometriosis, unexplained infertility, and women attempting pregnancy late in their reproductive years. In recent years, IVF has also been used to allow for preimplantation genetic diagnosis (PGD) in order to prevent the transfer of an affected embryo, especially in couples with high risk of transmitting genetic diseases to their offspring and in women at high risk of chromosomal abnormalities due to age or recurrent abortion.

Both *in vitro* fertilization and intracytoplasmic sperm injection (ICSI) use basically the same steps: controlled ovarian hyperstimulation, followed by oocyte collection,

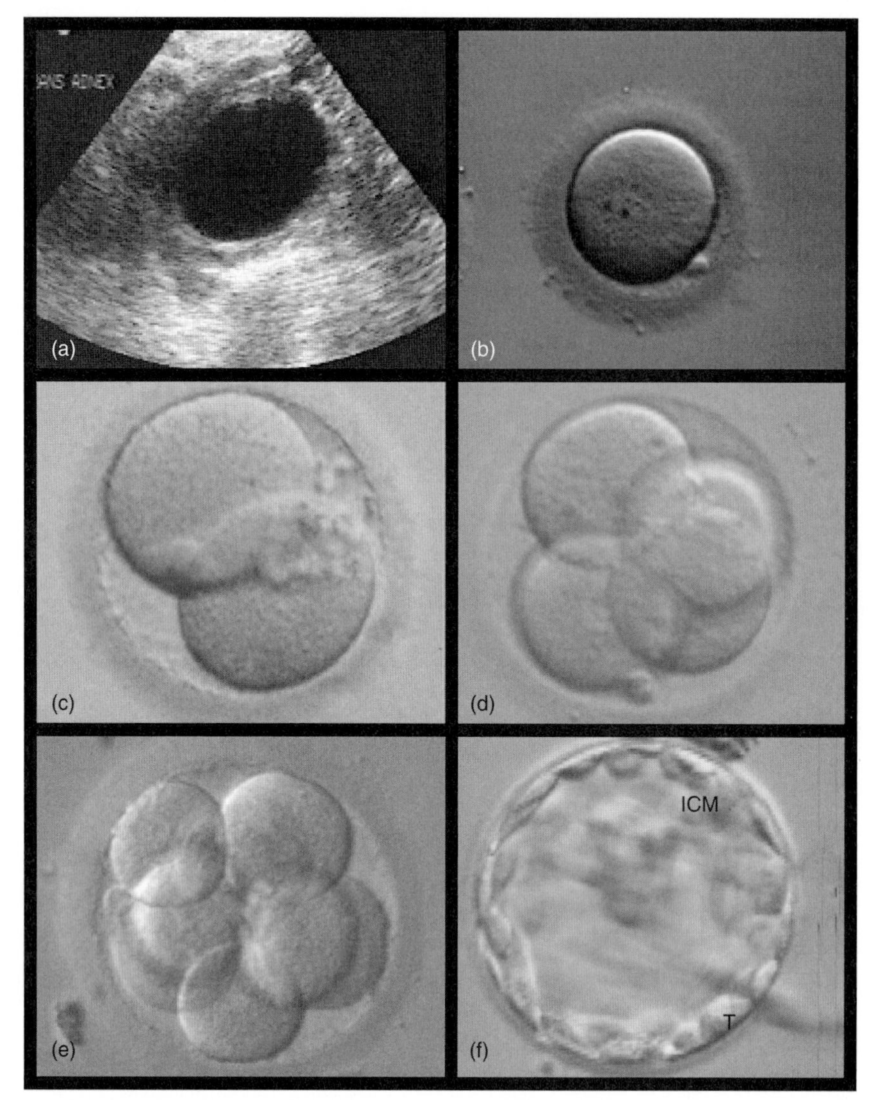

Figure 1 *In vitro* preimplantation development of the human embryo (a) Ultrasound image of a preovulatory follicle, prior to oocyte pick-up. (b) Metaphase-2 oocyte, denudated of surrounding cumulus oophorus. (c) Early-cleavage embryo; two blastomeres are seen 26 h after *in vitro* fertilization. (d) Early-cleavage embryo; four blastomeres are seen 48 h after *in vitro* fertilization. (e) Eight-cell embryo, 72 h after *in vitro* fertilization. (f) Blastocyst seen 96 h after *in vitro* fertilization. Two distinct cell types are seen: the trophoectoderm (T), which originates the placenta, and the inner cell mass (ICM), which originates the embryo.

in vitro fertilization and *in vitro* culture (see **Figure 1**), embryo transfer, and luteal phase support. The difference between IVF and ICSI is basically the fertilization process that takes place in the laboratory. In IVF, the oocytes are co-incubated in a culture medium with approximately 50 000 sperm cells to allow for spontaneous fertilization to take place. In ICSI, a single sperm is injected into an oocyte. Since only one sperm is necessary to perform ICSI, its indications have been gradually expanded to include nearly all men with serious infertility, including many who would previously have been considered not suitable for IVF.

A comprehensive review of ART can be obtained in a report of a WHO meeting, *Current Practices and Controversies in Assisted Reproduction* (Vayena *et al.*, 2002).

Inequality in Access to Fertility Treatments/ART

Throughout the world, the availability of infertility services is the result of public health policies associated with a variety of socioeconomic and political factors. Wide disparities exist in the availability, quality, and delivery of infertility services within developed countries, but most of all between developed and developing countries. Apart from infertile couples living in Israel and in Scandinavian countries, relatively few of the world's infertile population have complete, equitable access to the full range of infertility treatment at affordable levels. Even in wealthy countries such as Japan and the USA, access to IVF is marked by high disparity, and inequality in the access to treatment due to high cost. This is especially true in the access to modern reproductive technologies such as IVF.

Although a very large number of infertile couples have had children with the help of ART, with an estimated 3 million IVF-babies born to date (Adamson *et al.*, 2006; International Committee for Monitoring Assisted Reproductive Technology *et al.*, 2006), these treatments are not currently available to the majority of infertile couples worldwide. Furthermore, the mere knowledge that these treatments are available but beyond financial reach adds stress to infertile couples. The fact that the major cause for secondary infertility in poor communities is tubal damage due to pelvic inflammatory diseases, for which ART is the only treatment available, is also a source of anger and desolation.

Most countries do not consider infertility a disease or a relevant public health issue. While excessive population growth has concentrated public health initiatives for the past 30 to 40 years, in most countries, infertility is a condition that affects individuals; hence, no special provision is currently available except in those countries where decreased population growth is a national interest. An example is Israel, where extensive infertility services are provided; however, observers of the Israeli political process suggest that the state's interest in enlarging its Jewish population is a crucial underlying factor in the government's willingness to sustain its costly infertility treatment policy (Birenbaum-Carmeli, 2004). With a few exceptions – most notably the USA, where a minor proportion of ART cycles are covered by either public or private health insurance – developed countries are increasingly recognizing infertility as a medical condition, which means including infertility treatments as part of national health policies. On the other hand, the same governments may also deny access to specific groups of people to certain infertility services. For example, in Germany, the utilization of donor gametes is restricted; access to infertility services may be denied to lesbians, as well as to single and postmenopausal women. These restrictions as well as other national regulations, such as the new Italian law prohibiting embryo cryopreservation, have led to what has been referred to as "reproductive tourism" (Spar, 2005), in which couples or individuals with enough economic resources travel to other countries to receive treatment when they can not get the services they require at home.

Because most societies in the developing world are considered to be in need of population control, infertility is rarely acknowledged as a serious public health problem. Consequently, the treatment of infertility is rarely incorporated into programs of family planning in developing countries (Hamberger and Janson, 1997).

In developing countries, barriers to infertility treatment can be divided into three main categories: accessibility, economic cost, and cultural/social factors. It is not unusual for all three to be present simultaneously, creating an almost insurmountable obstacle to adequate reproductive health care.

In many regions of the world, conditions affecting accessibility to any form of medical treatment include lack of knowledge as to where to go, poor infrastructure, lack of electric power, and shortage of public transportation to the nearest clinic or hospital, among others. In other regions, such as in Costa Rica, the lack of access to certain treatments (i.e., ART) is the result of a legal prohibition imposed by the Supreme Court of that country.

Economic barriers include the high costs associated with more advanced techniques such as hormone assays, sperm analysis, hysterosalpingography, or laparoscopic procedures. Furthermore, even if diagnosis is made, modern reproductive technology is generally not available. For the majority of women, the cost of such treatments is out of reach, and this inability to receive appropriate infertility treatment has far-reaching consequences.

Cultural and social differences in belief can cause Western-style infertility treatments to appear disturbing and even threatening. Furthermore, religion may play a role in access to fertility treatment in both the developing world and the industrialized West. For example, in Latin America, the Catholic Church applies considerable moral pressure on legislators and the public to prevent or restrict access to ART.

As described previously, the rates of tubal and male factor infertility are often significantly higher in the developing world than they are in the developed world. These problems are virtually impossible to overcome without the use of ART. Therefore, the lack of availability and affordability of ART means that the majority of infertile couples have essentially no access to effective treatment at all (Nachtigall, 2006). **Table 6** shows population growth rates and the relative contributions of public and private, "out-of-pocket" payments to overall health expenditures for selected developed and developing countries. In general, people in less developed countries, even those from upper-middle income countries, have to contribute more from their own financial resources to health care than people from high-income countries (the USA is somewhat of an exception in this respect). Since infertility treatment is usually not provided for from public funds even in these upper-middle-income countries, few couples in these countries are able to afford the additional out-of-pocket expense for such treatment.

Several studies have shown that the number of couples seeking treatment, and therefore the number of procedures, greatly increases once the service is offered by health providers, demonstrating that the need for infertility care is underestimated. The financial burden is the only reason for many couples to delay or refrain from seeking medical help (Hughes and Giacomini, 2001).

In summary, one of the key questions is whether infertility should be recognized as a medical problem. If it is recognized as a medical condition, then a publicly funded

Table 6 Proportion of public versus private 'out-of-pocket' payments to overall health expenditures in selected high-income and upper-middle-income countries

Country	*Population growth (%)*	*Total health expenditure (% GDP)*	*Health expenditure*	
			Public[a]	*Private*[a]
United Kingdom	0.3	7.6	82.2	17.8
Denmark	0.3	8.4	82.4	17.0
Sweden	0.1	8.7	85.2	14.8
Belgium	0.2	8.9	71.7	28.3
France	0.5	9.6	76.0	24.0
USA	0.9	13.9	44.4	55.6
Argentina	1.1	9.5	53.4	46.6
Brazil	1.2	7.6	41.6	58.4
Chile	1.1	7.0	44.0	56.0
Mexico	1.2	6.1	44.3	55.7
Japan	0.2	8.0	77.9	22.1
Israel	1.6	8.7	69.2	30.8
Egypt	2.0	3.9	48.9	51.1

[a]As % of total health expenditure.
GDP = gross domestic product.

system should enforce solidarity between the fertile and infertile populations, as it does between the sick and the healthy for a number of medical conditions, and provide access to affordable diagnostic and therapeutic procedures for the majority of infertile couples.

See also: Female Reproductive Function; Gender Aspects of Sexual and Reproductive Health; Gynecological Morbidity; Male Reproductive Function; Reproductive Ethics: New Reproductive Technologies; Sexual and Reproductive Health: Overview; Sexually Transmitted Infections: Overview.

Citations

Adamson GD, de Mouzon J, Lancaster P, Nygren KG, Sullivan E, and Zegers-Hochschild F (2006) World collaborative report on in vitro fertilization, 2000. *Fertility and Sterility* 85: 1586–1622.
Bambra CS (1999) Current status of reproductive behaviour in Africa. *Human Reproduction Update* 5: 1–20.
Birenbaum-Carmeli D (2004) 'Cheaper than a newcomer': On the social production of IVF policy in Israel. *Sociology of Health and Illness* 26: 897–924.
Covington SN (1995) The role of the mental health professional in reproductive medicine. *Fertility and Sterility* 64: 895–897.
Faddy MJ, Gosden RG, Gougeon A, *et al.* (1992) Accelerated disappearance of ovarian follicles in mid-life: Implications for forecasting menopause. *Human Reproduction* 7: 1342–1346.
Fonttis AA, Napolitano R, Borda C, *et al.* (2004) Successful pregnancy and delivery after delaying the initiation of progesterone supplementation in a postmenopausal donor oocyte recipient. *Reproductive Biomedicine Online* 9: 611–613.
Hamberger L and Janson PO (1997) Global importance of infertility and its treatment: Role of fertility technologies. *International Journal of Gynecology and Obstetrics* 58: 149–158.
Hughes EG and Giacomini M (2001) Funding *in vitro* fertilization treatment for persistent subfertility: The pain and the politics. *Fertility and Sterility* 76: 431–442.
International Committee for Monitoring Assisted Reproductive Technology, *et al.* (2006) World collaborative report on *in vitro* fertilization, 2000. *Fertility and Sterility* 85: 1586–1622.
Larsen U (2000) Primary and secondary infertility in sub-Saharan Africa. *International Journal of Epidemiology* 29: 285–291.
Nachtigall RD (2006) International disparities in access to infertility services. *Fertility and Sterility* 85: 871–875.
Obermeyer CM (2005) The consequences of female circumcision for health and sexuality: An update on the evidence. *Culture, Health, and Sexuality* 7: 443–461.
Rutstein SO and Shah IH (2004) *Infecundity, Infertility, and Childlessness in Developing Countries*. ORC Macro and the World Health Organization: Calverton, MD: DHS Comparative Reports No. 9.
Schmidt L and Münster K (1995) Infertility, involuntary infecundity, and the seeking of medical advice in industrialized countries 1970–1992: A review of concepts, measurements and results. *Human Reproduction* 10: 1407–1418.
Smith S, Pfelfer SM, Collins JA, *et al.* (2003) Diagnosis and management of female infertility. *Journal of the American Medical Association* 290: 1767–1770.
Spar D (2005) Reproductive tourism and the regulatory map. *New England Journal of Medicine* 352: 531–533.
United Nations (1995) *Report of the International Conference on Population and Development*. Cairo, Egypt, 5–13 September 1994.
Vayena E, Rowe PJ, and Griffin PD (2002) *Report of a Meeting on "Medical, Ethical, and Social Aspects of Assisted Reproduction,"* Geneva, Switzerland, 17–21 September 2001. Geneva, Switzerland: World Health Organization.
World Health Organization (1994) *Health, Population, and Development.* (WHO Position Paper, International Conference on Population and Development). Geneva, Switzerland: World Health Organization.
World Health Organization (2006) *Reproductive Health Indicators: Guidelines for their Generation, Interpretation and Analysis for Global Monitoring*. Geneva, Switzerland: World Health Organization.
Zegers-Hochschild F, Nygren KG, Adamson GD, *et al.* (2006) The International Committee Monitoring Assisted Reproductive Technologies (ICMART) Glossary on ART Terminology. *Fertility and Sterility* 86: 16–19.

Further Reading

Berg BJ, Wilson JF, and Weingartner PJ (1991) Psychological sequelae of infertility treatment: The role of gender and sex-role identification. *Social Science and Medicine* 33: 1071–1080.
Collins JA and Rowe TC (1989) Age of the female partner is a prognostic factor in prolonged unexplained infertility: A multicenter study. *Fertility and Sterility* 52: 15–20.
Crosignani PG, Collins J, Cooke JD, *et al.* (1993) Recommendations of the ESHRE workshop on "Unexplained Infertility." *Human Reproduction* 8: 977–980.
Cundiff G, Carr BR, and Marshburn PB (1995) Infertile couples with a normal hysterosalpingogram. Reproductive outcome and its relationship to clinical and laparoscopic findings. *Journal of Reproductive Medicine* 40: 19–24.
Evers JL (2002) Female subfertility. *Lancet* 360: 151–159.
Imani B, Eijkemans MJ, te Velde ER, *et al.* (2002) A nomogram to predict the probability of live birth after clomiphene citrate induction of ovulation in normogonadotropic oligoamenorrheic infertility. *Fertility and Sterility* 77: 91–97.
Mulders AG, Eijkemans MJ, Imani B, *et al.* (2003) Prediction of chances for success or complications in gonadotrophin ovulation induction in normogonadotrophic anovulatory infertility. *Reproductive Biomedicine* 7: 170–178.
Olive DL and Pritts EA (2002) The treatment of endometriosis: A review of the evidence. *Annals of the New York Academy of Sciences* 955: 360–372; discussion 389–393, 396–406.
Practice Committee of the American Society for Reproductive Medicine (2004) Optimal evaluation of the infertile female. *Fertility and Sterility* 82(Suppl 1): S169–S172.

Rice JP, London SN, and Olive DL (1986) Reevaluation of hysterosalpingography in infertility investigation. *Obtetrics and Gynecology* 67: 718–721.
Tanahatoe SJ, Hompes PG, and Lambalk CB (2003) Investigation of the infertile couple: Should diagnostic laparoscopy be performed in the infertility work up programme in patients undergoing intrauterine insemination? *Human Reproduction* 18: 8–11.
Tummon IS, Asher LJ, Martin JS, *et al.* (1997) Randomized controlled trial of superovulation and insemination for infertility associated with minimal or mild endometriosis. *Fertility and Sterility* 68: 8–12.

Abortion

A Faúndes, State University of Campinas, Campinas, Brazil
F Alvarez, PROFAMILIA, Santo Domingo, Dominican Republic

Introduction

Abortion remains a major global health issue, with approximately 42 million of the 210 million pregnancies that occur each year ending in induced abortion; some 20 millions of these abortions are estimated to be unsafe (Sedgh *et al.*, 2007). Induced abortions occur because contraceptive methods are not widely accessible to those who need them and also because there is some risk of failure for all of these methods. Induced abortions occur in unsafe conditions primarily because in many countries the laws regarding voluntary pregnancy termination remain very restrictive, leading many women to seek clandestine, unsafe procedures. A woman's decision to have an abortion in countries with restrictive legislation may risk her to criminal prosecution and expose her to the dangers involved in unsafe procedures. Preventing unwanted pregnancies and unsafe abortion should be a priority among women's public health issues.

Defining Abortion

Abortion is defined as the termination of pregnancy before the fetus is able to survive independently. From the biological point of view, it is generally accepted that pregnancy begins with implantation. This is based on the understanding that pregnancy cannot exist independently from the woman who carries the embryo. In assisted reproduction, it is relatively easy to fertilize a woman's oocytes and to transfer one or more of these pre-embryos to the woman's uterus, but the women is only pregnant after confirmation of implantation.

Pregnancy termination before 37 completed weeks of gestation is defined as premature birth. The termination is considered an abortion (spontaneous or induced) if it occurs before the fetus has reached viability outside the uterus, with viability defined by the World Health Organization (WHO) as 22 completed weeks of gestation, or fetal weight of 500 g or more.

Induced abortion is defined by the Ethics Committee of the Fédération Internationale de Gynécologie et d'Obstétrique (International Federation of Gynecology and Obstetrics) (FIGO) as the termination of pregnancy using drugs or surgical intervention after implantation and before the embryo or fetus has become independently viable. The method or procedure used and the person who induces the abortion may vary enormously and determine whether the procedure is safe or unsafe for the woman's health.

WHO defines unsafe abortion as a procedure for terminating an unwanted pregnancy either by persons lacking the necessary skills or in an environment lacking the minimal medical standards, or both.

The Magnitude of Induced Abortion

A recent publication by Sedgh *et al.* (2007) estimated that a total of 42 million abortions were performed in 2003, down from 46 million estimated for 1995 using the same methods. The same study showed that 20% of all pregnancies, including miscarriages and stillbirths, terminate in abortion each year. This means that one out of every five pregnancies worldwide is voluntarily terminated annually, a statistic that illustrates its enormous dimensions. Expressed another way, the worldwide rate of induced abortion was approximately 29 per 1000 women between ages 15 and 44 years in 2003, down from 35 per 1000 in 1995. This means that globally one out of every 34 women within that age range has an abortion each year.

Regional Differences in Induced Abortion

The abortion rate in Eastern Europe was as high as 90 per 1000 women of childbearing age in 1995. This high number was influenced by a culture of small families and was compounded by the poor quality of contraceptive methods and inadequate access to them, while abortions have traditionally been free and easily accessible. Increased availability of modern, more effective contraceptive methods during the following decade may explain a rapid drop in the abortion rate to 44 per 1000 women of childbearing age in 2003.

The situation is the opposite in Western Europe, which has the lowest rate in the world, with 12 induced abortions per 1000 women between the ages of 15 and 44 years (Sedgh *et al.*, 2007).

Rates two to three times higher than those in Western Europe prevail in Latin America with 31 per 1000, and in Africa and Asia, both with an abortion rate of 29 per 1000. Abortion rates almost twice as high as those of Western Europe are observed in North America (21 per 1000) according to the most recent estimates (Sedgh *et al.*, 2007).

The country with the highest estimated abortion rate in 1995 was Viet Nam, which was close to 100 per 1000 women of childbearing age. The uncertainty of data makes it impossible to evaluate eventual changes during the period 1995–2003. In contrast, the estimated abortion rate for South-Central Asia, which includes some heavily populated countries such as India, Pakistan, and Bangladesh is the lowest of the Asian continent (27/1000).

Abortion rates in Latin America and the Caribbean appear to be relatively homogeneous, varying between 25 and 35 per 1000 women aged 15–44 years. Although Cuba had higher rates until the 1990s, there is evidence that its rate has fallen in recent years approaching the average for the region.

Great intracontinental variations are also apparent in Africa. The highest rate (39/1000) is found in the Eastern region, from Ethiopia and Somalia in the north to Mozambique and Zimbabwe in the south (Sedgh *et al.*, 2007).

In contrast, the rates in Northern Africa, comprising Algeria, Egypt, Libya, Morocco, Sudan, and Tunisia, are almost half the rate in the Eastern region (22/1000). Tunisia, one of the few countries in Africa where abortion is legal on broad grounds, has the lowest abortion rate in the region, at only seven per 1000 women of childbearing age, similar to that of the countries with the lowest rates in Western Europe.

Regional Differences in Occurrence of Unsafe Abortion

Most illegal abortions are performed under unsafe conditions, and most legal abortions are performed under safe conditions (although there are a number of exceptions to this general rule). Out of the 42 million induced abortions estimated for 2003, 22 million corresponded to generally safe legal abortions and 20 million to mostly unsafe illegal abortions (Sedgh *et al.*, 2007).

Safe and unsafe abortions are not homogeneously distributed throughout the different regions of the world, however. Developed countries make up more than 20% of the world's population, but only 2.5% of unsafe abortions occur in these countries. This incongruity occurs because abortions in developed countries are mostly legal and safe, whereas abortions in developing countries, with the exception of the People's Republic of China and a few other countries, are mostly illegal and unsafe. Like the People's Republic of China, India is also a densely populated country where abortion is legal on broad grounds, but in India a large proportion of induced abortions are still performed in unsafe conditions outside the official health-care system.

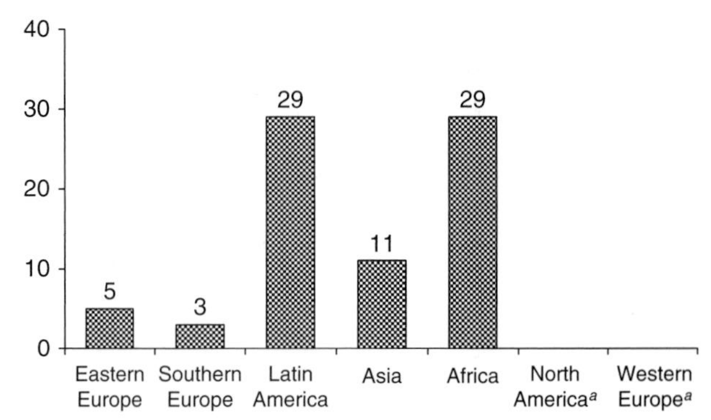

Figure 1 Unsafe abortion rate per 1000 women aged 15–44 years by region. Adapted from Sedgh G, Henshaw S, Singh S, Ahman E, and Shah IH (2007) Induced abortion: Estimated rates and trends worldwide. *Lancet* 370: 1338–1345. [a]: negligible

The most recent available estimates of unsafe abortion rates according to geographical region are shown in **Figure 1**. When subregions are considered, the highest rate is found in South America (29/1000), followed by Central America (18/1000), Southeastern Asia (16/1000) Eastern Africa and Northern Africa, both with a rate of 14/1000 women in fertile age. The rate of unsafe abortion in the more developed regions is close to nil.

Methods of Pregnancy Termination

Pregnancy can be terminated by a variety of procedures that may be safe or unsafe, mechanical or pharmacological, which can be carried out at any time during pregnancy. We will start describing the methods that are medically sanctioned as safe for pregnancy termination during the first and second trimesters of pregnancy, the period of pregnancy that fulfills the definition of abortion.

Surgical Abortion

All methods that involve the mechanical interruption of a pregnancy by a doctor or other trained health professional are classified as surgical abortions. The most traditional method of surgical abortion is dilatation and curettage (D&C). This procedure consists of dilating the cervix by either mechanical or pharmacological means, which allows the introduction of a sharp curette to scrape the walls of the uterine cavity. In recent times, D&C has been progressively replaced by aspiration of the uterine contents, which can be performed using either an electrical pump (EVA - electrical vacuum aspiration) or by simple manual vacuum aspiration (MVA). Several studies have shown that MVA is safer and less painful for the woman than D&C and far less expensive for the health system given its simplicity and lower complication rate.

Both D&C and vacuum aspiration can be used up to 12 weeks of pregnancy (up to 15 weeks for vacuum aspiration by specially trained providers). In very early pregnancy, up to 6 weeks of amenorrhea, vacuum aspiration can usually be carried out under local anesthesia without cervical dilatation and can be performed as an office procedure. Such a procedure is sometimes described as menstrual regulation to emphasize the possibility that the intervention may simply resolve a delay in menstruation rather than terminating a very early pregnancy, as is the situation in about 20% of cases since menstrual regulation procedures are typically done without carrying out a prior pregnancy test. Because of the absence of prior testing for pregnancy, menstrual regulation may be permissible in some countries (e.g., Bangladesh) that have restrictive abortion laws.

Beyond 14 weeks of gestation, the fetal parts are too large to be removed using D&C or vacuum aspiration, and a different procedure, referred to as dilatation and evacuation (D&E) is used. This procedure requires greater skill than a D&C and involves a higher risk to the woman. Greater cervical dilatation needs to be achieved to allow the use of a larger curette as well as forceps for the removal of fetal parts. Preparation of the cervix to allow its greater dilatation without causing cervical injury may involve use of either pharmacologic agents, such as prostaglandin analogs (discussed in more detail in the section titled 'Pharmacological Abortion'), or mechanical hydroscopic dilators, such as laminaria. Laminaria dilators, which are dried cones of a species of seaweed, are placed in the cervix where they absorb water from the surrounding environment and increase in diameter, thereby gently dilating the cervix. Greater dilatation can be achieved by the simultaneous use of several laminaria cones or by their sequential placement over time. Many providers find the use of ultrasound helpful during the actual D&E procedure, but its use is not essential.

In exceptional cases, a microcesarean section, using a vertical incision of the uterus, can be used. This procedure should be avoided because the uterine scar compromises the reproductive future of the woman in question.

Pharmacological Abortion

Pharmacological, or medical (nonsurgical), abortions may use a variety of regimens to terminate a pregnancy. Most commonly, regimens use a combination of the antiprogesterone mifepristone (formerly known as RU 486) and a prostaglandin analog. Mifepristone is a molecule very similar to progesterone that acts as an antagonist at the level of the progesterone receptor, thereby blocking the action of this ovarian hormone. When given alone, mifepristone is approximately 80% effective in inducing abortions up to 9 weeks amenorrhea. When a prostaglandin analog is given vaginally 24–48 h after oral administration of 200 mg of mifepristone, the effectiveness of this combination reaches up to 97%.

Although several prostaglandins have been used alone or in combination with mifepristone to induce abortion, the prostaglandin E1 analog, misoprostol, is most commonly used because it is stable at room temperature and equally or more effective and safe than other prostaglandins. Misoprostol alone, administered by the vaginal route at doses of 800 μg, is 85–90% effective in inducing abortion. It has the advantage of being effective at any gestational age and requires lower doses as pregnancy advances.

Methotrexate has also been used, in combination with misoprostol 800 μg vaginally between 3 and 7 days later, and repeated daily up to three times, with a success rate just above 90% to induce abortion within the first 9 weeks of pregnancy. No difference in effectiveness after oral or intramuscular administration of methotrexate has been found.

Misoprostol is also used to prepare the cervix before cervical dilatation for D&C or MVA. The most commonly used schedule is the administration of 400 μg orally or vaginally 3 h before the surgical procedure.

The later the abortion is performed, the greater the risk of complications and death (Cates *et al.*, 2003). Some 20 or 25% of all abortions are performed at 14 weeks or more of amenorrhea, and they represent a far greater problem than first-trimester abortions. In such cases, misoprostol offers the great advantage of being highly effective and safer than surgical methods.

Other methods, such as intra-amniotic injection of hypertonic saline or hyperosmolar urea; intra- or extra-amniotic administration of ethacridine; parenteral, intra-, or extra-amniotic administration of prostaglandin analogs; and intravenous or intramuscular administration of oxytocin, have been used in the past, but abandoned because of their higher rate and greater severity of complications.

The availability of drugs with the capacity to induce an abortion without the requirement of hospital facilities and operating rooms is already changing dramatically the resources required for safe abortions and reducing the risk run by women who self-induced their abortion in countries with restricted laws. Legal medical abortion can be carried out in outpatient clinics with appropriate emergency backup, and women can wait for the abortion in their homes, at least within the first 9 weeks of pregnancy. Several studies suggest that in countries with restrictive laws the availability of misoprostol appears to have reduced the rate of severe complications as compared with the use of other unsafe procedures described in the following section titled 'Methods of Unsafe Abortion.'

Methods of Unsafe Abortion

Unsafe abortions may be induced by the women themselves, by nonmedical persons, or by health workers using hazardous techniques in unhygienic conditions. A popular method of unsafe abortion has traditionally been the introduction of a solid, pointed object, such as a stick, wire, knitting needle, or stem, through the uterine cervix into the uterus. The intention is to rupture the membrane surrounding the gestational sac in order to induce abortion. It is easy to understand that the insertion of solid, pointed objects in unhygienic conditions by unskilled persons will often cause severe infection and, not uncommonly, perforation of the uterus and other pelvic organs.

A number of different potions, herbal teas, poisonous and harmful substances, such as bleach and hair dye, have also been used, either orally or vaginally. Over-the-counter or prescription medicines, sold illegally and in overdose by pharmacy salespersons, are also often used orally, by injection, and vaginally. Poisonous substances frequently intoxicate, and caustic substances cause genital tract injuries.

Another traditional method used mostly in Asia is to massage the pregnant woman's abdomen forcefully, or to apply blows or severe pressure to the abdomen by kneading. Women also report intentionally falling down stairs or jumping from heights in an attempt to abort. It is not difficult to understand that applying violence to the abdomen may cause internal injuries.

All these methods of unsafe abortion commonly result in incomplete abortions and infectious complications that may require emergency care to save the woman's life.

Complications and Consequences of Unsafe Abortion

Unsafe abortions are commonly associated with a range of complications that can lead to mortality, morbidity, and permanent disabilities, such as chronic pelvic pain, pelvic inflammatory disease, infertility, and genital fistula.

In contrast, when abortion procedures are performed appropriately by trained providers, complications rarely occur. In fact, the risk of death after a legal, safe abortion in the USA is ten times lower than the risk of death after childbirth (0.7 vs 7.0/100 000).

Acute Complications

The most frequent immediate complications of unsafe abortion are hemorrhage, infection, traumatic or chemical lesion of the genitals and other organs, and toxic reactions to products ingested or placed in the genital tract.

Hemorrhage, which can lead to acute anemia, shock, and death, frequently necessitates an emergency blood transfusion. In countries with a high prevalence of HIV and inadequate facilities for testing donated blood, blood transfusion may result in HIV infection. Bacterial infections may be introduced into the uterus during the maneuvers to procure an abortion and can be disseminated to the Fallopian tubes, the ovaries, and the abdominal cavity, causing pelvic inflammatory disease and peritonitis. Infections can also disseminate through lymphatic channels or through the blood, leading to sepsis and septic shock, which frequently ends in the woman's death.

Long-Term Sequelae

Women may suffer sequelae of unsafe abortion through two different mechanisms: the treatment required to prevent death may include the removal of the ovaries, the Fallopian tubes, and the uterus, rendering the woman infertile, or the consequences of infection or trauma to the reproductive tract may also lead to infertility or risk of ectopic tubal pregnancy. Ectopic pregnancy may be fatal if immediate access to treatment is not available.

Chronic inflammation of the internal genital organs or surgical scars may result in chronic pelvic pain, which can interfere with a woman's daily activities, including sexual intercourse. These chronic sequelae may have severe social consequences.

Maternal Mortality

It is estimated that roughly 13% of the 527 000 or more maternal deaths occurring worldwide every year are the result of complications of unsafe abortions (World Health Organization, 2004). This means that close to 70 000 women die every year following unsafely induced abortion. In many countries, abortion is the third or fourth most frequent cause of maternal death but in some, unsafe abortion is the most common cause of maternal death. Up to 98% of all abortion-related deaths occur in developing countries, where abortion is mostly illegal and therefore unsafe. Unsafe abortion-related mortality rates per 100 000 live births, by region, are presented in **Figure 2**.

Psychological Consequences

Reviews of the literature on the psychological consequences of induced abortion carried out over the last

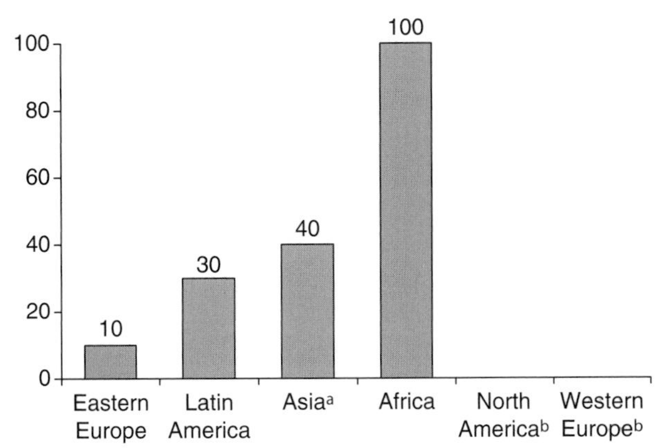

Figure 2 Mortality due to unsafe abortion per 100 000 live births according to geographical region. [a]Excluding Japan, [b]Negligible. Adapted from World Health Organization (2004) *Unsafe Abortion: Global and Regional Estimates of Incidence of Unsafe Abortion and Associated Mortality in 2000*, 4th edn. Geneva: World Health Organization.

30 years have found that adverse psychological sequelae occur in only a small percentage of women (Rogers *et al.*, 1989). More frequent and severe negative psychological effects have been described among women who are denied abortion and among the children who are born as a result (David, 2006).

Several studies have identified some factors associated with a greater risk of suffering psychological symptoms following induced abortion; these factors include having suffered the same symptoms prior to the abortion, having been under pressure to abort from partners or others, or, conversely, having experienced external cultural-religious pressure not to abort (Major *et al.*, 1998), or having terminated a wanted pregnancy because a genetic fetal defect was diagnosed. The unanimous conclusion of a panel of experts assembled by the American Psychological Association to examine legal abortion in the USA was that pregnancy termination has no negative psychological consequences for women who make the decision with no external pressure and that there was no evidence to support the purported existence of a so-called abortion trauma syndrome (American Psychological Association, 2005).

A more recent paper analyzed a cohort of women aged 15–25 years and controlled for a number of potentially confounding factors existing before abortion. The authors found a significantly greater risk of psychological disorder among women who had induced abortion than among those who had live births and those who were never pregnant (Fergusson *et al.*, 2006). They accept, however, that "an important threat to the study validity comes from the lack of information on contextual factors associated with the decision to seek an abortion." In fact, they controlled only for the first of several factors listed above and, consequently, did not invalidate the conclusions of the American Psychological Association. In summary, abortion may lead to psychological consequences if it is not the woman's free, informed decision, taken without external pressure either in favor or against pregnancy termination.

Economic Consequences

A safe abortion, carried out in a clinic or hospital environment, costs less than the average delivery, while the cost of caring for the complications resulting from an unsafe abortion is far greater and has put a heavy burden on the health system of many developing countries. The care of women with complications resulting from unsafe abortions can consume a large proportion of hospital beds, operating theatre time, medical equipment, antibiotics, intravenous fluids, blood and blood products, disposable supplies, and skilled human resources. For instance, the cost of treating a patient for complications resulting from an unsafe illegal abortion was nine times higher than the cost of a safe abortion and five times higher than the cost of delivery care in the main hospital of Maputo, Mozambique, in 1993. Use of antibiotics was 100 times greater, blood transfusions 16 times more frequent, and the duration of hospitalization was 15 times longer in women treated for complications resulting from unsafe abortion than for women who underwent in-hospital pregnancy termination (Hardy *et al.*, 1997).

Association Between Legal Status and the Consequences of Abortion

Legal Status of Abortion in the World

A recent review of abortion legislation in the world (Center for Reproductive Rights, 2007) shows that the law permits abortion to save the life or to preserve the physical health of the pregnant woman in 63% of the 190 countries or territories with a population that exceeds one million. In addition, in 47% of these countries the law explicitly permits abortion for the preservation of the woman's mental health. In 37% of the countries, abortion is also allowed for socioeconomic reasons, and in 29% (39.3% of the world's population) women do not need any justification to obtain a legal abortion, at least during the first 12 weeks of pregnancy. Abortion is permitted only to save a woman's life or prohibited altogether in 69 countries inhabited by 26% of the world's population (Center for Reproductive Rights, 2007).

Abortion laws tend to be more restrictive in developing countries. While in 87% of developed countries abortion is permitted upon request or for socioeconomic reasons, only 15% of developing countries have laws that are equally permissive. Laws tend to be more restrictive in Latin America and Africa than in other world regions (**Table 1**).

Table 1 Grounds on which abortion is permitted in 190 countries[a]

	55 more developed countries		*135 less developed countries*		*All 190 countries*	
	Number	*%*	*Number*	*%*	*Number*	*%*
To save the woman's life	54	98	132	99	186	98
To preserve her physical health	52	95	70	52	122	63
To preserve her mental health	50	91	39	29	89	47
Rape or incest	48	87	45	33	93	49
Fetal impairment	50	91	39	29	89	47
Socioeconomic reasons	48	87	20	15	68	36
Upon request	42	76	13	10	55	29

[a]Adapted from Center for Reproductive Rights (2007) *The World's Abortion Laws*. http://www.reproductiverights.org/pub_fac_abortion_laws.html (accessed February 2008).

In general, in addition to the circumstances in which abortion can be legally done, abortion laws establish other conditions that must be fulfilled, such as upper gestational age limits, the qualifications of the health professional(s) who can carry out abortions, and the facilities in which abortions can be performed, as well as requirements for informed consent, counseling, and waiting periods.

Legal Status and Abortion Rate

It is likely that the intention of legislators who sanction restrictive abortion laws is to inhibit its practice. Considering that the highest incidence of abortion is observed in countries of Eastern Europe and Viet Nam (60–100 per 1000 women of childbearing age) where abortion laws are rather liberal, and that countries in Latin America, where abortion is severely restricted, have moderate abortion rates (30–50), it would appear that the legislators in the latter group of countries have achieved their purpose. However, abortion rates are five to ten times lower in Western European countries, such as the Netherlands and Germany (6–10), where abortion is broadly permitted and easily accessible (Henshaw *et al.*, 1999). All in all, there are strong indications that legal prohibition is not an effective tool for reducing the incidence of abortion.

Abortion Law and Maternal Mortality

Whereas virtually all legal abortions are safe, the vast majority of illegal abortions are unsafe. The negative consequences of the criminalization of abortion were dramatically illustrated by the increase in maternal mortality observed in Romania following the prohibition of both abortion and contraception in November 1965. The abortion-related maternal mortality rate, which was below 20 per 100 000 live births in 1965, increased approximately eightfold to almost 150 per 100 000 live births between 1966 and 1988 (Stephenson *et al.*, 1992).

Determinants of Induced Abortion

Abortion is almost always the result of an unintended and unwanted pregnancy. A pregnancy may be unwanted for a variety of reasons present in most cultures: absence of a father, economic constraints, interference with life prospects, inability to provide good parenting, conflict with prevailing social norms, and health concerns.

The main determinants of unwanted pregnancy and resulting recourse to abortion are women's lack of power over their sexual activity; their lack of education, including inadequate and inaccurate knowledge of contraceptive methods; limited access to effective contraception; and the absence of social support for pregnant women and their children. As long as these conditions exist, a large number of unwanted pregnancies will occur, and, despite legal, moral, or religious prohibitions and sanctions, many will result in abortion. This is the situation in, for instance, Viet Nam and Eastern Europe, and to a lesser degree in the majority of Latin American and African countries. The low abortion rates in Western Europe are likely related to the greater access of the population to education in general and to sex education in particular, universal knowledge and access to contraception, greater power balance between the sexes, and greater social protection of motherhood, all of which contribute to reducing unwanted pregnancies and abortions, irrespective of the legality of and accessibility to abortion.

Knowledge of Contraception

The proportion of women who declared knowledge of at least one modern contraceptive method (hormonal methods, intrauterine devices, diaphragm, male or female condom, or surgical sterilization) varied from less than 50% in Chad to close to 98% in Kenya, Nepal, the Philippines, and Zambia, and almost 100% in developed countries as well as some developing countries such as Bangladesh, Brazil, and the Dominican Republic.

There are also large differences within each country between urban and rural residence and socioeconomic status. While 98% or more of women who have at least a secondary education in most developing countries declare that they know of at least one modern contraceptive method, only about 50% of women with no education have that knowledge.

Moreover, when a woman says that she knows of a method it may only mean that she has heard of the existence of that method; it is no guarantee that the knowledge she has is correct. She may believe, for example, that the IUD causes an abortion every month or that women who take the pill for a long period become sterile.

A study carried out among women living in a shanty town in Rio de Janeiro, Brazil, found that 23% of contraceptive pill users were using them incorrectly. Similarly, other studies have shown that adolescents and women with little education who attempt to use periodic abstinence for family planning have inaccurate knowledge of the menstrual cycle and the fertile period (Castaneda *et al.*, 1996). Even in the USA, a study found that close to one-third of female students and two-fifths of male students who were sexually active were unaware of the importance of leaving a space at the tip of the condom. One-third of both groups believed wrongly that Vaseline could be used with condoms (Crosby and Yarber, 2001). Some adolescents in both developed and developing countries are not even aware that girls can become pregnant the first time they have sexual intercourse.

The accuracy of contraceptive knowledge is positively related to education attainment, according to Demographic and Health Surveys (DHS) in a variety of countries. On the other hand, having a religious affiliation has been found to be associated with several misconceptions about how to use condoms correctly, according to a national sample of adolescents in the USA (Crosby and Yarber, 2001).

Distorted information on contraceptives disseminated by the media may also have an effect on the abortion rate, as occurred in the case of the sensationalist information published regarding the reported increase in the risk of adverse vascular effects associated with some new progestogens contained in so-called third-generation combined oral contraceptive pills. At least two studies showed that these reports coincided with a dramatic decrease in the use of oral contraceptive pills in the United Kingdom and Norway, and was followed by a sharp rise in unwanted pregnancies and abortion (Skjeldestad, 1997).

Access to Contraception

If a sexually active woman does not desire a child and has no access to family planning methods she is running a high risk of having an unwanted pregnancy and consequently of having an induced abortion. A study in Nepal found, for example, that for many women unsafe abortion was the only available method of fertility control.

The highest abortion rates have been observed in countries in Eastern and Central Europe that belonged to the former Soviet Union. Contraceptive prevalence in these countries was low because only high-dose contraceptive pills, associated with more side effects, were available, the quality of condoms and IUDs was poor, and there were legal restrictions to surgical sterilization. For many women who wanted to control their fertility, abortion was a more easily accessible option than contraception.

Unfriendly or inappropriate delivery systems and conflicting cultural values can create insurmountable barriers to obtaining contraception, affecting adolescents disproportionally in both developed and developing countries. Several studies carried out in developing countries, such as India and Tanzania, or in developed countries, such as Belgium and the USA, have found that many adolescents are either unaware that they have a right to request contraceptive services or are inhibited by cost, waiting times, embarrassment, and fear of gynecological examination (Silberschmidt and Rasch, 2001).

Effectiveness of Contraceptive Methods

No contraceptive method is 100% effective, and pregnancies resulting from contraceptive failure are often aborted. All methods have an intrinsic failure rate, but for the methods that depend strongly on the users' compliance such as the pill or behavioral methods, most of the pregnancies are caused by incorrect use by the woman or couple (Indian Council of Medical Research Task Force on Natural Family Planning, 1996). The contraceptive pill's effectiveness is close to 100% in controlled clinical studies, but the pregnancy rate in actual use can be as much as 6–8% per year of use in population-based studies. Knowledge of correct use, as well as user error, contribute to its decreased effectiveness outside of clinical trials.

Failure rates due to user error are negligible for methods that do not depend on user compliance, such as intrauterine devices and implants. The very high effectiveness of these methods in actual use makes their availability critical in the effort to reduce the number of unwanted pregnancies and abortions.

Gender Power Imbalance

Knowledge about and access to contraceptive methods is not enough if women do not have control over their use every time they have sexual relations. Coerced sex is far more common than has traditionally been acknowledged, because most studies are limited to the incidence of rape, defined as imposed sexual intercourse using force or the threat of force. The far more common cultural imposition of male so-called rights over a woman's body, ranging from sexual coercion (in exchange for maintaining a job or satisfying other personal needs) to the woman's sense of obligation to have sex with a stable partner, is usually not counted in these studies.

Prevalence of sexual violence varies from less than 10% to close to 40% of women of childbearing age, according to different studies. The variation between studies depends both on the social environment and on the definition of sexual violence. According to the United Nations definition, gender-based violence includes threatening and coercion, but many studies consider only direct physical imposition as sexual violence. The woman's cultural acceptance of her obligation to satisfy the sexual desires of a partner or husband above her own wishes and above the risk of unwanted pregnancy is seldom considered. This more subtle cultural imposition of unwanted sex and women's lack of power to negotiate use of protection against pregnancy are far more relevant than actual rape as a cause of unwanted pregnancy and abortion. A good illustration of this cultural conditioning is a study carried out among adolescent girls and boys in the USA, which showed that a large proportion believed that boys who had invested time or money in entertaining their female partners had the right to have sex with them. In South Africa, an overwhelming majority of adolescents had the same opinion.

In addition, studies also show that both adolescent and adult males believe that protection against pregnancy is the responsibility of women; therefore, men may often be an obstacle to contraceptive use. One-third of unwanted pregnancies among women who requested legal abortion in India could be attributed to the husband's unwillingness to use contraception (Banerjee *et al.*, 2001).

The review of several studies on the subject strongly suggests that male dominance in the decision to have sex and the lack of male responsibility with respect to contraceptive use contribute to the incidence of unintended pregnancy. This may be a major hurdle to reducing the incidence of abortion among adolescents, who are more resistant than adults to interventions against unwanted pregnancy and abortion.

Interventions to Reduce Induced Abortions

Abortions are the result of unwanted pregnancies; therefore, the logical strategy for decreasing the number of abortions is to help women avoid unwanted pregnancy. However, even after an unwanted pregnancy has occurred, the number of abortions can be reduced by providing the pregnant woman with the social support she needs to have a baby without sacrificing her plans for the future.

Prevention of unwanted pregnancy has proven to be the most effective strategy for decreasing the number of abortions. The low abortion rate in Western Europe coincides with conditions that favor the prevention of unwanted pregnancies. The high incidence of abortion in Latin America and Eastern Europe is a result of limited access to effective means for preventing unwanted pregnancy.

Family Planning

The 1994 International Conference on Population and Development (ICPD) Programme of Action urged all governments to "strengthen their commitment to women's health" and "deal with the health impact of unsafe abortion as a major public health concern." The conference participants pledged their commitment to reducing the need for abortion "through expanded and improved family planning services."

The call to reduce abortion through family planning is based on the global experience acquired over the decades since modern hormonal contraceptives and IUDs became widely available. In 1966, the Chilean government established family planning as an official component of the women's health-care program, providing free contraceptive services to 85% of the population through the national health service. Several authors confirmed a direct correlation between improved access to contraceptives and the dramatic decrease in the number of women with abortion complications who were admitted to public hospitals; these data are illustrated in **Figure 3**. Similarly, in Brazil, the number of abortion complications dealt with by the national health service dropped from approximately 345 000 in 1992 to 228 000 in 1998 after the rapid increase in the use of modern contraceptives.

Experience from many countries shows that when there is a rapid decrease in desired family size, contraceptive use

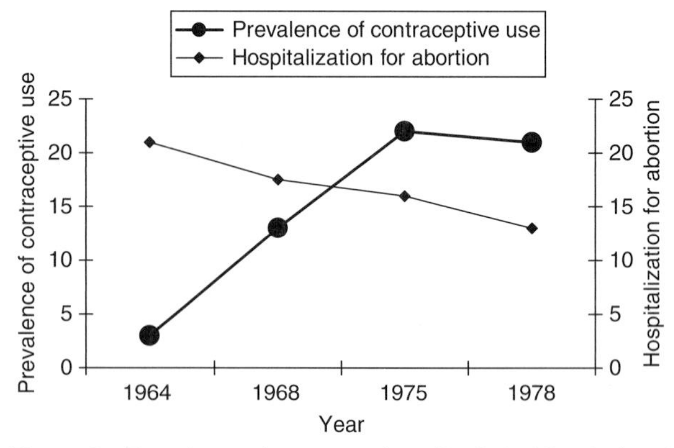

Figure 3 Prevalence of contraceptives (Statistical Service) and proportion of hospitalizations for complications of abortion related to all hospitalizations. Chile, 1964–78. Adapted from Maine D (1981) *Family Planning: Its Impact on the Health of Women and Children.* New York: Columbia University Center for Population and Family Health.

increases, but induced abortions may also increase if family planning access cannot satisfy the increased demand. Only when the unmet need for contraception has been met does the number of induced abortions invariably decline.

Sex Education

The majority of studies have shown that sex education programs of adequate duration, content, and methodology are effective in reducing pregnancy and abortion. Programs with greater effectiveness are those that promote responsibility, abstinence, and (for those who are sexually active) protection against pregnancy and sexually transmitted infections. Successful programs combine the promotion of mutual respect and gender power balance with knowledge of sexual and reproductive physiology and sexually transmitted infection, knowledge that is mostly lacking, particularly among young people in developing countries.

Contrary to the fears of some people, sex education does not increase sexual activity or promote early sexual experimentation; on the contrary, it may encourage adolescents to postpone the initiation of sexual activity, thus reducing the proportion of adolescents who have sexual relations.

The countries with the lowest abortion rates are those that have broad-based, progressive sex education programs in their schools and nearly universal school attendance. The Netherlands and several Scandinavian countries fulfilling these conditions have the lowest teen abortion and teen pregnancy rates. This has been achieved through implementation of comprehensive sex education; confidential, high-quality family planning services; and access to a wide selection of contraceptive methods. Age of first intercourse has remained unchanged in these countries, but contraceptive use has increased dramatically. For instance, a study in Finland showed that, in 1997, 87% of sexually active 16-year-old girls had used a condom during their last intercourse. These data suggest that sex education is an effective intervention for reducing gender power imbalance and promoting mutual respect concerning sexual decisions, a basic element in the efforts to reduce unwanted pregnancy and abortion.

In contrast, countries that have a high abortion rate either have no sex education programs or have local programs that are narrow in scope. The emergency created by the HIV epidemic has stimulated the acceptance of sex education, but the lack of both adequately trained personnel and strong political support reduces the effectiveness of the few programs that are being implemented.

Other Interventions

Young, unmarried women may become pregnant as a result of a subconscious or conscious desire to have a child. They may be pleased to be pregnant in the expectation that, with the support of a partner or family, their motherhood will not interfere with continuing their education or career. Since their partners may disappear when they learn of the pregnancy and family support may be lacking, these women end up choosing pregnancy termination as the only viable alternative. Studies show that women who find themselves in this situation might have continued their pregnancies if they had received more support.

Support is needed not only from the family but also from society as a whole. Society must ensure, for instance, that a pregnant adolescent does not have to terminate her studies in order to have a baby if she wishes to continue her education. Otherwise, she will have to decide between having a baby but discontinuing her studies or aborting as the only means of achieving her objectives for the future. An adult pregnant woman needs support to keep her job, needs to receive appropriate care during pregnancy, delivery, and the postpartum period, and facilities enabling her to continue her career while breastfeeding and taking care of her baby. Without that support, she may have to decide between aborting and losing a job that is incompatible with pregnancy, delivery, breastfeeding, and child care.

Interventions to Reduce Unsafe Abortions and Their Consequences

Even if all the interventions described above were implemented, some abortions would still occur as long as people are unable to fully anticipate what will happen in their lives and as long as contraceptive methods are not 100% effective. Therefore, other than diminishing the number of induced pregnancy terminations, unsafe abortion can be reduced by making safe all abortions that will still occur in the foreseeable future.

Decriminalizing or Extending Legal Grounds for Abortion

As described earlier, most legal abortions are safe while the vast majority of illegal abortions are unsafe. Consequently, the most effective approach to reducing unsafe abortions is to eliminate the laws that penalize the voluntary termination of pregnancy.

The reduction of unsafe abortion and its consequences following the liberalization of abortion laws has been documented in several countries. For instance, during the 3 years after the approval of the Abortion Act (1982 and 1984), which extended the legal grounds for abortion and greatly facilitated free access to safe abortion in England and Wales, no deaths related to induced abortion were registered, compared with 75–80 deaths per triennium before the act was passed (Stephenson *et al.*, 1992).

A 50% reduction in the abortion-related maternal death rate was observed in the state of New York, USA, during the 2 years after the 1970 liberalization of abortion laws. A dramatic decrease in abortion-related maternal mortality was also observed in the whole of the USA following the Supreme Court's 1973 *Roe v. Wade* decision (Cates *et al.*, 2003).

We have already described the negative health effect of the criminalization of abortion, dramatically illustrated by the increase in maternal mortality after the prohibition of both abortion and contraceptives in Romania in late 1965. Even more dramatic was the reduction in abortion-related maternal mortality after the decriminalization of abortion and contraception following the fall of the Romanian leader, Nicolae Ceausescu, in December 1989: the maternal mortality ratio fell from approximately 150 per 100 000 live births in the previous year to fewer than 50 per 100 000 live births 2 years after decriminalization (Stephenson *et al.*, 1992).

Improving Access to Legal Abortion

The experience of countries such as India, South Africa, and Zambia has shown that the decriminalization of abortion does not guarantee easy access to safe procedures. In India and Zambia, and (to a lesser degree) South Africa, the majority of women still do not have access to safe abortion services. In these countries, it is essential to accelerate the process of adapting and upgrading the public health system to meet the demand for voluntary pregnancy termination.

In a much larger number of countries, the restricted laws permit abortion under specific circumstances, but women who fulfill these requirements still may not have access to safe pregnancy termination and may have to resort to highly unsafe, back-street abortions. In these countries, access to safe abortion greatly depends on how liberally or restrictively the law is interpreted. A crucial role is played by obstetricians/gynecologists, since ultimately they are the ones who decide whether or not to perform the abortion, and by the legal profession, whose function it is to interpret the law.

Improving access to safe abortion to the full extent permitted by law is an important means for reducing the rates and consequences of unsafely induced abortion. Despite current efforts, a large number of women around the world who meet the requirements to have access to safe and legal abortion are still denied their right.

Access to Postabortion Care

The mortality rates and severity of short- or long-term complications of unsafe abortion may be strongly influenced by the quality of the care received after the abortion. Acute complications such as uterine perforation or the perforation or injury of other internal organs may be fatal if emergency care is not urgently provided. Infection will progress locally and systemically as long as the proper medical and surgical therapy is not given. The later and the less efficient the care received, the more severe the consequences are likely to be and the greater the risk of death. Therefore, it is recommended that the assessment of the woman's condition and the provision of postabortion care should be available on a 24-h basis and provided with the urgency demanded by the severity of the condition. Accordingly, the governments represented at the 1994 ICPD Conference agreed that "in all cases, women should have access to quality services for the management of complications arising from abortion."

Unfortunately, the prevalent discrimination against women who have induced abortions has frequently led to inadequate emergency care. Delay in receiving care can be attributable to lack of services within a reasonable distance of the woman's residence or lack of roads or means of transportation, but may also be the result of social, economical, religious, or legal restrictions on abortion. Several studies carried out in countries with restrictive abortion laws have found that, in some hospitals, particularly those run by the Catholic Church, women suspected of having had an induced abortion were turned away. In Chile, many women were jailed after being denounced by personnel at the hospital where they received postabortion care, effectively inhibiting women from seeking the medical assistance they needed. Thus, access to postabortion care may be limited by providers in at least two ways: by direct rejection or by threat of prosecution.

Postabortion care should not, however, be limited to the emergency situation, but should take into consideration the long-term needs of women, including counseling and services for the prevention of later unwanted pregnancy and abortion. Several decades ago, it was shown already that women who have had an abortion are at a much higher risk of aborting the next pregnancy than those who never had an abortion, and more recent studies have shown the effectiveness of postabortion contraception in reducing future unplanned pregnancies and abortions (Johnson *et al.*, 2002).

Conclusions

While there exists universal agreement on the need to reduce the number of abortions and their consequences, many governments in developing and developed countries have still not fully adopted the necessary policies to promote practices that could prevent abortion. The deaths of thousands of women that result from unsafe conditions in which abortion is practiced in the poorer regions of the world are preventable, as the causes of abortions and the interventions that could reduce their numbers and the severity of their consequences are well known. It is the obligation of international organizations and of all of those dedicated to the health and well-being of women to promote the implementation of the measures that would decrease the number of abortions and increase women's access to safe abortion services and to the means to avoid unwanted pregnancy (World Health Organization 2003). We expect the current trend toward a reduction in the number of abortions to continue, but it depends on the public health community and donors realizing that access to family planning information and services is still very limited, particularly in Africa and the more vulnerable communities in Latin America and parts of Asia, and deserves more attention than it has received in recent years. There has also been a trend toward liberalization of abortion laws in some large countries, while more restricted laws have been imposed in a few countries. A better understanding of the ineffectiveness of restrictive laws in reducing the number of abortions and the heavy consequences on health and well-being, unfairly borne almost exclusively by poor, young, and rural women, should progressively lead the world community to take the logical steps to reduce this human drama and public health burden: improving education and services in sexuality and contraception; greater social support to women who want to have a baby, but are abandoned by their partner and family; and more liberal abortion laws broadly implemented without social, economic, or age limitations. Great progress in all of these areas should take place in the next seven years to achieve the Millennium Development Goal of reducing, by the year 2015 from the 1990 baseline, by three-quarters the maternal mortality ratio in the world.

See also: Family Planning/Contraception; Maternal Mortality and Morbidity; Reproductive Ethics: Perspectives on Contraception and Abortion; Reproductive Rights; Sexual Violence; Violence Against Women.

Citations

American Psychological Association (2005) *APA Briefing Paper on the Impact of Abortion on Women*. http://www.apa.org/ppo/issues/womenabortfacts.html.

Banerjee N, Sinha A, Kriplani A, Roy KK, and Takkar D (2001) Factors determining the occurrence of unwanted pregnancies. *National Medicine Journal of India* 14: 211–214.

Castaneda X, Garcia C, and Langer A (1996) Ethnography of fertility and menstruation in rural Mexico. *Social Science and Medicine* 42: 133–140.

Cates W, Grimes DA, and Schulz KF (2003) The public health impact of legal abortion: Thirty years later. *Perspectives on Sexual and Reproductive Health* 35: 25–28.

Center for Reproductive Rights (2007) *The World's Abortion Laws*. http://www.reproductiverights.org/pub_fac_abortion_laws.html (accessed February 2008).

Crosby RA and Yarber WL (2001) Perceived versus actual knowledge about correct condom use among US adolescents: Results from a national study. *Journal of Adolescent Health* 28: 415–420.

David HP (2006) Born unwanted, 35 years later: The Prague study. *Reproductive Health Matters* 14: 181–190.

Fergusson DM, Horwood LJ, and Ridder EM (2006) Abortion in young women and subsequent mental health. *Journal of Child Psychology and Psychiatry* 47: 16–24.

Hardy E, Bugalho A, Faúndes A, Duarte GA, and Bique C (1997) Comparison of women having clandestine and hospital abortions: Maputo Mozambique. *Reproductive Health Matters* 9: 108–115.

Henshaw SK, Singh S, and Haas T (1999) The incidence of abortion worldwide. *International Family Planning Perspectives* 25 (supplement): 30–38.

Indian Council of Medical Research Task Force on Natural Family Planning (1996) Field trial of Billings ovulation method of natural family planning. *Contraception* 53: 69–74.

Johnson BR, Ndhlovu S, Farr SL, and Chipato T (2002) Reducing unplanned pregnancies and abortion in Zimbabwe through post-abortion contraception. *Studies in Family Planning* 33: 195–202.

Maine D (1981) *Family Planning: Its Impact on the Health of Women and Children*. New York: Columbia University Center for Population and Family Health.

Major B, Richards C, Cooper ML, Cozzarelli C, and Zubek J (1998) Personal resilience, cognitive appraisals, and coping: An integrative model of adjustment to abortion. *Journal of Personality and Social Psychology* 74: 735–752.

Rogers JL, Stoms GB, and Phifer JL (1989) Psychological impact of abortion: Methodological and outcomes summary of empirical research between 1966 and 1988. *Health Care for Women International* 10: 347–376.

Sedgh G, Henshaw S, Singh S, Ahman E, and Shah IH (2007) Induced abortion: Estimated rates and trends worldwide. *Lancet* 370: 1338–1345.

Silberschmidt M and Rasch V (2001) Adolescent girls, illegal abortions and "sugar-daddies" in Dar es Salaam: Vulnerable victims and active social agents. *Social Science and Medicine* 52: 1815–1826.

Skjeldestad FE (1997) Increased number of induced abortions in Norway after media coverage of adverse vascular events from the use of third-generation oral contraceptives. *Contraception* 55: 11–14.

Stephenson P, Wagner M, Badea M, and Serbanescu F (1992) Commentary: The public health consequences of restricted induced abortion. Lessons from Romania. *American Journal of Public Health* 82: 1328–1331.

World Health Organization (2003) *Safe Abortion: Technical and Policy Guidance for Health Systems*, 1st edn. Geneva, Switzerland: World Health Organization.

World Health Organization (2004) *Unsafe Abortion: Global and Regional Estimates of Incidence of Unsafe Abortion and Associated Mortality in 2000*, 4th edn. Geneva, Switzerland: World Health Organization.

Further Reading

AbouZahr C and Ahman E (1998) Unsafe abortion and ectopic pregnancy. In: Murray CJL and Lopez AD (eds.) *Health Dimensions of Sex and Reproduction: The Global Burden of Sexually Transmitted Diseases, HIV, Maternal Conditions, Perinatal Disorders, and Congenital Anomalies*, pp. 266–296. Cambridge, MA: Harvard University

Baulieu EE (1985) Contragestion by antiprogestin: A new approach to human fertility control. Ciba Foundation (ed.). *Proceedings of Ciba Foundation Symposium 115. Abortion: Medical Progress and Social Implications*, pp. 192–210. London: Pitman

Faúndes A and Barzelatto J (eds.) (2006) *The Human Drama of Abortion: A Global Search for Consensus,* 1st edn. Nashville, TN: Vanderbilt University Press.

Heise LL, Pitanguy J, and Germain A (1994) *Violence Against Women: The Hidden Health Burden*. World Bank Discussion Papers. Washington, DC: The World Bank.

Kirby D (2001) *Emerging Answers: Research Findings on Programs to Reduce Teen Pregnancy*. Washington, DC: National Campaign to Prevent Teen Pregnancy.

Maine D (1981) *Family Planning: Its Impact on the Health of Women and Children*. New York: Columbia University Center for Population and Family Health.

Nordic Family Planning Associations (1999) *The Nordic Resolution on Adolescent Sexual Health and Rights*. Presented at the International Conference on Population and Development (ICPD)+5 Forum, The Hague, February 1999.

Prada E, Kestler E, Sten C, Dauphinee L, and Ramírez L (2005) *Aborto y atención post-aborto en Guatemala: Informe de profesionales de la salud e instituciones de salud Informe ocasional*. New York: The Guttmacher Institute.

Tamang AK, Shrestha N, and Sharma K (1999) Determinants of induced abortion and subsequent reproductive behavior among women in three urban districts of Nepal. In: Mundigo AI and Indriso C (eds.) *Abortion in the Developing World*, pp. 167–190. New Delhi, India: Vistar Publications.

Trussell J (1998) Contraceptive efficacy. In: Hatcher RA, Trussell J, Stewart F, *et al.* (eds.) *Contraceptive Technology,* 17th edn., pp. 779–844. New York: Ardent Media.

United Nations (1995) *Report of the International Conference on Population and Development*. Cairo, September 5–13. New York: United Nations.

Warriner IK and Shah IH (eds.) (2006) *Preventing Unsafe Abortion and its Consequences. Priorities for Research and Action.* New York: The Guttmacher Institute.

Weeks A (ed.) (2007) Misoprostol for Reproductive Health: Dosage Recommendations. *International Journal of Gynecology & Obstetrics*, 99 (Supplement 2).

Maternal Mortality and Morbidity

L Say, World Health Organization, Geneva, Switzerland
R C Pattinson, Kalafong Hospital, Pretoria, Gauteng, South Africa

Introduction

Pregnancy and childbirth are among the most significant periods in women's lives, both physically and emotionally. Women normally adapt to the physiological changes of pregnancy and are capable of delivering healthy babies with full recovery following delivery. However, some pregnancies may pose risks to the health of women and/or their babies, leading to unwanted consequences in varying severity – ranging from pelvic lacerations due to obstetric trauma to death. At an individual level, it is not usually predictable which women will experience either a minor or a life-threatening complication, although certain characteristics such as age, number of previous pregnancies and their outcomes, health status, and family history are associated with some of the negative outcomes.

At the population level, socioeconomically disadvantaged women are more likely to experience negative outcomes of pregnancy and childbirth. Substantial differences in terms of the type and extent of the complications exist between women living in more developed countries and those living in developing countries and among subgroups characterized by social, economic, geographical, and ethnic features within the same country. The reasons underlying these differences include physical vulnerability in terms of poor nutrition and lifestyle conditions, type of the health care available, and the extent to which women can access this care or, in most cases, a mixture of all.

The patterns of complications of pregnancy and childbirth reflect the capacity of the health system in a population, including the capacity of the health system to address the needs of the most vulnerable segments of the population. For example, hemorrhage is the most common direct cause of maternal deaths in most developing country settings, but few women die due to this condition in developed countries because it is preventable and treatable. So, both the prevalence of hemorrhage and the probability of dying from it are much lower in countries with more developed health systems. In this context, the main causes of maternal deaths are rare conditions that the medical profession is not able to prevent (such as preeclampsia) or treat (such as complications of a congenital heart disease) adequately. Understanding the causes of maternal deaths, the patterns of morbidities, the characteristics of the groups affected most, and health system failures is essential to determine where to concentrate efforts to provide improvements in a population. Such information is also an indicator of the broader issues of social inclusion, women's status and rights, and socioeconomic development in the society.

Public health challenges remain at conceptual, measurement, and implementation (of effective interventions) levels, particularly in countries with less developed health systems. Developed countries are faced with a new challenge of sustaining the health system such that mortality and morbidity rates do not deteriorate as resources are shifted to other areas or training of health professionals ignores common problems.

Concepts and Definitions

Maternal Death

The World Health Organization (WHO) in the International Classification of Diseases and Related Health Problems (ICD-10) (1992) defines a maternal death as:

> The death of a woman while pregnant or within 42 days of termination of pregnancy, irrespective of the duration and site of the pregnancy, from any cause related to or aggravated by the pregnancy or its management but not from accidental or incidental causes.

This definition allows examination of maternal deaths according to their causes as direct and indirect deaths. Direct obstetric deaths are those resulting from obstetric complications of the pregnant state (pregnancy, delivery, and the postpartum period), from interventions, omissions, incorrect treatment, or from a chain of events resulting from any of the above. Deaths due to hemorrhage, preeclampsia/eclampsia, or those due to complications of anesthesia or cesarean section are, for example, classified as direct obstetric deaths.

Indirect obstetric deaths are those resulting from previous existing disease or diseases that developed during pregnancy and were not due to direct obstetric causes, but were aggravated by physiologic effects of pregnancy. For example, deaths due to aggravation of an existing cardiac or renal disease are indirect obstetric deaths.

Accurate identification of the causes of maternal deaths to be able to understand the extent to which they are due to direct or indirect obstetric causes, or due to accidental or incidental events, is not always possible, particularly in settings where deliveries occur mostly at home and/or information registration systems are not adequate. In these instances, the classical ICD definition of maternal death will not be useful. A concept of pregnancy-related death included in ICD-10 incorporates maternal deaths due to any cause. With this concept, any death during pregnancy, childbirth, or the postpartum period is a maternal death even if it is due to accidental or incidental causes.

Complications of pregnancy or childbirth can lead to death after the postpartum period (42 days following delivery) has passed. Deaths due to aggravation of chronic conditions, such as cardiac diseases, may happen at a relatively late stage. In addition, increasingly available modern life-sustaining procedures and technologies help more women survive adverse outcomes, but may also delay death. Despite having been caused by pregnancy-related events, these deaths do not appear as maternal deaths in registration systems that use only the conventional definition. An alternative concept of late maternal death is included in ICD-10 to capture delayed deaths (those due to the complications of pregnancy and childbirth that occur later than the completion of 42 days after termination of pregnancy up to one year) (**Table 1**).

These alternative definitions, particularly the concept of pregnancy-related death, allow identification and measurement of maternal deaths in settings where accurate causes of deaths are not known by means of reliable information registration systems. In these instances, determination of maternal mortality levels mostly depend on *ad hoc* household surveys, where relatives of a woman who has died at reproductive age are asked about her pregnancy status at the time of the death. In settings where established routine registration systems with correct attribution of causes of death exist, it is possible to identify maternal deaths that fit into the classical ICD definition of maternal death.

Table 1 Alternative definitions of maternal death in ICD-10

Pregnancy-related death – the death of a woman while pregnant or within 42 days of termination of pregnancy, irrespective of the cause of death
Late maternal death – the death of a woman from direct or indirect obstetric causes more than 42 days but less than 1 year after termination of pregnancy

Maternal Morbidity

Defining morbidity as an outcome is more difficult than defining death in general. Conceptualization of maternal or obstetric morbidity is particularly difficult as compared to defining morbidities in other areas of medicine because of the wide range of pregnancy outcomes, each with varying severity. Definitions of specific negative outcomes differ in different settings or for different purposes. For example, severe postpartum hemorrhage, one of the main causes of maternal deaths, is defined as blood loss greater than 500, 1000, or 1500 ml in different settings. One reason for this is that the amount of blood loss that could lead to death would differ according to a woman's preexisting health, in particular her hemoglobin status, or her probability of receiving timely and adequate care. An anemic woman is more likely to die with less blood loss from postpartum hemorrhage than a woman with normal hemoglobin level. Other definitions of hemorrhage exist for instances when it is not feasible to accurately identify the level of blood loss. These refer to the need for a specified amount of blood transfusion or to a defined level of drop in hemoglobin level. Similarly, for another serious pregnancy outcome, preeclampsia, the cut-off levels to define high blood pressure differ in different definitions. Finally, some conditions, such as urinary incontinence, are difficult to define, as a distinction between normal physiological symptoms due to pregnancy and its pathology may not be clear-cut.

It is, however, essential to measure maternal morbidity as well as the deaths. Monitoring maternal morbidity has some major advantages over monitoring maternal deaths. Maternal morbidity occurs much more frequently than maternal deaths; hence information on maternal care can be more rapidly gathered and analyzed, allowing for more rapid feedback and corrective intervention. The survival of the woman means that the woman can be interviewed to identify whether the health system failed or not. This is of particular value in women who have been referred from

Table 2 Range of pregnancy outcomes

1. Death
2. Near miss (severe acute maternal morbidity)
3. Severe morbidity
4. Minor obstetric morbidity
5. Sociocultural difficulties around major life event
6. Normal physiological experiences and adaptation of pregnancy, lactation, and postpartum period

Adapted from Bewley S, Wolfe C, and Waterstone M (2002) Severe maternal morbidity in the UK. In: MacLean AB and Neilson JP (eds) *Maternal Morbidity and Mortality*, pp. 132–146. London: RCOG.

one institution to another, and especially for assessing care at the primary health-care level. This information on care provided at the primary level is often not available in the investigation of maternal deaths.

In countries with low numbers of maternal deaths, the causes of the deaths are often peculiar to the particular case. For example, congenital heart disease is an important contributor to maternal death in the United Kingdom, but congenital heart disease is a rare condition. Information gained from analyzing deaths due to congenital heart disease, although useful, is limited to those very few people with the condition. Hence, lessons learned cannot be generalized to the whole population.

Several useful approaches to conceptualize maternal morbidity and to categorize the wide range of pregnancy outcomes exist. One approach classifies maternal morbidity based on the hierarchy of the severity of pregnancy outcomes. In this concept, the spectrum of the pregnancy outcomes ranges from death to normal physiological experiences, as shown in **Table 2**. The conditions that fall into categories 2–4 fit into the definition of maternal morbidity.

Maternal Death

Measurement

Although widely accepted standardized definitions of maternal mortality exist, as described above, it is difficult to accurately measure the level of maternal mortality in a population for several reasons. First, it is difficult to identify maternal deaths precisely. Particularly in settings where deaths are not reported comprehensively through routine registration systems, the death of a woman of reproductive age may not be recorded. Second, even if it is recorded, her pregnancy status may not be known and, third, even if the pregnancy status is known, where a medical certification of cause of death does not exist, correct attribution of death as a maternal death cannot be done. This means that even in settings where routine registration of deaths is in place, without special attention to enquiring about the causes of deaths, maternal deaths can go underreported. Chang and colleagues (2003) in their report of routine surveillance of maternal deaths in the USA during 1991–99, estimated that the true number of deaths related to pregnancy might increase 30–150% with active surveillance.

Where routine registration with correct cause–attribution of deaths does not exist, determination of the levels of maternal mortality relies on *ad hoc* population surveys. These surveys require very large sample sizes because, despite being unfairly high in some parts of the world, maternal deaths are relatively rare events in epidemiological terms. Even very large sample sizes produce estimates of maternal mortality levels with wide confidence intervals. Alternative methods to reduce the requirement for large sample sizes have been developed, but the problem of the large uncertainty margins remains.

Public health officials use statistical measures of maternal deaths primarily so that comparisons can be made over time or between areas. However, another aspect related to maternal death analysis is to detect where health systems fail. For this, only a large representative sample is needed. This is particularly useful in developing countries where inadequacies in data collection mean rates are not possible to calculate. Confidential enquiries into maternal deaths (CEMD) are a good example of this approach and are defined by the World Health Organization (2004) as:

> A systematic multidisciplinary anonymous investigation of all or a representative sample of maternal deaths occurring at an area, region (state) or national level, which identifies the numbers, causes and avoidable or remediable factors associated with them. Through the lessons learnt from each woman's death, and through aggregating the data, confidential enquiries provide evidence of where the main problems in overcoming maternal mortality lie and an analysis of what can be done in practical terms, and highlight the key areas requiring recommendations for health sector and community action as well as guidelines for improving clinical outcomes.

Measures of maternal mortality

The maternal mortality ratio (MMR) is the most frequently used statistical measure to evaluate maternal mortality. This measure refers to the number of maternal deaths during a given time period per 100 000 live births during the same time period. It is a measure of the risk of death once a woman has become pregnant. Other, less commonly used measures include the maternal mortality rate and lifetime risk of maternal death, as shown in **Table 3**.

Data sources and collection methods

Ideally, data for both the numerator (number of maternal deaths) and the denominator (number of live births) should be obtained by direct counting through routine registration systems with correct attribution of the causes

Table 3 Statistical measures of maternal mortality

Maternal mortality ratio (MMR) – number of maternal deaths during a given time period per 100 000 live births during the same time period
Maternal mortality rate – number of maternal deaths in a given time period per 100 000 women of reproductive age during the same time period
Lifetime risk of maternal death – the probability of dying from a maternal cause during a woman's reproductive lifespan

of deaths as determined by medical certification. For reasons mentioned earlier, this does not exist in many countries. Even if such systems exist, underreporting is usually a problem and identification of the true numbers of maternal deaths requires additional specific investigations into the causes of deaths, even in developed countries. A specific example for such investigation is the CEMD for the United Kingdom, which was initiated in 1928.

Alternatively, data are collected on an *ad hoc* basis using a variety of methods including direct household surveys, surveys using sisterhood methodology, reproductive-age mortality studies (RAMOS), and censuses. Derived estimates are based on respondents' accounts of maternal deaths in the household or among their sisters (sisterhood methodology) or studies of deaths among women of reproductive age (RAMOS methodology). Questions investigate deaths of women during pregnancy, childbirth, or the defined postpartum period, thus pregnancy-related deaths. Main methodologies used for estimating maternal mortality levels are shown in **Table 4**.

Global Levels and Determinants

The difficulties in measuring the true extent of maternal mortality, particularly in less developed countries, are well-established. Making international comparisons is also difficult due to the variety of methodologies used to determine maternal mortality in different settings and the different definitions used by different methodologies. It is, however, clear that a vast difference between more developed and developing countries exists in levels of maternal mortality. WHO, together with the United Nations Children's Fund (UNICEF), the United Nations Population Fund (UNFPA) and The World Bank, uses a methodology to calculate global estimates of maternal mortality in which countries are classified according to the availability of data from different sources and a modeling technique for those with no available reliable data is used. The 2007 WHO publication of maternal mortality estimates refers to levels of maternal mortality in 2000 and shows an MMR of 450 in developing country settings and 9 in more developed regions. As shown in **Figure 1**, among the developing regions of the world, the highest maternal mortality levels were in Africa (820 maternal deaths per 100 000 live births) and the lowest in Latin America (130 maternal deaths per 100 000 live births).

Countries in sub-Saharan Africa had very high levels of maternal mortality. For example, MMR was estimated as 2100, 1800, and 1500 in Sierra Leone, Niger, and Chad, respectively. In Afghanistan, 1800 women were estimated to have died from pregnancy-related causes per 100 000 live births. In contrast, maternal mortality ratios hardly exceed 9 per 100 000 live births in developed countries (**Figure 2**).

In addition to wide between-country differences, maternal mortality levels vary between subpopulations within countries, both more developed and developing. Ethnic group differences in the USA are among the most-cited examples of differentials of maternal mortality in developed settings. Anachebe and Sutton (2003) reported a fourfold disparity in the risk for maternal death among black women as compared to white women for 1987–96 in this country. In Germany, Razum and colleagues (1999) found that immigrant women were more likely to die from pregnancy-related causes as compared to their native counterparts. Women living in the most deprived areas of England had a 45% higher death rate as compared with women in the most affluent areas according to the CEMD 2004 in the United Kingdom for the years 2000–2002 (Lewis, 2004). Similar trends exist in developing countries. An analysis of nationally representative survey data from 10 developing countries (Graham *et al.*, 2004) showed the consistent increase in probability of dying from maternal causes with increased poverty.

Between-country differences in maternal mortality levels reflect the differences in socioeconomic development of countries as well as the capacity and functioning of health systems to deliver effective interventions as needed. Within-country differences show that certain subpopulations are disadvantaged through a variety of characteristics. These characteristics make them vulnerable to maternal complications (e.g., poor living conditions and nutrition) as well as less able to seek and obtain appropriate health care (e.g., education, health-care access, quality of care) for the prevention or management of these complications.

Causes of Maternal Deaths

As mentioned in the previous sections, maternal deaths are caused either by conditions specific to pregnancy, childbirth, and the postpartum period and/or their management (direct causes) or by preexisting medical conditions (indirect causes). Direct causes include hemorrhage, hypertensive disorders of pregnancy (preeclampsia, eclampsia), sepsis, obstructed labor, uterine rupture, obstetric embolism, ectopic pregnancy, complications of abortion, cesarean section, or anesthesia, and other less frequent conditions.

Indirect maternal deaths may occur due to chronic conditions such as cardiac or renal disease, or suicide

Table 4 Methodologies used for estimating maternal mortality

Methodology	*Characteristics*
Routine registration systems (vital registration)	Direct counting of maternal deaths (and live births) within routine registration. Even where coverage is complete and all deaths are medically certified, in the absence of active case finding, maternal deaths may be missed or misclassified • Confidential inquiries are used to evaluate the extent of misclassification and underreporting
Direct household survey	Where vital registration data are not appropriate for the assessment of cause-specific mortality, household surveys provide an alternative • Surveys using direct estimation are expensive and require large sample sizes to provide a statistically reliable estimate • Even with large sample sizes, uncertainty margins are wide – difficult to monitor trends
Sisterhood method	Obtains information by interviewing respondents about the survival of all their adult sisters – reduces high sample-size requirements, but • the problem of wide uncertainty margins remains • provides a retrospective rather than a current estimate (refers to some 35 years back, with a midpoint around 12 years before the survey) • originally developed version (indirect sisterhood method) is not appropriate for use in settings where fertility levels are low (total fertility rate <4) or where there has been substantial migration or other causes of social dislocation • the Demographic and Health Surveys use a variant of the sisterhood approach (direct sisterhood method). This relies on fewer assumptions than the original method but requires larger sample sizes and the analysis is more complicated. Obtained estimates refer to an approximate point estimate some three to four years before the survey. The problem of wide confidence intervals remains, hence the limitation of monitoring trends
Reproductive-age mortality studies (RAMOS)	Involves identifying and investigating the causes of all deaths of women of reproductive age in a defined area/population • Multiple and varied sources of information are used to identify deaths of women of reproductive age; no single source identifies all the deaths • Interviews with household members and health-care providers and reviews of facility records are used to classify the deaths as maternal or otherwise • Properly conducted, the RAMOS approach provides the most complete estimation of maternal mortality (in the absence of reliable routine registration systems), but can be complicated and time-consuming to undertake, particularly on a large scale
Verbal autopsy	Means to assign cause of death through interviews with family or community members where medical certification of cause of death is not available • Reliability and validity for assessing cause of death in general and identifying maternal deaths in particular has not been established • May fail to correctly identify a group of maternal deaths, particularly those occurring early in pregnancy (ectopic, abortion-related), those in which the death occurs some time after the termination of pregnancy (sepsis, cardiac disease), and indirect causes of maternal death (malaria, HIV/AIDS)
Census	A census of the whole population that: • could include questions on deaths in the household in a defined reference period • could include more detailed questions that would permit the identification of maternal deaths on the basis of time of death relative to pregnancy (verbal autopsy) and • eliminates sampling errors, thus allows trend analysis (weaknesses of verbal autopsy should be borne in mind)

due to pregnancy-related depression and psychosis. Non-pregnancy-related infections, including malaria and tuberculosis, are usually also regarded as indirect causes of maternal deaths. Deaths due to acquired immune deficiency syndrome (AIDS) are also often included as indirect deaths in the non-pregnancy-related infections category. Women rarely die due to AIDS alone, but rather due to a concomitant disease such as tuberculosis, *Pneumocystis carinii* pneumonia, cryptococcal meningitis, and other common diseases, such as malaria and community-acquired pneumonia, that are made more severe by the underlying condition of AIDS. In areas with high HIV prevalence, it is more appropriate to classify all women with respect to their HIV status and their stage of disease rather than have AIDS as a specific primary obstetric cause of maternal death. Analysis of women with or without HIV infection and those with or without AIDS can then be performed and the full impact of the disease assessed. When AIDS is classified within the indirect causes, this may conceal the actual cause of death, for example, a septic abortion. (When a woman with AIDS dies, there is a tendency to report only AIDS and not other conditions directly responsible for the death but imminently treatable such as a septic abortion.)

Although not regarded as maternal deaths within the classic definition, accidental causes including trauma and

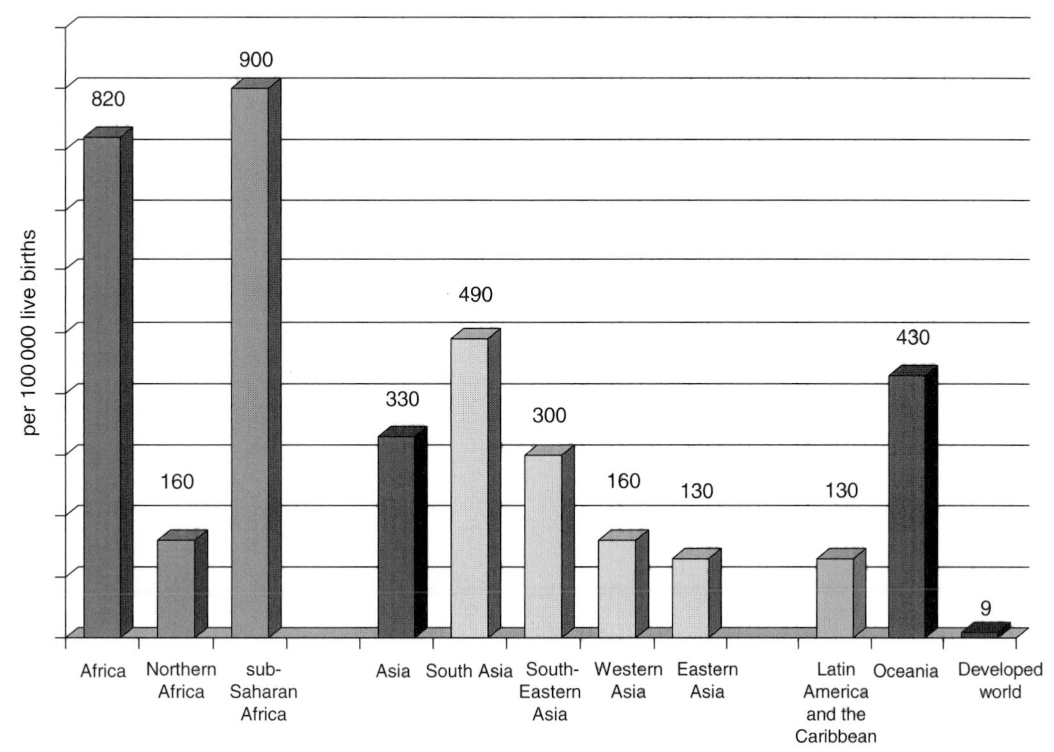

Figure 1 Maternal mortality according to world regions. Adapted from WHO/UNICEF/UNFPA/The World Bank (2007) *Maternal Mortality in 2005: Estimates Developed by WHO, UNICEF and UNFPA*. Geneva: World Health Organization.

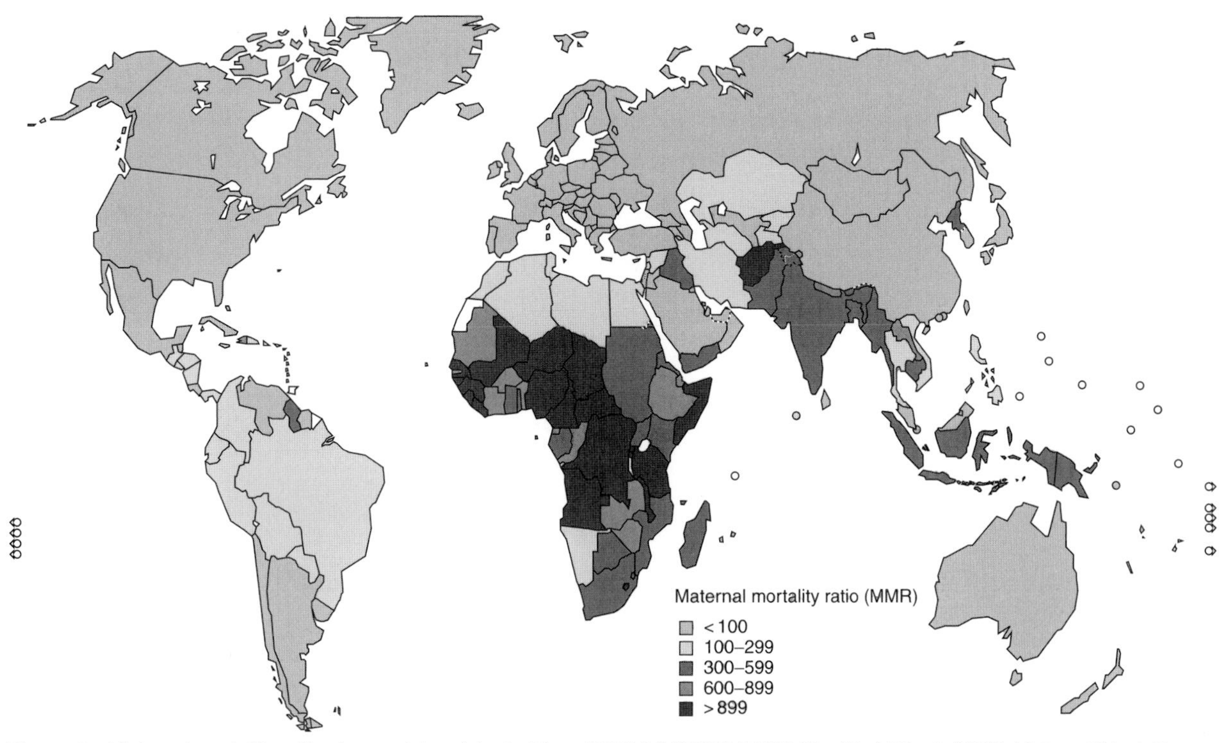

Figure 2 Maternal mortality ratios in countries. Adapted from WHO/UNICEF/UNFPA/The World Bank (2007) *Maternal Mortality in 2005: Estimates Developed by WHO, UNICEF and UNFPA*. Geneva: World Health Organization.

homicide are increasingly becoming important in cause distributions of maternal deaths in developed countries, where mortality due to other causes is reduced with provision of effective health care.

The patterns of the causes of maternal deaths vary according to the world regions due to the disease profiles in the regions as well as the extent of the development of health systems. An analysis of the causes of maternal

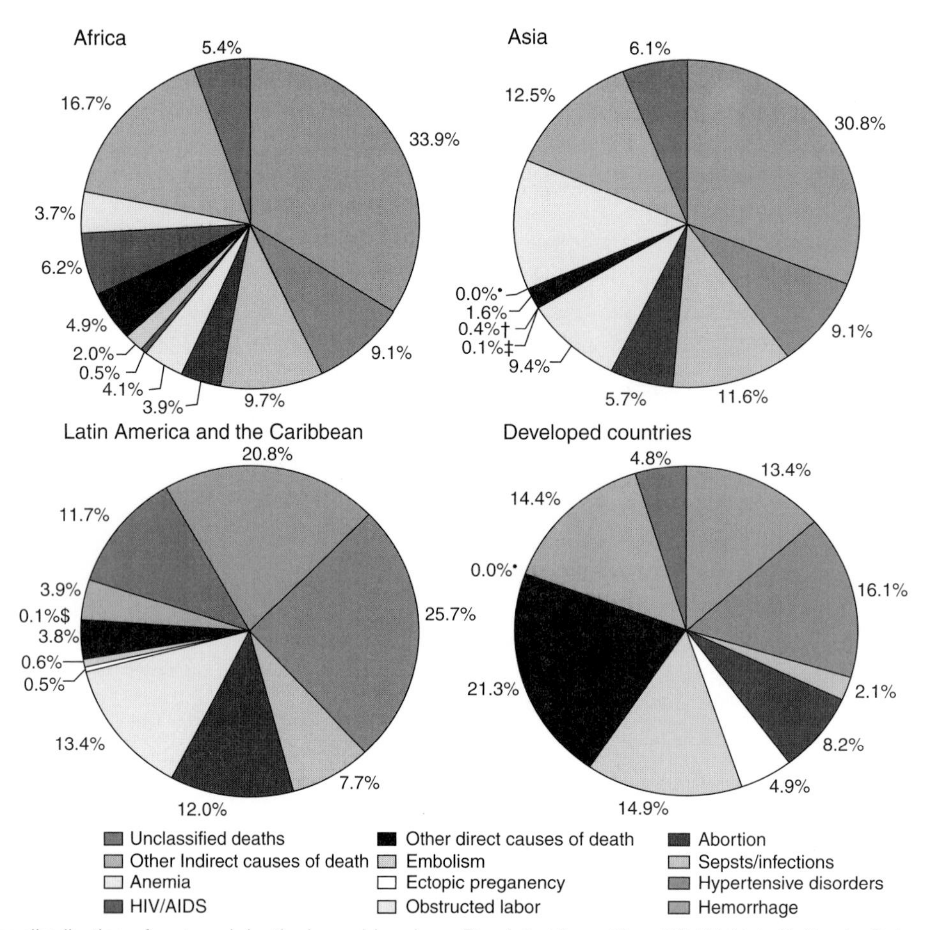

Figure 3 Cause distribution of maternal deaths in world regions. Reprinted from Khan KS, Wojdyla D, Say L, Gulmezoglu AM, and Van Look PFA (2006). WHO analysis of causes of maternal death: A systematic review. *Lancet* 367: 1066–1074, Copyright 2006, with permission from Elsevier.

deaths in the broad world regions (Khan *et al.*, 2006) shows that hemorrhage is the leading direct cause of death in Africa (33.9%) and Asia (30.8%). In Latin America and the Caribbean, hypertensive disorders including pre-eclampsia and eclampsia are responsible for the largest number of deaths (25.7%). In more developed country settings, deaths due to a group of direct obstetric causes including complications of anesthesia are the biggest contributors to maternal deaths (21.3%) (**Figure 3**). Deaths related to infections, in particular HIV/AIDS, anemia, and abortion are the region-specific characteristics of distribution of maternal deaths in Africa, Asia, and Latin America and the Caribbean, respectively.

Another emerging pattern in the cause distribution of maternal deaths in developed countries with low maternal mortality ratios is the dominance of causes indirectly related to pregnancy, particularly worsening of preexisting medical conditions. The contribution of indirect conditions was 59% to all deaths in the United Kingdom CEMD for the years 2000–2002 (Lewis, 2004). In the USA where MMR is slightly higher than in many European countries, one-third of maternal deaths during the 1990s were due to indirect morbidities including cardiomyopathy, complications of anesthesia, cerebrovascular accidents, and pulmonary or neurological problems, as reported by Chang and colleagues (2003). The shift from direct causes to indirect causes associated with chronic conditions is due to the decrease of deaths caused by direct morbidities as more developed health systems are able to prevent and treat these conditions.

Maternal (Obstetric) Morbidity

Severe Acute Maternal Morbidity (Near Miss)

The concept of severe acute maternal morbidity (SAMM) or “near miss” describes the severest type of maternal morbidity, defined by Mantel and colleagues (1998) as: “A very ill pregnant or recently delivered woman who would have died had it not been that luck and good care was on her side.”

This concept is relatively new in maternal care, but is increasingly becoming an important indicator of pregnancy-related risks and the quality of maternal health care. Particularly where maternal deaths are becoming less frequent or the geographic area is small, investigation of the low numbers of maternal deaths gives information relevant only to the limited number of cases. Examination of the cases that almost died, however, provides richer information on the quality of the health care in relation to common major morbidities in a particular setting.

Maternal deaths in developed countries are now rare, and the factors that surround the death are often peculiar to the event and are not generalizable. This does not mean that pregnancy is a safe condition in developed countries. Waterstone and colleagues (2001) reported a severe obstetric morbidity rate of 12.0 per 1000 births, and a severe morbidity to mortality ratio of 118:1 in the South East Thames region in London, United Kingdom. Contrary to what would be expected, the common causes of maternal morbidity are not the same as the common causes of maternal mortality in developed country settings. A review of only maternal deaths would neglect the important life-threatening other obstetric emergencies, which are common causes of morbidity in London, such as antepartum hemorrhage, postpartum hemorrhage, and preeclampsia. The most common cause of severe morbidity was severe hemorrhage at 6.7 per 1000 births in the London area. This is not unique to London, as the most common cause of near misses in Scotland (Brace *et al.*, 2004) was also hemorrhage. In comparison, there were only seven deaths in over 2 million deliveries (3.3 deaths per million deliveries) due to hemorrhage in the United Kingdom CEMD for the years 1997–99 (Lewis, 2001).

By looking at maternal deaths only, developed countries might be in danger of overlooking other major causes of morbidity in obstetric care. This might have already happened in the United Kingdom, as in the next triennium (2000–2002), according to the CEMD, there were 17 deaths (8.5 deaths per million deliveries) due to hemorrhage, a 140% increase (Lewis, 2004). A systematic analysis of severe morbidity would have kept hemorrhage as a priority condition, and the increase in deaths due to hemorrhage may potentially have been prevented. The new challenge in developed countries is to ensure that the gains made are sustained. Morbidity analysis will help in ensuring this, as resources will not be diverted away and health workers' training will not neglect conditions such as hemorrhage.

Operational definitions to identify SAMM cases include those according to existence of specified conditions (disease-specific); use of specific interventions or management techniques (management-based); and existence of organ failure (organ system-based) (**Table 5**).

Prevalence of SAMM varies according to which definition is used. Rates vary between 0.80–8.23% with disease-specific and between 0.38–1.09% with organ-system-based definitions (Say *et al.*, 2004). Rates are lower (0.01–2.99%) with management-based definitions. As expected, and similar to MMR, prevalence of SAMM varies between less developed and more developed countries. In less developed countries, 4–8% of pregnant women who deliver in hospitals experience SAMM when diagnosis is made on the basis of specific diseases. This rate is around 1% when organ-system failure is considered. In more developed country settings, the rates are around 1% with disease-specific and 0.4% with organ-system-based criteria, respectively.

In-depth reviews of SAMM cases in terms of the health care that the cases received give detailed information on how particular types of cases are managed and where the gaps and weaknesses are in the process of care.

Table 5 Definitions and criteria for severe acute maternal morbidity (SAMM)

Definition	*Identification*
Disease-specific – existence of severe forms of common conditions	• Severe hemorrhage, severe preeclampsia, eclampsia, sepsis, severe dystocia, other direct morbidities (requiring cesarean section, hysterectomy, and/or blood transfusion) (Prual *et al.*, 2000) • Severe hemorrhage, severe preeclampsia, eclampsia, hemolysis, elevated liver enzymes and low platelet count syndrome (HELLP) sepsis, severe sepsis, uterine rupture (Waterstone *et al.*, 2001)
Management-specific – use of specific interventions	• Admission to intensive care • Emergency hysterectomy
Organ-system failure-based – existence of severe organ dysfunction or failure	Organ systems affected and signs of organ failure (specified criteria for each system) • Cardiac – pulmonary edema, cardiac arrest • Circulatory – hypovolemia • Respiratory – intubation/ventilation requirement • Immunological – sepsis • Renal – oliguria • Hepatic – jaundice (with preeclampsia) • Metabolic – ketoacidosis, hypoglycemic coma, thyroid crisis • Coagulation – thrombocytopenia • Cerebral – coma (Mantel *et al.*, 1998)

The use of a combined measure of mortality and morbidity also provides a crude indicator of the quality of care. One approach to construct such a measure is to divide the number of the cases identified as severe morbidity by those of deaths (morbidity-to-mortality ratio). The drawback of this approach is that although severe morbidity and mortality are expected to rise and fall equally (in the same setting), they may be independent. A second approach is to calculate the ratio of maternal deaths to the sum of maternal deaths and SAMM cases (mortality index, or MI). This measure gives the proportion of women with SAMM who die.

Morbidity-to-mortality ratio is typically 10-fold higher in studies undertaken in developed countries as compared to those in developing countries. For example, MI was calculated as 5–8 and 49 in South African and Scottish studies, respectively, that used the same case-identification criteria (Say *et al.*, 2004).

Major Morbidities

Major conditions that lead to SAMM and/or maternal death are classified under this category and include obstetric hemorrhage (antepartum, postpartum), preeclampsia/eclampsia, pregnancy-related infections (mainly postpartum sepsis), thromboembolism, and obstructed labor (by leading to postpartum hemorrhage, sepsis, and uterine rupture) as direct maternal morbidities. Complications of abortion, when performed under unsafe conditions, constitute a major cause of maternal death through hemorrhage, infection, or uterine rupture. Other direct major morbidities are complications of ectopic pregnancy, anesthesia, or cesarean section.

Major morbidities also include indirect causes of maternal deaths, such as those that are based on preexisting medical conditions (e.g., chronic hypertension, cardiac or renal disease, diabetes, asthma, anemia) or suicide due to pregnancy-related depression, as well as non-pregnancy-related infections during pregnancy, such as AIDS, malaria, and tuberculosis.

Obstetric hemorrhage

Hemorrhage is the major cause of maternal deaths in many settings. It can cause up to one-third of maternal deaths in settings with less-developed health systems. Because of the variety of definitions and difficulties in assessing the blood loss involved, the incidence of hemorrhage cannot be determined precisely. Severe hemorrhage was reported in 3.05% of women delivering live babies in six African countries and 0.67% of deliveries in the United Kingdom. Approximately 3% and 0.3% of these cases died in the African countries and the United Kingdom, respectively (Prual *et al.*, 2000; Waterstone *et al.*, 2001).

Hemorrhage may occur during pregnancy, either antepartum (due to placenta previa or placental abruption) or, most commonly, postpartum (due to uterine atony, retained placenta, inverted or ruptured uterus, or cervical, vaginal, or perineal laceration). The majority of cases occur immediately after delivery (immediate postpartum hemorrhage) and the most common cause is uterine atony, that is, the failure of the uterus to contract after delivery. Risk factors for uterine atony include preeclampsia, prolonged labor, high parity, induction of labor, high doses of halogenated anesthetics, prior postpartum hemorrhage, and large or multiple fetuses. However, as many as two-thirds of women who have postpartum hemorrhage do not have any of these risk factors. Late (or secondary) postpartum hemorrhage occurs after the first 24 h of delivery and can be due to retained placenta, infection, or trophoblastic tumors.

Postpartum hemorrhage is traditionally defined as the loss of more than 500 ml of blood after completion of the delivery, although the accurate measurement of the amount of the blood loss has been problematic. When measured objectively, 500 ml of blood loss is observed in approximately half of normal vaginal deliveries. Clinically estimated loss of 500 ml or more, however, is considered as significant since clinical estimates of blood loss are found to be about half of the actual loss (objectively measured) (Pritchard *et al.*, 1962). This definition is particularly relevant to developing country settings, where women are more vulnerable to negative consequences of blood loss due to limitations of the delivery care they usually receive or to preexisting severe anemia. In developed country settings, blood loss of 1500 ml or more is usually defined as severe hemorrhage.

Other practical definitions of severe hemorrhage also exist since accurate measurement is not always feasible and clinical estimations are not reliable. Drop of a specified quantity in hemoglobin or hematocrit levels or the amount of blood transfused are alternative practical definitions of severe hemorrhage. Operational definitions used in two classic studies of severe morbidity in developing and developed country settings are shown in **Table 6**.

Severe hemorrhage can lead to SAMM or death in a very short period if not managed properly. Management depends on the cause and the severity of the bleeding. For example, hemorrhage due to uterine atony is managed by

Table 6 Operational definitions of severe hemorrhage

Prual et al., 2000 (seven countries in Western Africa)	*Waterstone et al., 2001 (United Kingdom)*
Leading to blood transfusion; OR Leading to hospitalization for $\geq$ 4 days; OR Leading to hysterectomy; OR Leading to death	Clinically estimated blood loss of $\geq$1500 ml; OR Peripartum hemoglobin drop of $\geq$4 g/dl; OR Acute transfusion of $\geq$4 units of blood

stopping the bleeding using uterotonics, physical compression techniques, uterine artery ligation, or hysterectomy, as well as replacement of fluid, clotting factors, and blood that have been lost due to the bleeding. Additional treatment will involve removal of retained fragments in the case of retained placenta, suturing in case of lacerations, and hysterectomy or repair in case of uterine rupture.

Postpartum hemorrhage due to uterine atony can be prevented in most instances. An intervention package involving administration of uterotonic agents, controlled cord traction, and uterine massage, referred to as active management of the third stage of labor, was shown to be effective in reducing the incidence of postpartum hemorrhage. Active management is therefore preferred over expectant management, which involves waiting for signs of placental separation and allowing for spontaneous delivery of the placenta.

Other measures to prevent hemorrhage include administering a uterotonic agent alone, reducing the incidence of obstructed labor by timely intervention as needed, minimizing the trauma due to instrumental delivery, and detection and treatment of anemia during pregnancy.

Hypertensive disorders of pregnancy (preeclampsia/eclampsia)

Pregnancy-specific hypertensive conditions, in particular preeclampsia/eclampsia, are among the leading causes of SAMM and maternal deaths. Four types of such conditions complicate pregnancy (**Table 7**).

As seen in **Figure 3**, 9–29% of maternal deaths are caused by hypertensive disorders in world regions. Hypertension complicates approximately 5% of all and 11% of first pregnancies, respectively. Preeclampsia constitutes half of these cases (2–3% of all pregnancies and 5–7% of first pregnancies). Between 0.8–2.3% of women with preeclampsia develop eclampsia (Magpie Trial Collaborative Group, 2002). Similar to hemorrhage, the likelihood of dying from preeclampsia or eclampsia is much higher in developing country settings than in more developed ones.

Table 7 Hypertensive disorders of pregnancy

Chronic hypertension – high blood pressure before the 20th week of pregnancy or, if only diagnosed during pregnancy, persisting after the sixth week after delivery
Gestational hypertension – development of hypertension after the 20th week of pregnancy without other signs of preeclampsia
Preeclampsia/eclampsia – development of hypertension after the 20th week of pregnancy with other signs of preeclampsia; eclampsia is the occurrence of seizures in women with preeclampsia
Preeclampsia superimposed on chronic hypertension – development of new signs associated with preeclampsia after the 20th week of pregnancy in women with chronic hypertension

Around 5% of women with eclampsia and 0.4% with preeclampsia die in developing country settings. These figures are as low as 0.7% and 0.03%, respectively, in more developed countries.

Preeclampsia is a multisystem disorder usually characterized by sudden onset of hypertension and proteinuria during the second half of pregnancy. The basic causes are changes in the endothelium, which is an intrinsic part of all blood vessels. Preeclampsia, therefore, has the potential of affecting any organ system. The nature of the signs and symptoms varies according to which organs are affected. In addition to hypertension, proteinuria, oliguria, cerebral or visual disturbance, pulmonary edema, cyanosis, thrombocytopenia, or signs of impaired liver function can be detected in a preeclamptic woman. It is, however, proteinuria and hypertension that are the defining features of preeclampsia.

A variety of definitions of preeclampsia with different specifications for both features is used, particularly for research purposes. Clinically:

- hypertension is defined as blood pressure of equal to or greater than 140 mmHg systolic or 90 mmHg diastolic in two consecutive measurements and occurring after the 20th week of pregnancy
- proteinuria is defined as a protein concentration of more than 500 mg/l in a random specimen of urine or a protein excretion of more than 300 mg per 24 h
- eclampsia is the occurrence of seizures in a woman with preeclampsia.

Women with first pregnancy, extremes of age, obesity, multiple pregnancy, preexisting hypertension or renal disease, who have had preeclampsia or eclampsia in a previous pregnancy, and with a family history of preeclampsia are more at risk of developing preeclampsia.

Identification of reduced placental perfusion or vascular resistance dysfunction by Doppler ultrasonography or high levels of urinary kallikrein may be useful in diagnosing preeclampsia during early pregnancy, but widespread use of these techniques, particularly in resource-poor settings, is limited due to technological requirements. Regular screening of blood pressure, urinary protein, and fetal size during antenatal care visits is currently the only strategy for early diagnosis of preeclampsia. High-risk women should be screened more frequently between 24 and 30 weeks of pregnancy, when the onset of the disease is most common.

If the hypertension is severe (blood pressure $\geq$170 mmHg systolic or $\geq$110 mmHg diastolic), treatment should include antihypertensive drugs. The effectiveness of antihypertensive treatment in mild to moderate hypertension has not been established. Magnesium sulfate is the drug of choice for preeclampsia to prevent eclampsia, as well as for eclampsia to reduce recurrence of convulsions and maternal death.

Studies on the effectiveness of a range of nutritional (vitamins, antioxidants) or non-nutritional (antiplatelet agents, calcium) strategies to prevent preeclampsia have provided no conclusive evidence. The only recommended strategy is the use of low-dose aspirin in high-risk nulliparous women and women with poor obstetric history.

Sepsis

Sepsis, in particular postpartum sepsis, is an important cause of maternal death in developing countries, but is now very rare in more developed countries where historically most of the deaths were due to sepsis. The invention and widespread use of antibiotics and recognition of the importance of clean delivery practices with skilled delivery attendants brought a sharp decline in the number of maternal deaths, in particular those caused by postpartum sepsis, during the first half of the 20th century. Currently, less than 1% of deliveries in developed country settings are complicated with sepsis, which usually results from rare hospital infections.

Sepsis is a systematic response to infectious agents or their byproducts. It develops most frequently from urinary tract infections, chorioamnionitis, postpartum fever (due to pelvic infection, surgical procedures, endometritis), or septic abortion. The infectious agents are either endogenous (i.e., they already exist as part of the normal flora of the woman's genital tract or as an existing infectious agent) or exogenous (i.e., acquired from outside sources such as deliveries or abortions taking place under unhygienic conditions) and most frequently include *Streptococcus*, Gram-negative bacteria, gonococcus, chlamydia, herpes simplex, and the organisms causing bacterial vaginosis. The effects of malaria and HIV/AIDS in areas where these infections are frequent are also increasingly recognized, although the mechanisms by which these infections cause sepsis are not clearly identified.

The usual signs of infection, such as fever, tachycardia, and leukocytosis at the early stages, if not treated appropriately, convert to severe sepsis with signs of multiple organ effects, such as hypothermia, hypotension, encephalopathy, oliguria, and thrombocytopenia. This can further lead to septic shock, which is a highly lethal syndrome in both developing and developed countries.

Management of sepsis involves treatment of the causative agent with appropriate antibiotics and interventions targeted to deal with presenting signs, such as hypovolemia, encephalopathy, or clotting problems.

Sepsis is preventable by:

- treatment of existing infection with appropriate antibiotics;
- prophylactic use of antibiotics for cesarean section, for women with preterm pre-labor rupture of membranes, and for high-risk women (women with previous spontaneous preterm delivery, history of low birthweight, pre-pregnancy weight less than 50 kg, or bacterial vaginosis in the current pregnancy); and
- clean delivery/pregnancy termination practices (infection control), including hand hygiene of providers and use of sterile, preferably disposable, supplies and equipment.

Obstructed labor

Obstructed labor is the failure of labor to progress due to mechanical obstruction. The source of the problem may be the mother (such as a contracted pelvis or tumor causing obstruction), the baby (such as a large or abnormal baby or abnormal position, presentation, or lie), or both. Most frequently, the problem is related to a relative size discrepancy between a normal baby and the pelvis of a healthy mother. In women experiencing their first pregnancy, obstructed labor usually leads to decreasing uterine activity and prolonged labor with the possibility of infection, postpartum hemorrhage, and vesicovaginal fistula formation, whereas in multiparous women uterine activity may continue to the point of uterine rupture.

Poor progress of labor may be due to obstruction, cervical dystocia, inefficient uterine activity, or a combination of these. The clinical diagnosis is usually based on poor progress of labor despite adequate uterine activity, together with visible signs of obstruction such as molding of the fetal skull. The definition of obstructed labor generally involves the length of labor (such as >12 or >18 h, or second stage >2 h), although other definitions involving clinical signs (such as uterine ring, pre-rupture, second stage transverse lie) are also used. According to the unpublished data from the WHO database of maternal mortality and morbidity, prevalence (with the definition of labor lasting >18 h) is, in general, less than 1% in developed country settings and may be up to 5% in less developed countries.

Black race, short maternal height, and maternal obesity have been suggested as potential risk factors, although there are no agreed criteria for their use to predict obstructed labor. In settings where female genital mutilation is common, scarring caused by the procedure may cause obstructed labor. Women with previous obstructed labor are at risk of experiencing future obstruction.

In settings with ready access to safe cesarean section, the management of obstructed labor is straightforward and related maternal mortality is extremely rare, although perinatal and maternal morbidity may occur. However, when cesarean section is not readily accessible, obstructed labor is a major cause of maternal mortality and severe morbidity by leading to uterine rupture, postpartum hemorrhage, sepsis, and obstetric fistula. Uterine rupture is a serious complication of obstructed labor, particularly in less developed settings. In developed countries, uterine rupture is typically seen in women with scarred uterus due to previous cesarean section.

The widely accepted interventions for preventing obstructed labor are the correction of breech presentation by external cephalic version at term and cesarean section. When cesarean section is not available or unsafe, symphysiotomy is a life-saving intervention for both the mother and the baby, although it has long been regarded as an unacceptable operation due to the perceptions of complications.

Venous thromboembolism

Venous thromboembolism refers to two related conditions – venous thrombosis and pulmonary embolism – that affect pregnant women. Pulmonary embolism arises from venous thrombosis, which is the process of clotting within the veins. It represents the leading cause of maternal deaths in developed country settings, despite being a rare event with a prevalence of less than 1%. Major risk factors for venous thromboembolism in pregnant women are shown in **Table 8**.

Venous thrombosis predominantly occurs in the legs and presents with clinical signs of pain, discomfort, swelling, and tenderness in the affected leg. Because problems such as swelling of the legs and discomfort are common in normal pregnancies, the clinical diagnosis of deep venous thrombosis is difficult. Only 10% of pregnant women who have such symptoms are diagnosed with deep venous thrombosis. It is, however, important to objectively diagnose the condition, because approximately one-fourth of women with untreated deep venous thrombosis develop pulmonary embolism.

In the majority of the cases where the thrombus is in the legs, the abovementioned clinical signs precede pulmonary embolism, but in others, where the thrombus is in the pelvic veins, the woman is usually asymptomatic until pulmonary embolus occurs. The most common clinical signs of pulmonary embolism are breathlessness, chest pain, cough, collapse, and hemoptysis. Advanced techniques such as compression ultrasonography and sophisticated computerized tomography scanners are used to diagnose deep venous thrombosis and pulmonary embolism, respectively.

The exact strategy to manage venous thromboembolism during pregnancy is still debated, but treatment generally includes the use of appropriate anticoagulants, mostly heparin. In deep venous thrombosis, the use of anticoagulants decreases the risk of developing embolism to less than 5%. Limitation of activity and use of elastic stockings are the recommended supportive measures. Anticoagulant use should be continued following delivery, but the evidence on the optimum length of use is limited. Current practice is to continue for at least 6 weeks. Thromboembolism can recur; thus, pregnant women with a previous event should be given prophylactic anticoagulants.

Table 8 Risk factors for venous thromboembolism

Age older than 35 years
Cesarean section
Operative vaginal delivery
Body mass index (BMI) >29
Thrombophilia
History of deep venous thrombosis or pulmonary thromboembolism
Gross varicose veins
Existing infection
Preeclampsia
Immobility
Medical conditions such as cardiac problems

Other major morbidities

A summary of other contributors of severe maternal morbidity and mortality not examined above is shown in **Table 9**. A final category that requires special attention is the effect of non-pregnancy-related infections – namely those of AIDS and malaria. These are classified among the indirect causes of maternal deaths as a subcategory of non-pregnancy-related infections.

The effect of AIDS

The most significant evidence on the contribution of AIDS to maternal deaths comes from the CEMD in South Africa (Pattinson, 2003, 2006). The inquiry attributed 17.0% and 20.1% of all maternal deaths to AIDS in 1999–2001 and 2002–2004, respectively. Moreover, the increase of deaths in hemorrhage, sepsis, and other non-pregnancy-related infections compared to the previous years was also suggested to be due to the effect of AIDS.

The mechanisms by which AIDS causes maternal deaths have not been clearly identified. There is some evidence from developing countries that pregnancy may be accelerating the disease, particularly in late stages. The presence of AIDS increases the severity of the complications of pregnancy and delivery, including ectopic pregnancy, spontaneous abortion, bacterial pneumonia, urinary tract infections, opportunistic infections (such as tuberculosis) and other infections, preterm delivery, hemorrhage, and sepsis that might eventually cause death. It appears that both the effects of infection and its complex interaction with related medical and social conditions that affect pregnancy are possible explanations of increased adverse pregnancy outcomes due to AIDS.

The effect of malaria

The proportion of maternal deaths attributable to malaria ranges between 2.9–17.6% in community-based studies in Africa (Brabin and Verhoeff, 2002), although the causal pathways between the infection and death, as in HIV, have not been established. The contribution of malaria to maternal deaths might be through its association with preeclampsia, eclampsia, hemorrhage, sepsis, and anemia. It has also been suggested that comorbidity with HIV increases malaria-related maternal deaths. In addition, pregnant women are more susceptible to infection than nonpregnant women are, and preventive measures should be a part of antenatal care where malaria is endemic.

Minor Morbidities

Obstetric trauma

Conditions that do not lead to serious maternal morbidity or death but cause long-term disability are included under this category. Sequelae of obstetric trauma, such as episiotomies or lacerations that are inappropriately managed and obstetric fistula, can cause lifelong suffering in varying degrees.

Lacerations involving the anal muscle and sphincter may cause rectal incontinence and uterine prolapse by weakening the support provided by the pelvic diaphragm. High parity, episiotomy, and cesarean section increase

Table 9 Other major maternal morbidities

		Epidemiology	
Condition	*Definition/notes*	*Developed countries*	*Developing countries*
Direct			
Ectopic pregnancy	Implantation of pregnancy outside the uterus – 95% in the Fallopian tubes	1.6–2.2% of all pregnancies; 6–8% of pregnancy-related deaths[a]	0.4–2.3% of all live births; 1.2–3.5% of pregnancy-related deaths (African countries)[a]
Anesthesia – complications	Failed intubation, aspiration of gastric contents, local anesthetic toxicity or high epidural block	1.6% of pregnancy-related deaths (USA)[b]	3.9% of pregnancy-related deaths (South Africa)[c]
Amniotic fluid embolism	Entering of amniotic fluid into circulation	1 of 20 000 deliveries; 7.5–10% of pregnancy-related deaths (USA)[b]	(All embolism) 6% of pregnancy-related deaths (South Africa)[c]
Complications of abortion	Difficult to estimate the burden – cause severe morbidity and death through hemorrhage, sepsis, uterine rupture; misclassification under these conditions is frequent	Negligible	Depends on legal, cultural, religious factors – 3.9% of pregnancy-related deaths (South Africa)[c]
Indirect			
Preexisting medical disease	Cardiac disease due to congenital anomalies, cardiomyopathy, aneurysm, myocardial infarction; cancer; renal disease; cerebrovascular disease (CVA)	Percentage of pregnancy-related deaths: cardiac disease: 8% (USA)[b], 14% (UK)[d] cancer: 9% (UK) CVA: 5% (USA)[b]	Percentage of pregnancy-related deaths: cardiac disease: 3% (South Africa)[c] preexisting maternal disease: 9% (South Africa)[c]
Suicide	Mostly due to postpartum depression – with the improvements in maternal health care and reductions in deaths from direct causes, increasingly becoming a significant contributor to maternal deaths	1.8–15% of all maternal deaths (studies from USA, Sweden, UK)[e]	1.8–15% of all maternal deaths (studies from India, Bangladesh, Zimbabwe)[e]
Anemia-related	Death due to anemia may occur due to its direct effect or secondary to adverse events, such as hemorrhage	Negligible	35–75% of pregnant women have anemia, 15–20% severe anemia (hemoglobin <8 g/dl);[f] significant contributor to maternal deaths in Asia
Coincidental – accidents, homicide	Homicide rate for pregnant women reported to be twice the rate for nonpregnant women	0.8–24.6% of all maternal deaths (studies from USA, Sweden)[e]	2% of all maternal deaths (one study in Bangladesh)[e]

[a]Goyaux N, Leke N, Keita N, and Thonneau P (2003) Ectopic pregnancy in African developing countries. *Acta Obstetricia et Gynecologica Scandinavica* 82: 305–312.
[b]Danel I, Berg CJ, Johnson JH, and Atrash H (2003) Magnitude of maternal morbidity during labor and delivery: United States, 1993–1997. *American Journal of Public Health* 93: 631–634.
[c]Pattinson RC (2003) *Saving Mothers 1999–2001: Second Report on Confidential Enquiries into Maternal Deaths in South Africa*. Pretoria, Africa: Government Printer.
[d]LewisG (ed.) (2004) *Why Mothers Die 2000–2002: The Confidential Enquiries Into Maternal Deaths in the United Kingdom*. London: RCOG Press.
[e]Shadigian E and Bauer ST (2005) Pregnancy-associated death: A qualitative systematic review of homicide and suicide. *Obstetrical & Gynecological Survey* 60: 183–190.
[f]van den Broek N (2002) The contribution of anaemia In: MacLean AB and Neilson JP (eds.) *Maternal Morbidity and Mortality*, pp. 79–88. London, United Kingdom: RCOG Press.

the risks of fecal or flatus incontinence. Urinary incontinence occurs with deep lacerations of the anterior vaginal compartment following spontaneous, forceps, or vacuum delivery.

Obstetric fistula (vesicovaginal or rectovaginal) occurs due to prolonged obstructed labor when the genital tract is compressed between the fetal head and the bony pelvis. Prolonged pressure may result in necrosis with subsequent vesicovaginal fistula formation, causing an uncontrollable leakage of urine from the vagina. Rarely, vesicocervical fistula occurs when the anterior cervical lip is compressed against the symphysis pubis. Rectovaginal fistula occurs in a similar way to vesicovaginal fistula. In this case, holes form between the tissues of the vagina and rectum, leading to uncontrollable leakage of feces. Some of these fistulas may heal spontaneously but, more often, special repair is necessary. Despite appropriate repair, Murray and colleagues (2002) reported persistence of urinary incontinence in 55% of cases in Ethiopia.

Psychological/psychiatric problems

Pregnancy may be associated with a level of anxiety reflecting fears surrounding delivery and the health of the baby, which, in some women, can lead to clinically meaningful levels of anxiety and depression. Sustained, high levels of anxiety may cause preterm delivery, prolonged delivery, and higher levels of delivery complications. Such anxiety also increases the risk of developing postpartum depression, which is the commonest cause of suicide among postpartum women, followed by postpartum psychosis. Suicide in the postpartum period is among the leading causes of maternal death, particularly in developed countries (**Table 9**).

Conclusion

Pregnancy and childbirth may have negative outcomes ranging from minor conditions to more serious morbidities and even death. Among all maternal deaths, 99% occur in developing parts of the world, where maternal morbidities are also more prevalent. The classic major severe morbidities of pregnancy contribute to a smaller part of the relatively few maternal deaths in developed countries because these conditions are preventable and treatable. In these settings, rare conditions, such as obstetric embolism, worsening of preexisting medical conditions, and depression-associated suicide have become the largest contributors of maternal deaths. In developing countries, the major direct morbidities are still the main causes of maternal deaths, together with infections such as AIDS and malaria or anemia. The patterns of maternal mortality and morbidity reflect the level of development of the health systems to be able to cope with severe consequences of major maternal morbidities. Public health challenges of maternal health remain to:

- accurately measure the extent of the mortality and morbidity and establish the patterns of both,
- address the weaknesses of health systems at both technical and functional levels, and
- ensure inclusion of populations that are most vulnerable in the delivery of services.

Disclaimer

L Say is a staff member of WHO. The author alone is responsible for the views expressed in this publication and they do not necessarily represent the decisions a policies of WHO.

See also: Abortion; Female Reproductive Function; Gynecological Morbidity; Sexual and Reproductive Health: Overview; Women's Mental Health Including in Relation to Sexuality and Reproduction.

Citations

Anachebe NF and Sutton MY (2003) Racial disparities in reproductive health outcomes. *American Journal of Obstetrics and Gynecology* 188: S37–S42.

Bewley S, Wolfe C, and Waterstone M (2002) Severe maternal morbidity in the UK. In: MacLean AB and Neilson JP (eds.) *Maternal Morbidity and Mortality*, pp. 132–146. London, United Kingdom: RCOG Press.

Brabin BJ and Verhoeff F (2002) The contribution of malaria. In: MacLean AB and Neilson JP (eds.) *Maternal Morbidity and Mortality*, pp. 64–78. London, United Kingdom: RCOG Press.

Brace V, Penney G, and Hall M (2004) Quantifying severe maternal morbidity: A Scottish population study. *BJOG: An International Journal of Obstetrics and Gynaecology* 111: 481–484.

Chang J, Elam-Evans LD, Berg CJ, *et al.* (2003) Pregnancy-related mortality surveillance: United States, 1991–1999. *Maternal Mortality and Morbidity Weekly Report* 52: 1–8.

Danel I, Berg CJ, Johnson JH, and Atrash H (2003) Magnitude of maternal morbidity during labor and delivery: United States, 1993–1997. *American Journal of Public Health* 93: 631–634.

Goyaux N, Leke N, Keita N, and Thonneau P (2003) Ectopic pregnancy in African developing countries. *Acta Obstetricia et Gynecologica Scandinavica* 82: 305–312.

Graham W, Fitzmaurice AE, Bell JS, and Cairns JA (2004) The familial technique for linking maternal death with poverty. *Lancet* 363(9402): 23–27.

Khan KS, Wojdyla D, Say L, Gulmezoglu AM, and Van Look PFA (2006) WHO analysis of causes of maternal death: A systematic review. *Lancet* 367(9516): 1066–1074.

Lewis G (ed.) (2001) *Why Mothers Die 1997–1999: The Confidential Enquiries Into Maternal Deaths in the United Kingdom.* London, United Kingdom: RCOG Press.

Lewis G (ed.) (2004) *Why Mothers Die 2000–2002: The Confidential Enquiries Into Maternal Deaths in the United Kingdom.* London, United Kingdom: RCOG Press.

Magpie Trial Collaborative Group (2002) Do women with preeclampsia, and their babies, benefit from magnesium sulphate? The Magpie Trial: A randomised placebo-controlled trial. *Lancet* 359: 1877–1890.

Mantel GD, Buchmann E, Rees H, and Pattinson RC (1998) Severe acute maternal morbidity: A pilot study of a definition for a near-miss. *British Journal of Obstetrics and Gynaecology* 105: 985–990.

Murray C, Goh JT, Fynes M, and Carey MP (2002) Urinary and faecal incontinence following delayed primary repair of obstetric genital fistula. *BJOG: An International Journal of Obstetrics and Gynaecology* 109: 828–832.

Pattinson RC (ed.) (2003) *Saving Mothers 1999–2001: Second Report on Confidential Enquiries into Maternal Deaths in South Africa.* Pretoria, South Africa: Government Printer.

Pattinson RC (ed.) (2006) *Saving Mothers 2002–2004: Third Report on Confidential Enquiries into Maternal Deaths in South Africa.* Pretoria, South Africa: Government Printer.

Pritchard JA, Baldwin RM, and Dickey JC (1962) Blood volume changes in pregnancy and the puerperium, 2. Red blood cell loss and changes in apparent blood volume during and following vaginal delivery, cesarean section and cesarean section plus total hysterectomy. *American Journal of Obstetrics and Gynecology* 84: 1271–1282.

Prual A, Bouvier-Colle MH, De Bernis L, and Breart G (2000) Severe maternal morbidity from direct obstetric causes in West Africa: Incidence and case fatality rates. *Bulletin of the World Health Organization* 78: 593–602.

Razum O, Jahn A, Blettner M, and Reitmaier P (1999) Trends in maternal mortality ratio among women of German and non-German nationality in West Germany 1980–1996. *International Journal of Epidemiology* 28: 919–924.

Say L, Pattinson RC, and Gulmezoglu AM (2004) WHO systematic review of maternal morbidity and mortality: The prevalence of severe acute maternal morbidity (near miss). *Reproductive Health* 1(1): 3.

Shadigian E and Bauer ST (2005) Pregnancy-associated death: A qualitative systematic review of homicide and suicide. *Obstetrical & Gynecological Survey* 60: 183–190.

van den Broek N (2002) The contribution of anaemia. In: MacLean AB and Neilson JP (eds.) *Maternal Morbidity and Mortality*, pp. 79–88. London, United Kingdom: RCOG Press.

Waterstone M, Bewley S, and Wolfe C (2001) Incidence and predictors of severe obstetric morbidity: Case-control study. *British Medical Journal* 322: 1089–1094.

WHO/UNICEF/UNFPA/The World Bank (2007) *Maternal Mortality in 2005: Estimates Developed by WHO, UNICEF, UNFPA and The World Bank.* Geneva, Switzerland: World Health Organization.

World Health Organization (1992) *International Statistical Classification of Diseases and Related Health Problems, Tenth Revision (ICD-10).* Geneva, Switzerland: World Health Organization.

World Health Organization (2004) *Beyond the Numbers: Reviewing Maternal Deaths and Complications to Make Pregnancy Safer.* Geneva, Switzerland: World Health Organization.

Further Reading

Critchley H, MacLean A, Allan PL and Walker J (eds.) (2003) *Preeclampsia.* London, United Kingdom: RCOG Press.

Cunningham FG, Leveno KJ, Bloom SL, *et al.* (eds.) (2005) *Williams Obstetrics,* 22nd edn. New York, NY: McGraw-Hill.

MacLean AB and Neilson JP (eds.) (2002) *Maternal Morbidity and Mortality.* London, United Kingdom: RCOG Press.

Sexual Violence

R Jewkes and L Dartnall, Medical Research Council, Pretoria, South Africa

Introduction

Sexual violence is endemic. More than one in five women report a lifetime experience of sexual assault from an intimate partner, while up to one-third of girls report their first sexual experience was forced (Jewkes *et al.*, 2002a). It is both a profound public health problem and human rights violation, while the short- and long-term consequences of it limit the potential of victims/survivors to achieve an optimum standard of health and well-being. Sexual violence violates multiple fundamental human rights, including the right to life, equality, liberty, security of persons, the right to be free from all forms of discrimination, and the right not to be subjected to torture or other cruel, inhuman, or degrading treatment or punishment. Despite this, it is one of the least researched of all forms of gender-based violence.

Sexual Violence: Definition

The World Report on Violence and Health defines sexual violence against women as:

> any sexual act, attempt to obtain a sexual act, unwanted sexual comments or advances, or acts to traffic, or otherwise directed against, women's sexuality, using coercion (i.e., psychological intimidation, physical force, or threats of harm), by any person regardless of relationship to the victim, in any setting, including but not limited to home and work. (Jewkes *et al.*, 2002a: 14a)

Sexual violence can include a range of acts, including rape in marriage and dating relationships; rape by strangers; rape as a weapon of war; sexual harassment (including demands of sex for jobs or school grades); sexual abuse of the mentally ill or impaired; sexual abuse of children; forced marriage or cohabitation, including marriage of children; sexual torture; and forced prostitution and trafficking in women for sexual exploitation. It also includes forced abortion; denial of the right to use contraception or protect self from disease; and acts of violence against women's sexuality such as female genital cutting and social virginity inspection. Sexual violence is predominantly experienced by women and girls, although it can be directed against men and boys.

Prevalence of Sexual Violence

Sexual Assault

Research on sexual violence is critical in highlighting the extent of the problem, and strengthening public health services and prevention programs. In no country has there been a study of the prevalence of perpetration of sexual assault of intimate partners and other women or girls in a randomly selected sample of men from the community. There are therefore many unanswered questions about patterns of perpetration; however, research from North America and South Africa seems to suggest that men who rape are very likely to do so first during their teenage years and that repeat perpetration is very common (White and Smith, 2004). In all countries, only a very small proportion of men who are sexually violent are ever punished by law. The extent to which convicted sex offenders resemble the total population of sexual violent men is not known.

Globally, there has been more research conducted with women documenting their experience as victims/survivors, and most of the data available describe the prevalence of this. Differences in measures, definitions, and social meanings as well as underreporting limit data accuracy. Sexual violence is particularly stigmatized. In some settings victims are ostracized, abandoned, and even killed, and very commonly lesser manifestations of stigma limit disclosure and help-seeking. These include shame, blame, denial, fear of the consequences, fear of retribution, and a sense of inevitability that disclosure will not be helpful. It is estimated that at most one in ten sexual assaults are reported to the police and in many settings the proportion is much lower. Despite these limitations, available evidence provides valuable insights into the epidemiology of sexual violence.

A global insight into the prevalence of sexual violence across the early and middle years of the life span is provided by the World Health Organization's (WHO) Multi-Country Study on Women's Health and Domestic Violence against Women (Garcia Moreno *et al.*, 2005). In the study, randomly selected samples of women aged 16–49 years from ten different countries, often from rural and urban areas, were interviewed to create a data set of more than 24 000 women. The prevalence findings are summarized in **Table 1**.

Sexual violence in childhood, involving inappropriate acts of exposure, or touching, of genitalia, as well as penetration, is very common. It is experienced by both girls (most commonly) and boys, but globally very much less is known about the prevalence, consequences, and context of abuse of boys than girls. Much of it occurs in homes, where male family members are often perpetrators, but children are also often sexually abused by respected older people in the community. The WHO study found a prevalence of child sexual abuse of between 1 and 20%.

In many countries, schools are also places where girls are at considerable risk. In developing countries, school-age girls may be forced by poverty into transactional sexual relationships with teachers or other older men in order to raise the money for school fees. Educational institutions are also settings in which sexual harassment and rape by teachers or fellow pupils occurs. In South Africa, the 1998 Demographic and Health Survey included questions about experience of rape before the age of 15 years and found that school teachers were the largest group of child rape perpetrators, responsible for 32% of the disclosed child rapes (Jewkes *et al.*, 2002b). Research in Africa describes high levels of apathy among officials about this, with a lack of information among pupils and parents and a reluctance to believe girls who make allegations. Teachers are generally unwilling to report colleagues' sexual misconduct and often see the behavior of boys as a normal part of growing up.

Research has increasingly shown that girls' first sexual intercourse is often forced, and the younger the age at which this occurs, the greater the likelihood of it being forced (**Table 1**). Forced first sex is also particularly common in settings where women's sexuality is more constrained and where sexual dominance by men is more strongly culturally sanctioned. The highest prevalence of first forced sex found in the WHO study, for example, was 30% in rural Bangladesh.

Intimate partner violence, usually defined as violence perpetrated by a current or ex-husband or boyfriend, encompasses both physically and sexually violent acts, which are usually located within a relationship that is characterized by a range of controlling and emotionally abusive dynamics. The prevalence of sexual intimate partner violence varies between settings and has been found to be as high as 44% (**Table 1**). In most settings, women are more at risk of being forced into unwanted sexual acts by an intimate partner than any other type of perpetrator.

Rape and attempted rape by acquaintances or strangers occurs in all settings with a prevalence of victimization as high as 12% (**Table 1**). Usually there is one perpetrator, but in some settings rape by more than one man commonly occurs and may even be ritualistic. A study from South Africa, where one-third of rapes reported to the police involve more than one perpetrator, found that 14% of young men disclosed in interviews having participated in group rape of a woman (Jewkes *et al.*, 2006). Throughout history, rape has been used as a weapon of war. In recent years, for example, between 23 200 to 45 600 Kosovar Albanian women are believed to have been raped in the 1998–99 conflict with Serbia, and very high rates of gang rape have been reported in the Democratic Republic of Congo, frequently resulting in genital fistulae.

Table 1 Prevalence of sexual coercion against women and children across the life cycle

	Sexual abuse in childhood before age 15 years (best estimate)[a]		*Forced first sex*		*Sexual violence by an intimate partner*			*Nonpartner sexual violence since age 15 years*	
Site	*N*	%	*N*	%	*N*	*Ever* %	*Current* %	*N*	%
Bangladesh, city	1602	7.4	1369	24.1	1373	37.4	20.2	1603	7.6
Bangladesh, province	1527	1.0	1326	29.9	1329	49.7	24.2	1527	0.5
Brazil, city	1172	11.6	1051	2.8	940	10.1	2.8	1172	6.8
Brazil, province	1473	8.7	1234	4.3	1188	14.3	5.6	1472	4.6
Ethiopia, province	3014	7.0	2238	16.6	2261	58.6	44.4	3016	0.3
Japan, city	1361	13.8	1116	0.4	1276	6.2	1.3	1368	3.5
Namibia, city	1492	21.3	1357	6.0	1367	16.5	9.1	1498	6.4
Peru, city	1414	19.5	1103	7.3	1086	22.5	7.1	1414	10.3
Peru, province	1837	18.1	1560	23.6	1534	46.7	22.9	1837	11.3
Samoa	1640	1.8	1317	8.1	1204	19.5	11.5	1640	10.6
Serbia and Montenegro, city	1453	4.2	1310	0.7	1189	6.3	1.1	1453	3.9
Thailand, city	1534	8.9	1051	3.6	1048	29.9	17.1	1534	6.1
Thailand, province	1280	4.9	1029	5.3	1024	28.9	15.6	1280	2.6
United Republic of Tanzania, city	1816	12.2	1557	14.3	1442	23.0	12.8	1816	11.5
United Republic of Tanzania, province	1443	9.5	1287	16.6	1256	30.7	18.3	1443	9.4

From Garcia-Moreno C, Jansen HEF, Ellsberg M, Heise L, and Watts CH (2005) *WHO Multi-country Study on Women's Health and Domestic Violence Against Women: Initial Results on Prevalence, Health Outcomes and Women's Responses.* Geneva: World Health Organization.

[a]In those sites where anonymous reporting was not linked to the individual questionnaire, the best estimate is the highest prevalence given by either of the two methods of data collection used in the multicountry study, i.e., face-to-face and anonymous report.

Sexual Trafficking

The *Report of the Special Rapporteur on the Human Rights Aspects of the Victims of Trafficking in Persons, Especially Women and Children* (Huda, 2006), defines trafficking in persons as:

> the recruitment, transportation, transfer, harbouring or receipt of persons, by means of the threat or use of force or other forms of coercion, of abduction, of fraud, of deception, of the abuse of power or of a position of vulnerability or of the giving or receiving of payments or benefits to achieve the consent of a person having control over another person, for the purpose of exploitation. Exploitation shall include, at a minimum, the exploitation of the prostitution of others or other forms of sexual exploitation, forced labour or services, slavery or practices similar to slavery, servitude or the removal of organs.

Evidence suggests that each year hundreds of thousands of women and girls throughout the world are bought and sold into prostitution and slavery. It is purportedly one of the fastest growing organized crimes.

Trafficked women and children are often trapped in slave-like conditions, with identification papers removed. In many cases, they may be led to believe that they have accepted domestic or waitressing work but they are taken to brothels, their passports confiscated, and they may be beaten, locked up and promised freedom only after earning their purchase price, travel and visa costs through sex work. This lucrative business has large profits and relatively few risks and has led to global networks of traffickers, often with mafia-style operations.

Harmful Traditional Practices

In many countries, there are traditional practices that are harmful to women and children, such as early and forced marriage, the offering of girls to temples to live a life of servitude often including prostitution, and giving girls to fetish shrines to serve as domestic and sexual slaves. Female genital cutting, which encompasses the removal of part, or all, of the external female genital tissue is estimated to effect 130 million girls worldwide and can lead to serious health problems. There has, however, been a shift in public opinion to traditional harmful practices, with an increasing recognition that they are violations of young girls' rights and integrity, which must be stopped.

Sexual Violence Against Men

Men and boys are recognized as potential victims of sexual assault, but globally there has been little research on the frequency with which this occurs and the contexts. It is recognized that both men and women may coerce young men and boys into sex, but those forcing older men are invariably male. There is some evidence that the meaning of acts of coercion of men may vary according to whether the perpetrator is a man or a woman, but coercion of young men into sex by women needs to be better understood and more widely recognized as a problem (Marston, 2005).

Circumstances in which coercion of men and boys occurs include child sexual abuse, forced sexual initiation, sexual harassment and coerced sex in schools, sexual harassment in families and from close relatives, and sexual violence experienced in police cells, prisons, during crime and during war. Rape of male prostitutes is also a well recognized and common problem (Jewkes *et al.*, 2002a). In some settings forced heterosexual initiation of boys is disclosed in surveys nearly as frequently as that of girls, with up to one-third of boys reporting this. However, generally the prevalence of coercion of boys and men is much lower than that reported for women (Jewkes *et al.*, 2002a). Sexual assault of men is also subject to underreporting, and this is particularly influenced by social expectations that normal men are heterosexual and always open to sex with women, and fears that a man may be considered gay if raped by a man.

Understanding Perpetration of Sexual Assault

Sexual violence is a subcategory of gender-based violence; in other words, it is inherently rooted in the lower power that women have in a society compared to men. Within societies there is a range of social expectations, behavioral norms, and values that stem from this, which vary between settings. They often include expectations that women and girls will be under the control of men; that men have sex as a right within marriage; that women should not have sexual desires or a right to sexual pleasure; that sexually aroused men cannot control themselves; and that women and girls are responsible for ensuring that they do not arouse inappropriate sexual desires in men. A nexus of power and control lies at the heart of explanations of why men rape. Rape is an act through which men communicate to their victim about their powerfulness, as well as one through which they may communicate with themselves, through the experience of power in the act of rape. Thus, rape may be an act of punishment and an expression of holding power over a victim, and it can also be an act through which perpetrators affirm to themselves a sense of power that they may not feel in other respects in their lives.

Globally, the base of research on risk factors for sexual assault perpetration has been narrow, and the great majority of what is known comes from North America and South Africa. Nonetheless, many of the findings from these settings point to considerable commonality in underlying factors. In North America, a multifactorial model known as the Confluence Model (Malamuth *et al.*, 1991) was developed to explain the interconnections among factors associated with sexual assault; it has been particularly influential in the field. The model has two paths through which factors are believed to operate to influence sexual assault perpetration and each has been empirically shown to independently predict perpetration, but also to work synergistically such that men who score highly on both paths are more likely to be sexually coercive.

The first path has been dubbed hostile masculinity and at the heart of this path are feelings of insecurity and defensiveness in relationships toward women, with sex being an act of power and dominance rather than love. Sexually violent men have been shown to be more likely to be hostile toward women, to perceive women as hostile to them, to mistrust women's affective expression, and thus be more likely to interpret assertiveness in women's interactions with them as hostility (Malamuth *et al.*, 1991; Abbey *et al.*, 2004a). The factors associated with or indicators of hostile masculinity are a high level of acceptance of interpersonal violence, relative isolation from women, and perceptions of sex as an act between adversaries. Sexually violent men, especially men who coerce sex more than once, are more likely to lack empathy for their victims, lack remorse (both features of psychopathy), and consider victims to be responsible for rape (Abbey *et al.*, 2004a).

The second path is the sexual promiscuity/impersonal sex path. At the heart of this are ideas of sex as a game to be won or for physical gratification, and not for emotional closeness or intimacy. This path builds on the observation that men are more likely to rape if they have experienced a range of trauma in childhood and this is seen to influence their later sexual practices and preferences. The experience of trauma in childhood is postulated to reduce the ability of men to form loving and nurturing attachments, and thus results in an orientation to impersonal sexual relationships rather than sex in the context of emotional bonding and short-term sex-seeking strategies. Malamuth *et al.* (1991) originally emphasized experience of childhood sexual abuse and witnessing violence between parents in childhood, but other authors have found an association with a wider range of experience of abuse (Knight and Sims-Knight, 2003; Jewkes *et al.*, 2006). Experience of trauma in childhood increases the likelihood of engagement in delinquent behavior, and rape is often perpetrated in the context of peer-related acts of delinquency. Sexually violent men are more likely to have more sexual partners than other men. Malamuth *et al.* (1991) argue that men who develop a relatively high emphasis on sexuality, particularly with sexual conquest, as a source of peer status and self-esteem, may use various means, including coercion, to induce girls into sexual acts. In this regard, peer pressure to have sex may also encourage some men to force sex (Jewkes *et al.*, 2006).

A number of rape researchers have argued that some important factors are missing from the Confluence Model. Alcohol plays a role in a high proportion of rapes (Abbey *et al.*, 2004b). It is a situational factor (Abbey *et al.*, 2004b) but may have a more complex role. Sexual assaults are very frequently associated with alcohol consumption. Alcohol has psychopharmacological effects of reducing inhibitions (like some drugs, notably cocaine), clouding judgment and enabling a greater focus on the short-term benefits of forced sex. It may also act as a cultural time-out for antisocial behavior. Thus, men are more likely to act violently when drunk because they do not feel they will be held accountable for their behavior (Abbey *et al.*, 2004b). Some forms of group sexual violence are associated with alcohol drinking and here drinking alcohol forms part of the group bonding with collectively reduced inhibitions and individual judgment ceded to that of the group.

The Confluence Model provides an understanding of factors associated with rape perpetration that overwhelmingly emphasizes individual characteristics. Research from outside the discipline of psychology suggests that there are several social factors that are important (social and economic status is one such factor), but unfortunately information on the association of these social factors with rape perpetration is limited. Some researchers (e.g., Bourgois, 1995) have argued that in the face of poverty and unemployment, men lack paths through which they may demonstrate success and may rape in an effort to reclaim, however transiently, a sense of powerfulness. Others have observed that in some research, men who rape have come from relatively more privileged backgrounds and were relatively more powerful than the women they raped (Duvvury *et al.*, 2002; Jewkes *et al.*, 2006). In these circumstances, their acts are better explained as stemming from an exaggerated sense of sexual entitlement. These explanations are not necessarily contradictory, because sexual assault occurs in a range of different contexts and may at different times have different roots.

A unifying idea between them is the location of sexual violence within a framework of ideas about manhood, which is developed and sustained not just at an individual, but also at a community or societal level. Socially constructed ideas about manhood are used by individual men in their self-evaluations of success and in development of their expectations and ideas about the appropriateness of particular approaches to and patterns of behavior. In South Africa, researchers have demonstrated a cluster of closely correlated male behaviors, including sexual and physical violence against women, having many sexual partners, transactional sex, and alcohol abuse, which link sexual conquest, risk taking, and violence with rape (Jewkes *et al.*, 2006). These are dimensions of a particular model of masculinity, which has been shown to change in response to an intervention that builds gender equity and ideas of sexual and social responsibility.

One of the community or societal level influences on sexual violence is the extent to which it is perceived to be an act that is socially stigmatized and liable to be punished. This is very closely linked to the status of women. Rape is one of the few acts of violence where the victims (usually women or girls) are very likely to be blamed equally or more severely than the perpetrator. This reflects the social position of women. In settings with greater gender equity, social understandings of sexual violence are more unequivocal, raping is stigmatized, and sanctions for perpetration are perceived as likely and feared. At an individual level, women and children are more likely to be victims of sexual violence if they are of lower social status overall, as well as with respect to men who rape. Men often perceive that they are less likely to face consequences if their victim is of lower social status or in some way socially devalued (for example, a mentally ill person or a sex worker). Rape by a nonpartner is often opportunistic and women and girls of lower socioeconomic status are more often in situations of vulnerability, for example, when walking on open land because they lack the resources to use transport or when living somewhere that can be more easily broken into.

Health Consequences of Sexual Assault

Sexual violence has a range of health consequences, both short-term and long-term. The most important immediate consequences are usually symptoms of psychological distress, experienced as shock, fear, and feelings of helplessness; hyperarousal caused by concerns for personal safety; and high levels of anxiety. This may be accompanied or followed by depression and, if prolonged, may develop into posttraumatic stress disorder (PTSD). Rape survivors not infrequently attempt suicide. In the longer term, especially where there has been repeated violations and where it has occurred in childhood, substance abuse is more common in sexual violence survivors. The psychological impact also includes fundamental changes in victims'/survivors' perception of self, which affect other areas of their lives, notably their relationships with men. Many victims/survivors struggle to have sexual relationships with men. As a group, they are more likely to engage in sexual risk taking including starting sexual relationships earlier, and having more partners, older partners, transactional sex and more unsafe sex. They are also more likely to have violent partners.

There are many important physical consequences of sexual violence. Pregnancy may occur in the absence of contraception and is particularly stressful if termination is not an option. Sexually transmitted infections may be acquired and, if untreated, can result in infertility, pelvic pain, and pelvic inflammatory disease. In high-prevalence settings, the transmission of HIV during sexual assault is a major concern for victims/survivors. Genital and other injuries may also occur in the course of sexual violence. Very violent rape, particularly gang rape, has been associated with very severe injuries including the development of genital fistulae and, if the woman is pregnant, pregnancy loss. The injuries associated with rape are at times fatal, these are known as rape homicides. In South Africa, rape is suspected in 16% of female homicides.

Ethics of Research on Sexual Violence

All studies face the challenge of ensuring that the research is of high quality, benefits are maximized, and the potential for doing harm through the research is kept to a minimum. The very sensitive nature of sexual violence poses a unique set of challenges, and a range of ethical and safety issues need to be considered and addressed prior to embarking on such research. In addition to the established codes of practice related to ethics in research, a number of specific issues need to be considered. The sensitive nature of the research makes it particularly important that every effort is made to ensure that benefits outweigh the possibility of harm, and for this reason it is important that the benefits and use of research on sexual violence are defined and assessed before research is commenced. The research must be methodologically sound and in every respect built on current experience and best practice. This is particularly important with respect to the approaches to data collection and questionnaire design. Where data are gathered from survivors of sexual violence, a provision should be made to make psychological support available to those disclosing violence. Research team members may experience vicarious trauma and they must be carefully selected and receive relevant and sufficient training and ongoing support. Both the research participants and the team may have their safety endangered by conducting research on sexual violence. Particular consideration needs to be given to ensuring confidentiality of the research, as well as of individuals included in studies.

Involving children in sexual violence research is important if we are to adequately understand their needs. However, special care and thought is needed beforehand; experts on research with children need to be consulted to ensure that methods used are appropriate and that all possible risks have been considered. Parents or care givers need to be consulted about the research and there may be mandatory reporting requirements that need to be followed if abuse is disclosed.

Research with men as perpetrators also entails a particular set of considerations. When conducted well, research on rape presents opportunities for men to confront the meaning of acts of sexual violence and this can play a constructive role in helping them to face up to what they have done and ultimately change their self-evaluation. However, there is a risk with less rigorous research of collusion with perpetrators of violence, which potentially could reinforce rape-supportive attitudes, or even encourage repeated acts of violence. Researchers may find the imperative of remaining neutral and detached very challenging in the face of accounts of perpetration of assault. Studies of rape perpetration need to be conducted in a manner that ensures that there is a balance between protecting the confidentiality of research participants while not suppressing information that could impact on the safety of others.

The World Health Organization has published ethics guidelines for research on domestic violence, with trafficked women and for research on sexual violence in emergencies. These documents are particularly useful resources for researchers planning work on sexual violence.

Responding to Sexual Assault in the Health Sector

Victims/survivors of sexual assault have a set of health needs that health services have to be in a position to address. Psychological support is of great importance and victims/survivors need to start to feel safe, regain a sense of control, and to recognize the psychological symptoms that they experience for what they are and to know that very often they are self-limiting. Women of reproductive age may require emergency contraception or termination of pregnancy. All victims/survivors need treatment for possible sexually transmitted infections, and in all but the lowest prevalence settings prophylaxis against HIV is also needed. Injuries sustained may require treatment and in many countries there is an established procedure for examinations and documentation of injuries for legal purposes. This is often accompanied by the collection of forensic evidence. Mental health services are needed to help victims/survivors who have longer-term mental health needs.

In most countries, health services for victims/survivors have been relatively neglected. Studies describing post-rape health services around the world highlighted important gaps in many countries (Christofides *et al.*, 2005). These include inequitable service provision, as many of the countries have first-rate services in only a small number of centers, with trained sensitive staff, good community participation, clinical protocols, psychological support, and good follow-up after the initial contact. There is a very substantial gap in most countries between the best and the normal (or worst) facilities and a need for great improvement to the majority of services. In many countries, basic clinical care after rape is deficient in important respects, particularly psychological support is often not provided and treatment for sexually transmitted infections is often inadequate. There is considerable variation between countries and within countries as to whether sexual violence patients are seen by designated service providers or generalist staff. The latter group are commonly almost completely unprepared for caring for sexual assault victims/survivors, because undergraduate medical and nursing courses generally devote very little, if any, time to teaching about gender and gender-based violence. Generalist staff may see rape victims/survivors infrequently, making it less likely that they would make special efforts to learn more about their care unless they were particularly motivated for other reasons. The attitudes of staff toward sexual assault victims/survivors have a very important impact on the quality of care provided (Christofides *et al.*, 2005). Staff often hold the same judgmental attitudes toward victims/survivors that are prevalent in the general population, and this poses a barrier to providing good care. Facilities for caring for victims/survivors after rape are often deficient in important respects, and the best services have a dedicated suite with washing facilities and a private waiting area.

In an attempt to improve the health sector response to sexual violence internationally, the World Health Organization has produced guidelines for medicolegal care for victims of sexual violence (World Health Organization, 2004). These are a valuable tool that can be used to improve health services by providing health-care providers with the knowledge and skills necessary to understand the needs and concerns of victims of sexual violence, provide quality health services, and carry out accurate and ethical collection of forensic evidence. These can most effectively be used within a framework of a national health sector policy on sexual violence.

Preventing Sexual Violence

Primary prevention of sexual violence is a substantially neglected area, but one that needs to be prioritized. There is a considerable need for research, particularly to understand perpetration, and to develop and evaluate approaches to prevent this. Prevention needs to be addressed both through individual-level interventions and community- or society-level interventions that seek to change social norms and the acceptability of sexual violence. There are two individual-level interventions that have been rigorously evaluated and shown to reduce perpetration. The first is the Safe Dates Programme (Foshee *et al.*, 2004), which was developed to address sexual and physical violence occurring in dating relationships. It was shown to be effective in a North Carolina (USA) school population in reducing sexual and physical violence in dating relationships 4 years after the intervention. The Stepping Stones Programme (Jewkes *et al.*, 2002c) is a gender transformative HIV prevention program that seeks to change risk taking and antisocial models of masculinity and through doing this promote sexual health as well as building gender equity. It has been shown to be effective in rural South African youth with an effect measured 2 years after intervention. These interventions show that it is indeed possible to change men's violent behavior. They require further study, adaptation, and testing in other settings. Many other interventions are used, and these need to be subjected to the same level of evaluation in order to create a knowledge base for evidence-based rape prevention.

Social norms related to the status of women and the acceptability of rape are also important. In this regard, general interventions to improve the status of women would be expected to impact on sexual violence. Legislation is needed to provide an appropriate legal framework for defining and responding to sexual violence, policy is needed in government departments, and governments need to show a political commitment to efforts to eradicate sexual violence. The responsiveness of police to victims when they report sexual assault, as well as that of health and other services, sends a powerful message about the way in which a society views the crime. This in turn will have an impact on general acceptability, and through this its prevalence. A key part of a comprehensive public health response to sexual violence necessarily involves ensuring that a society learns to speak with one voice in showing no tolerance for sexual violence.

See also: Abortion; Cultural Context of Reproductive Health; Gender Aspects of Sexual and Reproductive Health; Gynecological Morbidity; Violence Against Women; Women's Mental Health Including in Relation to Sexuality and Reproduction.

Citations

Abbey A and McAuslan P (2004a) A longitudinal examination of male college student's perpetration of sexual assault. *Journal of Consulting and Clinical Psychology* 72: 747–756.

Abbey A, Zawacki T, Buck PO, Clinton AM, and McAuslan P (2004b) Sexual assault and alcohol consumption: What do we know about their relationship and what types of research are still needed? *Aggression and Violent Behaviour* 9: 271–303.

Bourgois P (1995) In *Search of Respect: Selling Crack in El Barrio*. New York: Cambridge University Press.

Christofides N, Jewkes R, Webster N, Penn-Kekana L, Abrahams N, and Martin L (2005) "Other patients are really in need of medical attention": The quality of sexual assault services in South Africa. *Bulletin of the World Health Organization* 83: 495–502.

Duvvury N, Nayak MB, Allendorf K, *et al.* (2002) Links between masculinity and violence: Aggregate analysis. In: *Men, Masculinity and Domestic Violence in India.* Washington, DC: The International Center for Research on Women.

Foshee VA, Bauman KE, Ennett ST, Fletcher Linder G, Benefeld T, and Suchindran C (2004) Assessing the long term effects of the Safe Dates Programme and a booster effect in preventing and reducing adolescent dating violence victimisation and perpetration. *American Journal of Public Health* 94: 619–624.

Garcia-Moreno C, Jansen HEF, Ellsberg M, Heise L, and Watts CH (2005) *WHO Multi-country Study on Women's Health and Domestic Violence Against Women: Initial Results on Prevalence, Health Outcomes and Women's Responses*. Geneva, Switzerland: World Health Organization.

Huda S (2006) *UN Report of the UN Special Rapporteur on Trafficking: Report of the Special Rapporteur on the Human Rights Aspects of the Victims of Trafficking in Persons, Especially Women and Children*. http://www.glow-boell.de/media/en/txt_rubrik_2/SR_Trafficking_Report_to_UN_2007.pdf (accessed November 2007).

Jewkes R, Sen P, and Garcia-Moreno C (2002a) Sexual violence. In: Krug EG, *et al.* (eds.) *World Health Report on Violence and Health*, pp. 148–181. Geneva, Switzerland: World Health Organization.

Jewkes R, Levin J, Bradshaw D, and Mbananga N (2002b) Rape of girls in South Africa. *Lancet* 359: 319–320.

Jewkes R, Nduna M, and Jama N (2002c) *Stepping Stones*. A training manual for sexual and reproductive health, communication and relationship skills, 2nd edn. Pretoria, South Africa: Medical Research Council.

Jewkes R, Dunkle K, Koss MP, *et al.* (2006) Rape perpetration by young, rural South African men: Prevalence, patterns and risk factors. *Social Science and Medicine* 63: 2949–2961.

Knight RA and Sims-Knight JE (2003) The developmental antecedents of sexual coercion against women: Testing alternative hypotheses with structural equation modelling. *Annals of the New York Academy of Sciences* 989: 72–85.

Malamuth NM, Sockloskie RJ, Koss MP, and Tanaka JS (1991) Characteristics of aggressors against women: Testing a model using a national sample of college students. *Journal of Consulting and Clinical Psychology* 59: 670–681.

Marston C (2005) Pitfalls in the study of sexual coercion: What are we measuring and why? In: Jejeebhoy S, Shah I and Thapa S (eds.) *Sex Without Consent. Young People in Developing Countries.* London: Zed Press.

White JW and Smith PH (2004) Sexual assault perpetration and reperpetration: From adolescence to young adulthood. *Criminal Justice and Behaviour* 31: 182–202.

World Health Organization (2004) *Guidelines for Medico-legal Care for Victims of Sexual Violence*. Geneva, Switzerland: World Health Organization.

Further Reading

Bachar K and Koss MP (2001) Closing the gap between what we know about rape and what we do. In: Edleson J and Renzetti C (eds.) *Handbook of Research on Violence Against Women*, pp. 117–142. Newbury Park, CA: Sage.

Jejeebhoy S, Shah I, and Thapa S (eds.) (2005) *Sex Without Consent. Young People in Developing Countries*. London, UK: Zed Press.

Jewkes R and Abrahams N (2002) The epidemiology of rape and sexual coercion in South Africa: An overview. *Social Science and Medicine* 55: 153–166.

Koss MP (1993) Detecting the scope of rape. A review of prevalence research methods. *Journal of Interpersonal Violence* 8: 198–222.

Leach F (2003) *An Investigative Study of the Abuse of Girls in African Schools*. Department for International Development: Educational Papers No. 54, DFID, August 2003.

Malamuth N (2003) Criminal and non-criminal sexual aggressors. Integrating psychopathy in a hierarchical-mediational confluence model. *Annnals of the New York Academy of Sciences* 989: 33–58.

World Health Organization (2001) *Putting Women First: Ethical and Safety Guidelines for Research on Domestic Violence Against Women*. Geneva, Switzerland: World Health Organization (WHO/FCH/GWH/01.1. http://whqlibdoc.who.int/hq/2001/WHO_FCH_GWH_01.1.pdf).

World Health Organization (2003) *WHO Ethical and Safety Recommendations for Interviewing Trafficked Women*. Geneva, Switzerland: World Health Organization.

World Health Organization (2007) *WHO Ethical and Safety Recommendations for Researching, Documenting and Monitoring Sexual Violence in Emergencies*. Geneva, Switzerland: World Health Organization.

Relevant Websites

http://www.un.org/womenwatch/daw/vaw/SGstudyvaw.htm – The Secretary-General's in-depth study on all forms of violence against women.

http://www.svri.org – Sexual Violence Research Initiative.

http://www.who.int/gender/violence/who_multicountry_study/en/index.html – WHO multi-country study on women's health and domestic violence against women: Initial results on prevalence, health outcomes and women's responses, World Health Organization, Geneva.

Violence Against Women

C García-Moreno, World Health Organization, Geneva, Switzerland

Introduction

Violence against women is a major public health problem and an abuse of women's human rights. It is pervasive worldwide, although its prevalence varies between sites, as do the patterns and forms it takes. Violence against women is an important risk factor for women's health, resulting in a wide range of negative outcomes for women's health and well-being, including their sexual and reproductive health.

Violence against women can take many forms, including: physical, sexual, and emotional abuse by an intimate partner; rape and sexual assault whether by a partner, acquaintance, or stranger, or in the context of war; sexual abuse during childhood; trafficking for purposes of sex or forced labor; forced prostitution; female genital mutilation, child marriage, and other harmful traditional practices; and murders in the name of honor or related to dowry. Violence against women is also referred to as gender-based violence (GBV) because it is closely linked to gender inequality and to the social norms that perpetuate women's and girls' subordinate status in society.

This chapter focuses on intimate partner violence (also known as domestic violence), and sexual violence including during conflict and displacement, while also touching on child sexual abuse, trafficking of women and female genital mutilation, both because they are common forms of violence experienced by girls and women globally and because of their particular impact on sexual and reproductive health. While recognizing that the health consequences of violence are far-ranging and include, importantly, mental health, injuries, and other physical health problems, this chapter focuses on the sexual and reproductive health aspects.

How Widespread Is Violence Against Women?

A growing number of studies worldwide are documenting how common violence against women is. The majority of the population-based studies so far have focused on intimate partner violence, particularly physical and sexual abuse (few studies so far also include emotional abuse), and less so on rape (by all perpetrators) and other forms of sexual abuse. Trafficking of women, violence during conflict and war, and other forms of violence against women remain understudied and not so well documented.

Intimate Partner Violence

A review (to 1999) of population-based studies from around the world found that between 10 and 69% of women reported being physically abused by an intimate male partner at some point in their life (Heise and Garcia-Moreno, 2002). This violence is usually accompanied by sexual and emotional violence, and studies are beginning to collect data on these other forms of abuse. **Table 1** summarizes existing data on the prevalence of physical violence by partners.

Until recently, data on this problem, while valid, have been difficult to compare across countries due to methodological differences, for example, in sample size, measurement of violence, age, and characteristics of those interviewed.

The World Health Organization's (WHO's) Multi-Country Study on Women's Health and Domestic Violence was designed to document the magnitude and nature of violence that women experienced, with a focus on intimate partner violence (its risk and protective factors, association with health outcomes, and women's responses to such violence), with comparable methods across countries. Over 24 000 women were interviewed in 15 sites in 10 countries: Bangladesh, Brazil, Ethiopia, Japan, Namibia, Peru, Samoa, Serbia, Thailand, and the United Republic of Tanzania (comparable data are now also available from Equatorial Guinea, the Maldives, and New Zealand). The study found that the lifetime prevalence of physical intimate partner violence was between 13 and 61%, with most sites reporting between 23 and 49%. The lifetime prevalence of intimate partner sexual violence was between 6 and 59%. Overall, between 15 and 71% of women reported physical or sexual violence, or both, in their lifetime (Garcia-Moreno *et al.*, 2006). Emotional abuse was also measured, asking about the presence and frequency of acts such as being insulted or made to feel bad, belittled, or humiliated in front of others, or threats to hurt someone you loved. Controlling behaviors were also measured and included: keeping a woman from seeing her friends, restricting contact with her family, insisting on knowing where she is at all times, expected to ask permission to seek health care, etc. Although there is less agreement about what constitutes emotional abuse, making it hard to determine its prevalence, women often report this as a most disempowering and devastating aspect of abuse by an intimate partner. Controlling behaviors by an intimate partner in the WHO study were

Table 1 Prevalence of physical assaults on women by a male partner[a]

Country or area	*Year of study*	*Region covered*	*Sample size*	*Study population*	*Age (years)*	*Proportion of women physically assaulted by a partner last 12 months*	*Ever*
Africa							
Ethiopia	2002	Meskanena Woreda	2261	III	15–49	29	49
Kenya	1984–87	Kisii District	612	V	>15	42[b]	
Malawi[c]	2005	National	3546			30	
Namibia	2003	Windhoek	1367	III	15–49	16	31
South Africa	1998	Eastern Cape	396	III	18–49	11	27
	1998	Mpumalanga	419	III	18–49	12	28
	1998	Northern Province	464	III	18–49	5	19
	1998	National	10 190	II	15–49	6	13
Uganda	1995–96	Lira and Masaka	1660	II	20–44		41[b]
United Republic of Tanzania	2002	Dar es Salaam	1442	III	15–49	15	33
	2002	Mbeya	1256	III	15–49	19	47
Zambia	2001–02	National	3792	III	15–49	27	49
Zimbabwe	1996	Midlands Province	966	I	>18	17[d]	
Latin America and the Caribbean							
Barbados	1990	National	264	I	20–45		30[e,f]
Brazil	2001	Sao Paulo	940	III	15–49	8	27
	2001	Pernambuco	1188	III	15–49	13	35
Chile	1993	Santiago Province	1000	II	22–55		26[b]
	1997	Santiago	310	II	15–49	23	
	2004[g]	Santa Rosa	422	IV	15–49	4	25
Colombia	1995	National	6097	II	15–49		19[b]
	2000	National	7602	III	15–49	3	44
Dominican Republic	2002	National	6807	III	15–49	11	22
Ecuador	1995	National	11 657	II	15–49	12	
El Salvador	2002	National	10 689	III	15–49	6	20[b]
Guatemala	2002	National	6595	VI	15–49	9	
Honduras	2001	National	6827	VI	15–49	6	10
Haiti	2000	National	2347	III	15–49	21	29
Mexico	1996	Guadalajara	650	III	>15	27	
	1996[g]	Monterrey	1064	III	>15	17	
	2003	National	34 184	II	>15	9	
Nicaragua	1995	Leon	360	III	15–49	27	52
	1997	Managua	378	III	15–49	33	69
	1998	National	8507	III	15–49	13	30
Paraguay	1995–96	National	5940	III	15–49	10	
	2004	National	5070	III	15–44	7	19
Peru	2000	National	17 369	III	15–49	2	42
	2001	Lima	1019	III	15–49	17	50
	\2001	Cusco	1497	III	15–49	25	62
Puerto Rico	1995–96	National	4755	III	15–49		13[h]
Uruguay	1997	National	545	IIi	22–55		10[f]

Continued

Table 1 Continued

Country or area	Year of study	Region covered	Sample size	Study population	Age (years)	Proportion of women physically assaulted by a partner last 12 months	Ever
North America							
Canada	1993	National	12 300	I	>18	3[d,f]	29[d,f]
	1999	National	8356	III	>15	3	8[j]
USA	1995–96	National	8000	I	>18	1[e]	22[e]
Asia and Western Pacific							
Australia	1996	National	6300	I		3[a]	8[b,d]
	2002–03	National	6438	III	18–69	3	31
Bangladesh	1992	National (villages)	1225	II	<50	19	47
	1993	Two rural regions	10 368	II	15–49	42[b]	
	2003	Dhaka	1373	III	15–49	19	40
	2003	Matlab	1329	III	15–49	16	42
Cambodia	1996	Six regions	1374	III	15–49	16	
	2000	National	2403	III	15–49	15	18
China	1999–00	National	1665	II	20–64	15	
India	1998–99	National	90 303	III	15–49	10	19
	1999	Six states	9938	III	15–49	14	40
	2004[g]	Lucknow	506	IV	15–49	25	35
	2004[g]	Trivandrum	700	IV	15–49	20	43
	2004[g]	Vellore	716	IV	15–49	16	31
Indonesia	2000	Central Java	765	IV	15–49	2	11
Japan	2001	Yokohama	1276	III	18–49	3	13
Maldives	2006	National	1732	III	15–49	6	18
New Zealand	2002	Auckland	1309	III	18–64	5	30
	2002	North Waikato	1360	III	18–64	34	
Papua New Guinea	2002	National, rural villages	628	IIIi		67	
Philippines	1993	National	8481	IV	15–49	10	
	1998	Cagayan de Oro City and Bukidnon	1660	II	15–49	26	
	2004[g]	Paco	1000	IV	15–49	6	21
Republic of Korea	2004	National	5916	II	20 –	13.2	20.7
Samoa	2000	National	1204	III	15–49	18	41
Tajikistan[k]	2005	Khatlon region	400	I	17–49	19	36
Thailand	2002	Bangkok	1048	III	15–49	8	23
	2002	Nakonsawan	1024	III	15–49	13	34
Viet Nam	2004	Ha Tay province	1090	III	15–60	14	25
Europe							
Albania	2002	National	4049	III	15–44	5	8
Azerbaijan	2001	National	5533	III	15–44	8	20
Finland	1997	National	4955	I	18–74	7	
Finland	2005	National	4464	I	18–74	6	
France	2002	National	5908	II	>18	3	9[l]
Georgia	1999	National	5694	III	15–44	2	5

Germany	2003	National	10 264	III	16–85	23[d]	
Lithuania	1999	National	1010	II	18–74		42[b,d,m]
Netherlands	1986	National	989	I	20–60		21[e]
Norway	1989	Tronheim	111	III	20–49	18	
Norway	2003	National	2143	III	20–56		627
Republic of Moldova	1997	National	4790	III	15–44	8	15
Romania	1999	National	5322	III	15–44	10	29
Russian Federation	2000	Three provinces	5482	III	15–44	7	22
Former Yugoslav Republics of Serbia and Montenegro	2003	Belgrade	1189	III	15–49	3	23
Sweden	2000	National	5868	III	18–64	4[h]	18[h]
Switzerland	1994–96	National	1500	II	20–60	6[f]	21[f]
	2003	National	1882	III	>18	10	
Turkey	1998	East/South-East Anatolia	599	I	14–75		58[e]
Ukraine	1999	National	5596	III	15–44	7	19
United Kingdom of Great Britain and Northern Ireland	1993[g]	North London	430	I	>16	12[e]	30[e]
	2001	National	12 226	I	16–59	3	19[n]
Eastern Mediterranean							
Egypt	1995–96	National	7123	III	15–49	13	34
	2004[j]	El-Sheik Zayed	631	IV	15–49	11	11
Israel	1997	Arab population	1826	II	19–67	32	
West Bank and Gaza Strip	1994	Palestinian population	2410	II	17–65	52	

Key to: *Study population*
I: All women
II: Currently married/partnered women
III: Ever married/partnered women
IV: Women with a pregnancy outcome
V: Married women; half with pregnancy outcome, half without
VI: Women who had a partner within the last 12 months

[a]Source for all countries or areas, unless noted is adapted from Ellsberg M and Heise L (2005) *Researching Violence Against Women: A Practical Guide for Researchers and Activists.* Washington, DC: PATH/WHO; United Nations (2006) World report on violence and children. *Secretary-General's Report.* New York: United Nations.
[b]During current relationship.
[c]Pelser E, Gondwre L, Mayamba C, Mhango T, Phiri W, and Burton P (2005) Intimate partner violence: Results from a national gender-based study in Malawi. Crime and Justice Statistical Division, National Statistical Office.
[d]Although sample included all women, rate of abuse is shown for ever married/partnered women (number not given).
[e]Sample group included women who had never been in a relationship and therefore were not in exposed group.
[f]Physical or sexual assault.
[g]Publication date (field work dates not reported).
[h]Rate of partner abuse among ever married/partnered women recalculated from author's data.
[i]Nonrandom sampling methods used.
[j]Within the last five years.
[k]Haar RN (2005) Violence against women in marriage: A general population study in Khatlon Oblast, Tajikistan, baseline survey conducted by the NGO Social Development Group. In Secretary General's study on violence against women, United Nations document A/61/122/Add.1. Tajikistan: United Nations.
[l]Since the age of 18.
[m]Includes threats.
[n]Since the age of 16.

found to be associated with perpetration of physical and sexual abuse. The study confirmed that there is wide variation in prevalence both between and within countries. The difference is particularly striking when looking at violence in the past year, with women in developing countries generally having a higher prevalence.

Intimate Partner Violence during Pregnancy

Partner violence often persists during pregnancy, with negative consequences for both maternal and infant health. Studies from the USA, Canada, and Europe have found prevalence of violence during pregnancy to be between 3.4 and 11%. Studies from developing countries have found that 4 to 32% of women reported have been subjected to physical violence in pregnancy (Campbell *et al.*, 2004). In the WHO Multi-Country Study, the prevalence of physical abuse during pregnancy, among ever-pregnant women, ranged from 4 to 12% in most sites. Abuse during pregnancy often involves blows to the abdomen, which may have serious consequences for both the mother and the infant. Overall, it would appear that this violence is a continuation of ongoing violence, with a small but varying percentage of women, depending on the site, reporting that this abuse started during the pregnancy. The evidence suggests that, in some settings, pregnancy may offer protection, with violence decreasing during this time, while in others violence may increase (or start) as a result of pregnancy.

Sexual Violence, Including during Conflict and Displacement

Sexual violence is a global problem that until recently has remained hidden. It happens predominantly to women and girls, but boys and men are also sexually assaulted. The lifetime risk of attempted or completed rape is up to 20% for women (Jewkes *et al.*, 2002). Legal definitions may vary, but rape is usually defined as the nonconsensual penetration of the vagina, mouth, or anus, by a penis. When an object other than the penis is used, the term "assault" is usually employed.

There is increasing concern with the violence that women and children, primarily girls, experience during conflict and displacement. While exact estimates of the magnitude of such violence are difficult to determine, it has been documented in Bosnia, Colombia, Darfur in Sudan, the Democratic Republic of Congo, Kosovo, and Rwanda, to name a few places (McGinn, 2001; Amnesty International, 2004). Abductions, sexual servitude, and violent rape by armed actors have also been reported. In situations of conflict and displacement, women may be exposed to rape and sexual abuse during the flight, on arrival and in camps, and after the conflict due to increased societal disruption and presence of weapons. Services are difficult to find in these situations, making things even more difficult, and women may be forced to trade sex for food or money.

Child Sexual Abuse and Forced First Sex

Women and girls are most at risk of sexual violence from people they know, whether partners or other family members, boyfriends, neighbors, acquaintances, and, less frequently, strangers. Precise estimates are difficult to give since sexual violence, particularly during childhood, remains a highly stigmatized and taboo subject in most societies. However, studies from around the world show that approximately 20% of women and 5 to 10% of men report having been sexually abused as children (World Health Organization and International Society for Prevention of Child Abuse and Neglect, 2006). In the WHO Multi-Country Study the most commonly reported perpetrators of this were family members, particularly male family members other than fathers and stepfathers, although strangers were also an important category. Abuse during childhood has been found to be associated with abuse in later life (Fergusson *et al.*, 1997; Coid *et al.*, 2001). It is also associated with many unhealthy outcomes, particularly behavioral and psychological problems, low self-esteem and depression, and with high, sexual risk-taking behaviors, such as increased number of partners and increased use of alcohol and other substances.

Forced sexual initiation is also a common occurrence. The WHO Multi-Country Study confirmed that a substantial proportion of young women reported their first experience of sexual intercourse as coerced or forced, which is consistent with studies from other countries, such as Uganda (Koenig *et al.*, 2004) and Ghana (Glover *et al.*, 2003). This was more likely to be the case the younger the reported age of the first sexual encounter (**Figure 1**). Coerced sex has been linked with lower use of modern contraception and of condom at last intercourse, more unwanted pregnancies, and more genital tract symptoms among young girls in Uganda (Koenig *et al.*, 2004).

Trafficking of Women

This form of violence is hard to document, particularly as it is an illegal practice, often carried out by bands of organized crime. Several organizations collect data on human trafficking, and while there is no agreement on what is the best estimate, there is agreement that it affects hundreds of thousands of people, particularly women and children, who are trafficked across borders in many parts of the world. Often this is for purposes of sex and prostitution, and these women are at increased risk of violence, sexually transmitted infections (STIs), and mental health problems (Zimmerman, 2003).

Female Genital Mutilation

Other forms of violence against women include harmful practices such as female genital mutilation (FGM). FGM is a global concern (World Health Organization, 2008). WHO estimates that about 100–140 million women have

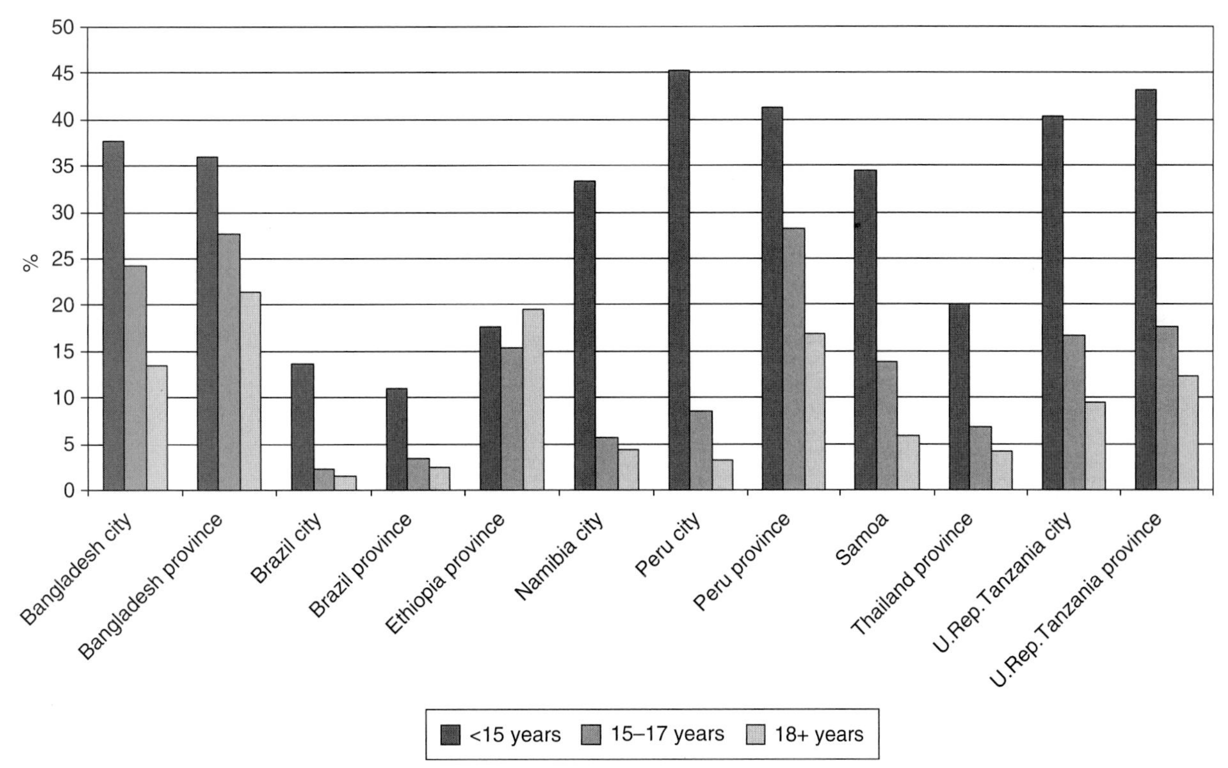

Figure 1 Percentage of women reporting forced first sexual intercourse by site and by age at time of first sexual experience. Reproduced from Garcia-Moreno C, *et al.* (2005) *WHO Multi-Country Study on Women's Health and Domestic Violence. Initial Results on Prevalence, Health Outcomes and Women's Responses.* Geneva, Switzerland: World Health Organization.

been subjected to FGM in 28 countries in Africa as well as among immigrants in Australia, New Zealand, Canada, Europe, and the USA. It appears that FGM is also practiced in some countries of Asia, especially among certain populations in India, Indonesia, and Malaysia. The practice is being reported in the Middle East, particularly in Northern Saudi Arabia, Southern Jordan, Iraq, and Yemen. It has been estimated that approximately 3 million girls are mutilated each year (Yoder *et al.*, 2004). The prevalence of FGM varies from country to country, and also varies between different ethnic groups within each country. For example, the prevalence is above 90% in Djibouti, Egypt, Guinea-Conakry, Mali, and Somalia, while the prevalence is only 5% in Niger and Uganda. The most severe form of FGM, Type III, involves excision of the labia minora and labia majora and suturing of the vaginal opening (with only a small opening left for urinating and menstruating). Since FGM procedures are generally carried out under unhygienic conditions, they commonly result in short- and long-term complications and sequelae.

The Sexual and Reproductive Health Consequences of Violence

Violence against women is associated with a wide range of negative health outcomes (Resnick *et al.*, 1997; Plichta and Falik, 2001; Campbell *et al.*, 2002), including injuries, mental health problems, and adverse effects on sexual and reproductive health (**Figure 2**). The latter include: unwanted pregnancies and STIs, including HIV/AIDS, gynecological problems (Wijma *et al.*, 2003), and abortion (Holmes *et al.*, 1996). Genital fistulae can also result from violent rape, and this is particularly common in some conflict settings.

There are both direct and indirect pathways leading to sexual and reproductive ill-health. Rape and sexual assault, for example, can directly result in unwanted pregnancy and STIs, including HIV. In addition, violence and fear of violence make it difficult to use contraception and to negotiate condom use and safe sex, thereby also leading to unwanted pregnancy and STIs. Sexual abuse during childhood has been associated with high-risk sexual behavior during adolescence and later in life, including an increased number of partners and early and unprotected sex, and use of alcohol and drugs – all factors that are associated with a higher risk of HIV infection.

Violence against women is associated with HIV and AIDS in a variety of ways. Violence by an intimate partner and fear of violence affect the opportunities for women to protect themselves and to request safer sex practices including condom use, and can also act as a barrier to HIV testing. Rape by an infected person can be responsible for HIV transmission or lead to other STIs and tears and lacerations, which increase the likelihood of HIV infection.

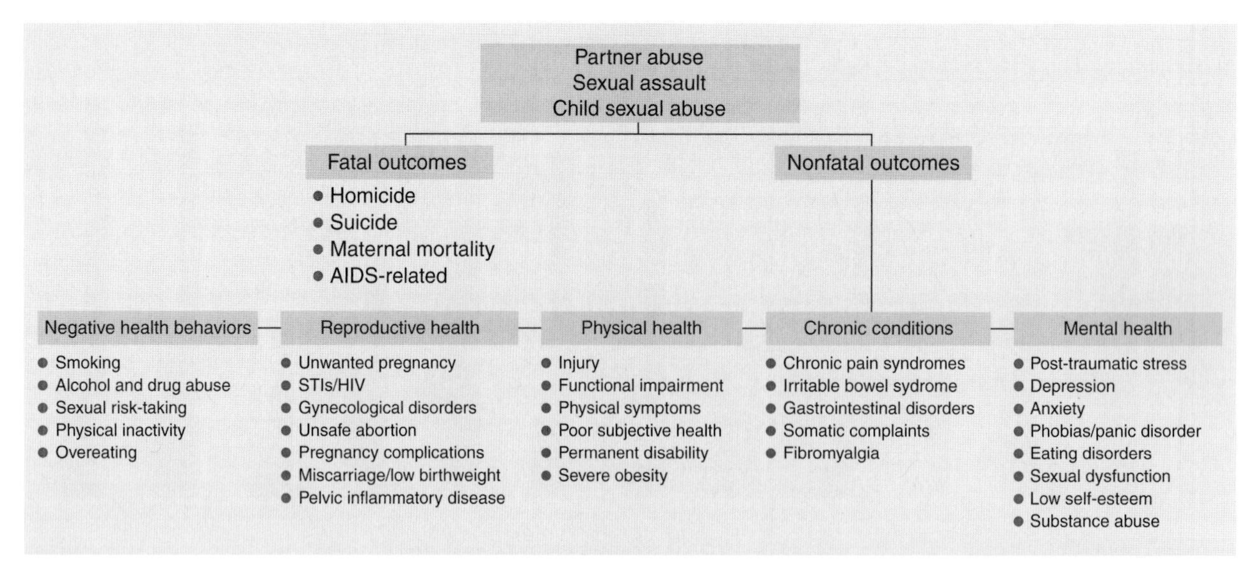

Figure 2 Health outcomes found to be associated with violence against women. Reproduced from Heise L, *et al.* (1999) *Ending violence against women*. Population Reports Series L, No. 11. Baltimore, MD: Johns Hopkins University School of Public Health.

Violence also interferes with women's ability to access care and treatment. Lastly, violence can be an outcome of taking an HIV test and of disclosure of a positive serostatus (Maman *et al.*, 2000; Dunkle *et al.*, 2004).

Intimate partner abuse often persists during pregnancy (although, as stated previously, for some women this may be a protected time during which the violence is reduced). Abuse during pregnancy has been associated with premature delivery, second and third trimester bleeding, low birthweight, risk behaviors such as smoking and substance abuse during pregnancy, and late entry into prenatal care.

FGM, particularly the more severe form, is associated with recurrent urinary tract infections, dyspareunia (pain during sexual intercourse), and genital ulcers. A recent study in six African countries found that, compared to women without FGM, women who had Type III genital mutilation were significantly more likely to experience cesarean section, postpartum hemorrhage, and prolonged hospitalization after delivery. Babies of women with FGM were more likely to require resuscitation and to be stillborn or suffer neonatal death (WHO Study Group on Female Genital Mutilation and Obstetric Outcome, 2006).

Responding to Violence Against Women

Preventing violence from occurring in the first place (i.e., primary prevention) is a public health priority, and the health sector can play an important role in gathering evidence and advocating for this. However, early childhood interventions (community-based and school-based), the media, and other approaches to challenge social norms and promote behavior change among men, may be better suited for this (**Figure 3**).

Health-care services, particularly for sexual and reproductive health, have an important role to play in secondary and tertiary prevention by identifying women who are suffering intimate partner violence as early as possible and contributing to prevent its recurrence and mitigate its effects on women's health (and lives) and that of their children. Most women come into contact with sexual and reproductive health services at some point in their lives, either for family planning, postabortion care, antenatal care, postpartum care, or treatment for STIs, and these contacts provide an opportunity for early identification and referral. These opportunities are, however, often lost because of health providers' lack of training, lack of time, and fear of offending women, among other constraints. Wijma *et al.* (2003), for example, documented that despite the high prevalence of physical, sexual, and emotional abuse among women attending gynecological clinics in five Nordic countries, most victims of abuse were not identified by their gynecologists. This may mean that abused women do not get the care they need. Since violence affects both women's health and the relevance and effectiveness of the care received, it is important that health-care providers understand and identify the problem as they may knowingly or unknowingly impair women's ability to deal with this violence, offer inappropriate care, or put women at risk.

Health-care providers need to help women in abusive relationships to assess their risk and do a safety plan, and ensure that they have access to other services as needed. They also need to document the information in ways that can be used in court if a woman desires to pursue this option, while maintaining confidentiality and privacy. Similarly, with rape and sexual assault, there is a need to ensure that any provider can provide at least the initial management, including treatment of injuries, preservation of forensic evidence, prevention of unwanted pregnancies and STIs,

Figure 3 Examples of information/advocacy campaigns against violence against women. Copyright (C) 2007 Men Can Stop Rape, Inc. Photography by Lotte Hansen.

referral for pregnancy termination where legal, and psychosocial support (World Health Organization, 2003).

More needs to be done to educate physicians, nurses, midwives, and other primary health-care providers on gender equality and equity issues and on violence, whether this is a focus of their work or not. The most promising approaches in this regard are those that use a systems approach that goes beyond training individual providers. Profamilia, the family planning association in the Dominican Republic, provides an example of such an approach, in which attention to gender-based violence was systematically integrated throughout all of the organization's services and at all levels (Population Council, 2006). Typically, these programs address all elements of care including both clinical and psychosocial support and establish partnerships with nongovernmental organizations or other service providers to ensure referral is possible. Others have attempted to provide all types of service in one location, usually in a hospital setting, as is the case with the One Stop Centers in Malaysia and other countries.

Programs need to be adapted to the specific context and the realities of health systems in different parts of the world. Everywhere, sexual and reproductive health-care providers must rise to the challenge of responding to women's needs. Recognizing how common violence against women is and its impact on women's health and lives is an important first step.

Conclusion

Violence against women is an important determinant of women's sexual and reproductive health and a public health concern in all parts of the world. Health-care providers, particularly those who care for women, need to understand the nature of the problem, its dynamics and impact on the health of women and that of their families, and what they can do to mitigate its consequences and provide the care and support that women need.

Disclaimer

C García-Moreno is a staff member of WHO. The author alone is responsible for the views expressed in this publication and they do not necessarily represent the decisions a policies of WHO.

See also: Abortion; Gender Aspects of Sexual and Reproductive Health; Gynecological Morbidity; Reproductive Rights; Sexual Violence; Women's Mental Health Including in Relation to Sexuality and Reproduction.

Citations

Amnesty International (2004) *Lives blown apart: Crimes against women in times of conflict: Stop violence against women*. ACT 77/075/2004. http://web.amnesty.org/library/Index/ENGACT770752004 (accessed January 2008).

Campbell J, Garcia-Moreno C, and Sharps P (2004) Abuse during pregnancy in industrialized and developing countries. *Violence against Women* 10: 770–789.

Campbell J, Jones AS, Dienemann J, *et al.* (2002) Intimate partner violence and physical health consequences. *Archives of Internal Medicine* 162: 1157–1163.

Coid A, Petruckevitch A, Feder G, *et al.* (2001) Relation between childhood sexual and physical abuse and risk of revictimisation in women: A cross-sectional survey. *Lancet* 358: 450–454.

Dunkle K, Jewkes R, Brown HC, *et al.* (2004) Gender-based violence, relationship power and risk of HIV infection among women attending antenatal clinics in South Africa. *Lancet* 363: 1415–1421.

Ellsberg M and Heise L (2005) *Researching Violence against Women: A Practical Guide for Researchers and Activists*. Washington, DC: PATH/WHO.

Fergusson DM, Horwood LJ, and Lynskey MT (1997) Childhood sexual abuse, adolescent sexual behaviours and sexual revictimization. *Child Abuse and Neglect* 21: 789–803.

Garcia-Moreno C, Jansen HAFM, Ellsberg M, Heise L, and Watts C (2005) *WHO Multi-Country Study on Women's Health and Domestic Violence. Initial Results on Prevalence, Health Outcomes and Women's Responses*. Geneva, Switzerland: World Health Organization.

Garcia-Moreno C, Jansen HFAM, Ellsberg M, Heise L, and Watts C (2006) Prevalence of intimate partner violence: Findings from the WHO multi-country study on women's health and domestic violence. *Lancet* 368: 1260–1269.

Glover EK, Bannerman A, Pence BW, *et al.* (2003) Sexual health experiences of adolescents in three Ghanaian towns. *International Family Planning Perspectives* 29: 32–40.

Heise L, Ellsberg M, and Gottemuller M (1999) *Ending violence against women*. Population Reports Series L, No. 11. Baltimore, MD: Johns Hopkins University School of Public Health.

Heise L and Garcia-Moreno C (2002) Intimate partner violence. In: Krug EG, Dahlberg LL, Mercy JA, Zwi AB, and Lozano R (eds.) *World Report on Violence and Health*, pp. 87–121. Geneva, Switzerland: World Health Organization.

Holmes MM, Resnick HS, Kilpatrick DG, and Best CL (1996) Rape-related pregnancy: Estimates and descriptive characteristics from a national sample of women. *American Journal of Obstetrics and Gynecology* 175: 320–324.

Jewkes R, Sen P, and Garcia-Moreno C (2002) Sexual violence. In: Krug EG, Dahlberg LL, Mercy JA, Zwi AB and Lozano R (eds.) *World Report on Violence and Health*, pp. 149–181. Geneva, Switzerland: World Health Organization.

Koenig MA, Zablotska I, Lutalo T, *et al.* (2004) Coerced first intercourse and reproductive health among adolescent women in Rakai, Uganda. *International Family Planning Perspectives* 30: 156–163.

Maman S, Campbell J, Sweat DC, and Gielen CA (2000) The intersections of HIV and violence: Directions for future research and interventions. *Social Science and Medicine* 50: 459–478.

McGinn T (2001) Reproductive health of war-affected populations: What do we know? *International Family Perspectives* 26: 174–180.

Pelser E, Gondwre L, Mayamba C, Mhango T, Phiri W, and Burton P (2005) *Intimate partner violence: Results from a national gender-based study in Malawi.* Crime and Justice Statistical Division, National Statistical Office.

Plichta SB and Falik M (2001) Prevalence of violence and its implications for women's health. *Women's Health Issues* 11: 244–258.

Population Council (2006) *Living up to their name: Profamilia takes on gender-based violence (Quality/Calidad/Qualite)*. New York, NY: Population Council.

Resnick HS, Acierno R, and Kilpatrick DG (1997) Health impact of interpersonal violence. 2: Medical and mental health outcomes. *Behavioural Medicine* 23: 65–78.

United Nations (2006) *Ending Violence against Women. From Words to Action. Study of the Secretary-General*. New York, NY: United Nations.

Wijma B, Schei B, Swahnberg K, *et al.* (2003) Emotional, physical and sexual abuse in patients visiting gynaecology clinics: A Nordic cross-sectional study. *Lancet* 361: 2107–2113.

World Health Organization (2003) *Guidelines for Medico-legal Care of Victims of Sexual Violence*. Geneva, Switzerland: World Health Organization.

World Health Organization (2008) *Interagency statement on the elimination of female genital mutilation*. Geneva, Switzerland: World Health Organization.

World Health Organization and International Society for Prevention of Child Abuse and Neglect (2006) *Preventing Child Maltreatment: A Guide to Taking Action and Generating Evidence*. Geneva, Switzerland: World Health Organization and International Society for Prevention of Child Abuse and Neglect.

WHO Study Group on Female Genital Mutilation and Obstetric Outcome (2006) Female genital mutilation and obstetric outcome: WHO collaborative prospective study in six African countries. *Lancet* 367: 1835–1841.

Yoder PS, Abderrahim N, and Zhuzhuni A (2004) *Female Genital Cutting in the Demographic and Health Surveys: A Critical and Comparative Analysis*. Calverton, MD: ORC Macro.

Zimmerman C (2003) *The Health Risks and Consequences of Trafficking in Women and Adolescents: Findings from a European Study*. London, United Kingdom: London School of Hygiene and Tropical Medicine.

Further Reading

Ellsberg M (2005) Violence against women and the Millennium Development Goals: Facilitating women's access to support. *International Journal of Gynecology and Obstetrics* 94: 325–332.

Guedes A (2004) *Addressing Gender-based Violence from the Reproductive Health/HIV Sector. A Literature Review and Analysis*. Washington, DC: The Population Technical Assistance Project (POPTECH Publication Number 04–164–020). http://www.igwg.org (accessed January 2008).

Murphy CC, Schei B, Myhr TL, and Dumont J (2001) Abuse: A risk factor for low birth weight? A systematic review and meta-analysis. *Canadian Medical Association* 164: 1567–1572.

Ramsay J, Rivas C, and Feder G (2005) *Interventions to Reduce Violence and Promote the Physical and Psychosocial Well-being of Women who Experience Partner Violence: A Systematic Review of Controlled Evaluations*. Final Report. London, United Kingdom: Barts and the London Queen Mary's School of Dentistry.

Spitz AM, Goodwin MM, Koenig L, *et al.* (2000) Special Issue: Violence and reproductive health. *Maternal and Child Health Journal* 4(2): 77–154.

United Nations (2006) *World report on violence and children. Secretary-General's Report.* New York, NY: United Nations.

World Health Organization (2005) *Addressing violence against women and achieving the Millennium Development Goals*. Geneva, Switzerland: World Health Organization.

Relevant Websites

http://www.ippfwhr.org/programs/program_gbv_e.asp – International Planned Parenthood Federation, Western Hemisphere Region (IPPF/WHR).
http://www.rhrc.org – Reproductive Health Response in Conflict Consortium.
http://www.stoprapenow.org – Stop Rape Now, UN Action on Sexual Violence in Conflict.
http://www.un.org/womenwatch/daw/vaw/SGstudyvaw – UN Division for the Advancement of Women, Violence against Women.
http://www.who.int/gender – WHO Department of Gender, Women and Health (GWH).
http://www.who.int/violence_injury_prevention – WHO Department of Violence and Injury Prevention and Disability (VIP).

Fetal Growth Retardation: Causes and Outcomes

S Sayers, Menzies School of Health Research Charles Darwin University, Darwin, Northern Territory, Australia
P A L Lancaster, Australian Health Policy Institute, University of Sydney, New South Wales, Australia

Introduction

Ideally, fetal growth retardation should be one of the main national indicators of pregnancy outcome, together with maternal, fetal, neonatal, and infant death rates. Instead, because information about the duration of pregnancy is not available in many developing countries, low birthweight is recommended as the key outcome by the World Health Organization (WHO).

Among the estimated 130–140 million births in the world each year, about 20 million infants (15%) weigh less than 2500 g (**Table 1**). In developed countries, low birthweight infants are more likely to be born preterm (less than 37 weeks' gestation) rather than growth retarded but in developing countries low birthweight is predominantly due to fetal growth retardation.

The likelihood that a significant proportion of chronic diseases in adult life, such as diabetes, hypertension, and coronary heart disease, in both developed and developing countries may be due to fetal growth retardation has given new impetus to improving our knowledge of its causes and prevention. The possible mechanisms whereby derangements of physiological and metabolic pathways cause fetal growth retardation have been extensively studied for many decades. Observations of variations in size at birth around the world, the impact on pregnant women of 'natural experiments' such as the siege of Leningrad and the Dutch famine in the Second World War, and the outcomes of growth-retarded fetuses after birth have all stimulated further epidemiological, clinical, and laboratory studies of fetal growth retardation. The rapid increase in birthweight by several hundred grams in Japan in the decades after the Second World War provided strong evidence that a significant change in maternal risk factors was responsible. Improved living conditions and better maternal nutrition seemed the most likely reasons.

Although more than 90% of the world's births occur in developing countries, the proportion of low birthweight is not evenly distributed across these regions. About half the total births are in South Asia, mainly India and neighboring countries, and in East Asia and the Pacific. However, low birthweight is four times more common in South Asia (29%) than in East Asia (7%) (**Table 1**). Across Africa and the Middle East, where there are more than a third of

Table 1 Estimates of low birthweight in regions and countries of the world

Country/region	*Estimated annual number of births (thousands)*	*Percentage of infants with low birthweight (less than 2500 g)*
Industrialized countries	10 848	7
Developing countries	120 128	16
Least developed countries	28 258	19
World	133 449	15
Sub-Saharan Africa	28 715	14
Eastern and Southern Africa	13 575	13
West and Central Africa	15 140	15
Middle East and North Africa	9743	15
South Asia	37 077	29
East Asia and Pacific	29 820	7
Latin America and Caribbean	11 651	9
CEE/CIS	5595	9

Reproduced from The United Nations Fund for Children (UNICEF) (2006) *The state of the world's children 2007. Women and children. The double dividend of gender equality.* New York, NY: UNICEF (derived from the published statistical tables)

the world's births, 13–15% are low birthweight, which is double that in the industrialized countries and East Asia. Within individual countries, many studies have consistently shown that mothers in higher socioeconomic groups are less likely to have an infant of low birthweight.

Concepts about the Relationship between Birthweight and Gestational Age

Gestational age indicates the duration of pregnancy, dated from the first day of the last menstrual period, whereas birthweight is the final achieved weight of the infant on completion of the pregnancy. Estimates of gestational age may be doubtful in a considerable proportion of pregnancies, either because of uncertainty about the date of the last menstrual period, abnormalities of fetal growth during pregnancy, or because mothers attend late or not at all for their antenatal care. Although gestational age increases continuously throughout pregnancy, by convention, gestational age is frequently referred to as being in the first, second, or third trimester rather than reported as completed single weeks.

While birthweight is readily measured, assessing gestational age accurately has proved to be much more difficult. It was not until the early twentieth century that the Finnish pediatrician, Arvo Ylppö, first proposed that small infants at birth should be designated as premature, based on a birthweight of 2500 g or less (Ylppö, 1919).

Initially, fetal growth during pregnancy was assessed indirectly by serial measurements of maternal weight, fundal height of the uterus, and abdominal girth. Various maternal hormonal and biochemical parameters, and fetal X-rays, were used to assess fetal maturity and well-being.

It was only in the 1960s that direct visualization of the fetus became possible when the evolving technique of clinical ultrasound was first used on pregnant women. Measurements of the biparietal diameter of the fetal skull were made in large numbers of pregnant women whose menstrual cycles had been regular, of normal duration, and in which the date of the last menstrual period was reliably known. These measurements at successive stages of pregnancy were then used to establish population values against which gestational age could be reliably assessed. As ultrasound could be safely used in early pregnancy, serial measurements could then be made at intervals to determine whether the pattern of fetal growth was normal or otherwise.

Birthweight percentile charts were introduced to assess whether an individual infant with known birthweight and gestational age was appropriately grown (Lubchenco *et al.*, 1976). Terms such as small-for-dates and large-for-dates became widely used to describe infants whose size at birth deviated from the normal value, especially those who were deemed to be growth-retarded.

Concurrently, it was increasingly recognized that low birthweight/premature infants are not a homogeneous group but rather a group in which some are born preterm and others are low birthweight because of poor fetal growth (Gruenwald, 1974).

For those who are less familiar with concepts related to growth from conception to birth at term some 270 days later, a better understanding of fetal growth can be gained by comparing this with the features of the postnatal growth of newborn infants from birth to adult size and maturity. Here, an individual's progress can be more readily observed directly and changes in height, weight, head circumference, and other indices can be measured serially, then compared with population normal values. Unless these concepts are well understood, the mainly unseen growth of the fetus remains mysterious to many!

As in postnatal life, a change in growth velocity or a drop to a lower percentile during intrauterine growth often suggests an adverse outcome. For example, a fetus consistently on the 75th percentile in serial measurements during pregnancy may be adversely affected by the onset of maternal hypertension in the third trimester of pregnancy and have a much reduced birthweight, say, on the 50th percentile. Such an infant will not usually be considered as growth-retarded because the birthweight is well above the cut-off weight designating small-for-gestational age. Yet this infant will be liable to some of the adverse outcomes of other more obviously growth-retarded infants.

The severity of the eventual outcome is influenced by the duration in pregnancy of the underlying cause and by the degree of fetal distress during the peripartum period, which can reflect a compromised supply of essential nutrients to the fetus.

Terminology and Definitions

No other topic in perinatal medicine has been so bedeviled by confusing terminology and definitions as small size at birth and fetal growth retardation. Similar problems are confronted in defining hypertensive disease of pregnancy, when terms such as pregnancy-induced hypertension, toxemia of pregnancy, preeclampsia, preexisting or gestational hypertension are often used interchangeably for the same condition. Many clinicians, nurses, and other health professionals tend to cling tenaciously to terms first used during their period of training.

WHO defines newborn infants with low birthweight as the percentage of liveborn infants with birthweight less than 2500 g in a given time period. Low birthweight may be subdivided into very low birthweight (less than 1500 g) and extremely low birthweight (less than 1000 g).

Preterm (less than 37 weeks' gestation), term (37–41 completed weeks). and post-term (42 weeks or more) describe the duration of pregnancy. In practice, the statistical definitions of the lower limits for gestational age and birthweight are often modified by some legal requirement within each country. The terms prematurity and post-maturity have been superseded by these more concise, less ambiguous definitions. Dysmaturity, which described an infant's abnormal appearance at birth and suggested a compromised nutrient supply-line *in utero*, has also been discarded.

Accurate information on fetal growth retardation is difficult to obtain in populations in which gestational age is not reliably known. Fetal growth during pregnancy is an ongoing, dynamic process and it is preferable to be specific that it is the fetus that is affected rather than using the more nebulous term intrauterine growth retardation. 'Fetal growth restriction' seems to imply some external, uterine cause. 'Intrauterine malnutrition' may often be the true underlying cause of fetal growth retardation, and emphasizes an important preventable cause, especially in developing countries, but other causes may be apparent after birth. Fetal growth retardation may have an onset early in pregnancy, for example, in fetuses with multiple malformations or congenital infections, or become apparent in the later stages of pregnancy, especially in twins and in other multiple pregnancies with more than two fetuses.

'Small- or large-for-gestational age' are terms preferred for infants who deviate most from normal birthweight. These are often abbreviated in clinical practice to the easier-to-say small- or large-for-dates. Selecting which percentile should be used to designate these extremes has proved to be problematical. Obviously, birthweight and other parameters of growth may be due to genetic factors or to some pathological cause. As these measurements broadly have a normal distribution around mean values, some infants in the lower percentile range are small because their parents, usually the mother, have small stature. In a statistical sense, the distribution of percentiles around the median value at each week of gestational age, rather than around the mean, is more appropriate. Birthweights do not have a normal distribution about the mean because relatively more infants are growth-retarded than are large-for-gestational age. In general, the greater the deviation from the median or mean, the more likely is a severe nutritional or other pathological cause. Nevertheless, some genetically small infants have birthweights in the lower percentiles but they often will have a low risk of complications associated with fetal growth retardation.

Birth Weight for Gestational Age Reference Standards

Accurate data sets needed to generate birthweight–gestational age percentile charts are difficult to produce and maintain. Just as there has been much debate about terms and definitions, there is often also disagreement about the most appropriate standards to use. National or regional standards based on nutritionally deprived mothers and infants fail to highlight such preventable problems (Lancaster, 1989).

Although birthweight is a simple and easy measurement, the quality of birthweight measurements may be affected by the time interval from birth. Weight should be measured preferably within the first hour of life before significant postnatal weight loss has occurred. Many observers show digit preference by recording birthweight rounded to 0 or the nearest 100 g value. National statistics are usually obtained from the birthweight data of viable newborns born in hospital and community clinics. However, in developing countries many babies are born at home and sometimes the proportion of babies weighed at birth is low. Hence, available data are likely to come from health institutions servicing a small privileged minority. There is lack of international uniformity in assessing signs of life and viability and, apart from birth size, local factors such as place of delivery, medical costs, maternity benefits, and burial costs may affect the decision to define a birth as liveborn.

Accurate gestational age estimations are essential, particularly if the common misclassification of small growth-retarded babies as preterm babies is to be avoided. Gestational age is estimated by a variety of methods according to the resources available. In developed populations, the estimated gestational age is based on maternal recall of the last menstrual period (LMP), often in association with an early fetal ultrasound measurement. In areas of limited resources, LMP recall and fundal height measurements are more likely to be used. Postnatal clinical examinations are used in the absence of or to verify estimations derived from antenatal sources of information. Nevertheless, gestational age estimations are not infrequently unreliable. Large data sets with adequate numbers to produce percentiles in each gestational age interval are likely to have gestational age misclassifications resulting in major biases. Smaller, more accurate hospital-based data sets may fail to have adequate numbers in the lower gestational age intervals and require supplementation from other data sets. Furthermore, to remain representative of the population, it is recommended that data sets should be updated at least every 10 years due to changing populations, lifestyles, and living conditions. Understandably, populations with limited resources may not be able to produce, analyze, and maintain large, accurate representative data sets.

As most data contributing to representative data sets are collected after birth, all curves up to 36 weeks of gestational age are, by necessity, taken from abnormal material. These problems argue for the use of birthweight percentile charts on the basis of both ultrasonography and birthweight data.

In many countries, a sizeable proportion of infants with a birthweight between the 3rd and 10th percentile do not have a pathological cause and may be small for familial reasons. Fetal growth retardation is then best defined as a birthweight below the 3rd percentile for each week of gestational age, recognizing that others may use the 5th or 10th percentile, or sometimes 2 standard deviations below the mean. In practice, especially in developing countries where fetal growth retardation is much more common, using the 10th percentile as the cut-off is more appropriate. This approach is based on assumptions that the 10th percentile is a valid cut-off point for identifying infants at higher risk and that the increased risk is constant over different gestational age intervals. However, for some fetal growth charts, the 10th percentile cut-off in all gestational age intervals underestimates higher risk while others overestimate it (Boulet *et al.*, 2006). More complex attempts to develop birthweight for gestational age standards have included three-dimensional contour lines of birthweight–gestational age distributions by race, sex, and metropolitan status (Hoffman *et al.*, 1974) and individualized growth assessments based on predicted second-trimester growth trajectories. Customized birthweight charts improve identification of fetal growth retardation associated with an increased risk of stillbirth and neonatal death but by their nature have limited application.

Separate percentile charts have been recommended for subpopulations if there is a unique and intrinsic advantage such as perinatal survival rates associated with the gestational age–specific categories. As girls have better survival than boys at a given weight for gestation, separate charts according to gender are biologically justified. However, there is little justification for separate charts based on racial/ethnic differences likely to be secondary to socioeconomic and lifestyle behaviors rather than fundamental biological differences (Roberts and Lancaster, 1999) An example of a birthweight for gestational age percentile chart is shown in **Figure 1**.

Causes of Fetal Growth Retardation

Traditionally, the causes of fetal growth retardation are subdivided into maternal, placental, and fetal (**Table 2**). Those of maternal and placental origin relate to insufficient gas exchange or delivery of nutrients to the fetus to allow it to thrive *in utero.* Causes affecting the fetus are familial, genetic, and chromosomal abnormalities, or

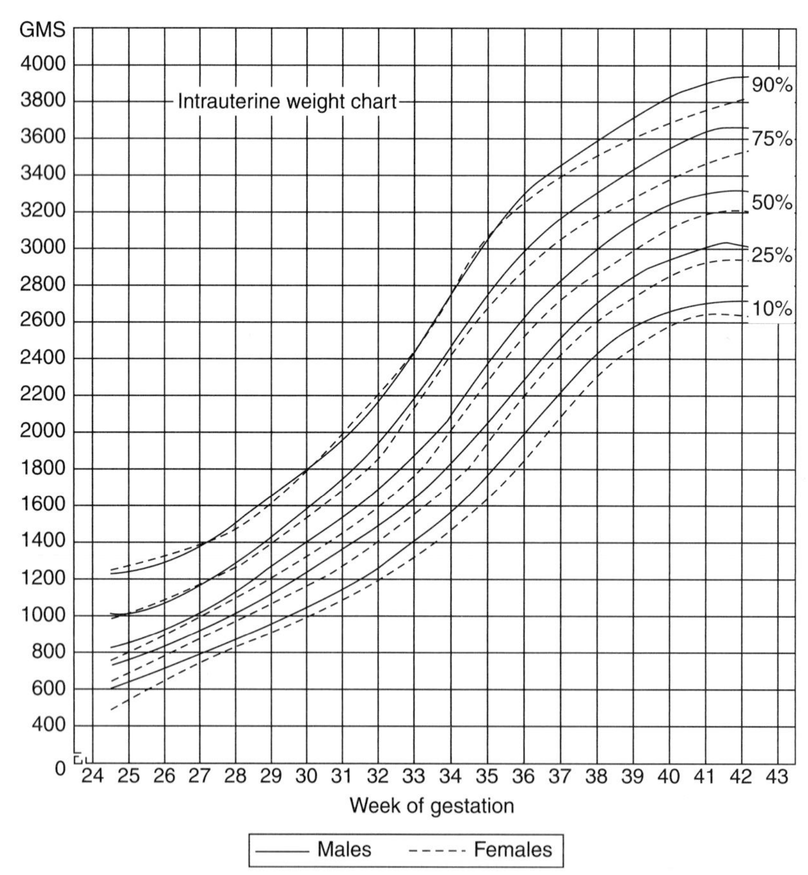

Figure 1 An example of a birthweight for gestational age percentile chart. The weights of liveborn Caucasian infants at gestational ages 24–42 weeks graphed as percentiles and published for the first time. Reproduced from Lubchenco L, Hansman LO, Dressler C, and Boyd M (1963) Intrauterine growth as estimated from liveborn birth-weight data 24–42 weeks of gestation. *Pediatrics* 32: 793–800.

Table 2 Maternal, placental, and fetal factors related to fetal growth retardation

Maternal	*Placental*	*Fetal*
Undernutrition: low pre-pregnant weight, short stature, poor gestational weight gain	Abnormal uteroplacental vasculature: abnormal trophoblast invasion, multiple infarcts, abruptio placentae	Congenital: Trisomy 21, Trisomy 18, Trisomy 13, Turner's syndrome, various chromosomal deletions
Malaria	Malaria	
Substance abuse and drug ingestion: smoking, alcohol, cocaine, phenytoin, heroin, warfarin	Damage due to thrombophilias: homocysteinuria	Infections: syphilis, cytomegalovirus, toxoplasmosis, rubella, hepatitis B, herpes simplex virus, human immunodeficiency virus
Hypertensive conditions: chronic hypertension, pregnancy-induced hypertension, preeclampsia, eclampsia	Chronic villitis	Multiple pregnancies
Chronic maternal disease: diabetes mellitus, renal disease, hypoxic lung disease, cyanotic heart disease, hemoglobinopathies	Placental mesenchyme dysplasia	
Prior history of fetal growth retardation: sibling, maternal		
High altitude		

Modified from Lin C and Santoloaya J (1998) Currents concepts of fetal growth restriction: Part 1. Causes, classification and pathophysiology. *Obstetrics and Gynecology* 92: 1044–1055.

congenital infections. For approximately 30% of cases of fetal growth retardation, no cause can be identified.

Maternal poverty and malnutrition are the prime causes of growth retardation in the fetus. If retarded fetal growth is exacerbated by poor living conditions, lack of maternal education, and inadequate nutrition in childhood, successive generations are likely to be affected, making it more difficult to interrupt this vicious cycle. When these underlying causes are widely recognized, the economic benefits of reducing low birthweight in developing countries will be unarguably clearer (Alderman and Behrman, 2006).

The major cause of fetal growth retardation in developing populations is maternal undernutrition. This may occur in combination or singly as low pre-pregnant weight due to inadequate nutritional status before conception, short stature primarily due to undernutrition and infection during childhood, and low gestational weight gain (GWG) due to inadequate diet during pregnancy.

The classical natural experiment on the effects of maternal undernutrition on birthweight relate to the Dutch famine of 1945 when a strict food embargo occurred during the last 6 months of World War II. The restriction of caloric intake of pregnant women to approximately 30% of the recommended daily allowance was associated with an 8–9% decrease in birthweights of babies (Smith, 1947).

However, it is the combination of a low pre-pregnant weight and a poor gestational weight gain that confers the greatest risk of delivering a growth-retarded baby, so for undernourished women in drought or other times of nutritional stress the effects of low pre-pregnant weight are even greater. Even for well-nourished women the calculated relative risk for fetal growth retardation is approximately two times higher if gestational weight gain is less than 7 kg (Kramer, 1987).

Malaria is a major determinant of fetal growth retardation in developing populations where there is endemic disease. In sub-Saharan Africa, one in four women has evidence of placental infection at the time of delivery. In areas of high malaria transmission, malaria-associated low birthweight approximately doubles if women have placental malaria, the greatest effect occurring in primigravid women.

In developed populations, cigarette smoking is a major cause of fetal growth retardation, with smoking increasing the risk up to 3.5-fold. Smoking effects are dose-related and may be dependent on the time of smoking in pregnancy, with effects increasing with maternal age. Other factors that, like smoking, affect oxygen-carrying capacity include hemoglobinopathies, maternal cyanotic heart disease, and severe maternal hypoxic lung disease.

In developed populations, hypertensive conditions such as pregnancy-induced hypertension, particularly if associated with proteinuria and/or preeclampsia, and chronic hypertension, are commonly responsible for fetal growth retardation. The mechanisms are unclear but high salt intake, low dietary calcium, and changes in the renin-angiotensin system have all been implicated.

Other maternal causes of fetal growth retardation include the pregnancies of growing adolescents in whom nutrients are used preferentially to promote maternal growth at the expense of the fetus. High-altitude living, associated with diminished oxygen saturation and maternal circulatory adjustments, limits fetal growth.

Fetal growth retardation is up to 10 times more frequent in twin deliveries because these and other multiple pregnancies can be affected by overcrowding, poor placental implantation, and twin-to-twin transfusion. The risk of fetal growth retardation is increased with primiparity and grandmultiparity and, in the absence of other specific pregnancy complications, there is a strong tendency for fetal growth retardation to be repeated in subsequent births. Intergenerational effects also occur with mothers born with fetal growth retardation more likely to have fetal-growth-retarded babies.

Alcohol consumption affects fetal growth and, separate from the fetal alcohol syndrome abnormalities, heavy maternal alcohol consumption, cocaine, methamphetamine or heroin use, and medications such as warfarin and phenytoin have all been linked to fetal growth retardation.

Abnormalities of placental structure and function are major causes of fetal growth retardation in developed populations. The placentae of fetal-growth-retarded births are frequently smaller, thinner, and lighter. Disease processes may be due to absent or incomplete uteroplacental vascular adaptation with increased placental resistance. Chronic inflammatory lesions of the placental villi may be present and vascular blockage producing placental infarcts may occur. Increased susceptibility to blood coagulation due to maternal blood-clotting disorders, such as factor V Leiden mutation and hyperhomocysteinuria, may contribute to thrombus and infarct pathology. When there is an accumulation of placental abnormalities leading to placental insufficiency over a length of time, fetal growth retardation is the likely outcome.

The fetal origins of fetal growth retardation relate to genetic and infective causes. Disorders due to an increased number of chromosomes (e.g., trisomy 21 – Down syndrome, trisomy 13, and trisomy 18) and chromosomal reductions (e.g., Turner's syndrome) and deletions (e.g., Cri du Chat syndrome) are all associated with fetal growth retardation. Single gene effects are rare but autosomal recessive metabolic diseases are consistently reported to be associated with smaller babies. Polygenic effects may be expressed through a wide spectrum of congenital malformations associated with fetal growth retardation.

The congenital infections of rubella, toxoplasmosis, cytomegalovirus, herpes simplex virus, varicella-zoster, and human immunodeficiency virus have all been implicated with fetal growth retardation. The mechanisms of reduced growth are thought to be due to infection of the vascular endothelium and limitation of cell multiplication.

Like so many other aspects of human pregnancy and care of the newborn, much has been learned about fetal growth by testing hypotheses experimentally in other species. For example, in the classic studies of the 1930s, heavy, powerful shire horses were crossbred with much smaller Shetland ponies. It was shown that maternal size was a crucial factor in limiting the size of the fetus at birth, thereby enabling easier pelvic descent and avoiding disproportion and obstructed labor and delivery (Walton and Hammond, 1938). This was subsequently confirmed by indirect studies in humans and became known as 'maternal constraint' of fetal growth (Ounsted and Ounsted, 1973).

Consequences of Fetal Growth Retardation

The disadvantages of fetal growth retardation continue over the entire life course. In early life there are increased fetal, neonatal, and infant death rates. There is poor postnatal growth and impaired neurodevelopment and immune function during childhood. Growing epidemiological evidence suggests that fetal responses to a deprived intrauterine environment may underlie the development of many chronic adult diseases like type 2 diabetes, hypertension, and cardiovascular disease.

Fetal growth retardation is associated with higher stillbirth rates, neonatal mortality and morbidity, and infant mortality rates in preterm, term, and post-term infants (Yu and Upadhyay, 2004). Compared to appropriately grown neonates in the neonatal period, fetal-growth-retarded babies are more prone to develop asphyxia and, as a result, meconium aspiration syndrome is more common. Many growth retarded infants have a characteristic appearance soon after birth of dry peeling skin and deficient subcutaneous tissue. Hypothermia (low temperature) is likely in these babies because of the greater surface area relative to weight and depleted subcutaneous fat and brown fat stores. Up to 25% may develop hypoglycemia (low blood sugar), partly due to diminished glycogen stores and poor adipose tissue. Hypocalcemia (low serum calcium levels) is more common. In approximately 15% of fetal-growth-retarded babies, polycythemia (abnormally large number of red blood cells in the circulatory system) occurs in response to chronic intrauterine hypoxia and this condition adds to the hypoglycemia and hyperbilirubinemia seen in fetal growth retardation. Asphyxia also results in a diversion of blood flow from mesenteric vessels to the brain and, together with the hyperviscosity accompanying polycythemia, contributes to the development of necrotizing enterocolitis (**Figure 2**).

Growth-retarded children have a high incidence of failure to thrive. Many fetal-growth-retarded babies, particularly those born at term, experience some catch-up growth. However, prospective studies that carefully define fetal growth retardation indicate that despite partial catch-up growth during the first 1 to 2 years of life fetal-growth-retarded babies continue to have a smaller body size compared with children who were not growth retarded in fetal life. By age 17–19 years, studies from both developed and developing populations show fetal-growth-retarded subjects about 5 cm less in height and 5 kg less in weight than those who were not fetal-growth-retarded (Sayers *et al.*, 2007). Babies born both fetal-growth-retarded and

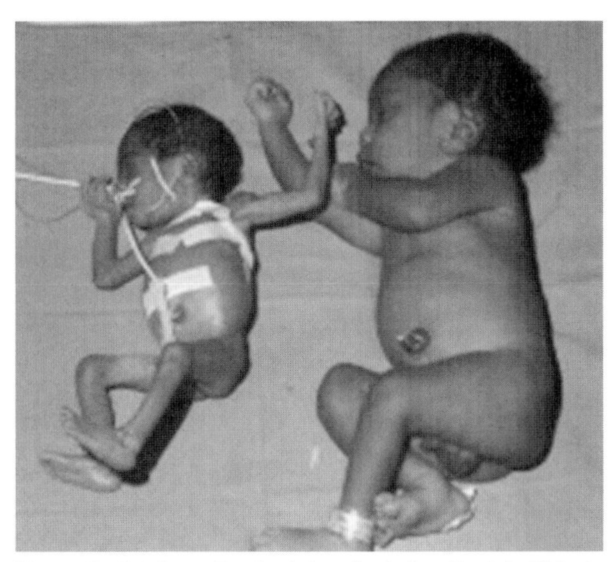

Figure 2 Fetal growth retardation. Australian Aboriginal infants of the same gestational age: the smaller baby is severely growth retarded.

preterm appear to have limited or no catch-up during the first years. Also, while studies are not conclusive, there is a trend for babies with symmetrical fetal growth retardation (i.e., the whole body is proportionally smaller) to have poorer growth outcomes compared to asymmetrical fetal growth retardation.

Compared to children who were appropriately grown babies, children who were fetal-growth-retarded exhibit more neurodevelopmental impairments including cerebral palsy, cognitive defects, clumsiness, behavioral problems associated with attention deficit disorders, and hyperactivity (Yanney and Marlow, 2004). School failure is more likely among fetal-growth-retarded children.

Impairment of immunological functions has been reported in a small number of studies, showing evidence of poor lymphocyte function and decreased IgG (immunoglobulin) production. More extensive studies are needed to confirm that fetal growth retardation lowers immunity and resistance to infection.

There is now increasing epidemiological evidence of the relationship of poor intrauterine growth (mostly represented by low birthweight) to later chronic adult disease associated with premature adult mortality. Initial observations linking lower birthweights in old birth records in Hertfordshire, UK, to adult coronary heart disease mortality have lead to numerous studies in different populations relating lower birthweights to later coronary heart disease, hypertension, type 2 diabetes, insulin resistance, and dyslipidemia (Barker, 2004).

The concept of early life events affecting later health and disease is now well established but the mechanisms involved are not clear. In a nutritionally deprived environment survival is enhanced by limited growth of the fetus accompanied by protection of brain growth. This occurs through dynamic physiological processes involving enzyme systems and endocrine pathways with alterations of blood supply from abdominal and musculoskeletal systems to the brain. These adaptations may lead to permanent structural and physiological effects.

Developmental plasticity (the phenomenon by which one genotype can give rise to a range of different physiological or morphological states in response to different environmental conditions during development), the thrifty phenotype, and predictive adaptations to postnatal life secondary to maternal cues may explain these phenomena. Whatever the mechanism, a mismatch between the intrauterine and postnatal nutritional environment appears to be fundamental to the development of disease (Gluckman *et al.*, 2007). The highest risks of adverse metabolic and cardiovascular outcomes occur when low birthweight is associated with higher child or adult body mass. Rapid postnatal growth of babies is associated with an enhanced risk for obesity, diabetes, hypertension, cardiovascular disease, and osteopenia in later life.

The finding of higher risk with the interaction of low birthweight and higher child and adult body mass has important implications. This nutritional profile is characteristic of populations undergoing nutritional transition. Countries with rapid economic development experience shifts in diet and physical activity that favor overnutrition which in combination with low birthweight increases the risk of chronic disease. Furthermore, in infancy, compensatory catch-up growth has been encouraged but it may be more prudent to avoid rapid postnatal weight gain, thereby balancing immediate gains with consideration of the future nutritional environment over the whole life span.

Prevention of Fetal Growth Retardation

Ideally, on a global scale, prevention should focus on successfully breaking the cycle of generations of growth-retarded babies by implementing major public health measures that aim to end poverty and the desolate socioeconomic conditions that contribute to child and maternal starvation and infections. However, the existing primary prevention programs are based mainly on specific health-care interventions. Systematic reviews of randomized controlled trials (RCTs) are the most objective way to evaluate the effectiveness of these health-care interventions in fetal growth retardation. However, methodological issues concerning rigorous randomization, reliable assessment of gestational age, and adequate sample size and confounders have meant fewer studies from developing populations are included in these reviews.

An overview of 65 RCTs of nutritional interventions during pregnancy for the prevention or treatment of fetal growth retardation, based on searches for Cochrane systematic reviews and RCTs published up to 2002, reports a systematic review of 13 trials of balanced protein energy supplementation in pregnancy, of which six reported fetal growth retardation. Data from these six trials on more than 4000 women showed that balanced protein energy supplementation given to women during pregnancy reduced the risk of fetal growth retardation by 30% (Merialdi *et al.*, 2003). The infants born to mothers receiving balanced protein energy supplementation during pregnancy (where protein accounts for $<25\%$ of the total energy content) tended to be heavier than those born to nonsupplemented mothers. Although the change in mean birthweight was small, it translated into relatively large reductions in the rate of fetal growth retardation. For example, a Gambian study that obtained an increase in mean birthweight of 136 g showed a reduction in low birthweight by 35%. In addition to these RCTs, an important nonrandomized study of supplementation during two consecutive pregnancies and the intervening lactation period showed even higher increases in mean birthweight. This suggests that the likelihood of improving birthweight in the chronically undernourished would be greater with longer-standing interventions.

High-protein supplementation has proved disappointing with one trial suggesting that high protein supplementation during pregnancy may be harmful and should be avoided. Single micronutrient supplementation interventions have not been shown to affect size at birth. Methodological concerns overshadow the findings that magnesium supplementation is reported to reduce fetal growth retardation by 30% in three trials of 1700 women and that the apparent beneficial effects of calcium supplementation may be related to an effect on gestation.

The lack of positive findings related to whether single nutrition interventions to improve fetal growth retardation raises questions about the likely effectiveness of interventions conducted for a few months during pregnancy and suggests that longer, more sustained interventions focusing on maternal preconception health and nutrition may be more effective.

Smoking cessation has been shown to reduce the occurrence of low birthweight. Observational data suggest that even partial reduction in smoking can have positive effects on birthweight. In general, behavioral strategies have been more effective in reducing smoking than just giving advice and increases in mean birthweights have been highest in groups who were most successful in stopping smoking.

A systematic review of antimalarial chemoprophylaxis use in endemic areas included 11 trials with data available on 3000 women. Although the studies were of variable designs, antimalarial chemoprophylaxis, overall, was associated with higher maternal hemoglobin levels and increases in birthweights. The effects were more marked in primigravida who are known to be most susceptible to infection. More research is needed to determine the optimum timing and combinations of drugs administered.

Management of Fetal Growth Retardation

The first step in the management of growth-retarded fetuses is to identify those fetuses that are truly at risk for adverse outcomes. This requires exclusion of small fetuses that are normally grown. It is also useful to identify those in whom fetal growth retardation is due to an underlying condition in which management will not greatly alter the outcome (e.g., chromosomal abnormalities, viral infections).

Currently, there is a lack of known therapeutic interventions to improve placental function and stimulate fetal growth so the management of fetal growth retardation relies on careful monitoring of fetal well-being and the planned delivery of the fetus when the risk of fetal death is greater than that of neonatal death and morbidity.

Antenatal surveillance of fetal well-being is resource dependent. In developing populations, assessment of fetal well-being may be confined to the monitoring of standardized fundal height measures, fetal movements, and listening to the fetal heart. Serial ultrasound measurements of biparietal diameter of the fetal skull and similar measurements of the abdomen as well as femur length, if available, are used to estimate fetal growth rates but need at least two measures at least 2 weeks apart.

Cardiotocography (CTG), or fetal heart rate analysis in response to movement and contractions, is used widely. However, difficulties of interpreting accelerations and decelerations and in reducing low birthweight changes often occurring only when urgent delivery is needed to prevent fetal death, limit its application. Similarly, a poor score in the composite biophysical profile including CTG, ultrasound measures of amniotic fluid volume recorded as the amniotic fluid index, fetal tone, movement, and breathing activity may also occur only when delivery is urgent. There is no evidence to show use of CTG or the biophysical profile improves perinatal outcomes. Doppler ultrasound measurements can be obtained from the umbilical artery, middle cerebral artery, ductus venosus, and uterine arteries. Umbilical artery Doppler studies have been well evaluated in RCTs and interventions based on the absent or reverse end-diastolic flow in the umbilical artery in fetal growth retardation pregnancies and in pregnancies with preeclampsia are associated with a 30% reduction in perinatal mortality. Combinations of fetal biometry, heart rate patterns, arterial and venous Doppler, and biophysical variables allow the most comprehensive evaluation of fetal well-being.

The most important aspect of prenatal management is the timing and mode of delivery individualized for the fetus at risk. There are no published RCTs on the optimum timing or mode of delivery. For the term fetus, timing of delivery is directly based on evidence of fetal distress or maternal disease but, for pregnancies of 25–32 weeks' gestation, each day gained *in utero* may improve survival. Early delivery of growth-retarded preterm infants results in higher liveborn rates at the expense of higher neonatal death rates. Intrapartum hypoxia is to be avoided and may necessitate cesarean delivery. Delivery should occur when immediate resuscitation and stabilization can be done by trained personnel.

Educational Programs

In teaching health-care practitioners in developing countries about concepts of fetal growth, and about deviations from the normal pattern of growth, it may be instructive to compare fetal growth with the more readily observable and better-understood features of postnatal growth and maturity through infancy and childhood. Starting with birth rather than conception, known causes of abnormal growth, whether genetic or environmental, can be compared and contrasted to enhance knowledge of those factors that affect growth of the embryo and fetus.

See also: Female Reproductive Function; Infant Mortality/Neonatal Disease; Perinatal Epidemiology.

Citations

Alderman H and Behrman JR (2006) Reducing the incidence of low birth weight in low-income countries has substantial economic benefits. *The World Bank Research Observer* 21: 25–48.

Barker DJP (2004) The developmental origins of adult disease. *Journal of American College of Nutrition* 23: 588S–595S.

Boulet SL, Alexander GR, Salihu HM, Kirby RS, and Carlo WA (2006) Fetal growth risk curves: Defining levels of fetal growth restriction by neonatal death risk. *American Journal of Obstetrics and Gynecology* 195: 151–157.

Gluckman PD, Hanson MA, and Beedle AS (2007) Early life events and their consequences for later disease: A life history and evolutionary perspective. *American Journal of Human Biology* 19: 1–19.

Gruenwald P (1974) Pathology of the deprived fetus and its supply line. Ciba Foundation Symposium 27. *Size at Birth*, pp. 3–26. Amsterdam, the Netherlands: Elsevier.

Hoffman HJ, Stark CR, Lundin FE Jr, and Ashbrook JD (1974) Analysis of birth weight, gestational age, and fetal viability, U.S. births, 1968. *Obstetrical & Gynecological Survey* 29: 651–681.

Kramer MS (1987) Determinants of low birth weight: Methodological assessment and metanalysis. *Bulletin of the World Health Organization* 659: 663–737.

Lancaster PA (1989) Birth weight percentiles for aborigines? *Medical Journal of Australia* 151: 489–490.

Lin C, and Santoloaya J (1998) Current concepts of fetal growth restriction: Part 1. Causes, classification and pathophysiology. *Obstetrics and Gynecology* 92: 1044–1055.

Lubchenco LO, Scaffer AJ and Markowitz M (eds.) (1976) *The High Risk Infant. Vol. 14. Major Problems in Clinical Pediatrics.* Philadelphia, PA: WB Saunders.

Merialdi M, Carroli G, Villar J, *et al.* (2003) Nutritional interventions during pregnancy for the prevention or treatment of impaired fetal growth: An overview of randomized controlled trials. *Journal of Nutrition* 133: 1626S–1631S.

Ounsted M and Ounsted C (1973) Familial factors. In: *On Fetal Growth Rate: Its Variations and Consequences (Clinics in Developmental Medicine)* 46, pp. 57–71. London, United Kingdom: Spastics International Medical Publications.

Roberts CL and Lancaster P (1999) National birth weight percentiles by gestational age for twins born in Australia. *Journal of Paediatrics and Child Health* 35: 278–282.

Sayers S, Singh G, Mackerras D, and Halpin S (2007) Growth outcomes of Australian Aboriginal children who were born with intrauterine growth retardation at term gestation. *Paediatric and Perinatal Epidemiology* 21: 411–417.

Smith C (1947) Effects of maternal undernutrition upon the newborn infant in Holland (1944–45). *Journal of Pediatrics* 30: 229–243.

United Nations Fund for Children (2006) The State of the World's Children 2007. *Women and Children. The double dividend of gender equality.* New York, NY: UNICEF.

Walton A and Hammond J (1938) The maternal effects on growth and conformation in Shire horse–Shetland pony crosses. *Proceedings of Royal Society of London. Series B, Biological Science* 125: 311–335.

Yanney M and Marlow N (2004) Paediatric consequences of fetal growth restriction. *Seminars in Fetal and Neonatal Medicine* 9: 411–418.

Ylppö A (1919) Das wachstum der frühgeborenen von der gebert bis zum schulalter. [The growth of prematures from birth to school age.] *Zeitschrift für Kindeheilkunde* 24: 111–178.

Yu VYH and Upadhyay A (2004) Neonatal management of the growth-restricted infant. *Seminars in Fetal and Neonatal Medicine* 9: 403–409.

Further Reading

Alberry M and Soohill P (2007) Management of fetal growth restriction. *Archives of Diseases of Childhood Fetal Neonatal Edition* 97: F62–F67.

Barker DJP (1994) *Mother, Babies and Diseases in Later Life*. London, United Kingdom: BMJ Publishing Group.

Gulmezoglu AM, de Onis M, and Villar M Jr (1997) Effectiveness of interventions to prevent or treat impaired fetal growth. *Obstetrical & Gynecological Survey* 52: 39–148.

Kramer MS (2003) The epidemiology of adverse pregnancy outcomes: An overview. *The Journal of Nutrition* 133(5S): S1592–S1599.

O'Brien PMS, Wheeler T and Barker DJP (eds.) (1999) *Fetal Programming Influences on Development and Disease in Later Life.* London, United Kingdom: RCOG Press.

Salafia CM, Minior VK, Pezzullo JC, Popek EJ, and Rosenkrantz TS (1995) Intrauterine growth restriction in infants of less than thirty-two weeks' gestation: Associated placental pathologic features. *American Journal of Obstetrics and Gynecology* 173: 1049–1057.

United Nations (2000) Administrative Committee on Coordination, Sub-Committee on Nutrition, Nutrition Policy Paper, No. 18, September 2000. Geneva, Switzerland: UNICEF.

World Health Organization (2006) *Promoting Optimal Fetal Development. Report of Technical Consultation.* Geneva, Switzerland: World Health Organization.

Infant Mortality/Neonatal Disease

P H Wise, Stanford University, Stanford, CA, USA

Introduction

The infant mortality rate has long been one of the most accessible and highly utilized public health statistics. This prominent public presence is due to its relative simplicity as well as to the social meaning that both the public health community and the general public convey to its interpretation. First, infant mortality implies tragedy, as the innocence and inherent promise of infancy imbues any measure of infant death with deep emotional content. Each infant death, regardless of cause, is perceived as a tragedy, as a 'shame.' Second, infant mortality has long been viewed as being linked to the adequacy of social conditions and patterns of social equity. This reflective capacity shifts the meaning of the infant mortality rate to the political arena, extending it from merely the tragic to the unjust. In this manner, the tragic content of infant death – as being a 'shame' – is coupled with its use as a social mirror – as being 'shameful.'

While this dual nature, tragedy and social mirror, has long defined the power of the infant mortality rate, it has also generated deep tensions in how the infant mortality rate is perceived by different disciplines and, more importantly, in how efforts to reduce infant mortality have been shaped and implemented in the real world. Those who view infant death as primarily a tragedy have directly focused on the infant mortality rate through the development and implementation of largely technical solutions. Those who perceive the infant mortality rate as an indicator of social conditions have tended to use the infant mortality rate as an advocacy tool and to elevate fundamental social processes as the primary determinants of health.

History

Although the struggle to find meaning in the death of an infant child has likely existed since parents began to value the promise of birth, the effort to measure infant mortality was largely rooted in the social reform movements of eighteenth-century Europe. Newly developed vital statistics systems and an expanding capacity to calculate population-based statistics provided an analytic foundation for a series of public reports documenting striking social inequalities in infant and young child mortality.

However, the response to these documented inequalities in infant mortality was characterized by a profound tension between those who stressed technical interventions and those who advocated for broader political reform. An important group of reformers, including Florence Nightingale and Edwin Chadwick, focused on unhealthy sanitary conditions as the primary cause of elevated mortality and stressed technical improvements such as improved sewage systems, clean water, and better housing. Consequently, the need for sanitary reform could be advocated without advancing the social claims of the affected communities. This technical approach has largely been adopted by the medical disciplines. However, another group of reformers, including Frederich Engels, saw the infant mortality rate as reflecting the deeper social fabric of an industrializing Europe. This group of reformers focused less on technical responses than on social ones, emphasizing economic and political reforms. Here the emphasis was on social rather than civil engineering.

Despite these longstanding tensions between technical and social frameworks, infant mortality is responsive to both technical and social influences. When highly efficacious technical interventions exist, then access to these interventions, often socially defined, will tend to dominate the distribution of infant outcomes. When technical efficacy is low or nonexistent, then access falls away and the determinants of underlying risk, particularly social conditions, will tend to dominate the distribution of infant outcomes. Therefore, the availability of efficacious technical interventions does not necessarily reduce the importance of social factors but rather alters the mechanisms by which they are ultimately expressed as patterns of infant death. Accordingly, the interaction of the technical and social determinants of infant mortality are best viewed as intensely dynamic and serve as the basis for examining global patterns of infant mortality and their specific causes.

Measurement of Infant Mortality

Definitions

An infant is defined as a liveborn child less than 1 year of age. The 'infant mortality rate' is defined as the number of deaths to liveborn infants who die before reaching 1 year of age divided by the total number of live births. This is usually reported as the number of infant deaths per 1000 live births. Generally, the infant mortality rate is calculated for a given calendar year, with all infant deaths occurring in that year included, even if some of these infants may have been born in the prior calendar year.

The infant mortality rate is best viewed as a composite of a series of component rates, each with its own spectrum of causation. The infant mortality rate can be stratified into two component rates: the 'neonatal mortality rate,' defined as the number of deaths to liveborn infants who die before reaching 28 days of age; and the 'postneonatal mortality rate,' defined as the number of deaths to liveborn infants who die between 28 days and 1 year of age. The sum of the neonatal mortality rate and the postneonatal mortality rate equals the infant mortality rate. This disaggregation of the infant mortality rate is helpful, as the age at death tends to be associated with different distributions of causes.

Sources of Data

In many settings, particularly industrialized parts of the world, infant mortality rates are generally derived from vital registration systems. Population-based death registries provide figures for the numerator, while birth registries generate the denominator of infant mortality rate calculations. However, in many parts of the world, accurate vital registration systems do not exist. Infant mortality rates in these settings are usually derived from surveys of selected, representative communities. Approximately half of all infant deaths in the world are identified through direct vital registration systems; the other half is estimated on the basis of surveys. Even where well-developed registries exist, differences in definition and childbearing practices can complicate the comparison of infant mortality rates. Of particular concern is the classification of deaths occurring to extremely premature newborns, as such deaths may be considered fetal deaths in many areas. Such differences can have a major impact on international comparisons when extreme prematurity accounts for a substantial portion of infant mortality.

Global Patterns of Infant Mortality

It is estimated that, in 2004, the world's infant mortality rate was approximately 50 deaths per 1000 live births (see **Table 1**). The highest rates in the world are reported in West-Central Africa (109 deaths per 1000 live births in 2004), although some areas ravaged by war and other humanitarian emergencies can also report extremely high rates. In contrast, the infant mortality rate for wealthy industrialized countries such as those of Western Europe was approximately 5 deaths per 1000 live births; the figure for Japan was approximately 3 deaths per 1000 live births in 2004.

Most areas of the world have witnessed major declines in infant mortality over the past several decades. However, the largest percentage decreases have occurred in areas already characterized by relatively low rates, particularly in Japan, Singapore, and Western Europe. This has had the effect of widening longstanding inequalities between materially wealthy countries and their poor counterparts. These inequalities have been exacerbated recently with a relative slowing in the annual improvements in infant mortality in materially poor areas of the world.

In general, recent reductions in infant morality rates have occurred in both neonatal and postneonatal mortality. However, in most high-income countries, postneonatal mortality rates have fallen more dramatically than neonatal rates, thereby increasing the portion of infant mortality that is accounted for by deaths in the neonatal

Table 1 Infant mortality rate, number of infant deaths, and percentage of global infant deaths by region and country development status, 2004

	Infant mortality rate	*Number of infant deaths*	*Percentage of global infant deaths*
Region			
Sub-Saharan Africa	102	2 882 826	40.1%
South Asia	67	2 482 484	34.5%
East Asia and Pacific	29	868 028	12.1%
Middle East and North Africa	44	423 280	5.9%
Latin America and Caribbean	26	303 524	4.2%
Central/Eastern Europe and Commonwealth of Independent States	32	178 240	2.5%
USA and Canada	7	30 578	0.4%
Western Europe	4	13 260	0.2%
Australia and New Zealand	5	1520	0.0%
All other countries	6	6158	
World Total	54	7 189 898	100%
Development status			
Developing countries	59	7 060 117	99.2%
Least developed countries	98	2 726 654	38.3%
Industrialized countries	5	54 195	0.8%
World Total	54	7 189 898	100%

Source: United Nations Children's Fund (2006) The State of the World's Children 2006. Excluded and Invisible. New York, NY: UNICEF.

period. For example, approximately two-thirds of the infant mortality rates for England and the United States are due to neonatal deaths.

Although the sub-Saharan African region contributes only 20% of all births in the world, it accounts for some 40% of all infant deaths. South Asia accounts for almost 35% of global infant deaths, with India alone accounting for approximately 22% of the global toll. Of the approximately 12% contributed from the East Asian and Pacific region, China accounted for about half of these deaths. In contrast to these major contributions, Western Europe and the USA contributed 0.6% of all infant deaths in the world, even though they accounted for some 6% of all births.

Components of Infant Mortality

The first year of life is characterized by dynamic changes in physiology and environmental and social exposures as the infant grows and develops. This results in a rapidly changing epidemiology of mortality. Moreover, the underlying social and health conditions in a population will dramatically alter the relative distribution of different causes of infant death.

Overall, it is estimated that the majority of infant deaths in the world are related to three general categories: (1) prematurity, birth asphyxia, and other obstetrical complications, (2) congenital anomalies, and (3) a variety of infectious causes. However, the causes of infant mortality vary considerably with different social and economic characteristics. Communities characterized by low material resources not only have dramatically elevated infant mortality rates but the ultimate causes of death are far more likely to be associated with pregnancy or birth complications, malnutrition, and infectious diseases. Populations of greater material resources, on the other hand, generally have infant mortality rates that are primarily due to problems noted at birth, such as prematurity and birth defects.

Recent estimates suggest that approximately half of infant mortality occurs during the neonatal period and half in the postneonatal period. This distribution can vary considerably, however, particularly when major differences exist in the reporting of early neonatal deaths. In general, neonatal mortality is dominated by conditions exhibited at birth or shortly thereafter. Postneonatal mortality tends to reflect delayed effects of birth defects and patterns of infectious causes, often with strong interactions with malnutrition. However, the causes of neonatal and postneonatal mortality overlap considerably, particularly where infectious determinants of mortality are highly prevalent.

The clinical causes of infant mortality should be seen as the ultimate pathophysiologic expression of a cascade of social and biologic interactions. Nevertheless, these direct causes provide important insights into the mechanisms of risk and potential opportunities for effective interventions. Detailed information on the clinical determinants of infant mortality is largely unavailable in many areas of the world. Therefore, global estimates regarding clinical causation are based on a mix of national data systems and selected surveys designed to be representative of large populations of neonatal deaths.

Neonatal Mortality

Neonatal deaths, those occurring in the first 28 days of life, are estimated to number approximately 4 million throughout the world annually. This accounts for more than one-third of all child deaths under 5 years of age. This suggests that the actual risk of mortality during the neonatal period is some 30 times higher than that during the remaining period between 29 days and 5 years of age. This high concentration of mortality is even more extreme in high-income countries, where the portion of young child deaths under 5 years of age occurring in the neonatal period approaches two-thirds.

Even within the neonatal period, mortality is heavily skewed to the first days of life. It is estimated that some three-quarters of all neonatal deaths occur in the first week of life, with up to half of these occurring in the first 24 h after birth.

Low birthweight

Low birthweight implies reduced fetal growth prior to delivery either through shortened gestation, growth retardation, or both. Generally defined as a newborn weighing less than 2500 g (approximately 5.5 lbs) at birth, low birthweight can elevate the risk of death from a variety of metabolic, nutritional, and infectious processes. It is estimated that low birthweight occurs in approximately 15% of all newborns, almost 20 million infants annually throughout the world, with the largest contribution coming from South Asia. This 15% of births, however, accounts for some 60–80% of all neonatal deaths.

Low birthweight is not a direct clinical cause of neonatal mortality *per se* as it can reflect a variety of specific pathophysiologic problems. Rather, low birthweight is merely a useful indicator of problem pregnancies and the risk of mortality during the neonatal period. Generally, low birthweight is the result of either shortened length of gestation or inadequate fetal growth for any given length of gestation. Shortened length of gestation of clinical significance is usually termed prematurity, while seriously reduced growth per unit of gestation is often labeled intrauterine growth retardation.

Prematurity

Globally, the most common direct clinical cause of neonatal death is prematurity, generally defined as a gestational age of less than 37 weeks (a normal pregnancy is approximately 40 weeks of gestation). It is estimated that some 27% of all neonatal deaths are associated with premature birth. However, in countries with relatively low neonatal mortality, this contribution is generally substantially higher. In addition, the risks associated with prematurity rise sharply as gestational age falls; in many settings, mortality for a newborn of 26 weeks gestation is some 100 times that of an infant born at 36 weeks of gestation. The problems premature infants face are multiple and reflect the immaturity of organ systems and immunity. Of special note are respiratory difficulties, metabolic problems such as an inability to feed or maintain body temperature, brain injuries due to hemorrhage or a lack of oxygen, and sepsis due to inadequate immune response to bacteria and other infections. The causes of prematurity are heterogeneous and poorly understood. However, infections or inflammation of the genitourinary tract, placental abnormalities, and maternal complications such as severely elevated blood pressure that may require delivery have all been associated with preterm birth.

Intrauterine growth retardation

Intrauterine growth retardation (IUGR) refers to seriously reduced fetal weight for its gestational age. Generally defined as a birthweight less than the 10th percentile of predicted fetal weight for its gestational age, IUGR is associated with elevated neonatal morbidity and mortality. The condition can be caused by a variety of pathways, including inadequate maternal–fetal circulation, intrauterine infections, and congenital anomalies such as trisomy 21 and trisomy 18. However, IUGR can also reflect inadequate maternal nutrition, both before and during pregnancy. IUGR elevates the risk of neonatal morbidity and mortality, particularly if growth retardation is severe or associated with other serious perinatal conditions. There are also some recent data suggesting that children born with even moderate IUGR can experience subtle cognitive or metabolic difficulties later in life.

Asphyxia

Asphyxia, a potentially catastrophic reduction in the supply of oxygen to the fetus during labor and delivery, is also a major contributor to newborn mortality. This can occur due to a variety of causes but often is associated with prolonged or obstructed labor or with difficult fetal presentations such as the breech position. Although asphyxia can cause death *in utero* (stillbirth), it contributes to neonatal mortality through the birth of an infant whose neurologic and metabolic status is severely compromised.

Effective interventions to prevent asphyxia include enhanced prenatal care, the improved management of labor and delivery, and basic resuscitative measures for hypoxic newborns. These strategies have been implemented as part of improved training for traditional birth attendants and midwives, who provide the largest portion of obstetrical and newborn care in most parts of the world. It is estimated that approximately half of asphyxia-related deaths could be prevented with available interventions and techniques.

Pneumonia and sepsis

Pneumonia, both with and without accompanying sepsis (serious bloodborne infection), also accounts for a major portion of neonatal mortality. Accounting for as much as one-quarter of neonatal deaths, pneumonia can kill through either interference with adequate oxygenation or through acting as a source for bloodborne, systemic infection. Although it is difficult to distinguish the specific infectious agent of any individual case, known etiologies of neonatal pneumonia include a variety of viral infections and bacterial agents, including group B streptococci and *Escherichia coli.*

Tetanus

Tetanus accounts for approximately 5% of all neonatal deaths worldwide. Tetanus is caused by the toxin-producing bacteria *Clostridium tetani* and, in newborns, is usually due to infection of the umbilical cord stump. Highly effective vaccines to prevent tetanus have made it extremely rare in industrialized countries. However, neonatal tetanus is still a major problem in areas where access to immunization is inadequate and where poor hygiene is practiced during the delivery process.

The importance of birthweight and causal interactions

It is common for neonatal deaths to be the result of multiple pathologic conditions, and sharp distinctions between causal groupings should be made with caution. For example, prematurity predisposes newborns to a variety of conditions, including severe respiratory problems and sepsis. Similarly, birth defects can often be associated with terminal conditions, such as pneumonia or dehydration.

The neonatal mortality rate is best viewed as a composite of a number of component rates, each with its own relationship to social forces and clinical pathways. Because birthweight is so closely tied to the infant risk and is relatively easy to measure, it has become a useful means of disaggregating infant mortality rate into different constituent arenas of potential prevention.

In general, elevations in neonatal mortality rates can be stratified into two components: birthweight distribution and birthweight-specific mortality. Birthweight distribution reflects the relative birth rates of infants at different birthweights, and particularly at birthweights that convey a high risk of morbidity and mortality. For example, an elevation in neonatal mortality could be due to elevated birth rates of low-birthweight infants. Alternatively, elevated neonatal mortality can be caused by differences in birthweight-specific mortality, generally defined as the risk of mortality for any given birthweight group. Here, an elevated neonatal mortality rate is due not to differences in the birth rate of low-birthweight newborns (birthweight distribution) but rather to differences in the risk of death once a low-birthweight infant is born (birthweight-specific mortality).

Although a helpful epidemiologic technique, the deeper utility of birthweight stratification rests on the general ability to tie these component rates to different arenas of causation and opportunities for constructive intervention. Birthweight-specific mortality rates have been most closely associated with the obstetrical management of high-risk deliveries and the provision of care to high-risk newborns once they are born. Birthweight distribution, on the other hand, is largely determined by prenatal influences and the general health of women.

Birthweight distribution effects

In large measure, the association of birthweight distribution and neonatal mortality is due to the birth rate of low birthweight infants, and particularly very low birthweight infants (<1500 g). This is because in most settings the risk of death rises sharply at the two tails of birthweight distribution, particularly at the lower end as birthweight declines below 1500 g. Accordingly, caution should be used when considering broad birthweight groupings, such as low birthweight (<2500 g), as they can be dominated by the relatively large number of infants falling near the upper boundaries of these categories, precisely the infants least likely to experience significant clinical problems. Similarly, comparisons of the mean or average birthweight of a general population of births will almost always reflect differences in the weights of infants born near the mean, which in most instances will have no bearing on morbidity or mortality. It should also be recognized that birthweight is generally used as a proxy for gestational age, since in large populations birthweight is usually far more accurately assessed than the duration of pregnancy itself. Although low birthweight can reflect growth retardation influences, the relationship between low birthweight and mortality is generally due to prematurity, which reflects the impact of immature organ system development secondary to being born too soon.

Reducing neonatal mortality through the improvement of birthweight distribution is heavily shaped by our ability to reduce very low birthweight and extreme prematurity. There is a large and diverse literature regarding risk factors for very low birthweight and prematurity. These risks include demographic factors such as parity and maternal age, social influences such as income and housing, maternal behaviors such as tobacco use, and a variety of preexisting medical and obstetrical conditions. However, the specific contribution of these documented risks to elevations in prematurity is difficult to assess, and together they account for only a relatively small portion of premature and very low-birthweight births. In response, research efforts have been focused on both a broader array of psychosocial influences as well as a deeper understanding of the pathophysiologic mechanisms of premature birth.

While the impact of absolute material deprivation on birth outcome is well recognized, there is growing interest in the potential impact of relative deprivation and social discrimination on the risk of prematurity. In addition, the organization and capacities of neighborhoods and community support systems and other broader 'social capital' influences could also play a role in mediating the ultimate expression of individual characteristics and behaviors on birth outcomes. Although the precise mechanisms by which these psychosocial effects operate remain unclear, there are suggestions that alterations in hormonal or prostaglandin pathways that help control the duration of pregnancy may be involved. These interactions in turn, may be influenced by genetic or epigenetic predispositions (changes to DNA that do not involve alterations in base pair sequencing). Similarly, there is evidence that a variety of infectious conditions may increase the risk for prematurity. Here too, socially defined exposures and biologic predisposition may be highly interactive in shaping patterns of elevated risk for prematurity.

Birthweight-specific mortality effects

Once an infant is born at a given birthweight, there are a variety of factors that can influence its relative chances of survival. For prematurity, access to high-quality obstetrical management of labor and delivery can be crucial in determining infant outcomes. Even more crucial is the quality of care the infant receives once born. In industrialized countries, most of such care has been organized into regional referral networks in which critically ill newborns are cared for in medical facilities with special expertise in neonatal intensive care. In such industrialized settings, the major reductions in neonatal mortality documented over the past several decades have been due to improvements in birthweight-specific mortality associated with improvements in obstetrical and neonatal care.

Birthweight-specific mortality rates also reflect differences in the rate of severe birth defects, as many such children are born at normal birthweights. Accordingly, any intervention that reduces the risk that a child will be born with such a condition can affect birthweight-specific mortality. For example, the addition of folic acid to the diet of women prior to and early in pregnancy can reduce the risk of neural tube defects, a set of conditions that elevate the risk of neonatal and infant mortality. Similarly, the use of prenatal ultrasound or other screening technologies to identify affected fetuses, can, if abortion is an option, reduce the birth rate of infants with severe birth defects.

Neonatal mortality and women's health

Major improvements in infant and particularly neonatal mortality will depend in some large measure on improving the health of young women. Particularly where prematurity, low birthweight, and birth defects account for a large portion of infant mortality, efforts to improve neonatal survival will likely require reducing the impact of complex and often longstanding maternal risks, such as chronic medical conditions, adverse maternal behaviors, poor prior obstetrical history, inadequate nutrition, and severe social needs. Moreover, a comprehensive approach recognizes that women's health extends beyond childbearing and that access to all forms of health services, including contraception and primary health care, are essential. Prenatal care, therefore, should always be viewed as merely a component, albeit an important component, of comprehensive women's health care over a lifetime.

Postneonatal Mortality

Although there are no sharp distinctions between the causes of neonatal and postneonatal mortality, postneonatal mortality patterns tend to be less reflective of birth outcomes and more the result of the direct interaction of the developing infant with the material and social environment. Accordingly, postneonatal mortality has long been tied to materially poor social conditions and tends to account for a larger portion of infant mortality in areas of the world characterized by severe poverty. Moreover, malnutrition is a major factor in predisposing infants for illness and death from virtually all causes. In this manner, inadequate familial resources and breast-feeding practices will affect the risk of acquiring and succumbing to a variety of infectious agents. Indeed, inadequate nutrition courses through the entire epidemiology of infancy and represents a central mechanism by which poverty shapes the risk of mortality during the first year of life.

Pneumonia

Pneumonia is a major cause of postneonatal mortality, accounting for an estimated 25% of deaths in this age group. Pneumonia in the postneonatal period tends to be caused by a different spectrum of infectious agents than that in the neonatal period. This is important because two major causes of pneumonia among older infants are *Streptococcus pneumoniae* and *Haemophilus influenzae*, two bacterial agents for which effective vaccines have been developed. Pneumonia in this age group often presents as fever and respiratory distress. As for neonatal pneumonias, the rapid initiation of appropriate antibiotic therapy can prove highly effective when bacterial infections are present. Other important causes include viruses, such as the respiratory syncytial virus, which can occur in major seasonal outbreaks.

The development of vaccines for *Streptococcus pneumoniae* and *Haemophilus influenzae* disease represents a major opportunity to reduce infant and young child illness and death from pneumonia. The use of these vaccines in the USA, Europe, and other industrialized countries has dramatically reduced illness and mortality from these causes over the past two decades. New global strategies to make these vaccines accessible to all infants and children could help ensure that similar reductions occur throughout the world.

Diarrhea

Diarrhea accounts for a significant portion of infant deaths, particularly among those in the postneonatal period. Although diarrhea during infancy can be due to a variety of causes, infectious etiologies account for the vast majority of diarrhea-related deaths in this age group. A variety of viral agents, including rotavirus, account for the majority of diarrhea-related infant deaths. Bacterial infections such as *Salmonella*, typhoid fever, cholera, and shigella as well as protozoan infections such as amebiasis and giardia can also cause severe diarrhea, particularly as part of local outbreaks. Generally, the risk of death from diarrhea is due to dehydration secondary to severe fluid and electrolyte loss. Therefore, regardless of the specific infectious etiology, the replacement of fluid and electrolytes during the acute phase of the diarrheal illness is highly effective in preventing mortality. In addition, newly developed rotavirus vaccines offer some promise in preventing a major portion of severe diarrhea in this age group throughout the world.

Birth defects

Birth defects account for an important component of postneonatal death. Although an important portion of birth defects are due to known genetic, toxic, nutritional, or infectious causes, most are of complex or unknown origin. Nevertheless, there are a variety of effective interventions to reduce mortality from birth defects. Adequate maternal nutrition including folic acid and iodine, vaccination against rubella, elimination of toxic exposures including maternal alcohol intake and teratogenic medications, and access to prenatal diagnosis and termination of pregnancy when the fetus is severely affected are among the most effective means of preventing serious birth defects. In addition, mortality from birth defects can be reduced through comprehensive monitoring and response to health problems in affected infants once born.

Malaria

Although malaria is not among the most common threats to infants around the world, it does account for a significant portion of deaths in several areas, particularly sub-Saharan Africa. Indeed, more than 90% of all malaria deaths in the world occur in Africa, which account for almost 1 in 5 deaths to young children in this region. In these endemic and hyperendemic areas, malaria parasites are present in approximately 10% of all infants by the time they reach 3 months of life, and some 90% by 1 year of age. Mortality from malaria is concentrated in young children.

There are effective interventions to reduce infant mortality from malaria. These include general efforts to eliminate mosquito breeding sites as well as the use of insecticide-impregnated mosquito nets for sleeping. In addition, confined spraying of dichlorodiphenyltrichloroethane (DDT) in homes in endemic areas has also been recently recommended. Long avoided because of environmental damage and possible direct human toxicity, the persistent if not growing threat of malaria to young children has generated calls for reinstituting purposeful DDT-based strategies in heavily affected areas. New therapeutic agents have also been developed that can be crucial in preventing death in children once they are infected with malarial parasites.

Acquired Immune Deficiency Syndrome (AIDS)

Infants infected with the human immunodeficiency virus (HIV) are at elevated risk of mortality due to a variety of infections. Virtually all infants acquire HIV through what has been termed vertical transmission, that is infection from the mother, which can occur during gestation, labor, and delivery, or through breastfeeding. In areas of high HIV prevalence, therefore, infant deaths due to AIDS can account for a substantial percentage of all infant deaths.

It has been shown that a major portion of HIV infection in infants can be prevented with the appropriate use of antiretroviral therapies during pregnancy, labor and delivery, and infancy. Different protocols have proven effective, and new strategies that attempt to reduce the complexity and cost of such interventions are undergoing active investigation. Yet, provision of these effective medications remains profoundly inadequate in many areas. Indeed, this problem reflects the general lack of access to antiretroviral therapy in general and to women in particular. In addition, recommendations to discourage breastfeeding by women with HIV infection have proven controversial in some areas, due to the lack of appropriate, alternative sources of nutrition and the lack of available treatment of women infected with HIV.

Conclusion

In settings of profound material deprivation, the purpose of health interventions is fundamentally to uncouple poverty from its implications for health. In this manner, technical strategies are directed at improving health outcomes even if poverty persists. The potential to accomplish this relative uncoupling in the arena of infant mortality has been estimated to be substantial, with approximately half of neonatal deaths and up to two-thirds of postneonatal deaths considered preventable with available technical capabilities. The burden, therefore, remains equitable and efficient provision to communities that would benefit from such capabilities, a burden that while concerned with technical solutions is inherently social in origin.

Strategies to address infant mortality in general and inequalities in infant mortality in particular must ensure the provision of efficacious interventions to all those in need. However, this strategy, while essential, must be tied to broader approaches that respond to the fundamental challenges of material deprivation, oppressive social structures, especially for women, and inadequate global commitment. In this manner, a comprehensive approach to reducing infant mortality must consider issues of power as well as those of technical intervention and recognize ultimately that the dual struggles – the struggle for medical progress and the struggle for social justice – will always be inextricably linked.

See also: AIDS: Epidemiology and Surveillance; Fetal Growth Retardation: Causes and Outcomes; Mother-to-Child Transmission of HIV; Perinatal Epidemiology.

Further Reading

Bang AT, Bang RA, and Reddy HM (2005) Home-based neonatal care: Summary and applications of the field trial in rural Gadchiroli, India (1993 to 2003). *Journal of Perinatology* 25: s108–s122.

Bryce J, *et al.* (2005) WHO estimates of the causes of death in children. *Lancet* 365: 1147–1152.

Darmstadt GI, *et al.* (2005) Evidence-based, cost-effective interventions: How many newborn babies can we save? *Lancet* 365: 977–988.

Jones G, *et al.* (2003) How many child deaths can we prevent this year? *Lancet* 362: 65–71.

Lawn JE, *et al.* (2005) Four million neonatal deaths: When? Where? Why? *Lancet* 365: 891–900.

United Nations Children's Fund (2006) *State of the World's Children 2006. Excluded and Invisible.* New York, NY: UNICEF.

Wise PH (2003) The anatomy of a disparity in infant mortality. *Annual Review of Public Health* 24: 341–362.

SECTION 3
REPRODUCTIVE CANCERS

Cancer Screening

M Hakama and A Auvinen, University of Tampere School of Public Health, Tampere, Finland

Introduction

The primary purpose of screening for cancer is to reduce mortality. Besides the effect on the length of life, screening also has other important consequences, including a burden on economic resources and implications for the quality of life. Screening usually implies an increase in health expenditure. The effects on the quality of life of screened subjects can be both positive and negative.

Cancers Suitable for Screening

Cancer is characterized both by an insidious onset and by an improved outcome when it is detected early. For a cancer to be suitable for screening, therefore, the natural history of the disease should include a phase without symptoms during which the cancer can be detected by a screening test earlier than by ordinary clinical diagnosis after the subject has experienced symptoms. The outcome of treatment following diagnosis during this detectable, preclinical phase (DPCP) (Cole and Morrison, 1978) should also be better than following clinical detection. Sometimes the screening program may reduce morbidity or improve the quality of life. For example, a mammography program may increase the number of women undergoing surgery because of overdiagnosis; early diagnosis increases the duration of sickness, but mammography enables breast-conserving operations, which improves the quality of life and produces less morbidity than the treatment of more advanced disease.

The objective of screening is to reduce the burden of disease. The impact of disease covers both death and morbidity, that is, detriment of the disease for the patient while alive, including reduced well-being and loss of functional capacity. The prognosis should be better following early treatment of screen-detected disease than of clinically detected disease. If a disease can be successfully treated after it manifests clinically, there is no need for screening. Screening should not be applied for untreatable diseases. Cancer is always a potentially lethal disease and the primary goal of treatment is saving the patient's life. Thus, a reduction of mortality is the most important indicator of the effectiveness of screening.

The disease should be common enough to justify the efforts involved in mounting a screening program. Disease in the preclinical phase is the target of detection. The frequency of preclinical disease in the population to be screened depends on the incidence of clinical disease and on the length of the detectable preclinical phase. Screening for disease is a continuous process. The prevalence of disease at initial screening may be substantially different from the prevalence at subsequent screens. This variation in yield is not due to the length of the preclinical phase, that is, the basic biological properties of the disease only, but is simply a consequence of the length of intervals between screening rounds, or the screening regimen.

The screening test must be capable of identifying disease in the preclinical phase. For cancer, the earliest stages of the preclinical phase are thought to occur in a single cell, and they remain beyond identification. The detectable preclinical phase starts when the disease becomes detectable by the test and ends when the cancer would surface clinically. The length of the DPCP is called the sojourn time (Day and Walter, 1984). The sojourn time is not constant, but varies from case to case. It can be described with a theoretical distribution. Factors influencing the sojourn time include (1) the biological characteristics of the disease (growth and progression rate) influencing the time from start of detectability to diagnosis at symptomatic stage, (2) the screening test and the cut-off value or criteria for positivity, used for identifying early disease (detectability threshold), and (3) behavioral factors and health care affecting the end of DPCP, or the time of diagnostic confirmation of the disease.

The occurrence of disease in a population is measured in terms of incidence or mortality rates. Stomach cancer is still common in middle-aged populations in many countries, but the risk is rapidly decreasing, which contrasts with a substantial increase for cancer of the breast, another disease occurring at a relatively early age in many countries. In contrast, the incidence of prostate cancer has been very low under 65 years of age, and the increase in incidence is mainly attributable to diagnostic activity. Therefore, the number of life-years that can potentially be saved by a (successful) screening program is small.

If screening postpones death, screening increases the prevalence of disease, that is, the number of people living who have been diagnosed with it at some time in the past. Prevalence may also increase as a consequence of earlier diagnosis or overtreatment, so the prevalence of disease and the survival of cancer patients are not appropriate indicators of the effect of screening.

Effects of screening on the quality of life and cost are in principle considered in decisions on whether or not to screen. In practice, however, public health policies related to cancer screening are invariably initiated, run, and evaluated by their effect on mortality. In fact, there is no agreement on how to apply criteria other than mortality to policy decisions about screening. Such a decision would require evidence on the magnitude of effect on these criteria, and agreement on how to weigh the benefits and harms in the different dimensions of death, quality of life, and cost.

Benefits and Harms of Screening

Independently of effectiveness, screening has adverse effects. The effectiveness of screening is related to the degree to which the objectives are met. As indicated above, the purpose of screening is to reduce disease burden and the main goal is the reduction of deaths from the disease. If the treatment of screen-detected disease has fewer or less serious adverse effects than the treatment of clinical disease, then the iatrogenic morbidity for the patient is decreased. If late-stage disease is avoided, highly debilitating effects can be reduced. Costs can be saved if treatment and follow-up of early disease requires fewer resources than clinically detected cancer. A correct negative test also has a beneficial effect in terms of reassurance for those without disease.

Because screening requires preclinical diagnosis of disease, the period of morbidity is prolonged by the interval between diagnosis at screening and the hypothetical time when clinical diagnosis would have occurred if the patient had not been screened. This interval is called the lead time.

Screen-positive cases are confirmed by the standard clinical diagnostic methods. Many screen-detected cases are borderline abnormalities, some of which would progress to clinical disease, while some would not progress even if untreated. Cervical cancer screening results in detection of early intraepithelial neoplasia (premalignant lesions or *in situ* carcinomas), not all of which would progress to (fatal) cancer, even without treatment (IARC, 2005). Occult intraductal carcinomas of the breast (IARC, 2002), occult papillary carcinomas of the thyroid gland (Furihata and Maruchi, 1969), or occult prostatic cancer (Hugosson *et al.*, 2000) may fulfill the histological criteria for malignancy, but would remain indolent clinically. Any screening program will disclose such abnormalities, which are indistinguishable from a case that would progress into clinical disease during the person's lifetime, if not subjected to early treatment. Therefore, one of the adverse effects of screening is overdiagnosis, that is, detection of indolent disease and its unnecessary treatment (overtreatment), which results in anxiety and morbidity that would be avoided without screening and which is unnecessary for achieving the goals of screening.

A false-negative screening test (a person with disease that is not detected by the test) provides undue reassurance and may result in delayed diagnosis and worse outcome of treatment. In such cases, the effect of screening is disadvantageous.

Screening tests are applied to a population without recognized disease. In addition to the abnormal or borderline diagnoses, therefore, there will be false-positive screening results (a person without disease has a positive test), and these can cause anxiety and morbidity. The test itself may carry a risk. For example, screening for breast cancer is based on mammography, which involves a small radiation dose. The small risk of breast cancer induced by irradiating a large population should be compared with the benefits of mammography.

Validity of Screening

The validity of screening indicates the process or performance of screening and consists of two components: sensitivity and specificity. Sensitivity is an indicator of the extent to which preclinical disease is identified and specificity describes the extent to which healthy individuals are so identified. Predictive values are derived from those indicators and they describe the performance from the point of view of the person screened (the screenee).

The purpose of the screening test is to distinguish a subset of the population with a high probability of having unrecognized disease from the rest of the population, with average or low risk. Sensitivity is the proportion of persons who have a positive test among those with the disease in the DPCP (**Table 1**). Sensitivity is a basic performance measure because it indicates the proportion of early disease that is identified by screening. Specificity is the proportion of persons with a negative test among all those screened who are disease-free. Specificity is a basic measure of the disadvantages of a test, since poor specificity results in high financial costs and adverse effects due to false-positive tests. Both the sensitivity and the specificity of a screening test are process indicators. A screening program based on a valid test may nevertheless fail in its objective of reducing mortality in the screened population.

Table 1 Validity of screening test

	Disease in DPCP[a]	
Screening test	*Present*	*Absent*
Positive	a	b
Negative	c	d

Sensitivity = a/(a+c)
Specificity = d/(b+d)
[a]DPCP, detectable preclinical phase.

As with clinical diagnostic procedures, the screening test is not always unambiguously positive or negative, and classification depends on the subjective judgment of the individual who is interpreting the test. The test result is often quantitative or semi-quantitative rather than dichotomous (positive or negative). Serum prostate-specific antigen (PSA) concentration is measured on a continuous scale and the results of Pap smear for cervical cancer were originally given on a 5-point scale of increasing degree of suspected malignancy (I, normal; II, benign infection; III, suspicious lesion; IV, probably malignant; V, malignant). The repeatability and validity of any classification are imperfect, because of both subjective interpretation and other sources of variation in the screening test. Because of this ambiguity, the cut-off point on the scale classifying the population in terms of screening positives and negatives can be selected at varying levels. The selection of the cut-off point crucially affects the test performance. Definition of a cut-off level influences both sensitivity and specificity in an opposite, counterbalancing fashion: a gain in sensitivity is inevitably accompanied by a loss in specificity.

Selection of a particular cut-off point will fix a particular combination of specificity and sensitivity. Several approaches have been proposed to select the cut-off point to distinguish best the high-risk group from the average-risk population. The simplest approach is to accept that cut-off which minimizes the total proportion of misclassification, which is equivalent to maximizing the sum of sensitivity and specificity. However, these two components of validity have different implications, which cannot be directly compared. Sensitivity is mainly related to the objective of screening and specificity to the adverse effects. When the total number of misclassified cases is minimized, sensitivity and specificity are considered of equal importance. This may be problematic, because it implies that false negatives and false positives have a similar impact. The importance of sensitivity relative to specificity implies a weighted sum of misclassification as the basis to find a correct cut-off point. However, there are no objective weightings for sensitivity and specificity. Selection of a particular combination for validity components, that is, sensitivity and specificity, always involves value judgment.

This does not mean that any combination of specificity and sensitivity is acceptable. It depends on the test and the disease to be screened: for Pap smear some false positives are regarded as acceptable. This is because the yield is of primary importance and confirmation of the diagnosis is regarded, not necessarily correctly, as relatively reliable, noninvasive, and inexpensive. False-positive diagnoses pose a problem in breast cancer screening, because some degree of abnormality is common in the breast, but confirmation of a positive test is rather expensive. Low specificity is problematic because of the potentially high cost of diagnostic confirmation and exceeding the capacity of clinical diagnostic services with screen-positive cases.

Episode validity describes the ability of the screening episode to detect disease in the DPCP and to identify those who are healthy. Attempts to confirm the diagnosis after a positive screening test may fail to identify the disease and a (true-positive) case may thus be labeled as a false-positive screening test. For example, the PSA test has a high sensitivity (Stenman *et al.*, 1994), but biopsy may fail to identify the malignant lesion. Therefore, many cases that are in the DPCP will not be diagnosed during the screening episode. The difference between test sensitivity and episode sensitivity is obvious for a screening test that is independent of the biopsy-based confirmation process. Screening for cervical cancer is based on exfoliated cells and the lesion where the malignant cells originated may not be detected at colposcopic biopsy. Even a biopsy for a breast cancer seen on screening mammography may fail to contain the malignant tissue.

Program validity is a public health indicator. Program sensitivity is related to the yield of the screening program, the detection of disease in DPCP in the target population. The program specificity indicates the correct identification of subjects free of the disease in the target population. Program validity depends on the screening test, confirmation of the test, attendance, the screening interval, and the success of referral for diagnostic confirmation of screen-positive cases.

Predictive Values of a Screening Test

Estimates of sensitivity are derived from the screening program itself. The most immediate indication of sensitivity is the yield, or cases detected at screen. The rate of detection is insufficient because it does not as such consider the total burden that stems from the cancers detected both at screen and clinically in between the screens if repeated. The relationship between these interval cancers and screen-detected cancers is not recommended as a measure of sensitivity because the cases that would not surface clinically if the population were not screened (overdiagnosis) cause a bias. Cancers diagnosed in nonattenders (those invited but not attending) at screening and in an independent control population contribute in estimating and understanding the screening validity. Methods are available that allow unbiased and comparable estimation of test, episode, and program sensitivity (IARC, 2005; Hakama *et al.*, 2007).

For the screenee, it is important to know the consequences of the result of a screening test and episode (including also the subsequent examinations). These can be described by the predictive values of a test and episode (**Table 2**). The positive predictive value (PPV) is the proportion of persons who do have unrecognized (preclinical) disease among those who have a positive test or episode. The negative predictive value (NPV) is the proportion of those who are free from the target condition

Table 2 Predictive values of a screening test

	Disease	
Screening	*Present*	*Absent*
Positive	a	b
Negative	c	d

Positive predictive value = a/(a+b)
Negative predictive value = d/(c+d)

Table 3 The effect of preclinical prevalence on the predictive value assuming specificity = sensitivity = 95%

	Predictive value (%)	
Prevalence in DPCP[a] *(%)*	*Positive*	*Negative*
10	68	99.4
1	16	99.9
0.1	2	99.99

[a]DPCP, detectable preclinical phase.

among those with a negative episode. The predictive values depend on the validity of the test and the episode and on the prevalence of the disease in the DPCP (**Table 3**). High predictive values require valid screening and diagnostic tests. Particularly for a rare disease, PPV is usually low and the majority of positive screening tests will then occur among those who do not have the disease. In contrast, if the prevalence is low, the NPV is high, that is, a negative test gives a very high probability of absence of the disease. Many of those who attend a screening program are seeking reassurance that they do not have the disease. In practice, this is the most frequent benefit of screening for rare diseases, and it emphasizes the importance of high specificity.

Evaluating the Effect of Screening

An effective program shows an impact in the process indicators, that is, intermediate end points in screening. Such proxy measures reflect the performance of the program, but are only indirect indicators of the ultimate goal, which is mortality reduction. To be effective, mammographic screening for breast cancer must provide sufficient coverage of the target population, identify preclinical breast cancer reliably, and lead to the detection of early cancers that are more curable than those diagnosed without screening. An evaluation of the effect of a screening program based on process indicators alone is inadequate: these are necessary but not sufficient requirements for effectiveness, because an ineffective program may still produce favorable changes in process indicators.

The screening test detects disease in the DPCP and the yield depends on the prevalence of unrecognized disease. Prevalence depends on the length of DPCP. For many cancers, the length of DPCP correlates with the prognosis: fast-growing cancers with a short DPCP have a poor prognosis. Screening detects a disproportionate number of slow-growing cancers compared with normal clinical practice, especially when pursuing a high sensitivity. Therefore, screen-detected disease tends to have a more favorable survival than clinically detected disease, because the cancers are selected to be more slow-growing than clinically detected ones. The bias introduced by this selection is called length bias (Feinleib and Zelen, 1969). Length bias cannot be directly estimated and adjustment for it is cumbersome. Therefore, study designs and measures of effect that are free from length bias should be used to assess the effectiveness of screening. They are based on the total target population. Randomized screening trials evaluating mortality outcome are free from bias caused by overdiagnosis, length bias, and lead time.

Lead time (Hutchison and Shapiro, 1968) is the amount of time by which the diagnosis of disease is brought forward as a result of screening compared with the absence of screening. By definition, an effective screening program gives some lead time, since earlier diagnosis is a requirement for achieving the goals of screening. The maximum lead time is equivalent to the length of the DPCP, or sojourn time. Therefore, even if screening does not postpone death, survival from the time of diagnosis is longer for a screen-detected case than for a clinically detected case. Comparison of survival between screen-detected and symptom-detected patients is therefore biased unless it can be corrected for lead time. Such corrections remain crude at best, and survival is not a valid indicator of the effectiveness of screening.

Length bias and lead time are theoretical concepts of key importance in screening. Empirical assessment of length bias and lead time can provide valuable information on the natural history of the disease and the effects of screening (Day *et al.*, 1984). These sources of bias, in addition to overdiagnosis, make indicators such as survival unsuitable for evaluating the effectiveness of screening programs.

Evaluation of the effectiveness of screening should therefore be made in terms of the outcome, which for cancer is mortality. However, screening programs also affect morbidity and, more broadly, quality of life. Such effects should also be taken into account in screening decisions although they should not be mixed with process indicators of the program. For example, fertility may be maintained after treatment for a screen-detected precursor lesion of cervical cancer, and breast-conserving surgery with less cosmetic and functional impairment may be used for screen-detected breast cancer compared with clinically detected cancer. Such procedures may reduce physical invalidity and adverse mental effects, and thereby improve the quality of life. However, the process measures, for example, the number of conisations of the cervix or breast-conserving surgery or their proportion of all surgical procedures, are invalid indicators of effect, because of potential overdiagnosis. Overdiagnosis increases the proportion of cases with favorable features because healthy individuals are misclassified as cases of disease.

A randomized preventive trial with mortality as the end point is the optimal and often the only valid means of evaluating the effectiveness of a screening program. Cohort and case-control studies are often used as a substitute for trials. Most evidence on the effectiveness of screening programs stems from comparisons of time trends and geographical differences between populations subjected to screening of different intensity. These nonexperimental approaches remain crude and insensitive, however, and do not provide a solid basis for decision making.

Several biases may distort comparability between attenders and nonattenders at screening. The most obvious is the 'healthy screenee' effect: an individual must be well enough to attend if he or she is to participate. Furthermore, a patient with, for example, a previous diagnosis or who is already under close medical surveillance due to a related disorder has no reason to attend a program aimed at detection of that disease. Elimination of bias in a nonexperimental design is likely to remain incomplete. In randomized trials, previously diagnosed cancer is an exclusion criterion, that is, prevalent cases are excluded and randomization with analysis based on the intention-to-screen principle guarantees comparability.

The randomized screening effectiveness trials with mortality as end point provide a solid basis for a routine screening program that is run as a public health policy. Such mass screening programs should be evaluated and monitored. Intervention studies without a control group (also called demonstration projects or single-arm trials) and other nonexperimental designs (cohort and case-control studies) have been proposed for this type of evaluation, but each approach has inherent biases. A randomized approach with controls must be considered the gold standard, to be adopted if possible.

All routine screening programs are introduced gradually. Extension from the initial stage requires time and planning and involves more facilities. In several countries, use of Pap smears developed from being very infrequent to become part of normal gynecological practice or public health service within about 10 years. It follows that a screening program can be introduced as a public health policy in an experimental design, with comparison of screened and unscreened groups formed by random allocation. Under such circumstances, provision of screening can be limited to a randomly allocated sample of the population, instead of a self-selected or haphazardly selected fraction of the population. As long as the resources for the program are only adequate for a proportion of the population, it is ethically acceptable to randomize, because screening is not withheld from anybody. There is, *a priori*, an equal chance for everybody in the target population to receive the potential benefit of the program and to avoid any adverse effects of the program. In this context, the equipoise (lack of firm evidence for or against an intervention), which is an ethical requirement for conducting a randomized trial, gradually disappears as evidence is accrued within the program. For those planning public health services, this will provide the most reliable basis for accepting or withholding new activities within the services.

Organizing a Screening Program

Screening is a chain of activities that starts from defining the target population and extends to the treatment and follow-up of screen-detected patients. A screening program consists of several elements that are linked together.

Different cancer screening programs consist of different components. In general, they can be outlined as eight distinct steps, divided into four components.

Population component:

1. Definition of target population.
2. Identification of individuals.
3. Measures to achieve sufficient coverage and attendance, such as personal letter of invitation.

Test execution component:

4. Test facilities for collection and analysis of the screen material.
5. Organized quality control program for both obtaining screen material and its analysis.

Clinical component:

6. Adequate facilities for diagnosis, treatment, and follow-up of patients with screen-detected disease.

Coordination component:

7. A referral system linking the screenee, laboratory (providing information about normal screening tests), and clinical facility (responsible for diagnostic examinations following an abnormal screening test and management of screen-detected abnormalities).
8. Monitoring, quality control, and evaluation of the program: availability of incidence and mortality rates for the entire target population, and separately for both attenders and non-attenders.

Routine screening can be divided into opportunistic (spontaneous, unorganized) and organized screening (mass screening, screening program). The major differences are related to the level of organization and planning, systematic nature, and scope of activity. The components described previously are characteristics of an organized screening program. Most of them are not found in opportunistic screening.

Spontaneous screening frequently focuses on high sensitivity at the cost of low specificity. This is due to several factors including economic incentives (fee-for-service) and risk-averse behavioral models (avoiding neglect, fear of litigation). However, more emphasis on specificity would be consistent with the primary ethical responsibilities of the physician under the Hippocratic oath: First, do no harm. For a high-technology screening program, such as mammography for breast cancer, low specificity results in high cost and frequent adverse effects. Furthermore, in countries with limited resources for the follow-up of screen-positive cases, screening competes for scarce resources that could be used for the treatment of overt disease with worse outcomes.

Major organizational considerations of any screening program are the age range to be covered and screening interval. For example, in Western populations with similar risk of disease and available resources, cervical cancer screening policies range from annual testing from the start of sexual activity to a smear every 5 years from age 30 years to age 55 years. Hence, the difference in the cumulative number of tests over a lifetime varies from 6 to more than 60, that is, 10-fold.

In general, only organized screening programs can be evaluated. An organized program with individual invitations can prevent excessively frequent screening and overuse of the service, and is therefore less expensive. If screening appears ineffective, an organized screening program is also easier to close than a spontaneous activity. Therefore, organized screening programs should be recommended over opportunistic screening.

Cancer Screening by Primary Site

The following sections on cervical, breast, and colorectal cancer screening describe cancer screening with proven effectiveness.

Cervical Cancer Screening

Cervical cancer is the second most common cancer among women worldwide, with 529 000 new cases in 2008 and 275 000 deaths. The great majority of the burden is in developing countries, with highest incidence rates in Africa and Latin America.

Natural history

Cervical cancer is thought to develop gradually, through a progression of a series of precursor lesions from mild abnormality (atypia) into more aberrant lesions (dysplasia) and eventually malignant changes (initially *in situ*, then microinvasive and finally invasive carcinoma). Human papillomavirus (HPV) infection is a common early event. Most infections are cleared spontaneously within 6–12 months or less. Early precursor lesions seem to be a consequence of persistent infection with oncogenic HPV types. High-grade neoplasia occurs rarely without persistent HPV infection and viral load also predicts the probability of progression. Regression of lesions occurs commonly at early stages and the rate of progression is likely to increase during the process, with accumulation of abnormalities. The duration of the detectable, preclinical phase has been estimated at as long as 12–16 years.

Screening with cervical smears

Screening for cervical cancer is based on detecting and treating unrecognized disease, primarily premalignant lesions to prevent their progression into invasive carcinoma. Traditional techniques are based on cytological sampling of cells from the cervix. The classical smear is based on cytological assessment of exfoliated cervical cells from the transformation zone, where the squamous epithelium changes into columnar epithelium. The sample is collected from the vaginal part of the cervix using a spatula and from the endocervix with a brush or swab. Sampled cells are fixed on a glass slide for evaluation with microscope.

There are several classifications for cytological and histopathological abnormalities of the cervix. The present Bethesda 2001 system consists of two classes of benign atypical findings (atypia thought to occur as reactive change related to infection; atypical squamous cells of unknown significance [ASCUS], or a result that cannot exclude higher grade lesion, ASC-H) and two classes of more aberrant changes (low- and high-grade squamous intraepithelial lesion, LSIL and HSIL, respectively). The diagnostic classification of preinvasive changes is based on cervical intraepithelial neoplasia (originally CIN1–3, corresponding to mild, moderate, and severe dysplasia [including carcinoma *in situ*], modified by combining mild, moderate, and *in situ* into high-grade CIN).

Diagnostic assessment requires colposcopic examination (microscopic visualization with 6- to 40-fold magnification) for assessment of morphological features of the cervix. Histologic assessment is based on colposcopy-directed punch or cone biopsy.

The incidence of cervical cancer is very low in the first year following a negative smear and increases gradually, returning to baseline at about 10 years. The risk is still 50% lower at 5 years, compared with unscreened women. The risk of cancer decreases further with the number of consecutive negative tests.

Effectiveness of screening

The objective of cervical cancer screening is to reduce both cervical cancer incidence and mortality. A successful screening program detects early, preinvasive lesions during the preclinical detectable phase and is able to reduce deaths by preventing the occurrence of invasive cancer.

No randomized trials were conducted to evaluate the mortality effects of cervical cancer screening at the time when it was introduced. There is, however, evidence for substantial incidence and mortality reduction from non-randomized studies conducted in several countries when screening was introduced. In such studies, a screened population is identified at the individual level. The screened and unscreened women may, however, not be comparable in terms of disease risk, which can induce selection bias. Several such studies have been summarized by IARC (2005). In British Columbia, Canada, a screening program was introduced in 1949. During 1958–1966, the incidence of cervical cancer among 310 000 women screened at least once was well below the rates preceding the screening project (standardized incidence ratio (SIR) 0.16, 13 cases), while 230 000 unscreened women showed no decrease (SIR 1.08, 67 cases). In Finland, a population-based cervical cancer screening program was started in 1963 and evaluation of more than 400 000 women covered showed effectiveness of 60% at 10 years in terms of incidence reduction. A Norwegian study with 46 000 women invited for screening showed approximately 20% lower cervical cancer incidence and mortality among participants than among the reference population.

Several case-control studies have been carried out in Denmark, Latin America, Canada, South Africa, and other countries to evaluate the efficacy of cervical cancer screening. Most have shown odds ratios in the range of 0.3–0.4 for invasive cervical cancer associated with ever versus never having had a screening test, but the results are prone to selection bias (IARC, 2005).

In ecological analyses without individual data, introduction of screening has been associated with a reduction in cervical cancer incidence and mortality. Major differences in the timing of introduction and the extent of cervical cancer screening between Nordic countries with similar baseline risk has allowed assessment of the effect of screening on cervical cancer incidence and mortality. Adoption of screening as a public health policy has been followed by a sharp reduction in cervical cancer occurrence. Comparison between counties in Denmark with and without cervical cancer screening showed that both incidence of and mortality from the disease were lower by a third in the areas with organized mass screening. Conversely, in an area where organized screening had been discontinued, increased incidence of invasive cancer was found. An analysis of 15 European countries was also consistent with a 30–50% reduction in cervical cancer incidence related to organized cervical cancer screening. Similar reductions in cancer incidence and/or mortality have been reported after launching screening programs in England and Wales, the USA, and Australia.

Screening programs

The recommended interval between screens has ranged from 1 to 5 years. Most screening programs start from 18–30 years of age and are discontinued after age 60–70 years. The most intensive screening protocols with an early start and frequent testing involve 10 times as many tests over a woman's lifetime as the most conservative approaches. Organized programs with large, population-based target groups tend to be least intensive, but nevertheless able to produce better results than less organized efforts, with ambitious screening regimens but very incomplete coverage of the target population and a weaker link among the program components (testing, diagnosis, and treatment). In some programs, the frequency of screening has been modified according to the screening result, either starting at annual screening, for example, and increasing the interval after negative results, or conversely, offering initially a longer, 3–5-year interval that is shortened if there is any abnormality.

Adverse effects of screening

Overdiagnosis of preinvasive lesions (i.e., detection and treatment of changes that would not have progressed into malignancy) appears common in cervical cancer screening, as only a small proportion of preinvasive lesions would develop into a cancer, even if left untreated. The cumulative risk of an abnormal screening test is relatively high compared with lifetime risk of cancer in the absence of screening (10–15% or higher versus approximately 3%). Treatment has several adverse effects. Excisional treatments are associated with pregnancy complications, including preterm delivery and low birthweight. Hysterectomy, quite obviously, leads to loss of fertility.

Other screening tests

Direct visualization of the cervix has been evaluated as a screening method that does not require sophisticated technology or highly trained personnel. Unaided visual inspection (also known as downstaging) can identify bleeding, erosion, and hypertrophy. Low-level magnification (× 2–4) has not been shown to improve performance of visual inspection. Use of acetic acid in visual inspection results in white staining of possible neoplastic changes and improves test sensitivity. Lugol's solution (or Schiller's iodine test) stains neoplastic epithelium yellow and it appears to have similar or better performance than acetic acid. Visual inspection with iodine solution or acetic acid offers an effective and affordable screening approach. With trained personnel and quality control a 25% reduction in cervical cancer mortality can be achieved (Sankaranarayanan *et al.*, 2007).

Commercially available HPV tests are based on nucleic acid hybridization and are able to identify more than 10 different HPV types. Results of trials comparing effectiveness of HPV testing with cytological smears indicate that HPV screening is likely to be at least as effective as screening based on smears, but is also likely to have more adverse effects including lower specificity. Etiology-based screening may label cervical cancer as a sexually transmitted infection, despite common non-sexual transmission of the virus. This may reduce the acceptability of screening and reduce participation. One hazard is the stigmatization of the woman carrying the virus, regardless of the route of transmission (e.g., infidelity of the woman's partner), especially in cultures where the norms guard female sexuality more strictly than that of men.

In summary, the effectiveness of cytological smears in cervical cancer screening has never been established with current, methodologically stringent criteria. However, there is extensive and consistent evidence showing that a well-organized screening program will reduce both the incidence of and mortality from invasive carcinoma.

In the future, HPV vaccination has the potential to influence profoundly the conditions in which screening operates, and possibly to reduce the demand for cervical cancer screening by lowering the risk of the disease. This may take at least one generation to achieve.

Breast Cancer Screening

Breast cancer is the most common cancer among women and it accounts for a fifth of all cancers among women worldwide. Approximately 1 383 000 new cases occurred in the world in 2008. The number of breast cancer deaths in 2008 was estimated at 458 000. Increasing incidence rates have been reported in most populations. The highest incidence rates have been reported in high-income countries, in North America, Europe, as well as in Australia and New Zealand. Yet, in terms of numbers of cases, the burden of breast cancer is comparable in rich and poor countries of the world.

Natural history

In a synthesis of seven reports on breast cancer as an autopsy finding, the median prevalence of invasive breast cancer at autopsy was 1.3%. The mean sojourn time has been estimated at 2–8 years. Sojourn times tend to be longer for older ages and may depend on the histological type of breast cancer.

Mammography screening

In screening, the primary target lesion is early invasive cancer, but ductal carcinoma *in situ* is also detected with a frequency of about 10–20% that of invasive cancer.

Mammography is X-ray imaging of the breast with a single or two views read by one or two radiologists. The screen-positive finding is a lesion suspicious for breast cancer, appearing typically as an irregular, starlike lesion or clustered microcalcifications. Two views are likely to increase detection rates by approximately 20%, with most benefit for detection of small cancers among women with radiologically dense breasts. In some screening programs, two views are used only at first screening, with only one view (mediolateral oblique) subsequently. Similarly, double reading appears to increase both the recall rate and detection of breast cancer by some 10%. Currently, digital mammography is replacing film technology.

Screen-positive findings result in the recall of 2–15% of women for additional mammographic examinations, with biopsy in approximately half of these women. Detection rates vary between populations of 3–11 per 1000, corresponding to PPVs of 30–70% for biopsied women. Sensitivity has been estimated at 60–70% in most randomized trials and close to that also in service screening. Specificity (the proportion of true negatives among all negative screens) has been reported at 95–98% in most studies, with lower figures in the initial (prevalence) screen than in subsequent screens.

Effectiveness of screening

Twelve randomized trials have evaluated mortality reduction in mammography screening (**Table 4**). In the age range 50–69 years, the trials have shown relatively consistent mortality reduction in the range of 20–35%. Excluding the Canadian trial with women 50–59 years, which showed no benefit, estimates of the number of women needing to be screened (NNS) to prevent one death from breast cancer has ranged from 1000 to 10 000 women at 10 years.

The Health Insurance Plan trial in New York was the first randomized screening trial. It had approximately 60 000 women randomized pair-wise. The subjects were aged 40–64 years at entry. Four screening rounds with 1-year intervals were performed using two-view mammography and clinical breast examination (CBE). CBE was also used in the control arm. Causes of death were evaluated by a committee unaware of the allocation. Breast cancer mortality was reduced in the screening arm by 20% (5.5 vs. 6.9 per 100 000). In the analysis with the longest follow-up, the mortality difference persisted after 18 years of follow-up.

Several cluster-randomized screening trials have been carried out in Sweden. In the Kopparberg trial with three screening rounds, an 18% reduction in breast cancer mortality was reported after 20 years of follow-up. The Östergötland study showed 13% lower breast cancer mortality in the screening arm after four screening rounds and 17 years of follow-up. The Malmö trial with 22 000 women aged 45–69 years reported an 18% reduction in breast cancer mortality for the screening group after 19 years. In the Stockholm trial, nearly 40 000 women aged 40–64 years were enrolled and breast cancer mortality was 12% lower in the screening group after 15 years of observation.

Two mammography screening trials have been conducted in Canada. Both were started simultaneously, one

Table 4 Randomized trials evaluating mortality effects of mammography screening

Reference	*Setting*	*Sample size*	*Age range*	*Follow-up (years)*	*Mortality[a] (10^{-5})*
Shapiro *et al.*, 1988	Greater NY	60 995	40–64	18	23/29
Andersson and Janzon, 1997	Malmö	42 283	45–70	19	45/55
	Malmö	17 793	43–49	9	26/38
Tabár *et al.*, 2000	Kopparberg	56 448	40–74	20	27/33
Nyström *et al.*, 2002	Östergotland	76 617	40–74	17	30/33
Alexander *et al.*, 1999	Edinburgh	52 654	45–64	13	34/42
Miller *et al.*, 2002	Canada	50 430	40–49	13	37/38
Miller *et al.*, 2000	Canada	39 405	50–59	13	50/49
Moss *et al.*, 2006	UK	160 921	39–41	11	18/22
Nyström *et al.*, 2002	Stockholm	60 117	40–64	15	15/17
Bjurstam *et al.*, 2003	Gothenburg	51 611	39–59	13	23/30
Hakama *et al.*, 1997	Finland	158 755	50–64	4	16/21

[a]Screening/control.

enrolling volunteers aged 40–49 years (Canadian National Breast Screening Study (CNBBS1) 50 000 women) and the other 50–59 years (NBBS2 40 000 women). Intervention was annual two-view mammography and CBE for 5 years in both trials, with no intervention for the control arm in the younger group and CBE only in the older cohort. No reduction in breast cancer mortality was found in either age group in analyses covering 13 years of follow-up.

The Edinburgh trial used cluster randomization based on general practices, with more than 44 000 women aged 45–64 years at entry. The screening interval was 12 or 24 months and covered four screening rounds. After 13 years of follow-up, breast cancer mortality was 19% lower in the screening arm.

The randomized trials have been criticized for methodological weaknesses (Olsen and Gøtzsche, 2001), especially for incorrect randomization and postrandomization exclusions leading to lack of comparability between the trial arms. In a systematic review excluding studies with possible shortcomings, only two trials were finally evaluated and no benefit was demonstrated (Olsen and Gøtzsche, 2001). It was also argued that breast cancer mortality is not a valid end point for screening trials. However, the validity of these criticisms has been rebutted by several investigators and international working groups. The dismissal of studies based on mechanistic evaluation of technical criteria of questionable relevance has been considered inappropriate.

Service screening

National breast cancer screening programs are ongoing in several European countries. They are organized either regionally or nationally and use guidelines and quality assurance systems for both radiology and pathology. Common features are target groups at ages 50–69 years and 2-year intervals. In several Northern European countries, participation around 80% has been achieved with recall rates of 1–8%. PPVs have been in the range of 5–10% and detection rate of invasive cancer generally has been at 4–10 per 1000 in the initial screen and 2–5 per 1000 in subsequent rounds (IARC, 2002). In the evaluation of effectiveness, nonrandomized approaches have been used (with the exception of Finland where screening was introduced using a randomized design). In such studies, definition of controls and estimation of expected risk of death is problematic. Extrapolations of time trends and comparisons between geographical areas have been used in evaluation of effectiveness. Yet, comparability between screened and nonscreened groups in such studies is questionable. Also, exclusion criteria are not as strict as in randomized trials and exclusion of prevalent cancers at baseline has not always been possible.

In Finland, the female population was divided into a screening group and a control group based on birth year. The women were aged 50–59 years at entry and 89 000 women born on even years were invited to mammography screening every second year, while 68 000 born on odd years were not invited but served as a reference group. Participation was 85%. Refined mortality was used, that is, breast cancers diagnosed before the start of the screening program were excluded. Overall, a 24% reduction in breast cancer mortality was reported at 5 years, which did not reach statistical significance.

In the Netherlands, a statistically significant reduction in breast cancer mortality was reported following introduction of mammography screening. The target age group was 50–69 years and participation reached 80%. Compared to rates before screening, the reduction in mortality was 19% at 11 years of follow-up (85.3 vs. 105.2 per 100 000). No similar decrease was found for women in older age groups, suggesting that the reduction was attributable to screening and not to improvement in treatment.

In an evaluation of mammography screening in Sweden, mortality from breast cancers diagnosed during the screening period was evaluated in seven counties between 1978 and 1990. A statistically significant 32% reduction in refined breast cancer mortality was observed following introduction of screening in counties with

10 years of screening and an 18% reduction in areas with shorter screening periods. A more recent assessment showed a 26% mortality reduction after 11 years among 109 000 women with early screening, compared with areas with later introduction of screening.

In England and Wales, breast cancer mortality decreased by 21% after introduction of mammography screening for women aged 50–69 years compared to that expected in the absence of screening (predicted from underlying trend). The estimated reduction in breast cancer mortality gained by screening was 6%, while the rest was attributed to improvements in treatment.

A Danish study showed a significant 25% reduction in nonrefined breast cancer mortality within 10 years after introduction of screening in Copenhagen for the age group 50–69 years compared with earlier rates and control areas.

No substantial effect on breast cancer mortality at population level was shown in an early analysis of a screening program in Florence, Italy. During the first 9 years of the program, breast cancer mortality declined by a mere 3%, which was attributed to the relatively low coverage of the program (60%).

In a comparison of Swedish counties that introduced mass screening with mammography in 1986–87 with those starting after 1992, only a nonsignificant 10% mortality reduction from breast cancer was found at 10 years among women aged 50–69 years at entry. Among women aged 70–74 years, the mortality reduction was even less (6%).

The results from nonrandomized evaluation are consistent with a mortality reduction obtained in screening trials and suggest that mortality reduction is achievable when mammography screening is applied as public health policy, though it may be somewhat less than the average effect of approximately 25% seen in randomized trials. Further, improvement in quality of life can be gained by early diagnosis, allowing a wider range of treatment options, with the possibility of avoiding radical surgery (and possibly adjuvant chemotherapy).

Adverse effects

The extent of overdiagnosis and subsequent unnecessary treatment of lesions that would not have progressed may not be as large for breast cancer screening as for several other cancer types. The estimates of overdiagnosis have been in the range of 3–5%.

Preinvasive cancer (*in situ* carcinoma) is detected at screening with a frequency of approximately 1 per 1000 screens and is commonly treated surgically, even if all cases would not progress to cancer. Mammography causes a small radiation dose (1–2 mGy) to the breast, which can be expected to increase breast cancer risk. The excess risk is, however, likely to remain very small (in the range of a 1–3% or less increase in the relative risk).

Other screening tests

Digital mammography has been adopted recently. It appears to yield a higher detection rate than conventional film mammography, but correspondingly, specificity seems lower. It remains unclear if the higher detection rate also increases overdiagnosis. No studies have evaluated the effect of digital mammography on breast cancer mortality.

Studies of magnetic resonance imaging among high-risk groups have suggested higher sensitivity compared with mammography, but no randomized trials have compared its effect on mortality with mammography. An advantage is avoidance of exposure to ionizing radiation. Yet, it is also more expensive and time-consuming.

CBE consists of inspection and palpation of the breasts by a health professional to identify lumps or other lesions suspicious for cancer. No randomized trials have evaluated the effectiveness of CBE alone, but it was included in the intervention arm of the Health Insurance Plan (HIP), Canadian, and Edinburgh trials. It may increase the sensitivity of screening if combined as an ancillary test in a mammography screening program.

Breast self-examination (BSE) has been evaluated as a resource-sparing option for early detection of breast cancer. A randomized trial in Shanghai, People's Republic of China, showed no reduction in breast cancer mortality following instruction in BSE. A similar conclusion was reached also in a trial carried out in the former Soviet Union.

Several randomized trials have shown that mammography screening can reduce mortality from breast cancer and evidence from studies evaluating service screening indicates a similar or slightly smaller effect at population level. The age group with most benefit is 50–69 years. The methodological limitations of the studies are not severe enough to justify ignoring their results.

Colorectal Cancer Screening

Colorectal cancer ranks as the second most common cause of cancer death. There were some 1 233 000 new cases in 2008, with an estimated 608 000 deaths.

Natural history

The majority of colorectal carcinomas are thought to arise from benign precursor lesions (adenoma). Adenoma (particularly those with a diameter of up to 1 cm, or dysplasia) and early carcinoma make up the principal target of screening. The duration of the detectable preclinical phase has been estimated at 2–6 years.

Screening tests

Several screening methods are available for colorectal cancer screening, including fecal occult blood testing (FOBT) and endoscopic examination (sigmoidoscopy or colonoscopy) as well as radiographic examination (double-contrast barium enema).

FOBT is based on detection of hemoglobin in stools using guaiac-impregnated patches, where an oxidative reaction (pseudoperoxidase activity) results in color change, which is detectable on inspection. The most commonly used test, Hemoccult II, is not specific to human blood. Other tests are also available that immunologically detect human hemoglobin, but they are also more expensive. Rehydration (adding water to the specimen) can be used to increase the detection rate, but this also leads to more false-positive results. For screening, two specimens are usually obtained on 3 consecutive days. Dietary restrictions (avoiding red meat, vitamin C, and nonsteroidal anti-inflammatory drugs) and combination of tests may increase specificity, but can also reduce acceptability.

The detection rate of carcinoma with FOBT has been 0.2–0.5% for biennial screening. Sensitivity is considerably lower for detection of polyps, which do not bleed as frequently as cancers. The PPV has been 10% for cancer and 25–50% for adenoma.

Effectiveness of FOBT

Two-year screening intervals have been most widely used and most studies have targeted the age groups 45–75 years. Three randomized trials evaluating incidence and mortality have been reported (**Table 5**). They show a consistent 6–18% reduction in mortality with biennial screening. A systematic review showed a 16% (95% confidence interval [CI] 7–23%) reduction in mortality. The NNS to prevent one colorectal cancer death was estimated as less than 1200 at 10 years, given two-thirds participation.

In the Nottingham (UK) trial, more than 150 000 subjects aged 45–74 years were recruited between 1981 and 1991 and randomized by household. No reduction in colorectal cancer incidence was shown, but mortality was 19% lower in the screening arm after a median 12 years of follow-up.

In the Danish Funen study, approximately 62 000 subjects aged 45–75 years were enrolled in 1985. Randomization was performed in blocks of 14 persons (with spouses always in the same arm). The mortality reduction was 11% after a mean follow-up of 14 years. No decrease in colorectal cancer incidence was observed.

In the Minnesota (USA) trial, 46 551 volunteers aged 50–80 years were recruited between 1975 and 1977. After a mean follow-up of 15 years, a mortality reduction of 33% was shown in the annual screening group and 21% in the biennial screening group, compared with the control arm. The incidence of colorectal cancer was also approximately 20% lower in the screened groups.

Service screening

Provision of colorectal cancer screening for the population has been tested in several countries. In France, a 16% mortality reduction was reported in a population of 90 000 offered screening, compared with control districts in the same administrative area. The incidence of colorectal cancer was similar in both populations during 11 years of follow-up.

Finland was the first country to launch a population-based FOBT screening program, in 2004. It was introduced using individual randomization, but the effects on colorectal cancer incidence and mortality have not been evaluated yet.

Adverse effects of screening

FOBT is safe, but a positive result requires further diagnostic examinations such as endoscopy, which cause inconvenience, rare complications (e.g., perforation), and costs. Some overdiagnosis is likely, because not all precursor lesions would advance to cancer. Yet, the morbidity related to the removal of polyps is very moderate.

Other screening tests

Flexible sigmoidoscopy covers approximately 60 cm of the distal colon, where roughly half of all colorectal cancers occur. Any adenomas detected can be removed during the procedure. Compliance with sigmoidoscopy as a screening investigation has only been 50% or less. The detection rate is higher than with FOBT, suggesting higher sensitivity. Three case-control studies and a cohort study have suggested lower mortality from colorectal carcinoma following sigmoidoscopy, as well as lower incidence. These studies do not provide evidence as strong as that from

Table 5 Randomized trials evaluating mortality effects of colorectal cancer screening based on fecal occult blood testing (FOBT)

Reference (setting)	*Sample size*	*Age range*	*Follow-up (years)*	*Mortality (RR)*[a]
Mandel *et al.*, 1999 (Minnesota, USA)	46 551[b]	50–80	15	0.67 (0.51–0.87)
Scholefield *et al.*, 2002 (Nottingham, UK)	152 850	45–74	11	0.87 (0.78–0.97)
Kronborg *et al.*, 2004 (Fynen, Denmark)	61 933	45–75	11	0.89 (0.78–1.01)
			14	0.82 (0.69–0.97)

[a]RR, ratio of mortality rates in screening arm and no-screening arm.
[b]Three arms, annual and biennial screening with control.

randomized intervention trials, because selection bias and other systematic errors may affect the results. Therefore, the mortality reduction achievable with sigmoidoscopy remains unclear. A population-based randomized trial is ongoing in Norway with 20 000 subjects aged 50–64 years, comparing one sigmoidoscopy with no intervention. It is expected to provide important new information.

Screening colonoscopy has the advantage of covering the entire colon, but the procedure is expensive, bears substantial discomfort and has the potential for complications such as perforation (reported in 1–2 patients per 10 000). No trials have evaluated the effectiveness of screening colonoscopy.

Recently, fecal DNA analysis has been introduced as a new option for colorectal cancer screening, but no studies assessing its effectiveness have been conducted.

In summary, FOBT has been shown to decrease mortality from colorectal cancer in several randomized trials. It appears to be an underutilized opportunity for cancer control. Other screening modalities are also available, but there is currently no solid evidence for their effectiveness.

Screening for Oral and Liver Cancer

Some evidence for effectiveness is available for oral cancer and liver cancer screening, but it is not as well established as for cervical, breast, and colorectal cancers.

Oral cancer is among the most common cancers in some areas of the world, largely due to the habit of tobacco chewing. Globally, more than 270 000 cases are detected annually, primarily in developing countries. A recent cluster-randomized trial of visual inspection for oral cancer demonstrated a 20% reduction in mortality among more than 190 000 subjects (Sankaranarayanan *et al.*, 2005).

Liver cancer is the sixth most common cancer in the world, with nearly 750 000 new cases in 2008. In terms of cancer deaths it ranks third, with nearly 700 000 deaths annually. Serum alpha-fetoprotein (AFP) and ultrasound have been used as a screening test for hepatocellular cancer. Two randomized trials have been carried out in the People's Republic of China, both among chronic carriers of hepatitis B virus, who are at high risk of liver cancer. The smaller study found a nonsignificant 20% reduction associated with 6-monthly AFP tests among 5500 men in Qidong county (Zhang *et al.*, 2004). Another trial involved 18 000 people and showed a one-third mortality reduction at 5 years with 6-monthly AFP and ultrasonography (Chen *et al.*, 2003).

The following section, 'Lung Cancer Screening,' describes cancer screening with evidence against effectiveness.

Lung Cancer Screening

Lung cancer is the most common cancer in many countries. Mortality rates are very similar to incidence due to its very poor prognosis. In 2008, the global number of cases was 1.61 million and there were 1.38 million deaths.

Natural history, diagnosis, and treatment

The target lesion for lung cancer screening is early, resectable (stage 1) carcinoma. Diagnostic examinations may include high-resolution computerized tomography (CT), positron-emission tomography (PET), transthoracic needle biopsy, and thoracotomy. A conclusive diagnosis of early lung cancer is based on biopsy.

Screening with chest X-rays, with or without sputum cytology

The plain chest X-ray was evaluated as a screening test in several randomized screening trials in the 1960s and 1970s. Commonly, it was combined with cytological assessment of exfoliative cells that are most commonly detected in squamous cell carcinoma.

Small nodules are easily missed in chest X-rays and sensitivity is low, with specificity above 90%. Detection rates have been 0.1–0.8% and positive predictive value 40–60%. False-negative results in sputum cytology are common among patients with lung cancer; its sensitivity is also regarded as inferior to that of chest X-rays.

Effectiveness of screening

Four randomized trials of lung cancer screening with chest X-rays have been conducted, but only one compared chest X-ray screening against no intervention (until the end of the 3-year study period, **Table 6**). One compared chest X-ray and sputum examination offered within the trial against a recommendation to have such tests. The other two trials assessed the impact of chest X-ray alone with chest X-ray and sputum cytology. The interval between rounds ranged between 4 months and 1 year. The detection rate of lung cancer at baseline examination was 0.1–0.8%. In a meta-analysis, lung cancer mortality was increased by 11% with the more intensive screening.

Adverse effects of screening

A false-positive test leads to invasive diagnostic procedures. Overdiagnosis has been estimated as 15% based on long-term follow-up in one of the trials.

Other screening methods

In spiral low-dose CT, the screen-positive finding is a noncalcified nodule, usually at least 1 cm in diameter. For smaller lesions, follow-up examinations may be needed to define whether the nodule is growing.

Table 6 Randomized trials assessing mortality effect of lung cancer screening based on chest X-ray with or without sputum cytology

Reference (setting)	*Study subjects*[a]	*Follow-up (years)*	*Mortality rate per 1000 (no. of deaths)*[a]
Studies comparing chest X-ray during vs. after study period			
Kubik *et al.*, 1990 (Czechoslovakia)	3172+3174	6 (15)	6.0 vs. 4.5 (247 vs. 216)
Studies comparing chest X-ray with sputum cytology vs. chest X-ray alone			
Melamed *et al.*, 1984 (Sloane-Kettering, USA)	4968+5072	8	2.7 vs. 2.7
Levin *et al.*, 1982 (Johns Hopkins, USA)	5161+5225	8	3.4 vs. 3.8
Marcus *et al.*, 2000 (Mayo Clinic, USA)	4618+4593	20	4.4 vs. 3.9 (337 vs. 303)

[a]Intervention vs. control.

No randomized trials have been done, so the effect on mortality has not been established. Adverse effects due to false-positive results appear common but the extent of overdiagnosis has not been established. In the USA, the National Lung Screening Trial is comparing annual chest radiography with annual low-dose CT among 50 000 smokers.

To conclude, chest X-rays have been shown *not* to reduce mortality from lung cancer. Opportunities provided by novel radiological technology have been eagerly advocated for screening, but currently there is no evidence on the effectiveness of screening based on spiral CT.

Neuroblastoma Screening

Neuroblastoma is a tumor of the sympathetic nervous system in children, with an overall annual incidence rate of approximately 1 per 100 000 under the age of 15 years. Peak incidence is during the first year of life, and occurrence decreases after that. The natural course is highly variable. Early-stage disease, occurring mainly in young children, has a very favorable prognosis, while diagnosis at an advanced stage (and usually above 1 year of age) is associated with poorer survival. There is a subgroup of tumors with the potential to disappear spontaneously or mature into a benign tumor (ganglioneuroma), even in the presence of metastases. Hence there is obvious potential for overdiagnosis. Screening is possible using urine tests for metabolites of the neuronal transmitters, catecholamines (homovanillic acid and vanillylmandelic acid), which are secreted by most (60–80%) tumors. Studies in Germany, Canada, and Japan have compared screened and unscreened cohorts to evaluate the effects of screening. Screening has led to a two- to sixfold *increase* in the recorded incidence of neuroblastoma, and an increase at young ages has not been counterbalanced by a reduction at older ages. No reduction in mortality or in the occurrence of advanced disease has been demonstrated. Screening for neuroblastoma is therefore not recommended, even if no randomized trials have been conducted. Screening was recently discontinued in Japan.

The follow section, 'Prostate Cancer Screening,' including other cancers, describes cancer screening without sufficient evidence for or against effectiveness.

Prostate Cancer Screening

The recorded incidence of prostate cancer has increased rapidly in the past 10–15 years in most industrialized countries and it is currently the most common cancer among men in several countries. Globally, 903 000 prostate cancers were diagnosed in 2008, with 258 000 deaths from it.

Natural history

The target lesion in prostate cancer screening is early invasive prostate cancer. The natural course of prostate cancer is highly variable, ranging from indolent to highly aggressive. Premalignant lesions such as prostatic intraepithelial neoplasia (PIN) exist, but are not strongly predictive of prostate cancer and are not considered as an indication for treatment. Latent prostate cancer is a common autopsy finding. It has been detected in more than 10% of men dying before the age of 50 years, and much more frequently in older men. The common occurrence of indolent prostate cancer is a clear indication for potential overdiagnosis. The mean lead time in prostate cancer has been estimated as 6–12 years.

Prostate cancer screening based on PSA

PSA is a serine protease secreted by the prostate and it is usually found in low concentrations in serum, with levels increased by prostate diseases such as benign prostatic hyperplasia, prostatitis, or prostate cancer.

The specificity of PSA has been estimated as approximately 90%. Detection rates have ranged from less than 2% to above 5%, depending on the cut-off and population. In serum bank studies, where no screening has been offered, but baseline PSA levels have been used to predict subsequent incidence of prostate cancer, sensitivity has been estimated as 67–86% at 4–6 years.

Effectiveness of PSA screening

Mortality analysis has been published from only one, relatively small, randomized trial. A study carried out in Quebec, Canada, randomized 46 500 men aged 45–80 years in 1988 with two-thirds allocated into the screening arm (Labrie *et al.*, 2004). Compliance with screening was only

24%. Screening interval was 1 year, with PSA cut-off 3 ng/ml. By the end of 1999, the mean length of follow-up was 10 years among unscreened men and 7 years in the screened group. When analyzed by trial arm (intention-to-screen analysis), no reduction in prostate cancer mortality was seen.

Two large randomized trials are being carried out, one in Europe and the other in the USA. The European Randomized trial of Screening for Prostate Cancer (ERSPC) has eight centers in the Netherlands, Finland, Sweden, Italy, Belgium, Spain, Switzerland, and France. It has recruited a total of more than 200 000 men aged 50–74 years. The first mortality analysis is planned in 2010.

In the USA, the Prostate, Lung, Colorectal and Ovary screening trial (PLCO) recruited 76 705 men aged 55–74 years in the prostate screening component between 1993 and 2001. Both serum PSA and digital rectal examination are used as screening tests. No mortality results are available yet.

Four case-control studies have evaluated prostate cancer screening, but they have not given consistent results. At any rate, the evidence from such nonrandomized studies should be seen as tentative.

Several ecological studies and time-series analyses have been published, correlating the frequency of PSA testing (or the incidence of prostate cancer as a surrogate for PSA testing) with prostate cancer mortality. The results have been inconsistent. Given the shortcomings inherent in these approaches, the findings do not have the potential to provide valid conclusions.

Adverse effects of screening

Overdiagnosis is potentially a major problem in prostate cancer screening. It has been estimated that 30–45% of the cancers detected by screening would not have been diagnosed in the absence of screening during the man's expected lifetime. Overdiagnosis leads to unnecessary treatment of prostate cancer, which has several major adverse effects, including high rates of erectile dysfunction and urinary incontinence with surgery, as well as urinary incontinence and irritation of the rectum or bladder (chronic radiation cystitis and proctitis) with radiotherapy.

Other screening tests

The effect of digital rectal examination (DRE) as a screening test on death from prostate cancer has not been evaluated in randomized studies. Five case-control studies have yielded inconsistent results, which is thought to be due to the low sensitivity of DRE for the detection of early disease.

In summary, no effect on mortality from prostate cancer with screening by serum PSA or other means has been established. Screening should be limited to randomized trials. Randomized trials are on-going and should provide important evidence.

Screening for other cancers

Ovarian cancer is among the 10 most common cancers among women. In 2008, 225 000 new cases were diagnosed globally and there were 140 000 deaths from ovarian cancer.

The natural history of ovarian carcinoma is not well understood. It is unclear how commonly cancers develop from benign or borderline lesions to malignant disease, relative to carcinoma arising *de novo*. Similarly, the duration of the detectable preclinical phase remains unknown. Screening tests include transvaginal or transabdominal ultrasound for imaging and serum CA-125 as a biochemical marker. There is no evidence on the effectiveness of ovarian cancer screening in terms of mortality reduction. Preliminary results obtained from nonrandomized studies are not encouraging: the sensitivity is low and false-positive findings are common.

Cutaneous melanoma incidence has been increasing rapidly in most industrialized countries and it now ranks among the 10 most common cancers in several European countries. There are approximately 200 000 new cases diagnosed annually in the world. Some of the increase may be due to more active case finding and changes in diagnostic criteria. Mortality has not shown a similar increase. Survival is favorable in the early stages. A substantial proportion of melanomas (approximately a fifth) arise from atypical moles (naevi). Visual inspection can be used to identify early melanomas (or premalignant lesions). Diagnostic assessment requires a skin biopsy. No randomized trials have been conducted to evaluate the effect of screening on melanoma mortality.

Gastric carcinoma is the third most common cancer in the world with 990 000 cases in 2008. With nearly 740 000 deaths annually, it is the second most common cause of cancer death globally. Fluoroscopic imaging (photofluorography) and endoscopy have been used in screening for stomach cancer. As no randomized trials have been reported, there is not sufficient evidence to allow sound evaluation of effectiveness or to make recommendations concerning screening.

Cancer Screening Guidelines

Several international and national organizations have given recommendations for cancer screening as public health policy (**Table 7**). They have been based on a variety of approaches from expert opinion and consensus development conferences to more objective methods of evidence synthesis. There is some degree of consistency between the guidelines, but also differences. For some organizations, the rationale for evaluation has been strictly based on evidence for effectiveness, with the goal of assessing whether there is sufficient high-quality research to justify screening. Other organizations, notably the American Cancer Society, have adopted a more

Table 7 Cancer screening recommendations by various organizations

Organization	*Cervical cancer*	*Breast cancer*	*Colorectal cancer*	*Prostate cancer*
WHO	Recommended without details	Every 2–3 years at ages 50–69 years	Mortality can be reduced; no clear recommendation	Not recommended
Union for International Cancer Control	Pap smear every 3–5 years from age 20–30 years until 60 years or later	Mammography every 2 years at ages 50–69 years	FOBT every 2 years at ages above 50 years	No evidence
European Union	Start at ages 20–30 years, screening interval 3–5 years, discontinue at age 60 years or later	Mammography every 2–3 years at ages 50–69 years	FOBT every 1–2 years at ages 50–74 years	Not recommended
Canadian Taskforce on Preventive Health Care	Pap test every year from age 18 years or start of intercourse; frequency every 3 years after two negative smears; discontinue at age 69 years	Mammography every 1–2 years at ages 50–59 years	Insufficient evidence	Not recommended
U.S. Preventive Services Taskforce	Pap smear at least every 3 years among sexually active women until age 65 years	Mammography every 1–2 years at ages 50–69 years	Yearly FOBT after age 50 years; sigmoidoscopy as alternative	Not recommended
American Cancer Society	Pap test annually since age 18 years or start of intercourse; less frequently after 2–3 negative smears	Mammography annually from age 40 years	Annual FOBT and sigmoidoscopy after age 50 years; alternatively, colonoscopy every 10 years or barium enema every 5–10 years	Annual DRE and PSA from age 50 years

Table 8 Summary of evidence for cancer screening

		Efficacy		*Effectiveness*	
Primary site	*Screening method*	*Nonrandomized*	*Randomized*	*Randomized trial*	*Service screening*
Cervical cancer	Pap smear		NA[a]	NA	0–80%
	Visual inspection			NA	NA
	HPV testing		NA	NA	NA
Breast cancer	Mammography		35%	15–25%	6–20
Colorectal cancer	Fecal occult blood		24	15	NA
	Sigmoidoscopy		NA	NA	NA
	Colonoscopy		NA	NA	NA
Lung cancer	Chest X-ray $\pm$ sputum cytology		None		NA
	Low-dose CT		NA	NA	NA
Prostate cancer	Serum PSA		NA	NA	NA
	Digital rectal exam	None	NA	NA	NA
Oral cancer	Visual inspection		20%		NA
Liver cancer	Serum AFP $\pm$ ultrasound		20–33%		NA
Ovarian cancer	Ultrasound with CA-125		NA	NA	NA

[a]NA, not available.

ideological approach, with a bias in favor of screening (a low threshold for advocating screening). Similarly, medical specialty societies tend to adopt screening recommendations relatively eagerly (not included in the table).

The role of the organization or the task of the working group also affects the outcome, so that groups with more responsibility for planning health-care services tend to apply more stringent evaluation criteria than those without such responsibility. Countries with publicly financed health-care systems tend to be more conservative than countries with systems based on fee-for-service health-care systems.

Conclusion

In summary, establishing the benefits of screening usually requires evidence of a significant reduction in mortality from large randomized trials. Such a knowledge base exists for cervical, breast, and colorectal cancer

(**Table 8**). The evidence is more limited for oral and liver cancer. Screening tests exist for numerous primary sites of cancer, but either their effectiveness has not been adequately evaluated, or a lack of effectiveness has been demonstrated. Even after randomized trials have shown efficacy (typically in specialist centers with volunteer subjects), the introduction of mass screening requires that pilot studies should first be done to demonstrate feasibility. After an organized screening program has been implemented, continuous evaluation is required to ensure that the benefits are maintained. Ideally, this is achieved by introducing a mass screening program with a randomized design, that is, comparing subjects randomly allocated to early entry with those covered only later. This approach is particularly important when the effect demonstrated in trials of efficacy is small. It is also highly recommended when an established technique is being replaced with a new one.

See also: Breast Cancer; Cervical Cancer; Endometrial Cancer; Ovarian Cancer; Prostate Cancer; Worldwide Burden of Gynecologic Cancer: The Size of the Problem.

Citations

Alexander FE, Anderson TJ, Brown HK, *et al.* (1999) Fourteen years of follow-up from the Edinburgh randomised trial of breast-cancer screening. *Lancet* 353: 1903–1908.

Andersson I and Janzon L (1997) Reduced breast cancer mortality in women under age 50: Updated results from the Malmo mammographic screening program. *Journal of the National Cancer Institute Monograph* 22: 63–67.

Bjurstam N, Björneld L, Warwick J, *et al.* (2003) The Gothenburg breast screening trial. *Cancer* 97: 2387–2396.

Chen JG, Parkin DM, Chen QG, *et al.* (2003) Screening for liver cancer: Results of a randomised controlled trial in Qidong, China. *Journal of Medical Screening* 10: 204–209.

Cole P and Morrison AS (1978) Basic issues in cancer screening. In: Miller AB (ed.) *Screening in Cancer*, UICC Technical Report Series, vol. 40, pp. 7–39. Geneva, Switzerland: UICC.

Day NE and Walter SD (1984) Simplified models for screening: Estimation procedures from mass screening programmes. *Biometrics* 40: 1–14.

Day NE, Walter SD, and Collette B (1984) Statistical models of disease natural history: Their use in the evaluation of screening programmes. In: Prorok PC and Miller AB (eds.) *Screening for Cancer I – General Principles on Evaluation of Screening for Cancer and Screening for Lung, Bladder and Oral Cancer,* UICC Technical Report Series, vol. 78, pp. 55–70. Geneva, Switzerland: UICC.

Feinleib M and Zelen M (1969) Some pitfalls in the evaluation of screening programs. *Archives of Environmental Health* 19: 412–415.

Furihata R and Maruchi N (1969) Epidemiological studies on thyroid cancer in Nagano prefecture, Japan. In: Hedinger CE (ed.) *Thyroid Cancer,* UICC Monograph Series, vol. 12, p. 79. Berlin: Springer-Verlag.

Hakama M, Auvinen A, Day NE, and Miller AB (2007) Sensitivity in cancer screening. *Journal of Medical Screening* 14: 174–177.

Hakama M, Pukkala E, Heikkilä M, and Kallio M (1997) Effectiveness of the public health policy for breast cancer screening in Finland: Population based cohort study. *British Medical Journal* 314(7084): 864–867.

Hugosson J, Aus G, Becker C, *et al.* (2000) Would prostate cancer detected by screening with prostate specific antigen develop into clinical cancer if left undiagnosed. *British Journal of Urology* 85: 1978–1984.

Hutchison GB and Shapiro S (1968) Lead time gained by diagnostic screening for breast cancer. *Journal of the National Cancer Institute* 41: 665–681.

IARC (2002) *Breast Cancer Screening. IARC Handbooks of Cancer Prevention* vol. 7. Lyon, France: IARC Press.

IARC (2005) *Cervix Cancer Screening. IARC Handbooks of Cancer Prevention* vol. 10. Lyon, France: IARC Press.

Kronborg O, Jorgensen OD, Fenger C, and Rasmussen M (2004) Randomized study of biennial screening with a faecal occult blood test: Results after nine screening rounds. *Scandinavian Journal of Gastroenterology* 39: 846–851.

Kubik A, Parkin DM, Khlat M, Erban J, Polak J, and Adamec M (1990) Lack of benefit from semi-annual screening for cancer of the lung: Follow-up report of a randomized controlled trial on a population of high-risk males in Czechoslovakia. *International Journal of Cancer* 45: 26–33.

Labrie F, Candas B, Cusan L, *et al.* (2004) Screening decreases prostate cancer mortality: 11-year follow-up of the 1988 Quebec prospective randomized controlled trial. *Prostate* 59: 311–318.

Levin ML, Tockman MS, Frost JK, and Ball WC (1982) Lung cancer mortality in males screened by chest X-ray and cytologic sputum examination: A preliminary report. *Recent Results in Cancer Research* 82: 138–146.

Mandel JS, Church TR, Ederer F, and Bond JH (1999) Colorectal cancer mortality: Effectiveness of biennial screening for fecal occult blood. *Journal of the National Cancer Institute* 91: 434–437.

Marcus PM, Bergstralh EJ, Fagerstrom RM, *et al.* (2000) Lung cancer mortality in the Mayo Lung Project: Impact of extended follow-up. *Journal of the National Cancer Institute* 92: 1308–1316.

Melamed MR, Flehinger BJ, Zaman MB, *et al.* (1984) Screening for early lung cancer. Results of the Memorial Sloan-Kettering study in New York. *Chest* 86: 44–53.

Miller AB, To T, Baines CJ, and Wall C (2000) Canadian National Breast Screening Study-2: 13-year results of a randomized trial in women aged 50–59 years. *Journal of the National Cancer Institute* 92: 1490–1499.

Miller AB, To T, Baines CJ, and Wall C (2002) The Canadian National Breast Screening Study-1: breast cancer mortality after 11 to 16 years of follow-up. A randomized screening trial of mammography in women age 40 to 49 years. *Annals of Internal Medicine* 137: 305–312.

Moss S, Cuckle H, Evans A, Johns L, Waller M, and Bobrow L (2006) Effect of mammographic screening from age 40 years on breast cancer mortality at 10 years' follow-up: A randomised controlled trial. *Lancet* 368: 2053–2060.

Nyström L, Andersson I, Bjurstam N, Frisell J, Nordenskjöld B, and Rutqvist LE (2002) Long-term effects of mammography screening: Updated overview of the Swedish randomised trials. *Lancet* 359: 909–919.

Olsen O and Gøtzche PC (2001) Cochrane review on screening for breast cancer with mammography. *Lancet* 358: 1340–1342.

Sankaranarayanan R, Esmy PO, Rajkumar R, *et al.* (2007) Effect of visual screening on cervical cancer incidence and mortality in Tamil Nadu, India: A cluster-randomised trial. *Lancet* 370: 398–406.

Sankaranarayanan R, Ramadas K, Thamas G, *et al.* (2005) Effect of screening on oral cancer mortality in Kerala, India: A cluster-randomised controlled trial. *Lancet* 365: 1927–1933.

Scholefield JH, Moss S, Sufi F, Mangham CM, and Hardcastle JD (2002) Effect of faecal occult blood screening on mortality from colorectal cancer: Results from a randomised controlled trial. *Gut* 50: 840–844.

Shapiro S, Venet W, Strax P, and Venet L (1988) Current results of the breast cancer screening randomized trial: The Health Insurance Plan (HIP) of Greater New York Study. In: Day NE and Miller AB (eds.) *Screening for Breast Cancer*, pp. 3–15. Geneva, Switzerland: International Union Against Cancer.

Stenman UH, Hakama M, Knekt P, Aromaa A, Teppo L, and Leinonen J (1994) Serum concentrations of prostate specific antigen and its complex with α1-ACT 0–12 years before diagnosis of prostate cancer. *Lancet* 344: 1594–1598.

Tabár L, Vitak B, Chen HH, *et al.* (2000) The Swedish Two-County Trial twenty years later. Updated mortality results and new insights from long-term follow-up. *Radiologic Clinics of North America* 38: 625–651.

Zhang BH, Yang BH, and Tang ZY (2004) Randomized controlled trial of screening for hepatocellular carcinoma. *Journal of Cancer Research and Clinical Oncology* 131: 417–422.

Worldwide Burden of Gynecologic Cancer: The Size of the Problem

R Sankaranarayanan and J Ferlay, International Agency for Research on Cancer, Lyon, France

Introduction

The most commonly used indicators of cancer burden are incidence, mortality and prevalence. Cancer incidence is expressed as the absolute number of new cases occurring in a defined population in one year, or as a rate in terms of number of new cases per 100 000 people per year. Cancer mortality is also expressed as the absolute number of people who die from a specific cancer during a given year or as a rate in terms of cancer deaths per 100 000 people per year. Prevalence pertains to the absolute number, and relative proportion, of individuals affected by the disease during a defined period in a population. Other more complex indices of disease burden used include 'person-years of life lost', which is defined as how many years of normal life are lost due to deaths from a given disease; and 'disability-adjusted life-years lost', which attempts to give a numerical score to the years lived with a reduced quality of life between diagnosis and death.

The estimation of cancer burden in terms of incidence, mortality and prevalence is necessary when formulating cancer control policies and planning health services. It enables an assessment of the demands of care and the implementation of strategies to provide health-care services to reduce disease burden and relieve suffering. It helps to define priorities for preventive, diagnostic, therapeutic and palliative care services and to evaluate the outcomes of targeted interventions in relation to costs and resource inputs. The comparison of disease burden between different populations over time enables the causal definition of hypotheses and the critical evaluation of the underlying risk factors, thereby providing potential opportunities for prevention.

In order to present the global gynecological cancer burden in the year 2002, this chapter describes the incidence, mortality and prevalence of cancers of the uterine cervix (International Classification of Disease [ICD]-10 code C53), uterine body (corpus uteri, ICD-10 C54), ovary and other uterine adnexa (ICD-10 C56, C57.0-4), together with the incidence of vagina (ICD-10 C51), vulva (ICD-10 C52) and choriocarcinoma (ICD-10 C58) for 18 world regions. It briefly discusses the sources and methods of estimation and data validity and the implications for health services worldwide in general, and those of the developing countries in particular.

Indices of Cancer Burden

'Incidence' provides an indication of the average risk of developing a cancer in a population (Ferlay *et al.*, 2002; Parkin *et al.*, 2005). When expressed as an absolute number of cases per year, it reflects the load of new patients diagnosed in a given region. The risk of disease in different populations within countries, regions, ethnicities or over different time periods can be compared using incidence. The impact of prevention strategies based on reducing or eliminating exposure of populations to disease-causing risk factors (e.g. tobacco use or infection with human papillomavirus [HPV], etc.) or early detection and treatment of precancerous lesions (e.g. cervical cancer screening) is measured in terms of reduction in incidence.

'Mortality' is the product of fatality and incidence, and provides an unambiguous measure of the outcome of disease (Ferlay *et al.*, 2002; Parkin *et al.*, 2005). 'Fatality' refers to the proportion of incident cancer cases that die. Mortality rates measure the average risk of dying from a given cancer in a given population, whereas fatality reflects the probability of an individual with cancer dying from it. Mortality rates are influenced by the trends in incidence rates as well as by the natural history of the disease, the efficacy of treatment interventions and of health services delivery. It is inappropriate to use mortality rate as a proxy measure of cancer incidence when comparing different populations, since it assumes equal survival/fatality in the populations compared, which is rarely true. The survival time of a cancer patient refers to the time duration between the diagnosis and death.

'Prevalence' refers to the number of persons who have had cancer diagnosed at some time in the past in a defined population, during a given period of time (Ferlay *et al.*, 2002; Parkin *et al.*, 2005). The prevalence of cancer cases diagnosed within 1, 3 and 5 years is likely to be relevant to the different phases of cancer management, such as initial treatment (1-year prevalence) and treatment of residual/recurrent disease and clinical follow-up (3- or 5-year prevalence).

Sources and Methods of Estimation of Data

The data discussed in this chapter are essentially derived from the comprehensive global cancer statistics published by the International Agency for Research on Cancer

(IARC) of the World Health Organization (WHO) (Ferlay *et al.*, 2002). The database, which is regularly updated, is available on internet at the CANCER mondial website (http://www-dep.iarc.fr) and provide the best possible estimates of the global cancer burden.

A brief description of how the cancer statistics have been compiled is useful to understand the validity, limitations and applications of the data in national, regional and global contexts. The methods of assembling these data have been described elsewhere (Ferlay *et al.*, 2002; Parkin *et al.*, 2001, 2005; Pisani *et al.*, 2002). The building-blocks of the global data are the best and most up-to-date information on incidence, mortality and survival in a given country, and are available from the Descriptive Epidemiology Group of IARC.

Population-based cancer registries produce cancer incidence data by systematically collecting information on a continuing basis on all new cancer cases diagnosed by all means (histological or clinical) in a defined resident population in a given geographical region (Armstrong, 1992). However, many countries do not have cancer registries and many populations, particularly in sub-Saharan Africa, certain regions of Asia (e.g. Central Asia) and Latin America (e.g. Central America) are not covered. Where available, cancer registries might cover the entire geographical regions of a country (national registries; e.g. Singapore, Oman, Finland, Costa Rica) or populations living in some regions in a country (e.g. Shanghai municipal area in the People's Republic of China, Kyadondo county in Uganda, Bas-Rhin department in France, Lima city in Peru). It is estimated that 16% of the world population (comprising 64% of developed and 5% of developing country populations) were covered by population-based cancer registries in 1995. The 2007 volume of Cancer Incidence in Five Continents (CI5) contains comprehensive and comparable incidence data from 300 populations in 60 countries for about the year 2000.

For this chapter, incidence rates were estimated, in order of priority, from:

- national incidence data
- national mortality data, with an estimation of incidence using sets of regression models that are specific for site, sex and age, derived from local cancer registry data (incidence plus mortality)
- local incidence from one or more regional cancer registries within a country
- frequency data from hospital registries or pathology registers.

If there are no data, the country-specific rates are calculated from neighbouring countries for which estimates could be made.

For cancers of the vagina, vulva and placenta, estimates for 18 world regions, as defined by the United Nations Population Division, were calculated by the population weighting average of the age-specific incidence rates observed in local cancer registries in the area.

We derived the mortality data from the different countries' death registration systems. The data are produced according to the underlying cause of death, which are available through WHO. However, the completeness, quality, accuracy and validity of mortality data available from different countries are highly variable. For instance, in 1995, only 29% of the world population was covered by death registration. In most developing countries, particularly in sub-Saharan Africa and certain areas of Asia and Latin America, the coverage of populations by death registration is grossly incomplete, and the mortality rates reported from these regions are low and unreliable. For countries where mortality data were of poor quality or unavailable, they were estimated from incidence, using survival data specific to a country or region.

We estimated the prevalence data from incidence and population-based cancer survival data produced by cancer registries by following-up all the incident cancer cases. Population-based survival rates indicate the average prognosis from a given cancer in a given region and reflect the efficiency of the local health services. Such survival data are widely available for developed countries, but for only limited regions in a few developing countries. The sources of population survival data include the population-based cancer registries under the Surveillance, Epidemiology and End Results (SEER) program, which covers 13% of the USA population (Ries *et al.*, 2004); the EUROCARE-3 project, which provides survival data for registries in several European countries (The EUROCARE Working Group, 2003); and the 'Cancer Survival in Developing Countries monograph', which provides survival data for selected populations in the People's Republic of China, Cuba, India, the Philippines and Thailand (Sankaranarayanan *et al.*, 1988, 1996), Singapore (Chia *et al.*, 2001) and first results from Uganda (Gondos *et al.*, 2004) and Zimbabwe (Gondos *et al.*, 2005).

From the above description, it is evident that the estimates of cancer burden for different countries vary in accuracy, depending on the extent and validity of the data available for each country. For several developing countries from where no data are available (e.g. Cambodia, Democratic Republic of Congo) our estimate was derived from neighbouring countries.

2002 Estimates of Gynecological Cancer Burden

It has been estimated that there were 10.9 million new cancer cases (all cancers excluding non-melanoma skin cancer) in both sexes in the world around the year 2002: 5.8 million cases in men and 5.1 million cases in women (Ferlay *et al.*, 2002; Parkin *et al.*, 2005).

Table 1 Cancers of the uterine cervix, uterine body, ovary, etc. Incident cases, deaths and 5-year prevalence in 18 world regions in 2002.

	Cervix			*Uterine body*			*Ovary*			*All*		
	Cases	*Deaths*	*5-year prev.*	*Cases*	*Deaths*	*5-year prev.*	*Cases*	*Deaths*	*5-year prev.*	*Cases*	*Deaths*	*5-year prev.*
World	492 800	273 200	1 409 200	198 600	50 200	775 400	204 200	124 700	538 400	895 600	448 100	2 723 000
More developed countries	83 400	39 500	309 900	136 300	29 100	557 400	96 700	62 200	262 300	316 400	130 800	1 129 600
Less developed countries	409 400	233 700	1 099 300	62 300	21 100	218 000	107 500	62 500	276 100	579 200	317 300	1 593 400
Eastern Africa	33 900	27 100	57 200	2400	800	8600	4700	3300	10 400	41 000	31 200	76 200
Middle Africa	8200	6600	13 900	700	200	3000	1100	800	2600	10 000	7600	19 500
Northern Africa	8100	6500	14 000	1500	600	5200	1800	1300	4200	11 400	8400	23 400
Southern Africa	7600	4400	13 100	600	200	2100	1000	600	2200	9200	5200	17 400
Western Africa	20 900	16 700	35 700	1400	500	5100	3600	2500	7900	25 900	19 700	48 700
Caribbean	6300	3100	18 400	1600	800	5400	800	400	2000	8700	4300	25 800
Central America	17 100	8100	49 300	2400	1000	8600	4000	1900	10 100	23 500	11 000	68 000
South America	48 300	21 400	139 200	10 600	3200	34 900	12 700	6100	31 500	71 600	30 700	205 600
Northern America	14 600	5700	58 200	51 500	6300	223 200	25 100	16 000	74 400	91 200	28 000	355 800
Eastern Asia	61 100	31 300	191 900	20 200	4700	80 000	30 600	15 000	88 000	111 900	51 000	359 900
South-eastern Asia	42 500	22 500	132 500	9100	3200	32 400	16 800	9200	44 300	68 400	34 900	209 200
South central Asia	157 700	86 700	446 100	13 100	5400	45 600	32 500	22 800	83 500	203 300	114 900	575 200
Western Asia	4400	2100	13 700	4100	1900	13 800	4000	2400	10 200	12 500	6400	37 700
Eastern Europe	30 800	17 100	107 700	29 600	10 100	111 500	23 600	15 200	56 700	84 000	42 400	275 900
Northern Europe	5600	2800	21 100	10 500	2200	39 500	10 500	7100	24 700	26 600	12 100	85 300
Southern Europe	10 600	4100	40 900	15 800	3600	61 600	11 600	6400	32 000	38 000	14 100	134 500
Western Europe	12 700	5600	49 200	21 000	4500	86 700	17 600	12 100	48 400	51 300	22 200	184 300
Oceania	2000	800	6500	2000	400	7500	1700	1000	4500	5700	2200	18 500

Table 2 Cancer of the vulva, vagina and choriocarcinoma: estimated incident cases and age-standardized incidence rates (world) per 100 000 (all ages) in 18 world regions (2002).

	Cancer						
	Vulva (C51)		*Vagina (C52)*		*Choriocarcinoma (C58)*		*All*
Area	*Cases*	*Rate*	*Cases*	*Rate*	*Cases*	*Rate*	*Cases*
Eastern Africa	0.5	0.80	0.3	0.34	0.5	0.46	1.3
Middle Africa	0.2	0.56	0.1	0.31	0.1	0.18	0.3
Northern Africa	0.2	0.35	0.2	0.30	0.0	0.04	0.5
Southern Africa	0.2	0.78	0.1	0.25	0.1	0.40	0.3
Western Africa	0.4	0.57	0.4	0.50	0.3	0.33	1.1
Caribbean	0.2	1.09	0.1	0.75	0.0	0.03	0.4
Central America	0.4	0.87	0.3	0.53	0.1	0.13	0.8
South America	2.4	1.46	1.1	0.68	0.5	0.26	4.0
Northern America	4.0	1.63	1.2	0.47	0.1	0.08	5.4
Eastern Asia	2.3	0.26	1.7	0.20	1.3	0.15	5.2
South-Eastern Asia	1.2	0.55	0.6	0.26	1.2	0.43	3.0
Southern Asia	2.6	0.46	4.1	0.70	1.0	0.13	7.7
Western Asia	0.4	0.61	0.2	0.24	0.2	0.16	0.8
Eastern Europe	4.8	1.61	1.3	0.47	0.2	0.10	6.2
Northern Europe	1.5	1.58	0.4	0.43	0.0	0.04	1.9
Southern Europe	2.4	1.45	0.5	0.36	0.0	0.05	2.9
Western Europe	2.7	1.33	0.8	0.45	0.0	0.05	3.5
Oceania	0.3	1.51	0.1	0.45	0.0	0.07	0.4
World	26.9		13.2		5.8		45.9

There were 6.7 million cancer deaths, 3.8 million in men and 2.9 million in women. There were 24.6 million persons with a diagnosis of cancer within 5 years from the initial diagnosis, of which 13.0 million were women.

The burden of uterine body, uterine cervix and ovarian cancer worldwide, in developing and developed countries and for 18 world regions is given in **Table 1**. The incidence rates and number of incident cases in 2002 of vaginal cancer, vulva cancer and choriocarcinoma are given in **Table 2**. Developed countries include those in North America and Europe (including all of Russia), Australia, New Zealand and Japan. All other countries constitute 'developing countries'.

New cancers of uterine body, cervix, ovary, vagina, vulva, and choriocarcinoma together constituted 942 000 cases, accounting for 18.6% of all incident cancers in women in the world. They accounted for 22.1% of all new cancer cases among women in developing countries compared to 14.5% of all new cases among women in developed nations. Of the total 2.9 million cancer deaths worldwide among women, gynecological cancer (excluding vagina, vulva and placental malignancies) accounted for 15.3% deaths; of the total 5-year prevalent cases, gynecological cancer accounted for 20.9% of cases.

Cancer of the Uterine Cervix

Cancer of the cervix is the second most common cancer among women worldwide, with an estimated 493 000 new cases and 273 000 deaths (**Table 1**) in the year 2002. Eighty-three per cent of new cases and 85% of deaths from cervical cancer occur in developing countries, where it is the most common cancer among women in many regions. Being the most common gynecological cancer in the developing world, it accounts for two-thirds of cases and continues to be a serious health problem. Of cervical cancers worldwide, 80–95% are squamous cell carcinomas (Parkin *et al.*, 2002).

There is an eight-fold variation in the incidence rates of cervical cancer worldwide. The highest incidence rates are observed in sub-Saharan Africa, Melanesia, Latin America and the Caribbean, South-Central Asia, and South-East Asia (**Figures 1** and **2**). One-third of the cervical cancer burden in the world is experienced in South-Central Asia (**Table 1**). The incidence is generally low in developed countries, with age-standardized rates less than 14 per 100 000 women (**Figures 1** and **2**). The low risk in these countries is due to effective screening programs. Before the introduction of screening programs, the incidence rates in most developed countries were similar to those found in developing countries today (Gustafsson *et al.*, 1997). Rates lower than 7 per 100 000 women are also observed in the countries of the Middle East and in the People's Republic of China.

Age-adjusted cervical cancer mortality rates exceed 15 per 100 000 women in most developing regions of the world, with rates as high as 35 per 100 000 in Eastern Africa (**Figure 2**). The high mortality rates are due to advanced clinical stage at presentation and to the fact that a significant proportion of patients do not receive or

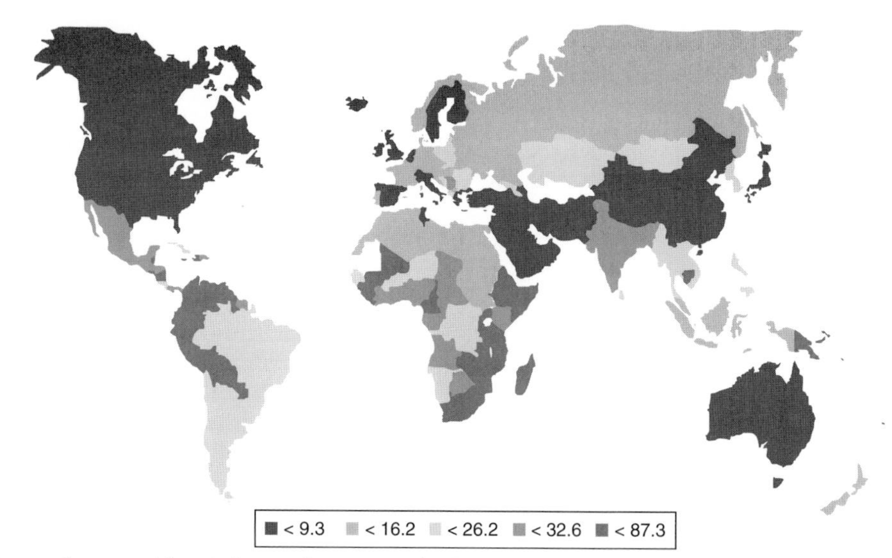

Figure 1 Incidence of cancer of the uterine cervix: age-standardized rates (world) per 100 000 (all ages).

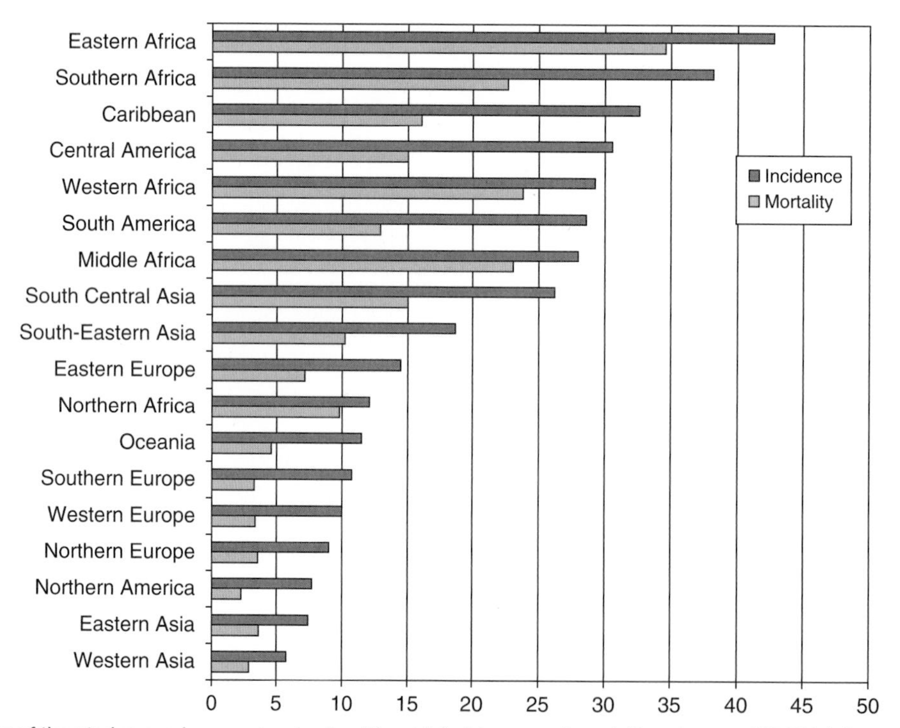

Figure 2 Cancer of the uterine cervix: age-standardized (world) incidence and mortality rates per 100 000 (all ages) in 18 world regions.

complete prescribed courses of treatment, due to deficiencies in treatment availability, accessibility and affordability in many developing countries. Mortality rates seldom exceed 5 per 100 000 women in developed countries with successful screening programs (**Figure 2**).

Higher survival is observed in populations in developed countries as compared to selected populations in developing countries from where survival data are available (**Table 3**). Five-year survival in the SEER registries, Europe and Singapore are considerably higher than those reported from sub-Saharan Africa, whereas survival rates are fair in India, the People's Republic of China and Thailand. The 5-year prevalence exceeds 1.4 million in the world, with 1 million of these cases being in the developing world (**Table 1**).

Cancer of the Uterine Body

Cancer of the uterine body, which is predominantly a cancer of postmenopausal women, has a similar geographic distribution to ovarian cancer and accounted for 199 000 new cases (3.9% of cancers in women) and 50 000 deaths (1.7% of all cancer deaths in women) worldwide in 2002 (**Table 1**). It is the most common gynecological cancer in developed countries, which account for two-thirds of the global burden. Around 90% of the uterine body cancers are endometrial adenocarcinomas; papillary serous carcinoma, clear cell carcinomas, papillary endometrial carcinoma and mucinous carcinoma account for the remaining cases. The highest incidences are observed in North America, with rates exceeding 20 per 100 000 women, whereas in Europe incidence varies between 11 and 14 per 100 000 women. (**Figures 3** and **4**). Incidence rates in South America are intermediate, whereas rates are low in Southern and Eastern Asia (including Japan) and most of Africa (less than 4 per 100 000). More than 90% of cases occur in women aged 50 years and over. Cancer of the uterine body has a much more favorable prognosis than ovarian and cervical cancers, with 5-year survival rates around 80% in developed countries and 70% in the developing countries (**Table 2**). The 5-year prevalence is around 775 000 cases worldwide.

Table 3 Five-year relative survival (1990)

Region/country	*Uterine body*	*Cervix uteri*	*Ovary*
USA (Surveillance, Epidemiology and End Results program)	83	72	42
English registries	74	64	31
French registries	73	68	38
Italian registries	76	67	37
Finland	81	66	35
Singapore	68	57	51
India, Mumbai		51	
India, Chennai		60	
China, Shanghai	77	52	44
Thailand, Chiang Mai	69	68	45
Thailand, Khon Kaen	79	57	36
Philippines, Rizal		29	
Uganda, Kampala		18	16
Harare, Zimbabwe: black		30	38

Ovarian cancer

Fallopian tube cancers and extraovarian primary peritoneal cancers are considered along with ovarian cancers because their biology and clinical characteristics are similar to ovarian cancer, although they are much less common. Ovarian cancer ranks as the sixth most common cancer in women and accounted for 204 000 cases and 125 000 deaths worldwide around 2002 (**Table 1**). It constitutes 4.0% of all female cancers and 4.2% of cancer deaths in women. It is the second most common gynecological cancer, accounting for 18.8% of all gynecological cancers in developing countries and 28.7% in developed countries. Developed countries account for half the worldwide burden of ovarian cancer. In the developed countries, more than 90% of ovarian cancers are epithelial in origin, while the remaining are constituted by germ cell tumors (2–3%) and sex cord-stromal tumors (5–6%). Germ cell tumors account for 10–15% of ovarian cancers in Asian and African populations. Dysgerminoma accounts for more than 70% of germ cell tumors, whereas granulosa tumors constitute the most common sex cord-stromal tumor. The vast majority of epithelial ovarian cancers are diagnosed in postmenopausal women, whereas germ cell tumors occur in young women of child-bearing potential, who are often in their twenties.

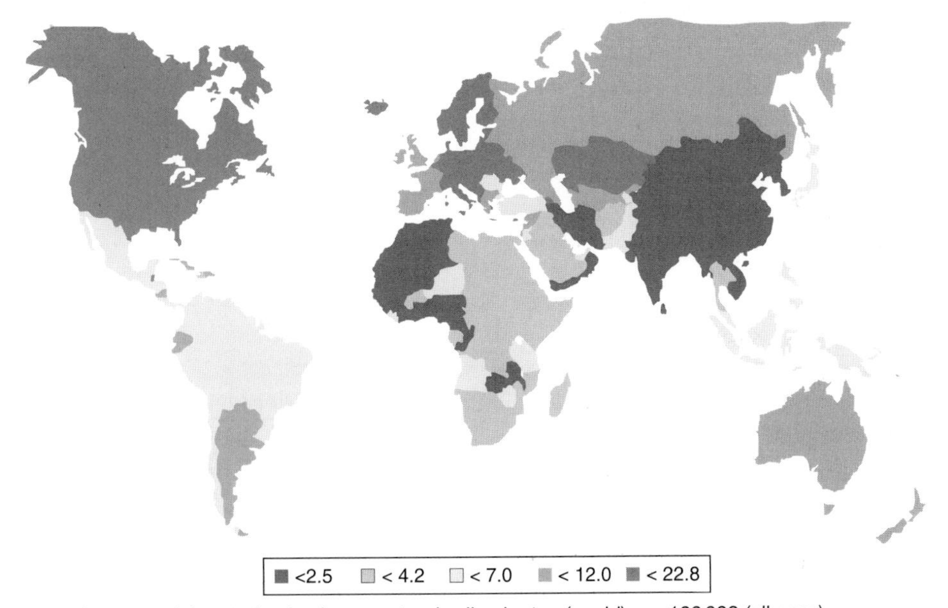

Figure 3 Incidence of cancer of the uterine body: age-standardized rates (world) per 100 000 (all ages).

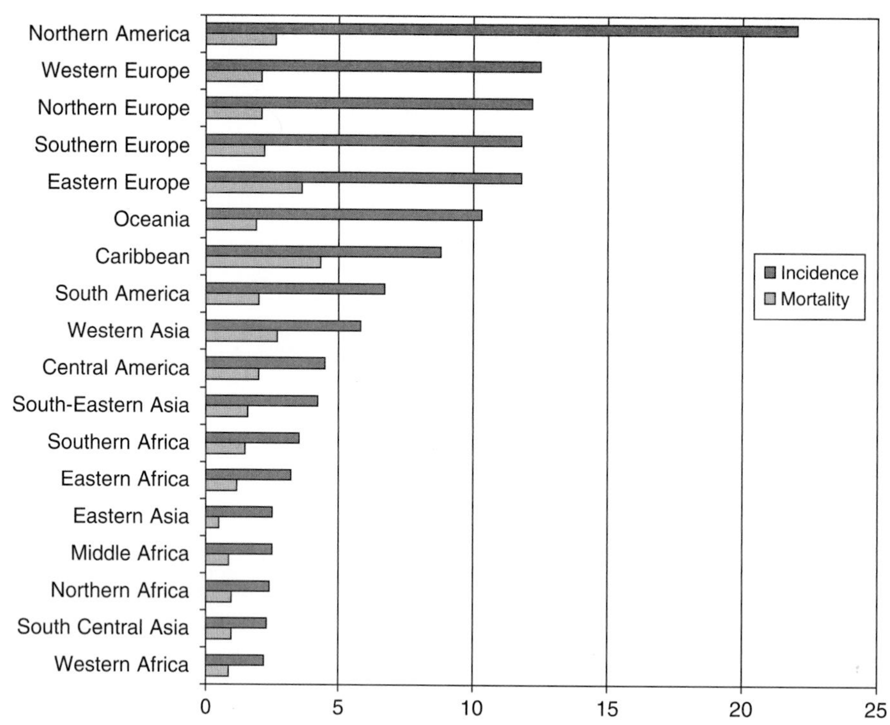

Figure 4 Cancer of the uterine body: age-standardized (world) incidence and mortality rates per 100 000 (all ages) in 18 world regions.

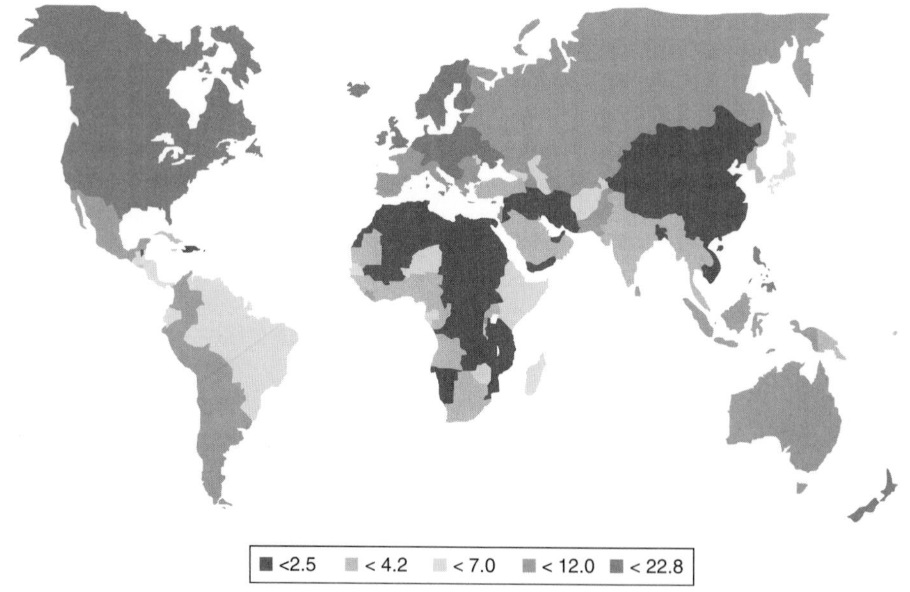

Figure 5 Incidence of ovarian cancer: age-standardized rates (world) per 100 000 (all ages).

Incidence rates are highest in developed countries (**Figure 5**), with rates in these areas exceeding 10 per 100 000 (**Figure 6**), except for Japan (6.4 per 100 000). The incidence in South America (7.7 per 100 000) is relatively high compared to many regions in Asia and Africa. The incidence rate of ovarian cancer has been slowly increasing in many developed countries over the last two decades.

The high proportion of deaths in relation to new cases reflects the much less favorable prognosis of ovarian cancer compared to cancers of uterine body and cervix (**Table 3**). Ovarian cancer mortality rates exceed 5 per 100 000 women in developed countries (**Figure 6**). The 5-year survival from ovarian cancer is mostly less than 50% (**Table 2**) and the 5-year prevalence is 538 000 cases worldwide.

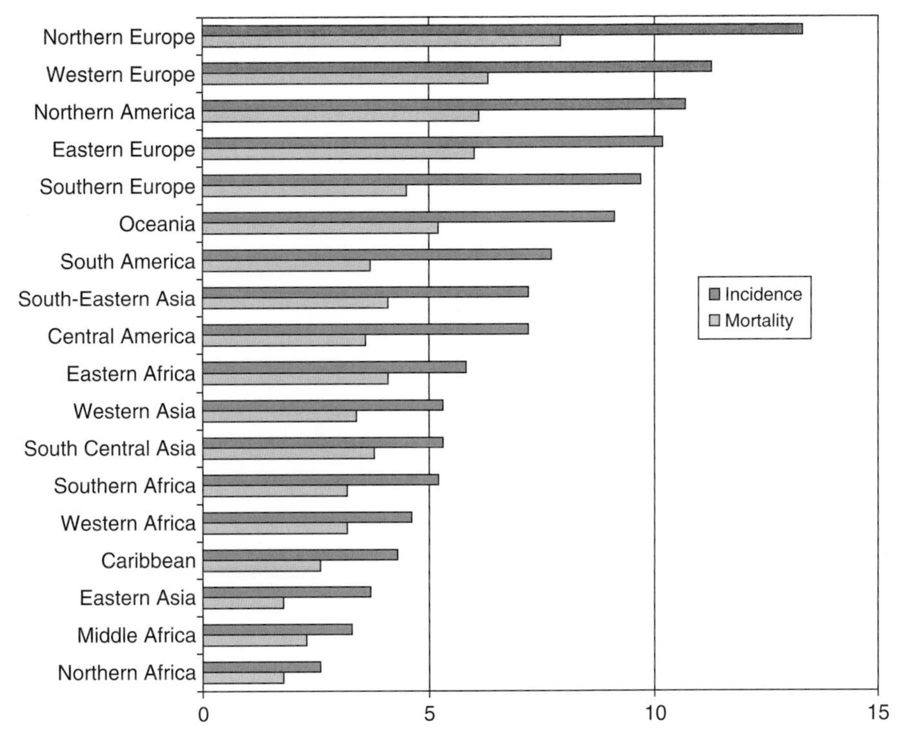

Figure 6 Cancer of the ovary: age-standardized (world) incidence and mortality rates per 100 000 (all ages) in 18 world regions.

Other Gynecological Cancers

Cancer of the vagina is rare, constituting less than 2% of gynecological cancers. It accounted for 13 200 cases worldwide around 2002 (**Table 3**); of these, 9000 occurred in developing countries. The incidence rates do not exceed 0.8 per 100 000 women in any world region. More than three-quarters of cases occur in women older than 60 years. Except for clear-cell carcinomas associated with maternal diethylstilbestrol (DES) exposure, vaginal cancer is rare in women younger than 40 years. Half of the tumors occur in the upper third of vagina.

Cancer of the vulva constitutes 3% of gynecological cancers. It accounted for 26 800 cases, of which 15 700 occurred in developed countries. The incidence rates exceed 1.5 per 100 000 women in North America, South America and Europe, whereas the rates are less than 1.0 per 100 000 in developing countries (**Table 3**). More than 50% of the cases are diagnosed in women over the age of 70 years; incidence rates peak in women aged 75 years and above. Two-thirds or more of the vulva cancers occur in the labia majora.

Choriocarcinoma is a rare cancer constituting 0.6% of all gynecological cancers. It accounted for 5800 cases worldwide, of which 5400 occurred in developing countries. Incidence rates in different regions vary 10-fold: age-standardized incidence rates of choriocarcinoma ranged from 0.04 in Southern Africa and Northern Europe to 0.43 per 100 000 women in South East Asia. An incidence rate of 1.7 per 100 000 women has been reported from Viet Nam (Altieri *et al.*, 2003).

Discussion

The estimates – the best possible information on global cancer burden – are more or less accurate for the different regions, depending on the extent and validity of the data available from those regions. The data indicate striking variations in the burden of different cancers, due to variations in exposure to suspected and established risk factors and wide disparity and inequality in health-care infrastructure and accessibility between developed countries and the medically underserved less developed countries across the world.

Among the gynecological cancers, cervical cancer offers a great potential for prevention, early detection and cure. Yet despite the slowly declining incidence rates, it remains the leading cause of cancer death for women in many developing countries due to lack or inefficiency of existing screening programs combined with a high background prevalence of HPV infection and high parity (Sankaranarayanan *et al.*, 2001; IARC handbooks of cancer prevention Volume 10, 2005; Bosch and de Sanjose, 2003). A low burden of disease is experienced in developed countries with screening programs and in countries with strict religious regulation of sexual behavior and a high prevalence of male circumcision (e.g. countries of the Middle East).

Persistent infection with one or more of the high-risk types of HPV types is now accepted as a necessary cause of cervical cancer and HPV DNA can be identified in more than 95% of cervical cancers (Bosch and de Sanjose, 2003; Munoz *et al.*, 2003). The high risk associated with high parity and oral contraceptive use seems to be due to the hormonal influences that maintain the transformation zone on the ectocervix, facilitating exposure to HPV and other cofactors. A lower risk of HPV infection in circumcised than in uncircumcised males, with a corresponding lower incidence of cervical cancer in their female partners, has been reported (Castellsague *et al.*, 2002).

The central role of HPV infection in the causation of cervical neoplasia has led to efforts to evaluate prophylactic vaccination in preventing infection and cervical intraepithelial neoplasia (CIN) and evaluating HPV detection and typing in cervical screening. Large-scale HPV vaccine efficacy trials have been undertaken with protection from the development of high-grade CIN as the primary endpoint. The results of these trials indicate that vaccinated women were protected against persistent infection with HPV 16 and 18 and CIN induced by these viral types (Koutski *et al.*, 2002; Harper *et al.*, 2004; Villa *et al.*, 2005). As a result of these trials, two preventive HPV vaccines are being licensed in an increasing number of countries around the world, generally for vaccination of young girls prior to first intercourse and, in a few countries, also for vaccination of young boys.

Most HPV infections are transient but some women with persistent infections have an increased risk of developing high-grade CIN, one-third to a half of which will progress to cervical cancer over a period of 10–15 years. Prevention of cervical cancer by early detection and treatment of CIN currently offers the most cost-effective, long-term strategy for cervical cancer control (Sankaranarayanan *et al.*, 2001; IARC handbooks of cancer prevention Volume 10, 2005). Cytology screening is a viable option for medium-resourced countries only if they have the ability to meet the requirements for testing and follow-up of positive screening results, such as trained human resources, supplies, mechanisms for delivery of samples and results, laboratory infrastructure, and the needed financial resources.

A critical appraisal of reasons for the failure or suboptimal performance of cytology screening has led to the evaluation of alternative testing approaches such as visual inspection with acetic acid (VIA), visual inspection with Lugol's iodine (VILI) and HPV testing. Currently, several studies are addressing the prospects of introducing HPV testing in mass screening programs in developed countries (Franco, 2003), whereas recent studies in developing countries indicate that VIA and VILI are accurate screening tests to detect CIN and early invasive cancer (Sankaranarayanan *et al.*, 2005). Results from a randomized intervention trial in India comparing VIA, cytology and HPV testing indicate that all tests had similar detection rates of CIN 2–3 lesions (Sankaranarayanan *et al.*, 2005a). Findings from the ongoing trials on the impact on disease burden are important for further conclusive evaluation of the HPV and visual screening tests in reducing disease burden (Sankaranarayanan *et al.*, 2004, 2005b). A commendable global effort in the context of reducing disease burden in recent years has been the studies of the Alliance for Cervical Cancer Prevention (ACCP) supported by the Bill & Melinda Gates Foundation (Alliance for Cervical Cancer Prevention (ACCP), 2004).

A number of studies in developed countries indicate that the incidence of cervical adenocarcinoma is increasing while the overall incidence of cervical cancer is declining (Vizcaino *et al.*, 1998; Smith *et al.*, 2000). Although the impact of human immunodeficiency virus (HIV) infection on the risk of progression of CIN to invasive cancer is not clear, the high risk of HPV infection and CIN in HIV-positive women demands active surveillance of these women with pelvic examination and screening.

Estrogenic stimulation produces cellular growth and glandular proliferation in the endometrial epithelium of the uterine body, which is balanced by the cyclical maturational effects of progesterone. Increased risk of endometrial cancer, associated with factors such as oral intake of estrogens (without progestins), early menarche, late menopause, low parity, extended periods of anovulation and obesity can be related to chronic unopposed estrogenic stimulation, leading to abnormal proliferation and neoplastic transformation of endometrial tissue. The association between long-term use of tamoxifen and endometrial cancer has been attributed to its estrogen agonist properties, and women receiving tamoxifen should be monitored carefully. Diabetes mellitus and hypertension have also been associated with increased risk. High parity, late age at last birth, long-term use of combined oral contraceptives and physical activity have a protective effect. Increasing trends among postmenopausal women have been observed in developed countries; in premenopausal and perimenopausal women, endometrial cancer is relatively rare, and where long-term data are available, incidence is decreasing (Bray *et al.*, 2005). The continuing increases in obesity and decreases in fertility, however, forewarn that endometrial cancer, as a postmenopausal disease, will become a more important public health problem in future years.

Most ovarian cancers are diagnosed in advanced clinical stages with poor prospects of long-term survival, particularly in developing countries. Early detection and treatment should decrease mortality from this disease. Screening methods currently undergoing evaluation include transvaginal ultrasound scanning of the ovaries and measurement of the tumour-marker cancer antigen 125 (CA 125) (Kramer *et al.*, 1993; Jacobs *et al.*, 1999). Routine screening for asymptomatic ovarian cancer is not currently recommended (U.S. Preventive Services Task Force. Screening for Ovarian Cancer: Brief Evidence Update. May 2004). There has been a significant improvement in the 5-year survival rates over the last decades in developed countries due to a better definition of patterns of spread of the disease, improvements in surgical techniques and supportive care and more effective chemotherapy regimes (Jemal *et al.*, 2003).

The relatively stable incidence of vulva cancer, despite a steady increase in the diagnosis of vulva intraepithelial neoplasia (VIN), suggests that effective treatment of VIN has prevented a significant increase in the incidence of invasive cancer in developed countries. An association between HPV infection and vaginal cancer has been reported (IARC monographs on the evaluation of carcinogenic risks to humans, 1995). Rapid growth, a high frequency of metastases in the lungs, vagina, pelvis, liver and brain, as well as hemorrhage, make choriocarcinoma a medical emergency for which chemotherapy is highly effective. Treatment has dramatically improved the outcome for this disease that had previously resulted in the death of 60% of patients with localized and 90% with disseminated disease.

Early clinical diagnosis and multidisciplinary treatment involving surgery, radiotherapy and chemotherapy are important in reducing mortality from gynecological cancers. Early clinical diagnosis and prompt treatment improves outcome from gynecological cancers. The death rates from cervical cancer had began to decline before the implementation of cytology screening programs, possibly due to improved awareness, early clinical recognition of disease and prompt treatment, among other factors (Wingo *et al.*, 1995; Ponten *et al.*, 1995). Survival rates suggest that early gynecological cancers have an excellent prognosis even in low- and medium-resource settings (Sankaranarayanan *et al.*, 1988, 1996; Chia *et al.*, 2001). Day-to-day translation of cancer management plans requires adequate health-care infrastructure for diagnosis and treatment as well as adequate trained human and financial resources. The health services infrastructure for diagnosis and management of cancers is poorly developed in many developing countries, and particularly in sub-Saharan Africa (Stewart and Kleihues, 2003). For instance, histopathology services are not available or are very limited in more than 20 African countries where cancer surgery facilities are also extremely limited (Stewart and Kleihues, 2003). In 32 countries in Africa with populations of >157 million, no radiotherapy services were available in the late 1990s (Levin *et al.*, 1999). There were only some 155 radiotherapy machines in the entire African continent, which was less than the number of radiotherapy machines in Italy alone (Levin *et al.*, 1999). The availability of radiotherapy services in many Asian countries also seems to be suboptimal (Tatsuzaki and Levin, 2001). Trained human resources in cancer diagnosis and management are distinctly inadequate in many regions. Training programs to generate such resources either do not exist or are strikingly limited in many countries.

Hospital cancer registry data from five premier cancer hospitals in different regions of India indicated that 12–37% of cervical cancer patients were not prescribed, or did not complete, the prescribed treatment (National Cancer Registry Programme, 2002).

Conclusion

Approximately one out of six cancer cases among women in the world is a gynecological cancer. Whereas cervical cancer is the most common gynecological cancer in developing countries, endometrial cancer is the most common in developed countries. Although mortality from cervical cancer is potentially entirely avoidable by current technologies, it still accounts for half of the global gynecological cancer burden due to lack of effective screening in low- and medium-resource countries. Implementation of current developments in screening has the potential to dramatically reduce the burden of this cancer. Prophylactic vaccination holds great promise for cervical cancer prevention in the future. The treatment outcomes in gynecological cancer can be improved by early clinical detection and appropriate treatment. The differences in the outcome of cancer treatment across the world are due to vast disparities in health service infrastructures, human resources, service delivery and accessibility to services. A significant proportion of patients, in many countries, are unable to access and avail themselves of (or complete) preventive, diagnostic and therapy services because of inadequate health-care services and financing. Creating awareness among the public and professionals is conducive for early detection. Advocacy and the political will to invest in the development of human resources and health-care infrastructure hold the key for effective gynecological cancer control and reducing the burden of disease in several regions of the world. A National Cancer Control Programme (NCCP), as advocated by WHO, provides a suitable and realistic framework to improve cancer preventive and management services in general (World Health Organization, 2002).

Practice Points

- Documenting the primary site of cancer, accurate clinical staging and treatment in medical records helps to quantify cancer burden and progress in cancer control.
- Documenting the underlying cause of death accurately in the death certificates helps to evaluate the outcome of cancer treatment and cancer control.
- Regular auditing of medical records and facilitating data collection by cancer registries can improve accuracy and validity of cancer statistics.

Research Points

- Establishing population-based cancer registries in countries/regions with no or inadequate coverage will improve the accuracy and validity of data on global cancer burden.

- Estimating population-based cancer survival data will help to evaluate and improve the efficiency of cancer health services in medically underserved countries/regions.

See also: Breast Cancer; Cancer Screening; Cervical Cancer; Endometrial Cancer; Ovarian Cancer; Prostate Cancer.

Citations

Alliance for Cervical Cancer Prevention (ACCP) (2004) *Planning and Implementing Cervical Cancer Prevention and Control Programs: A Manual for Managers*. Seattle: ACCP.

Altieri A, Franceschi S, Ferlay J, *et al.* (2003) Epidemiology and aetiology of gestational trophoblastic diseases. *Lancet Oncology* 4: 670–678.

Armstrong BK (1992) The role of the cancer registry in cancer control. *Cancer Causes and Control* 3: 569–579.

Bosch FX and de Sanjose S (2003) Human papillomavirus and cervical cancer burden and assessment of causality. *Journal of the National Cancer Institute Monographs* 31: 3–13.

Bray F, Dos Santos Silva I, Moller H, and Weiderpass E (2005) Endometrial cancer incidence trends in Europe: Underlying determinants and prospects for prevention. *Cancer Epidemiology, Biomarkers and Prevention* 14: 1132–1142.

Castellsague X, Bosch FX, Munoz N, *et al.* (2002) Male circumcision, penile human papillomavirus infection, and cervical cancer in female partners. *New England Journal of Medicine* 346: 1105–1112.

Chia KS, Du WB, Sankaranarayanan R, *et al.* (2001) Population-based cancer survival in Singapore, 1968 to 1992: An overview. *International Journal of Cancer* 93: 142–147.

Ferlay J, Bray F, Pisani P, and Parkin DM (2002) *GLOBOCAN.* Cancer Incidence, Mortality and Prevalence Worldwide. IARC Cancer Base No. 5 Version 2.0, Lyon: IARC Press, 2004.

Franco EL (2003) Primary screening of cervical cancer with human papillomavirus tests. *Journal of the National Cancer Institute Monographs* 31: 89–96.

Gondos A, Brenner H, Wabinga H, *et al.* (2005) Cancer survival in Kampala, Uganda. *British Journal of Cancer* 92: 1808–1812.

Gondos A, Chokunonga E, Brenner H, *et al.* (2004) Cancer survival in a Southern African urban population. *International Journal of Cancer* 112: 860–864.

Gustafsson L, Ponten J, Bergstrom R, and Adami HO (1997) International incidence rates of invasive cervical cancer before cytological screening. *International Journal of Cancer* 71: 159–165.

Harper DM, Franco EL, and Wheeler CM (2004) Efficacy of a bivalent L1 virus-like particle vaccine in prevention of infection with human papillomavirus types 16 and 18 in young women: A randomized controlled trial. *Lancet* 364: 1757–1765.

IARC handbooks of cancer prevention Volume 10 (2005) *Cervix cancer Screening*. Lyon: IARCPress.

IARC monographs on the evaluation of carcinogenic risks to humans (1995) *Human Papillomaviruses. IARC Monographs*, Vol. 64, Lyon: IARCPress.

Jacobs IJ, Skates SJ, Mac Donald N, *et al.* (1999) Screening for ovarian cancer: A pilot randomized controlled trial. *Lancet* 353: 1207–1210.

Jemal A, Murray T, Samuels A, *et al.* (2003) Cancer Statistics 2003. *CA: A Cancer Journal for Clinicians* 50: 5–26.

Koutski LA, Ault KA, and Wheeler CM (2002) A controlled trial of a human papillomavirus type 16 vaccine. *New England Journal of Medicine* 347: 1645–1651.

Kramer BS, Gohagen J, Prorok PC, and Smart C (1993) A National Cancer Institute sponsored screening trial for prostatic, lung, colorectal, and ovarian cancers. *Cancer* 71: 589–593.

Levin CV, El Gueddari B, and Meghzifene A (1999) Radiation therapy in Africa: Distribution and equipment. *Radiotherapy & Oncology* 52: 79–84.

Munoz N, Bosch FX, de Sanjose S, *et al.* (2003) Epidemiologic classification of human papillomavirus types associated with cervical cancer. *New England Journal of Medicine* 348: 518–527.

National Cancer Registry Programme (2002) *Five-Year Consolidated Report of the Hospital Cancer Registries* 1994–1998. *An Assessment of the Burden and Care of Cancer Patients*. New Delhi: Indian Council of Medical Research, 90–95.

Parkin DM, Bray FI, and Devessa SS (2001) Cancer burden in the year 2000. The global picture. *European Journal of Cancer* 37: S4–S66.

Parkin DM, Bray F, Ferlay J, and Pisani P (2005) Global cancer statistics, 2002. *CA: A Cancer Journal for Clinicians* 55: 74–108.

Parkin DM, Whelan SL, Ferlay J, *et al.* (eds.) (2002) *Cancer Incidence in Five Continents,* Vol. VIII. IARC Scientific Publication No.155. Lyon: IARCPress.

Pisani P, Bray F, and Parkin DM (2002) Estimates of the worldwide prevalence of cancer for 25 sites in the adult population. *International Journal of Cancer* 97: 72–81.

Ponten J, Adami HO, and Bergstrom R (1995) Strategies for global control of cervical cancer. *International Journal of Cancer* 60: 1–26.

Ries LAG, Eisner MP, Kosary CL, *et al.* (eds.) (2004) *SEER cancer statistics review, 1975–2001.* Bethesda, MD: National Cancer Institute. Available at: http://seer.cancer.gov/csr/1975_2001/ Accessed December 22, 2004.

Sankaranarayanan R, Black RJ and Parkin DM (eds.) (1988) *Cancer Survival in Developing Countries, IARC Scientific Publications No.15.* Lyon, IARCPress.

Sankaranarayanan R, Budukh A, and Rajkumar R (2001) Effective screening programs for cervical cancer in low- and middle-income developing countries. *Bulletin of the World Health Organization* 79: 954–962.

Sankaranarayanan R, Nene BN, Dinshaw KA, *et al.* (2005a) A cluster randomised controlled trial of visual, cytology and HPV screening for cancer of the cervix in rural India. *International Journal of Cancer* 116: 617–623.

Sankaranarayanan R, Gaffikin L, Jacob M, *et al.* (2005b) A critical assessment of screening methods for cervical neoplasia. *International Journal of Gynecology & Obstetrics* 89: S4–S12.

Sankaranarayanan R, Rajkumar R, Theresa R, *et al.* (2004) Initial results from a randomized trial of cervical visual screening in rural south India. *International Journal of Cancer* 109: 461–467.

Sankaranarayanan R, Swaminathan R, and Black RJ (1996) Global variations in cancer survival. *Cancer* 78: 2461–2464.

Smith HO, Tiffany MF, Qualis CR, *et al.* (2000) The rising incidence of adenocarcinoma relative to squamous cell carcinoma in the United States: A 24-year population-based study. *Gynecologic Oncology* 78: 97–105.

Stewart BW and Kleihues P (2003) *World Cancer Report*. Lyon: IARC Press.

Tatsuzaki H and Levin CV (2001) Quantitative status of resources for radiation therapy in Asia and Pacific region. *Radiotherapy & Oncology* 60: 81–89.

The EUROCARE Working Group (2003) EUROCARE-3: Survival of cancer patients diagnosed 1990-94 - results and commentary. *Annals of Oncology* 14: v61–v118.

U.S. Preventive Services Task Force (2004) Screening for Ovarian Cancer: Brief Evidence Update. May 2004. Agency for Healthcare Research and Quality. Rockville, MD. http://www.ahrq.gov/clinic/3rduspstf/ovariancan/ovcanup.htm.

Villa LL, Costa RR, and Petta CA (2005) Prophylactic quadrivalent human papillomavirus (types 6,11,16 and 18) L1 virus-like particle vaccine in young women: A randomized double-blind placebo-controlled multicentre phase II efficacy trial. *Lancet Oncology* 6: 271–278.

Vizcaino AP, Moreno V, Bosch FX, *et al.* (1998) International trends in the incidence of adenocarcinoma and adenosquamous carcinoma. *International Journal of Cancer* 75: 536–545.

Wingo PA, Tong T, and Bolden S (1995) Cancer statististics 1995. *CA: A Cancer Journal for Clinicians* 45: 8–30.

World Health Organization (2002) *National Cancer Control Programmes: Policies and Managerial Guidelines*, 2nd edition. Geneva: World Health Organization.

Cervical Cancer

L C Zeferino and S F Derchain, State University of Campinas, Campinas, Brazil

Introduction

According to the World Health Organization (WHO), around 270 000 deaths occur worldwide every year due to cervical cancer. Of these, almost 40 000 occur in more-developed regions of the world, encompassing Europe, North America, Japan, New Zealand and Australia, containing 1200 million inhabitants. More than 230 000 deaths occur in less-developed regions of the world, inhabited by around 5200 million individuals. The incidence and mortality rates of this neoplasia are not uniformly high throughout the less-developed world. The countries of Northern Africa, the Middle East and the People's Republic of China have rates similar to those of European countries. On the other hand, sub-Saharan Africa, most of Central and Southern Asia, and South and Central America have higher incidence and mortality rates for cervical cancer (International Agency for Research on Cancer, 2005). Therefore, the stage of development of a region is not the determinant factor in the incidence or the mortality rates of cervical cancer, although there is an association. What appears to characterize the less-developed regions, even those in which the incidence of cervical cancer is low, is that mortality rates are closer to the incidence rates (International Agency for Research on Cancer, 2005).

A meta-analysis of social inequality and the risk of cervical cancer found that both cervical infection by human papillomavirus (HPV), which is linked to both male and female sexual behavior, and access to adequate cervical cancer screening are closely linked to the higher incidence rates of cervical cancer observed in different socio-economic groups, and that the importance of these factors may vary from one geographical region to another (Parikh *et al.*, 2003).

Since the 1960s, a significant reduction in the incidence of cervical cancer has been observed in countries with an organized screening program (Bray *et al.*, 2002; Peto *et al.*, 2004). Nevertheless, some regions have presented an increasing incidence rate of cervical cancer, and rates in other regions have remained stable (Vizcaino *et al.*, 2000).

Squamous cell carcinoma and its precursor lesions are more frequent than glandular neoplasias. Cervical adenocarcinoma corresponds to less than 10% of invasive carcinomas; however, the incidence of cervical adenocarcinoma has increased in recent decades, particularly among young women (Chan *et al.*, 2003; Bulk *et al.*, 2005). This relative and absolute increase in the incidence of cervical adenocarcinoma may be partly due to a modification in risk factors or cofactors, and to the importance of the cytological diagnosis of atypical glandular cells, as defined by the Bethesda system (Krane *et al.*, 2001; Ruba *et al.*, 2004).

HPV

There is much evidence demonstrating that HPV is a primary risk factor for cervical cancer and that it plays a central role in its carcinogenesis (Walboomers *et al.*, 1999). Nevertheless, HPV is not in itself sufficient to lead to the development of a cervical carcinoma, since the majority of HPV infections regress spontaneously with no clinical manifestations (Sellors *et al.*, 2000). Up to the present time, around 200 types or subtypes of HPV have been identified, of which around 40 types have already been isolated from the anal–genital epithelium. Of these, 15 types may be considered oncogenic (high risk) and another three types may be considered as probably oncogenic. The most frequent oncogenic HPV types in patients with cervical carcinoma are 16, 18, 45, 31, 33, 52, 58 and 35— a similar HPV spectrum to that found in the general female population (Franco *et al.*, 2003).

The proportion of women infected with HPV varies according to the woman's age and among different populations. It is estimated that 15–40% of young, sexually active women and 3–10% of women aged 35 years or older are infected with HPV (Jacobs *et al.*, 2000). Prevalence is the result of a balance between infection and re-infection of HPV versus spontaneous regression; a process that is very intense among young women. Follow-up of these women shows that up to 80% may present HPV infection within a relatively short period of time (Brown *et al.*, 2005).

If the majority of HPV infections regress spontaneously and if virtually all cases of invasive cervical carcinoma are associated with oncogenic HPV, women with persistent HPV infection are indeed those at risk of developing the precursor lesions and invasive cervical cancer (Schlecht *et al.*, 2003; Franco *et al.*, 1999). Infection by HPV16 persists for longer, which partially explains its greater prevalence and would give it a selective advantage in triggering cancer (Ho *et al.*, 1995; Woodman *et al.*, 2001).

Cofactors of Risk for Cervical Cancer

The fact that HPV is a prerequisite for invasive cervical cancer implies that the other risk factors may be considered to be cofactors rather than independent factors.

These cofactors would act by influencing HPV acquisition, its persistence, and the development and progression of a neoplastic lesion. The environmental cofactors that have been most studied and that would influence cervical carcinogenesis are high parity, use of oral contraceptives, co-infection with other sexually transmitted agents, and diet. Cofactors related to the host may include immune response, genetic susceptibility, genetic variations of the human leukocyte antigen, and polymorphisms of the TP53 gene. Viral factors include HPV type and variant, viral load and viral integration in the genome of the host cell (Tortolero-Luna and Franco, 2004). Smoking seems to promote early cervical carcinogenic events by increasing the duration of oncogenic HPV infections and decreasing the probability of clearing oncogenic infections (Giulian *et al.*, 2002). Women who are smokers have an increased risk of developing squamous cell carcinoma but not adenocarcinoma of the cervix, and the risk of squamous cell carcinoma increases in current smokers with the number of cigarettes smoked per day and with younger age at starting smoking (International Collaboration of Epidemiological Studies of Cervical Cancer, 2005).

Screening Tests and Their Efficacy

Screening tests are not intended to be diagnostic, but they are used to distinguish apparently unaffected women from those who may have the disease. Ideally, their results require confirmation. The strategy of screening policies is to offer a test within a broad-reaching program; therefore, the test must be easy to perform, cheap and innocuous.

Sensitivity and specificity are very important parameters for evaluating the accuracy of the screening test. Sensitivity refers to the percentage of women with the disease in whom the test was positive, while specificity is the percentage of women who do not have the disease and who tested negative.

Conventional and Liquid-based Cytology

Conventional cervical cytology is the worldwide screening test for cervical cancer. There are several terminologies for reporting precursor lesions of the cervix (**Table 1**). The Bethesda system, created in 1988 for cervical cytological reports, includes separate diagnoses for squamous and glandular cells, as well as analysis of the adequacy of the specimen.

Samples of cells from the transformation zone of the uterine cervix may be collected by doctors, nurses or other trained health workers who prepare the smear. Based on the authors' experience, each cytotechnician is able to read between 60 and 100 cytology tests/day. The cytopathologists should be responsible for supervision and for preparing final reports. Training cytotechnicians is a lengthy procedure and this is an obstacle that must be overcome in low-resource settings. Quality control procedures, which should ideally consist of both internal and external quality control, are required for evaluation of the performance of the cytopathology laboratory as well as the performance of the health workers responsible for taking the samples (Miller *et al.*, 2000).

Despite the proven effectiveness of cervical cytology screening in reducing the incidence of cervical cancer through organized programs, the accuracy of conventional cervical cytology has been discussed recently. The levels of sensitivity and specificity of cervical cytology vary greatly. A meta-analysis on the accuracy of cervical cytology, in which data from 59 studies were combined, reported estimates of sensitivity and specificity ranging

Table 1 Summary of the systems of terminology used for cervical neoplasia

Modified Papanicolaou class[*,a]	*World Health Organization*[b]	*Richart*[c]	*Bethesda system*[d]
Class III	Mild dysplasia	CIN 1	Low-grade SIL
	Moderate dysplasia	CIN 2	High-grade SIL
	Severe dysplasia	CIN 3	High-grade SIL
Class IV	Carcinoma in situ	CIN 3	High-grade SIL
Class V	Micro-invasive carcinoma	Invasive carcinoma	Squamous cell carcinoma
	Invasive carcinoma		
			Atypical squamous cells
			Endocervical adenocarcinoma in situ
			Adenocarcinoma
			Atypical glandular cells

CIN, cervical intra-epithelial lesion; SIL, squamous intraepithelial lesion.
*Class I, absence of atypical or abnormal cells; Class II, atypical cytology, but no evidence of malignancy.
[a]Papanicolaou GN (1954) *Atlas of Exfoliative Cytology*. Boston: Commonwealth Fund University Press.
[b]Riotton G, Christopherson WM, and Lunt R (1973) *Cytology of the Female Genital Tract. International Histological Classification of Tumours No. 8.* Geneva: World Health Organization.
[c]Richart RM (1973) Cervical intraepithelial neoplasia. In: Sommers SC (ed.) *Pathology Annual*, pp. 301–328. New York: Appleton-Century-Crofts.
[d]Solomon D, Davey D, Kurman R, *et al.* (2002) The 2001 Bethesda system: Terminology for reporting results of cervical cytology. *JAMA* 287: 2114–2119.

Table 2 Accuracy of conventional cytology in detecting cervical intra-epithelial neoplasia (CIN) 2–3 lesions and invasive cancer in selected cross-sectional studies in developing countries (test positive threshold of low-grade squamous intra-epithelial lesion)

Author, year of publication, country of study	*No. of participants*	*Sensitivity, % (95% CI)*	*Specificity, % (95% CI)*
University of Zimbabwe/JHPIEGO, 1999, Zimbabwe	2092	44 (37–51)	91 (89–92)
Denny *et al.*, 2000, South Africa	2885	78 (67–87)	95 (94–96)
Wright *et al.*, 2000, South Africa	1352	61 (46–74)	96 (94–97)
Denny *et al.*, 2002, South Africa[†]	2754	57 (46–67)	96 (95–97)
Cronjé *et al.*, 2003, South Africa	1093	48 (38–60)	96 (94–97)
Sankaranarayanan *et al.*, 2004a, India[*,†]	22.663	61 (56–66)	95 (94–95)

CI, confidence interval.
*Pooled results of five studies.
[†]Alliance for Cervical Cancer Prevention study.

from 11% to 99% and 14% to 97%, respectively (Fahey *et al.*, 1995). A systematic study reported sensitivity and specificity ranging from 30% to 87% and 86% to 100%, respectively (Nanda *et al.*, 2000). More recent cross-sectional studies from developing countries have reported sensitivity varying from 44% to 78% and specificity varying from 91% to 99% (**Table 2**). These data suggest that the variability in accuracy is related to the quality of cervical cytology, as sensitivity is dependent on the method and not on disease prevalence. However, cervical cytology can be carried out with a high degree of accuracy, even in developing countries.

Some developing countries, particularly countries in Latin America, have introduced cytology screening programs over the last 30 years. However, there is little evidence of the success of these screening programs in reducing the burden of cervical cancer (Zeferino *et al.*, 2005). The problems identified were the suboptimal performance of cytology, lack of quality control, and inefficiency in the follow-up and treatment of detected cases (Sankaranarayanan *et al.*, 2001; Robles *et al.*, 1996; Gage *et al.*, 2003; International Agency for Research on Cancer, Cancer, 2005). The target women and screening interval adopted in the majority of these countries are similar to those recommended in developed countries, but the predominant screening modality is opportunistic.

Liquid-based cytology (LBC) was developed in an attempt to improve the performance of conventional cytology. In this technique, the cell sample is transferred to a liquid preservative solution, and the glass slide is prepared in the laboratory. The LBC advantages are borderline and include higher sensitivity, more representative transfer of cells from the collection device to the glass slide, improvement of specimen adequacy and greater microscopic readability. The cell suspension remaining after preparation of the smear can be used for additional testing procedures (Karnon *et al.*, 2004). Analyses that have been carried out recently in developed countries offer greater certainty with respect to the potential cost-effectiveness of LBC compared with conventional cervical cytology (Karnon *et al.*, 2004). Nevertheless, the results of such analyses cannot be applied to developing countries where conventional cytology is usually cheaper than LBC, and the conclusions reached from these analyses may be less favorable than those with LBC. Although there are some diagnostic advantages, the impact of LBC on reducing cancer mortality remains to be established (Sankaranarayanan *et al.*, 2005).

HPV Testing

If cervical cancer and its precursor lesions are HPV-related lesions, HPV testing could potentially be of use as a screening tool. Hybrid capture is the most tested DNA-HPV detection technique; however, polymerase chain reaction and enzyme immunoassay have also been tested (Bulkmans *et al.*, 2004). A positive characteristic of HPV-DNA testing is its higher sensitivity for detecting high-grade squamous intra-epithelial lesions compared with conventional cytology. Nevertheless, the specificity of HPV testing is lower than that of cytology, particularly in younger women, and this represents an important restriction to the adoption of this technique as a single screening tool (Cuzick, 2002).

In fact, these cross-sectional studies evaluated the tests in a single screening and therefore failed to provide information on the changes in the performance of the test during successive screenings. In a longitudinal evaluation of the test performance, the sensitivity should increase as screening cytology is repeated routinely because an initially false-negative test could be a true positive test in the next screening without resulting in any harm to the woman, given that the precursor lesions take a considerable time to progress (Holowaty *et al.*, 1999). Conversely, implementation of a low-sensitivity screening test will result in many healthy women being reported with the disease due to false-positive results, and more healthy women will be added to this group at each new screening. This problem is accentuated after successive screenings because the prevalence of the lesions decreases progressively, thereby leading to a decrease in the positive predictive value of the test. Hence, following many screenings, the burden of the false-positive tests could be large, resulting in much unnecessary anxiety and other morbidities, as well as an increase in the cost of screening. Even in developed countries, HPV testing should not be used as a screening tool in women under 30 years of age (Wright *et al.*, 2004). Although the sensitivity of HPV testing is higher, there is no evidence that it is any more efficient at reducing the mortality rate from cancer than cytology (Holmes *et al.*, 2005).

According to recent cross-sectional studies carried out in developing countries, the sensitivity of HPV testing for detecting cervical intra-epithelial neoplasia (CIN) 2 and CIN 3 lesions varied from 62% to 78%, while specificity varied from 91% to 99% (**Table 3**). In the Alliance for Cervical Cancer Prevention (ACCP) cross-sectional studies carried out in India, the sensitivity of HPV testing in detecting CIN 2–3 lesions and invasive cancer varied from 46% to 81% and specificity varied from 92% to 95% (Sankaranarayanan *et al.*, 2004b). These data suggest that HPV testing in developing regions could present lower sensitivity than that observed in developed countries.

Considering the restrictions in access to health services and the cultural issues that exist in some populations in low-resource settings, the vaginal self-sampling HPV test could be an alternative screening tool that may help to overcome these obstacles. However, this means of collection has shown lower sensitivity than direct sampling of cervical cells carried out by health workers (Wright *et al.*, 2000; Franco, 2003).

Currently, cost-effectiveness is one of the major concerns of health authorities in the planning of public health programs. In countries such as the UK, the Netherlands, France and Italy, HPV-DNA testing has the potential to improve health benefits at a reasonable cost compared with current screening policies (Holmes *et al.*, 2005; Kim *et al.*, 2005). Nevertheless, these analyses of the cost-effectiveness of HPV testing in developed countries cannot be transferred to developing countries where conventional cytology is usually cheaper (Syrjanen *et al.*, 2005).

VIA, VIAM and VILI

Visual inspection with acetic acid (VIA) involves naked-eye inspection of the cervix under bright light conditions at least 1 min after the application of 3–5% diluted acetic acid. The test can be carried out by nurses or midwives, who check for acetowhite areas in the uterine cervix. Uniform criteria for reporting VIA results are still to be established, but a positive result is based on the appearance of well-defined, densely opaque, acetowhite lesions in the transformation zone. It is a simple, inexpensive test that does not require a laboratory infrastructure. Providers can be trained in 5–10 days.

Recent selected cross-sectional studies, including the ACCP studies, reported sensitivity of VIA varying from 52% to 79% and specificity varying from 49% to 88% (**Table 4**). These sensitivity data are similar to those for cervical cytology, while the specificity data are closer to those for HPV testing. VIA is inefficient in detecting lesions located in the cervical canal of the uterus.

However, the major logistic advantage of the visual tests is the immediate availability of results, allowing diagnostic investigation and treatment at the same screening appointment. Nevertheless, the lower specificity of these tests means that a large number of women will undergo investigation, although most of them could be negative for neoplasia or may not have any clinically significant lesion. An ACCP study carried out in Thailand assessed the safety, acceptability and feasibility of combining VIA testing and immediate cryotherapy of VIA-positive cases. At the 1-year follow-up visit, the VIA-negative rate was 94.3%, (Royal Thai College of Obstetricians and Gynaecologists (RTCOG) CCJHPIEGO Corporation Cervical Cancer Prevention Group, 2003), but if specificity is low, many healthy women may have been treated unnecessarily, thereby falsely improving the rate of control.

VIAM is VIA with low-level magnification (2–4X) using a hand-held device to inspect the cervix. There is no evidence that this procedure improves the performance of the naked-eye visualization test (Sankaranarayanan *et al.*, 2004c).

VILI is visual inspection using Lugol's iodine solution that stains glycogen stored in cervical epithelial cells. The neoplastic or immature squamous metaplastical epithelium stores less glycogen than the normal mature squamous epithelium. The yellow-coloured changes associated with a positive VILI test result are recognized more easily by health workers than the acetowhite changes associated with VIA. The ACCP studies, involving almost 50 000 women, found more than 2% of

Table 3 Accuracy of human papillomavirus testing in detecting cervical intra-epithelial neoplasia (CIN) 2–3 lesions and invasive cancer in selected cross-sectional studies

Author, year of publication, country of study	*No. of participants*	*Sensitivity, % (95% CI)*	*Specificity, % (95% CI)*
Womack *et al.*, 2000, Zimbabwe	2140	81 (74–86)	62 (59–64)
Denny *et al.*, 2000, South Africa	2885	73 (62–82)	92 (91–93)
Wright *et al.*, 2000, South Africa	1352	84 (71–92)	83 (80–85)
Franco, 2003, review of 13 studies	2754	66–100	61–96
Sankaranarayanan *et al.*, 2004b, India†	18 085	68 (61–74)	94 (93–94)

CI, confidence interval.
†Alliance for Cervical Cancer Prevention study.

Table 4 Accuracy of visual inspection with acetic acid in detecting cervical intra-epithelial neoplasia (CIN) 2–3 lesions and invasive cancer in selected cross-sectional studies

Author, year of publication, country of study	*No. of participants*	*Sensitivity, % (95% CI)*	*Specificity, % (95% CI)*
University of Zimbabwe/JHPIEGO, 1999, Zimbabwe	2148	77 (70–82)	64 (61–66)
Denny *et al.*, 2000, South Africa	2885	67 (56–77)	84 (92–85)
Belinson *et al.*, 2001, People's Republic of China	1997	71 (60–80)	74 (71–76)
Denny *et al.*, 2002, South Africa*,‡	2754	70 (59–79)	79 (77–81)
Cronjé *et al.*, 2003, South Africa	1093	79 (69–87)	49 (45–52)
Sankaranarayanan *et al.*, 2004d, India†,‡	54 981	79 (77–81)	86 (85–86)
Gontijo *et al.*, 2005, Brazil*	684	52 (31–74)	79 (31–74)
Bastos *et al.*, 2005, Brazil*	809	54 (51–58)	88 (86–91)

CI, confidence interval.
*Estimates for CIN 2–3 lesions only.
†Pooled results of 11 studies from Jaipur, Kolkata, Mumbai and Trivandrum in India, and Burkina Faso, Congo, Guinea, Mali and Niger in Africa.
‡Alliance for Cervical Cancer Prevention study.

Table 5 Accuracy of screening tests for the detection of high-grade squamous intra-epithelial lesions in women aged 30–65 years, Mumbai, India[a]

Test	*Sensitivity (95% CI)*	*Specificity (95% CI)*	*PPV (%)*	*NPV (%)*
VIA	59.7 (45.8–72.4)	88.4 (87.4–89.4)	7.0	99.3
VIAM	64.9 (51.1–77.1)	86.3 (85.2–87.4)	6.5	99.4
VILI	75.4 (62.2–85.9)	84.3 (83.1–85.4)	6.5	99.6
Cytology	57.4 (43.2–70.8)	98.6 (98.2–99.0)	37.8	99.4
HPV testing	62.0 (47.2–75.4)	93.5 (92.6–94.3)	12.1	99.4

PPV, positive predictive value; NPV, negative predictive value; VIA, visual inspection with acetic acid; VIAM, visual inspection with acetic acid using low-level magnification; VILI, visual inspection with Lugol's iodine; HPV, human papillomavirus; CI, confidence interval.
[a]Shastri SS, Dinshaw K, Amin G, et al. (2005) Concurrent evaluation of visual, cytological and HPV testing as screening methods for the early detection of cervical neoplasia in Mumbai, India. *Bulletin of the World Health Organization* 83: 186–194.

high-grade lesions or worse. They reported sensitivity varying from 78% to 98% and specificity varying from 73% to 91% (Sankaranarayanan *et al.*, 2004d). These data indicate that VILI is a more sensitive test than VIA.

One of the most important concerns in assuring good-quality screening with VIA or VILI is adequate preparation of the examiners. To achieve this goal, the ACCP prepared a comprehensive and hands-on training program for examiners in low-resource settings, and recently reported on their accomplishments (Blumenthal *et al.*, 2005).

Concurrent Screening Test Assessment

In Mumbai, India, Shastri *et al.* (2005) carried out a concurrent evaluation of visual, cytological and HPV testing as screening methods, using a cross-sectional study in which each screening test was carried out independently by trained workers with high-school education. The tests were carried out in women aged 30–65 years in the following order: specimen collection for cytology; specimen collection for HPV testing; VIA; VIAM; and VILI. Colposcopy was carried out before VILI, and was repeated afterwards in order to perform punch biopsies on suspicious areas of the cervix. This study design is very suitable for comparing the performance of these five methods because all methods were carried out in the same women. The sensitivity was similar with the five methods; nevertheless, cytology showed the highest specificity, followed by HPV testing, while VIA, VIAM and VILI showed a similar, lower specificity (**Table 5**).

Organization of Cervical Cancer Screening Programs

The success of cervical cancer screening is shown by its ability to reduce the incidence of cervical cancer and its resultant mortality. The high-grade intra-epithelial lesion is the direct precursor of invasive squamous carcinoma, and the best strategy for achieving success in reducing the incidence and mortality rates is the detection and treatment of this lesion. To achieve a reduction in the mortality rate, cervical cancer screening must achieve three goals: provide high coverage of the population at risk; provide women with an accurate screening test as part of high-quality services; and ensure that women with positive test results are properly managed (Bradley *et al.*, 2005). To achieve these goals, program policies have to define who is to be screened and treated, how often screening should occur, which screening test and treatment options should be used, and where and by whom screening should be carried out, as well as ensuring sufficient political and financial investment.

In many developed and less-developed countries, cervical cancer screening has been carried out as part of an obstetrical or gynecologic consultation, or in family-planning services. Other institutions have introduced cervical screening for women attending a hospital for any type of health care (Pinotti *et al.*, 1981). In these conditions, younger women are screened more frequently than older women, and the frequency of screening is dependent on medical appointments. This modality of screening is referred to as 'opportunistic'. Under this system, some women are overscreened, while others have no access whatsoever to health services and are therefore excluded from any screening program. The impact of this screening modality on the incidence of cervical cancer is null or very limited (International Agency for Research on Cancer, Cancer, 2005).

Ideally, cervical cancer screening should be organized as an integrated system capable of recruiting, screening and guaranteeing the treatment of detected cases according to the age group, screening method and screening interval established. Health services should be very accessible and capable of providing coverage for the entire target population group. A system for the collection of health data and a structure of quality assurance should be set up. This modality of screening is referred to as an 'organized program' and is capable of reducing the incidence of invasive cervical cancer (Laara *et al.*, 1987).

Several strategies with respect to target women, screening interval and therapeutic approaches have been evaluated for application in resource-poor settings. The use of these strategies will inevitably result in some cases of disease being missed; however, in the long term, they will lead to maximum reduction in cervical-cancer-related morbidity and mortality in relation to the resources available. One ethical issue that should be mentioned is that the screening strategy should not divert resources from other more important health-care programs because, in these circumstances, offering screening may end up reducing the overall level of health in a community.

Age and Screening Interval

The screening test that has been used worldwide is conventional cytology, and most guidelines recommend this test. The burden of low-grade lesions is very high in women under 25 years of age (D'Ottaviano-Morelli *et al.*, 2004), and the majority of these lesions regress, resulting in a normal smear within a 2-year period (Holowaty *et al.*, 1999). Detection and treatment of such lesions increases the cost and the morbidity rate without any proven reduction in incidence rates. Recommendations are usually to initiate screening at 20–25 years of age since, in the case of younger women, the risk of harm may be greater than any potential benefit, and invasive cervical cancer is rare before 30 years of age. A good reference guide would be to take the age at the beginning of the rise in the incidence of cervical cancer rates and initiate screening 5 years prior to this age, which would be around 30 years of age (Miller *et al.*, 2000).

The optimal screening interval is one that provides the most favourable ratio between the degree of disease control and the cost of screening or the available resources. Understanding the natural history of cervical cancer, particularly the duration of precursor lesions, may help determine the frequency of screening. The most important precursor lesion, CIN 3, may either regress (Östör, 1993) or take a long time to progress to invasive carcinoma (Gustafsson and Adami, 1989). The most common screening interval adopted in developed countries is every 3 years.

The ACCP consensus recommends that where resources are limited, programs should focus on screening women between the ages of 30 and 49 years at least once in their lifetime (World Health Organization, International Agency for Research on Cancer-IARC, 1986). There is evidence that even once-in-a-lifetime screening can result in a 19% reduction in the incidence of cervical cancer (Goldie *et al.*, 2001). One cost-effective option is to have three screenings throughout a woman's lifetime, 5 years apart, beginning at age 35 or 40 years (Goldie *et al.*, 2001). Screening every 3 years has a similar impact to annual screening, and screening every 10 years can also reduce mortality. This schedule of screening is feasible in middle or more developed regions (World Health Organization, International Agency for Research on Cancer-IARC, 1986).

It should be emphasized once again that the impact of cervical screening is very limited, and valuable resources may be wasted, if younger women are included in the target population and if a 3-year screening interval or less is implemented but with low coverage rate. Therefore, the principal goal to be achieved in cervical cancer screening is that of high coverage of the established target population. After reaching this goal, the target population may be progressively increased, extending coverage to other age groups and repeating screening at shorter intervals.

When to Stop Screening

The incidence rate of cervical cancer is higher in older women around 50–60 years of age (D'Ottaviano-Morelli *et al.*, 2004). Ideally, some countries recommend stopping screening and following-up women at 60 or 65 years of age if they have tested normal at two or three regular tests over the previous 10 years. If they have never been screened or have not been screened for many years, they should be encouraged to have at least two tests; if both are negative, there is no need for further screening (International Agency for Research on Cancer, Cancer, 2005). If resources are limited, a single test should be performed on these women. If resources are very limited, women in this age group should not be included in the screening program because they tend to be poor attenders, their smears tend to be of poorer quality so long after the menopause, and cervical neoplasia often appears as a clinically invasive carcinoma.

Follow-up and Treatment of Detected Abnormalities

In low-resource settings, therapeutic services may be unavailable, inaccessible or inadequately linked to screening services. Colposcopy with cervical biopsy is the standard procedure for the diagnosis of cervical neoplasia, but this procedure is not generally available for the care of women with suspicious cervical lesions. To deal with this limitation, some strategies should be developed.

Immediate colposcopy and biopsy is the general consensus for the diagnosis of high-grade cytological abnormalities. Women with a low-grade lesion or those with undetermined cytological diagnoses should have repeat cytology, and only those with persistently positive results should undergo colposcopy. However, the problem with this conservative approach is low attendance as it requires multiple visits to health services. Women in rural areas, those living further away from the health centres or those with family or work responsibilities may require different approaches. It may be preferable to treat these cases immediately, depending on the available resources and facilities.

Excisional methods for treating precursor lesions, such as the loop electrosurgical excision procedure, are effective and provide tissue specimens for histological examination (Martin-Hirsch *et al.*, 2004), but this method may only be available in middle-resource settings or higher.

Cryotherapy has been in use for more than 40 years and there is sufficient positive evidence about its safety, effectiveness and acceptability. Anesthesia is not required. The equipment is simple, inexpensive and requires few consumable supplies and no electricity. This method should not be used for lesions extending into the endocervical canal. As an ablative procedure, however, no tissue is available for histological examination, but a punch biopsy specimen can be taken prior to the procedure (Jacob *et al.*, 2005).

Obviously, in settings where there is no possibility of histological confirmation of the lesion, deciding to treat every woman who had an abnormal screening test would result in a proportion of the women screened being treated correctly while others would be overtreated. In a study carried out in 35–65-year-old women from South Africa who had undergone no previous screening, Denny *et al.* (2000) reported that 14.4%, 11.7%, 3.0% and 3.4% of screened women with no disease would be overtreated based on VIA, HPV (cut-off 1 RLU), HPV (cut-off 10 RLU) and cytology, respectively.

Recruiting the Women

Developed countries have achieved high coverage and a more regular screening interval by using a regular call–recall system every 3 or 5 years to recruit target women, and this is an important component of an organized screening program. In the UK, there were specific regions and populations in which the screening programs were unable to prevent the appearance of new cases, principally among younger women. Coverage of preventive action was found to be below 50%, leading to the implementation in 1988 of a periodic call-in system for women. Coverage practically doubled, reaching 85% of the target population, and the incidence of cervical cancer began to decline in young women (Bulk *et al.*, 2005).

Less-developed countries may not have an adequate health information system with names and addresses for the entire population, thereby making the call–recall system unfeasible, and this is an important limitation to introducing an adequate, high-coverage screening program with an established target population and screening interval. Alternative sources of population data may include population registries, electoral roles and medical records.

Success in delivering a screening program requires a good understanding of, and attention to, behavioral factors, including communication about cervical cancer and the screening process, the psychological consequences of participating and the issues that affect participation. When making a decision about participation in screening, women should ideally take into account their own understanding of cervical cancer and their perception of risk of acquiring it, in addition to their understanding of the benefits of screening. The main determinants of participation include factors such as the organization of the health-care system; women's perceptions, vulnerability, anxiety and fear about cervical cancer; concurrent family difficulties and education level. All these issues and factors have to be dealt with (International Agency for Research on Cancer, Cancer, 2006).

The absence of a health information system in less-developed countries may be partially compensated by involving volunteers, social workers or community health workers in recruiting women for screening, and by ensuring that women with positive test results have access to proper diagnosis and treatment. Other strategies should include: developing community partnerships to listen to and learn from the community, thereby enhancing the appropriateness and quality of services; developing culturally appropriate messages and educational material; and identifying effective ways of encouraging women to complete diagnosis and treatment regimens (Agurto *et al.*, 2005).

Prospects for Controlling Cervical Cancer

Currently, studies are ongoing in developing countries to evaluate the efficacy and cost-effectiveness of a single round of visual, cytology and HPV screening in reducing cervical cancer incidence and mortality, as well to assess the determinants of participation in screening and the effectiveness of information education programs.

Recent research on the safety and efficacy of candidate prophylactic vaccines against HPV have shown very promising results, with nearly 100% efficacy in preventing the development of persistent infections and precancerous cervical lesions (Harper *et al.*, 2004; Villa *et al.*, 2005). Ongoing clinical studies are expected to provide further evidence on the vaccines that should be available for general use within a few years. Although the future seems bright on the HPV vaccine front, policy makers are strongly cautioned to avoid scaling back cervical cancer screening because screening will be needed for several decades.

Conclusion

To reduce mortality, cervical cancer screening must achieve three goals: provide high coverage of the target population; provide an accurate screening test; and ensure that women with positive test results are properly managed. A high level of participation and good-quality cytology can be achieved in low-resource settings. After high coverage has been achieved, the target population may be progressively extended to include other age groups and to repeat screening at shorter intervals. VIA and VILI have been shown to be viable alternatives to traditional cytology in the lowest-resource settings, but these methods require careful monitoring. Although the sensitivity of HPV testing is higher than that of cytology, there is no evidence that it is any more efficient at reducing the mortality rate from cancer than cytology. Cryotherapy is a safe and effective method that is acceptable to women and can be delivered by a range of health-care providers, not necessarily physicians. Application of special strategies with respect to the target population, screening interval and therapeutic approaches for resource-poor settings may result in some cases of the disease being missed; however, in the long term, they will lead to maximum reduction in cervical-cancer-related morbidity and mortality in relation to the resources available. The main determinants of a woman's participation include factors such as the organization of the health-care system; women's perceptions, vulnerability, anxiety and fear about cervical cancer; concurrent family difficulties; and education level.

Practice Points

- Oncogenic HPV infection is a prerequisite for invasive cervical cancer.
- Results and conclusions of studies on cost-effectiveness of screening tests carried out in more-developed countries cannot be applied to less-developed countries.
- HPV-DNA testing provides higher sensitivity, but lower specificity, for the diagnosis of high-grade squamous intra-epithelial lesions compared with conventional cytology.
- The major logistic advantage of visual tests is the immediate availability of results.
- VIA provides similar sensitivity but lower specificity than conventional cytology.
- Cytology screening can be initiated 5 years prior to the age at which the incidence of cervical cancer starts to rise.
- There is evidence that even once-in-a-lifetime screening between the ages of 30 and 49 years can reduce the incidence of cervical cancer.
- There are no reasons to continue screening women older than 65 years with a previous normal smear in the last 10 years.

Research Agenda

- The efficacy and cost-effectiveness of a single round of visual, cytology and HPV screening in reducing cervical cancer incidence and mortality.
- The determinants of participation in screening and the effectiveness of information and education.
- The long-term safety and efficacy of prophylactic vaccines against HPV.
- Psychosocial outcomes of HPV testing.
- Cost-effectiveness of HPV testing as primary screening test.
- Cost-effectiveness of HPV testing as an adjunct to cervical cytology.

Acknowledgments

The authors wish to acknowledge the collaboration of Elsevier Ireland Ltd, who kindly gave their permission to reproduce **Tables 2**, **3** and **4** of this chapter from the *International Journal of Gynaecology and Obstetrics* (International Agency for Research on Cancer, Cancer, 2005).

See also: Cancer Screening; Female Reproductive Function; Worldwide Burden of Gynecologic Cancer: The Size of the Problem.

Citations

Agurto I, Arrossi S, White S, *et al.* (2005) Involving the community in cervical cancer prevention programs. *International Journal of Gynecology & Obstetrics* 89 (Suppl 2): S38–S45.

Bastos JFB, Derchain SFM, Sarian LO, *et al.* (2005) Aided visual inspection with acetic acid (VIA) and HPV detection as optional screening tools for cervical cancer and its precusor lesions. *Clinical & Experimental Obstetrics & Gynecology* 32: 225–229.

Belinson JL, Pretorius RG, Zhang WH, *et al.* (2001) Cervical cancer screening by simple visual inspection after acetic acid. *Obstetrics & Gynecology* 98: 441–444.

Blumenthal PD, Lauterbach M, Sellors JW, *et al.* (2005) Training for cervical cancer prevention programmes in low-resource settings: Focus on visual inspection with acetic acid and cryotherapy. *International Journal of Gynecology & Obstetrics* 89 (Suppl 2): S30–S37.

Bradley J, Barone M, Mahe C, *et al.* (2005) Delivering cervical cancer prevention services in low-resource settings. *International Journal of Gynecology & Obstetrics* 89 (Suppl): S21–S29.

Bray F, Sankila R, Ferlay J, *et al.* (2002) Estimates of cancer incidence and mortality in Europe in 1995. *European Journal of Cancer* 38: 99–166.

Brown DR, Shew ML, Qadadri B, *et al.* (2005) A longitudinal study of genital human papillomavirus infection in a cohort of closely followed adolescent women. *Journal of Infectious Diseases* 191: 182–192.

Bulk S, Visser O, Rozendaal L, *et al.* (2005) Cervical cancer in the Netherlands 1989–1998: Decrease of squamous cell carcinoma in older women, increase of adenocarcinoma in younger women. *International Journal of Cancer* 113: 1005–1009.

Bulkmans NW, Rozendaal L, Snijders PJ, *et al.* (2004) POBASCAM, a population-based randomized controlled trial for implementation of high-risk HPV testing in cervical screening: Design, methods and baseline data of 44,102 women. *International Journal of Cancer* 20: 94–101.

Chan PG, Sung HY, and Sawaya GF (2003) Changes in cervical cancer incidence after three decades of screening US women less than 30 years old. *Obstetrics & Gynecology* 102: 765–773.

Cronjé HS, Parham GP, Cooreman BF, *et al.* (2003) A comparison of four screening methods for cervical neoplasia in a developing country. *American Journal of Obstetrics & Gynecology* 188: 395–400.

Cuzick J (2002) Role of HPV testing in clinical practice. *Virus Research* 89: 263–269.

Denny L, Kuhn L, Pollack A, *et al.* (2000) Evaluation of alternative methods of cervical cancer screening in resource poor settings. *Cancer* 89: 826–833.

Denny L, Kuhn L, Pollack A, *et al.* (2002) Direct visual inspection for cervical cancer screening: An analysis of factors influencing test performance. *Cancer* 94: 1699–1707.

D'Ottaviano-Morelli MG, Zeferino L, Cecatti JG, *et al.* (2004) Prevalence of cervical intraepithelial neoplasia and invasive carcinoma based on cytological screening in the region of Campinas, Sao Paulo, Brazil. *Cadernos de Saúde Publica* 20: 153–159.

Fahey MT, Irwig L, and Macaskill P (1995) Meta-analysis of Pap test accuracy. *American Journal of Epidemiology* 141: 680–689.

Franco EL (2003) Primary screening of cervical cancer with human papillomavirus tests. *Journal of the National Cancer Institute Monographs* 31: 89–96.

Franco EL, Schlecht NF, and Saslow D (2003) The epidemiology of cervical cancer. *Cancer J* 9: 348–359.

Franco EL, Villa LL, Sobrinho JP, *et al.* (1999) Epidemiology of acquisition and clearance of cervical human papillomavirus infection in women from a high-risk area of cervical cancer. *Journal of Infectious Diseases* 180: 1415–1423.

Gage JC, Ferreccio C, Gonzales M, *et al.* (2003) Follow-up care of women with an abnormal cytology in a low-resource setting. *Cancer Detection and Prevention* 27: 466–471.

Giulian AR, Sedjo RL, Roe DJ, *et al.* (2002) Clearance of oncogenic human papillomavirus (HPV) infection: Effect of smoking (United States). *Cancer Causes and Control* 13: 839–846.

Goldie SJ, Kuhn L, Denny L, *et al.* (2001) Policy analysis of cervical cancer screening strategies in low resource settings: Clinical benefits and cost-effectiveness. *JAMA* 285: 3107–3115.

Gontijo RC, Derchain SFM, Montemor EB, *et al.* (2005) Pap smear, hybrid capture II, and visual inspection in screening for uterine cervical lesions. *Cadernos Saúde Publica* 21: 141–149.

Gustafsson L and Adami HO (1989) Natural history of cervical neoplasia: Consistent results obtained by an identification technique. *British Journal of Cancer* 60: 132–141.

Harper DM, Franco EL, Wheeler C, *et al.* (2004) Efficacy of a bivalent L1 virus-like particle vaccine in prevention of infection with human papillomavirus types 16 and 18 in young women: A randomised controlled trial. *Lancet* 364: 1757–1765.

Ho GY, Burk RD, Kein S, *et al.* (1995) Persistent genital human papillomavirus infection as a risk factor for persistent cervical dysplasia. *Journal of the National Cancer Institute* 87: 1365–1371.

Holmes J, Hemmett L, and Garfield S (2005) The cost-effectiveness of human papillomavirus screening for cervical cancer. A review of recent modelling studies. *European Journal of Health Economics* 6: 30–37.

Holowaty P, Miller AB, Rohan T, and To T (1999) Natural history of dysplasia of the uterine cervix. *Journal of the National Cancer Institute* 91: 252–258.

International Agency for Research on Cancer (2005) *Cancer Database, Cancer Mondial, Globocan 2002.* Available at http://www-dep.iarc.fr/.

International Agency for Research on Cancer, Cancer (2005) *IARC Handbooks of Cancer Prevention. Cervical Cancer Screening*, pp. 117–162. Lyon: International Agency for Research on Cancer.

International Collaboration of Epidemiological Studies of Cervical Cancer (2006) Cervical carcinoma and tobacco smoking: Collaborative reanalysis of individual data on 13,541 women with carcinoma of the cervix and 23,017 women without carcinoma of the cervix from 23 epidemiological studies. *International Journal of Cancer* 118: 1481–1495.

Jacob M, Broekhuizen FF, Castro W, *et al.* (2005) Experience using cryotherapy for treatment of cervical precancerous lesions in low-resource settings. *International Journal of Gynecology & Obstetrics* 89: S13–S20.

Jacobs MV, Walboomers JMM, Snijders PJF, *et al.* (2000) Distribution of 37 mucusotropic HPV types in women with cytologically normal cervical smears: The age-related patterns for high-risk and low-risk types. *International Journal of Cancer* 87: 221–227.

Karnon J, Peters J, Platt J, *et al.* (2004) Liquid-based cytology in cervical screening: An updated rapid and systematic review and economic analysis. *Health Technology Assessment* 8: 1–78.

Kim JJ, Wright TC, and Goldie SJ (2005) Cost-effectiveness of human papillomavirus DNA testing in the United Kingdom, The Netherlands, France, and Italy. *Journal of the National Cancer Institute* 97: 888–895.

Krane JF, Granter SR, Trask CE, *et al.* (2001) Papanicolaou smear sensitivity for the detection of adenocarcinoma of the cervix: A study of 49 cases. *Cancer* 93: 8–15.

Laara E, Day NE, and Hakama M (1987) Trends in mortality from cervical cancer in the Nordic countries: Association in the organised screening programmes. *Lancet* 1: 1247–1249.

Martin-Hirsch PL, Paraskevaidis E, and Kitchner H (2004) *Surgery for Cervical Intraepithelial Neoplasia.* Chichester: John Wiley and Sons, Cochrane review. The Cochrane Library Issue 1.

Miller AB, Nazeer S, Fonn S, *et al.* (2000) Report on consensus conference on cervical cancer screening and management. *International Journal of Cancer* 86: 440–447.

Nanda K, McCrory DC, Myers ER, *et al.* (2000) Accuracy of the Papanicolaou test in screening for and follow-up of cervical cytologic abnormalities: A systematic review. *Annals of Internal Medicine* 132: 810–819.

Östör AG (1993) Natural history of cervical intraepithelial neoplasia: A critical review. *International Journal of Gynecological Pathology* 12: 186–192.

Papanicolaou GN (1954) *Atlas of Exfoliative Cytology.* Boston: Commonwealth Fund University Press.

Parikh S, Brennan P, and Boffetta P (2003) Meta-analysis of social inequality and the risk of cervical cancer. *Int J Cancer* 105: 687–691.

Peto J, Gilham C, Fletcher O, *et al.* (2004) The cervical cancer epidemic that screening has prevented in the UK. *Lancet* 364: 249–256.

Pinotti JA, Brenelli HB, Moragues JD, *et al.* (1981) Preventive obstetrics and gynecology programme: Pilot plan for integrated medical care. *Bulletin of the Pan American Health Organization* 15: 104–112.

Richart RM (1973) Cervical intraepithelial neoplasia. In: Sommers SC (ed.) *Pathology Annual*, pp. 301–328. New York: Appleton-Century-Crofts.

Riotton G, Christopherson WM, and Lunt R (1973) *Cytology of the Female Genital Tract. International Histological Classification of Tumours No. 8.* Geneva: World Health Organization.

Robles SC, White F, and Peruga A (1996) Trends in cervical cancer mortality in the Americas. *Bulletin of the Pan American Health Organization* 30: 290–301.

Royal Thai College of Obstetricians and Gynaecologists (RTCOG) CCJHPIEGO Corporation Cervical Cancer Prevention Group (2003) Safety, acceptability, and feasibility of a single-visit approach to cervical-cancer prevention in rural Thailand: A demonstration project. *Lancet* 361: 814–820.

Ruba S, Schoolland M, Allpress S, *et al.* (2004) Adenocarcinoma in situ of the uterine cervix: Screening and diagnostic errors in Papanicolaou smears. *Cancer* 102: 280–287.

Sankaranarayanan R, Budukh A, and Rajkumar R (2001) Effective screening programmes for cervical cancer in low- and middle-income developing countries. *Bulletin of the World Health Organization* 79 (Suppl): 954–962.

Sankaranarayanan R, Thara S, Anjali S, *et al.* (2004a) Accuracy of conventional cytology: Results from a multicentre screening study in India. *Journal of Medical Screening* 11: 77–84.

Sankaranarayanan R, Chatterji R, Shastri S, *et al.* (2004b) Accuracy of human papillomavirus testing in the primary screening of cervical neoplasia: Results from a multicentre study in India. *International Journal of Cancer* 112: 341–347.

Sankaranarayanan R, Shastri SS, Basu P, *et al.* (2004c) The role of low-level magnification in visual inspection with acetic acid for the early detection of cervical neoplasia. *Cancer Detection and Prevention* 28: 345–351.

Sankaranarayanan R, Basu P, Wesley R, *et al.* (2004d) Accuracy of visual screening for cervical neoplasia: Results from an IARC multicentre study in India and Africa. *International Journal of Cancer* 110: 907–913.

Sankaranarayanan R, Gaffikin TL, Jacob M, *et al.* (2005) A critical assessment of screening methods for cervical neoplasia. *International Journal of Gynecology and Obstetrics* 89(Suppl 2): S4–S12.

Schlecht NF, Platt RW, Duarte-Franco E, *et al.* (2003) Human papillomavirus infection and time to progression and regression of cervical intraepithelial neoplasia. *Journal of the National Cancer Institute* 95: 1336–1343.

Sellors JW, Mahony JB, Kaczorowski J, *et al.* (2000) Prevalence and predictors of human papillomavirus infection in women in Ontario, Canada. Survey of HPV in Ontario Women (SHOW) Group. *CMAJ* 163: 503–508.

Shastri SS, Dinshaw K, Amin G, *et al.* (2005) Concurrent evaluation of visual, cytological and HPV testing as screening methods for the early detection of cervical neoplasia in Mumbai, India. *Bulletin of the World Health Organization* 83: 186–194.

Solomon D, Davey D, Kurman R, *et al.* (2002) The 2001 Bethesda system: Terminology for reporting results of cervical cytology. *JAMA* 287: 2114–2119.

Syrjanen K, Naud P, Derchain S, *et al.* (2005) Comparing PAP smear cytology, aided visual inspection, screening colposcopy, cervicography and HPV testing as optional screening tools in Latin America. Study design and baseline data of the LAMS study. *Anticancer Research* 25: 3469–3480.

Tortolero-Luna G and Franco EL (2004) Epidemiology of cervical, vulvar, and vaginal cancers. In: Gershenson DM, Mcguire WP, Gore M, *et al.* (eds.) *Gynecologic Cancer: Controversies in Management*, pp. 3–30. Philadelphia: Elsevier Churchill Livingstone.

University of Zimbabwe/JHPIEGO Cervical Cancer Project (1999) Visual inspection with acetic acid for cervical cancer screening: Test qualities in a primary-care setting. *Lancet* 353: 869–873.

Villa LL, Costa RL, Petta CA, *et al.* (2005) Prophylactic quadrivalent human papillomavirus (types 6, 11, 16, and 18) L1 virus-like particle vaccine in young women: A randomised double-blind placebo-controlled multicentre phase II efficacy trial. *Lancet Oncology* 6: 271–278.

Vizcaino AP, Moreno V, Bosch FX, *et al.* (2000) International trends in incidence of cervical cancer: II. Squamous-cell carcinoma. *International Journal of Cancer* 86: 429–435.

Walboomers JM, Jacobs MV, Manos MM, *et al.* (1999) Human papillomavirus is a necessary cause of invasive cancer worldwide. *Journal of Pathology* 189: 12–19.

Womack SD, Chirenje ZM, Blumenthal PD, *et al.* (2000) Evaluation of a human papillomavirus assay in cervical screening in Zimbabwe. *BJOG: An International Journal of Obstetrics and Gynaecology* 107: 33–38.

Woodman CB, Collins S, Winter H, *et al.* (2001) Natural history of cervical human papillomavirus infection in young women. *Lancet* 357: 1831–1836.

World Health Organization, International Agency for Research on Cancer-IARC (1986) WHO: IARC working group on evaluation of cervical cancer screening programmes. Screening for squamous cervical cancer: Duration of low risk after negative results of cervical cytology and its implication for screening policies. *British Medical Journal* 293: 659–664.

Wright TC Jr, Denny L, Kuhn L, *et al.* (2000) HPV-DNA testing of self-collected vaginal samples compared with cytologic screening to detect cervical cancer. *JAMA* 283: 81–86.

Wright TC, Schiffman M, Solomon D, *et al.* (2004) Interim guidance for the use of human papillomavirus DNA testing as an adjunct to cervical cytology for screening. *Obstetrics & Gynecology* 103: 304–309.

Zeferino LC, Pinotti JA, Jorge JP, *et al.* (2006) Organization of cervical cancer screening in Campinas and surrounding region, Sao Paulo State, Brazil. *Cadernos de Saúde Publica* 22: 1909–1914.

Endometrial Cancer

Y Sonoda and R R Barakat, Memorial Sloan-Kettering Cancer Center, New York, NY, USA

Introduction

Endometrial carcinoma is the most common gynecologic cancer in the USA. According to the American Cancer Society statistics, there will be 40 880 new cases of uterine corpus cancer in 2005 (Jemal *et al.*, 2005). The lifetime risk of being diagnosed with endometrial cancer is approximately 2.7%, and the median age at diagnosis 65 years. Endometrial cancer is commonly thought of as a 'curable cancer', primarily owing to the fact that almost three-quarters of cases are diagnosed before disease has spread outside the uterus. The overall 5-year survival for all stages is 86%, with disease confined to the uterus having a 97% 5-year survival.

Despite the advances that have been made in other cancers, both the annual incidence of and the death rate associated with endometrial cancer appear to be rising. According to American Cancer Society statistics, an estimated 37 000 cases of endometrial cancer were diagnosed in 1985. The number of deaths has more than doubled to 7310 in 2005 compared with 2900 in 1985. Although the incidence of endometrial cancer is higher in white women than in black women in the USA, mortality is higher in blacks. Endometrial cancer is one of the cancers in which the discrepancy in survival between blacks and whites is the largest (Hill *et al.*, 1996; Ries *et al.*, 1997). This discrepancy is attributable in part to the higher proportion of black women who present with advanced-stage disease, as well as a higher frequency of aggressive, more lethal histologic subtypes such as serous cancers. However, because endometrial cancer is more common among white women, most epidemiologic studies have included few black women.

Few studies have conducted centralized pathology review to determine histologic type or analyzed results separately by histologic type. Very little is known about risk factors for serous tumors; they do not seem to be related to the hormonal factors important in endometrioid tumors (Westhoff *et al.*, 2000). The etiology and epidemiology of endometrial cancer will be reviewed along with methods of screening and strategies for prevention.

Epidemiology

Rates of endometrial cancer among American women are among the highest in the world. There is tremendous geographic variation, with nearly a five-fold differential in rates from high- to low-risk countries, supporting the notion of a strong influence of environmental factors. Many of these factors have received extensive epidemiologic attention. A number of risk factors have been well defined, including increased risks associated with nulliparity, early age at menarche, late age at menopause, obesity, and long-term use of estrogen-replacement therapy (ERT). By contrast, women who have had multiple births, as well as users of oral contraceptives and cigarette smokers, are at decreased risk. Many of these factors can be tied to the hypothesis of the causal factor being an unopposed estrogen, whereby exposure to estrogens unopposed by progesterone or synthetic progestins leads to increased mitotic activity of endometrial cells, increased DNA replication errors, and somatic mutations resulting in malignant phenotype (Akhmedkhanov *et al.*, 2001).

Although most cases of endometrial carcinoma occur sporadically, a hereditary form has been identified. Endometrial cancer is the most common extracolonic malignancy of the autosomal dominant hereditary non-polyposis colorectal cancer (HNPCC) syndrome. The identifiable risk of endometrial cancers is so great in this patient population that surveillance is recommended (Smith *et al.*, 2005).

Risk Factors

Endometrial cancers can arise in normal, atrophic, or hyperplastic endometrium and have been classified as two types (Kurman *et al.*, 1994). Type I cancers are the more common form and are associated with increased levels of circulating estrogen. These tumors usually begin as endometrial hyperplasia and progress to cancer. They tend to occur at a younger age and are less aggressive. Histologically, these are the lower-grade endometrioid adenocarcinomas that constitute 75–80% of cases. The type II cancers are the higher-grade, more aggressive forms of cancer that tend to arise spontaneously. Histologically, these encompass the serous, clear cell, adenosquamous, and grade 3 adenocarcinomas. They tend to arise in the older patient and do not have an estrogen-related precursor.

Epidemiologic studies have focused on the risk factors for endometrial cancers of the type I variant where estrogen is felt to be the driving factor in development. These will be discussed in this chapter Risk factors for the type II cancers are less well established.

Estrogen-Alone and Estrogen-With-Progestin Replacement

The increased use of estrogen-replacement therapy in the late 1970s led to a major epidemic of endometrial cancers in the USA, which stimulated a series of investigations to assess specific relationships of risk with patterns of usage. The first report was published in 1975 by Smith *et al.* In a case-control study, the authors compared 317 patients with endometrial cancer to an equal number of matched controls with other types of gynecologic neoplasms. There were 152 patients with endometrial cancer who used estrogen, compared with 54 patients in the control group. The risk of endometrial cancer was 4.5 times greater among women exposed to estrogen. Multiple studies showed strong relationships (10- to 20-fold relative risks) with long-term usage (Herrinton and Weiss, 1993). In most studies, cessation of use was associated with a relatively rapid decrease in risk. After controlling for other known risk factors for endometrial cancer, the risk appeared highest in patients without obesity and hypertension, two characteristics typically associated with endometrial cancer. Epidemiologic studies have generally shown that the excess risk of endometrial cancer associated with estrogens alone can be counteracted by the addition of a progestin; this, however, appears to be dependent on the progestin being given for at least 10 days each month (Archer, 2001; Pike *et al.*, 1997).

The Postmenopausal Estrogen Progestin Interventions (PEPI) Trial illustrated the effects of estrogen and progestin on the endometrium. This trial involved nearly 600 women, who were randomized to placebo, conjugated equine estrogen daily, or one of three combinations of conjugated equine estrogen with progestins (either medroxprogesterone acetate or micronized progesterone). Patients were monitored with annual endometrial biopsies. The patients taking combined estrogen and progestin developed endometrial hyperplasia at a similar rate to those in the placebo arm (Effects of hormone replacement therapy on endometrial histology in postmenopausal women. The Postmenopausal Estrogen/Progestin Interventions (PEPI) Trial, 1996). Patients taking estrogen alone had a higher incidence of endometrial hyperplasia, including atypical hyperplasia, which is felt to be a precursor to type I endometrial cancer.

More recently, an analysis of the patients involved in the Women's Health Initiative (WHI) Study, which evaluated the impact of combined progestin and estrogen therapy compared with placebo, did not demonstrate an increased risk of endometrial cancer. The WHI study involved over 16 000 women and had an average follow-up of 5.6 years. Women randomized to combined hormonal replacement therapy with estrogen and progestin had a hazard ratio for endometrial cancer of 0.81(95% confidence interval (CI) 0.48–1.36) compared with those taking placebo. There was no significant increase in risk of endometrial cancer associated with use of combined hormonal replacement therapy (Anderson *et al.*, 2003).

Tamoxifen

Tamoxifen is a synthetic estrogen antagonist used in the treatment and prevention of breast cancer. It has an estrogenic rather than an antiestrogenic effect on the endometrium. Its potential association with endometrial cancer was reported in 1987 by Killackey *et al.*, in their report of three breast cancer patients receiving tamoxifen who developed endometrial cancer. Since that time, multiple studies have evaluated the increased risk of endometrial cancer with tamoxifen use.

Fornander *et al.* (1989) reviewed the frequency of new primary cancers in 1846 patients with breast cancer randomized to receive tamoxifen (40 mg/day for 2–5 years) versus placebo. With a median follow-up of 4.5 years, they reported a 6.4-fold increase in the relative risk of endometrial cancer in the tamoxifen group compared with the placebo group. The authors of the National Surgical Adjuvant Breast and Bowel Project (NSABP) P-1 trial (Fisher *et al.*, 1998) noted that, in the group on tamoxifen, the average annual rate of developing endometrial cancer was 2.30 per 1000 women. This was compared with an average annual rate of 0.91 per 1000 women in the placebo group. Thus, patients on tamoxifen had a 2.53 times greater risk of developing invasive endometrial cancer. The increase in endometrial cancer was seen almost exclusively in women over the age of 50 years. This trial noted that the rate of breast cancer development in women deemed at risk for breast cancer was reduced by nearly 50% in patients taking tamoxifen. The authors concluded that despite the side effects that patients on tamoxifen have, its use as a breast cancer preventive agent is appropriate in patients at increased risk. A recent review of the NSABP trials revealed that tamoxifen use might be associated with an increased rate of uterine sarcoma. Fortunately, this type of highly malignant tumor occurred at a very low rate of 0.17 per 1000 women in those patients treated with tamoxifen (Wickerham *et al.*, 2002).

Obesity

One of the strongest identified risk factors for endometrial cancer is that of obesity, with studies suggesting a three- to four-fold differential between women in the highest and lowest quartiles of weight or body mass index (BMI) (Goodman *et al.*, 1997; Swanson *et al.*, 1993; Terry *et al.*, 1999; Weiderpass *et al.*, 2000). Among postmenopausal women, this association is believed to be largely due to the conversion of androstenedione to estrone by aromatase in adipose tissue (Akhmedkhanov *et al.*, 2001). Obese women have also been shown to have lower levels of sex-hormone-binding globulin, which might result in a higher amount of bioavailable estrogen (Potischman *et al.*, 1996). In premenopausal women, obesity-associated anovulation may be more central, e.g. in association with polycystic ovarian syndrome.

Some studies suggest that body fat distribution might have an independent effect, with high risk seen for women whose weight distributes abdominally rather than peripherally (Swanson *et al.*, 1993). Whether associations with body fat distribution are reflective of unique hormonal profiles has yet to be resolved. Relationships with timing of obesity and weight cycling are also under scrutiny.

Higher BMI might also place a patient at higher risk of death from endometrial cancer. The Cancer Prevention Study II followed a prospective cohort of 495 477 women for 16 years. The relative risk of dying from endometrial cancer was 6.25-fold higher in those women with the greatest BMI ($\geq$40 kg/m^2) compared with women considered to be of normal weight (Calle *et al.*, 2003).

Diabetes and Hypertension

Diabetes and hypertension are two conditions that have been linked to endometrial cancer. Such conditions are commonly found in obese patients, which could potentially account for the increased risk. Diabetes, and in particular non-insulin dependent diabetes, is associated with a state of hyperinsulinemia. Hyperinsulinemia has been associated with a hyper-estrogenic state including an increase in steroid production (Poretsky and Kalin, 1987), stimulation of the conversion of testosterone to estradiol (Garzo and Dorrington, 1984), and suppression of the circulating concentrations of sex-hormone-binding globulin (Nestler *et al.*, 1991).

In an Italian case-control study, Parazzini *et al.* (1999) reviewed reported history of diabetes among 752 patients with endometrial cancer and 2606 controls. A total of 132 (17.6%) cases reported a history of diabetes compared with only 116 (4.5%) controls. There was no association with diabetes diagnosed under age 40 years with endometrial cancer compared with an odds ratio (OR) of 3.1 (95% CI 2.3–4.2) for diabetes diagnosed at age ≥40 years. The authors concluded that a history of diabetes, in particular non-insulin dependent diabetes, is associated with a higher risk of endometrial cancer, independent of such other risk factors as obesity.

The association between diabetes and endometrial cancer has not been universally identified in all studies, however. Shoff *et al.* (1998) examined the relation of diabetes to the risk of endometrial cancer on the basis of BMI. Subjects were categorized as not overweight, overweight, and obese on the basis of BMI of <29.1, 29.1–31.9, and >31.9 respectively. The risk of endometrial cancer was not significantly greater in non-overweight and overweight women who reported diabetes when compared with non-overweight, non-diabetic subjects. Obese, diabetic women were at increased risk for endometrial cancer. The authors concluded that diabetes conferred no additional risk of endometrial cancer in women who are neither overweight nor obese.

In a report from the Iowa Women's Health Study, the association between self-reported diabetes and incidental endometrial cancer was examined in 24 664 postmenopausal women. The relative risk of endometrial cancer was not significantly increased in those subjects with diabetes compared with those without. There did appear to be an increased diabetes-associated risk in patients who were of higher BMI. The authors concluded that diabetes is associated with a modestly increased risk for endometrial cancer (Anderson *et al.*, 2001).

The association of hypertension and endometrial cancer is also not well defined. In a Swedish study of over 4000 patients, women with hypertension had a significantly higher BMI than those without. After adjusting for this difference in BMI, history of hypertension was not associated with endometrial cancer. When evaluating obese patients alone, there was an increased risk of endometrial cancer in patients with a BMI of 30 or greater. However, contrary to this, Soler *et al.* (1999) found hypertension to be associated with endometrial cancer risk even when adjustment was made for BMI.

Parity

By interrupting the continued stimulation from estrogen, pregnancy confers protection from endometrial cancer. Women who are nulliparous are twice as likely to develop endometrial cancer as women who have delivered a child. The high progesterone levels during pregnancy might be responsible for the protective effect.

The data from the Iowa Women's Health Study were analyzed to determine if reproductive factors affected risk of endometrial cancer (McPherson *et al.*, 1996). As expected, the mean gravidity of patients with endometrial cancer was significantly less than the controls (2.6 versus 3.5, $P < 0.0001$). An Italian study also demonstrated the protective effects of parity. A total of 752 women with endometrial cancer were compared with 2606 women without for the role of reproductive factors on the risk of endometrial cancer. Endometrial cancer risk decreased with one, two, and three or more births. The risk of endometrial cancer also decreased with induced abortions. Timing of last birth also seemed to affect endometrial cancer risk. Patients with the last birth 10–19 years before and <10 years before had an OR of 0.6 (95% CI 0.4–0.9) and 0.3 (95% CI 0.1–0.9), respectively, for developing endometrial cancer when compared with women who reported their last birth as being more than 20 years before (Parazzini *et al.*, 1998).

Menstrual Factors

Specific menstrual factors have been associated with an increased risk of endometrial cancer. In a population-based, case-controlled study in a low-risk population of women in Shanghai, People's Republic of China, early menarche age and later menopausal age were associated with an increased risk of endometrial cancer. There was a clear dose–response relation between endometrial cancer risk and years of menstruation. After adjusting for gravidity, early age at menarche, late age at natural menopause, and total length of ovulation span were associated with increased endometrial cancer risk. Both completed and incomplete pregnancy was associated with a protective effect against endometrial cancer. This protective effect seemed to increase with the total number of pregnancies (Xu *et al.*, 2004).

In the Iowa Women's Health Study, the inverse relationship between age at menarche and endometrial cancer incidence was apparent. Women with menarche at age 15 years or older had a third of the risk of developing endometrial cancer compared with women who began menstruating at age 10 years or younger. Those subjects with menopause at age 55 years or older had a relative risk of endometrial cancer 1.87 times (95% CI 1.12–3.09) that of women who had undergone menopause before age 45 years. When comparing span of ovulations, those subjects in the highest quantile of years of ovulation had over three times the endometrial cancer risk as those in the lowest (McPherson *et al.*, 1996).

In addition to menstrual factors that expose women to prolonged periods of ovulation, chronic anovulation as in patients with polycystic ovary syndrome also confers an increased risk of endometrial cancer. These patients are hyperandrogenic, which results in a constant stimulation of the endometrium from the peripheral conversion of androgens to estrogen. However, these patients do not ovulate; thus, they lack the normal progesterone secretion of the corpus luteum. Such women are at risk for the development of estrogen-driven endometrial cancers (Coulam *et al.*, 1983; Schmeler *et al.*, 2005).

Diet

Dietary fat has been associated with an increased risk of endometrial cancer. Littman *et al.* (2001) performed a case-controlled study using telephone interviews and questionnaires to examine the associations of dietary fat and plant foods with endometrial cancer. In all, 679 incident cases of endometrial cancer were compared with 944 controls. The OR for endometrial cancer was estimated after adjusting for age, county, energy intake, hormone use, smoking, and BMI. Women with the highest percent of kilocalories per day from fat were at significantly higher risk (OR 1.8; 95% CI 1.3–2.6) of developing endometrial cancer. Saturated and monosaturated fats were the main contributors to risk. Conversely, consumption of fruits and vegetables was inversely associated with endometrial cancer risk.

In a case-control study of dietary associations and endometrial cancer, Potischman *et al.* (1993) studied the dietary intake of 399 cases of endometrial cancer and 296 controls. The subjects in the highest quantile of caloric intake had a modest increase in risk compared with subjects in the lowest quantile (OR 1.5; 95% CI 0.9–2.5). After adjusting for other risk factors, high intakes of fat were associated with increased risk. Saturated fats and oleic acid had the highest risk. Intake of complex carbohydrates, specifically breads and cereals, was associated with a reduced risk. Similar conclusions have also been reached in other case-controlled studies (Goodman *et al.*, 1997) and support the role of diet as a contributing factor to endometrial cancer.

Genetic Predisposition

It has been established through a number of studies that an increased risk of endometrial cancer exists among individuals with a family history of the disease, particularly if the diseases among the family members were diagnosed at young ages. Subjects with a family history of colon cancer also have been documented as being at an increased risk of endometrial cancer, supporting a role for the dominantly inherited HNPCC gene.

The hereditary non-polyposis colorectal cancer (HNPCC) syndrome is caused by mutations in certain DNA mismatch repair genes (MSH2, MLH1, PMS1, PMS2, MSH6). Patients who have these genetic mutations carry a lifetime risk as high as 80% for the development of colorectal cancer (Aarnio *et al.*, 1999). In addition to this, females with this syndrome carry a lifetime risk of 40–60% for the development of endometrial cancer (Aarnio *et al.*, 1999; Watson and Lynch, 2001; Dunlop *et al.*, 1997). Patients who fulfill the modified Amsterdam criteria for this syndrome should be referred for genetic counseling. The criteria are as follows: three family members affected with an HNPCC-associated cancer (colorectal cancer, cancer of the endometrium, ovary, small bowel, ureter or renal pelvis), one being a first-degree relative of the other two, one diagnosed before age 50, and cases must occur in at least two successive generations.

In a recent report by Lu *et al.* (2005) reviewing the presenting cancer in 223 families that met Amsterdam criteria, 117 women were identified as having dual primary cancers. Sixteen women had synchronous colon and endometrial/ovarian cancers diagnosed simultaneously. Of the remaining 101 patients, there were 52 (51%) patients who presented with a gynecologic cancer as the first cancer. Forty-six of these patients presented with endometrial cancer as the first cancer with a median age of 45 years (range 27–61 years). The authors stressed that female patients whose disease belongs to this syndrome are at equal or higher risk of developing endometrial cancer than colon cancer, and that endometrial cancer as well as ovarian and colon cancer may be the initial cancer diagnosis. Given that these patients are also at risk for the development of ovarian cancer, prophylactic hysterectomy and bilateral salpingo-oophorectomy may be appropriate in this population.

Prevention

There are several protective factors for the development of endometrial cancer. These factors are linked to the protective effect of reducing the estrogenic stimulation on the endometrium, which is the causative factor for type I cancers.

Oral Contraceptives

Combined oral contraceptive use has been shown to decrease the risk of endometrial cancer. This is likely due to the suppression of endometrial proliferation by the progestin component of oral contraceptives. This protective effect may be a function of duration of use. Patients with 12–23 months of oral contraceptive use have a 40% reduction in risk of endometrial cancer, whereas subjects with at least 10 years of use have a 60% reduction (Herbst, 1994).

In a population-based Swedish study of postmenopausal women aged 50–74 years, 709 cases of endometrial cancer were compared with 3368 controls. Any oral contraceptive use was found to decrease the risk of endometrial cancer by 30%. The use of the progestin-only pill seemed to lower risk further but lack of statistical power precluded a more detailed analysis. Three or more years of combined oral contraceptive use conferred a 50% reduction in endometrial cancer risk whereas use of 10 or more years resulted in an 80% reduction (Weiderpass *et al.*, 1999).

Studies have been conducted to determine if the different progestin content of the combined oral contraceptive pill has an effect on endometrial cancer risk. Voight *et al.* (1994) conducted personal interviews with 316 women diagnosed with endometrial cancer. Compared with 501 controls, they determined that the relative risk of endometrial cancer did not differ according to the progestin potency of the contraceptive pill used. The reduced risk was seen only in women who used combined oral contraceptives for 5 years or more and who were not long-term users of unopposed postmenopausal estrogens.

Smoking

Epidemiologic studies have demonstrated an inverse relationship between smoking and endometrial cancer. There are several biological mechanisms for this relationship. Smoking is associated with a lower endogenous level of estrogen (Baron *et al.*, 1990), possibly by altering its metabolism by the liver (Michnovicz *et al.*, 1986). Smoking has also been associated with an earlier age at menopause, one of the commonly quoted risk factors for endometrial cancer.

In an Italian case-control study, 726 patients with endometrial cancer and under age 75 years were compared with 1452 controls. Smoking history was obtained using a structured questionnaire administered by trained interviewers. Information obtained pertained to smoking status, total duration of smoking, and average number of cigarettes per day. Compared with never smokers, the relative risk of endometrial cancer was 0.8 (95% CI 0.7–1.1) in current smokers and 0.6 (95% CI 0.4–0.9) in ex-smokers. The number of cigarettes smoked per day and the duration of smoking habit were inversely related to endometrial cancer risk (Parazzini *et al.*, 1995).

The effect of smoking on hormonal risk factors was studied by Newcomer *et al.* (2001) in a population-based case-control study in the USA. Current smoking was associated with a relative risk of 0.8 (95% CI 0.6–1.0) compared with never smokers. When the patients in the highest quartile of BMI were considered, the endometrial cancer risk was higher among non-smokers when compared with smokers. The risk of endometrial cancer in postmenopausal patients on HRT was higher among non-smokers when compared with smokers. Smoking was also associated with a lower risk of endometrial cancer in patients with non-insulin-dependent diabetes. These findings suggested that smoking moderates the risk of endometrial cancer among patients felt to be at the highest risk.

A recent review of the Nurses' Health Study demonstrated that, compared with never smokers, the multivariate relative risk of endometrial cancer was significantly lower among both current and past smokers. This suggests that the effects of smoking on endometrial cancer risk are long lasting. This is the first prospective study to demonstrate a significantly lower risk of endometrial cancer in both past and current smokers (Viswanathan *et al.*, 2005).

Needless to say, although smoking seems to be associated with a lower endometrial cancer risk, this does not warrant its use for endometrial cancer prevention given the other major health risks to which it has been linked.

Physical Activity

Studies evaluating physical activity and endometrial cancer risk have shown a weak to moderate inverse relationship between the two. Assessment of physical activity between these studies varied, making comparisons therefore difficult.

A population-based study from Norway of over 24 000 women evaluated the effect of physical inactivity on endometrial cancer risk. Participants were asked to grade both their recreational and occupational activity according to specified definitions. Physical inactivity was found to be a major risk for endometrial cancer that was independent of BMI (Furberg and Thune, 2003).

The Netherlands Cohort Study on Diet and Cancer (Schouten *et al.*, 2004) involved 62 573 women who were followed over a 10-year period. A total of 226 cases of endometrial cancer were identified. The study revealed a 46% reduction (relative risk 0.54; 95% CI 0.34–0.85) in risk of endometrial cancer in women who were physically active for more than 90 minutes per day compared with those active less than 30 minutes per day. Similar conclusions resulted from a case-control study from Sweden in which 709 cases of endometrial cancer were compared with 3368 controls. Leisure-time physical activity was measured by patient questionnaire and occupational physical activity was classified by experts based on the subjects' reported occupations. The authors concluded that both occupational and leisure-time physical activity might reduce the risk for postmenopausal endometrial cancer (Moradi *et al.*, 2000). Hypothetically, the inverse relationship might be due to a reduction in obesity or a reduction in the serum estrone levels that may be associated with increased physical activity (Cauley *et al.*, 1989).

Screening for Endometrial Cancer

According to the American Cancer Society guidelines (Smith *et al.*, 2005) for the early detection of cancers, there is no proven role for screening asymptomatic women who are at average risk for endometrial cancer. This also applies to those women at somewhat increased risk due to such factors as history of unopposed estrogen therapy, tamoxifen therapy, late menopause, nulliparity, infertility, or failure to ovulate. The American Cancer Society screening guidelines for endometrial cancer, which were written by an expert panel after a thorough review of the literature, recommend that when these women reach menopause they are informed about the risks and symptoms of endometrial cancer so that they can in turn alert their physicians. The panel concluded, however, that screening *should* be considered in those women at highest risk for developing endometrial cancer. This includes those who harbor mutations in the genes associated with the HNPCC syndrome, those who are at greater likelihood of being a mutation carrier, and women with a suspected autosomal predisposition to colon cancer in the absence of genetic testing. Women belonging to an HNPCC kindred are at an increased risk of endometrial cancer with a lifetime risk of 42 to 60% (Aarnio *et al.*, 1999; Watson and Lynch, 2001; Dunlop *et al.*, 1997). Such women can be offered annual screening, beginning at age 35 years, after they have been counseled properly about the risks, benefits, and limitations of screening. Several modalities of screening have been studied.

Outpatient Endometrial Biopsy

Endometrial biopsy is the most common technique employed for obtaining endometrial tissue for histologic evaluation. A meta-analysis of 39 studies compared the accuracy of the different endometrial sampling devices in the detection of endometrial cancer and atypical hyperplasia. The detection rates for endometrial cancer using the Pipelle in postmenopausal and premenopausal women were 99.6% and 91%, respectively. The Vabra aspirator resulted in sensitivities of 97.1% and 80% in postmenopausal and premenopausal women, respectively. The sensitivity for detecting atypical hyperplasia was also significantly better for the Pipelle than for the Vabra device and lavage. The authors concluded that endometrial biopsy with the Pipelle is superior to the other available devices and its accuracy was higher in postmenopausal women (Dijkhuizen *et al.*, 2000). In a series of patients with known endometrial cancer, Pipelle biopsy detected 54 of the 65 cases for a sensitivity of 83% (SD ± 0.5%). Of the eleven patients with false-negative results, five had tumors limited to a polyp and three had disease localized to <5% of the surface area (Guido *et al.*, 1995).

Issues of inadequate access to the endometrial cavity or sampling error can limit the clinical significance of a negative result. Several large studies have demonstrated that endometrial biopsy might produce insufficient tissue for diagnosis. In a study of 801 asymptomatic perimenopausal and postmenopausal women, endometrial biopsy was performed prior to initiating HRT. The biopsy yielded insufficient tissue for diagnosis in 195 (24.5%) of women and identified only one case of endometrial cancer. The authors concluded that screening endometrial biopsy was not justified in this group of women (Archer *et al.*, 1991). Other large studies employing office biopsy have also demonstrated a high rate of insufficient tissue. In a study of postmenopausal women with breast cancer, 96 (36%) of 268 Pipelle biopsies were considered insufficient for diagnosis (Duffy *et al.*, 2003).

Ultrasound

There are many reports on the use of ultrasound as a non-invasive modality to evaluate the presence of cancer in patients with postmenopausal bleeding. Its use as a screening modality for endometrial cancer in asymptomatic patients has been less well studied. Using a 6-mm threshold, Fleischer *et al.* (2001) reported on the use of transvaginal ultrasonography in postmenopausal asymptomatic patients. This study screened 1926 patients by transvaginal ultrasound; 93 were found to have an endometrial thickness greater than 6 mm. Only 42 (45%) patients with an abnormal endometrial thickness had an aspiration biopsy, which revealed one case of endometrial adenocarcinoma. Of 1833 patients with an endometrial thickness of ≤6 mm, 1750 patients underwent an endometrial biopsy revealing one endometrial cancer and four cases of atypical hyperplasia. The sensitivity and specificity for an endometrial thickness of ≤6 mm were 17% and 98%, respectively. The negative predictive value was >99%, but the positive predictive value of an endometrial thickness >6 mm was only 2%. The authors of the study emphasized that the low sampling rate of the patients with a thickened endometrium did not allow an accurate assessment of positive predictive value.

Langer *et al.* (1997) evaluated the use of transvaginal ultrasound for the screening of asymptomatic patients before or during treatment with estrogen or estrogen with progesterone. Concurrent endometrial biopsy and sonographic results were available for 577 examinations among this population. Using a threshold value of 5 mm for endometrial thickness, transvaginal ultrasonography had a sensitivity and specificity of 90% and 48%, respectively. The positive predictive value was only 9% for detecting any abnormality. The authors concluded that, because of the high false-positive rate, ultrasonography is not a practical screening tool in asymptomatic women, regardless of whether they are on estrogen-replacement therapy. Other studies have also demonstrated that ultrasound should not be considered a screening tool in the asymptomatic patient (Gambacciani *et al.*, 2004).

In addition to patients on HRT, transvaginal ultrasound has been studied in a prospective manner as a screening test for postmenopausal patients with breast cancer treated with tamoxifen. Using an endometrial thickness cutoff of 10 mm, 52 asymptomatic patients with a thickened or morphologically suspect endometrium underwent hysteroscopy with dilatation and curettage. This yielded only one case of endometrial cancer. There were 20 patients in this study who reported vaginal bleeding, and two endometrial cancers were found. From these data, the authors concluded that screening by transvaginal ultrasound is not warranted in asymptomatic tamoxifen-treated patients (Gerber *et al.*, 2000).

Mass Screening

Although mass screening is not recommended in the general population, its utility has been studied in certain groups felt to be at increase risk for this disease. Owing to the increased risk of endometrial cancer associated with tamoxifen use, the utility of screening this population has been studied in a prospective manner. Love *et al.* (1999) reported on 357 asymptomatic women treated with tamoxifen for a mean of 66 months compared with 130 controls who were screened with transvaginal ultrasound. All abnormal ultrasounds were investigated further with outpatient hysteroscopy. Women on tamoxifen had a greater mean endometrial thickness (7.3 mm versus 2.5 mm); 41% of the tamoxifen-treated patients were found to have a thickened endometrium, compared with none in the controls. Either outpatient hysteroscopy or dilatation and curettage was performed in these patients with thickened endometrium, and 46% had atrophic endometrium. The remaining patients all had benign findings and the authors concluded that transvaginal ultrasound has a high false-positive rate and is a poor screening tool. They also concluded that the low frequency of significant findings suggests that screening in asymptomatic women is not worthwhile.

Others have also examined the role of screening in the tamoxifen-treated population. Fung *et al.* (2003) used transvaginal ultrasound and color-flow Doppler to screen 304 breast cancer patients on adjuvant tamoxifen therapy. An endometrial thickness of up to 9 mm was considered to be normal. Six cases of endometrial cancer were detected, and all presented with abnormal vaginal bleeding. The specificity was 60.4% and the positive predictive value was only 1.4% for an endometrial cancer. These authors also deemed ultrasound surveillance in asymptomatic women on tamoxifen not to be useful.

More invasive screening with endometrial biopsy was performed by Barakat *et al.* (2000) who screened 159 breast cancer patients on tamoxifen for a 2-year period. Fourteen patients (12.6%) underwent a dilatation and curettage for an abnormal biopsy and one patient was found to have complex hyperplasia. The authors concluded that the utility of endometrial biopsy for screening in this group was limited.

Conclusion

Endometrial cancer remains a common problem, especially in Western countries. Fortunately, the majority of these cancers are the type I variant and associated with a favorable prognosis. Multiple epidemiologic studies have been conducted to identify the specific risk and protective factors associated with this disease. These factors seem to relate either directly or indirectly to the estrogenic stimulation of the endometrium. Genetic risk, in particular the HNPCC syndrome, has been identified and places affected individuals at very significant risk.

Because the majority of patients will present with symptoms of irregular bleeding, women should be counseled to bring this to their physicians' attention. Screening tests such as sonogram and office biopsy have been studied; however, in the asymptomatic female at average or increased risk, the utility of these tests remains in question. In fact, they might lead unnecessarily to more expensive and invasive testing. Patients with or at high risk for HNPCC are the only group that the American Cancer Society recommends be offered annual screening with endometrial biopsy beginning at age 35 years. It should be stated that there are no data to support this recommendation, and this is only based on expert opinion. Proper patient education about the risks and early warning signs of this disease is encouraged for all women.

Practice Points

- When employing hormone replacement therapy (HRT) in menopausal females with an intact uterus, at least 10 days of progestins should be prescribed with each cycle.
- Women should be counseled on the early signs of endometrial cancer when they reach menopause.
- Women taking tamoxifen should be counseled on the increased risk of endometrial cancer and the early warning signs of endometrial cancer.
- Women with a hereditary risk for endometrial cancer should be counseled on available screening methods as well as on prophylactic hysterectomy.

Research Agenda

- To improve our knowledge of the exact risk factors associated with endometrial cancer.
- To identify the genetic mechanism of endometrial cancer development.
- To determine the risk factors associated with the type II endometrial cancers.
- To identify an accurate and effective screening method for endometrial cancer in the general, asymptomatic population.

See also: Cancer Screening; Female Reproductive Function; Worldwide Burden of Gynecologic Cancer: The Size of the Problem.

Citations

Aarnio M, Sankila R, Pukkala E, *et al.* (1999) Cancer risk in mutation carriers of DNA-mismatch-repair genes. *International Journal of Cancer* 81: 214–218.

Akhmedkhanov A, Zeleniuch-Jacquotte A, and Toniolo P (2001) Role of exogenous and endogenous hormones in endometrial cancer. Review of the evidence and research perspectives. *Annals of the New York Academy of Sciences* 943: 296–315.

Anderson KE, Anderson E, Mink PJ, *et al.* (2001) Cancer Epidemiology, Biomarkers & Prevention. *Cancer Epidemiology, Biomarkers & Prevention* 10: 611–616.

Anderson GL, Judd HL, Kaunitz AM, *et al.* (2003) Women's Health Initiative Investigators. Effects of estrogen plus progestin on gynecologic cancers and associated diagnostic procedures: The Women's Health Initiative randomized trial. *JAMA* 290: 1739–1748.

Archer DF (2001) The effect of the duration of progestin use on the occurrence of endometrial cancer in postmenopausal women. *Menopause: The Journal of the North American Menopause Society* 8: 245–251.

Archer DF, McIntyre-Seltman K, Wilborn WW Jr., *et al.* (1991) Endometrial morphology in asymptomatic postmenopausal women. *American Journal of Obstetrics & Gynecology* 165: 317–320.

Barakat RR, Gilewski TA, Almadrones L, *et al.* (2000) Effect of adjuvant tamoxifen on the endometrium in women with breast cancer: A prospective study using office endometrial biopsy. *Journal of Clinical Oncology* 18: 3459–3463.

Baron JA, La Vecchia C, and Levi F (1990) The antiestrogenic effect of cigarette smoking in women. *American Journal of Obstetrics & Gynecology* 162: 502–514.

Calle EE, Rodriguez C, Walker-Thurmond K, and Thun MJ (2003) Overweight, obesity, and mortality from cancer in a prospectively studied cohort of U.S. adults. *New England Journal of Medicine* 348: 1625–1638.

Cauley JA, Gutai JP, Kuller LH, LeDonne D, and Powell JG (1989) The epidemiology of serum sex hormones in postmenopausal women. *American Journal of Epidemiology* 129: 1120–1131.

Coulam CB, Annegers JF, and Kranz JS (1983) Chronic anovulation syndrome and associated neoplasia. *Obstetrics & Gynecology* 61: 403–407.

Dijkhuizen FP, Mol BW, Brolmann HA, and Heintz AP (2000) The accuracy of endometrial sampling in the diagnosis of patients with endometrial carcinoma and hyperplasia: A meta-analysis. *Cancer* 89: 1765–1772.

Duffy S, Jackson TL, Lansdown M, *et al.* (2003) The ATAC adjuvant breast cancer trial in postmenopausal women: Baseline endometrial subprotocol data. *BJOG: An International Journal of Obstetrics and Gynaecology* 110: 1099–1106.

Dunlop MG, Farrington SM, Carothers AD, *et al.* (1997) Cancer risk associated with germline DNA mismatch repair gene mutations. *Human Molecular Genetics* 6: 105–110.

Effects of hormone replacement therapy on endometrial histology in postmenopausal women. The Postmenopausal Estrogen/Progestin Interventions (PEPI) Trial (1996) The Writing Group for the PEPI Trial. *JAMA* 275: 370–375.

Fisher B, Costantino JP, Wickerham DL, *et al.* (1998) Tamoxifen for prevention of breast cancer: Report of the National Surgical Adjuvant Breast and Bowel Project P-1 Study. *Journal of the National Cancer Institute* 90: 1371–1388.

Fleischer AC, Wheeler JE, Lindsay I, *et al.* (2001) An assessment of the value of ultrasonographic screening for endometrial disease in postmenopausal women without symptoms. *American Journal of Obstetrics & Gynecology* 184: 70–75.

Fornander T, Rutqvist LE, Cedermark B, *et al.* (1989) Adjuvant tamoxifen in early breast cancer: Occurrence of new primary cancers. *Lancet* 1(8630): 117–120.

Fung MF, Reid A, Faught W, *et al.* (2003) Prospective longitudinal study of ultrasound screening for endometrial abnormalities in women with breast cancer receiving tamoxifen. *Gynecological Oncology* 91: 154–159.

Furberg AS and Thune I (2003) Metabolic abnormalities (hypertension, hyperglycemia and overweight), lifestyle (high energy intake and physical inactivity) and endometrial cancer risk in a Norwegian cohort. *International Journal of Cancer* 104: 669–676.

Gambacciani M, Monteleone P, Ciaponi M, Sacco A, and Genazzani AR (2004) Clinical usefulness of endometrial screening by ultrasound in asymptomatic postmenopausal women. *Maturitas* 48: 421–424.

Garzo VG and Dorrington JH (1984) Aromatase activity in human granulosa cells during follicular development and the modulation by follicle-stimulating hormone and insulin. *American Journal of Obstetrics & Gynecology* 148: 657–662.

Gerber B, Krause A, Muller H, *et al.* (2000) Effects of adjuvant tamoxifen on the endometrium in postmenopausal women with breast cancer: A prospective long-term study using transvaginal ultrasound. *Journal of Clinical Oncology* 18: 3464–3470.

Goodman MT, Hankin JH, Wilkens LR, *et al.* (1997) Diet, body size, physical activity, and the risk of endometrial cancer. *Cancer Research* 57: 5077–5085.

Goodman MT, Wilkens LR, Hankin JH, Lyu LC, Wu AH, and Kolonel LN (1997) Association of soy and fiber consumption with the risk of endometrial cancer. *American Journal of Epidemiology* 146: 294–306.

Guido RS, Kanbour-Shakir A, Rulin MC, and Christopherson WA (1995) Pipelle endometrial sampling. Sensitivity in the detection of endometrial cancer. *Journal of Reproductive Medicine* 40: 553–555.

Herbst AL (1994) OCs and genital tract malignancies. *Dialogues in Contraception* 4(3): 5–7.

Herrinton LJ and Weiss NS (1993) Postmenopausal unopposed estrogens. Characteristics of use in relation to the risk of endometrial carcinoma. *Annals of Epidemiology* 3: 308–318.

Hill HA, Eley JW, Harlan LC, Greenberg RS, Barrett RJ, and Chen VW (1996) Racial differences in endometrial cancer survival: The black/white cancer survival study. *Obstetrics & Gynecology* 88: 919–926.

Jemal A, Murray T, Ward E, *et al.* (2005) Cancer statistics, 2005. *CA: A Cancer Journal for Clinicians* 55: 10–30.

Killackey MA, Hakes TB, and Pierce VK (1987) Endometrial adenocarcinoma in breast cancer patients receiving antiestrogens. *Cancer Treatment Report* 69: 237–238.

Kurman RJ, Zaino RJ, and Norris HJ (1994) Endometrial carcinoma. In: Kurman RJ (ed.) *Blaustein's Pathology of the Female Genital Tract*, 4th edn., pp. 439–486. New York: Springer-Verlag.

Langer RD, Pierce JJ, O'Hanlan KA, *et al.* (1997) Transvaginal ultrasonography compared with endometrial biopsy for the detection of endometrial disease. Postmenopausal Estrogen/Progestin Interventions Trial. *New England Journal of Medicine* 337: 1792–1798.

Littman AJ, Beresford SA, White E (2001) The association of dietary fat and plant foods with endometrial cancer (United States). *Cancer Causes and Control* 12: 691–702.

Love CD, Muir BB, Scrimgeour JB, Leonard RC, Dillon P, and Dixon JM (1999) Investigation of endometrial abnormalities in asymptomatic women treated with tamoxifen and an evaluation of the role of endometrial screening. *Journal of Clinical Oncology* 17: 2050–2054.

Lu KH, Dinh M, Kohlmann W, *et al.* (2005) Gynecologic cancer as a 'sentinel cancer' for women with hereditary nonpolyposis colorectal cancer syndrome. *Obstetrics & Gynecology* 105: 569–574.

McPherson CP, Sellers TA, Potter JD, Bostick RM, and Folsom AR (1996) Reproductive factors and risk of endometrial cancer. The Iowa Women's Health Study. *American Journal of Epidemiology* 143: 1195–1202.

Michnovicz JJ, Hershcopf RJ, Naganuma H, Bradlow HL, and Fishman J (1986) Increased 2-hydroxylation of estradiol as a possible mechanism for the anti-estrogenic effect of cigarette smoking. *New England Journal of Medicine* 315: 1305–1309.

Moradi T, Weiderpass E, Signorello LB, Persson I, Nyren O, and Adami HO (2000) Physical activity and postmenopausal endometrial cancer risk (Sweden). *Cancer Causes and Control* 11: 829–837.

Nestler JE, Powers LP, Matt DW, *et al.* (1991) A direct effect of hyperinsulinemia on serum sex hormone-binding globulin levels in obese women with the polycystic ovary syndrome. *Journal of Clinical Endocrinology and Metabolism* 72: 83–89.

Newcomer LM, Newcomb PA, Trentham-Dietz A, and Storer BE (2001) Hormonal risk factors for endometrial cancer: Modification by cigarette smoking (United States). *Cancer Causes and Control* 12: 829–835.

Parazzini F, La Vecchia C, Negri E, *et al.* (1999) Diabetes and endometrial cancer: An Italian case-control study. *International Journal of Cancer* 81: 539–542.

Parazzini F, Negri E, La Vecchia C, *et al.* (1998) Role of reproductive factors on the risk of endometrial cancer. *International Journal of Cancer* 76: 784–786.

Parazzini F, La Vecchia C, Negri E, Moroni S, and Chatenoud L (1995) Smoking and risk of endometrial cancer: Results from an Italian case-control study. *Gynecologic Oncology* 56: 195–199.

Pike MC, Peters RK, Cozen W, *et al.* (1997) Estrogen-progestin replacement therapy and endometrial cancer. *Journal of the National Cancer Institute* 89: 1110–1116.

Poretsky L and Kalin MF (1987) The gonadotropic function of insulin. *Endocrine Reviews* 8: 132–141.

Potischman N, Swanson CA, Brinton LA, *et al.* (1993) Dietary associations in a case-control study of endometrial cancer. *Cancer Causes and Control* 4: 239–250.

Potischman N, Swanson CA, Siiteri P, and Hoover RN (1996) Reversal of relation between body mass and endogenous estrogen concentrations with menopausal status. *Journal of the National Cancer Institute* 88: 756–758.
Ries LAG, Kosary CL, Hankey BF, Miller BA, Harras A and Edwards BK (eds.) (1997) *SEER Cancer Statistics Review, 1973–1994.* Bethesda MD: National Cancer Institute NIH Pub. No. 97-2789.
Schmeler KM, Soliman PT, Sun CC, Slomovitz BM, Gershenson DM, and Lu KH (2005) Endometrial cancer In young, normal-weight women. *Gynecologic Oncology* 99: 388–392.
Schouten LJ, Goldbohm RA, and van den Brandt PA (2004) Anthropometry, physical activity, and endometrial cancer risk: Results from the Netherlands Cohort Study. *Journal of the National Cancer Institute* 96: 1635–1638.
Shoff SM and Newcomb PA (1998) Diabetes, body size, and risk of endometrial cancer. *American Journal of Epidemiology* 148: 234–240.
Smith RA, Cokkinides V, and Eyre HJ (2005) American Cancer Society Guidelines for the Early Detection of Cancer, 2005. *CA: A Cancer Journal for Clinicians* 55: 31–44; quiz 55–56.
Smith DC, Prentice R, Thompson DJ, and Herrmann WL (1975) Association of exogenous estrogen and endometrial carcinoma. *New England Journal of Medicine* 293: 1164–1167.
Soler M, Chatenoud L, Negri E, Parazzini F, Franceschi S, and la Vecchia C (1999) Hypertension and hormone-related neoplasms in women. *Hypertension* 34: 320–325.
Swanson CA, Potischman N, Wilbanks GD, *et al.* (1993) Relation of endometrial cancer risk to past and contemporary body size and body fat distribution. *Cancer Epidemiology, Biomarkers & Prevention* 2: 321–327.
Terry P, Baron JA, Weiderpass E, *et al.* (1999) Lifestyle and endometrial cancer risk: A cohort study from the Swedish twin registry. *International Journal of Cancer* 82: 38–42.
Viswanathan AN, Feskanich D, De Vivo I, *et al.* (2005) Smoking and the risk of endometrial cancer: Results from the Nurses' Health Study. *International Journal of Cancer* 114: 996–1001.
Voigt LF, Deng Q, and Weiss NS (1994) Recency, duration, and progestin content of oral contraceptives in relation to the incidence of endometrial cancer (Washington, USA). *Cancer Causes and Control* 5: 227–233.
Watson P and Lynch HT (2001) Cancer risk in mismatch repair gene mutation carriers. *Familial Cancer* 1: 57–60.
Weiderpass E, Adami HO, Baron JA, Magnusson C, Lindgren A, and Persson I (1999) Use of oral contraceptives and endometrial cancer risk (Sweden). *Cancer Causes and Control* 10: 277–284.
Weiderpass E, Persson I, Adami H-O, and Magnusson C (2000) Body size in different periods of life, diabetes mellitus, hypertension, and risk of postmenopausal endometrial cancer (Sweden). *Cancer Causes and Control* 11: 185–192.
Westhoff C, Heller D, Drosinos S, and Tancer L (2000) Risk factors for hyperplasia-associated versus atrophy-associated endometrial carcinoma. *American Journal of Obstetrics & Gynecology* 182: 506–508.
Wickerham DL, Fisher B, Wolmark N, *et al.* (2002) Association of tamoxifen and uterine sarcoma. *Journal of Clinical Oncology* 20: 2758–2760.
Xu WH, Xiang YB, Ruan ZX, *et al.* (2004) Menstrual and reproductive factors and endometrial cancer risk: Results from a population-based case-control study in urban Shanghai. *International Journal of Cancer* 108: 613–619.

Breast Cancer

U Veronesi, European Institute of Oncology, Milan, Italy
P Boyle, International Agency for Research on Cancer, Lyon, France

Glossary

pTNM Tumor-node-metastasis classification system.
- p refers to pathological examination;
- T refers to the size of the tumor;
- N describes whether or not the cancer has spread to nearby lymph nodes, and if so, how many;
- M shows whether the cancer has spread (metastasized) to other organs of the body;

$pT1_{,mic}$ Refers to primary tumor size and degree of local extension; mic indicating microinvasive cancer.
pN0(i+) No regional lymph node metastasis histologically, positive isolated tumor cells.
Her2/neu An oncogene often expressed in breast cancer.
BRCA-positive Result of genetic testing for breast and ovarian cancer indicating a high-risk subject.
ipsilateral carcinoma Carcinoma located on the same breast.
EORTC European Organisation for Research and Treatment of Cancer.
Ki67 The Ki-67 biomarker is a proliferation index that is detected by immunohistochemical staining.

Introduction

Breast cancer is one of the major public health problems. Every year some 1.2 million new cases of breast cancer are diagnosed and some 400 000 women die from it. Worldwide, some 50 million women are living after a previous breast cancer experience. Approximately 200 000 deaths from breast cancer occur in developed countries and 200 000 in developing countries. In Europe, there were an estimated 370 000 new cases and 130 000 deaths in 2004. Mortality rates rose from 1951 to about 1990 but have fallen since then in most European countries, noticeably in the UK (**Figure 1(a)**), but mortality rates in Central and Eastern European countries have been rising (**Figure 1(b)**). Although rates in Hong Kong and Japan have been lower than those in Europe, they have also been increasing (**Figure 1(b)**). Rates in North and South America have been similar to those in Western Europe (**Figure 1(c)**).

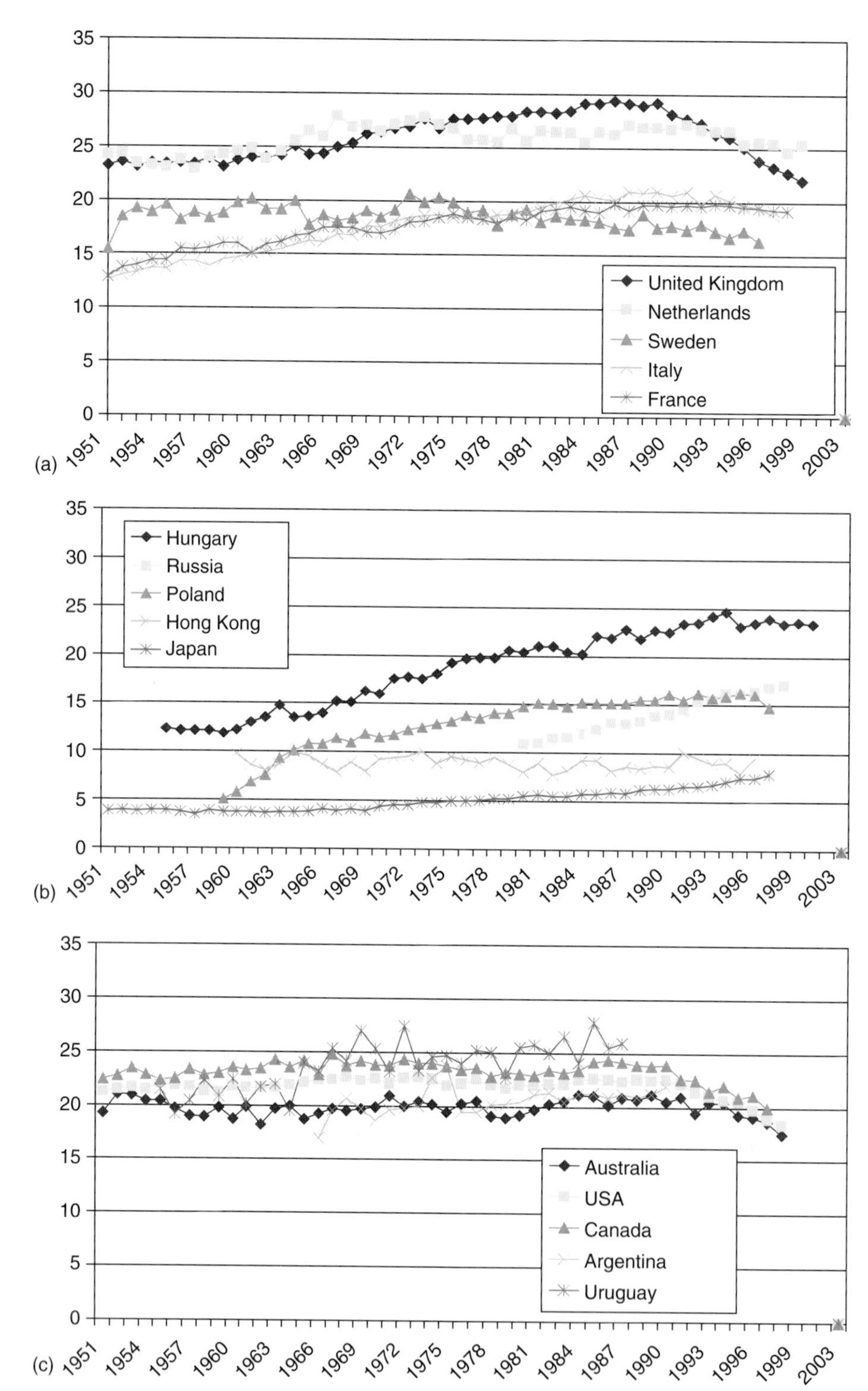

Figure 1 Breast cancer mortality, 1951–2000: All ages, age-standardized rates per 100 000. From Veronesi U, Boyle P, Goldhirsch A, Orecchia R, and Viale G (2005) Breast cancer. *Lancet* 365: 1727–1741.

The decline in mortality rates in Western Europe, Australia, and the Americas may be due to widespread mammographic screening, good diagnosis, and increased numbers of women receiving the best treatment, including hormonal drugs (Ferlay *et al.*, 2004).

Causes

Family History

In countries where breast cancer is common, the lifetime excess incidence of breast cancer is 5.5% for women with one first-degree relative who has had breast cancer and 13.3% for women with two. Eight out of nine women who develop the disease do not have an affected mother, sister, or daughter, however. Only 3–4% of women with breast carcinoma have a genetic mutation (*BRCA1* or *BRCA2*), although these are the most strongly associated risk factors: affected women have a 50–70% risk of developing breast carcinoma during their lifetime.

Pregnancy-Related and Hormone-Related Factors

Women who have their first full-term birth at an early age have a lifetime reduction in risk. Increased parity is associated with a long-term risk reduction, even when the age at first birth is controlled for; the additional, long-lasting protective effect of a young age at subsequent full-term pregnancies is not as strong as that for the first full-term pregnancy; a nulliparous woman has roughly the same risk as a woman who has her first full-term birth aged about 30 years; the risk is transiently increased after a full-term pregnancy. Long duration of lactation confers a small, additional reduction in risk after the age at pregnancy and the number of full-term pregnancies is controlled for. Breast cancer risk factors are set out in **Table 1**.

The decision not to breastfeed or a very short lifetime duration of breastfeeding, typical of women in developed countries, contributes substantially to the high incidence of breast cancer in these areas. The risk is significantly reduced by breastfeeding, in addition to the reduction for every birth. Breastfeeding practices can be modified and promoted usefully as a strategy to prevent breast cancer.

There is no pronounced excess risk of diagnosis in women 10 or more years after the cessation of oral contraceptive use. The cancers diagnosed in women who have used combined oral contraceptives tend to be less advanced clinically than those diagnosed in women who have never used them. The risk of breast cancer is raised in women using hormone-replacement therapy and increases with duration of use. This effect declines after cessation of hormone-replacement therapy and largely disappears after about 5 years; the benefits and risks associated with this hormone treatment should be taken into account.

After adjustment for known risk factors, induced abortion is not associated with an increased risk of breast cancer. Pregnancies that end as a spontaneous or induced abortion have been shown not to increase a woman's risk of developing breast cancer (Veronesi *et al.*, 2005).

Anthropometric Indices and Physical Activity

With pooled data from seven prospective studies (337 819 women and 4385 incident breast cancer cases in total) and after adjustment for reproductive, dietary, and other risk factors, the pooled relative risk of breast cancer per height increment of 5 cm was 1.02 (95% CI 0.96–1.10) in premenopausal women and 1.07 (1.03–1.12) in postmenopausal women. The body mass index showed substantial inverse and positive associations with the disease

Table 1 Risk factors in breast cancer

Factor	*Relative risk*	*High-risk group*
Age	>10	Elderly individuals
Geographical location	5	Developed countries
Breast density	>5	Extensive dense breast tissue visible on mammogram
Age at menarche	3	Before age 11 years
Age at menopause	2	After age 54 years
Age at first full pregnancy	3	First child after age 40 years
Family history	≥2	Breast cancer in first-degree relative
Previous benign breast disease	4–5	Atypical hyperplasia
Cancer in other breast	>4	Previous breast cancer
Socioeconomic group	2	Groups I and II
Body mass index		
Premenopausal	0.7	High body mass index
Postmenopausal	2	High body mass index
Alcohol consumption	1.07	7% increase with every daily drink
Exposure to ionizing radiation	3	Abnormal exposure to young girls after age 10 years
Breastfeeding and parity	Relative risk falls by 4.3% for every 12 months of breastfeeding in addition to a 7% reduction for each birth	Women who do not breastfeed
Use of exogenous hormones		
Oral contraceptives	1.2	Current users
Hormone replacement therapy	1.66	Current users
Diethylstilbestrol	2	Use during pregnancy

From Veronesi U, Boyle P, Goldhirsch A, Orecchia R, and Viale G (2005) Breast cancer. *Lancet* 365: 1727–1741.

in premenopausal and postmenopausal women, respectively. Height is an independent risk factor for breast cancer after menopause but not in premenopausal women.

In postmenopausal women not taking exogenous hormones, general obesity is an important predictor of breast cancer. In premenopausal women, weight and body mass index showed nonsignificant inverse associations with breast cancer.

Increased physical activity seems to be inversely related to the risk of breast cancer, although the findings are inconsistent. Physical activity and weight control can be recommended at present, although further research may highlight additional benefits.

Dietary Factors

A pooled analysis of eight prospective studies showed relative risks for an increment of 5% of energy intake were 1.09 for saturated fat, 0.93 for monounsaturated fat, and 1.05 for polyunsaturated fat, compared with equivalent energy intake from carbohydrates.

The Nurses' Health Study II (Cho *et al.*, 2003) showed that intake of animal fat, mainly from red meat, before menopause, was associated with an increased risk of breast cancer. To assess the risk of invasive breast cancer associated with total and beverage-specific alcohol consumption and to establish whether dietary and nondietary factors change such an association, data from six prospective studies were examined. Alcohol consumption correlated with breast cancer incidence in women. A reduction of consumption among women who drink alcohol regularly could reduce their risk of breast cancer. Cigarette smoking, frequently analyzed with alcohol consumption in etiological studies, does not seem to be related to risk.

Environmental Exposures

An increased risk of breast cancer in women exposed to ionizing radiation, particularly during puberty, has been widely accepted even with low-dose exposure. Environmental exposure to organochlorines has been examined as a potential risk factor for breast cancer. Based on current evidence, the association between risk and exposure to organochlorine pesticides and their residues seems to be small, if it exists at all. The combined evidence from five large USA studies that assessed the link between breast cancer risk and concentrations of 1,1-dichloro-2,2-bis(p-chlorophenyl) ethylene and polychlorinated biphenyls in blood plasma does not support such an association (Cuzick *et al.*, 2003).

Possibilities of Chemoprevention

The pharmacological prevention of cancer represents a comparatively novel field in clinical oncology, but it offers a very promising approach to reducing the burden of cancer and its incidence. In cardiology, it is common practice to treat subjects at higher risk for cardiovascular disease long before clinical evidence of the disease can be detected. This has made a definite contribution to lower mortality. A similar strategy could be adopted for cancer prevention in subjects 'at higher risk' (Hong and Sporn, 1997).

The peculiarity of carcinogenesis is that it is a multistep, multipath, and multifocal process, involving a series of genetic and epigenetic alterations that develop from genomic instability all the way to the final development of cancer. This is the key notion lying behind the rationale for intervention in the initial steps of the process, by employing natural or synthetic agents capable of delaying, arresting, or even reversing the pathogenesis of cancer. Since the process is generally very long (10–20 years, sometimes more), there is potentially a great deal of time to assess the true risk and intervene with nutrients and/or pharmacological agents that may interrupt the chain of molecular events long before the onset of clinical symptoms. This may prove of particular use where solid tumors are concerned, since they are often characterized by multifocality and metachronous growth. Recently, a number of compounds have been shown to be clinically effective in breast carcinoma, covering all three settings into which prevention may be typically divided. In primary prevention, the goal is to prevent the onset of the disease, selecting healthy cohorts who are at high risk because of their environment, lifestyle, or familial/genetic factors. Secondary prevention (screening) is aimed at detecting and treating persons with a premalignant condition or an *in situ* malignancy, thus blocking its evolution to an invasive cancer. Tertiary prevention is a term that can be applied to the protection of individuals who have previously been treated for cancer from developing a second primary tumor (**Figures 2** and **3**; **Table 2**).

Pathogenesis

Progression from Healthy Tissue to Invasive Carcinoma

In contrast to the position for adenocarcinoma of the colon, no definitive model of progression from the common benign proliferative lesions of the breast to invasive malignancy has been identified. Cytological or architectural dysplastic changes can be identified in various nonmalignant breast diseases, such as florid and columnar duct hyperplasia, adenosis, and papilloma, but their true precancerous potential remains undefined.

Atypical duct hyperplasia has been regarded as the missing link between healthy duct hyperplasia and low-grade, ductal neoplasia *in situ* (DIN). Morphological

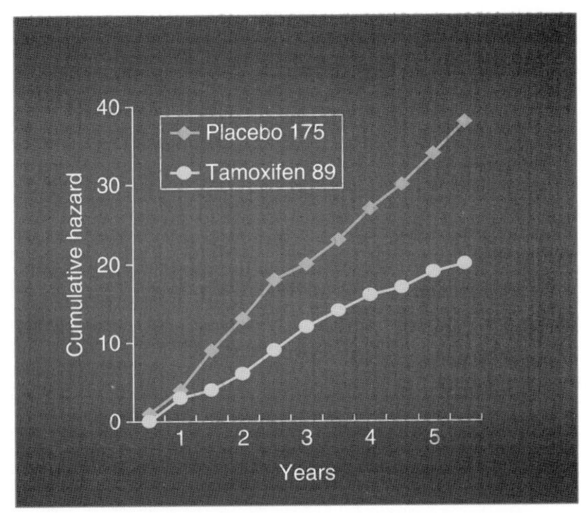

Figure 2 Cumulative incidence of breast cancer in the American Tamoxifen Cancer Prevention Trial. Fisher B, Costantino JP, Wickerham DL, *et al.* (1998) Tamoxifen for prevention of breast cancer: Report of the National Surgical Adjuvant Breast and Bowel project p-1study. *Journal of the National Cancer Institute* 90: 1371–1388.

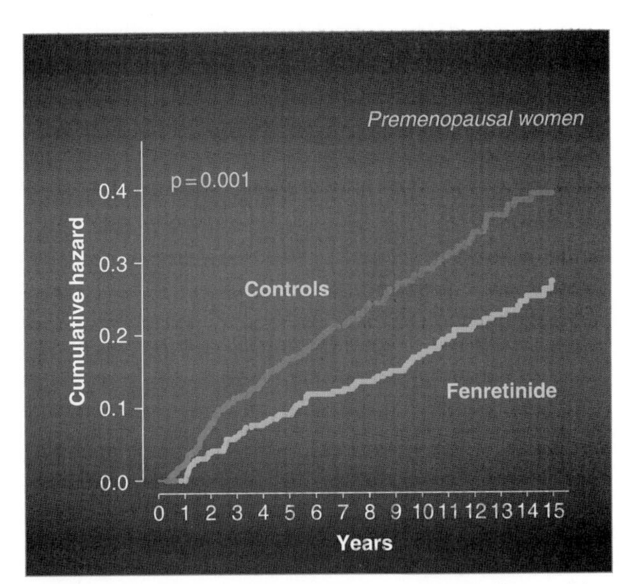

Figure 3 Fenretinide to prevent breast cancer (Italian study). From Veronesi U, Mariani L, Decensi A, *et al.* (2006) Fifteen-year results of a randomized phase III trial of fenritinide to prevent second breast cancer. *Annals of Oncology* 17: 1065–1071.

Table 2 Big killers: Pharmacoprevention

Breast	Tamoxifen, SERMs (raloxifene), fenretinide, aromatase inhibitors, NSAIDs, statins
Colon	NSAIDs, aspirin, vitamin D/calcium, allopurinol
Prostate	Finasteride, selenium E, vitamin E, green tea
Oral cavity	Vitamin A, NSAIDs
Lung cancer	Budesonide, NSAIDs
Cervix	α-Dfmo, indole-3-carbinol, curcumin

features of atypical duct hyperplasia, which are intermediates of those in healthy and malignant tissues, and the substantially raised risk for subsequent carcinoma in affected women, have been claimed as sufficient proof of a precancerous nature. However, genetic changes in atypical duct hyperplasia are identical to changes in fully developed DIN, which raises questions about whether atypical duct hyperplasia is a distinct entity from low-grade intraductal neoplasia.

Novel approaches such as gene-expression profiling will increasingly be used to ascertain the occurrence of true preneoplastic lesions in the breast. Precise identification of these precursor lesions will be vital for planning interventions in women at high risk of breast cancer and to assess the effectiveness of prevention trials.

Ductal lavage is currently undergoing investigation. In this procedure, luminal cells from the ductal tree are obtained by cannulation of the lactiferous ducts and gentle massage of the breast. Harvested cells can then be examined not only for morphological changes but also for the expression of early markers of cell transformation that will eventually be identified. This procedure is being tested for validation as an additional research instrument to identify patients at high risk of developing breast carcinoma (O'Shaughnessy *et al.*, 2002).

In view of the uncertainty of the occurrence of true preneoplastic lesions of the breast, the morphologically identifiable initial phases of neoplastic transformation remain *in situ* neoplasia, either ductal or lobular. However, this encompasses various lesions, ranging from low-grade to high-grade neoplasms, with remarkably different modes of presentation, histopathological features, genetic alterations, risk of relapse and of progression to invasive carcinoma.

To emphasize the non-life-threatening nature of *in situ* lesions and reduce any psychological effect caused by the use of carcinoma as a description, the term ductal intraepithelial neoplasia has been suggested to define these cell masses. This has been revised to encompass candidate preneoplastic lesions: flat epithelial atypia and atypical duct hyperplasia (Tavassoli and Devilee, 2003). The same procedure has been done for noninvasive - lobular neoplasms (atypical lobular hyperplasia and lobular carcinoma *in situ*), which have been classified into a three-tiered system of lobular intraepithelial neoplasia.

Invasion and metastasis are the hallmarks of fully developed breast carcinoma. Extensive histopathological examination of axillary sentinel lymph nodes by complete and serial sectioning at very close cutting intervals (e.g., at least 60 serial sections at 50-μm intervals, as used at the European Institute of Oncology, Milan, Italy) has greatly improved the detection rate of axillary lymph-node association (Veronesi *et al.*, 2003). Detection is strongly associated with the definitive features of the primary breast cancer, such as tumor size and type, occurrence of peritumoral vascular invasion, and multifocality.

Randomized trials (Veronesi *et al.*, 2003) have shown that the recorded number of patients with clinically overt axillary progression of breast cancer is much lower than expected, based on the false-negative rate of the sentinel lymph-node biopsy. This difference suggests that metastatic cells may not progress to clinical disease in all patients and that only some cells are able to sustain tumor progression, which is consistent with the hypothesis that the growth, progression, and clinical outcome of a cancer depend on the activation of tumorigenic stem/progenitor cells.

This redefinition of the malignancy of breast cancer is recognized in the new pTNM classification, whereby minimum nodal invasion (1 mm or less) is classed as $pT1_{mic}$ (indicating microinvasive cancer) and isolated tumor cells or tumor-cell clusters (0.2 mm or less) in the regional lymph node are no longer regarded as metastatic, and qualify as pN0(i+). These new classifications are designed to prevent overstaging of the disease and hence overtreatment for the patient.

Systemic adjuvant therapy is currently offered to patients according to selected clinicopathological features of the primary tumor, which include the status of estrogen and progesterone receptors and expression of human epidermal-growth-factor receptor 2 (HER2/*neu*); such treatment is undertaken independently of the axillary node status, with an equivalent survival benefit.

Quantification of tumor cells circulating in the blood of patients with breast cancer may also be a predictor of the duration of survival.

Gene-expression profiling of breast carcinoma has already shown that differential expression of specific genes is a more powerful prognostic indicator than traditional determinants such as tumor size and lymph-node status. These molecular assays now await clinical validation by prospective randomized trials before being introduced into routine clinical practice.

Diagnosis and Staging

Organized screening, education programs, and improved consciousness of the female population have substantially changed the type of patients seen nowadays compared with those a few decades ago. The revolution in diagnostic imaging over the past 20 years has also profoundly modified diagnostic strategies in breast cancer.

Diagnostic Procedures

Procedures commonly used for the diagnosis of breast cancer include mammography, ultrasonography, MRI (magnetic resonance imaging), and PET (positron emission tomography). Physical examination of the breast remains important, however, because a substantial minority (11%) of breast cancers are not seen on mammography.

Mammography remains the most important diagnostic tool in women whose breast tissue is not dense. After menopause, mammography is generally the best method to discover tiny, nonpalpable lesions. By contrast, ultrasonography is the most effective procedure to diagnose small tumors in women with dense breast tissue and to differentiate solid lesions from cystic lesions. Although mammography can identify suspicious microcalcifications, it is not good at distinguishing between breast densities and has difficulty in identifying certain lobular invasive carcinomas, Paget's disease of the nipple, inflammatory carcinoma, and particularly peripheral, small carcinomas.

MRI is mainly used as a problem-solving method after conventional diagnostic procedures. It is highly sensitive and is used mainly to screen high-risk, *BRCA*-positive patients. In dynamic, contrast-enhanced MRI, images are acquired before and after patients are given a substance to improve contrast when imaging the lesion. Malignant lesions are generally highly permeable, with rapid uptake and elimination of contrast, whereas benign lesions have slow-rising, persistent-enhancement kinetics. Although MRI has good diagnostic accuracy, the false-positive detection rate is still high and MRI findings should not be the sole indication for breast surgery.

PET is currently used to detect metastatic foci in any distant organ or to assess the status of axillary nodes in preoperative staging. However, PET may fail to identify low-grade lesions and tumors smaller than 5 mm.

The use of imaging techniques to detect unknown breast cancers in women who had no symptoms (i.e., screening) was inaugurated by the Health Insurance Plan of New York in the 1960s. In many randomized studies and population studies, mammography has been confirmed as the only screening test that can reduce breast cancer mortality if a large proportion of the population uses the procedure.

However, ultrasonography seems promising for women with dense breast tissue, such as those before menopause, and MRI has been valuable in the screening of women at high risk of breast cancer who are younger than 50 years.

Staging

The TNM (tumor-nodes-metastasis) system defines the extent of disease. It is the language used to compare different cases from various centers. With respect to the primary carcinoma (T), T1 can be divided into three subgroups (T1a, T1b, T1c), depending on the size of the primary lesion. However, with new subdivisions, most instances arise in one subcategory (e.g., T1c). In the era of computerized data analysis, classification is thought to be less necessary, whereas precise description of specific cases is regarded as essential and functional to the different needs of statisticians. Therefore, the T classification will probably be determined by a continuous metric description of the size (cm) of the carcinoma (e.g., T0.9, T2.4). The same system could apply to nodes (N) in which the numbers of involved and examined nodes will define the patient's condition (e.g., N2/18 for two lymph nodes with tumor out of 18 sampled).

Finally, we believe that the TNM system should rely more on biological characteristics (e.g., hormonal receptors, proliferation rates) and biomolecular aspects (e.g., gene expression profile) of tumors. The present biometric, anatomical description will probably be replaced by molecular staging.

Surgery

Once a diagnostic procedure indicates a tumor in the breast, cytological or histological confirmation is vital before further treatment is given. Cytology is effective in solid lesions, especially if sonographically guided. The histology of the lesion, which can be obtained by a core biopsy, is most useful for surgeons. This is the simplest method for palpable lesions that are easily reached, whereas a vacuum-assisted needle biopsy can obtain enough material for a good histological diagnosis in non-palpable or deep lesions (Burbank *et al.*, 1996). Excision biopsy a few days before definitive surgery is no longer done, because it creates a local anatomical distortion that makes conservative treatment difficult.

The sophisticated technique of sentinel lymph node biopsy provides knowledge about the condition of the axillary nodes without the need for dissection, when lymph nodes are not affected. Internal mammary nodes can also be easily reached during surgery, to complete the staging procedure. With respect to distant, occult metastases, PET will help identify occult foci of cancer cells anywhere in the body.

Breast conservation is the most popular treatment because most carcinomas have a restricted size and large primary tumors can be reduced in size by primary (neoadjuvant) chemotherapy before surgery. In most breast cancer centers, conservative surgery (lumpectomy) accounts for 75–85% of all operations. Total removal of the mammary gland (mastectomy) is required for multicentric invasive carcinoma, extensive intraduct carcinoma, inflammatory carcinoma, and large primary carcinomas that have not been sufficiently shrunk by neoadjuvant chemotherapy. Early recurrences or even a second ipsilateral carcinoma of restricted size can also be treated with conservative surgery.

Several options are available for reconstruction of the breast, from simple positioning of an expander to the use of musculocutaneous flaps (such as the thoracodorsal or abdominal flap, TRAM). One method becoming widely used is the skin-sparing mastectomy that conserves an extensive section of skin, as well as the more recent skin- and nipple-sparing mastectomy, which preserves the nipple-areolar complex.

Identical 5-year survival has been recorded in women with axillary dissection and in women who underwent axillary dissection only if the sentinel lymph node was affected by tumor, although other clinical trials of the long-term effect on survival are ongoing. The histological diagnosis of the sentinel node should be immediately available. The traditional frozen-section procedure (which takes three or four sections of the node) often fails to detect micrometastases. As a consequence, surgeons should completely and definitively examine the sentinel node during surgery, and accurately section the node (up to 60–80 sections) to avoid missing even very small micrometastases. In about 85% of cases in which a micrometastatic sentinel node is found, other axillary nodes are not implicated. Therefore, many surgeons now consider the option of simply monitoring patients carefully with ultrasonography and PET.

In situ lesions are mainly treated with local mammary resection. Since axillary metastases are rare, both lymph node dissection and biopsy are optional.

In situ neoplasia should not be incorporated in the TNM classification. Instead, it should be described with the new ductal-intraepithelial-neoplasia system proposed by Tavassoli (Tavassoli *et al.*, 2003).

Radiotherapy

Radiotherapy in Breast Conservation

The current standard of care for patients with early-stage breast cancer consists of breast-conserving surgery, followed by 5–6 weeks' postoperative radiotherapy. The need for radiotherapy in breast conservation is still debated. Some subgroups of patients may be expected to have a low risk of local recurrence, and radiotherapy could therefore be avoided. Attempts have been made to identify these populations, which might include individuals with small, low-grade tumors that are estrogen-receptor-positive, or elderly patients resected with wide tumor-free margins, but no subgroup has been identified that would be adequately treated by breast-conserving surgery alone.

Radiotherapy is used on the whole breast. Some data support the effectiveness of an additional dose applied to the tumor bed (i.e., boost irradiation) to reduce local recurrence. The EORTC study results suggest that the patients deemed to receive the greatest absolute benefit from boost doses are those younger than 50 years and at higher risk of local recurrence (large tumor size or positive or small tumor-free margins in the surgical specimen).

Different radiation treatment schedules with rapid fractionation have been used for years in centers in the UK and Canada. Results from a randomized trial support delivery of a reduced total dose in a shortened schedule (42.5 Gy in 16 fractions for 22 days) in patients with lymph-node-negative breast cancer treated by lumpectomy. A short schedule (20 fractions) with concurrent use of the boost dose is currently used at the European Institute of Oncology in Milan, Italy, after quadrantectomy. In patients younger than 48 years who receive an intraoperative boost dose of 12 Gy, a rapid course of external radiotherapy is used (13 fractions of 2.85 Gy each).

Partial breast irradiation

The rationale for partial breast irradiation (restricted to the excision site and adjacent tissues) instead of the conventional approach is based on the finding that most recurrences arise near the primary tumor location. Partial breast irradiation can be delivered by different techniques, such as low- or high-dose-rate brachytherapy (delivered interstitially or with an intracavitary balloon), conformal external-beam irradiation (including intensity-modulated radiotherapy), and intraoperative radiotherapy.

Intraoperative radiotherapy

ELIOT (ELectron Intra Operative Therapy) refers to the application of a high dose of radiation during surgical intervention, after removal of the tumor.

ELIOT is currently used in early-stage breast cancer as the only treatment at the European Institute of Oncology, Milan, Italy, and a prospective randomized trial is ongoing. Two miniaturized mobile-linear accelerators producing a variable range of electron energies are available. Apart from low costs, ELIOT is advantageous because it potentially overcomes problems related to the accessibility of radiotherapy centers after surgery and has a beneficial impact on the patient's quality of life. ELIOT does not irradiate the skin or the other breast, and irradiation to the lung and the heart is greatly reduced. ELIOT can also be used to give boost doses: one boost of 10–15Gy in an intraoperative session will extend surgery by just 10–20 min and reduces the time for external treatment by 2 weeks. The TARGIT (TARGeted Intraoperative radioTherapy) trial is based on the use of a low-energy radiography source to compare one fraction of radiotherapy with a conventional postoperative approach.

Radiotherapy for ductal carcinoma in situ

The role of radiotherapy in ductal carcinoma *in situ* (DCIS) managed with conservative treatment has been defined by results from three randomized trials. Addition of radiotherapy reduced the local recurrence rate by about 50%, with no effect on survival, and women with positive margins benefited the most. Despite these positive data, the best management of DCIS is still controversial. In fact, according to an analysis of a database of more than 25 000 patients treated between 1992 and 1999, almost half the women did not undergo postoperative radiotherapy after breast-conserving surgery.

Development in radiation techniques

Nowadays, the target volume can be tailored to individuals, which reduces the dose to the lung on the same side as the affected breast, and to the heart, the other breast, and surrounding soft tissue. Intensity-modulated beam arrangement ensures a more homogeneous dose delivery. The increasing use of optic or electronic devices (or both) to monitor organ motion and daily setup variations guarantees the accuracy and safety of the delivery system.

Radiotherapy in Locally Advanced Carcinoma

Breast-conserving surgery followed by radiotherapy can be offered to patients with locally advanced disease who respond to induction chemotherapy. A study of 340 patients given this combined treatment had a total locoregional recurrence rate of only 9% at 5-year follow-up. Breast reconstruction after mastectomy has become a standard procedure, and postmastectomy radiotherapy might represent an obstacle to good aesthetic results because of radiation-related fibrosis. The use of radiotherapy after mastectomy is still controversial.

Radiotherapy of Metastases

With metastases in the skeleton, short courses of irradiation can palliate symptoms and prevent fractures. Radiotherapy and diphosphonates can improve the efficacy of the treatment. Brain metastases or carcinomatous meningitis can be treated successfully: radiotherapy after complete surgical resection can substantially improve control. Patients with one brain metastasis who can be treated with more aggressive therapies, including surgery and high-precision radiotherapy, are especially challenging. Stereotactic radiotherapy is also used in other secondary tumor sites such as the liver, lung, and soft tissues.

Systemic Treatment

Treatment of Locally Advanced Disease

The presence of estrogen and progesterone receptors in tumor cells shown by immunohistochemical staining is a good predictor of endocrine-responsiveness. Staining for either receptor indicates a response to endocrine therapies. Chemotherapy is also effective in endocrine-responsive disease, but the chances of more extensive cell killing are lower for these tumors than for tumors that are unresponsive.

The distinction between disease lacking expression of steroid-hormone receptors and disease showing some presence of these receptors is associated with gene-expression profiling and with the clinical course. Recognition of such a distinction will require a fundamental shift away from reporting whether the receptor status is positive or negative, which is current practice in many laboratories, to the quantitative reporting of receptor determinations.

Overexpression of the epithelial growth factor receptor HER2/*neu* on tumor-cell membranes is a strong predictor for response to trastuzumab (Tripathy *et al.*, 2004). Overexpression of both steroid-hormone receptors and HER2/*neu* has been postulated as a condition for selective resistance to tamoxifen, but less so to aromatase inhibitors in postmenopausal women, and not to tamoxifen combined with suppression of ovarian endocrine function.

Adjuvant Treatments

Adjuvant systemic therapy is given to attempt eradication of micrometastatic disease, which may still be present in all patients with invasive breast cancer. It aims to reduce relapse and increase survival. Postoperative adjuvant therapies cannot be checked for efficacy except with respect to long-term outcomes in a randomized trial. In contrast, the efficacy of systemic treatment given either before surgery (i.e., primary treatments for operable or locally advanced breast cancer) or for metastases, should enable evaluation of the effect of treatment after short-term treatment.

Adjuvant systemic treatments are usually offered to reduce the risk of relapse. An expected 10-year survival below 90% would justify the use of adjuvant chemotherapy. The major concern about adjuvant cytotoxic treatments is that they are offered to a large proportion of patients who are either cured by local treatments or who might have their small risk of relapse reduced by endocrine drugs alone.

Selection of adjuvant treatments is based on the distinction between endocrine-unresponsive and endocrine-responsive breast cancer.

Patients with endocrine-responsive disease are offered adjuvant systemic therapy based on endocrine treatments. High risk of relapse (with metastatic lymph nodes in the operated axilla or vascular invasion) can justify some chemotherapy being given before endocrine treatment in adjuvant therapy.

Premenopausal women with endocrine-responsive disease are usually offered tamoxifen, with or without suppression of ovarian function. Use of cytotoxic drugs before endocrine therapy is recommended only if the risk of relapse is very high. However, the role of both ovarian function suppression and chemotherapy is still uncertain for many of these patients, and trials are in progress. Aromatase inhibitors, which need ovarian function suppression, represent an additional treatment choice.

Women with endocrine-responsive disease after menopause are usually offered endocrine therapy with tamoxifen and increasingly with aromatase inhibitors. Efficacy in reduction of recurrence and mortality well beyond 5 years' treatment (carryover effect) is the reason for its standard use. When prescribed, chemotherapy should be given before the start of tamoxifen treatment.

New alternatives for tamoxifen are available to treat postmenopausal women with endocrine-responsive disease after surgery, after 2–3 years of tamoxifen to complete standard duration, or after 5 years to further reduce risk of relapse (especially for patients at high risk of relapse) (Coombes *et al.*, 2004). These alternatives include nonsteroidal (anastrozole and letrozole) and steroidal (exemestane) aromatase inhibitors. The IBCSG (International Breast Cancer Study Group) trial 18–98 or BIG 1–98, which compares tamoxifen and letrozole alone or in the two possible sequences, recently provided data on greatly improved disease-free survival for postmenopausal patients with endocrine-responsive disease who received letrozole compared with those who received tamoxifen.

Neoadjuvant (Primary) Systemic Treatments

Systemic primary treatment is usually offered to patients with large primary tumors and aims to reduce tumor size for breast-conserving surgery. With such treatment, physicians can also induce regression of axillary node metastases and obtain knowledge on the responsiveness of the disease to treatment.

Endocrine therapies for patients with endocrine-responsive disease showed an improved outcome for aromatase inhibitors compared with that for tamoxifen.

Endocrine-unresponsive disease and high proliferation rates (e.g., Ki67 expressed in $\geq$20% of tumor cells) are important predictors of complete pathological response to six courses of primary chemotherapy. Disease-free survival is substantially longer for patients with endocrine-responsive disease than for patients who do not express steroid-hormone receptors, even though patients with endocrine-unresponsive disease are at least four times more likely to obtain a pathological complete remission after primary chemotherapy (Colleoni *et al.*, 2004).

Anthracyclines and taxanes are usually used for patients with both operable and locally advanced disease (Nowak *et al.*, 2004). Anthracycline-based primary chemotherapy has been reported to yield a large proportion of responses in small-sized tumors with a high proliferation index (Ki67) or grade, and with simultaneous overexpression of HER-2/*neu* and topoisomerase II, whereas mutation of *p53* has been associated with a reduced response rate to chemotherapy. Chemotherapy regimens that do not contain anthracycline (that have vinorelbine, platinum, and fluorouracil) were also reported to be effective, especially for patients with endocrine-unresponsive disease, with or without inflammatory features.

Conclusion

Breast cancer is the most common type of tumor in women in most parts of the world. Although stabilized in Western countries, its incidence is increasing in other continents. Prevention of breast cancer is difficult because the causes are not well known. We know of many risk factors such as nulliparity, late age at first pregnancy, little or no breastfeeding, which, however, are linked to the historic development of human society. On the contrary, a great effort is needed to improve early detection of the tumor. Screening programs among the female population should therefore be implemented. The early discovery of a small breast carcinoma leads to a very high rate of curability and entails very mild types of treatment, with preservation of the body image. Treatments are improving, but a strict interdisciplinary approach is essential. It is conceivable that in all countries specialized centers or units for breast cancer management should be set up.

See also: Cancer Screening; Female Reproductive Function; Ovarian Cancer; Worldwide Burden of Gynecologic Cancer: The Size of the Problem.

Citations

Burbank F, Parker SH, and Fogarty TJ (1996) Stereotactic biopsy: Improved tissue harvesting with the Mammotome. *American Surgeon* 62: 738–744.
Cho E, Spiegelman D, Hunter DJ, *et al.* (2003) Premenopausal fat intake and risk of breast cancer. *Journal of the National Cancer Institute* 95: 1079–1085.
Colleoni M, Viale G, Zahrieh D, *et al.* (2004) Chemotherapy is more effective in patients with breast cancer not expressing steroid hormone receptors: A study of preoperative treatment. *Clinical Cancer Research* 10: 6622–6628.
Coombes RC, Hall E, Gibson LJ, *et al.* (2004) Intergroup Exemestane Study. A randomized trial of exemestane after two to three years of tamoxifen therapy in postmenopausal women with primary breast cancer. *New England Journal of Medicine* 350: 1081–1092.
Cuzick J, Powles T, Veronesi U, *et al.* (2003) Overview of main outcomes in breast cancer prevention trials. *Lancet* 361: 296–300.
Ferlay J, Bray F, Pisani P, and Parkin DM (2004) *Globocan 2002: Cancer Incidence, Mortality and Prevalence Worldwide. IARC CancerBase No. 5, version 2·0*. Lyon, France: IARC Press.
Fisher B, Costantino JP, Wickerham DL, *et al.* (1998) Tamoxifen for prevention of breast cancer: Report of the National Surgical Adjuvant Breast and Bowel project p-1study. *Journal of the National Cancer Institute* 90: 1371–1388.
Hong WK and Sporn MB (1997) Recent advances in chemoprevention of cancer. *Science* 278: 1073–1077.
Nowak AK, Wilcken NR, Stockler MR, Hamilton A, and Ghersi D (2004) Systematic review of taxane-containing versus non-taxane containing regimens for adjuvant and neoadjuvant treatment of early breast cancer. *Lancet Oncology* 5: 372–380.
O'Shaughnessy JA, Ljung BM, Dooley WC, *et al.* (2002) Ductal lavage and the clinical management of women at high risk for breast carcinoma: A commentary. *Cancer* 94: 292–298.
Tavassoli FA and Devilee P (eds.) (2003) *World Health Organization Classification of Tumours. Pathology and Genetics of Tumours of the Breast and Female Genital Tract.* Lyon, France: IARC Press.
Tripathy D, Slamon DJ, Cobleigh M, *et al.* (2004) Safety of treatment of metastatic breast cancer with trastuzumab beyond disease progression. *Journal of Clinical Oncology* 22: 1063–1070.
Veronesi U, Paganelli G, Viale G, *et al.* (2003) A randomized comparison of sentinel-node biopsy with routine axillary dissection in breast cancer. *New England Journal of Medicine* 349: 546–553.
Veronesi U, Boyle P, Goldhirsch A, Orecchia R, and Viale G (2005) Breast cancer. *Lancet* 365: 1727–1741.
Veronesi U, Mariani L, Decensi A, *et al.* (2006) Fifteen-year results of a randomized phase III trial of fenritinide to prevent second breast cancer. *Annals of Oncology* 17: 1065–1071.

Ovarian Cancer

R J Edmondson, Queen Elizabeth Hospital, Tyne and Wear, UK
A R Todd, Northern Centre for Cancer Treatment, Newcastle General Hospital, Newcastle upon Tyne, UK

The term ovarian cancer is used to describe a range of tumors arising from the ovary. Many of these tumors arise from very different cell types and are biologically distinct. The commonest, and the most lethal, are the epithelial tumors, and this chapter will focus predominantly on these types of tumors. For a review of the other common types of ovarian tumor, the reader is referred to Sanusi *et al.* (2002).

Epidemiology

A consistent finding from the many epidemiological studies carried out is that the number of lifetime ovulations is associated with the risk of developing ovarian cancer. Factors that reduce ovulation, including pregnancy, breast-feeding, and use of the combined oral contraceptive pill, appear to confer a degree of protection. For instance, a single full-term pregnancy reduces the subsequent risk of developing ovarian cancer by half (relative risk 0.47) (Whittemore *et al.*, 1992). Although an early menarche and a late menopause may increase the number of ovulations, in fact many cycles at the beginning and end of menstrual life are nonovulatory, and this is borne out by the lack of evidence to support the hypothesis that early menarche and late menopause are associated with epithelial ovarian cancer (Edmondson and Monaghan, 2001).

Use of hormone replacement therapy appears to confer a small risk of developing ovarian cancer, although many of the trials to date have shown only marginal results. Other factors that may be protective include previous hysterectomy and previous sterilization.

The effect of infertility and of treatment for infertility on the subsequent risk of developing ovarian cancer has been widely debated. While true infertility may confer an increased risk, the use of infertility treatments has little or no effect on the subsequent risk of developing ovarian cancer.

Although the majority of ovarian cancers are sporadic, it is clear that some are related to a family history of either breast or ovarian cancer. It is now apparent that most of these familial cancers are related to germ line mutations of the breast cancer genes *BRCA1* or *BRCA2* (Venkitaraman, 2002). Other genes implicated in familial ovarian cancer include *MLH1* and *MSH2*, often referred to as the hereditary nonpolyposis colon cancer (HNPCC) phenotype. It is presumed that familial cancers share the same risk factors as sporadic cancers, but there is little evidence to support this.

Biology and Etiology

The biology of ovarian cancer remains poorly understood. This is partly because the disease is often diagnosed late in its progression, so that study of early invasive disease is difficult. Equally, the ovaries are not easily accessible, and study of early invasive or preinvasive lesions is impossible.

It is now believed that epithelial ovarian cancers arise from the ovarian surface epithelium (OSE), a single layer of poorly committed cells that lies in continuity with the peritoneum. The processes underlying malignant transformation are not completely understood, but great advances have been made in the last 10 years following the development of techniques to culture OSE cells in the laboratory. It is now clear that the first step toward malignant transformation is an increasing commitment toward an epithelial phenotype by these cells. This appears to occur more in crypts and inclusion cysts within the ovary, and it is now suggested that most ovarian cancers arise in these areas, although to date no definite premalignant lesions have been identified (Auersperg *et al.*, 2001).

This ability of the OSE cell to undergo a differentiation process toward a more epithelial phenotype may explain why epithelial ovarian cancers can display a range of morphological types including serous, mucinous, endometrioid, and clear-cell morphologies, although it is not clear why different tumors develop along these different pathways. It is interesting to note that ovarian cancer appears to be unique among tumors in having a differentiation process as the first part of the carcinogenetic process.

Ovarian cancer commonly spreads by the transcelomic route: metastases are usually seen in the omentum and on the peritoneum, including the peritoneal covering of the bowel. Lymphatic spread is seen in some cases, classically to the pelvic lymph nodes but also to the common iliac and para-aortic nodes.

Hematogenous spread is uncommon but can involve the liver and spleen as well as distant organs.

Ovarian cancer is often associated with ascites and pleural effusions; these often contain malignant cells.

Diagnosis and Treatment Pathway

Ovarian cancer is notorious for its late presentation. It has been called 'the silent killer' because the majority of women with ovarian cancer present very late, that is, they already have disseminated disease by the time they are first seen by a doctor. Diagnosis is often delayed because of the wide variety of symptoms experienced by these patients. Ovarian cancer can present as a result of symptoms or may be detected by screening.

Symptomatic Presentation

The range of symptoms associated with ovarian cancer is wide and nonspecific. Symptoms may be related to the presence of a pelvic mass, such as urinary frequency, pain, and constipation; or to other intra-abdominal disease including disease on or invading the bowel, such as rectal bleeding or altered bowel habit; or to the presence of ascites, leading to abdominal distension; or to general symptoms related to cancer, such as nausea, vomiting, anorexia, and cachexia. Although it is important for all physicians to have a high index of suspicion, given that the variety of nonspecific symptoms is so great, it is not surprising that diagnosis is often delayed and patients are often referred to general surgical teams and other specialists before referral to a gynecologist.

Pelvic examination is useful in identifying the presence of pelvic masses, but it does not provide information about the nature of such a mass, so imaging is required for further evaluation. Transvaginal ultrasound remains the most useful modality because it is cheap, well-tolerated, and provides useful information about the contents and nature of a pelvic mass or the presence of free fluid in the pelvis. Transvaginal ultrasound is often combined with transabdominal ultrasound to identify intra-abdominal metastases. Other imaging modalities include computerized tomography (CT) scan and magnetic resonance imaging (MRI), but overall these confer little advantage over ultrasound.

Since its identification in 1983 (Bast *et al.*, 1983), serum levels of CA 125 have been used widely in the diagnosis and monitoring of ovarian cancer. CA 125 is a protein produced by mesodermally derived tissues and for this reason is relatively nonspecific; CA 125 levels can also be raised in nonmalignant ascites and cardiac failure, as well as in pregnancy and benign conditions including endometriosis and fibroids. Furthermore, some ovarian cancers, particularly early-stage and mucinous tumors, do not produce CA 125.

Nevertheless, CA 125 remains the most useful tumor marker for ovarian cancer, and the test is commonly performed alongside carcinoembryonic antigen (CEA), levels of which help to differentiate ovarian tumors from colorectal malignancies.

There is increasing evidence that patients treated by specialist gynecological oncologists have a better prognosis than patients who undergo surgery by general gynecologists or general surgeons (Junor *et al.*, 1999). For this reason, there is much interest in devising mechanisms for improving preoperative diagnosis in order that patients who have ovarian cancer are referred to specialist gynecological oncologists prior to surgery. Using a combination of the patient's menopausal status, the results of ultrasound, and the CA 125 level, an algorithm has been devised (the 'risk of malignancy' index) that can predict the presence of invasive ovarian cancer (Jacobs *et al.*, 1996). Such algorithms have been adopted by many centers in the United Kingdom, with a reasonable degree of success (Kirwan *et al.*, 2002).

Screening

The concept of screening for ovarian cancer has attracted much interest, given that the disease tends to present clinically at such a late stage, when it has a very high fatality. Unfortunately, our knowledge of the natural history of the disease is limited, and there is no evidence to support the presence of a latent or premalignant phase of the disease. Both of these factors limit the ability to develop an effective screening program.

Efforts have nevertheless been made to develop a suitable screening test. Those that have attracted the most attention are transvaginal ultrasound and serum CA 125. In isolation, these tests tend to have low specificity, which would lead to large numbers of women without cancer being investigated with surgery. This has led to the development of a multimodal approach in which patients are initially screened with a CA 125 test and only subjected to ultrasound when the CA 125 measurement is abnormal. Pilot studies of this approach have been encouraging, and a large multicenter trial is currently under way in the United Kingdom to assess whether screening can be effective in reducing mortality from this disease.

Despite the lack of evidence to support the use of these tests in a mass screening program, they have tended to be offered to women who are thought to be at high risk of developing ovarian cancer, because of a strong family history. In this context, both CA 125 measurement and transvaginal ultrasound are often offered, on an annual basis.

Primary Management

Surgery

The primary treatment for epithelial ovarian cancer has traditionally been surgical. It is based on the principle of cytoreductive or debulking surgery. At the time of surgery, disease is often disseminated throughout the peritoneal cavity. This makes it impossible to perform a complete removal of the tumor with a clear margin of normal tissue, often considered to be the gold standard for surgical oncology. Several studies have demonstrated that patients' overall prognosis is improved if most of the disease is removed at surgery compared with patients in whom it is impossible to remove disease. This has led to the concept of optimal surgery, often defined as residual disease less than 1–2 cm in greatest diameter.

More recently, the concept of cytoreductive surgery has been questioned: is the altered outcome following cytoreductive surgery merely a reflection of the biology of the disease? That is, do patients who have had optimal cytoreductive surgery do better than patients who are suboptimally debulked, not because their disease has been removed but simply because their disease can be removed?

This question has not yet been fully answered, but there are interesting data emerging regarding the concept of ultraradical surgery. If one accepts the argument that outcome following surgery is principally related to the biology of the disease, then more aggressive surgery and removal of more disease should have no effect on overall outcome (Bristow and Berek, 2006). However, data from Eisenkop *et al.* (2003) have demonstrated that an ultraradical approach, that is, removing the maximal amount of disease possible, thereby achieving optimal debulking rates of up to 96%, is associated with a 5-year survival rate as high as 49%. It is important to note that these data are not from randomized studies, but then it is unlikely that a randomized study could or will ever be carried out to address this question.

Surgical intervention has also been used as an 'interval' procedure, that is, part way through a course of chemotherapy. Evidence from randomized studies suggests that for patients who are suboptimally debulked at initial surgery and respond to chemotherapy, there is a prognostic benefit from undergoing interval surgery followed by further chemotherapy (van der Burg *et al.*, 1995).

Laparoscopy has gained in popularity in the management of ovarian cancer, principally as a diagnostic tool, but also for prophylactic surgery in women at high risk of developing ovarian cancer as a result of a genetic predisposition.

Chemotherapy

While surgery is the primary treatment modality for ovarian cancer, chemotherapy is used in two settings in the initial management of this disease: first as adjuvant treatment following surgery, and second as neoadjuvant or primary treatment in those patients for whom surgery is inadvisable because of significant comorbidity or extensively disseminated disease.

The most effective form of chemotherapy is based on platinum drugs. It was first used in the 1970s in the combination of cisplatin, cyclophosphamide, and adriamycin (CAP). After a number of trials questioned whether this three-drug combination was necessary, the ICON-2 trial confirmed that single-agent carboplatin was indeed as effective as the combination, and caused significantly less toxicity (ICON, 1998). In the 1990s, the addition of paclitaxel to the regimen was found to improve efficacy still further in the GOG-111 and OV-010 trials (McGuire *et al.*, 1996; Piccart *et al.*, 2000). These trials both used the combination of cisplatin with paclitaxel, but a number of later trials comparing this regimen with carboplatin/paclitaxel have found the two to be equivalent in efficacy, with the carboplatin regimen producing significantly less toxicity. It is important to note that two other studies looking at the addition of paclitaxel to platinum-based regimens, ICON-3 and GOG-132, did not show a survival advantage from the addition of paclitaxel (ICON, 2002; Muggia *et al.*, 2000). One reason proposed to explain this is the extent of crossover to taxane-based regimens in the control groups of these studies. Another possible explanation is that paclitaxel may reduce the response to the platinum agent if given at the same time. Indeed, in the GOG-132 trial, patients treated with the combination regimen experienced a myeloprotective effect compared to those treated with single-agent cisplatin. It is suggested that this may be because paclitaxel was attenuating the effects of the platinum.

Proponents of this theory suggest that the optimal regimen may be sequential treatment with a platinum agent and a taxane rather than simultaneous treatment with the two drugs (Moss and Kaye, 2002). As well as comparing the efficacy of alternative platinum drugs, the use of alternative taxanes has been studied. The SCOTROC trial compared paclitaxel/carboplatin with docetaxel/carboplatin and found that the docetaxel regimen had a similar efficacy to paclitaxel/carboplatin with less severe neurotoxicity. However there was more neutropenia in this group (Vasey *et al.*, 2004). In addition, there are ongoing and recently closed trials of new combinations of agents such as the five-arm ICON-5 trial, which looked at gemcitabine, liposomal doxorubicin, and topotecan in various combinations and sequences, compared to standard carboplatin and paclitaxel.

The trials mentioned above have all looked at patients with advanced ovarian cancer. the ICON-1 and ACTION trials looked at the effect of adjuvant chemotherapy in high-risk early-stage ovarian cancer and found a survival advantage for adjuvant treatment in these patients also (ICON, 2003; Trimbos *et al.*, 2003).

In an effort to try to improve the outcome in patients with ovarian cancer, intraperitoneal chemotherapy and maintenance chemotherapy have also been investigated. Intraperitoneal chemotherapy has been extensively studied in the past. Three large randomized trials have found a survival benefit in women treated with intraperitoneal chemotherapy compared with those treated with intravenous chemotherapy. However, it is associated with high levels of gastrointestinal and systemic toxicity. A GOG Phase II study is ongoing to investigate whether less toxic regimens can be given with the same efficacy (Ozols, 2006). The use of maintenance chemotherapy has also been investigated, with extended cycles of induction chemotherapy and prolonged administration of a single chemotherapeutic agent, but no improvement in survival has been found (Ozols, 2006).

The current standard regimen following debulking surgery is therefore paclitaxel and carboplatin, given intravenously every 3 weeks for a total of six cycles. Combination chemotherapy is generally offered to patients with FIGO Stage Ic disease or worse.

Chemotherapy would usually be discussed with those patients with Stage Ia or Ib disease with poor prognostic features, in the form of single-agent platinum treatment. The CA 125 level is measured at the start of each cycle of chemotherapy. If this falls to within normal levels within the first three cycles, there is an improved prognosis. If, at the end of six cycles, the CA 125 has fallen but not completely normalized, consideration may be given to treating with a further two cycles of chemotherapy, although there is no published evidence showing that this is of benefit.

Neoadjuvant chemotherapy is currently used in patients who are not fit to undergo surgery, or in patients with extensive disease, to try to downstage the disease so that optimal debulking surgery can be performed. A recent study of neoadjuvant chemotherapy found a higher tumor resection rate and a survival advantage in women who were treated with chemotherapy in the first instance (Kuhn *et al.*, 2001). The use of neoadjuvant chemotherapy is therefore currently being investigated in a randomized UK study (CHORUS) in which patients who are healthy enough and suitable for either chemotherapy or surgery are randomized to primary treatment with one of these modalities. Those receiving primary surgery subsequently receive adjuvant chemotherapy, while those receiving primary chemotherapy (usually with carboplatin and paclitaxel) undergo interval debulking surgery midway through their chemotherapy course.

Management of Relapse

Over 75% of women with ovarian cancer have advanced disease when they are first diagnosed. While the disease is very chemosensitive and the majority of patients achieve complete remission following surgery and chemotherapy, cancer will recur in most patients (Ozols, 2006). For this reason, following completion of chemotherapy, patients are routinely followed up on a regular basis with physical examination and recording of their CA 125 level. In patients with recurrent ovarian cancer, the aim of treatment is palliation of symptoms rather than cure. There is debate about whether measuring CA 125 levels is of any value in the asymptomatic patient. Currently, the OVO5 study is investigating this by double-blinded estimation of CA 125 levels. At present, if a patient is symptomatic and/or shows a rapidly rising CA 125, relapse is confirmed by CT scan. The initiation of second-line treatment would usually be on the development of symptoms. However, the response to treatment is affected by tumor bulk and the performance status of the patient, and it may be that by waiting for symptoms to develop, tumor bulk has increased and performance decreased such that response is adversely affected. Equally, starting treatment too soon may cause significant toxicity in a patient who had previously been asymptomatic.

The choice of second-line agent depends largely on the woman's response to her original chemotherapy, the length of time since it was completed, and the toxicities experienced, as well as her current performance status and preference. Treatment for relapsed disease may continue intermittently for months or even years, and the effect that the treatment will have on quality of life is therefore even more important and must be carefully considered.

Patients who relapse more than 6 months after their initial platinum regimen are said to be platinum-sensitive and should be rechallenged with carboplatin with or without paclitaxel. The ICON-4 trial showed an improved 2-year survival after further treatment with the combination regimen compared with carboplatin alone (ICON/AGO, 2003), but toxicity is greater with the combination, and in the palliative setting this needs to be borne in mind. In patients who are platinum-resistant (relapse less than 6 months after their original platinum regimen) or platinum-refractory (had no response to the original platinum regimen), there is no standard second-line treatment. A number of treatments are known to have activity (Moss and Kaye, 2002). Liposomal doxorubicin and topotecan have been recommended by NICE for second-line treatment in this group of patients. Tamoxifen may also be considered (Williams and Simera, 2001). Paclitaxel should be considered if it was not used in the initial regimen prior to relapse. Other agents with activity include gemcitabine and oral etoposide. The response to second-line agents is poor – of the order of 15–20%. Therefore, patients should also be considered for entry into Phase I and II trials of investigational agents.

Although the treatment of relapsed ovarian cancer is essentially with chemotherapy, both surgery and radiotherapy do have a role to play. Surgery is used for the excision of bulky disease or for the removal of lesions causing specific symptoms. In addition, in the patient with a long disease-free interval, further debulking surgery may be an option. Ovarian cancers are generally not radiosensitive, and therefore radiotherapy is only indicated for the treatment of specific symptoms such as vaginal bleeding or pain.

Palliation

Given that 75% of women with ovarian cancer will die of their disease, it is important to recognize that most patients will require palliative input at some point in their care. In addition to pain, one of the most common problems encountered is bowel dysfunction, with symptoms such as nausea and vomiting, but in many cases it may develop into bowel obstruction.

Intermittent, partial obstruction can often be managed conservatively, and treatment centers on control of nausea and vomiting and promotion of bowel motility using antiemetics such as metoclopramide. Conservative management of complete obstruction utilizes antispasmodics and antisecretory drugs such as hyoscine butylbromide and octreotide, respectively. Surgery is generally reserved for those patients who have a discrete level of obstruction in whom a defunctioning or bypass operation can be performed. It is important to note that such defunctioning surgery does not increase longevity but merely acts as a palliative procedure and is associated with a high perioperative morbidity and mortality. Nevertheless, some patients will benefit from this, allowing them to be discharged from hospital and managed at home.

Prevention

Combined oral contraception has been shown to be protective against the development of ovarian cancer, and it has therefore attracted attention as a possible preventative treatment. Unfortunately, because of the increased risk of breast cancer in users of combined oral contraception, it cannot be recommended even for the population at greatest risk, namely those patients with mutations in the *BRCA* genes.

Similarly, prophylactic surgery is often considered for these high-risk patients. Salpingo-oophorectomy can be performed laparoscopically, as a day case, with minimal morbidity. The subsequent risk of developing ovarian cancer is hugely reduced, and, if performed before the menopause, removal of the ovaries and Fallopian tubes (salpingo-oophorectomy) has been shown to reduce the risk of developing breast cancer. There is still great controversy regarding the timing of surgery, particularly because surgery will render the woman menopausal and the use of HRT in this group is controversial (Griffiths *et al.*, 2005).

Conclusion

Epithelial ovarian cancer remains a significant health problem with incidence and mortality rates that have barely changed over the last 40 years, partly due to the late presentation of the disease. Nevertheless, some progress has been made, not least in the area of chemotherapy. The increase in number of active drugs has allowed patients to live longer with their disease, although ultimately the majority of patients will still succumb to the cancer.

Challenges for the future lie in developing our ability to predict chemosensitivity to the different agents available and therefore to be able to individualize therapies for patients. Longer-term strategies are to detect the disease earlier and ultimately, of course, to prevent the development of the cancer in the first place.

See also: Breast Cancer; Cancer Screening; Female Reproductive Function; Worldwide Burden of Gynecologic Cancer: The Size of the Problem.

Citations

Auersperg N, Wong A, Choi K, Kang S, and Leung P (2001) Ovarian surface epithelium: Biology, endocrinology, and pathology. *Endocrine Reviews* 22: 255–288.

Bast R, Klug T, and St John E (1983) A radioimmunoassay using a monoclonal antibody to monitor the course of ovarian cancer. *New England Journal of Medicine* 309: 883–887.

Bristow RE and Berek JS (2006) Surgery for ovarian cancer: How to improve survival. *Lancet* 367(9522): 1558–1560.

Edmondson R and Monaghan J (2001) The epidemiology of ovarian cancer. *International Journal of Gynecological Cancer* 11: 423–429.

Eisenkop SM, Spirtos NM, Friedman RL, Lin W-CM, Pisani AL, and Perticucci S (2003) Relative influences of tumor volume before surgery and the cytoreductive outcome on survival for patients with advanced ovarian cancer: A prospective study. *Gynecologic Oncology* 90: 390–396.

Griffiths S, Lopes A, and Edmondson R (2005) The role of prophylactic salpingo-oophorectomy in women who carry mutations of the BRCA genes. *Obstetrician and Gynaecologist* 7: 15–19.

ICON (1998) ICON 2: Randomized trial of single-agent carboplatin against three-drug combination of CAP (cyclophosphamide, doxorubicin and cisplatin) in women with ovarian cancer. *Lancet* 351: 1571–1576.

ICON (2002) Paclitaxel plus carboplatin versus standard chemotherapy with either single-agent carboplatin or cyclophosphamide, doxorubicin, and cisplatin in women with ovarian cancer: The ICON 3 randomized trial. *Lancet* 360: 505–515.

ICON (2003) International collaborative ovarian neoplasm trial 1: A randomized trial of adjuvant chemotherapy in women with early-stage ovarian cancer. *Journal of the National Cancer Institute* 95: 125–132.

ICON/AGO (2003) Paclitaxel plus platinum-based chemotherapy versus conventional platinum-based chemotherapy in women with relapsed ovarian cancer. *Lancet* 361: 2099–2106.

Jacobs IJ, Skates S, Davies AP, *et al.* (1996) Risk of diagnosis of ovarian cancer after raised serum CA 125 concentration: A prospective cohort study. *British Medical Journal* 313(7069): 1355–1358.

Junor E, Hole D, McNulty L, Mason M, and Young J (1999) Specialist gynecologists and survival outcome in ovarian cancer: A Scottish national study of 1866 patients. *British Journal of Obstetrics and Gynaecology* 106: 1130–1136.

Kirwan JMJ, Tincello DG, Herod JJO, Frost O, and Kingston RE (2002) Effect of delays in primary care referral on survival of women with epithelial ovarian cancer: Retrospective audit 10.1136/bmj.324.7330.148. *British Medical Journal* 324(7330): 148–151.

Kuhn W, Rutke S, Spathe K, Schmalfeldt B, *et al.* (2001) Neoadjuvant chemotherapy followed by tumor debulking prolongs survival for patients with poor prognosis in International Federation of Gynecology and Obstetrics Stage IIIC ovarian carcinoma. *Cancer* 92: 2585–2591.

McGuire MP, Hoskins WJ, Brady MF, *et al.* (1996) Cyclophosphamide and cisplatin compared with paclitaxel and cisplatin in patients with Stage III and Stage IV ovarian cancer. *New England Journal of Medicine* 334: 1–6.

Moss C and Kaye SB (2002) Ovarian cancer: Progress and continuing controversies in management. *European Journal of Cancer* 38: 1701–1707.

Muggia FM, Braly PS, Brady MF, *et al.* (2000) Phase III randomized study of cisplatin versus paclitaxel versus cisplatin and paclitaxel in patients with suboptimal Stage III or IV ovarian cancer: A Gynaecologic Oncology Group study. *Journal of Clinical Oncology* 18: 106–115.

Ozols RF (2006) Systemic therapy for ovarian cancer: Current status and new treatments. *Seminars in Oncology* 33 (2 Supp 6): S3–S11.

Piccart MJ, Bertelsen K, James K, *et al.* (2000) Randomized intergroup trial of cisplatin-paclitaxel versus cisplatin-cyclophosphamide in women with advanced epithelial ovarian cancer: Three year results. *Journal of the National Cancer Institute* 92: 699–708.

Sanusi F, Carter P, and Barton D (2002) Non-epithelial ovarian tumors. *The Obstetrician and Gynaecologist* 2: 37–39.

Trimbos JB, Vergote I, Bolis G, *et al.* (2003) Impact of adjuvant chemotherapy and surgical staging in early-stage ovarian carcinoma: European Organisation for Research and Treatment of Cancer-Adjuvant Chemotherapy in Ovarian Neoplasm Trial. *Journal of the National Cancer Institute* 95: 113–125.

van der Burg MEL, van Lent M, Buyse M, *et al.* (1995) and The Gynecological Cancer Cooperative Group of the European Organization for Research and Treatment of Cancer. The effect of debulking surgery after induction chemotherapy on the prognosis in advanced epithelial ovarian cancer. *New England Journal of Medicine* 332: 629–634.

Vasey PA, Jayson GC, Gordon A, *et al.* (2004) Phase III randomized trial of docetaxel-carboplatin versus paclitaxel-carboplatin as first-line chemotherapy for ovarian carcinoma. *Journal of the National Cancer Institute* 96: 1682–1691.

Venkitaraman AR (2002) Cancer susceptibility and the functions of BRCA1 and BRCA2. *Cell* 108: 171–182.

Whittemore AS, Harris R, Itnyre J, *et al.* (1992) Characteristics relating to ovarian cancer risk: Collaborative analysis of 12 US case-control studies. II. Invasive epithelial ovarian cancers in white women. *American Journal of Epidemiology* 136: 1184–1203.

Williams CJ and Simera I (2001) Tamoxifen for relapse of ovarian cancer (Review). *Cochrane Database of Systematic Reviews* 1: CD001034.

Relevant Websites

http://www.cancerresearchuk.org/ – Cancer Research UK.

http://www.nlm.nih.gov/medlineplus/ovariancancer.html – Medline Plus.

http://www.cancer.gov/cancertopics/types/ovarian – National Cancer Institute.

http://www.ukctocs.org.uk/ – U.K. Collaborative Trial of Ovarian Cancer Screening.

Prostate Cancer

G G Giles, Cancer Epidemiology Centre, Carlton, Victoria, Australia

Glossary

Aggressive tumor An abnormal growth or mass of tissue that has the potential to become malignant (see metastasis).
Androcline A steady age-related decline in testosterone levels in men.
Androgen A steroid hormone that stimulates or controls the development and maintenance of masculine characteristics.
Androgen deprivation Hormonal therapies used to reduce the production of androgens, including castration and drugs that are antiandrogens and antiadrenal androgens. Used in combination termed androgen blockade.
Apoptosis One of the main types of programmed cell death.
Benign prostatic hyperplasia The increase in size of the prostate in middle-aged and elderly men.
Brachytherapy Involves the placement of about 100 small seeds containing radioactive material directly into the tumor.
Carotenoid organic pigments Pigments that naturally occur in plants.
Digital rectal examination The examiner inserts a gloved, lubricated finger into the rectum to examine the adjoining prostate.
Dihydrotestosterone A biologically more active form of the male sex hormone testosterone.
Histologic Refers to thin sections of tissue when examined under a microscope by a pathologist.
Insulin-like growth factor Polypeptides with high similarity to insulin.
Lead time The length of time between when a disease becomes present in a person's body and its usual clinical presentation.
Metastasis The transfer of cancer from one organ or part to another organ or part not directly connected with it. Only malignant tumor cells have the capacity to metastasize.
Nonsteroidal anti-inflammatory drugs Drugs other than steroids that reduce pain, fever, and inflammation.
Prostatectomy Surgical removal of all (radical) or part of the prostate gland.
Prostate specific antigen (PSA) A protein (enzyme) produced by the cells of the prostate gland; functions to liquefy semen after ejaculation.
PSA velocity The rate of increase of PSA levels in the blood.
Randomized controlled trial Scientific procedure most commonly used in testing medicines or medical procedures; considered the most reliable form of scientific evidence.
Sexually transmitted infection Illness caused by an infectious agent that has a significant probability of transmission between humans by means of sexual contact.
Transurethral resection Commonly called a TURP, this is a surgical procedure performed when the tube (urethra) from the bladder to the penis is blocked by prostate enlargement.
Tumor An abnormal growth or mass of tissue also called a neoplasm. A tumor can be either malignant of benign.
Watchful waiting Also called active surveillance; refers to observation and regular monitoring without invasive treatment.
X-linked The X chromosome is larger than the Y chromosome and has many more genes. X-linked inherited diseases occur far more frequently in males than in females because they only have one X chromosome.

Introduction

Prostate cancer is the most commonly diagnosed male malignancy in developed countries. With age, the total prevalence of prostate tumors reaches very high levels. Only a few per cent of these become clinically significant with the potential to kill, and the challenge described by Boccon-Gibod (1996) is to distinguish tigers from pussycats; to identify the minority of lethal cancers from the majority of nonaggressive tumors. Answers to this challenge remain elusive. We do not know what causes lethal prostate cancer; little can be done to prevent it, and, until it is possible to tell the tigers from the pussycats, this is not likely to change.

Anatomy, Physiology, and Pathology

The prostate is a walnut-sized gland located at the base of the pelvis beneath the urinary bladder, surrounding the urethra (**Figure 1**). Its initial development is stimulated by testosterone from the fetal testes but it remains incompletely developed until puberty, when it grows to its adult

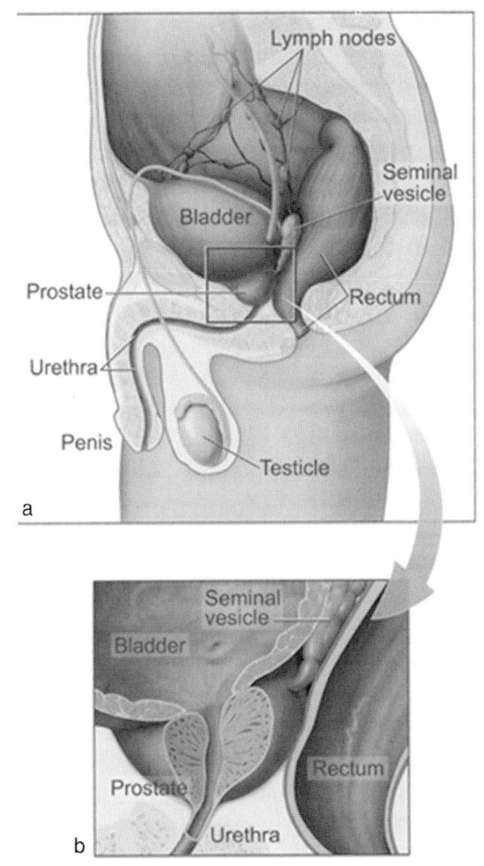

Figure 1 (a) The prostate and nearby organs. (b) The inside of the prostate, urethra, rectum and bladder. Source: http://www.cancer.gov/cancertopics/wyntk/prostate/allpages#ab3d4f20-6ab9-4428-9717-067035d2e691.htm.

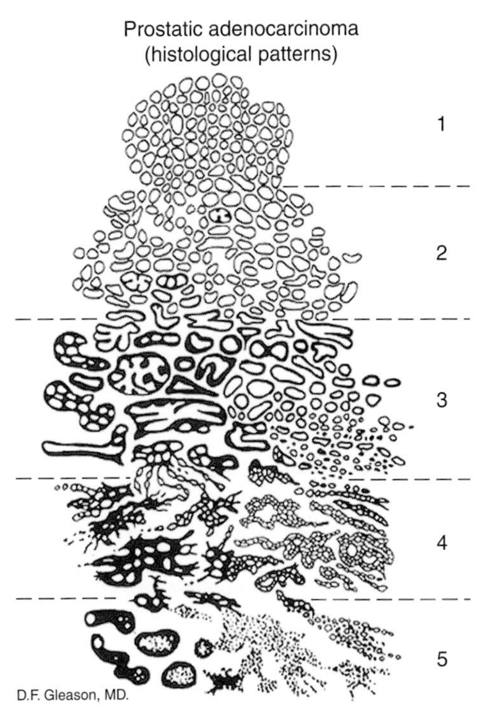

Figure 2 The Gleason grading system. From Gleason DF (1977) The Veteran's Administration Cooperative Urologic Research Group: Histologic grading and clinical staging of prostatic carcinoma. In: Tannenbaum M (ed.) *Urologic Pathology: The Prostate.* Philadelphia: Lea and Febiger.

size of about 20 g around age 20 years. The prostate develops 30–50 branched ducts that are lined with glandular epithelium and open into the prostatic urethra. Its function is to contribute up to 30% of the volume of semen with slightly alkaline secretions including prostaglandins, proteolytic enzymes, acid phosphatase, zinc, and citric acid, which help maintain sperm viability.

McNeal (1981) first identified the prostate's zonal anatomy. The peripheral zone, where 60–70% of tumors occur, contains about 70% of the total volume and is accessible via digital rectal examination (DRE). The central zone, containing 25% of the volume and the ejaculatory ducts, is where inflammatory processes such as prostatitis arise. The transitional zone, where 25% of tumors occur, contains only 5% of the volume. The anterior zone is fibromuscular.

As the hormonal milieu changes with middle age, the prostate grows, a condition termed benign prostatic hyperplasia (BPH). Due to the inelasticity of the prostate's fibrous outer capsule, increasing pressure is placed on the prostatic urethra, producing urinary symptoms. Depending on severity, these problems come to medical attention and in this context cancer is often diagnosed. BPH was once considered a risk factor for prostate cancer, but this view is no longer held; both conditions simply occur commonly in the aging prostate gland. A small proportion of cancers are found in tissue fragments from transurethral resections performed to relieve obstructive symptoms of BPH.

Virtually all prostatic tumors are adenocarcinomas arising from the glandular epithelium and most men will develop them if they live long enough. Although much research has sought tumor markers that would separate the tigers from the pussycats, this is still beyond reach. A tumor's potential lethality (aggressiveness) is currently based on its degree of spread beyond the capsule and microscopic inspection by a pathologist, who identifies the two largest tumor foci (primary and secondary patterns) and grades each from 1 to 5 according to the histologic features described by Gleason (Gleason, 1977) (see **Figure 2**).

Gleason grades for the two foci, which are rarely more than one grade apart, are added to give a sum from 2 to 10. Tumors with Gleason sums below 5 are not considered aggressive. Historically, tumors with Gleason sums 5–7 were considered low risk but tumors with a Gleason score of 7 have been added to the high-grade aggressive category (formerly Gleason scores 8–10). The significance of Gleason sum 7 tumors remains controversial, especially with respect to decisions about treatment. Approximately

30% of Gleason sum 7 tumors have a primary focus of grade 4 and these are considered to be more aggressive than those having a primary focus of grade 3. Some clinicians consider that any focus of Gleason grade 4 warrants clinical suspicion.

Diagnosis, Screening, and Treatment

Prostate cancer has no specific symptoms until advanced. Its size and the presence of any lumps are assessed by DRE and ultrasound; the diagnosis of cancer is aided by measurement of the serum levels of prostate specific antigen (PSA), a protease enzyme contributed by the prostate to seminal fluid, some of which permeates into the bloodstream. Originally used clinically to monitor cancer progression after treatment, PSA is now used for early detection. High serum PSA levels almost certainly indicate the presence of malignancy, but its use also leads to overdiagnosis, that is, the diagnosis of tumors that would never have progressed to clinically significant tumors during life, the so-called pussycats.

Suspicious PSA levels are followed up with ultrasound-guided needle biopsies to establish a histopathologic diagnosis. Prostate tumors are commonly small and multifocal, and there is an element of chance in whether a tumor can be adequately identified by biopsy. The number of biopsies used to establish a diagnosis has increased over time. The PSA threshold level used to prompt clinical investigation has also fallen, especially in the USA.

PSA testing for early prostate cancer is already being performed on an *ad hoc* or opportunistic basis in many countries. Randomized clinical trials (RCTs) are being conducted to test the efficacy of systematic PSA screening in reducing prostate cancer mortality in the whole population (Schroder 1994). Because many controls (men who were randomly assigned not to have a PSA test) in these RCTs have in fact received PSA tests as part of their normal community care, the power of the trials to provide definitive evidence of efficacy has been reduced, and this has lengthened the duration of the trials by some years (Gohagen *et al.*, 1994).

Screening, to be effective, requires a cheap, sensitive, and specific test, a good understanding of the natural history to identify a curable early form of the disease, and good evidence that treatment of this form of disease reduces mortality by comparison with treatment of disease diagnosed in the ordinary way at the onset of symptoms. Prostate cancer screening by PSA testing currently fails to satisfy these criteria.

Given the uncertainties that surround its biological heterogeneity and limited evidence of treatment benefits, prostate cancer management is complex. Men with low-grade, localized cancer and a low PSA may opt for watchful waiting, with regular repeat PSA tests to monitor changes, termed PSA velocity. Once PSA velocity reaches a certain point, decisions are made with respect to treatment. Radical treatments for localized disease are surgery (prostatectomy) or implantation with radioactive seeds to deliver localized radiation (brachytherapy). Advanced disease is treated by androgen deprivation/blockade, which can involve surgical or medical castration, and external beam radiation. Principal concerns are side effects, such as impotence, incontinence, and damage to the rectum and bladder neck, and the uncertainty posed by the limited evidence for substantial benefit of treatment.

Variation in Incidence and Mortality

Autopsy studies carried out by Yatani and others (1982) showed that the prevalence of microscopic, nonaggressive tumors (historically termed latent cancers) increased with age, affecting a majority of prostates by age 60 years, but varied little between populations, any variation in total incidence being due to differences in levels of invasive disease. Because of their high prevalence, the detection of nonaggressive cancers depends mainly on the intensity with which they are sought. The total incidence of prostate cancer is, therefore, highly sensitive to the medical-care setting.

Prior to PSA testing, a 30-fold international variation in prostate cancer incidence was observed, with high rates in Western countries, particularly the USA, and low rates in Asia. Although age-standardized incidence rates differed between countries, trends were stable for decades prior to PSA testing (see **Figure 3**). After PSA testing became widespread in the late 1980s, incidence more than doubled in some populations. This happened because nonaggressive tumors were in effect transferred from the pool of previously undiagnosed disease to tumors requiring clinical attention. The dramatic increase occurred earlier and rose

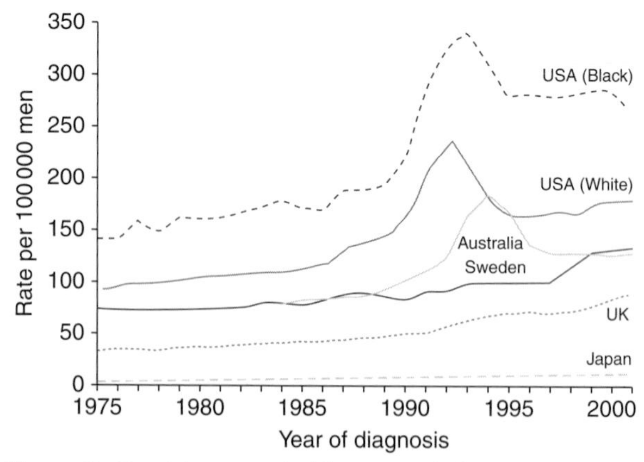

Figure 3 Prostate cancer incidence trends from a selection of countries to demonstrate the effect of the introduction of PSA testing.

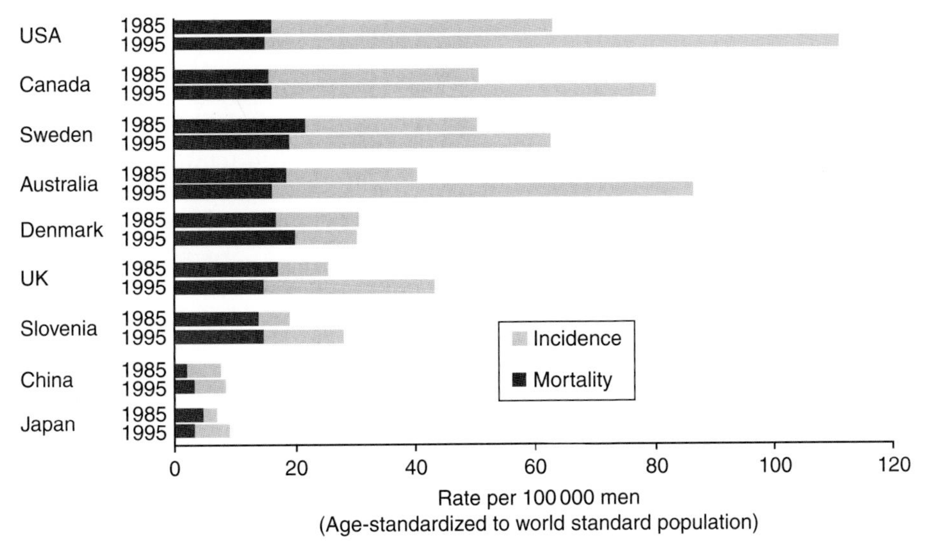

Figure 4 Age-standardized incidence and mortality rates for prostate cancer from selected countries prior to and after the widespread use of PSA testing.

higher in the USA than elsewhere. After peaking in 1992, incidence fell, but it has remained at a higher level than before. Countries where PSA testing is not common do not show this huge increase.

Figure 4 shows age-standardized prostate cancer incidence and mortality rates for a selection of countries centered on 1985 and 1995, before and after PSA testing. The remarkable features are the huge differences in incidence and minor differences in mortality. PSA testing was slow to emerge in the UK and is rare in Denmark, and this is reflected in the rates.

Against a long-term trend of slowly increasing prostate cancer mortality, some countries have experienced a small decrease that has sparked debate about the contribution of PSA testing. It is impossible, however, to distinguish possible early detection effects from simultaneous improvements in treating advanced cancers. Mortality has also fallen in countries without PSA testing and this argues against the impact of PSA testing on mortality. Such opportunistic testing is unlikely to have reduced mortality so quickly unless the lead time (advancement of diagnosis by the test) were extremely short, which is improbable. We will have to wait for randomized controlled trials (RCTs) to resolve this question.

Total prostate cancer survival has improved since the introduction of PSA testing, due mainly to the diagnosis of an increased proportion of nonaggressive tumors. Relative survival estimates at 5 years are virtually 100% for localized cancer and are around 33% for cancer with distant metastases.

Prostate cancer shows strong ethnic variation, with African-Americans, especially those of Jamaican descent, having the highest reported rates of incidence and mortality in the USA. It had long been considered that native Africans were at low risk for prostate cancer but recent surveys suggest that the incidence for thoroughly screened African populations may be as high as for African-Americans.

Japanese men have historically experienced a low incidence of prostate cancer, but this changed among Japanese migrants to the USA. Incidence rates for first-generation Hawaiian Japanese rose above those for Japan and the rates for the second generation rose even further. It is impossible to ascribe this to specific changes in environment because little is known about the environmental causes of prostate cancer. The fact that mortality rates also increased in the migrants argues that at least some of the change might be due to environmental factors.

Risk Factors for Prostate Cancer

Apart from age, family history, and race, few risk factors have been established for prostate cancer, especially for its lethal forms. It is thus not currently possible to prevent prostate cancer. Given the amount of research conducted to identify modifiable risk factors, this is disappointing. One problem has been the biological heterogeneity of prostate tumors. Another has been the dominance of reports in the literature from small case-control studies that, due to their retrospective design, inherent biases and poor statistical power have produced inconsistent findings concerning a large number of possible risk factors.

In the late twentieth century, the first reports began to appear from large cohort studies. Unfortunately, the prostate cancers in these studies have been diagnosed during the PSA era, and the majority of tumors are nonaggressive. Without rigorous efforts to identify risk factors for aggressive cancer and distinguish tigers from pussycats, progress in our understanding of prostate cancer will continue to be slow.

Family History and Genetic Factors

Having a first-degree relative with prostate cancer incurs a twofold to threefold risk and this increases with the number of relatives affected and earlier age at diagnosis. These observations are consistent with an inherited genetic predisposition. Analyses of families with multiple cases of prostate cancer have inferred dominant patterns of inheritance, as well as recessive and X-linked patterns. It is therefore considered that mutations in more than one gene might increase susceptibility to developing prostate cancer. Several loci have been identified, for example, on chromosomes 1, 17, 20, and X, but reports have been difficult to replicate. Putative prostate cancer susceptibility genes have also been reported, for example, *ELAC2, RNASEL*, and *MSR1*, but none has been generally accepted. The *BRCA2* gene associated with increased risk of breast cancer for women is a possible exception, as male *BRCA2* mutation carriers have increased risk of early-onset prostate cancer.

A region on chromosome 8q found in Icelandic families has been replicated in two other studies and may contribute to increased risk for African-Americans (Amundadottir *et al.*, 2006). The risk associated with this region appears regardless of severity, so the putative gene is likely to be associated with early events in carcinogenesis.

The search for prostate cancer genes may have been thwarted by tumor heterogeneity, since different genes may be involved in susceptibility to aggressive and nonaggressive tumors. There is evidence that men with a family history are more likely to have PSA testing and this will lead to increased diagnosis of nonaggressive tumors in families and make the search for susceptibility genes more difficult.

Association studies have been carried out for common polymorphisms in candidate genes known to be in pathways important to prostate development and function. Candidates have been selected from pathways such as steroid hormone metabolism, DNA repair, insulin, and insulin-like growth factors (IGFs) and response to infection and inflammation. For example, much has been published concerning genetic variants in the androgen receptor, 5-alpha reductase type 2, the vitamin D receptor, IGF-1, interleukins, toll-like receptors, and a number of cytochrome P450 enzymes. This approach has had a similar lack of success to the search for susceptibility genes. Seldom have initial reports generated from small, poorly designed studies been replicated by studies with greater statistical power. It is increasingly recognized that genetic association studies need to be substantially larger than previously thought. Some very large studies are under way, but an equivalent emphasis on tumor aggressiveness is also required if knowledge is to be advanced.

Hormones

Androgens are required for normal growth and function of the prostate. Testosterone and its active form dihydrotestosterone (DHT) bind to the androgen receptor and result in increased transcription and cell division. Some argue that androgens are permissive factors for prostate cancer, in that their presence is essential for other carcinogenic factors to play their part. Androgen deprivation in almost all forms leads to involution of the prostate, a fall in PSA levels, and apoptosis of prostate cancer and epithelial cells. Prostate cancer is termed hormone-dependent because, when advanced, it is initially responsive to androgens and can be controlled by surgical or chemical castration. Men with congenital 5-alpha reductase deficiency cannot convert testosterone to DHT and do not develop a normal prostate. It is often said that eunuchs do not develop prostate cancer but the evidence for this observation is slight.

It has long been reasoned that the hormone dependency of advanced disease might also influence incidence, the androgen hypothesis. Many case-control studies tested this hypothesis, with contradictory results. Early cohort study analyses supported the androgen hypothesis, because circulating levels of testosterone were associated with prostate cancer risk, while those of sex-hormone binding globulin (SHBG) were protective.

Early analyses failed to take into account tumor heterogeneity, combining all cancers as if they were the same, the majority of those being nonaggressive. Later studies that have stratified the analysis on Gleason score or tumor grade have reported that circulating levels of androgens are associated with increased risk of nonaggressive tumors but a reduced risk of aggressive tumors. This is consistent with the known function of androgens in maintaining prostate cellular differentiation.

Clearly, the relationship between androgens and prostate carcinogenesis is complex. Putting their role in the genesis of nonaggressive tumors to one side, the question remains of how and at what stage androgens might lose their protective influence on cellular differentiation and promote aberrant proliferation. A pertinent observation is that prostate cancer increases dramatically with age from the fifth decade, a time linked with the androcline and changes in hormonal milieu toward falling androgen and rising estrogen levels.

There is evidence that IGF-1 and its binding protein, IGFBP-3, are associated with risk, but findings from cohort studies are contradictory. A review of early studies reported a positive association with IGF-1 and a negative association with IGFBP-3 levels. Larger studies later found little evidence of association with IGF-1 and some have found positive associations with IGFBP-3, in direct contrast with the earlier observations. The latter, if real, may help explain a link with obesity. IGFBP-3 levels are reported to decrease with increasing levels of physical activity and are potentially modifiable.

Sexual Activity

The androgen hypothesis has also influenced research about sexual behaviors and markers of androgen influence, for example, age at starting shaving, sexual activity and marriage, the number of sexual partners including prostitutes, and the frequency of sexually transmissible infections (STIs). With the exception of STIs, these reports have lacked consistency. Some studies also report risk associated with diminished opportunity for sexual activity, for example, increased prostate cancer mortality for Roman Catholic priests, somewhat analogous to breast cancer in nuns. The case-control literature on these topics has its usual flaws but additionally lacks a standard approach to measuring sexual activity.

A large case-control study and a prospective cohort study using a standard questionnaire on ejaculatory frequency by decade of life reported similar findings, namely that men having high ejaculatory frequency, especially in early adulthood, reduced their risk by about one-third. The significance of the observation is not clear; hypotheses range from considerations of prostatic duct hygiene to more subtle hormonal feedback relationships between the prostate, testes, and the pituitary gland. The contrasting age-related relationships between androgen levels, ejaculatory frequency, and prostate cancer incidence may offer some insight.

Infection and Inflammation

There is general support for the hypothesis that infection might be causally related to prostate cancer. Self-reports of STIs are unreliable, however, and few studies have attempted to measure evidence of infection directly. Further, many common STIs in men are asymptomatic or insufficiently bothersome to come to medical attention and it is not possible to identify culpable organisms for a substantial proportion of those that do.

The findings concerning human papilloma virus infection are essentially null, an earlier positive association failing to be replicated by later studies. Studies have also failed to demonstrate any consistent association with herpes simplex viruses. There is limited serologic evidence that syphilis infection is associated with risk. On the other hand, a protective effect of serum positivity to *Chlamydia trachomatis* infection has been reported, consistent across different serotypes and demonstrating dose–response relationships. This raises the possibility that some STIs might be able to increase the prostate's immune response.

The idea that one consequence of infection, chronic inflammation, might contribute to prostate carcinogenesis is gaining ground. Inflammation in prostate specimens is common and foci of proliferative inflammatory atrophy (PIA) are commonly observed in the peripheral zone. Some consider PIA to be a precancerous lesion or a marker of an intraprostatic environment that is favorable to cancer development, because it is often found adjacent to areas of both intraepithelial neoplasia and cancer. Although not fully consistent, a number of studies have shown a protective effect of aspirin and nonsteroidal anti-inflammatory drugs on prostate cancer risk, particularly for advanced disease. Further evidence for the possible causal significance of inflammation arises from reports that variants in genes involved in the inflammatory response to infection, for example, the interleukin and toll-like receptor signaling pathways, are associated with risk. Intriguingly, some putative susceptibility genes, for example, *MSR1* and *RNASEL*, have roles in this pathway.

Dietary and Nutritional Factors

A large number of inconsistent dietary associations have been reported from ecological correlation studies and case-control studies. The measurement error associated with dietary assessment coupled with the biases inherent in case-control studies provides a strong rationale for focusing attention on reports from well-conducted cohort studies. Cohort studies (mostly North American) have reported positive associations with saturated fat and/or meat, dairy products (calcium) and zinc, and negative associations with tomato-based foods (lycopene), vitamin D, vitamin E, selenium, fish (omega-3 fatty acids), soy, isoflavones, and polyphenols. Cohort study findings are also not fully consistent, possibly because the majority of cancers in these studies are PSA-detected and advanced tumors are rare. Although there is a trend toward analysis by tumor grade or stage, statistical power is usually a limiting factor.

It is clear from cohort studies that fruit and vegetables are not associated with risk (Key *et al.*, 2004). Although a protective role for phytoestrogens has been promoted, there is little evidence that men who consume foods containing significant amounts are at lower risk. The evidence regarding cruciferous vegetable consumption is also poor. No effect of cruciferous vegetables was observed in either the European Prospective Investigation of Cancer (EPIC) or the Health Professionals Follow-Up Study (HPFS). EPIC had no information on tumor stage but in the HPFS there was evidence of a protective effect for disease confined to the prostate in men aged less than 65 years (Giovannucci *et al.*, 2003).

Although the data from different studies are not entirely consistent, tomato-based foods (especially tomato sauce), and their principal carotenoid, lycopene, may be protective and, more so, against advanced disease. Lycopene is concentrated in the prostate and has antioxidant ability twice that of beta-carotene and ten times that of vitamin E. Associations with other carotenoids are weaker than for lycopene but are similar in direction.

Micronutrients of interest for chemoprevention have been identified incidentally from RCTs having other principal outcomes. The Alpha Tocopherol Beta Carotene (ATBC) trial in male smokers showed an unexpected deficit in prostate cancer in the alpha tocopherol (vitamin E) arm but no effect of beta carotene (Albanes *et al.*, 1996). There is little evidence that vitamin E within the normal dietary range is protective. Similarly, a trial of selenium for skin cancer prevention also showed a deficit of prostate cancer. These agents, both of which have antioxidant properties, are now subject to RCTs. A concern for the RCTs is the outcome, which is unlikely to be aggressive cancer but a screen-detected endpoint such as elevated PSA or microscopic disease. The ATBC trial's other unexpected outcome was the increased risk of lung cancer in the beta carotene arm, a salutary example to those who would promote the supplementary consumption of micronutrients in high doses. It is matched by the finding that men who take high doses of zinc supplements put themselves at increased risk of prostate cancer.

Meat, particularly red meat, remains under suspicion, but it has been difficult to separate an effect of meat from that of fat, for which meat-based foods provide a major source. Some hold the view that the association with red meat might be related to carcinogenic by-products of cooking methods, particularly char grilling. There is little available evidence, and similar associations have not been observed with other grilled meats such as chicken or fish. Fish consumption appears to be protective against advanced prostate cancer, possibly as a rich source of marine fatty acids with known anti-inflammatory properties that might be relevant to prostate cancer prevention.

Cohort studies have yielded mixed results concerning fat, but those that do report associations have indicated them more consistently with advanced prostate cancer. Analyses by type of fat and for specific fatty acids have produced similarly inconsistent results, the strongest association being reported with alpha-linolenic acid. Although the evidence is inconclusive, several hypotheses have been advanced to explain an association with fat, including an effect on serum androgen levels and the possibility that certain types of fatty acids or their metabolites may initiate or promote carcinogenesis.

Dairy foods have been suspected for their fat and calcium content. It has been hypothesized that high calcium intake, possibly by lowering $1,25(OH)_2$ vitamin D levels, is associated with poorer differentiation in prostate cancer and thereby with fatal prostate cancer. Rigorous reviews and meta-analyses have concluded that the risk associated with dairy foods, if any, is very small.

Vitamin D is also manufactured by the skin when exposed to sunlight. Vitamin D is known to promote differentiation and impede proliferation of prostate cells and its deficiency is, therefore, hypothesized to decrease risk, but epidemiological studies are inconclusive.

Overweight and obesity are products of imbalance between energy intake and expenditure by way of physical activity. The evidence that body size and composition are associated with the risk of prostate cancer is inconsistent. Some inconsistencies are due to variations in design, power, and physical measurements, many studies relying on self-report rather than direct measurement. A comprehensive meta-analysis of the associations reported from cohort studies concluded that body mass index was weakly associated with prostate cancer risk (particularly for advanced tumors). Few studies have examined central obesity and prostate cancer risk, though there is growing recognition of its possible importance. One cohort study that made direct body measurements has shown waist circumference to be positively associated with the risk of aggressive disease.

Several prospective cohort studies have found that obese men are more likely to die from prostate cancer. Also, among men with prostate cancer, obese men are more likely to experience biochemical progression (rising PSA levels). It has been postulated that obesity might differentially influence the development of disease, reducing the risk of nonaggressive disease, while increasing the risk of aggressive disease.

Given the positive associations observed between measures of obesity and prostate cancer, it would be expected that physical activity would be protective against prostate cancer. Reviews have concluded that physical activity may be inversely associated with prostate cancer risk but the evidence is inconsistent, the magnitude of the risk reduction observed is small, and these reductions could be attributed to the studies' methodological limitations.

Other Factors

Alcohol

Reviews generally conclude that there is no association between low to moderate alcohol consumption and prostate cancer, but cannot exclude the possibility of an association with heavy drinking. There is a single case-control study report of a protective effect of red wine consumption, which, if real, may be related to the antioxidant properties of red wine pigments such as resveratrol.

Tobacco

Overall, epidemiological studies do not support an association between smoking and the risk of developing prostate cancer. There is evidence, though, that smoking is associated with increased prostate cancer mortality.

Radiation

The prostate is one of the few organs in which cancer is not associated with ionizing radiation, even at high doses. There is weak evidence of a negative association with ultraviolet light exposure, possibly via vitamin D synthesis.

Occupation

Although a large literature exists on occupational associations with prostate cancer, very little has been firmly established. The risks that have been observed may be the result of uncontrolled confounding with social class. In the PSA era, the degree of detection bias in occupational studies has increased further, because men who believe that they have been exposed to a possible carcinogen tend to seek PSA tests more frequently and are more likely to have a diagnosis made than colleagues.

Early studies focused on occupational exposures to cadmium. As the prostate concentrates zinc, it was considered that cadmium could replace zinc and be procarcinogenic, but follow-up studies of cadmium smelter and battery production workers failed to confirm early suggestions of increased risk.

Dioxin, a contaminant of a herbicide used in Viet Nam, is similar to many components of herbicides used in farming. A review of the linkage between dioxin and prostate cancer risk by the National Academy of Sciences reported that the available data on the association between dioxin exposure and prostate cancer was inconclusive. A number of studies have looked at the risks associated with farming, particularly related pesticide and herbicide exposures, but most have been ecologic and have failed to measure exposures at the individual level or to control for confounders.

Vasectomy

Studies have been inconsistent with respect to possible risks of prostate cancer following vasectomy. Reviews have concluded that virtually all studies had been deficient in their handling of bias that could account for the modest risks observed in the positive studies.

Conclusion

Prostate cancer has changed irrevocably since the widespread use of the PSA test for early detection. In many countries, incidence continues to increase, while in others early detection has yet to begin. Although many men now diagnosed with prostate cancer would have died with the disease rather than as a result of it, their diagnosis is just as meaningful in terms of public health services as would have been a diagnosis based on symptoms. The man-made epidemic of prostate cancer will continue to impact hugely on the use of health services not only with respect to initial diagnosis and treatment, but also in relation to the management of the long-term side effects of treatment, with their enormous potential to diminish quality of life.

There is little prospect of change until advances are made in the molecular taxonomy of these tumors, so that the tigers can be discriminated from the pussycats. Without such progress, the epidemic of prostate cancer will continue unabated.

See also: Cancer Screening; Male Reproductive Function.

Citations

Albanes D, Heinonen OP, Taylor PR, *et al.* (1996) Alpha-tocopherol and beta-carotene supplements and lung cancer incidence in the alpha-tocopherol, beta-carotene cancer prevention study: Effects of base-line characteristics and study compliance. *Journal of the National Cancer Institute* 88: 1560–1570.

Amundadottir LT, Sulem P, Gudmundsson J, *et al.* (2006) A common variant associated with prostate cancer in European and African populations. *Nature Genetics* 38: 652–658.

Boccon-Gibod L (1996) Significant versus insignificant prostate cancer – Can we identify the tigers from the pussy cats? *Journal of Urology* 156: 1069–1070.

Giovannucci E, Rimm EB, Liu Y, Stampfer MJ, and Willett WC (2003) A prospective study of cruciferous vegetables and prostate cancer. *Cancer Epidemiology, Biomarkers and Prevention* 12: 1403–1409.

Gleason DF (1977) The Veteran's Administration Cooperative Urologic Research Group: Histologic grading and clinical staging of prostatic carcinoma. In: Tannenbaum M (ed.) *Urologic Pathology: The Prostate.* Philadelphia, PA: Lea and Febiger.

Gohagan JK, Prorok PC, Kramer BS, and Cornett JE (1994) Prostate cancer screening in the prostate, lung, colorectal and ovarian cancer screening trial of the National Cancer Institute. *Journal of Urology* 152: 1905–1909.

Key TJ, Allen N, Appleby P, *et al.* (2004) Fruits and vegetables and prostate cancer: No association among 1104 cases in a prospective study of 130, 544 men in the European Prospective Investigation into Cancer and Nutrition (EPIC). *International Journal of Cancer* 109: 119–124.

McNeal JE (1981) The zonal anatomy of the prostate. *The Prostate* 2: 35–49.

Schroder FH (1994) The European Screening Study for Prostate Cancer. *Canadian Journal of Oncology* 4 (supplement 1): 102–105.

Yatani R, Chigusa I, Akazaki K, *et al.* (1982) Geographic pathology of latent prostatic carcinoma. *International Journal of Cancer* 29: 611–616.

Further Reading

Bostwick DG, Burke HB, Djakiew D, *et al.* (2004) Human prostate cancer risk factors. *Cancer* 101 (supplement 10): 2371–2490.

Boyle P, Severi G, and Giles GG (2003) The epidemiology of prostate cancer. *Urologic Clinics of North America* M30(2): 209–217.

Donn AS and Muir CS (1985) Prostatic cancer: Some epidemiological features. *Bulletin Cancer* 72: 381–390.

Friedenreich CM and Thune I (2001) A review of physical activity and prostate cancer risk. *Cancer Causes and Control* 12: 461–475.

Friedrich MJ (1999) Issues in prostate cancer screening. *Journal of the American Medical Association* 281: 1573–1575.

Godley PA (1999) Prostate cancer screening: Promise and peril – A review. *Cancer Detection and Prevention* 23: 316–324.

Platz EA (2002) Energy imbalance and prostate cancer. *Journal of Nutrition* 132 (supplement 11): 3471S–3481S.

Platz EA and Giovannucci E (2004) The epidemiology of sex steroid hormones and their signaling and metabolic pathways

in the etiology of prostate cancer. *Journal of Steroid Biochemistry and Molecular Biology* 92: 237–253.
Platz EA, De Marzo AM, and Giovannucci E (2004) Prostate cancer association studies: Pitfalls and solutions to cancer misclassification in the PSA era. *Journal of Cell Biochemistry* 91: 553–571.
Ross RK, Pike MC, and Coetzee GA (1998) Androgen metabolism and prostate cancer: Establishing a model of genetic susceptibility. *Cancer Research* 58: 4497–4504.
Schaid DJ (2004) The complex genetic epidemiology of prostate cancer. *Human Molecular Genetics* 13: R103–R121.
Shimizu H, Ross RK, Bernstein L, *et al.* (1991) Cancers of the prostate and breast among Japanese and white immigrants in Los Angeles County. *British Journal of Cancer* 63: 963–966.
Whittemore AS (1994) Prostate cancer. *Cancer Surveys* 19–20: 309–322.

Testicular Cancer

E Huyghe, Hôpital Ranguel, Toulouse, France

Introduction

Testicular tumors are germ cell tumors (GCT) in most cases (95% of the cases). All these tumors have a common origin and are usually grouped by the generic name 'testicular cancer (TC).' TC is a relatively rare cancer, but in young men it is the most frequent solid tumor. Its incidence has been increasing in almost all industrialized countries (Huyghe *et al.*, 2003). As a consequence, research is aimed to identify risk factors and underlying conditions involved in the dramatic increase of TC incidence. Age, ethnicity, country of origin, cohort of birth, and family history are clearly associated with TC incidence. Although evidence is not fully consistent, environmental factors may play a role in the development of this disease (Fenner-Crisp, 2000; Fisher, 2004). TC is characterized by an important improvement in survival and has become a model of a curable cancer, even for patients with advanced disease (Aareleid *et al.*, 1998; Germa-Lluch *et al.*, 2002). However, a group of poor prognosis patients still persists, in which new therapeutic protocols are currently investigated (Hendry, *et al.*, 2002). For the majority of patients, with the availability of effective treatment, attention has been turned to reduction of treatment morbidity (Joly *et al.*, 2002; Huyghe *et al.*, 2004). TC is often diagnosed several months after the onset of the first symptom, and therefore, measures for promotion of early diagnosis are investigated (Post and Belis, 1980; Moul *et al.*, 1990). In this chapter, I summarize these various aspects of TC pathology.

Histology

Histologically, two groups of TC must be distinguished (Mostofi and Sesterhenn, 1985; Mostofi *et al.*, 1994):

- germ cell tumors (GCT), which account for 94% to 97% of cases, arise from intratubular neoplasia (**Figure 1**) and are divided into the seminoma and the nonseminomatous germ-cell tumors (NSGCT) (**Table 1**)
- rare tumors that arise from nongerminal elements (Leydig cells, Sertoli cells, etc.).

In this chapter, I focus on GCT that represent the main public health concern.

Seminoma

Seminoma is the most common GCT type (accounting for 30–50% of GCT). At presentation, the testis is enlarged without loss of normal shape. On macroscopic specimen examination, the tumor is usually solitary with distinct borders. It is pink-white, smooth, and usually homogeneous. Microscopically, seminoma cells are uniform and have clear cytoplasm and well-defined cell borders. Age at presentation is generally several years older than for NSGCT tumors (30–40 years vs. 20–35 years). Three subtypes have been described: typical (85% of the cases), anaplastic (5% to 10%), and spermatocytic (2% to 10%). Initially, the anaplastic variety was thought to carry a worse prognosis than the typical variety; however, currently this difference is debated.

Spermatocytic seminoma usually presents later in the patient's life. Its prognosis is excellent and cases of metastasis are extremely rare.

Nonseminomatous Germ Cell Tumors

Embryonal carcinoma is an aggressive tumor that tends to metastasize early. Embryonal carcinoma is often associated with other cell types in the metastatic sites. Pure embryonal carcinoma represents 3–6% of GCT.

Choriocarcinoma is another aggressive tumor with a high potential to metastasize (lungs). Even a small intratesticular lesion may present with evidence of advanced distant metastases. Pure choriocarcinoma is extremely rare (1% of GCT), and it is often a component of a mixed tumor.

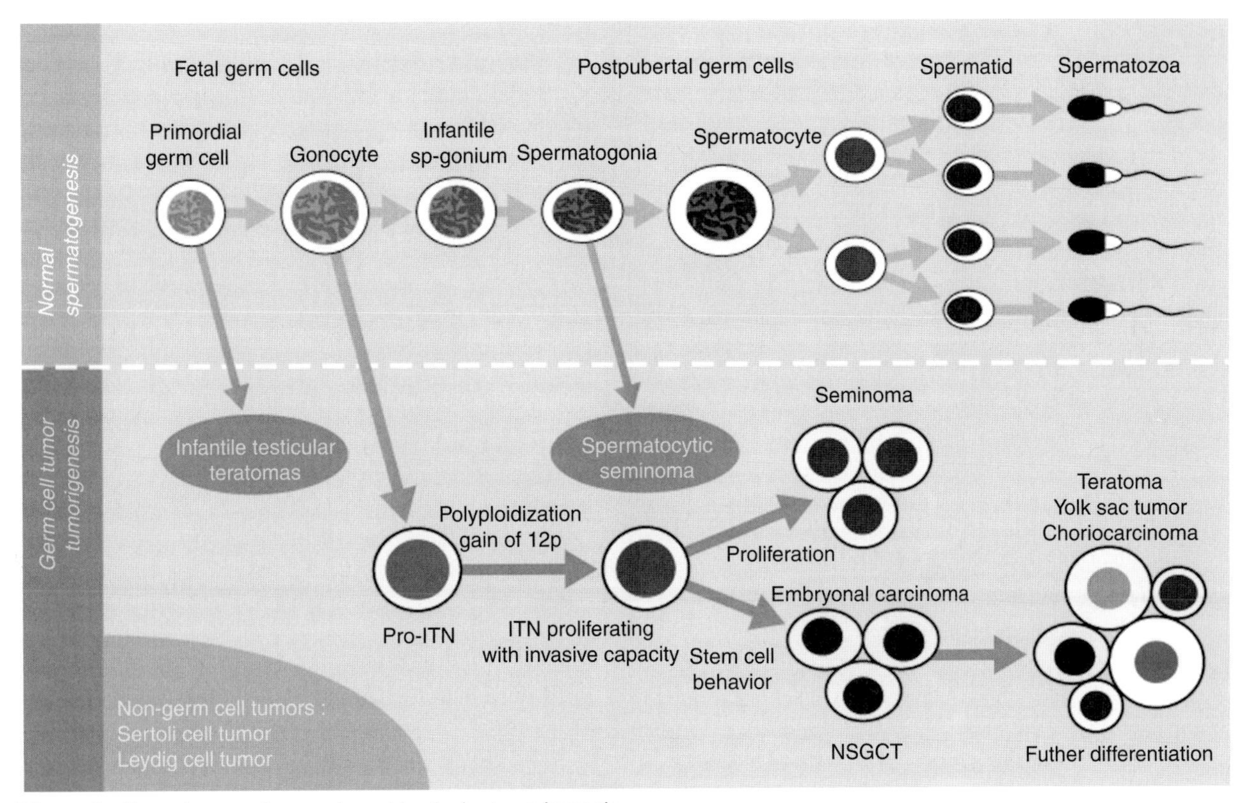

Figure 1 Normal spermatogenesis and testicular tumorigenesis.

Table 1 World Health Organization classification of testis cancer

Precursor tumors	*Intratubular germ cell neoplasia (CIS)*
Tumors of one histologic type (pure form)	Seminoma Spermatocytic seminoma Embryonal carcinoma Yolk sac tumor Polyembryoma Trophoblastic tumor (choriocarcinoma) Teratoma (mature, dermoid cyst, immature)
Tumors of more than one histologic type (mixed forms)	Specify types and estimate percentage

Teratoma contains more than one germ cell layer in various stages of maturation and differentiation. Mature elements seem like benign structures derived from normal ectoderm, endoderm, and mesoderm; however, in adults, mature teratoma must be considered as malignant because it has the ability to dedifferentiate into more aggressive forms. Immature teratoma consists of undifferentiated primitive tissues from each of the three germ cell layers.

Yolk sac tumor is the most common testis tumor of infants and children. In adults, it occurs almost exclusively in combination with other histological types.

Tumors of more than one histological type are considered a separate entity and are also known as mixed GCT; they make up approximately 60% of all GCTs. Components of embryonal carcinoma and seminoma are very common in mixed GCT.

Incidence

Approximately 52 000 new cases were reported worldwide in 2008. Wide geographical discrepancies exist between countries: the rates in developed countries are about six times higher than those in developing countries (Huyghe *et al.*, 2003). Age-standardized incidence rate ranges from less than 1/100 000 in Asia and Africa to more than 10/100 000 in Denmark (**Figure 2**) (Purdue *et al.*, 2005).

Trends in Incidence

Since the Second World War, TC incidence has been increasing in nearly all industrialized countries, especially in Europe, Northern America, and Oceania (Adami *et al.*, 1994; Moller *et al.*, 1995; McGlynn *et al.*, 2003; Richiardi *et al.*, 2004). The worldwide incidence rate of testicular cancer has doubled over the last 50 years. In Europe, strong differences exist between countries (**Figure 2**) and a gradient in the increase has been observed, the

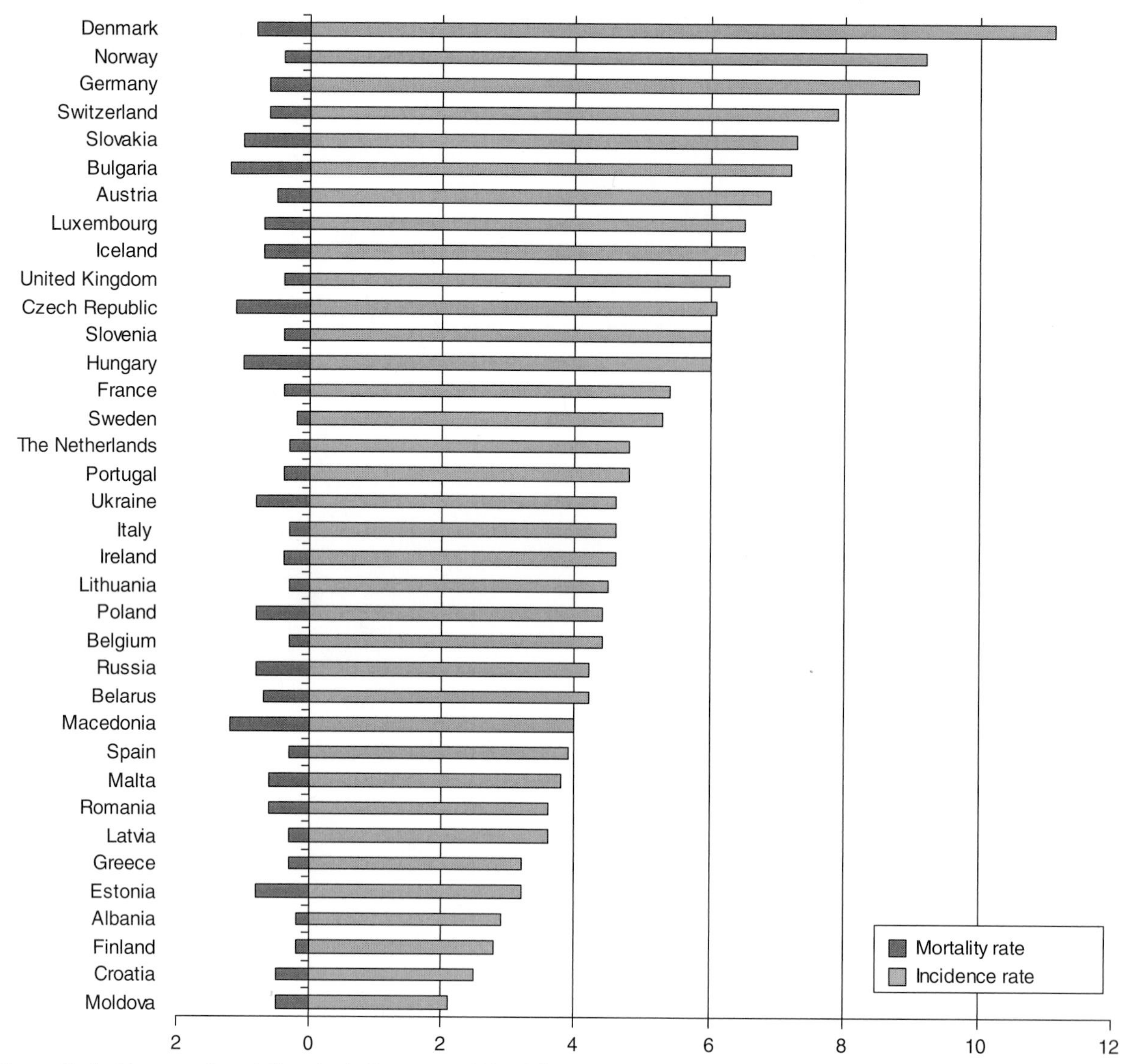

Figure 2 Incidence and mortality rates in Europe per 100 000 (data 1995–2000).

highest rate in Europe is centered in Denmark, Norway and Germany, and decreases progressively in a centrifugal way (Huyghe *et al.*, 2006). Recent data may indicate a trend to stabilization in the incidence rate in Denmark, Switzerland, and the USA (Levi *et al.*, 1990; Pharris-Ciurej *et al.*, 1999; Moller, 2001).

Birth Cohort Effect

A clear birth cohort effect (that is, the incidence rate varies with the cohort of birth) has been identified in TC (Bergstrom *et al.*, 1996; Zheng *et al.*, 1996). In Europe, during the last century, the risk of testicular cancer has been increasing regularly with successive birth cohorts; an interruption occurred in this rise in countries such as those in Scandinavia during the Second World War, followed by a new increase after the War (Moller, 1989; Bergstrom *et al.*, 1996). These epidemiological findings have been interpreted to support the hypotheses that TC may be secondary to an early exposure (maybe as early as embryonic or fetal life) to etiologic factors and that these factors may be present in the environment.

Age

Testicular cancer occurs typically in young adults (20 to 35 years), and it is the most frequent malignancy in this age group in several countries. So far, the risk of developing TC is predominant in young age groups (15–35 years) (Moller *et al.*, 1995). Overall, the highest incidence is noted in young adult males, making these neoplasms the most common solid tumors of men age 20 to 34 years and the second most common of men age 35 to

40 years. Seminoma is rare before age 10 years and after age 60 years; peak incidence occurs at 35 years. Embryonal carcinoma occurs predominantly between 25 and 35 years, choriocarcinoma more often between 20 and 30 years. Pure yolk sac tumors and teratoma are predominant lesions of childhood but frequently appear in combination with other elements in adulthood.

Racial Factors

The incidence of TC is highest in Europeans and European-derived populations, with age-standardized rates of between 2 to 10 per 100 000 World Standard Population (International Agency for Research on Cancer, 2002; Purdue *et al.*, 2005). In non-European populations the incidence rates are usually below 2 per 100 000. The Maori population of New Zealand is an exception, with an incidence rate of 7 per 100 000 (Wilkinson *et al.*, 1992). Several examples of variation in incidence among ethnic groups come from multi-ethnic countries: the incidence of TC in American blacks is approximately one-third that in American whites (but 10 times that in African blacks) (McGlynn *et al.*, 2005). In Israel, Jewish people have at least an eight-fold higher incidence than the non-Jewish people (Parkin and Iscovich, 1997). In Hawaii, the incidence in the Filipino and Japanese population is one-tenth that in the white population.

Occupation and Sociodemographic Characteristics

The level of evidence of a link between sociodemographic data and TC incidence rate is low, even if higher incidence rates have been noted in the upper and middle socioeconomic classes (Swerdlow *et al.*, 1991).

Significant relationship with the occurrence of TC has been found in men exposed to various environmental or occupational conditions, such as leather tanners, airframe repairers, policemen, gas and petroleum workers, carpet and textile workers, paper, plastic, and metal workers, and pesticide-exposed farmers. However, no definite conclusion can be ascertained, due to the long list of potential causative agents. Regarding the link between occupation of parents and the occurrence of TC in children, TC has been found to be associated with paternal occupations (metal workers, food and beverage service workers, painters and printing workers, food product workers, wood processors). Interestingly, no relationship was shown with maternal occupations (Pearce *et al.*, 1987; Swerdlow *et al.*, 1991; Van den Eeden *et al.*, 1991; Moller, 1997; Pollan *et al.*, 2001).

Laterality

Testicular cancer appears to be slightly more common in the right testis than in the left, similar to the slightly greater incidence of right-sided cryptorchidism. Approximately 2 to 3% of testicular tumors are bilateral, occurring either simultaneously or successively (Sokal *et al.*, 1980). Similar, rather than different, histology in the two testes predominates with bilateral tumors. A history of cryptorchidism (unilateral or bilateral) in nearly half these men is consistent with observations that bilateral dysgenesis occurs frequently in unilateral maldescent. Long-term surveillance of patients with a history of cryptorchidism or previous orchiectomy for GCT is mandatory.

Risk Factors

Cryptorchidism

About 7 to 12% of patients with TC have a prior history of cryptorchidism (the absence of one or both testes from the scrotum) (Schottenfeld *et al.*, 1980; Boyle and Zaridze, 1993). The exact incidence of cryptorchidism is unknown because of difficulties in defining cryptorchidism. Scorer estimated that approximately 4.3% of neonates, 0.8% of infants and children, and 0.7% of adults older than 18 years harbor a truly cryptorchid testis (Scorer, 1979). In patients with cryptorchidism, there is an estimated 2 to 32-fold increased risk of TC (Schottenfeld *et al.*, 1980; Herrinton *et al.*, 2003). Between 5% and 10% of patients with a history of cryptorchidism develop malignancy in the controlateral (normally descended) gonad (Berthelsen *et al.*, 1982). The basis for the relationship between cryptorchidism and testicular cancer remains unclear; it is hypothesized that both are different manifestations of a common condition that may result from genetic, lifestyle, and/or environmental factors. Orchiopexy that consists of descending surgically an undescended testis in scrotal position is mandatory in order to preserve fertility (for this purpose, it must be realized early in childhood, preferably before 3 years of age) and to facilitate clinical surveillance of these patients; however, there is no clear evidence that orchiopexy prevents TC (Strader *et al.*, 1988; Jones *et al.*, 1991; Cortes *et al.*, 2001; Herrinton *et al.*, 2003; Pettersson *et al.*, 2007).

Hormones

Changes in the testosterone–estrogen balance may contribute to the development of testicular tumor.

Several prenatal risk factors for germ cell cancer may be related with estrogenic disturbances (Moller and Skakkebaek, 1997; Wanderas *et al.*, 1998; Richiardi *et al.*, 2003):

- Low birth weight due to intrauterine growth retardation
- Low parity of mother
- Children born preterm
- High maternal age of the first-born boy.

Relative risk for testicular tumor in the sons of diethylstilbestrol-treated mothers ranges from 2.8 to 5.3% (Schottenfeld *et al.*, 1980).

Genetic Factors

Although a relatively higher incidence of testicular tumors has been reported in twins, brothers, and family members, genetic causative factors have not been clearly identified. Standardized incidence ratios for testicular cancer are 3.8 and 7.6 when a father and a brother have testicular cancer, respectively (Hemminki and Chen, 2006). However, the high familial risk may be the product of heritable causes or it may also result from shared environmental conditions. Nicholson and Harland (1995) reported that one-third of all testis cancer patients are genetically predisposed to disease, probably a homozygous (recessive) inheritance of a single predisposing gene. Certain genetic conditions are frequently observed in TC (alteration of the sex chromosome isochromosome of 12(i12p), intersex conditions) (Peltomaki, 1991; Peltomaki *et al.*, 1992; Lothe *et al.*, 1995; Bianchi *et al.*, 2002; Smiraglia *et al.*, 2002).

Diet

The number of studies examining diet and TC risk is still limited. Intake of milk, dairy products, and cheese has been found to be associated with TC risk by several authors. Cheese and dairy products contain a high level of fat, calcium, and proteins, and they also contain large amounts of the female hormones estrogen and progesterone. The impact of calcium or meat consumption is still debated. Other potential dietary TC risk factors are high saturated, animal, and high total fat intakes and increased intake of baked products (that include milk, sugar, and eggs). The critical periods could be prenatal and perinatal periods, but childhood and adolescence are also pointed out. Relatively late exposures have also been suspected: an association between a high fat consumption one year before diagnosis in one study and high dairy product intake two years before diagnosis in another study were associated with an increased risk of TC (Davies *et al.*, 1990, 1996; Bonner *et al.*, 2002).

Etiology and Pathogenesis

Experimental and clinical evidence supports the importance of congenital factors in the etiology of GCTs. Existence of a birth cohort effect and data on migrants in Sweden, which show that the relative risk of developing testicular cancer is 0.34 in Finnish migrants to Sweden compared with the Swedish male population, strengthens the hypothesis of a pre- and postnatal environmental exposure (Hemminki *et al.*, 2002; Ekbom *et al.*, 2003). The temporal decline of male fertility in humans and wildlife is paralleled by an increase in endocrine-dependent pathologies in the male, such as hypospadias, male infertility, cryptorchidism, and testicular cancer, collectively called Skakkebaek testicular dysgenesis (Skakkebaek *et al.*, 1998, 2001). The 'estrogen hypothesis' has been put forward as a possible explanation for these trends. The hypothesis is that *in utero,* the primordial germ cell may be altered by environmental factors, mainly manufactured substances with estrogen-like action (xenoestrogens) resulting in disturbed differentiation (Skakkebaek, 2002). *In vitro* studies suggest that some heavy metals also have an estrogen-like action (Chan *et al.*, 1983; Ragan and Mast, 1990). The estrogen-like action seems not to be the only mechanism of action of endocrine disruptors, however. Some endocrine disruptors are anti-androgenic, such as phthalates (Fisher, 2004). Even if a large body of animal experiments supports the endocrine disruption hypothesis, fewer data are available from human studies (Virtanen *et al.*, 2005; Bay *et al.*, 2006; Frederiksen *et al.*, 2007).

Diagnostic Delay

Patients who present with advanced disease (stage III) generally have a much poorer prognosis than do those with disease confined to the testis or those with regional nodal involvement only. Diagnostic delay, which is defined as the time elapsing from the onset of tumor symptoms to the diagnosis of the tumor, has a prognostic value (Huyghe *et al.*, 2007). Diagnostic delay up to six months is not uncommon and seems to be related to survival. It may be correlated to patient factors such as ignorance, denial, and fear, as well as physician factors such as misdiagnosis. As almost half of patients continue to present with metastatic disease, the need clearly exists for population education through campaigns of information on TC symptoms and natural history and programs such as those advocating testicular self-examination (TSE). Only through widespread public health techniques will the knowledge of testicular tumors be promulgated so that diagnosis can occur earlier. Persistent physician-related delay of treatment pleads for continuing education and awareness campaigns. TSE is recommended by the American Medical Association, the American Urological Association, and the American Cancer Society as a regular monthly practice, arguing that TSE is the best way of detecting early TC, especially in men at high risk of developing TC. On the contrary, the United States Preventive Services Task Force and the National Cancer Institute consider that TC screening by TSE would not result in an appreciable decrease in mortality but would lead to numerous unnecessary diagnostic procedures. Although TSE is inexpensive and relatively easy to teach, its acceptance by young males and regular practice remains low in the USA and Europe (Sheley *et al.*, 1991; Singer *et al.*, 1993; Brenner *et al.*, 2003).

Signs and Symptoms

The usual presentation is a lump, painless swelling, or hardness of the testis (60%). Description of an enlargement by patients with a previously small atrophic testis is not uncommon. In approximately 10% of patients, acute pain is the presenting symptom and, in 30%, patients may complain of sensation of fullness or heaviness in the inguinal region or scrotum. In a minority of patients (10%), infertility may be the presenting complaint (Raman *et al.*, 2005). On rare occasions (5–10%), patients continue to present with manifestations due to metastases such as a neck mass (supraclavicular lymph node metastasis); respiratory symptoms, such as cough or dyspnea (pulmonary metastasis); gastrointestinal disturbances, such as anorexia, nausea, vomiting, or hemorrhage (retroduodenal metastasis); lumbar back pain (bulky retroperitoneal disease involving the psoas muscle or nerve roots); bone pain (skeletal metastasis); central and peripheral nervous system manifestations (cerebral, spinal cord, or peripheral root involvement); or unilateral or bilateral lower-extremity swelling (iliac or caval venous obstruction or thrombosis). Finally, gynecomastia due to endocrine manifestation of these neoplasms is seen in a majority of Leydig cell tumors and about 5% of patients with testicular GCTs (Gabrilove and Furukawa, 1984; Mellor and McCutchan, 1989).

Physical Examination

Any firm, hard, or fixed area within the substance of the tunica albuginea should be considered suspicious until proved otherwise. In general, seminoma tends to expand within the testis as a painless, rubbery enlargement. Embryonal carcinoma may produce an irregular, rather than discrete, mass, although this distinction is not always easily appreciated.

Testicular tumors tend to remain ovoid (egg-shaped), being limited by the tough investing tunica albuginea. In 10 to 15% of patients, spread to the epididymis or cord may occur. A hydrocele may be present and increase the difficulty of appreciation of a testicular neoplasm. Ultrasonography of the scrotum is a rapid, reliable technique to exclude hydrocele or epididymitis and should be used if there is any suspicion of testicular tumor.

Natural History

All GCT in adults should be regarded as malignant. The only GCT that may be considered as benign is infantile teratoma. Generally, NSGCT are considered as more aggressive than seminoma, and at the time of diagnosis, approximately half of patients with NSGCT present with disseminated disease. Moreover, the growth rate of NSGCT tends to be high. Doubling times range from 10 to 30 days. A total of 85% of patients dying from NSGCT do so within 2 years and the majority of the remainder within 3 years (Aass *et al.*, 1990; Price *et al.*, 1990).

Seminoma has a slower course: approximately 75% of seminomas are confined to the testis at the time of clinical presentation, which represents a greater percentage than that among patients with NSGCT. Because of this indolent course, seminoma may recur from 2 to 10 years after apparently successful initial management, and a long follow-up is mandatory (Borge *et al.*, 1988; Chung *et al.*, 2002).

Staging

The American Joint Committee on Cancer (AJCC) staging for GCTs conceived a tumor, nodes, metastasis, and serum marker staging (TNMS) system (**Tables 2** and **3**).

Serum markers used in the AJCC staging are serum human chorionic gonadotropin (HCG), α-fetoprotein (AFP), and lactate dehydrogenase (LDH).

Treatment

'Radical' orchiectomy, consisting of removal of the whole testis by inguinal approach, is the first step of treatment, except in case of bilateral tumor or tumor occurring in a solitary testis that may be treated conservatively. Orchiectomy provides histologic diagnosis of local staging (pT) and local control of the neoplasm. Because more than half of patients with testicular tumors present with metastatic disease, further treatment after orchiectomy is usual. Modalities of treatment (surgery, radiotherapy, chemotherapy) depend on histology and staging. Not only multimodal therapy, but also the accuracy of clinical staging and the ability to recognize failure early are of first importance for treatment efficacy.

Prognosis

Seminoma

In stage I seminoma, the 3-year and the 5-year survival rates after treatment are 99% and 95%, respectively. In stages II and III, the 5-year survival rates range from 80% to 90%.

Nonseminomatous Germ Cell Tumor

The cure rate for patients with pathologically confirmed stage I NSGCT is roughly 95% with surgery alone. Patients with metastatic NSGCT disease classified as good or intermediate prognosis do well with standard chemotherapy, with response rates of 80–90%. Patients with advanced

Table 2 Testicular cancer staging system of the American Joint Committee on Cancer and the International Union Against Cancer

Primary tumor (T)

pTX: Primary tumor cannot be assessed (if no radical orchiectomy has been performed, TX is used)
pT0: No evidence of primary tumor (e.g., histologic scar in the testis)
pTis: Intratubular germ cell neoplasia (carcinoma in situ)
pT1: Tumor limited to the testis or epididymis and no vascular/lymphatic invasion
pT2: Tumor limited to the testis or epididymis with vascular/lymphatic invasion or tumor extended through the tunica albuginea with involvement of tunica vaginalis
pT3: Tumor invades the spermatic cord with or without vascular/lymphatic invasion
pT4: Tumor invades the scrotum with or without vascular/lymphatic invasion

Regional lymph nodes (N)

NX: Regional lymph nodes cannot be assessed
N0: No regional lymph node metastasis
N1: Lymph node mass 2 cm or less in greatest dimension or multiple lymph node masses, none more than 2 cm in greatest dimension
N2: Lymph node mass more than 2 cm but not more than 5 cm in greatest dimension or multiple lymph node masses, any one mass greater than 2 cm but not more than 5 cm in greatest dimension
N3: Lymph node mass more than 5 cm in greatest dimension

Distant metastases (M)

M0: No evidence of distant metastases
M1: Nonregional or pulmonary metastases
M2: Nonpulmonary visceral metastases

Serum tumor markers (S)	*LDH*	*HCG (mIU/mL)*	*AFP (ng/mL)*
S0	N	N	N
S1	< 1.5 X N	< 5000	< 1000
S2	1.5 – 10 X N	5000–50 000	1000–10 000
S3	> 10 X N	> 50 000	> 10 000

Table 3 Stage grouping for testis cancer according to the AJCC staging system

Stage grouping	*T*	*N*	*M*	*S*
Stage 0	pTis	N0	M0	S0
Stage I	T1–T4	N0	M0	SX
Ia	T1	N0	M0	S0
Ib	T2	N0	M0	S0
	T3	N0	M0	S0
	T4	N0	M0	S0
Ic	Any T	N0	M0	S1–S3
Stage II	Any T	Any N	M0	SX
IIa	Any T	N1	M0	S0
	Any T	N1	M0	S1
IIb	Any T	N2	M0	S0
	Any T	N2	M0	S1
IIc	Any T	N3	M0	S0
	Any T	N3	M0	S1
Stage III	Any T	Any N	M1	SX
IIIa	Any T	Any N	M1	S0
	Any T	Any N	M1	S1
IIIb	Any T	Any N	M0	S2
	Any T	Any N	M1	S2
IIIc	Any T	Any N	M0	S3
	Any T	Any N	M1a	S3
	Any T	Any N	M1b	Any S

disease classified as poor prognosis, however, have only a 50–60% therapeutic response. Therefore, more aggressive chemotherapy should be used in this population.

Trends in Mortality

Comparison of specific death rates from testis cancer revealed that in Eastern Europe, mortality decreased slower than in Western Europe, and that the decline in mortality occurred later (only since the late 1980s). In the USA, as in Western Europe, mortality has fallen by about 70% since the 1970s as a result of the introduction of modern cisplatin-based chemotherapies.

See also: Male Reproductive Function.

Citations

Aareleid T, Sant M, and Hédelin G (1998) Improved survival for patients with testicular cancer in Europe since 1978. EUROCARE Working Group. *European Journal of Cancer* 34(14 Spec. No): 2236–2240.
Aass N, Fossa SD, Ous S, Stenwig AE, Lien HH, Paus E, and Kaalhus O (1990) Prognosis in patients with metastatic non-seminomatous testicular cancer. *Radiotherapy and Oncology* 17: 285–292.

Adami HO, Bergstrom R, Möhner M, *et al.* (1994) Testicular cancer in nine northern European countries. *International Journal of Cancer* 59: 33–38.

Bay K, Asklund C, Skakkebaek NE, and Andersson AM (2006) Testicular dysgenesis syndrome: possible role of endocrine disrupters. *Best Practices & Research Clinical Endocrinology & Metabolism* 20: 77–90.

Bergstrom R, Adami HO, Möhner M, *et al.* (1996) Increase in testicular cancer incidence in six European countries: A birth cohort phenomenon. *Journal of the National Cancer Institute* 88: 727–733.

Berthelsen JG, Skakkebaek NE, von der Maase H, Sørensen BL, and Mogensen P (1982) Screening for carcinoma in situ of the contralateral testis in patients with germinal testicular cancer. *British Medical Journal (Clinical Research Edition)* 285(6356): 1683–1686.

Bianchi NO, Richard SM, Peltomäki P, and Bianchi MS (2002) Mosaic AZF deletions and susceptibility to testicular tumors. *Mutatation Research* 503: 51–62.

Bonner MR, McCann SE, and Moysich KB (2002) Dietary factors and the risk of testicular cancer. *Nutrition and Cancer* 44: 35–43.

Borge N, Fossa SD, Ous S, Stenwig AE, and Lien HH (1988) Late recurrence of testicular cancer. *Journal of Clinical Oncology* 6: 1248–1253.

Boyle P and Zaridze DG (1993) Risk factors for prostate and testicular cancer. *European Journal of Cancer* 29A: 1048–1055.

Brenner JS, Hergenroeder AC, Kozinetz CA, and Kelder SH (2003) Teaching testicular self-examination: Education and practices in pediatric residents. *Pediatrics* 111: e239–e244.

Chan WY, Chung KW, Bates JM Jr, LeBlanc M, Tease LA, Griesmann GE, and Rennert OM (1983) Zinc metabolism in testicular feminization and surgical cryptorchid testes in rats. *Life Sciences* 32: 1279–1284.

Chung P, Parker C, Panzarella T, *et al.* (2002) Surveillance in stage I testicular seminoma – Risk of late relapse. *Canadian Journal of Urology* 9: 1637–1640.

Cortes D, Thorup JM, and Visfeldt J (2001) Cryptorchidism: Aspects of fertility and neoplasms. A study including data of 1,335 consecutive boys who underwent testicular biopsy simultaneously with surgery for cryptorchidism. *Hormone Research* 55: 21–27.

Davies TW, Prener A, and Engholm G (1990) Body size and cancer of the testis. *Acta Oncology* 29: 287–290.

Davies TW, Palmer CR, Ruja E, and Lipscombe JM (1996) Adolescent milk, dairy product and fruit consumption and testicular cancer. *British Journal of Cancer* 74: 657–660.

Ekbom A, Richiardi L, Akre O, Montgomery SM, and Sparén P (2003) Age at immigration and duration of stay in relation to risk for testicular cancer among Finnish immigrants in Sweden. *Journal of the National Cancer Institute* 95: 1238–1240.

Fenner-Crisp PA (2000) Endocrine modulators: Risk characterization and assessment. *Toxicology and Pathology* 28: 438–440.

Fisher JS (2004) Environmental anti-androgens and male reproductive health: Focus on phthalates and testicular dysgenesis syndrome. *Reproduction* 127: 305–315.

Frederiksen H, Skakkebaek NE, and Andersson AM (2007) Metabolism of phthalates in humans. *Molecular Nutrition and Food Research* 51: 899–911.

Gabrilove JL and Furukawa H (1984) Gynecomastia in association with a complex tumor of the testis secreting chorionic gonadotropin: Studies on the testicular venous effluent. *Journal of Urology* 131: 348–350.

Germa-Lluch JR, Garcia del Muro X, Maroto P, *et al.* (2002) Clinical pattern and therapeutic results achieved in 1490 patients with germ-cell tumors of the testis: The experience of the Spanish Germ-Cell Cancer Group (GG). *European Urology* 42: 553–562; discussion 562–563.

Hemminki K and Chen B (2006) Familial risks in testicular cancer as aetiological clues. *International Journal of Andrology* 29: 205–210.

Hemminki K, Li X, and Czene K (2002) Cancer risks in first-generation immigrants to Sweden. *International Journal of Cancer* 99: 218–228.

Hendry WF, Norman AR, Dearnaley DP, Fisher C, Nicholls J, Huddart RA, and Horwich A (2002) Metastatic nonseminomatous germ cell tumors of the testis: Results of elective and salvage surgery for patients with residual retroperitoneal masses. *Cancer* 94: 1668–1676.

Herrinton LJ, Zhao W, and Husson G (2003) Management of cryptorchism and risk of testicular cancer. *American Journal of Epidemiology* 157: 602–605.

Huyghe E, Matsuda T, and Thonneau P (2003) Increasing incidence of testicular cancer worldwide: A review. *Journal of Urology* 170: 5–11.

Huyghe E, Matsuda T, Daudin M, *et al.* (2004) Fertility after testicular cancer treatments: Results of a large multicenter study. *Cancer* 100: 732–737.

Huyghe E, Plante P, and Thonneau PF (2006) Testicular cancer variations in time and space in Europe. *European Urolology* 51: 621–628.

Huyghe E, Muller A, Mieusset R, *et al.* (2007) Impact of diagnostic delay in testis cancer: Results of a large population-based study. *European Urolology* 52: 1710–1716.

International Agency for Research on Cancer (IARC) (2002) *Cancer Incidence in Five Continents* vol. 8. Lyon, France: IARC Scientific Publication.

Joly F, Héron JF, Kalusinski L, *et al.* (2002) Quality of life in long-term survivors of testicular cancer: A population-based case-control study. *Journal of Clinical Oncology* 20: 73–80.

Jones BJ, Thornhill JA, O'Donnell B, Kelly DG, Walsh A, Fennelly JJ, and Fitzpatrick JM (1991) Influence of prior orchiopexy on stage and prognosis of testicular cancer. *European Urology* 19: 201–203.

Levi F, Te VC, and La Vecchia C (1990) Changes in cancer incidence in the Swiss Canton of Vaud, 1978–87. *Annals of Oncology* 1: 293–297.

Lothe RA, Peltomäki P, Tommerup N, Fosså SD, Stenwig AE, Børresen AL, and Nesland JM (1995) Molecular genetic changes in human male germ cell tumors. *Laboratory Investigations* 73: 606–614.

McGlynn KA, Devesa SS, Sigurdson AJ, Brown LM, Tsao L, and Tarone RE (2003) Trends in the incidence of testicular germ cell tumors in the United States. *Cancer* 97: 63–70.

McGlynn KA, Devesa SS, Graubard BI, and Castle PE (2005) Increasing incidence of testicular germ cell tumors among black men in the United States. *Journal of Clinical Oncology* 23: 5757–5761.

Mellor SG and McCutchan JD (1989) Gynaecomastia and occult Leydig cell tumor of the testis. *British Journal of Urology* 63: 420–422.

Møller H (1989) Decreased testicular cancer risk in men born in wartime. *Journal of National Cancer Institute* 81: 1668–1669.

Møller H (1997) Work in agriculture, childhood residence, nitrate exposure, and testicular cancer risk: A case-control study in Denmark. *Cancer Epidemiology Biomarkers & Prevention* 6: 141–144.

Møller H (2001) Trends in incidence of testicular cancer and prostate cancer in Denmark. *Human Reproduction* 16: 1007–1011.

Møller H, Jørgensen N, and Forman D (1995) Trends in incidence of testicular cancer in boys and adolescent men. *International Journal of Cancer* 61: 761–764.

Møller H and Skakkebaek NE (1997) Testicular cancer and cryptorchidism in relation to prenatal factors: Case-control studies in Denmark. *Cancer Causes and Control* 8: 904–912.

Mostofi FK, Davis J Jr., and Rehm S (1994) *Tumours of the Testis.* Lyon, France: IARC Scientific Publication (111): 407–429

Mostofi FK and Sesterhenn IA (1985) Pathology of germ cell tumors of testes. *Progress in Clinical and Biological Research* 203: 1–34.

Moul JW, Paulson DF, Dodge RK, and Walther PJ (1990) Delay in diagnosis and survival in testicular cancer: Impact of effective therapy and changes during 18 years. *Journal of Urology* 143: 520–523.

Nicholson PW and Harland SJ (1995) Inheritance and testicular cancer. *British Journal of Cancer* 71: 421–426.

Parkin DM and Iscovich J (1997) Risk of cancer in migrants and their descendants in Israel: II. Carcinomas and germ-cell tumours. *International Journal of Cancer* 70: 654–660.

Pearce N, Sheppard RA, Howard JK, Fraser J, and Lilley BM (1987) Time trends and occupational differences in cancer of the testis in New Zealand. *Cancer* 59: 1677–1682.

Peltomäki P (1991) DNA methylation changes in human testicular cancer. *Biochimica et Biophysica Acta* 1096: 187–196.

Peltomäki P, Lothe RA, Børresen AL, Fosså SD, Brøgger A, and de la Chapelle A (1992) Chromosome 12 in human testicular cancer: Dosage changes and their parental origin. *Cancer Genetics and Cytogenetics* 64: 21–26.

Pettersson A, Richiardi L, Nordenskjold A, Kaijser M, and Akre O (2007) Age at surgery for undescended testis and risk of testicular cancer. *New England Journal of Medicine* 356: 1835–1841.
Pharris-Ciurej ND, Cook LS, and Weiss NS (1999) Incidence of testicular cancer in the United States: Has the epidemic begun to abate? *American Journal of Epidemiology* 150: 45–46.
Pollan M, Gustavsson P, and Cano MI (2001) Incidence of testicular cancer and occupation among Swedish men gainfully employed in 1970. *Annals of Epidemiology* 11: 554–562.
Post GJ and Belis JA (1980) Delayed presentation of testicular tumors. *Southern Medical Journal* 73: 33–35.
Price P, Hogan SJ, Bliss JM, and Horwich A (1990) The growth rate of metastatic nonseminomatous germ cell testicular tumours measured by marker production doubling time–II. Prognostic significance in patients treated by chemotherapy. *European Journal of Cancer* 26: 453–457.
Purdue MP, Devesa SS, Sigurdson AJ, and McGlynn KA (2005) International patterns and trends in testis cancer incidence. *International Journal of Cancer* 115: 822–827.
Ragan HA and Mast TJ (1990) Cadmium inhalation and male reproductive toxicity. *Review of Environmental Contamination and Toxicology* 114: 1–22.
Raman JD, Nobert CF, and Goldstein M (2005) Increased incidence of testicular cancer in men presenting with infertility and abnormal semen analysis. *Journal of Urology* 174: 1819–1822; discussion 1822.
Richiardi L, Askling J, Granath F, and Akre O (2003) Body size at birth and adulthood and the risk for germ-cell testicular cancer. *Cancer Epidemiology Biomarkers & Prevention* 12: 669–673.
Richiardi L, Bellocco R, Adami HO, *et al.* (2004) Testicular cancer incidence in eight northern European countries: Secular and recent trends. *Cancer Epidemiology Biomarkers & Prevention* 13: 2157–2166.
Schottenfeld D, Warshauer ME, Sherlock S, Zauber AG, Leder M, and Payne R (1980) The epidemiology of testicular cancer in young adults. *American Journal of Epidemiology* 112: 232–246.
Scorer CG (1979) Cryptorchidism: a renewed plea. *British Medical Journal* 1(6163): 616.
Sheley JP, Kinchen EW, Morgan DH, and Gordon DF (1991) Limited impact of testicular self-examination promotion. *Journal of Community Health* 16: 117–124.
Singer AJ, Tichler T, Orvieto R, Finestone A, and Moskovitz M (1993) Testicular carcinoma: A study of knowledge, awareness, and practice of testicular self-examination in male soldiers and military physicians. *Military Medicine* 158: 640–643.
Skakkebaek NE (2002) Endocrine disrupters and testicular dysgenesis syndrome. *Hormone Research* 57(Suppl 2): 43.
Skakkebaek NE, Rajpert-De Meyts E, Jørgensen N, *et al.* (1998) Germ cell cancer and disorders of spermatogenesis: An environmental connection? *Acta Pathologica Microbiologica et Immunologica Scandinavica* 106: 3–11; discussion 12.
Skakkebaek NE, Rajpert-De Meyts E, and Main KM (2001) Testicular dysgenesis syndrome: An increasingly common developmental disorder with environmental aspects. *Human Reproduction* 16: 972–978.
Smiraglia DJ, Szymanska J, Kraggerud SM, Lorhe RA, Peltomäki P, and Plass C (2002) Distinct epigenetic phenotypes in seminomatous and nonseminomatous testicular germ cell tumors. *Oncogene* 21: 3909–3916.
Sokal M, Peckham MJ, and Hendry WF (1980) Bilateral germ cell tumours of the testis. *British Journal of Urology* 52: 158–162.
Strader CH, Weiss NS, Daling JR, Karagas MR, and McKnight B (1988) Cryptorchism, orchiopexy, and the risk of testicular cancer. *American Journal of Epidemiology* 127: 1013–1018.
Swerdlow AJ, Douglas AJ, Huttly SR, and Smith PG (1991) Cancer of the testis, socioeconomic status, and occupation. *British Journal of Industrial Medicine* 48: 670–674.
van den Eeden SK, Weiss NS, Strader CH, and Daling JR (1991) Occupation and the occurrence of testicular cancer. *American Journal of Industrial Medicine* 19: 327–337.
Virtanen HE, Rajpert-De Meyts E, Main KM, Skakkebaek NE, and Toppari J (2005) Testicular dysgenesis syndrome and the development and occurrence of male reproductive disorders. *Toxicology and Applied Pharmacology* 207(2 Suppl): 501–505.
Wånderas EH, Grotmol T, Fosså SD, and Tretli S (1998) Maternal health and pre- and perinatal characteristics in the etiology of testicular cancer: A prospective population- and register-based study on Norwegian males born between 1967 and 1995. *Cancer Causes & Control* 9: 475–486.
Wilkinson TJ, Colls BM, and Schluter PJ (1992) Increased incidence of germ cell testicular cancer in New Zealand Maoris. *British Journal of Cancer* 65: 769–771.
Zheng T, Holford TR, Ma Z, Ward BA, Flannery J, and Boyle P (1996) Continuing increase in incidence of germ-cell testis cancer in young adults: Experience from Connecticut, USA, 1935–1992. *International Journal of Cancer* 65: 723–729.

Further Reading

Boisen KA, Kaleva M, Main KM, *et al.* (2004) Difference in prevalence of congenital cryptorchidism in infants between two Nordic countries. *Lancet* 363(9417): 1264–1269.
Bosl GJ and Motzer RJ (1997) Testicular germ-cell cancer. *New England Journal of Medicine* 337: 242–253.
Bray F, Richiardi L, Ekbom A, Pukkala E, Cuninkova M, and Møller H (2006) Trends in testicular cancer incidence and mortality in 22 European countries: Continuing increases in incidence and declines in mortality. *International Journal of Cancer* 118: 3099–3111.
Brenner H (2002) Long-term survival rates of cancer patients achieved by the end of the 20th century: A period analysis. *Lancet* 360(9340): 1131–1135.
Herrinton LJ, Zhao W, and Husson G (2003) Management of cryptorchism, risk of testicular cancer. *American Journal of Epidemiolology* 157: 602–605.
Horwich A, Shipley J, and Huddart R (2006) Testicular germ-cell cancer. *Lancet* 367(9512): 754–765.
Levi F, LaVecchia C, Boyle P, Lucchini F, and Negri E (2001) Western and eastern European trends in testicular cancer mortality. *Lancet* 357: 1853–1854.
Logothetis CJ (1988) Adjuvant chemotherapy for testicular cancer. *New England Journal of Medicine* 319: 53.
Oliver RT (2001) Trends in testicular cancer. *Lancet* 358(9284): 841.
Oliver RT, Mason MD, Mead GM, *et al.* (2005) Radiotherapy versus single-dose carboplatin in adjuvant treatment of stage I seminoma: A randomised trial. *Lancet* 366(9482): 293–300.
Wein AJ, Kavoussi LR, Novick AC, Partin AW, and Peters CA (2007) *Campbell – Walsh Urology*, 9th edn. Philadelphia, PA: Saunders, Elsevier.

SECTION 4

SELECTED ETHICAL AND OTHER GENERAL ASPECTS

Reproductive Ethics: New Reproductive Technologies

R A Ankeny, The University of Adelaide, Adelaide, South Australia, Australia

Introduction

Assisted human reproductive technologies (ARTs) have become increasingly common in the mid to late twentieth century in most developed countries. Technological and theoretical advances together with pharmaceutical developments and improved practical techniques taken from gynecology, genetics, urology, and associated medical specialties have combined to enable treatment for individuals or couples experiencing infertility due to a wide range of causes. The ART market is predicted to grow as it provides a possible solution to public health problems associated with declining birth rates in many countries, due both to increasing rates of unexplained infertility and lower fertility associated with delayed marriage and reproduction reflecting changes in work and lifestyle patterns. Although originally ARTs were available primarily in developed countries, as costs have decreased, some of the less complex technologies are now available in less technologically developed countries, particularly those with cultures that highly value biologically related children. However, ARTs bring with them a number of controversial ethical issues, including who should be permitted to access these technologies and for what indications, whether or not they are considered a regular part of medical care and hence covered by governmental or private insurance, and if they represent an effective public health strategy.

General Ethical Issues Associated with ARTs

There are a number of religious arguments typically raised against ARTs. Some Christian traditions claim that all ARTs are ethically impermissible because they separate the natural process of procreation from intercourse within marriage, which is seen as unnatural (for discussion, see Cahill and McCormick, 1987; Fisher, 1989). ARTs are also viewed as a threat to the concept of the family and to the dignity of human beings, particularly inasmuch as technology dominates the origin of the human being. Thus, many of these critics argue both that reproduction should not be interfered with (e.g., by using contraception) and that technology should not be used to intervene in, or help to achieve, reproduction. Other traditions emphasize the unnaturalness of ARTs. Some Confucian commentators claim that any non-conjugal reproduction weakens the blood ties between family members and leads to moral and social instability (Qiu, 2002). Aside from those writing from a particular religious perspective, there are other critics who do not oppose (married heterosexual) couples making autonomous decisions about reproduction, such as using contraception, but who do not view ARTs or other unnatural arrangements as morally permissible (e.g., Marquis, 1989). Still others consider ARTs in relation to environmental ethics, arguing that infertile couples should adopt since the world is already overpopulated and that the desire to have a biologically related child is selfish (for a related argument, see McKibben, 1998). According to this reasoning, individual desires to have children should give way to broader concerns about population health and control.

Further ethical issues are raised by the selection criteria that are used to determine who receives ARTs. In many countries and clinics, these technologies are only made available for married heterosexual couples. Access to ARTs by unmarried (or *de facto*) couples and lesbian or single women is much more restricted, the latter being considered an instance of social rather than medical infertility. There have also been concerns about postmenopausal women using ARTs and donor eggs to conceive, and many clinics have rules limiting the provision of ARTs to women over a particular age (typically early to mid-forties) in part out of concerns for the welfare of the future child and relatively low success rates (Hope *et al.*, 1995). Finally, particularly in public health-care systems where ARTs are freely available, screening criteria such as age are used to choose the best candidates (those with most likelihood of success) due to limited resources, or limits are placed on the number of cycles that can be undergone to attempt to achieve a pregnancy.

In some jurisdictions (e.g., most states in the USA), there is no requirement that insurance companies provide coverage for infertility treatments, which means that only the most affluent can afford to use ARTs. Where public funds are directed at research or treatment using ARTs, there are economic concerns about whether it is just to use considerable health resources to help a relatively small number of people conceive in what are typically overstretched public health-care systems. Some argue that resources might be better utilized for research into and prevention of the various causes of infertility and related population-based problems including environmental issues, rather than individualized clinical solutions. Further, there is considerable disquiet about support for

ARTs in populations where evidence shows they are unlikely to be successful, for example older women.

Feminist scholars have expressed concerns about ARTs placing additional psychological, economic, and physical pressures on women to produce biologically related children (Sherwin, 1992; Donchin, 1996). They cite the problematic case of using ARTs to treat primary male infertility. This requires women to undergo onerous fertility treatments involving hyperstimulation of the ovaries and a series of surgical procedures despite the women themselves not being infertile. Further, they argue that infertility itself is a socially defined and interpreted category, rather than a natural disease category (Sherwin, 1992) and one that has been reinforced by a largely male-dominated medical profession. Most of these commentators do not deny that many women wish to have biologically related children, but they emphasize that the social and economic pressures associated with ARTs often are ignored. Others argue that ARTs have not been sufficiently well assessed, particularly with regard to their potential long-term negative effects on women's health, especially due to the side effects of hyperstimulation (de Melo-Martin, 1998).

Artificial Insemination

Some of the simpler advanced reproductive technologies that do not require the involvement of medical professionals were traditionally adapted from animal husbandry for use in human reproduction, notably artificial insemination by donor (AID) or by husband/partner (AIH). The introduction of semen or concentrated specimens of spermatozoa into a woman's reproductive tract by noncoital means can be successfully performed with instruments as simple as a turkey baster (Wikler and Wikler, 1991). AI is sometimes coupled with use of hormones to stimulate ovulation at the time of insemination to maximize the chances of fertilization occurring, although these drugs are associated with some risks to the women involved.

In recent years, fears about donor health status, risk of infection (HIV and otherwise), and legal issues (such as establishing paternity) have caused most AI to be performed in medical clinics under a physician's supervision. Hence some critics note that this procedure has become unduly medicalized. In the past, some doctors avoided paternity issues by mixing sperm from several donors including the male partner, but recent advances in genetic technologies allow paternity testing using DNA and have resulted in clarification in many jurisdictions of the legal standing of children born from AI. Legal issues remain in some places, for instance with custody and adoption of AID children born to lesbian couples.

Some religious traditions do not consider AID to be ethically permissible as they hold it to be equivalent to adultery. Historically, there has typically been considerable stigma associated with AID, and often information about the biological father (or even the fact that the child was produced using AID) was not revealed to children born using AID. However, in many countries recent changes in the law or court decisions have established that children born of AID have the right to information about their biological fathers once they are 18 years of age. This change is due in part to long-term psychosocial studies that have shown that there are considerable benefits to disclosure, as is the case with adopted children (Blyth, 1998; McWhinnie, 2000; Ethics Committee of the American Society of Reproductive Medicine, 2004). Following these changes, the rate of anonymous donation for AID programs has decreased dramatically in many countries, apparently because anonymity cannot be guaranteed once the child becomes an adult. In contrast, in the USA, some clinics have reported increased donor rates, apparently among those men who would be happy to have contact with their biologically related children in the future.

In Vitro Fertilization and Embryo Transfer

The first successful birth using in vitro fertilization and embryo transfer (IVF-ET) occurred in 1978. The British scientists involved, Patrick Steptoe and Robert Edwards, drew on embryological studies done for over 20 years in mice, rabbits, and other animals. The procedure involved laparoscopic aspiration of an egg during a woman's natural cycle, followed by IVF using ejaculated sperm and transfer of the dividing embryo in its early stages into the woman's uterus, hence creating what became known as a test tube baby. More generally in IVF-ET, eggs are harvested and mixed in Petri dishes either with donor sperm or with sperm from the male partner (if primary male infertility is not thought to be at issue), typically using the healthiest sperm to facilitate fertilization. Eggs may be obtained from the female partner being treated or donated by another woman (e.g., in cases of premature ovarian failure, genetic abnormalities, or reduced egg production due to advanced maternal age). Most women undergoing IVF-ET also have controlled ovarian hyperstimulation prior to aspiration of eggs to increase the number of eggs available for IVF. Early fears that babies produced through IVF would be abnormal have not been substantiated. However, in many localities there is inadequate tracking of offspring and their potential health problems, including their future reproductive health.

Depending on the clinic, the country, and the putative cause of infertility in the couple, different numbers of fertilized embryos are created and transferred. Improved methods recently have created higher success rates both in terms of the number of viable embryos that can be created as well as the likelihood of embryo implantation following transfer (the latter remains the major technological barrier to successful pregnancies via IVF). The result has been numerous cases of multiple gestations; however, due to the increased risks of premature birth, low birthweight and other abnormalities associated with multiple births, selective reduction (i.e., termination) of one or more of the fetuses may be offered to the parents, which present difficult decisions, particularly for those opposed to termination. Consequently, many clinics have adopted more conservative approaches to the number of embryos transferred at any one time, and there are now laws or professional guidelines in some countries to limit the number of embryos that can be transferred during any one IVF cycle.

There are social, ethical, and legal issues associated with the status and disposition of embryos that are surplus to the needs of the individuals pursuing IVF. Excess or supernumerary embryos can be cryopreserved at a very early stage of development for later IVF cycles; most clinics allow limited storage but require destruction of embryos after a set period of time unless the embryos are donated to other infertile couples or under certain circumstances donated for stem cell or fertility research. In all of these cases, issues arise concerning the moral status of the embryo; some people who believe that the human embryo has the moral standing of a child or even adult human being (i.e., it is a person deserving of protection and respect) tend to favor donation to other infertile couples. Others who do not believe that personhood starts in these early stages are more likely to consider donating embryos for research or disposing of them.

One of the major ethical issues that has arisen with IVF relates to how clinics report success rates. Previously, clinics reported success in terms of pregnancies established, whereas clients typically were interested in the likelihood of taking home a baby and hence had unrealistic expectations. Standards have recently been developed by professional societies to standardize how this information is conveyed to clients, though some would claim that considerable empty rhetoric still surrounds the IVF industry and that clients are not always well informed.

IVF-ET can now be combined with preimplantation genetic diagnosis (PGD) techniques, which permit testing of embryos for genetic diseases and chromosomal abnormalities to allow selection and transfer only of unaffected embryos. This technique was originally developed as an alternative to prenatal diagnosis (and possibly termination for fertile couples with known genetic risks). However, there are debates surrounding what types of tests using PGD should be offered, for instance whether embryos should be screened for only severe, early-onset disease conditions or if later-onset conditions or nondisease conditions should be included (Robertson, 2003). The combination of PGD and human leukocyte antigen (HLA) typing (or tissue typing) allows couples to have an unaffected child who can serve as a donor for an ill sibling who has a medical condition requiring a hematopoietic stem cell transplant (preferably from a well-matched donor). The creation of these so-called savior siblings is viewed by some as morally well justified, particularly when the parents were already planning to have additional children (Fost, 2004).

IVF-ET theoretically also could be used to implant a cloned embryo. Human embryo clones could be created by taking an egg from a woman, extracting the nucleus, and leaving behind only the cytoplasm. The nucleus of a somatic (body) cell (from another person or the same woman who donated the original egg) is then extracted and inserted into the cytoplasm. Under the correct culture conditions, the entity created will start to divide normally, behaving like a naturally created embryo (i.e., one produced by the union of egg and sperm). The resulting embryo would have the same genetic material as that of the donor of the somatic cell nucleus, instead of its genome being the usual blend of two parents; thus, this technique could allow potential parents to avoid passing on certain types of genetic conditions. Most commentators are opposed to human reproductive cloning, where the intention is to produce a cloned human embryo to implant in a woman in order to produce a pregnancy and eventually a child. As there are concerns about safety and long-term effects of cloning based on animal models, reproductive cloning is explicitly legally prohibited in several countries and is not known to be currently in clinical use.

Other ARTs

Additional types of ARTs have been developed in the last 20 years. Gamete intrafallopian transfer (GIFT) involves placement of eggs (which have been removed from the follicles) together with sperm directly into the oviducts for fertilization and is used with women with Fallopian tube problems. After its development in the mid-1980s, this technique became very popular because it did not require sophisticated IVF culture systems and could be done in clinics with less ART expertise and without a full IVF laboratory. It also seemed to produce better results than IVF, perhaps because fertilization occurs in a natural environment; greater success rates also are due to the fact that patients are generally fertile except for blockage of the Fallopian tubes. Zygote intrafallopian transfer (ZIFT) is a less common technique that involves transfer of the zygote (a fertilized egg that has not yet divided) into the oviduct after IVF. The ethical issues associated with GIFT and ZIFT are similar to those related to IVF-ET, although these procedures avoid the creation of embryos that might subsequently not be implanted.

Intracytoplasmic sperm injection (ICSI) is a micromanipulation technique in which pregnancy is achieved by injecting a single spermatozoon directly into the cytoplasm of the egg. It is used to enhance fertilization rates for men who have reduced sperm counts or impaired sperm motility, with banked sperm (obtained prior to chemotherapy or radiation), or with sperm obtained through electroejaculation (e.g., in those with spinal cord injuries or the newly dead, the latter being ethically and legally problematic). The technique also can be combined with those allowing separation of male and female sperm to avoid birth of children of a particular sex, for example to avoid sex-linked genetic disease or for sex selection, which is also ethically controversial. The practice of sex selection is outlawed in a number of countries because it is considered to be a morally problematic and discriminatory choice (or potentially because it results in gender imbalance in the broader population). It remains popular in cultures where having a child of a particular sex, or having a balanced family (i.e., some boys and some girls), is an important social norm.

Conclusion

Although ARTs have progressed rapidly over the past 30–40 years, there are still many limitations to their effectiveness. In addition, the ethical issues presented by ARTs remain controversial, particularly with regard to their potential impacts on women and their health, the appropriate use of selection criteria to determine who is eligible for treatment, and the disposition of excess embryos. In addition, there are specific public health implications that arise from the development of various ARTs outlined above. These include whether their use is appropriate in terms of resource allocation, especially since inability to reproduce might be viewed by many as less important than other, life-threatening health problems, and which selection criteria should be utilized, particularly given low success rates in some populations. In addition, concern exists about whether prevention of infertility, including identification and elimination of its causes, is being neglected, especially among certain minority populations, given the widespread availability of ARTs to the economically privileged. The dominant view that considers infertility as an individual problem, rather than as one which is associated with broader population-level social and medical issues, undoubtedly will continue to influence research and policy with regard to reproduction; however, it warrants more active and explicit debate.

See also: Gender Aspects of Sexual and Reproductive Health; Infertility; Reproductive Rights.

Citations

Blyth E (1998) Donor assisted conception and donor offspring rights to genetic origins information. *International Journal of Children's Rights* 6: 237–253.

Cahill LS and McCormick RA (1987) The Vatican document on bioethics. *America* 156: 246–248.

de Melo-Martin I (1998) Ethics and uncertainty: In vitro fertilization and risks to women's health. *Risk: Health Safety & Environment* 9: 201–227.

Donchin A (1996) Feminist critiques of new fertility technologies. *Journal of Medicine and Philosophy* 21: 476–497.

Ethics Committee of the American Society of Reproductive Medicine (2004) Informing offspring of their conception by donor gametes. *Fertility and Sterility* 81: 527–531.

Fisher A (1989) *IVF: The Critical Issues*. Melbourne, Australia: Collins Dove.

Fost NC (2004) Conception for donation. *Journal of the American Medical Association* 291: 2125–2126.

Hope T, Lockwood G, and Lockwood M (1995) Should older women be offered in vitro fertilisation? The interests of the potential child. *British Medical Journal* 310: 1455–1456.

Marquis D (1989) Why abortion is immoral. *Journal of Philosophy* 86: 183–202.

McKibben B (1998) *Maybe One: A Personal and Environmental Argument for Single-Child Families*. New York: Simon & Schuster.

McWhinnie A (2000) Children from assisted reproductive technology: The psychological issues and ethical dilemmas. *Child Development and Care* 163: 13–23.

Qiu R-Z (2002) Sociocultural dimensions of infertility and assisted reproduction in the Far East. In: Vayena E, Rowe PJ and Griffin PD (eds.) *Current Practices and Controversies in Assisted Reproduction: Report of a Meeting on Medical, Ethical and Social Aspects of Assisted Reproduction, 17–21 September 2001*. Geneva, Switzerland: World Health Organization. www.who.int/reproductive-health/infertility/12.pdf (accessed November 2007).

Robertson JA (2003) Extending preimplantation genetic diagnosis: The ethical debate. *Human Reproduction* 18: 465–471.

Sherwin S (1992) *No Longer Patient: Feminist Ethics and Health Care*. Philadelphia, PA: Temple University Press.

Wikler D and Wikler NJ (1991) Turkey-baster babies: The demedicalization of artificial insemination. *Milbank Quarterly* 69: 5–40.

Further Reading

Cohen CB (ed.) (1996) commissioned by the National Advisory Board on Ethics in Reproduction (1996) *New Ways of Making Babies: The Case of Egg Donation*. Bloomington, IN: Indiana University Press.

Dooley D and Dalla-Vorgia P (2003) *The Ethics of New Reproductive Technologies: Cases and Questions*. Oxford, UK: Berghahn Books.

Hildt E and Mieth D (eds.) (1998) *In Vitro Fertilisation in the 1990s*. Aldershot, UK: Ashgate.

National Research Council (US) (1989) *Medically Assisted Conception: An Agenda for Research*. Washington, DC: National Academy Press.

Robertson JA (1994) *Children of Choice: Freedom and the New Reproductive Technologies*. Princeton, NJ: Princeton University Press.

Royal Commission on New Reproductive Technologies (1993) *Proceed with Care*. Ottawa, Canada: Minister of Government Services.

Seibel MM and Crockin SL (eds.) (1996) *Family Building Through Egg and Sperm Donation: Medical, Legal, and Ethical Issues*. Boston, MA: Jones and Bartlett.

Seibel MM, Bernstein J, Kiessling AA, *et al.* (eds.) (1993) *Technology and Infertility: Clinical, Psychosocial, Legal and Ethical Aspects*. New York: Springer-Verlag.

Warnock M (1985) *A Question of Life: The Warnock Report on Human Fertilisation and Embryology*. Oxford, UK: Basil Blackwell.

Reproductive Ethics: Ethical Issues and Menopause

W Rogers, Flinders University, Adelaide, South Australia, Australia

Introduction

The cessation of menstruation is a universal event for women who live long enough to reach their middle years. Menopause is the name given to this event, defined retrospectively as the last menstrual period. Discussion about menopause and its treatment raises a number of issues relevant to public health ethics. These include whether menopause is a disease, the role of exogenous estrogens alone or combined with progestogens in the treatment of menopausal symptoms (known as hormone replacement therapy, or HRT), and more recently, the role of HRT as preventative treatment for a range of disorders. In particular, there has been controversy over the relationship between HRT and cardiovascular disease in postmenopausal women. Specific ethical issues include the implications of defining menopause as a disease and its subsequent medicalization, the quality of information available for women to make informed decisions about HRT, the emphasis on decreasing individual risk factors through medication, and broader questions of research ethics.

Menopause

As women age, there is a decrease in ovarian follicular activity, with a decline in estradiol and progesterone. This leads to a rise in follicle stimulating hormone (FSH) and luteinizing hormone (LH). Initially, women may experience irregular menstrual cycles with changes in the amount of blood loss. This menopausal transition is known as the perimenopause, which starts at a median age of 45.5–47.5 years in Western women. The average time to the last menstrual period from the start of the perimenopause is 4 years. During the perimenopause, there is increased frequency of anovulatory cycles, increased menstrual irregularities, and decreased fertility (O'Connor and Kovacs, 2003).

The symptoms experienced by women as they go through menopause vary considerably, with significant cultural and geographic differences in both the nature and intensity of symptoms. Many women have minimal symptoms. The classic symptoms of lowered levels of estrogen are vasomotor instability leading to hot flushes, palpitations, and night sweats; and genitourinary symptoms such as vaginal dryness and urinary frequency. Other symptoms that have been described include changes in general quality of life and psychological well-being associated with mood swings, anxiety, depression, memory and sleep disturbances, decreased energy and changes in libido, and a range of other symptoms such as muscle and joint aches, headaches, and sensory disturbances. These symptoms may last for several years, preceding and after the last menstrual period.

Following menopause, women have low levels of estradiol and progesterone, and increased levels of LH and FSH. This altered hormonal milieu is an inevitable part of aging; the relationship between these hormonal changes and diseases, such as osteoporosis and cardiovascular disease that become more common with increasing age in women, remains the subject of some controversy.

Culture and Menopause

The symptoms that women experience at menopause and the meanings that they attach to menopause are profoundly affected by prevailing sociocultural influences such as attitudes toward aging and older women, and their status in society. Factors thought to play a role in women's experiences of menopause include culturally influenced behaviors, such as diet, smoking, and exercise; cultural attitudes toward and expectations of menopause, including the influence of medicalization; meanings assigned to menopause such as whether it is seen as normal, deviant, or an illness; previous health, including sexual and reproductive health; mother's experience of menopause; relationship with and attitudes of husbands/partners; attitudes toward women and child rearing; social supports; socioeconomic status; education; career; and religious beliefs (Melby *et al.*, 2005). Lifestyle choices influenced by culture, such as smoking and diet, can modify the underlying biology of menopause, with associated changes in symptomatology.

The reported frequency and importance of symptoms varies enormously. For example, women in rural India described loss of vision as their primary symptom of menopause (Singh and Arora, 2005), whereas women of low socioeconomic status in the USA described mood swings, sleeping and memory problems, and lower energy levels more frequently than hot flushes, but despite this, felt positive about menopause (Schnatz *et al.*, 2005). Recent results from the Study of Women's Health across the Nation (SWAN), which is the latest and largest of epidemiological studies of menopause, following on from the Massachusetts Women's Health Study and the Healthy Women Study, show significant variations between

American women. American Caucasian women reported significantly more symptoms than other menopausal USA women; African American women had the highest rates of vasomotor symptoms; and Chinese American and Japanese American women reported significantly fewer symptoms than Caucasians, African Americans, or Hispanics.

Despite all these variations, research from many locations shows that the majority of women pass through menopause with relatively little discomfort (Melby *et al.*, 2005).

Is Menopause a Disease?

Given the universal nature of menopause as part of women's reproductive cycle, there are strong reasons to consider menopause to be a natural transition, similar to puberty. There have been various challenges to this view, however, most famously by Wilson (1966) in his publication *Feminine Forever*. Wilson put forward the argument that the lowered levels of estrogen experienced by postmenopausal women should be considered a deficiency disease and treated by administration of exogenous estrogen. This argument draws upon a quantitative conception of disease as a variation from a statistical norm. In this quantitative approach to disease, population measurements of various physiological parameters such as blood pressure, blood sugar levels, or hemoglobin are used to develop descriptive statistics. Cut-off levels for 'normal' are then defined, often using two standard deviations from the mean as the cut-off, with any values lying outside of these defined as abnormal. There are various diseases defined in this way, such as diabetes, hypertension, and anemia. Wilson argued that, like levels of blood sugar in diabetes or thyroxine in thyroid disease, postmenopausal levels of estrogen lie outside the norm and are therefore a sign of disease. Diseases defined in this way as deviations from the norm can be cured by treatment aimed at returning the abnormal parameter to normal. Wilson proposed that the 'estrogen deficiency disease' of menopause should be cured by treatment with exogenous estrogens.

This conception of menopause as a statistically defined deficiency disease has been challenged on the grounds that physiological parameters should not be interpreted in isolation as the sole criterion for diagnosing disease. The context is important in determining whether a particular measurement is pathological. A low hemoglobin level, for example, might be due to pregnancy rather than anemia, or a diastolic blood pressure of 90 mmHg might be normal for a 75-year-old but pathological for a 20-year-old. The postmenopausal state is characterized by lower estrogen levels than during women's reproductive years, but the levels are normal for age, just as pre-pubertal girls also have 'low' estrogen levels that are normal for their age. According to this line of reasoning, it is wrong to declare menopause the start of an estrogen deficiency disease as hormone levels are normal for age; what is wrong is expecting a 60-year-old woman to have the same estrogen levels as a 30-year-old (Rogers, 1999).

Diseases can also be defined qualitatively, according to the way that physiological events or symptoms are experienced, with unpleasant states classified as diseases. This conception of disease can place the individual and her experiences at the center of the diagnostic process. As already discussed, however, symptoms attributed to menopause vary by culture, thus making it difficult to make a claim for menopause as a disease based upon a universal set of symptoms. In addition, when symptoms, events, or practices that are culturally unwelcome are redefined as diseases, this intertwines the medical with the moral and can bring under medical control areas of life that were previously taken for granted. This happened in nineteenth-century responses to menopause, which was said to be associated with moral insanity demonstrated by peevishness and fits of temper or self-absorption and exacerbated by reading novels, dancing, or going to the theater. The treatment involved rest and avoiding excitement, effectively excluding women from those activities which society deemed inappropriate for 'good' women. Cultural norms have changed, but it is possible to see the pressure to treat menopause with HRT as part of a wider cultural approach to women, most noted in wealthier countries, that values them primarily for youthful attributes, which can be potentially sustained by long-term ingestion of exogenous estrogens (Rogers, 1999).

There are significant implications for public health and for health-care provision if an inevitable physiological change such as menopause is defined as a disease. These implications range from reinforcing cultural stereotypes about the attractiveness of youth to creating expectations about unpleasant symptoms at menopause to creating demand for medical treatments. Many studies have found that naturally occurring menopause (as opposed to menopause caused by premature failure or removal of the ovaries) appears to have no major impact on women's health and behavior and can be self-managed. Conversely, there appears to be a positive association between believing menopause to be a problem and then experiencing those problems (Rogers, 1999).

Making Informed Decisions about Treatment for Menopause

Whatever view we take about whether or not menopause should be defined as a disease, a proportion of women experience significant symptoms which are severe enough to lead them to seek medical care. The ethical concept that is relevant in this situation is that of informed consent. There is a need for accurate information so that women can make informed decisions about treatment. At a minimum, this information should include:

- the nature of the treatment on offer (what it involves, time course, evidence about efficacy, etc.)
- the likely benefits
- the potential harms or side effects
- other options, including no treatment.

This ethical requirement to provide information is complicated by the current uncertainty surrounding HRT. It is well accepted that treatment with HRT is effective in decreasing the frequency and severity of vasomotor symptoms and in relieving vaginal dryness and urinary symptoms. There is less agreement about the long-term benefits and safety of HRT, although most authorities agree that there is an increased risk of breast cancer associated with use of HRT (Greiser *et al.*, 2005). Health-care professionals have an obligation to be familiar with the latest research evidence so that they are able to provide women with accurate and relevant information.

There has been considerable interest in the processes for making decisions about treatment for menopause, in part because it has been recognized that menopausal treatment is largely symptomatic, rather than aimed at preventing long-term disability or disease. This has led to research into and the creation of decision support aids to assist women in identifying the values that are important to them in making a decision about accepting medical treatment (Fortin *et al.*, 2003).

Hormone Replacement Therapy and Disease Prevention

One of the major ethical issues raised by menopause concerns the promotion and use of HRT as a preventative health-care measure, particularly in relation to osteoporosis and cardiovascular disease. With regard to osteoporosis, women lose approximately 30% of their bone mass in the 20 years following menopause, and have two to three times the risk of fractures compared with men of the same age. The association between osteoporosis and postmenopause led to the widespread promotion of HRT as a preventative treatment for osteoporosis. This promotion supported the view of menopause as an estrogen-deficiency disease and the desirability of taking HRT both to treat the 'disease' and to decrease the risk of fractures. The former benefits the individual patient, but the latter has the potential to be a significant public health intervention, given the morbidity and mortality associated with fractures in elderly women. Research in this area indicates that HRT has a consistent and favorable effect on bone density at all sites, and confers protection against fractures for the duration of treatment (Wells *et al.*, 2002). This protection wears off rapidly after HRT use ceases (Banks *et al.*, 2004). From a public health perspective, there are many risk factors for osteoporotic fractures apart from being postmenopausal that are amenable to physical, dietary, and nonhormonal therapies. In particular, HRT has no effect on physical fitness, which is highly associated with falls, the major cause of fractures (Uusi-Rasi *et al.*, 2005).

In relation to coronary heart disease (CHD), women prior to menopause have lower rates than men, but after menopause their rates rise to rates similar to those of men. This and other observations led to the hypothesis that estrogen has a protective effect on the cardiovascular system of women. During the 1980s and early 1990s, there were many observational studies suggesting that there was a substantial reduction in CHD risk in women who took estrogens. A 1992 meta-analysis showed that there was one-third less fatal heart disease in postmenopausal women taking HRT than in women not taking HRT. Based upon these calculations, it was argued that taking HRT at a population level would be an effective intervention, with a net gain in lives saved. This was deemed so despite increased risk of death from breast or uterine cancers, because CHD is more common than those cancers (Barrett-Connor, 2003). These arguments had a significant effect upon bodies such as the American College of Physicians, the American College of Obstetrics and Gynecology, and the American Heart Association, who issued recommendations that all postmenopausal women should be offered HRT to prevent heart disease. Offering HRT became an accepted indicator used in assessing quality of medical practice, and not offering HRT was considered unethical (Barrett-Connor, 2003).

Some commentators at the time argued that the observational data were potentially biased, as women taking HRT tended to be white, educated, upper middle class, and lean, all of which would reduce their risk of CHD independent of taking HRT. The hypothesis needed to be tested by a large-scale randomized controlled trial (RCT) to eliminate potential biases in the observational studies. There had been one RCT of estrogen in a secondary prevention trial with participants with known CHD in the 1960s, with male participants. The estrogen arm of this trial was stopped early because estrogen-treated men had an increased rate of thromboembolic events and myocardial infarction (Barrett-Connor, 2003). Twenty-three years later, the first RCT to evaluate whether estrogen plus progestin therapy reduced the risk for CHD events in postmenopausal women with established coronary disease started. This was the Heart and Estrogen/Progestin Replacement Study, known as HERS. Barrett-Connor provides a detailed description of these trials and commentary on their findings. In summary, HERS found no overall difference in the primary CHD outcomes between treated and nontreated groups, and there was excess mortality in the treated group in the first year of treatment. These results were unexpected, although consistent with the earlier trial in men and with emerging results from smaller trials. One of the main criticisms of these trials was that giving HRT to women with established CHD may be too late and that estrogen may be more effective for primary prevention.

The HERS trial was followed shortly after by a large primary prevention trial called the Women's Health Initiative or WHI. The WHI trial compared three preventative strategies (HRT, diet, and calcium supplements) on disease outcomes in healthy postmenopausal women. This was a complex trial with multiple arms, one arm of which (combined HRT in women with an intact uterus) was stopped early due to excess breast cancers. Initial analysis showed that none of the excess risks or benefits was large (all fewer than 10 events per 10 000 women per year). Overall, the numbers of breast cancers, cardiovascular events, and pulmonary emboli in the hormone-treated groups exceeded the fractures prevented and decreased occurrence of colonic cancers in those groups. The results of WHI, first published in the *Journal of the American Medical Association* in 2002 (Rossouw *et al.*, 2002), received extensive media coverage, caused widespread confusion and anxiety among women taking HRT, and led to reported decreases in the numbers of women taking HRT (Hoffmann *et al.*, 2005).

The debate about the meaning of these results continues. It has been argued that the WHI used the wrong estrogen and progestogen, used the wrong-aged women, had inadequate blinding, and had high rates of dropouts and that the results lacked generalizability. Klaiber *et al.*, for example, in their review focus upon the effects of cohort age and the use of a combined continuous regimen of HRT to argue that the design flaws in the WHI led to adverse conclusions about HRT, and that further research with different hormonal regimens and younger women are required (Klaiber *et al.*, 2005).

With our current state of knowledge, HRT is not recommended as a preventative therapy for osteoporosis or CHD, as the risks seem to outweigh the benefits. In addition, there are good public health reasons for advocating for population health approaches to the treatment of multifactorial diseases such as CHD and osteoporosis rather than using individual hormonal treatments. Finally, treatment with HRT can be expensive, not only in terms of the direct costs of the pharmaceutical agents, but also in terms of the associated costs of increased medical care, largely associated with breast and uterine diagnostic and treatment interventions. These costs fall after the first year of treatment but remain significant (Ohsfeldt *et al.*, 2004).

Research Ethics and Menopause

Treating menopause is a multimillion dollar business in Western countries, although profits have fallen following publicity about the results of the WHI trial. As well as prescribed medications, many women take complementary or alternative therapies, also at significant cost. Much of the research evidence that we have about HRT, in particular the observational studies that suggested that HRT had cardio-protective effects, was generated in trials funded by pharmaceutical companies. In her 2004 critique of the pharmaceutical industry, Angell warns us 'to question how reliable publications from industry-sponsored research really are' (Angell, 2004, p. 113). With respect to the treatment of symptoms at menopause, the research agenda has been dominated by pharmaceutical interventions. There was significant lobbying of medical practitioners to promote HRT prior to the results of the HERS and WHI trials through activities such as continuing medical education funded by pharmaceutical companies and visits by pharmaceutical representatives.

Ideally, research agendas should be set in partnership with those who are affected by symptoms, include a range of treatment alternatives, and lead to research that provides practical information for women. At present this has not been achieved. It is hoped that research projects in progress will lead to greater clarity about the place of HRT and other symptom-relief measures for menopausal women.

Conclusion

Menopause, as a universal feature of female aging, provides a major opportunity for public health interventions aimed at improving women's health. Many women require information and general advice about optimizing their health at this time, rather than pharmaceutical treatment with uncertain benefits and identifiable risks or complementary and alternative therapies of little proven benefit. There are high costs and potential health risks in adopting a pharmaceutical treatment model for menopause, and the opportunity for effective public health interventions may be lost. At present we lack sufficient information for women and their health providers to make informed decisions about the long-term use of HRT.

See also: Female Reproductive Function; Menopause; Women's Mental Health Including in Relation to Sexuality and Reproduction.

Citations

Angell M (2004) *The Truth about Drug Companies: How They Deceive Us and What to Do about It.* New York: Random House.

Banks E, Beral V, Reeves G, Balkwill A, Barnes I, and Million Women Study Collaborators (2004) Fracture incidence in relation to the pattern of use of hormone therapy in postmenopausal women. *JAMA* 291: 2212–2220.

Barrett-Connor E (2003) An epidemiologist looks at hormones and heart disease in women. *Journal of Clinical Endocrinology and Metabolism* 88: 4031–4042.
Fortin JM, Goldberg RJ, Kaplan S, *et al*. (2003) Impact of a personalized decision support aid on menopausal women – Results from a randomized controlled trial. *Annual Symposium Proceedings, AMIA Symposium:* 843.
Greiser CM, Greiser EM, and Doren M (2005) Menopausal hormone therapy and risk of breast cancer: A meta-analysis of epidemiological studies and randomized controlled trials. *Human Reproduction Update* 11: 561–573.
Hoffmann M, Hammar M, Kjellgren KI, *et al.* (2005) Changes in women's attitudes towards and use of hormone therapy after HERS and WHI. *Maturitas* 52: 11–17.
Klaiber EL, Voge W, and Rako S (2005) A critique of the Women's Health Initiative hormone therapy study. *Fertility and Sterility* 84: 1589–1601.
Melby MK, Lock M, and Kaufert P (2005) Culture and symptom reporting at menopause. *Human Reproduction Update* 11: 495–512.
O'Connor V and Kovacs G (2003) *Obstetrics, Gynaecology and Women's Health.* New York: Cambridge University Press.
Ohsfeldt RL, Gavin NI, and Thorp JM (2004) Medical care costs associated with postmenopausal estrogen plus progestogen therapy. *Value in Health* 7: 544–553.
Rogers W (1999) Menopause: Is this a disease and should we treat it? In: Donchin A and Purdy LM (eds.) *Embodying Bioethics*, pp. 203–219. Lanham, MD: Rowman & Littlefield Publishers, Inc.
Rossouw JE, Anderson GL, Prentice RL, *et al.* (2002) Risks and benefits of estrogen plus progestin in healthy postmenopausal women: Principal results from the Women's Health Initiative randomized controlled trial. *Journal of the American Medical Association* 288: 321–333.
Schnatz PF, Banever AE, Greene JF, *et al.* (2005) Pilot study of menopause symptoms in a clinic population. *Menopause* 12: 623–629.
Singh A and Arora AK (2005) Profile of menopausal women in rural north India. *Climacteric* 8: 177–184.
Uusi-Rasi K, Sievanen H, Heinonen A, *et al.* (2005) Determinants of changes in bone mass and femoral neck structure, and physical performance after menopause: A 9-year follow-up of initially peri-menopausal women. *Osteoporosis International* 16: 616–622.
Wells G, Tugwell P, Shea B *et al*; Osteoporosis Methodology Group and the Osteoporosis Research Advisory Group (2002) Meta-analyses of therapies for postmenopausal osteoporosis. V. Meta-analysis of the efficacy of hormone replacement therapy in treating and preventing osteoporosis in postmenopausal women. *Endocrine Reviews* 23: 529–539.
Wilson R (1966) *Feminine Forever.* New York: Evans and Co.

Further Reading

Barrett-Connor E, Grady D, and Stefanick ML (2005) The rise and fall of menopausal hormone therapy. *Annual Review of Public Health* 26: 115–140.
Dubey R, Imthurn B, Zacharia L, and Jackson E (2004) Hormone replacement therapy and cardiovascular disease: What went wrong and where do we go from here? *Hypertension* 44: 789–795.
Houck JA (2003) "What do these women want?": Feminist responses to *Feminine Forever,* 1963–1980. *Bulletin of the History of Medicine* 77: 103–132.
Krieger N, Löwy I, Aronowitz R, *et al.* (2005) Hormone replacement therapy, cancer, controversies, and women's health: Historical, epidemiological, biological, clinical, and advocacy perspectives. *Journal of Epidemiology and Community Health* 59: 740–748.
Lamy O, Krieg MA, Burckhardt P, and Wasserfallen JB (2003) An economic analysis of hormone replacement therapy for the prevention of fracture in young postmenopausal women. *Expert Opinion on Pharmacotherapy* 4: 1479–1488.
Prentice R, Langer R, Stefanick M, *et al.* for the Women's Health Initiative Investigators (2005) Combined postmenopausal hormone therapy and cardiovascular disease: Toward resolving the discrepancy between observational studies and the Women's Health Initiative clinical trial. *American Journal of Epidemiology* 162: 404–414.
The Women's Health Initiative Steering Committee (2004) Effects of conjugated equine estrogen in postmenopausal women with hysterectomy: The Women's Health Initiative randomized controlled trial. *Journal of the American Medical Association* 291: 1701–1712.
U.S. Preventive Series Task Force (2005) Hormone therapy for the prevention of chronic conditions in post-menopausal women: Recommendations from US Preventive Series Task Force. *Annals of Internal Medicine* 142: 855–860.

Relevant Websites

http://www.fwhc.org/index.htm – Feminist Women's Health Centre.
http://www.healthinsite.gov.au/topics/Menopause – Health Insite: Menopause.
http://www.nlm.nih.gov/medlineplus/menopause.html – Medline Plus Menopause.
http://www.nih.gov/PHTindex.htm – National Institutes of Health Menopausal Hormone Therapy Information.
http://www.nia.nih.gov/ResearchInformation/ScientificResources/SWANdescription.htm – SWAN Repository.
http://www.nhlbi.nih.gov/whi/ – Women's Health Initiative.

Reproductive Ethics: Perspectives on Contraception and Abortion

E Mulligan, Flinders University of South Australia School of Law, Adelaide, South Australia, Australia
M Ripper, University of Adelaide, Adelaide, South Australia, Australia

Introduction

The goals of public health include improving the health status of populations, augmenting the ability of parents to raise healthy children, and maximizing the opportunities for individual self-determination. Abortion and contraception are only two among many interventions that may contribute to achieving these public health goals. The ethical considerations that public health practitioners apply to the provision of abortion and contraception include respecting the autonomy of individuals, seeking to minimize harm, acknowledging competing interests, and appreciating the fairness of providing access to health services for everyone.

While there is extensive debate concerning the ethics of abortion and (to a lesser degree) all forms of fertility control, only those arguments that apply specifically to the field of public health are considered here. The same ethical principles underpin public health provision of all forms of fertility control. Using ethical grounds to distinguish between abortion and contraception suggests that women who are not yet pregnant (and therefore potentially contracepting) have a different moral status from those who are pregnant (and therefore potentially considering abortion). This distinction has become less meaningful following the widespread adoption of postcoital fertility control methods such as menstrual extraction, pharmaceutical interception of pregnancy prior to implantation, and the use of contraceptive devices that prevent implantation but not conception. The continuum between contraception and abortion technologies highlights the need for public health practitioners to rely on ethical principles that are applicable to both.

The Autonomy of Women

The most significant ethical considerations of fertility control center upon the autonomy of individual women and their authority to make decisions regarding reproduction, their future, and their health care. Forced sterilization, surreptitious contraception, and compelled abortion are all unethical because each denies the autonomy of women.

Family formation, pregnancy, and childbearing have an enormous impact on women's life chances. To be able to influence these events gives women the opportunity to enhance their own health and well-being. Women also seek to maximize the opportunities for the health and development of their children. For women who have very little opportunity to make decisions about their circumstances or their sexual relationships (such as those living with domestic violence, with drug dependency, or on the very margins of their society), fertility control may provide an opportunity to exercise self-determination and exert some influence over their future.

Respect for the autonomy of women requires their consent to all health interventions. In order to make informed decisions and provide consent, the risks of contraception, abortion, and childbearing must be understood. The risks associated with each of these vary according to the healthcare practices and products used and the social and economic context that prevails. Providing accurate information is an ethical and often a legal obligation. In many jurisdictions with a common law tradition, medical practitioners have a duty of care, and this duty includes providing warnings of the risks of treatment. Valid evidence concerning risk is required by both practitioner and patient to inform consent. There is a strong body of scientific evidence demonstrating the safety of the widely available forms of contraception as well as the safety of abortion.

Respect for autonomy also underlies the requirement for privacy. The concepts that a woman has a right to privacy and that decisions concerning abortion and contraception are private matters to be decided by a woman and her doctor were embodied in *Roe v. Wade* (1973), a USA Supreme Court decision that established access to abortion and identified the right to privacy as a facet of the USA Constitution. Privacy legislation adopted by many members of the Organisation for Economic Co-operation and Development (OECD) is derived from the Guidelines on the Protection of Privacy and Transborder Flows of Personal Information (Organisation for Economic Co-operation and Development, 1980). These guidelines assert that people may control disclosure of personal information and that they have a right of access to the records kept about them.

The right of patients to expect confidential health care has been consistently defended by medical practitioners and public health advocates. Practicing confidentiality acknowledges the autonomy of patients by respecting their decisions about who may know their secrets. It also encourages patients to trust health providers with sensitive information that may be important in providing care. This has long been recognized as a prerequisite for treating sexually transmitted infections and is necessary for the provision of all sexual and reproductive health services, including contraception and abortion. In particular, sexually active teenagers will forgo sexual health care where they fear that confidentiality will not be protected (Marks *et al.*, 1983; Zabin *et al.*, 1991; Thrall *et al.*, 2000).

Minimizing Harm

Another major ethical principle on which provision of contraception and abortion rests is the minimization of harm to women. Two questions arise from consideration of the ethical principle of nonmaleficence (do no harm). Is giving women decision-making authority over their sexuality and reproduction harmful or health-enhancing to women, and is fertility control harmful or health-enhancing for women?

Is Self-Determination Harmful to Women?

Public health requires attention to a wide range of factors that contribute to well-being. When contraception and abortion are accessible, women may choose to raise a smaller number of healthier children. For women, opportunities to gain and maintain employment, participate in public life, and contribute to decision-making forums all enhance health. Women's social participation is restricted by many factors. When child care is entirely the responsibility of mothers and employment conditions conflict with parental responsibilities, child rearing causes disadvantages for women seeking employment. When women are faced with discrimination, access to fertility control provides them with opportunities for self-development.

The ability to determine family size improves women's sense of well-being irrespective of the number of children they bear. The contribution that control over family size makes to women's well-being is illustrated by research such as that conducted by Hardee *et al.* (2004) in Indonesia.

Is Contraception or Abortion Harmful to Women?

The very low risk associated with contraception or abortion conducted using safe techniques and the contrasting higher risks associated with childbirth and unsafe abortion are ethically significant. They indicate that generally available contraception is a harm-minimization strategy and legally sanctioned abortion is the least harmful pregnancy outcome for women.

Consideration of potential harm to women has been important in framing legislation regulating abortion in many jurisdictions. Legislation may permit abortion when the risk to the pregnant woman from abortion is less than the risk of continuing the pregnancy (all pregnancies) or, more restrictively, when there is grave risk to the life of the pregnant woman if the pregnancy continues (few pregnancies). While legislation varies in the range of circumstances in which abortion is sanctioned, the underlying ethical principle embodied in these provisions is that the life of the pregnant woman is important and she will be permitted to take action that increases her chances of survival.

Assessment of risk has concerned legislators in regulating abortion in many jurisdictions. A concern that women may not act in their own best interests when choosing abortion has led to the substitution of another decision maker (often a judicial officer or a doctor) in some jurisdictions. The substitute decision maker is then called upon to decide whether the risks of abortion are acceptable in individual cases.

Where only unsafe abortion is available, women who seek abortion appear to act against their own best interests. The risk that women are prepared to face can be seen as a measure of the adversity that they expect in continuing an unwanted pregnancy. In societies in which the consequence of a pregnancy conceived outside marriage may include withdrawal of family support, social condemnation, or death for mother or child, women will choose abortion even when they are aware that the risk is high.

Because contraception is widely used and abortion is very common, it has been possible to collect data concerning the associated risks across diverse populations. A low frequency of complications and low rates of contraceptive failure resulting in a pregnancy have been established. There is strong evidence demonstrating that the risk of death, physical injury, mental injury, or reduction in future fertility is very low following abortions conducted by trained practitioners using appropriate facilities. The risk of death from abortion using surgical (Kulier *et al.*, 2005a) or medical (Kulier *et al.*, 2005b) procedures is much lower than the risk of childbirth or other commonly performed surgical procedures (such as cesarean section). Women have lower mortality rates from all causes following an abortion than women who have not been pregnant (Gissler *et al.*, 2004). There is no difference in the rates of mental illness between women who have had abortions and those who have not (Broen *et al.*, 2004). There is no association between abortion and an increased risk of breast cancer (American College of Obstetrics and Gynecology Committee on Gynecologic Practice, 2003). There is some evidence of a slightly increased risk of miscarriage or premature labor in subsequent pregnancies following either a miscarriage or an abortion (Henriet and Kaminiski, 2001; Sun *et al.*, 2003). Both miscarriage and premature labor are relatively common, and the increase in risk is small in comparison to the background rate.

There is a significant contrast between the health risk associated with abortions conducted by appropriately trained practitioners using sterile techniques and those conducted in circumstances that do not support safe methods. These differences have been mapped globally by the World Health Organization (2004) in a report that estimates the maternal death rate attributable to unsafe abortion to be 1:270 (**Figure 1**). The harm that restrictive abortion laws cause to women's health and morbidity are well documented, and world trends are toward liberalizing access to abortion (Cook *et al.*, 1999).

Competing Interests

Ethical discussion of fertility control may focus entirely on women, overlooking the role of men as reproductive beings with responsibility for their own sexual activity and contraception. Men's rights are most often discussed in relation to a right to paternity rather than in relation to their own responsibility to constrain their capacity to impregnate and make ethical decisions about paternity.

Two major areas in which third parties claim rights in relation to abortion are the rights of the fetus and the rights of men to fatherhood. All claims that a fetus has rights rest on the presumption of the moral personhood of the fetus. If personhood is accorded prior to birth, then a claim to the protection of that life follows. The moral status of the fetus is a matter of religious belief for some, and for others the stage of development or viability of the fetus (its ability to survive independently of its mother) is a marker of personhood. Many jurisdictions accord rights only to born

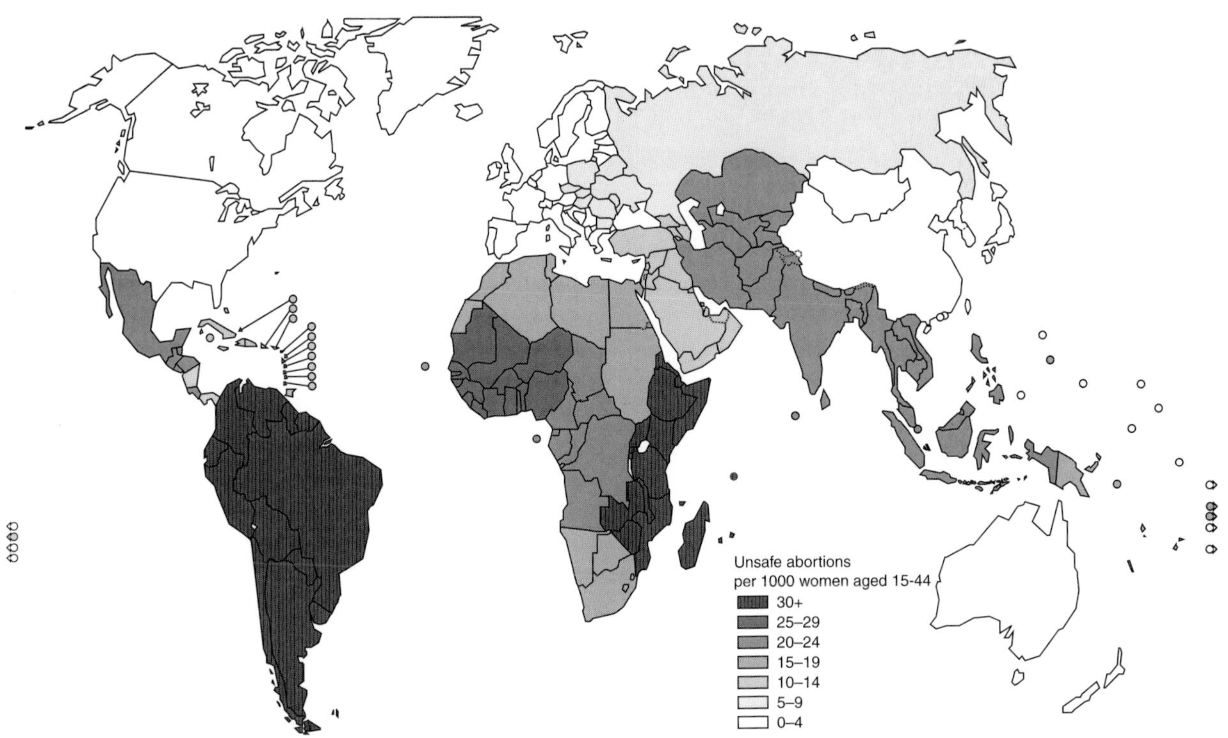

Figure 1 Estimated annual incidence of unsafe abortion per 1000 women aged 15–44 years, by United Nations subregions, 2000. Reproduced from the World Health Organization (2004) *Unsafe Abortion – Global and Regional Estimates of the Incidence of Unsafe Abortion and Associated Mortality in 2000,* 4th edn. Geneva: World Health Organization.

individuals, while others allow intervention to sustain fetal life against a pregnant woman's wishes and some outlaw abortion entirely because of the presumption of fetal personhood. When fetal (or embryonic) personhood is conceded, exercising rights on behalf of the fetus remains incompatible with the principles of the autonomy of, and nonmaleficence toward, the pregnant woman.

The ethical principle of proportionality is relevant in weighing competing rights claims, including male rights to paternity. The burden to a woman of continuing a pregnancy unwillingly is generally considered to far outweigh the benefit to the man in becoming a father. Claims by would-be fathers in a pregnancy that they have caused have been given very little legal or ethical standing worldwide. Although cases have been brought in USA and European courts by men seeking to prevent a woman from having an abortion, these rarely succeed, primarily because recognition of paternal rights contravenes the principles of respect for the woman's autonomy and informed consent. When the principle of proportionality is applied, the rights of the pregnant woman prevail.

Justice

Distributive justice is a key ethical principle that applies to the provision of social goods including public health services. Health services are an instrumental, rather than an absolute, good in that they are not good in and of themselves, but only insofar as they facilitate survival, human dignity, and full citizenship. The principle of distributive justice requires that health services be accessible to individuals according to need and within the context of resource availability. When there are barriers preventing access to contraception and abortion, distributive justice is compromised. Access to health care is often stratified by race, class, and region. This is also true for access to abortion and contraception. Many factors compromise access to fertility control, including material considerations, cost, availability, and religious or national policies.

Access to safe contraceptive and abortion services requires sufficient regulation of providers and manufacturers to ensure safe services. However, a highly regulated environment can compromise services. In some jurisdictions, regulatory mechanisms retard or restrict the distribution of contraceptives and medications and devices used for postcoital contraception and abortion. Currently this restriction exists in some countries in relation to the antiprogestin mifepristone (RU486) used in medical abortion despite this drug being included on the World Health Organization's Model List of Essential Medicines (2010).

Access to contraception and abortion is also compromised when services are under attack, and when service providers and/or patients are intimidated and stigmatized. The marginalization of abortion services from mainstream health provision produces further barriers to access. These include difficulties in recruiting, training, and sustaining a skilled workforce, which compromise the quality of services.

Stigma associated with abortion and contraception creates an environment in which normal requirements for duty of care by medical practitioners can be compromised. For example, where law or practice allows health workers conscientious exemption, those practitioners who decline to provide services still have an ethical obligation to refer the patient for these services elsewhere.

Conclusion

Contraception and abortion are ethical when they result from informed, voluntary decisions by individual women. They support the woman's agency and authority over her own life. They reflect the principle of bodily autonomy and maximize women's opportunities to be healthy. In situations in which it is possible to provide contraception and safe abortion, these minimize maternal and infant deaths and enhance the health and social well-being of women and children. It is ethical to provide fertility control services for these reasons.

Self-determination and access to contraception and safe abortion are not harmful to women. In contrast, many maternal deaths are caused by unsafe abortion and childbirth without health care. Access to all forms of fertility control, including contraception and safe abortion, is contested across the world, with negative consequences for public health.

Although there are competing claims over a pregnancy, these do not outweigh the ethical value accorded to the autonomy of women and the benefits that flow from self-determination.

The ethical principle of justice supports access to health services for all people. Equitable and confidential access to contraception and abortion are part of any comprehensive public health system.

Contraception and abortion are not the only means of supporting women's self-determination, ability to maximize their health potential, or their ability to raise healthy children. Education and economic independence are critical in supporting the well-being of women and their dependents.

See also: Abortion; Family Planning/Contraception; Gender Aspects of Sexual and Reproductive Health; Reproductive Ethics: New Reproductive Technologies; Reproductive Rights.

Citations

American College of Obstetrics and Gynecology Committee on Gynecology Practice (2003) AOG Committee Opinion. Induced abortion and breast cancer risk. *International Journal of Gynecology and Obstetrics* 83: 233–235.

Broen A, Moum T, Bodtker A, and Ekeberg O (2004) Psychological impact on women of miscarriage versus induced abortion: A 2 year follow-up study. *Psychosomatic Medicine* 66: 265–271.

Cook R, Dickens B, and Bliss L (1999) International developments in abortion law from 1988 to 1998. *American Journal of Public Health* 89: 579–586.

Gissler M, Berg C, Bouvier-Colle M, and Buekens P (2004) Pregnancy-associated mortality after birth, spontaneous abortion, or induced abortion in Finland, 1987–2000. *American Journal of Obstetrics and Gynecology* 190: 422–427.

Hardee K, Eggleston E, Wong E, Irwantoth, and Hull T (2004) Unintended pregnancy and women's psychological well-being in Indonesia. *Journal of Biosocial Sciences* 36: 617–626.

Henriet L and Kaminski M (2001) Impact of induced abortions on subsequent pregnancy outcome. *British Journal of Obstetrics and Gynaecology* 108: 1036–1042.

Kulier R, Fekih A, Hofmeyr G, and Campana A (2005a) Surgical methods for first trimester termination of pregnancy. *Cochrane Database of Systematic Reviews* 4.

Kulier R, Gulmezoglu A, Hofmeyr G, Cheng L, and Campana A (2005b) Medical methods for first trimester abortion. *Cochrane Database of Systematic Reviews* 4.

Marks A, Malizio J, Hoch J, Brody R, and Fisher M (1983) Assessment of health needs and willingness to utilize health care resources of adolescents in a suburban population. *Journal of Pediatrics* 102: 456–460.

Mulligan E and Braunack-Mayer A (2004) Why protect confidentiality in health information? A review of the research evidence. *Australian Health Review* 28: 48–55.

Organisation for Economic Co-operation and Development (OECD) (1980) *Guidelines on the Protection of Privacy and Transborder Flows of Personal Information.* Paris, France: OECD.

Roe v. Wade (1973) 410 U.S. 113.

Sun Y, Che Y, Gao E, Olsen J, and Zhou W (2003) Induced abortion and risk of subsequent miscarriage. *International Journal of Epidemiology* 32: 449–454.

Thrall J, McCloskey L, Ettner S, and Rothman E (2000) Confidentiality and adolescents' use of providers for health information and for pelvic examinations. *Archives of Pediatrics & Adolescent Medicine* 154: 885–892.

World Health Organization (2004) *Unsafe Abortion – Global and Regional Estimates of the Incidence of Unsafe Abortion and Associated Mortality in 2000,* 4th edn. Geneva, Switzerland: World Health Organization.

World Health Organization (2010) *16th WHO Model List of Essential Medicines.* Geneva, Switzerland: World Health Organization.

Zabin L, Stark H, and Emerson M (1991) Reasons for delay in contraceptive clinic utilization: Adolescent clinic and nonclinic populations compared. *Journal of Adolescent Health* 12: 225–232.

Further Reading

British Medical Association (1999) *The Law and Ethics of Abortion: BMA Views.* London: British Medical Association.

Cohen AL, Bhatnager J, Reagan S, *et al.* (2007) Toxic shock associated with *Clostridium sordellii* and *Clostridium perfringens* after medical and spontaneous abortion. *Obstetrics & Gynecology* 110: 1027–1033.

Hakim-Elahi E, Harold MM, Tovell MD, and Burnhill MS (1990) Complications of first trimester abortion: A report of 170,000 cases. *Obstetrics & Gynecology* 76: 129–135.

Henderson JT, Hwang AC, Harper CC, and Stewart FH (2005) Safety of mifepristone abortions in clinical use: A review of 195,000 cases. *Contraception* 72: 175–178.

Mulligan E (2005) Striving for excellence in abortion services. *Australian Health Review* 30: 468–473 (outcomes following 35,000 South Australian cases).

Reime B, Schücking BA, and Wenzlaff P (2008) Reproductive outcomes in adolescents who had a previous birth or an induced abortion compared to adolescents' first pregnancies. *BMC Pregnancy Childbirth* Jan 31: 8, 4.

Royal College of Obstetricians and Gynaecologists (2004) *The Care of Women Requesting Induced Abortion.* London: Royal College of Obstetricians and Gynaecologists.
Visintine J, Berghella V, Henning D, and Baxter J (2008) Cervical length for prediction of preterm birth in women with multiple prior induced abortions. *Ultrasound in Obstetrics and Gynecology* 31: 198–200.
Winikoff B, Sivin I, Coyaji K, *et al.* (1997) Safety, efficacy, and acceptability of medical abortion in China, Cuba, and India: A comparative trial of mefipristone-misoprostol versus surgical abortion. *American Journal of Obstetrics and Gynecology* 176: 431–437.
World Health Organization (2003) *Safe Abortion: Technical and Policy Guidelines for Health Systems.* Geneva, Switzerland: World Health Organization.
Zhou W, Gunnar LN, Moller M, and Olsen J (2002) Short-term complications after surgically induced abortions: A register-based study of 56117 abortions. *Acta Obstetricia et Gynecologica Scandinavica* 81: 331–336.

New Technologies: Ethics of Stem Cell Research

R A Ankeny, The University of Adelaide, Adelaide, Australia

Introduction

Human embryonic stem cell (hESC) research has been surrounded by considerable controversy in recent years, particularly with regard to the ethical issues associated with such research. The scientific facts are relatively straightforward. Stem cells are undifferentiated, unspecialized cells with the ability both to multiply for long periods and to differentiate into specific kinds of cells after being stimulated by chemical or other signals. This latter property means that stem cells could be transformed into specialized cells with specific functions, such as heart muscle cells, blood cells, or nerve cells. Hence promoters of stem cell research claim that it holds promise for the development of therapeutic interventions for a wide range of diseases in which cells have been damaged or destroyed (e.g., heart disease, diabetes, spinal cord injury, and Parkinson's disease). Patients with leukemia have been routinely treated with adult stem cells (ASCs) from the blood of compatible donors since the 1970s, and the success of this therapy is often cited as a precedent for future uses of hESCs. However, other types of stem cell therapies are highly experimental and are currently only used in animal models.

hESCs can be very difficult to isolate and establish in culture, but a number of stem cell lines have been successfully obtained from the inner cell mass of 4- to 6-day-old embryos (blastocysts). Cells are isolated at this early stage because it is thought that in later stages of development, they become more restricted in terms of what types of tissues the stem cells can become. A stem cell line contains cells that continue dividing without differentiating into specialized cells and which remain pluripotent (able to differentiate into any cell type with a few exceptions). Stem cell banks hold cell lines that have been produced under standardized conditions, which have been quality controlled, and provide them to basic and clinical researchers (Faden *et al.*, 2003). The UK Stem Cell Bank is the most prominent of these and is overseeing the International Stem Cell Initiative to track and characterize all stem cell lines worldwide.

Blastocysts used for creating hESCs typically are derived from frozen embryos that were originally created as part of in vitro fertilization (IVF) treatments with the intention of achieving successful pregnancies but that were never implanted. In many countries, there are hundreds of thousands of unused frozen embryos stored in assisted reproductive technology (ART) clinics. In some countries (e.g., Australia, Canada, and the United Kingdom), once couples have finished IVF, they can consent to donation of their surplus stored embryos for research purposes, with some limitations (Isasi and Knoppers, 2006). Elsewhere, excess ART embryos cannot be donated for any type of research (e.g., Italy) or creation of hESCs is explicitly banned (e.g., Germany, which does allow importation of lines created elsewhere for research purposes).

Ethical Debates About the Moral Status of the Embryo

For many, the main ethical issues associated with hESC research are related to the fact that an embryo must be destroyed to obtain stem cells, although research is underway regarding accessing stem cells without destroying the embryo. A central ethical question related to this is: What is the moral status of a human embryo? Beliefs regarding when personhood begins (that is, when a human being has moral status) differ. For instance, some religious traditions hold that personhood begins at conception, and thus hold hESC research (and all other forms of destructive embryo research) to be unethical (Doerflinger, 1999; Meilaender, 2001). We should not use embryos as a means to other ends, they claim, and hence should value them as human beings with moral status and full human rights, perhaps deserving of additional protections because they are especially vulnerable and dependent.

A similar argument can be made without relying on religion: Each of us began life as an embryo. If my life is worthy of respect and hence inviolable by virtue of my humanity, so was my life in earlier stages. Since we cannot define a precise time between conception and birth at which we become a person, we must hold embryos to possess the same inviolability as fully developed human beings (for the problems with this argument, see Sandel, 2004). Opponents of embryo research also argue that allowing such research might create a sort of slippery slope, which will lead to increased social toleration of loss of life for vulnerable persons (such as infants and the disabled), although there is limited empirical evidence about social attitudes to support such a claim.

Other more intermediate views hold that the stage at which embryos would be used for hESC research (less than 6 days old) is so early that the embryo cannot be argued to have moral status and hence ethically can be used for destructive research. The time at which personhood or moral status is said to occur differs, but even many religious traditions would nominate implantation or the formation of the primitive streak at around 14 days. These are said to be markers of an individualized human entity with the inherent potential to become a human person; up until this point, the moral status of the embryo is not that of a person, and its use for certain kinds of research aimed at alleviating illness and suffering can be ethically justifiable (and may in fact be preferable to simple destruction, as research is an acknowledgment of the value of the embryo; see Annas *et al.*, 1999). Nonetheless, the embryo should be given some protections, in recognition of its potential as a human being.

More liberal views claim that to attribute basic and equal human worth to an individual requires more than cells that have the potential to develop into a person (Robertson, 1999). They would cite a much later stage of development as the cut-off point, such as the time at which a termination would no longer be legally allowed under normal circumstances, which is much later than when destruction for hESC research would occur. In this view, there is no moral harm committed when an embryo is destroyed or not transferred into the uterus to attempt pregnancy. Some have argued that embryos have higher and lower moral status, depending on whether or not they are part of a parental project where there is a desire and intention to bring a pregnancy to term. These arguments typically do not imply that we can do anything we wish with embryos (such as cosmetic testing) as we need to give them special respect because of their source and their symbolic value, but it would allow research aimed at curing serious diseases.

The most liberal view holds that embryos are merely body parts until they can exist independently. Thus, they have no independent moral status and belong (as property does) to those persons who donated the parts needed to make them. Therefore, all decision making about disposition of embryos should reside with the couple who contributed the genetic materials, and we must respect these rights as we would any other property rights.

There also is a related debate over intentionality, and in particular whether it is ethically permissible to create embryos explicitly for research purposes (see Davis, 1995; Juengst and Fossel 2000; FitzPatrick, 2003), either by using IVF to join a sperm and an egg or via cloning (see the section titled 'Cloning' below). Some people who would find it ethically permissible to use already created, stored embryos for research oppose the creation of embryos explicitly for research purposes. Societies that permit or support IVF technologies might be argued to have implicitly endorsed the idea that there will be some wastage of embryos inherent in such procedures, but creating new embryos for research is more morally problematic for many. The Council of Europe's Oviedo Convention of 1997 prohibits the creation of human embryos for research purposes, but it has yet to be ratified by a majority of its member states. Others argue that the creation of embryos for research purposes is ethically permissible, as such embryos do not have the social significance associated with those created for reproduction and hence have a different moral status.

Cloning

The term cloning is used to describe a range of processes that involve genetic copying, ranging from copies of sections of DNA (genes) to whole organisms (such as plants). Ethical debates about cloning escalated in 1997 with the birth of Dolly the sheep, a whole organism produced using cloning techniques. Cloning is relevant to debates about hESC research because it has the potential to provide a source of tailored stem cells that might be particularly useful for individualized medical treatments. Human embryo clones could be created using somatic cell nuclear transfer (SCNT), which involves taking an egg cell from a woman in a similar way to how eggs are obtained for in vitro fertilization. The nucleus of the egg is extracted, leaving behind only the cytoplasm. A nucleus is then extracted from a somatic (body) cell of a person (who can be another person or the same woman who donated the original egg) and inserted into the cytoplasm.

Under the correct conditions, the entity created should start to divide as would an embryonic nucleus created naturally (i.e., by the union of egg and sperm). The resulting embryo that is formed has the same genetic material as that of the donor of the somatic cell nucleus, instead of its genome being a blend of the two parents as is usual, and its inner cell mass can be used to obtain stem cells as outlined above. The potential advantage of cloning via SCNT is that if the donor of the somatic cell nucleus is also the potential recipient, the resulting hESCs are less likely to be rejected by the recipient's immune system when transplanted.

Clones could also be created by splitting an embryo at early stages of cell division; by inserting an embryonic stem cell nucleus into the cytoplasm of an egg; or by stimulation of egg cells such that they become embryos without being fertilized by sperm (parthenogenesis), although many of these techniques are highly experimental and have only been used with animal models.

Ethical Debates on Cloning

In 2005 in a split vote, the United Nations adopted by 84 votes a nonbinding resolution to prohibit all forms of human cloning as they are "incompatible with human dignity and the protection of human life" (United Nations, 2005). However, numerous countries (34 voted against and 37 abstained) did not support the resolution because it failed to distinguish between reproductive and nonreproductive cloning. Typically, ethical responses to cloning split in terms of purposes for which cloning will occur, though many claim that this distinction is difficult to maintain and will be difficult to monitor in practice (Bowring, 2004). Most commentators are opposed to human reproductive cloning, where the intention is to produce a cloned human embryo to implant in a woman in order to produce a pregnancy and eventually a child. There do not seem to be particularly strong rationales for creating this type of human clone, and there are concerns about safety and long-term effects based on animal models; reproductive cloning is explicitly legally prohibited in several countries.

Nonreproductive cloning or SCNT is less controversial for those who are not opposed to hESC research, as the clone is created for research purposes and destroyed at an early stage. This type of cloning is permitted in some countries (e.g., The Republic of Korea and New Zealand) or is allowed with a license (e.g., the United Kingdom, China, and Australia). Some commentators argue that pursuing cloning research is an ethical imperative, so long as there is appropriate oversight, transparency, and accountability (Devolder and Savulescu, 2006). However, SCNT still requires the donation of eggs, and some ethicists have argued that women who are potential donors might be subject to coercion, including economic or workplace pressures (e.g., for those working in stem cell research labs), or that it would be difficult to foster truly informed consent (see Magnus and Cho, 2005). Many countries do not permit payment beyond reasonable medical expenses for egg donors, hence partially mitigating concerns about economic coercion.

Other Issues Related to hESC Research

Another key issue often raised in relation to hESC research is whether ASC (or somatic stem cell) research, which for many is much less morally problematic, might not be as useful (or more so). ASCs are undifferentiated cells found among differentiated cells in a tissue or organ; although their precise origin is unknown, they can renew themselves and it is thought that they can differentiate to yield the major specialized cell types of the tissue or organ, though perhaps not as many cell types as hESCs. Skeptics claim that ASCs are relatively rare, hard to find with available techniques, and difficult to culture outside of the body. Opponents of hESC research argue that ASC research holds as much if not more promise, and hence should be pursued instead of hESC (though they often conflate their ethical claims with scientific ones). Others would claim that there is no need to put the two types of research in opposition to each other, but that both lines should be pursued so long as there is reasonable evidence and expectation of benefit.

Some critics have argued that ART clinics may put pressure on women and couples to produce more embryos than are needed for IVF treatment in order to allow them to donate for research purposes. Consequently, many countries have laws that prohibit donating embryos created after a certain date, which usually reflects when legislation governing hESC research was enacted. However, there seems to be little evidence of such practices occurring, and many safeguards (e.g., cooling-off periods following consent to donation and limits on the number of embryos produced) exist to counter these concerns.

A final set of ethical concerns are associated with the commercialization of these technologies and who will have access to any beneficial therapeutics that might be developed. Some feminist critiques of ART note that these technologies can contribute to oppression; in the case of hESC research, altruistic donation of eggs or embryos could be undermined if these donations are used for commercialized research.

Conclusion

Although many have noted that there is considerable hype associated with the promises held out for stem cell research, there still has not been adequate debate about the broader issues associated with it (Dresser, 2005), including the implications for public health. For instance, as disability advocates among others have noted, it is unclear whether the relatively high costs associated with current research, let alone potential costs associated with any therapies which might be developed, reflect adequate assessment or debate regarding appropriate allocation of health-care resources. As with many areas of medicine, it can be argued that a focus on the prevention of relevant disease conditions might prove more beneficial to a larger number of people. In addition, given that initial costs are likely to be extremely high for stem cell-based treatment options, concerns exist about the possibility for inequities in access. Furthermore, there is no consensus on whether limitations should be placed on commercialized research or how any benefits from such research are likely to reach those who most need care. The patenting of stem cell products might well keep them inaccessible for all but the most advantaged. Stem cell banks which keep resources including stem cell lines in the public domain and accessible to researchers worldwide are one key step in making certain that at least some of the results of research stay in the public domain. hESC research is likely to remain fraught with controversy, and policies about governing it should be made through transparent, democratic processes. The disagreements between opponents and supporters of hESC research are unlikely to be resolved, and hence policy makers face a difficult task of finding ways to accommodate deeply held, conflicting views, particularly when formulating public health policies.

Citations

Annas GJ, Caplan A, and Elias S (1999) Stem cell politics, ethics and medical progress. *Nature Medicine* 5: 1339–1341.

Bowring F (2004) Therapeutic and reproductive cloning: A critique. *Social Science and Medicine* 58: 401–409.

Davis D (1995) Embryos created for research purposes. *Kennedy Institute of Ethics Journal* 5: 343–354.

Devolder K and Savulescu J (2006) The moral imperative to conduct embryonic stem cell and cloning research. *Cambridge Quarterly of Healthcare Ethics* 15: 7–21.

Doerflinger RM (1999) The ethics of funding embryonic stem cell research: A Catholic viewpoint. *Kennedy Institute of Ethics Journal* 9: 137–150.

Dresser R (2005) Stem cell research: The bigger picture. *Perspectives in Biology and Medicine* 48: 181–194.

Faden RR, Dawson L, Bateman-House AS, *et al.* (2003) Public stem cell banks: Considerations of justice in stem cell research and therapy. *Hastings Center Report* 33: 13–27.

FitzPatrick W (2003) Surplus embryos, nonreproductive cloning, and the intend/foresee distinction. *Hastings Center Report* 33: 29–36.

Isasi RM and Knoppers BM (2006) Mind the gap: Policy approaches to embryonic stem cell and cloning research in 50 countries. *European Journal of Health Law* 13: 9–25.

Juengst E and Fossel M (2000) The ethics of embryonic stem cells—now and forever, cells without end. *Journal of the American Medical Association* 284: 3180–3184.

Magnus D and Cho MK (2005) Issues in oocyte donation for stem cell research. *Science* 308:1747–1748.

Meilaender G (2001) The point of a ban: OR, how to think about stem cell research. *Hastings Center Report* 31: 9–16.

Robertson JA (1999) Ethics and policy in embryonic stem cell research. *Kennedy Institute of Ethics Journal* 9: 109–136.

Sandel MJ (2004) Embryo ethics—the moral logic of stem-cell research. *New England Journal of Medicine* 351: 207–209.

United Nations (2005) Ad hoc Committee on an International Convention against the Reproductive Cloning of Human Beings. http://www.un.org/law/cloning

Further Reading

Holland S, Lebacqz K, and Zoloth L (eds.) (2001) *The Human Embryonic Stem Cell Debate.* Cambridge, MA: MIT Press.

Panno J (2004) *Stem Cell Research: Medical Applications and Ethical Controversy.* New York: Facts on File.

Reid L, Johnston J, and Baylis F (eds.) (2006) Ethics and stem cell research: Shifting the discourse. *special issue: Journal of Bioethical Inquiry* 3(1–2): 1–121.

Ruse M and Pynes CA (eds.) (2003) *The Stem Cell Controversy: Debating the Issues.* Amherst, NY: Prometheus.

Waters B and Cole-Turner R (eds.) (2003) *God and the Embryo: Religious Voices on Stem Cells and Cloning.* Washington, DC: Georgetown University Press.

Gender Aspects of Sexual and Reproductive Health

J C Cottingham, World Health Organization, Geneva, Switzerland
T K S Ravindran, Achutha Menon Centre for Health Science Studies, Trivandrum, India

Definitions

Sexual and reproductive health was first defined at the International Conference on Population and Development (ICPD), held in Cairo in 1994. The term 'reproductive health' reflected a sharp departure from previous approaches to the question of how, and to what extent, humans reproduce. In reaction to the often aggressive population control programs launched in the 1960s and 1970s in Africa, Asia, and Latin America, the international women's movement and others called for programs that would both respect women's human rights and provide them with services for a range of issues beyond just birth control – going through childbirth safely, having a healthy, wanted child, and being free from sexually acquired reproductive tract infections.

The definition of reproductive health also pointed to the central importance of human sexual relationships, and the fact that they often represent an unequal exchange between women and men. This inequality, rooted in society's view of what a man or a woman should be (termed gender), plays a critical role in women's and men's sexual and reproductive health. Gender is defined as encompassing the different roles, rights, expectations, and obligations that culture and society attach to individuals according to whether they are born with male or female sex characteristics. Sex refers to the biological and physiological differences and genetic susceptibilities and immunities associated with being biologically male or female. Both sex and gender have an impact on, and are affected by, sexual and reproductive health.

Dimensions of Gender Considerations

The past two decades have seen an explosion of studies and interventions on the way in which gender differences have an impact on all dimensions of development. In the area of health – including sexual and reproductive health – it is generally accepted that the different characteristics of gender play a role in women's and men's vulnerability to exposure to disease and the outcome of that exposure and in their health-seeking behavior, their ability to access health information and services, their experience of health services, and the social and economic consequences of a health problem (Klugman and Theobald, 2006). Gender characteristics, or dimensions, include (1) the sociocultural norms and values attached to what women and men, girls and boys, do, that is, gender roles, (2) the differential access to control (including making decisions) over resources such as money, information, and education, (3) the influence that social institutions such as family, school, religious institutions, workplaces, and health services have in reinforcing gender norms, and (4) the impact of laws and policies, which often have gender-based inequalities written into them. For the purposes of this chapter, we will use these categories to highlight some of the major gender aspects of sexual and reproductive health. However, examination of gender dimensions in any area of inquiry must start with sex-disaggregated data.

The Burden of Sexual and Reproductive Ill-Health, by Sex

Estimates of the global burden of disease, which assesses the burden to the world's population from premature mortality and disability by using a combined measure, the disability-adjusted life year (DALY), indicate that sexual and reproductive ill-health contributes to 20% for women and 14% for men (see **Figure 1**). The elements included in the calculation are maternal and perinatal mortality and morbidity, reproductive cancers, HIV/AIDS, and sexually transmitted infections (STIs). Since only women get pregnant and give birth, the burden from maternal causes is unique to women, making the overall burden higher for them than for men. In addition, there are biological differences in, for instance, susceptibility to STIs. Biologically, women, and particularly young women, are at greater risk of infection through heterosexual sex (Zierler and Krieger, 1997). This is partly because of prolonged exposure to organisms when infected ejaculate is retained in the vagina, particularly for pathogens that produce discharge or are present in genital secretions, such as gonococci, chlamydia, trichomonads, and HIV (Howson *et al.*, 1996). Reproductive cancers are more numerous in women, such as cancers of the cervix and corpus of the uterus and ovarian and breast cancer, than in men (prostate and testicular cancer).

This situation already puts women at a disadvantage from a health point of view. In reality, the situation is probably even more unbalanced since the global burden estimates are limited in their definition of sexual and reproductive health. They do not include conditions such as obstetric fistula, urinary incontinence, uterine prolapse, menstrual disorders, non-sexually transmitted

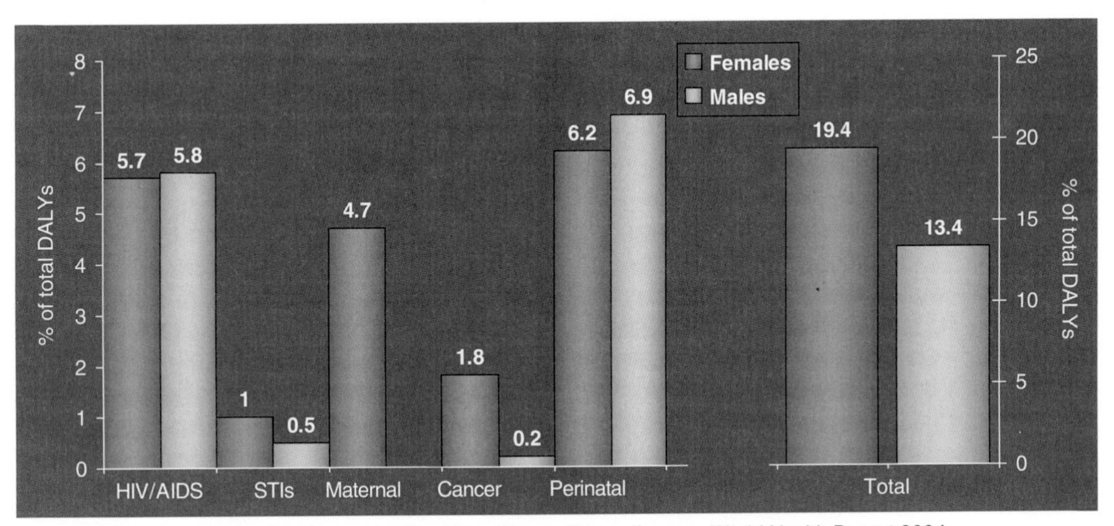

Figure 1 Global burden of sexual and reproductive ill-health conditions. Source: *World Health Report 2004*.

reproductive tract infections, female genital mutilation, stillbirth, infertility, and the sexual and reproductive health consequences of sexual abuse. Nor do the estimates capture the social and psychological effects of sexual and reproductive ill-health, some of which – like those associated with obstetric fistulae – may be significant (AbuoZahr and Vaughan, 2000). Therefore, it seems likely that, if and when estimates are made to include these additional elements, the burden for women would be considerably higher.

Norms and Values

Norms and values in any given society or community are a powerful influence in the shaping of gender roles and responsibilities. They define, from an early age, what is appropriate behavior, including sexual behavior, for men and women. In many parts of the world, masculinity is associated with early sexual activity, having many sexual partners and experiences, virility, and pleasure. Femininity, on the other hand, is frequently associated with passivity, virginity, chastity, and fidelity. The need for men to ensure that their progeny are really theirs has given rise, over the centuries, to the devising of myriad methods of maintaining women's chastity, from chastity belts to seclusion to partial or total covering of the body in public, with sometimes severe penalties for transgressors. In some parts of the world, this pattern has changed to a more affirmative vision of female sexuality and softer notions of masculine sexuality, with an accompanying affirmation that sexuality includes homosexuality as well as heterosexuality.

These norms and values attached to gender roles and sexuality have a considerable impact on sexual and reproductive health. For instance, in sub-Saharan Africa in 2003, young women aged 15–24 years were at least three times more likely to be infected with HIV than young men, as compared with the same numbers in 1985 (UNAIDS *et al.*, 2004). The major factors that contribute to this have been identified as the culture of silence around sexuality, exploitative transactional and intergenerational sex, and violence against women in relationships (UNAIDS *et al.*, 2004). In countries as different as Côte d'Ivoire (Painter *et al.*, 2007), India (Schensul *et al.*, 2006), and Mexico (Hirsch *et al.*, 2007), marriage – implying sexual exclusivity – may not afford protection for women from HIV or other STIs because of men's extramarital sexual behavior. At the same time, poverty may often push young – especially urban – women into early, premarital sexual relations, incurring a higher risk of contracting HIV (Hattori and Dodoo, 2007), but this does not increase men's vulnerability to HIV in marriage (Glynn *et al.*, 2003).

Masculine sexuality often incorporates elements of self-destructive risk, conquest, and even violence. This gender-based difference has an enormous impact on girls' and women's (and, to a lesser extent, boys' and men's) sexual and reproductive health. Data emerging on the extent of violence against women show that, in selected countries, lifetime prevalence of sexual violence by an intimate partner ranges between 6–59% (Garcia-Moreno *et al.*, 2006). The prevalence of sexual abuse before the age of 15 years varied from 1–21%, and an adult male family member was usually the perpetrator. Overall, more than 5% of women reported their first sexual experience as forced, with the prevalence ranging from 1% in Japan to 24–30% in Bangladesh, 17% in Ethiopia, 24% in Peru, and between 14–17% in Tanzania. Other population-based surveys indicate that between 3–23% of married young women (aged 15–24 years) in some developing countries (Cambodia, Colombia, Haiti, India, Nepal, Nicaragua, and Zambia) have ever experienced nonconsensual sex by a current or former spouse. In most settings, sexual coercion is initiated early in marriage, and in settings characterized by early and arranged marriage, especially at sexual initiation. Sex workers – the majority of whom are women, sometimes very young and trafficked into prostitution – are particularly vulnerable to forced sex and its attendant sexual and reproductive ill-health consequences since sex work is often illegal and sex workers considered legitimate targets of violence. Surveys in different continents show that up to 70% of sex workers report being raped, frequently by uniformed men (World Health Organization, 2006). Irrespective of the setting, gender power imbalances tend to underlie the persistence of forced sex – both marital and extramarital – among young women (Jejeebhoy and Bott, 2003). It is important to note that a small proportion of boys (see **Figure 2**) also undergo coerced first sex – often from older boys or men.

The sexual and reproductive health outcomes of such coercive and/or violent relationships include unwanted pregnancy and an associated risk of unsafe abortion (especially in countries where abortion is legally restricted), chronic pelvic pain, sexually transmitted infections including HIV, and painful intercourse. Having suffered sexual abuse, particularly in childhood, is also associated with increased risk-taking, such as alcohol and drug abuse, multiple sex partners, low self-esteem, and depression. Interventions to change gender-related inequalities, especially with regard to sexual behavior, may be one of the most important contributions to the improvement of sexual and reproductive health in the world.

Gender norms and values also contribute to early marriages and early childbearing, lack of knowledge about one's body, embarrassment around discussing contraception with partners, and silence around sexual and reproductive health problems and conditions, all of which can lead to delays in health-care seeking or unwillingness to disclose or discuss one's problem with a health provider (Roudi-Fahimi, 2006). Despite the considerable decline in early marriages in many parts of the

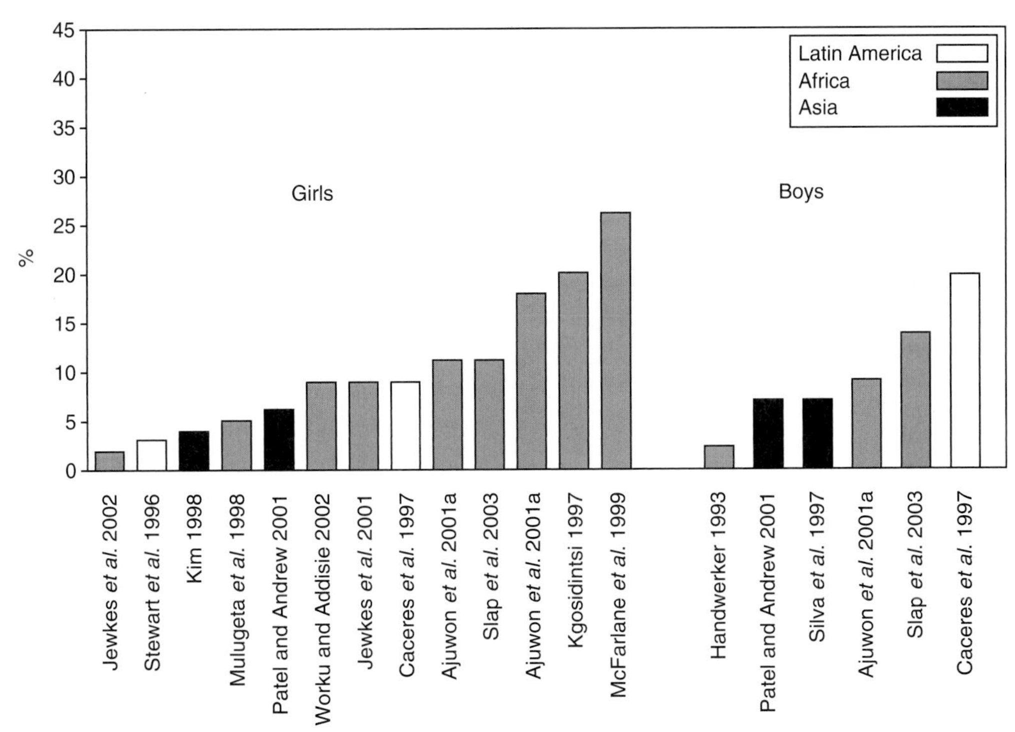

Figure 2 Percentage of young women and men (aged 10–24 years) reporting a coercive sexual relationship. Reproduced with permission from Jejeebhoy S and Bott S (2003) *Non-Consensual Sexual Experiences of Young People: A Review of the Evidence from Developing Countries*. New Delhi, India: Population Council.

world, between 30–45% of women are married or in union by age 18 years in much of sub-Saharan Africa, South-Central and Southeast Asia, and some parts of Latin America and the Caribbean (Mensch *et al.*, 2005). Among men, however, marriage during the teenage years is very rare, and marriage in the early 20s has also declined in most places over the past decade. Early marriage for women leads to early childbearing, which carries well-documented increased risks of death in pregnancy or childbirth, premature labor, complications during delivery, low birthweight, and a higher chance that the newborn will not survive (Senderowitz, 1995).

Female genital mutilation (FGM) – also called female circumcision or excision – is closely related to perceptions of femininity and female sexuality. It is often performed on young girls as a rite of passage into adulthood (although in some parts of the world it is practiced on girls soon after they are born, and sometimes on adult women). Current estimates indicate that between 100–140 million girls and women have undergone FGM (World Health Organization, 2000), and that 3 million are at risk of undergoing FGM each year (Yoder *et al.*, 2004). It is practiced in 28 countries across northern and central Africa, the Middle East, and some countries in Asia. It is also practiced by members of communities from these regions who have emigrated to Europe and North America. The practice of FGM is considered a violation of human rights, and a marker for serious gender-based inequalities.

The health consequences of FGM – particularly the more extreme types of mutilation – include hemorrhage, infection, chronic pelvic pain, infertility, painful urination, difficulties with menstruation and sexual intercourse, problems during childbirth, and death. A multicountry study found that the death rate among babies during and immediately after birth was between 15–55% higher (depending on the severity of the FGM) in mothers with FGM than in those without. It also found an increased risk of postpartum hemorrhage and cesarean section for the mothers with FGM (World Health Organization Study Group on Female Genital Mutilation and Obstetric Outcome, 2006).

Access to Resources

Gender roles have an impact on how and whether men and women have access to essential resources such as education, information, disposable income, and health services. All of these resources in their turn have an impact on sexual and reproductive health.

Although the gender gap in education has been closing (dramatically in some places) over the past decade, girls are still less likely to stay in school than boys in a fair number of countries. The link between education and (sexual and reproductive) health has been well-established, with many studies having documented greater use of contraception, antenatal care services, and

Table 1 Contraceptive prevalence rates by women's years of education, select Latin American countries (Demographic and Health Surveys data)

Country	*All*	*0 years of schooling*	*1–3 years of schooling*	*4–6 years of schooling*	*7–9 years of schooling*	*≥ 10 years of schooling*
Bolivia	30	12	23	31	43	53
Brazil	66	47	59	71	76	73
Colombia	65	53	61	65	73	73
Dom. Rep	50	38	47	51	49	57
Ecuador	44	18	37	43	50	61
El Salvador	47	37	42	55	51	64
Guatemala	23	10	24	42	60	60
Mexico	53	25	44	58	70	69
Peru	46	19	33	46	60	67

Modified from Martin TC and Juarez F (1995) The impact of women's education on fertility in Latin America: Searching for explanations. *International Family Planning Perspectives* 21: 52–57, 80.

skilled attendance at childbirth by women with higher levels of education (see **Table 1**). Uneducated and rural women have the highest unmet need for contraception and the least access to health services.

While men the world over tend to have more years of schooling than women, men also have needs for information and services that may not be easy for them to fulfill. For instance, they use health services less than women and so may be less exposed to health-specific information. Especially in the domain of sexuality, social norms may dictate that men 'know' what should be done, and they do not want to appear ignorant or foolish. Yet, fewer than one-third of men in many developing countries know that two ways of avoiding sexually transmitted infections are condom use and either abstinence or having only one, uninfected partner (Alan Guttmacher Institute, 2003).

Sexual relationships have an economic basis that determines the terms on which sex takes place. In marriages, women who are financially and socially dependent on men may not feel able to refuse sex or particular kinds of sex. While boys and men also go into prostitution, the vast majority of sex workers in the world are women, nearly always for economic reasons. If women have no control of financial resources, they may turn to prostitution as a seemingly easy way to earn money. Thus, the very act of attempting to gain access to economic resources can put women at risk of sexual ill-health. Because sex work is privately condoned but publicly condemned, women all over the world move in and out of situations where they are condemned for immorality while earning money, frequently to support their children. The power implicit in economic relationships also can lead young boys into exploitative situations with older men.

Economic dependency has an impact on whether or not women have access to food and health services they need during pregnancy, childbirth, and the postpartum period. A large proportion of women who become pregnant in developing countries are poorly nourished and frequently anemic. Anemia compromises immunity to infections and healing capacity, making childbirth much more risky than for well-nourished women. In addition, where women's status is low – as it is in many societies and families – family resources are unlikely to be mobilized to cover costs of emergency care and transportation that could save women from dying or becoming severely ill from pregnancy or birth-related causes. Where family resources are almost nonexistent, women's needs in childbirth are therefore not likely to be met. Compounded, as is often the case, by a weak health system that cannot provide essential health services such as pregnancy, delivery, and postpartum care free to poor women, this situation is a stark example of how gender and poverty work together to the detriment of women, with sometimes fatal consequences.

Gender Biases in Health Services

While it is difficult to generalize from a global point of view, there is evidence that many health service providers make different assumptions about their male and female patients. Health-care providers also have gender norms and values instilled. Their lack of understanding of the ways in which gender roles affect women's ability to access health services can lead to 'victim blaming,' for instance, when women do not seek timely antenatal care or emergency obstetric care. In societies where sexual relations outside marriage are condemned, pregnant unmarried women may be treated with disrespect, while unmarried men who have a sexually transmitted infection would not be treated in the same way.

Typically, sexual and reproductive health-care providers have little or no training in talking about sexuality and sexual relationships, nor are they prepared to listen to and treat people who fall outside of gender and related sexual norms, such as men who have sex with men, lesbians, transgender people, or even sex workers. For health-care providers – be they men or women – being aware of gender norms and values and the impact these have on women's and men's health (as described above) is an essential tool for improving the quality of care in sexual and reproductive health services. Numerous studies also show that the sex of the health-care provider plays a key role in a patient's satisfaction and whether or not he or she will continue to use a health service; generally speaking, both women and men prefer same-sex providers. Health services may also be organized in gender-insensitive ways. For example, user fees may be introduced for routine sexual and reproductive health commodities that women need, such as contraceptives, in situations where other essential medicines are provided free of charge.

At the health policy level, the central part played by gender inequalities in sexual and reproductive (ill-) health makes it imperative to give priority to services that only women need – such as emergency and other obstetric care, antenatal and postnatal care, abortion services, and repair of pregnancy-related problems such as obstetric fistulas and urinary incontinence. Examples from both developed and developing countries show that, where such services have been given high priority, the health of women (and of their infants) has improved enormously (Van Lerberghe and De Brouwere, 2001).

Laws and Policies

Gender inequalities are often captured and reinforced in laws and policies, some of which have a direct effect on sexual and reproductive health. Marriage law is one example. In many countries, both the legal minimum age and actual age of marriage, particularly for girls, are still very low, leading to early or very early childbearing with its attendant risks, as discussed above. The enactment of laws that prohibit early marriage, while not sufficient to halt the practice initially, has been shown to contribute to positive change and an increased protection of girls and women (UNICEF, 2001).

The nonrecognition in law of rape within marriage is a serious gap from which women suffer. Since the Declaration on Violence Against Women (1993) and the Fourth World Conference on Women in Beijing in 1995, many countries have enacted laws to criminalize the various forms of domestic violence against women, including marital rape, and to protect women's rights and safety. Creating such a legal environment contributes to changing the societal perception of women, so that violence becomes unacceptable behavior.

Despite many years of international attention to promoting women's autonomous decision making as central to their human rights, many countries still have in place laws or policies that require a husband's approval for a woman to use contraception, to have professional assistance during childbirth, or to have access to her own medical records. Other laws may allow only married women to obtain access to health services. These regulations and rules represent a barrier to women obtaining access to services that they need in order to remain healthy, which are either not required for men or are irrelevant since men do not use the services.

Perhaps the most serious example of gender inequalities reinforced by law is the restriction on access to safe abortion services. Worldwide, for every 1000 women of childbearing age, 29 are estimated to have an induced abortion each year – a rate that is very similar in developed and developing regions (Sedgh *et al.*, 2007). Nearly half of these abortions are illegal. Unintended/unwanted pregnancies occur because of women's and men's lack of access to, or lack of use of, appropriate contraceptive methods, method failure, sexual coercion or rape, and changing economic circumstances. Women will therefore always have need for safe abortion services, something that men will never need. Where laws severely restrict such services to cases where a woman's life is threatened, it has been clearly demonstrated that deaths and injury from the complications of unsafe abortion are much higher than in countries that provide legal access to abortion on broad grounds including on request (Berer, 2004). Providing safe abortion services is a cost-efficient way of contributing substantially to the reduction of maternal mortality and morbidity around the world.

Conclusion

The social norms and values that govern what women and men do and how they behave have an extensive impact on sexual and reproductive health. Biologically, women and men are differently susceptible to certain conditions such as sexually transmitted infections, but gender dimensions can significantly magnify such vulnerability through specific behaviors that are socially acceptable (or not acceptable), through access (or lack of it) to essential resources, through the ways in which institutions perceive and treat men and women, and through laws that may uphold gender inequalities or not protect one sex or the other against injustices.

For these reasons, sexual and reproductive health services have to be more than medical interventions. Gender-sensitive sexual and reproductive health services call for a mindset that takes into account not only the client's medical history but also the many ways in which gender affects his or her sexual and reproductive behavior. In serving men, services can help counteract negative aspects of masculinity contributing to risky behaviors; promote respect for every woman's right to self-determination in matters related to sexuality and reproduction; and support a woman's dignity, and the exercise of her rights.

For the past two decades, interventions to counteract these gender inequalities have ranged from the passing of laws to support an end to gender discrimination, often as part of the implementation of the Convention on the

Elimination of All Forms of Discrimination Against Women, to the training of health workers and policy makers in gender analysis and approaches. They have included programs to empower women through becoming financially independent and, more recently, programs specifically targeted to men on sexual and reproductive health and encouraging zero tolerance for violence. While some advances have been made, particularly in the widespread recognition of gender inequalities, such interventions constitute only some of the necessary elements on a long path toward lasting change in gender roles and relations.

Disclaimer

J C Cottingham is a staff member of WHO. The author alone is responsible for the views expressed in this publication and they do not necessarily represent the decisions a policies of WHO.

See also: Abortion; Reproductive Rights; Sexual and Reproductive Health: Overview; Sexual Health; Sexual Violence; Violence Against Women.

Citations

AbouZahr C and Vaughan P (2000) Assessing the burden of sexual and reproductive ill-health. *Bulletin of the World Health Organization* 78: 655–666.

Alan Guttmacher Institute (2003) *In Their Own Right: Addressing the Sexual and Reproductive Health Needs of Men Worldwide*. New York: Alan Guttmacher Institute.

Berer M (2004) National laws and unsafe abortion: The parameters of change. *Reproductive Health Matters* 12 (supplement 24): 1–8.

Garcia-Moreno C, Jansen H, Ellsberg M, *et al.* (2006) Prevalence of intimate partner violence: Findings from the WHO multi-country study on women's health and domestic violence. *Lancet* 368(9543): 1260–1269.

Glynn JR, Carael M, Buve A, *et al.* (2003) HIV risk in relation to marriage in areas with high prevalence of HIV infection. *Journal of Acquired Immune Deficiency Syndromes* 33: 526–535.

Hattori MK and Dodoo N-A (2007) Cohabitation, marriage and 'sexual monogamy' in Nairobi's slums. *Social Science and Medicine* 64: 1067–1078.

Hirsch JS, Meneses S, Thompson B, *et al.* (2007) The inevitability of infidelity: Sexual reputation, social geographies, and marital HIV risk in rural Mexico. *American Journal of Public Health* 97: 986–996.

Howson CP, Harrison PF, Hotra D and Law M (eds.) (1996) *In Her Lifetime: Female Morbidity and Mortality in Sub-Saharan Africa.* Washington, DC: National Academy Press.

Jejeebhoy S and Bott S (2003) *Non-Consensual Sexual Experiences of Young People: A Review of the Evidence from Developing Countries*. New Delhi, India: Population Council.

Klugman B and Theobald S (2006) Gender, health and development III: Engendering health research. *Progress in Development Studies* 6: 337–342.

Martin TC and Juarez F (1995) The impact of women's education on fertility in Latin America: Searching for explanations. *International Family Planning Perspectives* 21: 52–57,80.

Mensch BS, Singh S, and Casterline JB (2005) *Trends in the Timing of First Marriage Among Men and Women in the Developing World*. Working Paper No. 202. New York: Population Council.

Painter TM, Diaby KL, Matia DM, *et al.* (2007) Faithfulness to partners: A means to prevent HIV infection, a source of HIV infection risks, or both? A qualitative study of women's experiences in Abidjan, Côte d'Ivoire. *African Journal of AIDS Research* 6: 25–31.

Roudi-Fahimi F (2006) *Gender and Equity in Access to Health Care Services in the Middle East and North Africa*. Population Reference Bureau. http://www.prb.org/Articles/2006/Gender and Equity in Access to Health Care Services in the Middle East and North Africa.aspx (accessed November 2007).

Schensul SL, Mekki-Berrada A, Nastasi BK, *et al.* (2006) Men's extramarital sex, marital relationships and sexual risk in urban poor communities in India. *Journal of Urban Health* 83: 614–624.

Sedgh G, Henshaw S, Singh S, Åhman E, and Shah I (2007) Induced abortion: Estimated rates and trends worldwide. *Lancet* 370: 1338–1345.

Senderowitz J (1995) *Adolescent Health: Reassessing the Passage to Adulthood.* World Bank Discussion Paper No. 272. Washington, DC: The World Bank.

UNAIDS, UNFPA, and UNIFEM (2004) *Women and HIV/AIDS: Confronting the Crisis*. Geneva, Switzerland: UNAIDS.

UNICEF Innocenti Research Centre (2001) Early marriage, child spouses. *Innocenti Digest* 7.

Van Lerberghe W and De Brouwere V (2001) Of blind alleys and things that have worked: History's lessons on reducing maternal mortality. In: De Brouwere V and Van Lerberghe W (eds.) *Safe Motherhood Strategies: A Review of the Evidence*, pp. 7–33. Antwerp, Belgium: ITG press.

WHO Study Group on Female Genital Mutilation and Obstetric Outcome (2006) Female genital mutilation and obstetric outcome: WHO collaborative prospective study in six African countries. *Lancet* 367: 1835–1841.

World Health Organization (2000) *Female Genital Mutilation.* Fact Sheet No. 241. Geneva, Switzerland: World Health Organization.

World Health Organization (2006) *Violence Against Sex Workers and HIV Prevention.* Information Bulletin Series No. 3. Geneva, Switzerland: World Health Organization and The Global Coalition on Women and AIDS.

World Health Report (2004) *Changing History*. Geneva, Switzerland: World Health Organization.

Yoder PS, Abderrahim N, and Zhuzhuni A (2004) *DHS Comparative Reports No. 7: Female Genital Cutting in the Demographic and Health Surveys: A Critical and Comparative Analysis*. Calverton, MD: ORC Macro.

Zierler S and Krieger N (1997) Reframing women's risk: Social inequalities and HIV infection. *Annual Review of Public Health* 18: 401–436.

Further Reading

Berer M (ed.) (2004) Abortion law, policy and practice in transition. *Reproductive Health Matters* 12 (supplement 24).

Cottingham J and Myntti C (2002) Reproductive health: Conceptual mapping and evidence. In: Sen G, George A, and Ostlin P (eds.) *(2002) Engendering International Health: The Challenge of Equity.* Cambridge, MA: MIT Press.

Sen G, George A, and Ostlin P (2002) *Engendering International Health: The Challenge of Equity*. Cambridge, MA: MIT Press.

WHO, UNFPA, and UNICEF (1997) *Joint Statement on Female Genital Mutilation*. Geneva, Switzerland: World Health Organization.

World Health Organization (2005) *WHO Multi-Country Study on Women's Health and Domestic Violence Against Women: Initial Results on Prevalence, Health Outcomes and Women's Responses*. Geneva, Switzerland: World Health Organization.

World Health Organization (2005) *Integrating Gender into HIV/AIDS Programmes*. Geneva, Switzerland: World Health Organization.

Relevant Websites

http://www.who.int/gender/en/ – World Health Organization, Department of Gender, Women and Health.

http://www.who.int/reproductive-health/gender/index.html – World Health Organization, Gender and Reproductive Rights.

Reproductive Rights

J N Erdman and R J Cook, University of Toronto, Toronto, Ontario, Canada

Introduction

Reproductive rights recognize that the sexual and reproductive health of both women and men requires more than scientific knowledge or biomedical intervention. It requires recognition and respect for the inherent dignity of the individual. Reproductive rights refer to the composite of human rights that protect against the causes of ill-health and promote sexual and reproductive well-being. An international consensus definition of the term, adopted at the 1994 United Nations International Conference on Population and Development (ICPD) in Cairo, Egypt, states that:

> reproductive rights embrace certain human rights that are already recognized in national laws, international human rights documents, and other consensus documents. These rights rest on the recognition of the basic right of all couples and individuals to decide freely and responsibly the number, spacing, and timing of their children and to have the information and means to do so, and the right to attain the highest standard of sexual and reproductive health. It also includes their right to make decisions concerning reproduction free of discrimination, coercion, and violence, as expressed in human rights documents.
> (United Nations, 1994, paragraph 7.3)

While reproductive rights also address the health needs of men, they are more critical for women. A major burden of disease in women relates to their reproductive capacity and the unequal treatment of women because of their gender.

The Evolution of an International Consensus

Human rights are defined in domestic laws, national constitutions, and regional and international treaties. Human rights embody norms that guide clinical care, health policy making, and wider social interactions. Moreover, human rights serve as standards that enable individuals to hold governments and other institutions accountable for violations of the embodied norms.

Rights in the Service of Health, Welfare, and Development

Sexuality and reproduction were long governed by religious and political ideology, enforced through criminal law. Individual control over these matters was regarded as threatening to moral values, gender hierarchies, and family security. Birth control, abortion, and certain forms of sexual behavior were, and in many countries remain, defined as crimes against morality (Cook *et al.*, 2003).

Empirical evidence of the dysfunction and harmful health effects of punitive laws contributed to the adoption of policies that promoted individuals' interests in their own health and welfare. At the 1968 International Conference on Human Rights in Tehran, Iran, reproductive health became a recognized subject of international human rights. Governments affirmed that "[p]arents have a basic human right to determine freely and responsibly the number and spacing of their children" (United Nations, 1968, paragraph 16). Public health and welfare interests were recognized as best served by individuals' free and responsible decision making.

The health and welfare rationale garnered support in international development debates regarding fertility reduction. In 1974, the first World Population Conference in Bucharest, Romania, promoted the reformulated basic right of "all couples and individuals ... to decide the number and spacing of their children and to have the information, education and means to do so" as an effective tool to achieve population and development goals (United Nations, 1974, paragraph 14f). At the 1984 International Conference on Population in Mexico City, Mexico, this language was reaffirmed (United Nations, 1984).

Rights in Respect of the Inherent Dignity of the Individual

During the United Nations Decade for Women (1976–85), a growing international women's rights movement also advocated for the right to reproductive decision making. While adopting the language of Tehran, Bucharest, and Mexico City, the movement articulated a different rationale to support the right. In 1975, the World Conference of the International Women's Year affirmed that "[t]he human body, whether that of woman or man, is inviolable and respect for it is a fundamental element of human dignity and freedom" (United Nations, 1976, Art. 11). The right to decide matters related to sexuality and reproduction derives from this respect for women's dignity and freedom. The conception of sexual and reproductive health from the perspective of women as equal and autonomous individuals was reaffirmed at subsequent international women's conferences in Copenhagen, Denmark (1980) and Nairobi, Kenya (1985) (United Nations, 1980, 1985).

In 1979, the United Nations General Assembly adopted the *Convention on the Elimination of All Forms of Discrimination against Women* (the *Women's Convention*) (United Nations, 1979). The treaty codified the right to reproductive decision making as a legally enforceable right. Prior to the *Women's Convention*, recognition of the right was limited to international conference documents that bind governments politically but not legally.

Article 16.1 of the *Women's Convention* specifically guarantees to women, on the basis of equality with men, the:

> same rights to decide freely and responsibly on the number and spacing of their children and to have access to the information, education, and means to enable them to exercise these rights.

In the tradition of the international women's movement, the right is recognized as an important end in itself. It serves the sexual and reproductive health needs and preferences of individual women. It is a human right premised on the inherent dignity of the individual.

The 1993 United Nations World Conference on Human Rights in Vienna reflected the transformation of women's interests from the periphery to the center of the human rights discourse. Women's rights were recognized as "an inalienable, integral and indivisible part of universal human rights" (United Nations, 1993, paragraph 18). The Conference reaffirmed:

> on the basis of equality between women and men, a woman's right to accessible and adequate health care and the widest range of family planning services, as well as equal access to education at all levels.
>
> (United Nations, 1993, paragraph 41)

An International Consensus on Reproductive Rights

At the 1994 International Conference on Population and Development in Cairo, Egypt, a human rights perspective was adopted in the development context. In a paradigm shift, the *ICPD Programme of Action* departs from the instrumentality of rights in the service of demographically based targets. It promotes policies that respect women's inherent dignity, recognizing individual women's perceptions, needs, and circumstances as the basis for reproductive decision making. The *ICPD Programme of Action* expressly acknowledges that:

> [t]he empowerment and autonomy of women and the improvement of their political, social, economic, and health status is a highly important end in itself.
>
> (United Nations, 1994, paragraph 4.1)

As defined in the *ICPD Programme of Action*, reproductive rights:

> rest on the recognition of the basic right of all couples and individuals to decide freely and responsibly the number, spacing, and timing of their children and to have the information and means to do so . . . free of discrimination, coercion and violence.
>
> (United Nations, 1994, paragraph 7.3)

More broadly, reproductive rights include the "right to attain the highest standard of sexual and reproductive health," defined as:

> a state of complete physical, mental and social well-being . . . in all matters relating to the reproductive system and to its functions and processes.
>
> (United Nations, 1994, paragraph 7.2)

Reproductive health further encompasses sexual health and:

> a satisfying and safe sex life . . . the purpose of which is the enhancement of life and personal relations.
>
> (United Nations, 1994, paragraph 7.2)

The promotion of these rights became:

> the fundamental basis for government- and community-supported policies and programmes in the area of reproductive health, including family planning.
>
> (United Nations, 1994, paragraph 7.2)

and the normative standard against which to assess government action.

At the 1995 Fourth World Conference on Women in Beijing, People's Republic of China, governments reaffirmed their commitment to the *ICPD Programme of Action*, and extended the definition of reproductive rights to recognize that:

> [t]he human rights of women include their right to have control over and decide freely and responsibly on matters related to their sexuality . . . free of coercion, discrimination, and violence.
>
> (United Nations, 1995, paragraph 96)

The *Beijing Platform for Action* further recognizes that:

> [i]n most countries, the neglect of women's reproductive rights severely limits their opportunities in public and private life, including opportunities for education and economic and political empowerment. The ability of women to control their own fertility forms an important basis for the enjoyment of other rights.
>
> (United Nations, 1995, paragraph 97)

By 2004, over 90% of governments committed to the *ICPD Programme of Action* and the *Beijing Platform for Action* had adopted legislation or implemented policies in furtherance of the consensus objectives (United Nations, 2004).

Since 2000, the United Nations Millennium Development Goals (MDGs) have become the primary international development framework. The MDGs include: poverty eradication, improvement in education, promotion of gender equality and women's empowerment, reduction of child mortality and improving maternal health, and the combating of HIV/AIDS. In 2006, the United Nations General Assembly adopted "universal access to reproductive health by 2015" as an explicit target under the MDG framework (United Nations, 2006).

The Protection of Reproductive Rights as Human Rights

Reproductive rights are protected through a composite of human rights guaranteed in national laws, constitutions, and regional and international treaties. Their content and meaning are informed by the *ICPD Programme of Action* and *Beijing Platform for Action*. Article 12(2) of the Constitution of the Republic of South Africa, for example, guarantees everyone "the right to make decisions concerning reproduction" (Government of South Africa, 1996). Albania, Benin, and Mali adopted the *ICPD Programme of Action* definition of reproductive health into national law (Republic of Albania, 2002; Mali, 2002; Benin, 2003).

Professional and ministerial guidelines also ensure the protection of reproductive rights. The International Federation of Gynecology and Obstetrics, through its Committee for the Ethical Aspects of Human Reproduction and Women's Health, issued ethical guidelines on the duty of providers to respect women's rights (FIGO, 2003). The Brazilian Ministry of Health published *Guidelines for the Prevention and Care of the Consequences of Sexual Violence Against Women and Adolescents* to ensure women's access to appropriate legal abortion services when pregnancy results from rape (Brazil Ministry of Health, 1998).

Reproductive rights may be divided into three broad categories of rights: (1) rights to reproductive self-determination, (2) rights to sexual and reproductive health services, information, and education, and (3) rights to equality and nondiscrimination.

Rights to Reproductive Self-Determination

Rights to reproductive self-determination recognize women as autonomous agents with authority to make sexual and reproductive decisions free from interference, discrimination, coercion, and violence.

As explained by the Committee on Elimination of Discrimination against Women (CEDAW), the right of nondiscrimination obligates governments to:

> refrain from obstructing action taken by women in pursuit of their health goals . . . includ[ing] laws that criminalize medical procedures only needed by women and that punish women who undergo those procedures.
> (CEDAW, 1999, paragraph 14)

For example, several treaty-monitoring bodies have called on governments to review criminal laws that restrict access to abortion. Where evidence discloses that high rates of maternal deaths are related to restrictive laws, the Human Rights Committee regards the maintenance of such laws and the lack of access to safe, legal abortion services as a violation of a woman's right to life. The Human Rights Committee held the Peruvian government in violation of multiple rights under the International *Covenant on Civil and Political Rights* (the *Political Covenant*) (United Nations, 1966a), including freedom from cruel, inhumane, and degrading treatment, for denying a 17-year-old a legal abortion despite evidence of danger to her life and health following the diagnosis of an anencephalic fetus (*Karen Llontoy v. Peru*, 2003).

The *Protocol on the Rights of Women in Africa*, which supplements the *African Charter on Human and Peoples' Rights* (Organization of African Unity, 2003), explicitly requires governments to take all appropriate measures to:

> protect the reproductive rights of women by authorising medical abortion in cases of sexual assault, rape, incest, and where the continued pregnancy endangers the mental and physical health of the mother or the life of the mother or the foetus. (Art. 14(2) (c))

Since the 1990s, abortion law reform has been achieved in several countries. The Constitutional Court of Colombia recently liberalized a criminal prohibition to render abortion legal in cases of fetal malformation, rape, and risk to the life or health of the woman (Republica de Colombia, Corte Constitucional).

Rights to self-determination address interferences in decision making by both state and private actors, including husbands and parents. CEDAW advises that governments:

> should not restrict women's access to health services . . . on the grounds that women do not have the authorization of husbands, partners, parents, or health authorities.
> (CEDAW, 1999, paragraph 14)

Under the *European Convention for the Protection of Human Rights and Fundamental Freedoms* (Council of Europe, 1950), a UK law that permits women to terminate their pregnancies without spousal authorization was upheld as protective of women's right to privacy (*Paton v. United Kingdom*, 1980).

The Committee on the Rights of the Child, which oversees adherence to the *Convention on the Rights of the Child* (CRC) (United Nations, 1989) by state parties to the Convention, encourages governments to ensure access to health services in a manner consistent with the evolving capacities of the child, and not according to strict age limitations. Best practice guidelines issued by the UK Department of Health, in accordance with the CRC

General Comment on HIV/AIDS and the rights of the child (CRC, 2003a, paragraph 17), emphasize privacy rights and professional duties of confidentiality in the provision of adolescent sexual and reproductive health services (Department of Health, United Kingdom, 2004).

Governments are further obligated to ensure that women can exercise their rights to reproductive self-determination free from discrimination, violence, and coercion. Forced sterilization policies violate women's rights to bodily integrity and freedom from torture or other cruel, inhumane, or degrading treatment. The Committee on the Elimination of Racial Discrimination (CERD), which oversees adherence to the *Convention on the Elimination of All Forms of Racial Discrimination* (United Nations, 1965) by state parties to the Convention, encourages governments to investigate and punish perpetrators of coercive sterilization policies and to provide compensation for victims. Under a friendly settlement facilitated by the Inter-American Commission on Human Rights, the government of Peru recognized responsibility for violating rights to life, physical integrity, humane treatment, and equality in relation to the forced sterilization of an indigenous woman and her subsequent death following a substandard surgical operation (*Maria Mamerita Mestanza Chávez v. Peru*, 2003).

Recognizing that reproductive decision making occurs within broader constraints of family, community, and state, women's effective exercise of rights to reproductive self-determination depends on their full and equal status in civil, cultural, economic, political, and social life. The *Women's Convention* obligates governments to take all appropriate measures to eliminate social and cultural patterns and practices that perpetuate notions of women's inferiority. The *Convention on the Rights of the Child* requires governments to take measures to abolish traditional practices harmful to children's health, including child and forced marriage (Art. 24.3). Enacting and implementing a minimum legal age of marriage at 18 years, as recommended by both the CRC and CEDAW, enables young girls to delay childbirth and to decrease risks associated with premature pregnancy and childbirth (CRC, 2003b, paragraph 20; CEDAW, 1994, paragraph 36).

Rights to Sexual and Reproductive Health Services, Information and Education

As elaborated by the Committee on Economic, Social and Cultural Rights (CESCR), the right to health under the *International Covenant on Economic, Social and Cultural Rights* (United Nations, 1966b), encompasses an entitlement to the information, education, and means necessary to realize the highest attainable standard of sexual and reproductive health. This includes:

> access to family planning, pre- and post-natal care, emergency obstetric services and access to information, as well as to resources necessary to act on that information.
>
> (CESCR, 2000, Paragraph 14)

The right to health obligates governments to ensure that health facilities, goods, and services related to sexual and reproductive health are available in sufficient quantity, accessible to everyone without discrimination, acceptable (respectful of medical ethics and culturally appropriate), and of good scientific and medical quality (CESCR, 2000, Paragraph 12).

Rights to sexual and reproductive health services may thus require increased expenditure or the redistribution of public resources to improve efficient access to and quality of services. Pursuant to the right to life under the *Political Covenant*, the Human Rights Committee has required governments to adopt "positive measures" necessary to preserve life, including the provision of essential obstetric care to reduce high rates of avoidable maternal death. Under the country's constitutional rights to life and health, the Supreme Court of Venezuela required the government to reallocate budgets to provide access to antiretroviral therapy within the public system (*Cruz Bermudez et al.* v. Ministerio de Sanidad y Asistencia Social, 1999). The Constitutional Court of South Africa similarly required the reasonable provision of treatment to pregnant women with HIV/AIDS under the constitutional right to health (*Minister of Health v. Treatment Action Campaign*, 2002). Rights to health services further encompasses the right to effective participation in all political processes, including the setting of spending priorities and the evaluation of government budgets, which affect sexual and reproductive health.

In addition to timely and appropriate health care, the right to health extends:

> to the underlying determinants of health, such as ... access to health-related education and information, including on sexual and reproductive health.
>
> (CESCR, 2000, Art. 11)

The *Women's Convention* explicitly recognizes a right of:

> [a]ccess to specific educational information to help to ensure the health and well-being of families, including information and advice on family planning. (Art. 10(h))

Rights to Equality and Nondiscrimination

Reproductive rights are guaranteed to everyone without discrimination on the basis of marginalized status, including, for example, poverty, age, sex, race, ethnicity, disability, health or marital status, geography, and sexual orientation. CESCR recognizes that, where services are not provided on principles of substantive equality, states are in violation of the right to health (CESCR, 2000, Art. 12).

Substantive equality requires that governments do not impose burdens or withhold benefits on the basis of presumed group characteristics unrelated to an individual's health needs. CERD regards targeting for mandatory HIV testing on the basis of national origin or race as a form of discrimination.

Substantive equality further requires governments to recognize and address particular health needs of population subgroups. The CEDAW Committee recognizes that societal factors determinative of health status vary between women and men and among women themselves (CEDAW, 1999, Art. 6). Governments are obligated to address these differences. Under the *International Convention on the Elimination of All Forms of Racial Discrimination*, a woman's inability to access appropriate reproductive health-care services because of her race, color, descent, national, or ethnic origin violates the right of nondiscrimination.

Conclusion

Reproductive rights refer to a composite of human rights guaranteed in national laws, constitutions, and regional and international treaties that can be applied to protect against the causes of ill-health and promote sexual and reproductive well-being. These rights may be broadly divided into three categories: (1) rights to reproductive self-determination, (2) rights to sexual and reproductive health services, information, and education, and (3) rights to equality and nondiscrimination. Rights to reproductive self-determination recognize women as autonomous agents with the authority to make sexual and reproductive decisions free from interference, coercion, and violence. Reproductive rights also encompass entitlements to the means necessary, including health facilities, goods, services, information, and education, to realize the highest attainable standards of sexual and reproductive health. Reproductive rights further guarantee that all couples and individuals have the right to decide freely on matters related to their sexual and reproductive health, and the means to do so, without discrimination and on the basis of substantive equality. While reproductive rights are instrumental to achieving population, health, and development goals, they are important in themselves. Reproductive rights, as all human rights, are intended to protect the inherent dignity of the individual.

See also: Abortion; Gender Aspects of Sexual and Reproductive Health; Sexual and Reproductive Health: Overview; Sexual Health.

Citations

Benin (2003) *Law No. 2003–04 on Reproductive and Sexual Health.*

Brazil Ministry of Health (1998) *Norma Técnica de Prevenção e Tratamento dos Agravos Resultantes da Violência Sexual Contra Mulheres e Adolescentes*. Brasilía DF: Ministério da Saúde.

CEDAW (1994) *General Recommendation 21: Equality in Marriage and Family Relations*. A/49/38. New York: United Nations.

CEDAW (1999) *General Recommendation 24: Women and Health*. A/54/38/Rev.1. New York: United Nations.

CESCR (2000) *General Comment 14: The Right to the Highest Attainable Standard of Health*. E/C.12/2000/4. New York: United Nations.

Cook RJ, Dickens BM, and Fathalla MF (2003) *Reproductive Health and Human Rights: Integrating Medicine, Ethics and Law*. Oxford, UK: Clarendon Press.

Council of Europe (1950) *European Convention for the Protection of Human Rights and Fundamental Freedoms*. Strasbourg, France: Council of Europe.

CRC (2003a) *General Comment 3: HIV/AIDS and Rights of the Child*. CRC/GC/2003/1. New York: United Nations.

CRC (2003b) *General Comment 4: Adolescent Health and Development in the Context of the Convention on the Rights of the Child*. CRC/GC/2003/4. New York: United Nations.

Cruz Bermudez et al. v. Ministerio de Sanidad y Asistencia Social (MSAS) (1999) Cano No. 15789 (Supreme Court Venezuela).

Department of Health, United Kingdom (2004) *Best Practice Guidance for Doctors and Other Health Professionals on the Provision of Advice and Treatment to Young People Under 16 on Contraception, Sexual and Reproductive Health*. London: Department of Health, United Kingdom.

FIGO (2003) *Recommendations on Ethical Issues in Obstetrics and Gynecology by the FIGO Committee for the Ethical Aspects of Human Reproduction and Women's Health*. London: FIGO.

Government of South Africa (1996) *Constitution of the Republic of South Africa*. PCT 108 of 1996.

Karen Llontoy v. Peru (2003) Human Rights Committee 85th Session. Case No. CCPR/C/85/D/1153/2003.

Mali (2002) *Law No. 02–44 on Reproductive Health.*

María Mamerita Mestanza Chávez v. Peru (2003) *Petition 12.191*. Friendly Settlement (Inter-American Commission on Human Rights).

Minister of Health v. Treatment Action Campaign (TAC) (2002) Case CCT 8/02 (Constitutional Court of South Africa).

Organization of African Unity (2003) *Protocol to the African Charter on Human and Peoples' Rights on the Rights of Women in Africa*. CAB/LEG/66.6. Maputo: AOU.

Paton v. United Kingdom (1980) 3 E.H.R.R. 408 (European Commission of Human Rights).

Republica de Colombia, Corte Constitucional (2006) *Comunicado de Prensa Sobre la Sentencia Relativa al Delito de Aborto.*

Republic of Albania (2002) *Law No. 8876 on Reproductive Health.*

United Nations (1965) *International Convention on the Elimination of All Forms of Racial Discrimination*. A/6014. New York: United Nations.

United Nations (1966a) *International Covenant on Civil and Political Rights*. A/6316. New York: United Nations.

United Nations (1966b) *International Covenant on Economic, Social and Cultural Rights*. A/6316. New York: United Nations.

United Nations (1968) *Proclamation of Tehran, Final Act of the International Conference on Human Rights*. A/Conf.32/41. New York: United Nations.

United Nations (1974) *Report of the United Nations World Population Conference*. A/Conf.60/19. New York: United Nations.

United Nations (1976) *Report of the World Conference of the International Women's Year*. A/Conf.66/34. New York: United Nations.

United Nations (1979) *Convention on the Elimination of All Forms of Discrimination against Women*. A/Res/34/180. New York: United Nations.

United Nations (1980) *Report of the World Conference of the United Nations Decade for Women: Equality, Development and Peace*. A/Conf.94/35. New York: United Nations.

United Nations (1984) *Report of the International Conference on Population, 1984*. E/Conf.76/19. New York: United Nations.
United Nations (1985) *Report of the World Conference to Review and Appraise the Achievements of the United Nations Decade for Women: Equality, Development and Peace*. A/Conf.116/28. New York: United Nations.
United Nations (1989) *Convention on the Rights of the Child*. A/44/49. New York: United Nations.
United Nations (1993) *Vienna Declaration, World Conference on Human Rights*. A/Conf.157/24. New York: United Nations.
United Nations (1994) *Programme of Action of the International Conference on Population and Development*. A/Conf.171/13. New York: United Nations.
United Nations (1995) *Beijing Declaration and the Platform for Action, Fourth World Conference on Women*. A/CONF.177/20. New York: United Nations.
United Nations (2004) *Report of the Secretary-General on the review and appraisal of the progress made in achieving the goals and objectives of the Programme of Action of the International Conference on Population and Development*. E/CN.9/2004/3. New York: United Nations.
United Nations (2006) *Report of the Secretary-General on the work of the Organization*. A/61/1. New York: United Nations.

Further Reading

Center for Reproductive Law and Policy (2000) *Reproductive Rights 2000: Moving Forward*. New York: Center for Reproductive Law and Policy.
Center for Reproductive Rights (2002) *Bringing Rights to Bear: An Analysis of the Work of UN Treaty Monitoring Bodies on Reproductive and Sexual Rights*. New York: Center for Reproductive Rights.
Center for Reproductive Rights (various years) *Women of the World: Laws and Policies Affecting Their Reproductive Lives: East and South East Asia* (2005), *South Asia* (2004), *Eastern Central Europe* (2000), *Francophone Africa* (2000), *Anglophone Africa* (1997), *Latin America and the Caribbean* (1997). New York: Center for Reproductive Rights.
Chavkin W and Chesler E (eds.) (2005) *Where Human Rights Begin: Health, Sexuality, and Women in the New Millennium*. Princeton, NJ: Rutgers University Press.
Coliver S (ed.) (1995) *The Right to Know: Human Rights and Access to Reproductive Health Information*. London: Article 19.
Cook RJ and Dickens BM (2003) Human rights dynamics of abortion law reform. *Human Rights Quarterly* 25: 1–59.
Cook RJ and Ngwena C (2006) Women's access to health care: The legal framework. *International Journal of Gynecology and Obstetrics* 94: 216–225.
Corrêa S and Reichmann R (1994) *Population and Reproductive Rights: Feminist Perspectives from the South*. London: Zed Books.
Freedman LP and Isaacs SL (1993) Human rights and reproductive choice. *Studies in Family Planning* 24: 18–30.
Ngwena C (2004) Access to legal abortion: Developments in Africa from a reproductive and sexual health rights perspective. *South African Journal of Public Law* 19: 328–350.
Organization of African Unity (1981) *African Charter on Human and Peoples' Rights*. CAB/Leg/67/3/Rev.5. Addis Abbaba: Organization of African Unity.
Petchesky RP and Judd K (1998) *Negotiating Reproductive Rights: Women's Perspectives Across Countries and Cultures*. London: Zed Books.
Sen G, Germain A, and Chen L (1994) *Population Policies Reconsidered: Health, Empowerment and Rights*. Cambridge, MA: Harvard University Press.
United Nations Millennium Project (2005) Who's got the power? Transforming health systems for women and children. *Report of the Task Force on Child Health and Maternal Health*. New York: United Nations Development Programme.
Yamin AE (ed.) (2005) *Learning to Dance: Advancing Women's Reproductive Health and Well-Being for the Perspectives of Public Health and Human Rights*. Cambridge, MA: Harvard University Press.

Relevant Websites

http://www.acpd.ca – Action Canada for Population and Development.
http://www.astra.org.pl/ – ASTRA Network, Central and Eastern European Women's Network for Sexual and Reproductive Health and Rights.
http://www.aidslaw.ca – Canadian HIV/AIDS Legal Network.
http://www.reproductiverights.org – Center for Reproductive Rights.
http://www.dawnorg.org/ – Development Alternatives with Women for a New Era (DAWN).
http://www.womenslinkworldwide.org – Women's Link Worldwide.

Cultural Context of Reproductive Health

R Root, Baruch College, City University of New York, New York, NY, USA
C H Browner, University of California, Los Angeles, CA, USA

> The act of giving birth to a child is never simply a physiological act but rather a performance defined by and enacted within a cultural context.
>
> (Romalis, 1981: 6)

Introduction

The concept of reproduction is a cultural product rooted in the biomedical paradigms of human anatomy, physiology, disease, and clinical medicine. In societies and subcultures where biomedicine is not the only or dominant framework, anthropology has demonstrated in an abundance of ethnographic detail how understandings and practices relating to the body, sexuality, and reproduction are culturally patterned and mediated by political–economic and ideological processes linked to gender, class, race, and ethnicity that have substantive implications for public health. These patterns, in turn, shape prenatal and postpartum activities, management of labor and parturition, as well as the highly variable experiences of gender across cultures. In short, the study of human reproduction is an inherently anthropological one that benefits from the field's ethnographic methods, and utilises theoretical and applied models which link social processes and cultural practices with ethical reasoning, new technologies, and advanced knowledge of human biological evolution.

This chapter offers an overview of anthropological research on human reproduction: its scope, organizing concerns, and landmark publications. It begins with a brief history and provides a topical overview of contemporary approaches to the subject. The latter fall loosely, but by no means exclusively, into the traditional fields of biological, materialist, and sociocultural anthropology. The chapter concludes with a description of the prolific research underway on the politics of reproduction cross-culturally (Ginsburg and Rapp, 1995), the effects of biomedical and indigenous (also commonly referred to as lay or local) concepts on reproductive decision making, and the expanded use of reproductive technologies. These include techniques to increase infertile women's and couples' probabilities of conception; monitor pregnancy, often through ultrasound, and detect birth anomalies *in utero*, primarily with use of amniocentesis.

Within the discipline's broad scope, medical anthropology, encompassing the comparative and cross-cultural study of health, illness, health care, and the body, has focused most extensively and explicitly on issues of human reproduction. The anthropology of human reproduction explores distinct but overlapping foci: the comparative study of obstetrical events and the influence of ethnophysiology (local, lay, or indigenous understandings about bodily processes) on reproductive decision making; the meanings of the human body, male and female, in relation to society as a body politic (Lock and Kaufert, 1998); and the cultural implications of reproductive technologies on constructs of motherhood, personhood, and the social organization of medicine. Illustrative research includes works on cross-cultural perceptions of when human life begins and how the body and the fetus have become cultural objects at the intensely contested intersection of technology and politics (Morgan and Michaels, 1999).

History of Anthropological Research on Human Reproduction

Anthropology emerged in the late nineteenth century as a science of culture that aspired to explain and justify unilinear models of human physical and cultural evolution. The field's subsequent focus on cultures as systems of interlocking parts emphasized empirical and longitudinal documentation of human biological and cultural variation. Most early anthropological data on human reproduction were therefore embedded within densely detailed ethnographies of specific cultures. Montagu's (1949) analysis of Australian Aboriginal concepts of conception and fetal development and Malinowski's (1932) account of Trobriand Islanders' understandings and practices associated with reproduction were important, albeit rare, exceptions.

As anthropological interest shifted in the mid-twentieth century from documenting distinct cultures to the ethnological project of cross-cultural comparison, researchers began to systematically cull data from respective ethnographies to produce comparative surveys of reproductive understandings and practices. The results ranged from carefully detailed efforts to demonstrate broad theoretical principles, for example of the biologically or socially adaptive value of birth practices, to laundry-list accounts of reproductive customs around the world.

Anthropology's radical reformulation in the 1960s diversified both the corps of anthropologists and its theoretical concerns to produce among others, the fields of feminist, applied, and urban anthropology. Studies of, for example, reproductive behavior in preindustrial societies provided insights into how maternity care in the industrialized world might be improved (Jordan, 1993). Others analyzed the empirical bases of medicinal plants and other substances for reproductive activities, such as enhancing and limiting human fertility, and facilitating childbirth, postpartum recovery, and lactation. These new paradigms provided valuable information on the now-euphemized management of reproduction, but left largely unexamined the articulation of biological reproduction with less easily formulated political-economic processes.

Currently, anthropological research on human reproduction is focused in three broad areas: assessing the differential effects of culture, class, ethnicity, and race in shaping the social context of reproduction and reproductive health; identifying differences in the cognitive evaluation and subjective experience of reproductive technologies across cultures and the issues of ethical relativism they raise; deconstructing local discourses of reproductive rights in light of global health programs and human rights legislation. Cross-cultural research on male sexuality, concepts of paternity, and reproductive health practices, long underexplored, is an important part of anthropology's expanding research enterprise.

Biological Anthropology and the Study of Human Reproduction

Sexual reproduction emerged as the dominant logic of procreation in tandem with industrialization, and associated scientific paradigms of knowledge production, in late nineteenth century Europe and the USA. Today, it refers to the sexual production of human beings and the physiology of human reproductive and aging processes, including puberty, menstruation, sexuality, coitus, conception, pregnancy, gestation, parturition, infertility, abortion, menopause, and senescence. However, where biomedical paradigms construct normal against abnormal processes in each domain and presume a unitary meaning for the term reproductive success, biological anthropology, and its related fields of human evolutionary demography, behavioral ecology, paleoanthropology, and physical and primate anthropology, attempt to trace the diachronic and synchronic evolution of human and other primate species as the basis for understanding fertility, weaning, male and female physiology, sexuality, disease, and adaptation as dynamic factors in human reproduction. Other branches of anthropology investigate the relation between biological and social variables as well. For instance, to document the full range of human reproductive experience, medical anthropologists have identified physiological and psychological correlates of such reproductive events as infertility, spontaneous abortion, and prematurity, which are classified as disorders in the biomedical and bioscientific clinical literature.

Structural–Materialist Anthropology and the Study of Human Reproduction

The concept of reproduction is variably used in the anthropological literature, so that its divergent meanings as biological processes that produce human beings and social processes that reproduce social institutions and relations are often conflated. At the same time, the reproduction/production heuristic is a recurrent and important theme in the literature on reproduction, because it reflects a dialectic between the production of people and the production of material things (Collier and Yanagisako, 1987). Both materialist and structuralist anthropology take this dualism as a starting point, conceptualizing human reproduction in terms of the differential access that social groups have to the material and symbolic capital needed to sustain a society's members and reproduce its social institutions.

Studies in a social structuralist mode show that members of a given society may have reproductive goals that conflict with one another and that differential access to power and prestige correlate strongly with how these conflicts are mediated. For example, comparative analyses of societies where women are ascribed a lesser status demonstrate how gendered asymmetries are often structured by the material and social constraints of childrearing. Feminist frameworks further identify the limitations of purely biological approaches that situate reproduction outside its political and economic contexts. By exploring these links cross-culturally, scholars have been able to broaden the materialist–structuralist concept of reproduction to include biology and the entire set of social relationships associated with the sustenance of society's nonproducing members.

Sociocultural Anthropology and the Study of Human Reproduction

The proliferation of international public health efforts devoted to maternal and child health (MCH) following World War II drew scholarly attention to the importance of sociocultural analysis in the design and implementation of MCH programs. Further energized by 1960s feminist activism, ethnographers demonstrated that key constructs, such as menstruation, conception, fetal development, pregnancy, motherhood, and menopause, are socially mediated and therefore vary enormously across cultures.

Comparative studies of menstruation and menopause, for instance, reveal how variable aspects of women's reproductive cycles can determine their social status and social position. In many cultures, menstruation has polluting and negative connotations. Yet in others, a menstruating woman is at the height of her creative power. Analysis of the cultural construction of pregnancy likewise shows the myriad ways women become biological mothers, and the broader contexts within which families are formed, gender roles embodied and enacted, social networks leveraged, and the maternal role performed.

Anthropological research of human reproduction has identified culture-specific pressures entailed in fertility decision making. One commonality seen in many parts of the world is the relationship between a woman's status and the number of children, especially males, she bears. The social logic by which women may be pressured to be reproductively prolific also serves to convey the differential value societies place on children. In many agrarian societies, the material imperative for labor exerts overt and relentless reproductive pressures and is reflected in paradigms of maternity that glorify fertility, childbearing, and maternal roles. In Benin, for instance, Sargent (1982) found that among the Bariba, female virtue is displayed during parturition through stoicism, and thereafter through self-sacrifice for one's children. Elsewhere as well, women gain important prestige from maternal self-sacrifice. However, women are not universally valued primarily for their reproductive abilities. In many hunter-gatherer and hunter-horticultural societies, themes of motherhood and biological reproduction are far less central to cultural conceptions of the female than are women as sexual beings.

Pronatalist ideologies continue to shape reproductive patterns in low-income and postindustrial societies as well. In the former, many women in poor rural and urban areas face contradictory pressures as subsistence activities and high infant mortality require high birth rates even as the costs of childrearing soar. In the latter, unmarried status, childlessness, and one-child families were not accorded, and in many places continue to lack, the same cultural value as married, two-parent, larger family households. But currently, throughout much of Europe and parts of Asia, declining fertility is posing sharp challenges to these historical trends. Economic pressures, broadened opportunities for women beyond maternity and a resultant improvement in their social status, along with disaffection with conventional family structures and gender roles, are contributing to women's and couples' decisions to have one child or none at all. An important counterpoint to this decline is the medicalization of childbirth, which has increased opportunities for individuals who in the past did not fit the reproductive profile, due, for example, to age, sexual orientation, or biological sex; all of which has further diversified concepts of maternity and is the subject of considerable investigation.

Public Health and the Anthropology of Reproduction

Anthropological research reveals how studies of human reproduction are also studies of the relationship between science and social life. Its research questions are operationalized through studies, for example, of the routinization of ultrasonography and other prenatal screening tests that are increasingly part of prenatal care. Studies have shown that differential utilization of reproductive technologies across cultures and social groups can be explained in part by decision-making processes that are culturally patterned and rendered in widely variant sociopolitical and economic contexts. Rapp's (2000) work, for example, on amniocentesis reveals the importance of capturing the meanings of the procedure to women who choose it, and of assessing how social dynamics of including race and social class are embedded in their decision-making processes.

Analyzing the cultural patterning of obstetrical events is an effective approach for exploring how gender relations are organized cross-culturally, the variable nature of domestic power relations, some of the forces that shape ritual and health-seeking behavior, and the components of ethnomedical and other cognitive systems. Understanding the hierarchy of authority in the obstetrical domain also offers insight into how legitimate decision-making power is distributed in a given society and how the domain of authoritative knowledge is constituted cross-culturally. In any society, authoritative knowledge refers to rules that carry more weight than others because they explain the world better for the purposes in question, are associated with a stronger power base, or both (Davis-Floyd and Sargent, 1997). Research in the cross-cultural study of reproduction using this framework has cast light on the broad yet variable basis of biomedical authority, as well as other kinds of authority.

Cultural studies of reproductive decision making have aided the comparative study of ethnophysiology and contributed to the field of cognitive anthropology. A pathbreaking analysis of indigenous fertility regulating practices in seven societies (Newman, 1985) demonstrated complex relationships between culture-specific understandings of conception, gestation, and parturition and childbearing behavior. In the same vein, in-depth studies using semi-structured and open-ended interviews of women's decisions regarding contraceptives, choice of birth attendants, and abortion make it possible to identify how women employ biomedical concepts and technologies within local ethnophysiological frameworks. Whether negotiation between lay concepts of reproductive physiology and biomedical theories and practices produces confusing or pragmatic choices for women has been the subject of much comparative investigation.

Ethnographic methods have enabled researchers to detail women's reproductive decision-making processes, examine the social norms and relationships these decisions entail, and identify institutional structures within which they are rendered, along with women's consequent health-seeking behaviors. DeBessa's (2006) fieldwork among Brazilian women in low-income communities, for example, showed that patterns of incorrect contraceptive use, harmful health risks, and extreme reliance on surgical sterilization were largely a function of how they interpret biomedical perspectives on contraception against competing ethnophysiologies of fecundity, menstruation, and conception.

Importantly, research has also shown that biomedicine and lay concepts of reproductive physiology are not always competing, mutually exclusive, or antagonistic. Obermeyer's (2000) study of birth knowledge and practices in Morocco, using women's narratives of recent birth experiences, observations of medical encounters, and cultural norms of prescribed pregnancy behaviors, found that the traditional Moroccan ethnophysiology of birth is not inconsistent with the precepts of biomedicine; that in fact, eclecticism, flexibility, and pragmatism characterize many women's experiences and choices of traditional and modern practitioners during pregnancy and birth.

The importance of scholarship in the anthropology of reproduction to public health is illustrated by Nichter's

(1983) study of ethnophysiology and the folk dietetics of pregnancy in rural southern India. He showed that women's preference for smaller babies along with indigenous dietary concepts helped to explain why many pregnant women were reluctant to comply with the iron sulfate, tetanus toxoid, and vitamin therapy protocols recommended by local primary health care staff. Also important for public health is work on the reasons why women in low-income countries have been reluctant to use biomedical contraceptives despite their expressed desires to have fewer children. These studies employ cross-cultural surveys to determine women's (and occasionally men's) knowledge, attitudes, and practices with regard to biomedical birth control techniques.

Conclusion

Anthropological research on human reproduction has accelerated in the past two decades, enhanced by internet communications and ready access to scholarly networks, information databases, and research funding across academic, governmental, and nonprofit entities. Its trajectory has been one of expanding scope, coordination, and methodologies to generate new hypotheses on the interrelatedness of biological, political–economic, and sociocultural factors involved in human reproduction. Current research continues to explore the vast topography described above and to further detail how ethnicity and social class shape perceptions and behaviors within the reproductive domain. Studies are also increasingly concerned with the underexamined relationships between masculinity, male physiology, sexuality, and reproduction. Anthropologists will continue to deploy the field's distinctive multidisciplinary frameworks to provide important insights into the diverse ways that gender, technology, political-economy, and social organization sustain, threaten, and transform human reproduction.

See also: Abortion; Family Planning/Contraception; Female Reproductive Function; Gender Aspects of Sexual and Reproductive Health; Infertility; Menopause; Reproductive Ethics: Ethical Issues and Menopause; Reproductive Ethics: New Reproductive Technologies; Reproductive Ethics: Perspectives on Contraception and Abortion; Reproductive Rights.

Citations

Collier JF and Yanagisako SJ (eds.) (1987) *Gender and Kinship: Essays Toward a Unified Analysis.* Stanford: Stanford University Press.

Davis-Floyd R and Sargent C (1997) *Childbirth and Authoritative Knowledge: Cross-cultural Perspectives*. Los Angeles, CA: University of California Press.

de Bessa GH (2006) Ethnophysiology and contraceptive use among low-income women in urban Brazil. *Health Care for Women International* 27: 428–452.

Ginsburg F and Rapp R (eds.) (1995) *Conceiving the New World Order: The Global Politics of Reproduction.* Berkeley, CA: University of California Press.

Jordan B (1993) *Birth in Four Cultures: A Cross-cultural Investigation of Childbirth in Yucatan, Holland, Sweden, and the United States*. Prospect Heights, IL: Waveland Press.

Lock M and Kaufert PA (eds.) (1998) *Pragmatic Women and Body Politics.* New York: Cambridge University Press.

Malinowski B (1932) *The Sexual Life of Savages in Northwestern Melanesia*. London: Routledge and Kegan Paul.

Montagu MFA (1949) Embryology from antiquity to the end of the 18th century. *Ciba Symposia* 10: 994–1008.

Morgan LM and Michaels MW (eds.) (1999) *Fetal Subjects, Feminist Positions.* Philadelphia, PA: University of Pennsylvania Press.

Newman L (ed.) (1985) *Women's Medicine.* New Brunswick, NJ: Rutgers University Press.

Nichter M (1983) The ethnophysiology and folk dietetics of pregnancy: A case study from South India. *Human Organization* 42: 235–246.

Obermeyer C (2000) Pluralism and pragmatism: Knowledge and practice of birth in Morocco. *Medical Anthropology* 14: 180–201.

Rapp R (2000) *Testing Women, Testing the Fetus: The Social Impact of Amniocentesis in America*. New York: Routledge.

Romalis S (ed.) (1981) Childbirth: Alternatives to Medical Control, pp 3-32. Austin, TX: University of Texas Press.

Sargent C (1982) *The Cultural Context of Therapeutic Choice: Obstetrical Decisions Among the Bariba of Benin*. Dordrecht, Holland: D Reidel Publishing.

Further Reading

Browner CH (2000) Situating women's reproductive activities. *American Anthropologist* 102: 773–788.

Buckley T and Gottlieb A (eds.) (1988) *Blood Magic: The Anthropology of Menstruation.* Berkeley, CA: University of California Press.

Clarke AE (1998) *Disciplining Reproduction: Modernity, American Life Sciences, and 'the Problem of Sex'.* Berkeley, CA: University of California Press.

Ellison PT (2001) *On Fertile Ground: A Natural History of Human Reproduction*. Cambridge, MA: Harvard University Press.

Greenhalgh S (ed.) (1995) *Situating Fertility: Anthropological and Demographic Inquiry*. Cambridge, UK: Cambridge University Press.

Inhorn M and van Balen F (eds.) (2002) *Infertility Around the Globe: New Thinking on Childlessness, Gender and Reproductive Technologies.* Berkeley, CA: University of California Press.

Kay MA (ed.) (1981) *Anthropology of Human Birth.* Philadelphia, PA: FA Davis Co.

MacCormack CP (ed.) (1982) *Ethnography of Fertility and Birth.* London: Academic Press.

Martin E (1992) *The Woman in the Body: A Cultural Analysis of Reproduction*. Boston, MA: Beacon.

Mullings L and Wali A (2001) *Stress and Resilience: The Social Context of Reproduction in Central Harlem*. New York: Kluwer.

Newton N and Newton M (1972) Childbirth in cross-cultural perspective. In: Howells J (ed.) *Modern Perspectives in Psycho-obstetrics*, pp. 150–172. Edinburgh, UK: Oliver and Boyd.

Ragoné H and Twine FW (2000) *Ideologies and Technologies of Motherhood: Race, Class, Sexuality, Nationalism*. New York: Routledge.

Sievert LL (2006) *Studies in Medical Anthropology. Menopause: A Biocultural Perspective*. Piscataway, NJ: Rutgers University Press.

Whiteford M and Poland M (eds.) (1989) *New Approaches to Human Reproduction: Social and Ethical Dimensions.* Boulder, CO: Westview Press.

Women's Mental Health Including in Relation to Sexuality and Reproduction

P S Chandra and V N G P Raghunandan, National Institute of Mental Health and Neuro Sciences, Bangalore, India
V A S Krishna, Washington University School of Medicine, St. Louis, MO, USA

Introduction

It is now widely understood that women's well-being is multifactorial and is not only determined by biological factors and reproduction, but also by the effects of poverty, nutrition, stress, war, migration, and illness. Approaching mental health problems from a female perspective and mainstreaming it requires a broad framework of health for women that addresses mental health throughout the life cycle and in domains of both physical and mental health. First, we present evidence demonstrating that women disproportionately suffer from certain mental disorders and are more frequently subject to social issues that lead to mental illness and psychosocial distress. Subsequently, the chapter will deal with the nature and types of mental disorders in women; factors contributing to vulnerability; and specific issues such as poverty, migration, HIV, war, natural disasters, and pregnancy-related psychiatric problems. It will also describe landmark studies that have been conducted in the area and report interventions that have been attempted with a public health perspective.

Sex Differences and Mental Disorders

Which Mental Disorders Are More Common in Women?

Mental disorders affect women and men differently: some disorders are more common in women and some express themselves with different symptoms. Researchers are only now beginning to tease apart the contribution of various biological and psychosocial factors to mental health and mental disorders in both women and men.

The disability-adjusted life year data recently tabulated by The World Bank reflect these differences. Depressive disorders account for close to 30% of the disability from neuropsychiatric disorders among women, but only 13% of that among men. Conversely, alcohol and drug dependence accounts for 31% of neuropsychiatric disability among men, but accounts for only 7% of the disability among women. These patterns for depression and general psychological distress and substance-abuse disorders are consistently documented in many quantitative studies carried out in societies throughout the world (Murray and Lopez, 1996). **Table 1** describes the sex differences in prevalence of various psychiatric disorders.

Clinical Profile of Various Mental Disorders in Women

Depression

Epidemiological and clinical studies have consistently documented that depression across different cultures is about twice as common in women as in men. Research shows that before adolescence and late in life, females and males experience depression at about the same frequency. Because the gender difference in depression is not seen until after puberty and decreases after menopause, scientists hypothesize that hormonal factors are involved in women's greater vulnerability. In addition, the changing psychological status and role of women in society following puberty may place them in a vulnerable position in times of stress.

The manifestation of depression also tends to be different in women. They very often present with medically unexplained symptoms including vague aches and pains.

Table 1 Lifetime prevalence of DSM III psychiatric disorders by gender

	ECA[a]		*NCS*[b]	
Psychiatric disorder	*Men (%)*	*Women (%)*	*Men (%)*	*Women (%)*
Prevalence of any disorder	36	30	48	50
Schizophrenic disorders	1.2	1.7	–	–
Affective disorders	2.3	5.0	17.5	24.9
Major depression	2.6	7.0	12.9	20
Alcohol abuse and/or dependence	23.8	4.6	19.6	7.5
Drug abuse and/or dependence	7.7	4.8	11.6	4.8
Panic disorder	0.99	2.1	3	6.2
Agoraphobia	3.18	7.86	1	1.6
Social phobia	2.53	2.91	11	13
Obsessive compulsive disorder	2	3	1	2.6
Somatization disorder	0.02	0.23	–	–
Antisocial personality	4.5	0.8	–	–

[a]Data from Robins LN and Regier DA (1991) *Psychiatric Disorders in America: The Epidemiologic Catchment Area Study.* New York: The Free Press.
[b]Data from Kessler RC, Berglund PA, Demler O, *et al.* (2005) Lifetime prevalence and age-of-onset distributions of DSM-IV disorders in the National Comorbidity Survey Replication (NCS-R) *Archives of General Psychiatry* 62: 593–602.

Though severity has been reported to be similar in both genders, depression in women has been found to be associated with increased functional impairment and rates of suicide attempts are higher than in men. Women also tend to have onset of depression at an earlier age and often become symptomatic in mid-adolescence, while depression in men usually begins in their twenties. Longitudinal studies have shown that women develop more recurrent depression and the individual episodes last longer. Comorbid medical disorders such as thyroid problems, migraine, and rheumatologic disorders are particularly common. In addition, depression in women frequently coexists with other psychiatric disorders, particularly panic disorder and simple phobia. Other psychiatric conditions such as eating disorders and personality disorders are often associated with depression in women and add to functional impairment and diagnostic problems.

Anxiety disorders

Anxiety disorders include generalized anxiety disorder, panic disorder, phobias, and posttraumatic stress disorder (PTSD), with women outnumbering men for each of these illness categories. Women not only have a higher risk of developing posttraumatic stress disorders, but they are also more likely to develop long-term PTSD than males, with higher rates of co-occurring medical and psychiatric problems.

Schizophrenia

Schizophrenia is the most chronic and disabling of the mental disorders, affecting about 1% of women and men worldwide. The illness typically appears earlier in men, usually in their late teens or early twenties, while women are generally affected in their twenties or early thirties. Thus, schizophrenia starts later in women compared to men, with a second peak in the menopausal period. The later age of onset confers some protection to women as they are usually better socialized and have a better clinical outcome.

Though the disease is reportedly less severe in women, they may have more depressive symptoms, paranoia, and auditory hallucinations than men and tend to respond better to typical antipsychotic medications. A significant proportion of women experience increased symptoms during the postpartum period and may also have significant problems in bonding with the child. Despite the better clinical outcome, women with schizophrenia often have to face higher stigma and face major problems in assimilating with the mainstream. In addition, they are prone to abuse – both physical and sexual – which puts them at further risk for physical and mental health problems. Additional problems that women with serious mental illness face include problems related to parenting, sexuality, and being more prone to drug side effects such as extrapyramidal symptoms, endocrine side effects, and osteoporosis.

Dementias: Alzheimer's Disease

The main risk factor for developing Alzheimer's disease (AD) is increased age. Studies have shown that while the number of new cases of AD is similar in older adult women and men, the total number of existing cases is somewhat higher in women. Possible explanations include that AD may progress more slowly in women than in men, that women with AD may survive longer than men with AD, and that men, in general, do not live as long as women and die of other causes before AD has a chance to develop.

Caregivers of a person with AD are usually family members – often wives and daughters. The chronic stress often associated with the care-giving role can contribute to mental health problems; indeed, caregivers are much more likely to suffer from depression than the average person. Since women in general are at greater risk for depression than men, female caregivers of people with AD may be particularly vulnerable to depression.

Suicide

Although men are four times more likely than women to die by suicide, women report attempting suicide about two to three times as often as men. Self-inflicted injury, including suicide, ranks ninth out of the ten leading causes of disease burden for females worldwide. A recent study on causes of maternal mortality in the first year after childbirth in the UK reported suicide as being the most common reason for death among women within 1 year of childbirth.

Substance Use

Several studies have reported a marked difference in rates of all substance use, with men outnumbering women. However, complications related to substance use such as alcoholic liver disease, neurological problems including cognitive deficits, and sexual and reproductive consequences of substance use in women lead to substantial disability.

What Are the Factors That Contribute to Increased Vulnerability in Women to Mental Health Problems?

Life Stress and Mental Health Problems in Women

Serious adverse life events are clearly implicated in the onset of depression (**Table 2**). Most work investigating the relationship of stressful life events and major depression has largely or exclusively employed samples of women and few studies have examined sex differences with regard to stress and depression. However, initial research in this area has demonstrated that women are three times more likely

Table 2 Factors that contribute to increased vulnerability among women

Life stresses
Sexual abuse and coercion
Intimate partner violence
Economic determinants: poverty
Migration
War
Natural disasters
Reproductive health: menstrual cycle, pregnancy, and menopause
Hormonal and endocrine factors
Gender disadvantage and discrimination
Medical conditions
Self-esteem and body image issues

than men to experience depression in response to stressful events. Brown and Harris were among the first to systematically describe the relationship of life stress to depression and subsequent research has confirmed the role of stress in women's mental health.

One of the first research studies to look at the role of social factors in depression in women was published by George Brown and Tirril Harris in 1978 (Brown and Harris, 1978). This study, conducted in inner city London, delineated three important sets of factors in the causation and manifestation of depression among women. The factors included vulnerability factors (lack of a confiding and intimate relationship, three children under the age of 14 years at home, and loss of the mother before the age of 11years), provoking factors (stressful life events), and symptom formation factors (past history of depression, severe life event after the onset of depression, and any past loss). The study was among the first that described a conceptual and interactive model for the social and personal causes of depression in women.

Sexual Abuse and Sexual Coercion

Trauma experienced by women, such as childhood sexual abuse, adult sexual assault, and intimate partner violence, also have been consistently linked to higher rates of depression in women, as well as to other psychiatric conditions (e.g., posttraumatic stress disorder, eating disorders) and physical illnesses. Features of abuse that determine the nature and severity of mental health problems include the duration of exposure to abuse, use of force, and relationship to the perpetrator. In addition, cognitive styles such as low self-esteem and even related appraisals also have an important impact on how women cope with abusive and traumatic life situations. In the context of sexual assault, characteristics such as degree and nature of physical force and perceived fear of death or injury significantly affect psychological outcome.

Intimate Partner Violence

Depression is also highly prevalent among women who experience male partner violence, which is often repetitive and concealed. Depression and posttraumatic stress disorder, which have substantial comorbidity, are the most prevalent mental health sequelae of intimate partner violence. Depression in women experiencing intimate partner violence has also been associated with other life stressors that often accompany domestic violence, such as younger age at marriage, childhood abuse, daily stressors, many children, coercive sex with an intimate partner, and negative life events. While some women might have chronic depression that is exacerbated by the stress of a violent relationship, there is also evidence that first episodes of depression can be triggered by such violence.

Though most research focuses on physical and sexual abuse, the impact of psychological abuse on mental health is also evident. Most of the data point to the association of substance use (particularly alcohol use) in women experiencing intimate partner violence. Substance use has also been found to be a sequel to the experience of violence among women. A postulated explanation of substance use as an outcome of intimate partner violence is through posttraumatic stress disorder. Women with posttraumatic stress disorder might use drugs or alcohol to calm or cope with the specific groups of symptoms associated with posttraumatic stress disorder: intrusion, avoidance, and hyperarousal. Women can also begin to abuse substances through their relationships with men or from wanting to escape the reality of intimate partner violence. It is important to address and understand these complex relations between intimate partner violence, mental health, and behavior to make an accurate diagnosis and intervene in substance abuse problems. In addition to mental health problems, women who experience violence use medical and emergency services more often and are known to present to primary care with unexplained somatic symptoms.

Poverty

Poverty among women has been steadily increasing and it has been linked to mental health problems among women. This association was observed more than two decades ago by Brown and Harris (1978), and this link has been confirmed in recent studies. Affective disorders are common among both men and women living in poverty and it has been observed that more women live in poverty than men. Poverty among women increases their risk for exposure to traumatic life experiences such as physical and sexual victimization. These experiences at times serve as barriers to accessing mental health services. In addition, women in poverty find it difficult to access health care because of lack of insurance and transportation and inflexible jobs. Mental health problems, when untreated, can lead to severe disabilities for women in poverty. In addition, mental health problems, specifically depression, have been reported to contribute to significant economic burden in women living in poverty (Patel *et al.*, 2007).

Migration

Migration involves uprooting oneself fully or partially from the familiar, traditional community and relocating in a foreign land. Migration can be planned and voluntary for better prospects in life or it can be unplanned, unanticipated, and forced, as in the case of civic unrest, armed conflict, and violation of human rights. Migration always calls for making adaptations, as people traverse several interpersonal, cultural, language, ecological, and geographic boundaries. The sociocultural differences and scarcity of resources could lead to feelings of fear, isolation, alienation, and helplessness in migrants. Migration as such may not affect the mental health of the immigrants, but the process of migration and adaptation can cause increased stress and vulnerability to mental health problems. For instance, evidence from South Asian immigrant women in Canada suggests that these women are specifically at risk for mental health problems because of the rigid gender roles in the South Asian community that make smooth integration into the adopted country a challenge. Likewise, South Asian women residing in the UK were more likely to report anxiety and depressive symptoms compared to their male counterparts. Loss of social support, low social status, constraints in finances and in accessing health services were major stressors for immigrant women in Canada (Ahmad *et al.*, 2004). Research studies conducted on divergent ethnic and racial immigrant groups residing in Europe found that complicated grief and posttraumatic stress disorder were noted to be common psychiatric problems among refugees and asylum seekers.

War

Wars are known to cause immense human suffering in several ways. War results in a shortage of food, water, fuel, and electricity that are basic requirements of every human being. During war, many civilians are exposed to traumatic experiences such as shooting and shelling and also seeing dead and wounded people, witnessing and experiencing violence, and injuries to self and others. Experiences of separation and displacement from relatives and forced migrations are very common. In addition to loss of valuable human resources, war results in damage of infrastructure and depletion of natural resources. There is substantial evidence on the mental health consequences of war for individuals. High rates of posttraumatic stress disorder as well as depressive and anxiety disorders were documented even among civilians. Studies documented inconsistent but high prevalence rates of PTSD among women affected by war. In addition, the aftermath or the postwar sociopolitical conditions are likely to contribute to mental health problems in women. For instance, life in Afghanistan has been disrupted for the past 20 years because of social and political disturbances. The war had extraordinary health outcomes for Afghan women, and women in Afghanistan have been found to have significantly poorer mental health compared to men (Cardozo, 2005).

Natural Disasters

Natural disasters are out of human control but the consequences of natural disasters overlap with the consequences of war or combat. In both contexts, there is human suffering caused by damage to life, personal property, and infrastructure. Families are displaced and victims lose shelter. This is complicated further by immense shortages of food and drinking water. Several medical and psychological problems among the victims are major offshoots of natural disasters. A summary of research studies conducted between 1981 and 2004 in both developing and developed nations yielded consistent results with regard to the psychological consequences of disasters on women. PTSD and major depressive and anxiety disorders were the common mental health consequences. The gender of the victims predicted several post-disaster outcomes in many of these studies; consistently, women were more likely to be affected than men. For PTSD alone, the rates for women exceeded those for men by a ratio of 2:1. In a few studies, being married was found to be a risk factor because the severity of husbands' symptoms predicted the severity of wives' symptoms more than vice versa (**Figure 1**).

Clinical Interface of Women's Mental Health with Reproductive Health and Medical Disorders

Psychiatric Disorders in Relation to Pregnancy and the Postpartum Period

Studies on psychiatric disorders among pregnant women in community-derived samples have shown lifetime depression risk estimates between 10% and 25%, while studies that screened obstetric patients at random for depressive symptoms found that up to 20% of patients met criteria for a diagnosis of a depressive episode. Risk factors for depression during pregnancy include young age, low education, a large number of children, a history of child abuse, a personal or family history of mood disorder, and stressors such as marital dysfunction.

Marcus *et al.* (2003) found that one in five pregnant women experience depression, but few seek treatment. The stigma of having depression during pregnancy may prevent women from seeking active treatment because

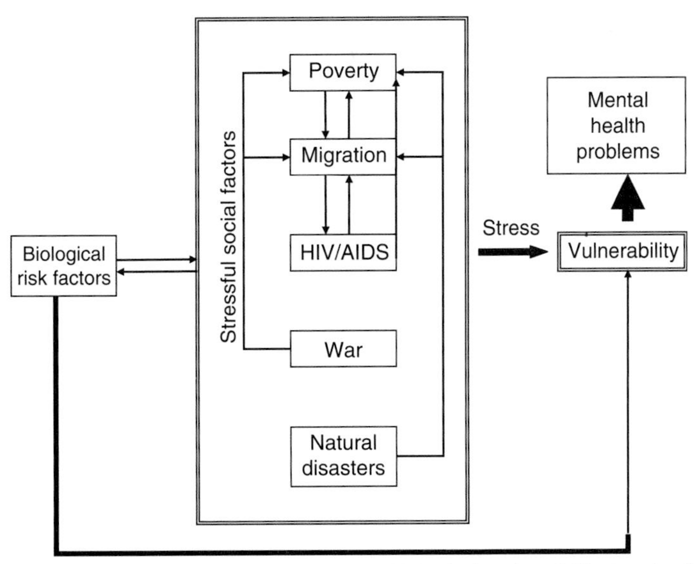

Figure 1 Hypothetical model demonstrating the interaction between biological and social factors leading to mental health problems among women.

they may feel guilty for suffering during what is supposed to be a happy period. The impact of untreated depression during pregnancy is known to have negative effects on both mother and child.

Major consequences of untreated maternal depression

In the mother

Mothers with depression may suffer from impaired social function, emotional withdrawal, and excessive concern regarding their future ability to parent. They are less likely to regularly attend obstetrical visits or have regular ultrasounds and show lack of initiative and motivation to seek help, and experience a negative perception regarding any potential benefit of obstetric services. Mothers suffering from depression are also more likely to smoke or use alcohol and have lower-than-normal weight gain throughout pregnancy because of diminished appetite. Severe depression also carries the risk of self-injurious, psychotic, impulsive, and harmful behaviors. Untreated depression may have associated obstetric complications, such as spontaneous abortion, bleeding during gestation, and spontaneous early labor.

In the child

Untreated maternal depression has also been associated with low birthweight in babies, babies small for their gestational age, preterm deliveries, perinatal and birth complications, admission to a neonatal care unit, and neonatal growth retardation. Neurobehavioral effects include reduced attachment, reduced mother–child bonding, and delays in offsprings' cognitive and emotional development. Lower language achievements and long-term behavioral problems may also be seen in some children whose mothers suffered from depression.

Postpartum psychiatric disorders

Postpartum psychiatric disorders are unarguably one of the most complex groups of disorders that encompass human experience. The joy of having a baby coupled with distress caused by impaired mental health can render the experience particularly traumatic to the mother, infant, and family. History taking in overburdened maternity wards usually does not allow for details about the mother's psychiatric history, let alone her current emotional status. In addition, short hospital stays and lack of follow-up make early recognition of emotional disorders difficult. The consequences of undiagnosed and hence untreated puerperal disorders can have negative consequences both on the mother and on the developing infant. Psychiatric disorders in the postpartum period include depression, anxiety-related disorders such as panic disorder and obsessive compulsive disorders, mother–infant bonding disorders and the relatively less common group of severe mental illness. In addition, women with preexisting psychiatric problems may also have worsening of symptoms in the postpartum period.

Epidemiology

Epidemiological surveys by Kendell *et al.* (1987) and Terp and Mortensen (1999) have established the incidence of postpartum psychosis as somewhat less than 1/1000 deliveries. Postpartum depression affects approximately 10–15% of all mothers in the developed world, while slightly higher rates have been reported from India (Patel, 2002) and South Africa. **Table 3** gives the prevalence rates based on studies done in several countries around the world.

Table 3 Prevalence rates for postpartum depression across various sites

Country/(state)	*Percentage (%)*
India/(Goa)	23
China	11.2
Japan	17
Arabia	15.8
Zimbabwe	16
South Africa	34.7
Australia	14–17
Western population[a]	10–15[a]

[a]Data from meta-analysis: O'Hara MW and Swain AM (1996) Rates and risk of postpartum depression: A meta-analysis. *International Review of Psychiatry* 8: 37–54.

The presence of depression in the last trimester of pregnancy is a strong predictor of postpartum depression. The preference for male children, deeply rooted in some societies, coupled with the limited control a woman has over her reproductive health may make pregnancy a stressful experience for some women. Local cultural factors are also pertinent in shaping maternal socioaffective well-being. For example, in rural China, mother-in-law conflict was reported in nearly one-third of young women who attempted suicide, as reported by Pearson (2002). Similar data on the role of mothers-in-law in domestic violence in pregnancy are available from different cultures, such as Hong Kong, India, The Republic of Korea, and Japan.

Severe mental illness in the postpartum period

Psychoses in the postpartum period have been classified as organic psychosis, schizophrenia, mania, or acute psychosis. Several studies report a relationship between puerperal psychosis and bipolar disorders. Some of the common clinical features of psychosis in the postpartum period include a polymorphic presentation, perplexity, confusion, emotional lability, and psychotic ideas related to the infant. In addition, preexisting severe mental illness can also worsen in the postpartum period. Severe mental illness after childbirth may raise several psychosocial issues, particularly related to safety of the mother and the infant. The Confidential Enquiry into Maternal Deaths 1997–1999, carried out in the UK, reported that psychiatric disorder, and suicide in particular, was the leading cause of maternal death in the first year after childbirth. The study has highlighted the need for routine assessment of preexisting and recent-onset psychiatric disorder in obstetric settings. It also emphasized the need for specialized clinical services for mothers with severe mental illness.

Depression

The presentation of this group of disorders can be heterogeneous. Mothers with chronic dysthymia, prepartum depression continuing into the puerperium, depression associated with recent adversity, and bipolar depression all fall into this common heading. Postnatal depression can have untoward effects on infant development, and depression may lead to reduced interaction and irritability misdirected at the child.

One of the most popular and widely used screening tools used for detection of postpartum depression is the Edinburgh Postnatal Depression Scale (EPDS) (**Table 4**) originally designed by Cox *et al.* (1987). This scale is available in several languages around the world. A high score on this ten-item self-rating questionnaire needs to be followed by an interview clarifying the symptoms of depression and comorbid psychiatric disorders. In addition to detecting depression, it is also important to explore the wider context, including the mother's life history, personality, and circumstances; the course of the pregnancy, including parturition and the puerperium; and relationships with the spouse, other children, family of origin and, especially, the infant. In addition to diagnosing depression and other disorders, one must identify vulnerability factors and the availability of support (**Table 5**).

Treatment and prevention of maternal mental disorders

Several studies have assessed the efficacy of different forms of treatment and prevention of postpartum depression. These include interpersonal psychotherapy, home visits by nurses, prenatal and postnatal classes, debriefing visits, and continuity of care models. A recent meta-analysis on the efficacy of psychosocial interventions in preventing postpartum depression has reported that interventions that target at-risk women, are individually based, and are done in the postpartum period rather than during pregnancy appear to show more benefit. Interpersonal psychotherapy has also been used in the treatment of postpartum depression with some efficacy.

Acute treatment of bipolar illness and psychosis is usually with psychotropic drugs including mood stabilizers. However, knowledge regarding the safety of antipsychotics and mood stabilizers in pregnancy and lactation is needed and second-generation antipsychotics may be safer. Lithium and other mood stabilizers are useful in treatment of bipolar disorders provided they are used with monitoring and also following discussions with the mother and the family. Antidepressants are indicated for moderate to severe depression, especially when biological functions are impaired or there is prominent suicidal ideation. Inpatient care may be indicated in more severe cases and requires specialized nursing and psychiatric care.

Postpartum psychosis has a recurrence rate of at least one in five pregnancies and mothers with a past history of puerperal or nonpuerperal psychosis have an enhanced risk. There is some evidence that prophylaxis, given immediately after delivery, reduces this risk.

Table 4 Edinburgh Postnatal Depression Scale (EPDS)

1. I have been able to laugh and see at the funny side of things.
   As much as I always could.
   Not quite so much now.
   Definitely not so much now.
   Not at all.
2. I have looked forward with enjoyment to things.
   As much as I ever did.
   Rather less than I used to.
   Definitely less than I used to.
   Hardly at all.
3. I have blamed myself unnecessarily when things went wrong.
   Yes, most of the time.
   Yes, some of the time.
   Not very often.
   No, never.
4. I have been anxious or worried for no good reason.
   No, not at all.
   Hardly ever.
   Yes, sometimes.
   Yes, very often.
5. I have felt scared or panicky for no very good reason.
   Yes, quite a lot.
   Yes, sometimes.
   No, not much.
   No, not at all.
6. Things have been getting on top of me.
   Yes, most of the time I haven't been able to cope at all.
   Yes, sometimes I haven't been coping as well as usual.
   No, most of the time I have coped quite well.
   No, I have been coping as well as ever.
7. I have been so unhappy that I have had difficulty sleeping.
   Yes, most of the time.
   Yes, sometimes.
   Not very often.
   No, not at all.
8. I have felt sad or miserable.
   Yes, most of the time.
   Yes, quite often.
   Not very often.
   No, not at all.
9. I have been so unhappy that I have been crying.
   Yes, most of the time.
   Yes, quite often.
   Only occasionally.
   No, never.
10. The thought of harming myself has occurred to me.
   Yes, quite often.
   Sometimes.
   Hardly ever.
   Never.

Cox JL, Holden JM, and Sagovsky R (1987) Detection of postnatal depression: Development of the 10-item Edinburgh Postnatal Depression Scale. *British Journal of Psychiatry* 150: 782–786.

Mother–infant bonding disorders that can occur as a consequence of psychiatric problems or infant-related issues are treated depending on the cause. Play therapy and baby massage under supervision or done in a graduated manner are often quite effective.

Table 5 Risk factors for postpartum depression

Low marital age
Low education
Marital discord and violence
Inadequate social support
Low socioeconomic status
Past depression
Personality vulnerability
Life events during pregnancy or near delivery
Complicated delivery
Adverse in-law relationship

Menstrual Cycle and Menopause

Mild mood changes in relation to the premenstrual phase have been reported commonly; however, significant mental health problems have been reported in only 3–5% of women. This usually occurs in the form of late luteal phase dysphoric disorders, which are mainly characterized by mood changes that significantly impair social, personal, and occupational functioning.

Mental health problems in menopause may be attributable to a combination of factors including hormonal, cognitive, and life-stage-related causes. Psychosocial factors including exit events, such as illness or death of spouse, retirement, and loneliness, may contribute significantly to mental health in the postmenopausal stage.

While the emphasis of the Women's Health Initiative (WHI) was more on the role of estrogen and progesterone on osteoporosis, breast cancer, and cardiovascular disease, the study also generated a large amount of data relevant to mental health. The WHIMS (Women's Health Initiative Memory Study) found a lack of evidence for the role of hormone therapy (estrogen alone or estrogen and progesterone) in protection against dementia among older women. The study also has important findings in the area of quality of life and lifestyle issues related to physical health, alcohol use, trauma, and panic disorders in older women. The study, which included Caucasian, African-American, and Asian women, was conducted in the USA and has important implications in the care of women over 65 years of age.

Reproductive Health Problems and Women's Mental Health

Infertility, female sterilization, and reproductive tract complaints have all been related to poor mental health in women. Infertility and the newer reproductive technologies are often fraught with uncertainties, leading to depression. Infertility in cultures where fertility and having children often determine the status of women has important implications for mental health. Vaginal discharge – both pathological (resulting from infections) and nonpathological – is often associated with symptoms of depression and anxiety. Women with somatic complaints and depression are known to present to clinics with a presenting complaint of vaginal discharge and need to be screened for mental health problems.

A recent study from West Africa (Coleman *et al.*, 2006), explored associations between depression and reproductive health conditions in 565 rural African women of reproductive age. The weighted prevalence of depression in the community was 10%, but more importantly, being depressed was significantly associated with widowhood or divorce, infertility, and severe menstrual pain.

Malignancies and Impact on Mental Health

Cancer of the cervix and breast cancer are the commonest cancers among women (the former in the developing world) and have several mental health implications. Mental health influences help-seeking, early detection, and participation in cancer screening programs. Subsequent to diagnosis, depressive disorders are common and may influence coping methods used in handling the illness. Studies done on women with breast cancer have emphasized the role of coping not only in the context of mental health but also in the progression of the disease.

HIV/AIDS

The efficacy of antiretroviral drugs has resulted in the decline in the incidence of AIDS cases in developed parts of the world but the proportion of women living with AIDS in resource-poor countries has been increasing steadily. Feelings of shame, guilt, fears related to stigma, death and dying, and concerns associated with childbearing and transmission of HIV to children are common sources of stress among women with HIV/AIDS. HIV-infected women face higher stigma and experience lesser social support than men. A high incidence of depressive symptoms and anxiety disorders among women with HIV has been found in recent studies. HIV-infected women were found to be four times more likely to report current major depressive disorder and anxiety symptoms compared to HIV-seronegative women. In the current context of antiretroviral treatment being available to a growing number of HIV-infected women even in the developing world, the recent literature has reported that depressive symptoms among women with HIV could lead to poor antiretroviral treatment utilization and adherence (Cook *et al.*, 2002). Early detection and treatment of mental health problems in HIV-infected women therefore may have an impact on help-seeking behaviors and medication compliance. In addition, psychosocial factors, such as poor socioeconomic conditions, race, and ethnicity, are also likely to be associated with mental health problems among women infected with HIV. For instance, HIV-infected women in the USA are disproportionately African-American or Latina and often live in poverty, and are, as a result, vulnerable to several social disadvantages. Understanding the psychological response to AIDS among women and the psychosocial context in which AIDS occurs is vital to prevent mental health problems and improve adherence to treatment.

Medical Disorders and Impact on Mental Health

Several medical conditions, especially endocrinological diseases (such as thyroid and parathyroid disorders) and collagen vascular disorders occur more commonly in women. These can cause mental health problems in two ways, one resulting from direct neuropsychiatric effects and the other as a consequence of the disability caused by these conditions. Pain and somatic symptoms in medical disorders can be worsened because of coexisting mood disorders, which are commoner among women.

Interventions

Primary Prevention

Health policies that incorporate mental health into public health and address women's needs and concerns in different life stages can be developed in numerous ways. Health promotion through public health initiatives related to education, men's attitudes toward women, gender discrimination, violence, safety, substance abuse, prenatal care, and regular health assessments of older women will help in preventing several mental health problems. Health policies must also face the challenge of formulating ethical but culturally sensitive responses to practices that are damaging to the emotional and physical health of women and girls (such as female genital mutilation, female infanticide, gender-specific abortion, and feeding practices that discriminate against girl children).

Secondary Prevention

Integrated health programs that address and handle the stigma of major mental illness, consequences of sexual or domestic violence, the consequences of gender discrimination, and the stress of poverty are an important part of public health. One of the more troubling mental health consequences of the general health status of communities is the effect on mothers of high infant and child mortality rates and high HIV infection rates affecting multiple family members across generations. Communication between health workers, physicians, and women patients (and often men as well) needs to be emphasized to facilitate disclosure of mental health issues by women to facilitate early detection. Training of nonmental health professionals in the use of simple screening tools to detect mental health problems such as the use of the General Health questionnaire or the EPDS (see the section titled 'Depression') will strengthen early detection efforts in primary care. Training and enhancing the competence of primary care physicians, mental health professionals, and health workers to detect and treat the consequences of domestic violence, sexual abuse, and psychological distress can also play an important role. Helplines for women in distress and suicide helplines have been shown to be effective for women seeking early and accessible help in the community. Women in most communities will need services near their homes and without causing inconvenience to child care and family.

Tertiary Prevention

Skilled clinicians as well as broader multidisciplinary programs in the community with links to hospitals are necessary to address the more distressing and difficult needs of women with serious psychiatric problems. These services should also be available in other medical centers, such as those dealing with oncology or HIV infection and in obstetric and gynecology clinics.

Although the social roots of many of these problems mean that they cannot be managed only with medical interventions, there is a need to strengthen the potential role of the health-care system. In addition, there should be increased consumer participation by women in formulating health-care policies and programs.

Conclusion

Women's mental health needs to be considered in the context of the interaction of physical, reproductive, and biological factors with social, political, and economic issues at stake. The multiple roles played by women, such as childbearing and child rearing, running the family, caring for sick relatives, and, in an increasing proportion of families, earning income, are likely to lead to considerable stress. The reproductive roles of women, such as their expected role of bearing children, the consequences of infertility, and the failure to produce a male child in some cultures, are examples of mechanisms that make women vulnerable to suffering from mental disorders. In addition, biological factors may play a major role, particularly in reproductive life events, such as pregnancy, the postpartum period, and menopause as well as in the clinical manifestations of various mental health problems.

Public health and social policies aimed at improving the social status of women are needed along with those targeting the entire spectrum of women's health needs. Efforts to improve and enhance social and mental health services and programs aimed at increasing the competence of professionals are also required.

See also: Female Reproductive Function; Gender Aspects of Sexual and Reproductive Health; Menopause; Reproductive Ethics: Perspectives on Contraception and Abortion; Sexual Violence; Violence Against Women.

Citations

Ahmad F, Shik A, Vanza R, *et al.* (2004) Voices of South Asian women. *Immigration and Mental Health* 40: 113–129.
Brown GW and Harris T (1978) *Social Origins of Depression. A Study of Psychiatric Disorder in Women.* London: Tavistock.
Cardozo BL (2005) Mental health of women in postwar Afghanistan. *Journal of Women's Health* 14: 285–293.
Coleman R, Morison L, Paine K, Powell RA, and Walraven G (2006) Women's reproductive health and depression: A community survey in the Gambia, West Africa. *Social Psychiatry and Psychiatric Epidemiology* 41: 720–727.
Cook JA, Cohen MH, Burke J, *et al.* (2002) Effects of depressive symptoms and mental health quality of life on use of highly active antiretroviral therapy among HIV-seropositive women. *Journal of Acquired Immune Deficiency Syndrome* 30: 401–409.
Cox JL, Holden JM, and Sagovsky R (1987) Detection of postnatal depression: Development of the 10-item Edinburgh Postnatal Depression Scale. *British Journal of Psychiatry* 150: 782–786.
Kendell RE, Chalmers JC, and Platz C (1987) Epidemiology of puerperal psychoses. *British Journal of Psychiatry* 150: 662–673.
Kessler RC, Berglund PA, Demler O, *et al.* (2005) Lifetime prevalence and age-of-onset distributions of DSM-IV disorders in the National Comorbidity Survey Replication (NCS-R). *Archives of General Psychiatry* 62: 593–602.
Marcus SM, Flynn HA, Blow FC, *et al.* (2003) Depressive symptoms among pregnant women screened in obstetrics settings. *Journal of Women's Health (Larchmont)* 12: 373–380.
Murray CJL and Lopez AD (eds.) (1996) *The Global Burden of Disease and Injury Series, vol. 1: A Comprehensive Assessment of Mortality and Disability from Diseases, Injuries, and Risk Factors in 1990 and Projected to 2020.* Cambridge MA: Harvard School of Public Health.
O'Hara MW and Swain AM (1996) Rates and risk of postpartum depression: A meta-analysis. *International Review of Psychiatry* 8: 37–54.
Patel V, Rodrigues M, and DeSouza N (2002) Gender, poverty, and postnatal depression: A study of mothers in Goa, India. *American Journal of Psychiatry* 159: 43–47.
Patel V, Chisholm D, Kirkwood BR, and Mabey D (2007) Prioritizing health problems in women in developing countries: Comparing the financial burden of reproductive tract infections, anaemia and depressive disorders in a community survey in India. *Tropical Medicine and International Health* 12: 130–139.
Pearson V (2002) Attempted suicide among young rural women in the People's Republic of China: Possibilities for prevention. *Suicide and Life-Threatening Behavior* 32: 359–369.
Robins LN and Regier DA (1991) *Psychiatric Disorders in America: The Epidemiologic Catchment Area Study.* New York: The Free Press.
Terp IM and Mortensen PB (1999) Post-partum psychoses: Clinical diagnoses and relative risk of admission after parturition. *British Journal of Psychiatry* 172: 521–526.

Further Reading

Benjamin L, Hankin L, and Abramson Y (2001) Development of gender differences in depression: An elaborated cognitive vulnerability-transactional stress theory. *Psychological Bulletin* 127: 773–796.
Carta MG, Bernal M, Hardoy MC, *et al.* (2005) Migration and mental health in Europe (The State of the Mental Health in Europe Working

Group: Appendix I): Review. *Clinical Practice and Epidemiology in Mental Health* 1: 13.
Fischback RL and Herbert B (1997) Domestic violence and mental health: Correlates and conundrums within and across cultures. *Social Science and Medicine* 45: 1161–1170.
Lumley J, Watson L, Small R, Brown S, Mitchell C, and Gunn J (2006) PRISM (Program of Resources, Information and Support for Mothers): A community-randomised trial to reduce depression and improve women's physical health six months after birth. *BMC Public Health* 6: 37.
Kornstein SG and Clayton AH (eds.) (2003) *Women's Mental Health: A Comprehensive Textbook.* New York: Guilford Press.
Mazure CM, Keita GP, and Blehar MG (2002) *Summit on Women and Depression: Proceedings and Recommendations.* Washington DC: American Psychological Association.
Miranda J and Green B (1999) The need for mental health services research focusing on poor young women. *The Journal of Mental Health Policy and Economics* 2: 73–80.
Oates M (2003) Perinatal psychiatric disorders – A leading cause of maternal morbidity and mortality. *British Medical Bulletin* 67: 219–229.
Patel V, Araya R, de Lima A, *et al.* (1999) Women, poverty and common mental disorders in four restructuring societies. *Social Science and Medicine* 49: 1461–1471.
Rahman A, Iqbal Z, Bunn J, Lovel H, and Harrington R (2004) Impact of maternal depression on infant nutritional status and illness: A cohort study. *Archives of General Psychiatry* 61: 946–952.
Watson M, Homewood J, Havilland J, *et al.* (2005) Influence of psychological response on breast cancer survival: 10-year follow-up of a population-based cohort. *European Journal of Cancer* 41: 1710–1714.

Relevant Websites

http://www.cdc.gov/hiv/ – Center for Disease Control and Prevention: HIV/AIDS.
http://www.who.int/mental_health/resources/gender/en/ – Gender and Women's Mental Health, World Health Organization.
http://www.womensmentalhealth.org/ – Massachusetts General Hospital Center for Women's Mental Health.
http://www.ncptsd.va.gov/ncmain/index.jsp – National Center for Posttraumatic Stress Disorder, United States Department of Veterans Affairs.
http://www.whi.org/ – Women's Health Initiative.
http://www.who.int/mental_health/media/en/67.pdf – Women's Mental Health: An evidence-based review, World Health Organization.

Subject Index

Notes

The index is arranged in set-out style with a maximum of three levels of heading. Major discussion of a subject is indicated by bold page numbers. Page numbers suffixed by T and F refer to Tables and Figures respectively. vs. indicates a comparison.

A

B

C

D

E

F

G

H

I

K

L

M

N

O

P

Q

R

T

U

V

W

Z